Carbon Nanomaterials

Second Edition

Advanced Materials and Technologies Series

Series Editor

Yury Gogotsi

Drexel University
Philadelphia, Pennsylvania, U.S.A.

Carbon Nanomaterials

Second Edition

Edited by

Yury Gogotsi and Volker Presser

CRC Press is an imprint of the
Taylor & Francis Group, an **informa** business

CRC Press
Taylor & Francis Group
6000 Broken Sound Parkway NW, Suite 300
Boca Raton, FL 33487-2742

First issued in paperback 2017

© 2014 by Taylor & Francis Group, LLC
CRC Press is an imprint of Taylor & Francis Group, an Informa business

No claim to original U.S. Government works

Version Date: 20130819

ISBN 13: 978-1-138-07681-5 (pbk)
ISBN 13: 978-1-4398-9781-2 (hbk)

Library of Congress Cataloging-in-Publication Data

Carbon nanomaterials / editors, Yury Gogotsi, Volker Presser. -- Second edition.
       pages cm -- (Advanced materials and technologies series)
    Includes bibliographical references and index.
    ISBN 978-1-4398-9781-2 (hardcover : alk. paper) 1. Nanostructured materials. 2. Carbon. I.
    Gogotsi, IU. G., 1961- II. Presser, Volker.

    TA418.9.N35C34 2014
    620.1'15--dc23   2013032517

Visit the Taylor & Francis Web site at
http://www.taylorandfrancis.com

and the CRC Press Web site at
http://www.crcpress.com

This book is dedicated to our families who have always been there for us.

In memory of John E. (Jack) Fischer (1939–2011)—a dedicated carbon scientist, excellent colleague, and friend.

Contents

Preface

Originating from humble beginnings when the term was coined in 1974 by Norio Taniguchi, nanotechnology has become by far the fastest growing area and the shooting star in materials science and engineering. Its impact can be seen in everyday products with improved or novel properties and, often as a buzz word, nanotechnology is widely present in the news and popular media. Typical examples range over a tremendously wide field, from high-performance computer chips and UV-blocking sun care cosmetics to nature's very own nanotechnology, including the adhesion of gecko feet even to smooth surfaces or the incredible strength of spider webs. It is only since the advent of high-resolution characterization techniques and advanced computer modeling that we have started to understand the mysteries and marvels of the "nano world" in such a way that we are able to capitalize on its unique phenomena and tailor material properties for making nanoscale devices. Many of the unique properties of nanomaterials, such as fluorescence or high electron mobility, may vanish for macroscale objects.

Carbon nanomaterials are widely used in commercial products. In mere numbers of scientific publications, the research on carbon nanomaterials, such as graphene, nanotubes, and fullerenes, exceeds by far all other fields of nanotechnology. Carbon nanotechnology has gained significant attention, energized by discoveries such as fullerenes, followed by carbon nanotubes, and, of course, the latest addition to the carbon family, graphene. Almost all recent discoveries relate back to the work that was done decades ago, such as the work on graphene from Hans-Peter Boehm in the 1960s or Eiji Osawa's pioneering work that predicted fullerenes in 1970. As is common for materials science, it takes years for materials development to attain a mature state and, after the first hype of a new discovery, the actual use and applicability of such materials start to become clear. However, it is noticeable how the speed of transferring the novel discoveries from the laboratory to scalable production and applications has increased significantly. For example, when looking at carbon nanotubes, manufacturing costs have been constantly decreasing during the past 20 years and today, large amounts of nanotubes are available for various applications. Currently, we see the same trend for graphene, yet at an even faster pace. Owing to the ability of carbon to occur in sp, sp^2, and sp^3 hybridization, carbon is truly the most versatile element in the periodic table with a large variety of allotropes and structures of various dimensionality, and it is exciting to see how new carbon materials with unique properties are discovered and explored for various applications—ranging from energy or gas storage, to catalysis, water treatment, medical implants, drug delivery, biofiltration, electronics, and many more fields.

While tutoring students and teaching graduate courses on nanostructured carbon materials, the need for a comprehensive, yet up-to-date, textbook on carbon nanotechnology became clear to us. The first edition of *Carbon Nanomaterials* from 2006 has become outdated because of the rapid developments in the field. Currently, there are a growing number of mostly topical books on carbons; however, the variety of carbon nanomaterials and their applications is insufficiently reflected in such compilations. By completely revising and extensively expanding the first edition of *Carbon Nanomaterials*, we hope to reflect the diversity of carbon nanotechnology regarding their synthesis and properties and also provide insights into the actual applications. We follow the successful approach of having leading experts in their respective fields author single chapters to provide a "first hand" experience to the reader and in this way we also provide an outlook on ongoing and future developments. We acknowledge that no single book can capture all aspects of carbon nanotechnology or provide a complete listing of all carbon nanomaterials. However, we believe that in the current compilation with 16 chapters, ranging from energy conversion and electronic applications, to water treatment and biomedicine, a representative cross-section is provided. The broad topic variety is tailored particularly to provide a basis for graduate students and scientists beginning to work in the field to be familiarized with the current topics of carbon nanotechnology. Following the pioneering words "there's plenty of room at the bottom" by Richard Feynman, we are looking forward to future developments and discoveries of novel carbon structures and new synthesis methods. Beyond the scope of the characterization and comprehensive understanding of the properties of carbon nanomaterials, we are particularly excited to see how material engineers will be able to create unique carbon-based devices.

The chapters from the 2nd edition have been either revised or represent completely new contributions. Examples for new chapters are the sections on carbon onions, nanodiamonds, graphene, and applications of carbon nanomaterials for biosensing and cell probes. With the rapid growth in the field of supercapacitors based on nanoporous carbons, a stand-alone chapter in the first edition has been the seed for the chapter's authors Francois Beguin and Elzbieta Frackowiak to write two books on that topic, namely, *Carbons for Electrochemical Energy Storage and Conversion Systems* (2010) and *Supercapacitors* (2013). With such broad coverage of that field, we are not covering this research topic in the 2nd edition.

We would like to acknowledge the help of many people who have assisted in preparing this book. This book is as much their accomplishment as it is ours. First, we are deeply grateful to our families and friends, for their support, patience, and understanding. We would also like to thank all contributing authors for taking time out of their busy work and teaching schedules to provide a comprehensive overview of their research on carbon nanomaterials. Our research associates, postdocs, students, and assistants did a fantastic job in performing some of the research described in this book and thus allowing us time to concentrate on the task of editing and revising the book. In particular, we would like to thank Dr. Jennifer S. Atchison for her tedious proofreading and helpful discussions. Also, the staff at Taylor & Francis, in particular, Allison Shatkin, helped immensely in preparing this volume for publication. Finally, we would like to acknowledge funding for our research on carbon nanomaterials, including the U.S. National Science Foundation (NSF), the U.S. Department of Energy (DOE), the Army Research Office, the German Federal Ministry of Research and Education (BMBF), and many private corporations. In particular, we would like to thank Dr. David Wesolowski (director of the FIRST Energy Frontier Research Center, ORNL) for his support and for providing an outstanding research environment.

Volker Presser
Saarbrücken, Germany

Yury Gogotsi
Philadelphia, Pennsylvania, USA

Editors

Yury Gogotsi is a distinguished university professor and trustee chair at the Department of Materials Science and Engineering of Drexel University, Philadelphia, Pennsylvania, USA. He also serves as the director of the A. J. Drexel Nanotechnology Institute. He earned his MS (1984) and PhD (1986) from Kiev Polytechnic and a DSc from the Ukrainian Academy of Sciences in 1995. He did his postdoctoral research in Germany supported by the Alexander von Humboldt Foundation, in Norway supported by NATO, and in Japan, supported by the Japan Society for the Promotion of Science (JSPS). His research group works on nanostructured carbons and other nanomaterials. He has coauthored more than 390 journal papers and is a Fellow of AAAS, MRS, ECS, and ACerS and a member of the World Academy of Ceramics.

Volker Presser is the leader of a BMBF Junior Investigator Group at the INM-Leibniz Institute for New Materials and assistant professor at Saarland University in Saarbrücken, Germany. He earned his doctoral degree in applied mineralogy from Eberhard Karls Universität in Tübingen, Germany in 2009. Being awarded the Feodor Lynen Research Fellowship from the Alexander von Humboldt Foundation, he joined Drexel University, Philadelphia, Pennsylvania, USA, as a postdoctoral research fellow in 2010 and continued his research as an assistant research professor until 2012. He has coauthored more than 70 journal papers. His research group works on carbon and carbide nanomaterials, mostly porous nanocarbons, for applications such as energy and gas storage and water purification.

Contributors

Marina Baidakova
Ioffe Physical-Technical Institute
St. Petersburg, Russia

P. Maarten Biesheuvel
Centre of Excellence for Sustainable
    Water Technology
Leeuwarden, the Netherlands

and

Department of Environmental
    Technology
Wageningen University
Wageningen, the Netherlands

Goknur C. Büke
Department of Materials Science and
    Engineering
Cankaya University
Ankara, Turkey

Yuriy Butenko
European Space Agency, ESTEC
Noordwijk, the Netherlands

Li Cao
Department of Chemistry
Clemson University
Clemson, South Carolina

Hui-Ming Cheng
Shenyang National Laboratory for
    Materials Science
Institute of Metal Research
Shenyang, China

Liming Dai
Case School of Engineering
Case Western Reserve University
Cleveland, Ohio

Sheng Dai
Oak Ridge National Laboratory
Oak Ridge, Tennessee

and

Department of Chemistry
University of Tennessee
Knoxville, Tennessee

Artur Dideikin
Ioffe Physical-Technical Institute
St. Petersburg, Russia

Svetlana Dimovski
Department of Materials Science and
    Engineering
Drexel University
Philadelphia, Pennsylvania

Bastian J. M. Etzold
Institute of Chemical Reaction Engineering
Friedrich-Alexander Universität
    Erlangen-Nürnberg
Erlangen, Germany

Pasquale F. Fulvio
Oak Ridge National Laboratory
Oak Ridge, Tennessee

and

Department of Chemistry
University of Tennessee
Knoxville, Tennessee

Yury Gogotsi
Department of Materials Science and
    Engineering
Drexel University
Philadelphia, Pennsylvania

Masoud Golshadi
Department of Microsystems Engineering
Rochester Institute of Technology
Rochester, New York

Joanna Gorka
Oak Ridge National Laboratory
Oak Ridge, Tennessee

and

Department of Chemistry
University of Tennessee
Knoxville, Tennessee

Pingang He
Department of Chemistry
East China Normal University
Shanghai, China

Matthew A. Hood
Department of Materials Science and
    Engineering
Drexel University
Philadelphia, Pennsylvania

Peng-Xiang Hou
Shenyang National Laboratory for Materials
    Science
Institute of Metal Research
Shenyang, China

Michael R. C. Hunt
Department of Physics
University of Durham
Durham, United Kingdom

Yair Korenblit
Department of Materials Science and
    Engineering
Georgia Institute of Technology
Atlanta, Georgia

Eric D. Laird
Department of Materials Science and
    Engineering
Drexel University
Philadelphia, Pennsylvania

Christopher Y. Li
Department of Materials Science and
    Engineering
Drexel University
Philadelphia, Pennsylvania

Chang Liu
Shenyang National Laboratory for Materials
    Science
Institute of Metal Research
Shenyang, China

Yamin Liu
Department of Chemistry
Clemson University
Clemson, South Carolina

Aurelio Mateo-Alonso
Institut für Organische Chemie und Biochemie
Albert-Ludwigs-Universität Freiburg
Freiburg im Breisgau, Germany

Richard T. Mayes
Oak Ridge National Laboratory
Oak Ridge, Tennessee

and

Department of Chemistry
University of Tennessee
Knoxville, Tennessee

Mohammed J. Meziani
Department of Chemistry
Clemson University
Clemson, South Carolina

and

Department of Chemistry and Physics
Northwest Missouri State University
Maryville, Missouri

Sebastian Osswald
Department of Physics
Naval Postgraduate School
Graduate School of Engineering and
    Applied Science
Monterey, California

Ge Peng
Department of Chemistry
Clemson University
Clemson, South Carolina

Slawomir Porada
Centre of Excellence for Sustainable
    Water Technology
Leeuwarden, the Netherlands

and

Department of Polymers and
    Carbon Materials
Wroclaw University of Technology
Wroclaw, Poland

Volker Presser
INM–Leibniz Institute for New Materials GmbH
and
Saarland University
Saarbrücken, Germany

Jieshan Qiu
Carbon Research Laboratory
Dalian University of Technology
Dalian, China

Sushant Sahu
Department of Chemistry
Clemson University
Clemson, South Carolina

Francesco Scarel
Freiburg Institute for Advanced Studies (FRIAS)
Albert-Ludwigs-Universität Freiburg
Freiburg im Breisgau, Germany

Michael G. Schrlau
Department of Mechanical Engineering
Rochester Institute of Technology
Rochester, New York

Jaganathan Senthilnathan
Department of Materials Science and
    Engineering
National Cheng Kung University
Taiwan, Republic of China

Lidija Šiller
School of Chemical Engineering and
    Advanced Materials
Newcastle University
Newcastle upon Tyne, United Kingdom

Ya-Ping Sun
Department of Chemistry
Clemson University
Clemson, South Carolina

Alexander Vul'
Ioffe Physical-Technical Institute
St. Petersburg, Russia

Albert van der Wal
Department of Environmental
    Technology
Wageningen University
Wageningen, the Netherlands

Masahiro Yoshimura
Department of Materials Science and
    Engineering
National Cheng Kung University
Taiwan, Republic of China

Gleb Yushin
Department of Materials Science and
    Engineering
Georgia Institute of Technology
Atlanta, Georgia

Mei Zhang
Case School of Engineering
Case Western Reserve University
Clcvcland, Ohio

Zongbin Zhao
Carbon Research Laboratory
Dalian University of Technology
Dalian, Chinae

1 Graphene
Synthesis, Properties, and Applications

Zongbin Zhao and Jieshan Qiu

CONTENTS

1.1 HISTORY AND DEVELOPMENT

Graphene, a two-dimensional (2D) and atomically thin crystal, consists of a single layer of carbon packed in a hexagonal lattice with a C–C distance of 0.142 nm.[1] Although zero-dimensional (0D) fullerene (C_{60}) and one-dimensional (1D) carbon nanotubes (CNTs) have been discovered since several decades, 2D materials graphene was discovered only recently. This discovery is attributed to Geim's group from Manchester University; the famous Scotch™ tape work published in *Science* in 2004 was regarded as a landmark for 2D crystal.[1] Since then, graphene has become one of the most exciting topics in fundamental research. The Nobel Prize in Physics for the year 2010 was awarded to Andre K. Geim and Konstantin S. Novoselov "who have made decisive contributions to the research of graphene."[2] However, pioneering work on graphene can be traced back to at least more than 60 years: P.R. Wallace predicted theoretically the unique electronic structure and the linear dispersion relation of graphene in 1947,[3] J.W. McClure formulated the wave equation for excitations in 1956,[4] and G.W. Semenoff, D.P. DiVincenzo, and E.J. Mele compared the equation with the Dirac equation in 1984.[5,6] Along with the progress in the theory of graphene, great efforts have been made for growth and identification of graphene in the community of surface science.[7,8] For example, John May grew and identified graphene on the surface of platinum as early as 1969,[7] and Jack Blakely and coworkers systematically investigated the formation of carbon layer on metals (Ni, Pt, Pd, Co, and Fe) from chemical vapor deposition (CVD). The observation of sheet-like graphitic material during arc discharge can be dated back to the 1990s. In 1992, Ebbesen et al.[9] reported that some sheet-like graphitic material was formed when they prepared multiwalled carbon nanotubes (MWCNTs) by helium arc discharge at low pressure. Because the yield of graphitic sheets was very low, and they always coexisted with other carbon allotropes, such as MWCNTs, it did not arouse much attention. In 1997, Y. Ando, X. Zhao, and coworkers observed petal-like graphite sheets, which were essentially stacked with graphene, deposited on the cathode surface by hydrogen arc discharge.[10,11] The number of graphene layers is two or three for the thinnest cases based on the observation in HRTEM micrographs.[10] In 2002, Wu et al. synthesize "carbon nanowalls," a similar structure with petal-like graphite sheets, by a microwave plasma-enhanced chemical vapor deposition (PECVD) method.[12] In fact, the sheet-like graphitic material, petal-like graphite sheets, and carbon nanowalls are all multiwalled graphene from the point of crystal structure.

Being the building block of graphite, graphene may not appear like a "new" material. Graphite is the most common and thermodynamically stable form of carbon. Its structure has been well-documented and consists of an ordered stacking of numerous graphene layers on top of each other. Each such layer is bonded by weak van der Waals force, which can be easily overcome, for example, by mechanical force as demonstrated by the excellent lubrication provided by graphite.[13,14] However, for a long time, the belief that a freestanding atomically thin carbon sheet would be thermodynamically unstable had discouraged further efforts in the research of graphene significantly.

Preparing graphene consisting of a few layers is quite simple and is encountered on a daily level while using a simple pencil. However, the isolation, identification, and characterization of single-layer graphene are not a trivial task because depending on the synthesis method, the monolayers are only the minority phase which are accompanied by thick flakes.[15] Geim's group was the first to isolate single-layer graphene from graphite and to identify its structure. In November 2005, the exciting electron transport measurement results reported by Geim et al. and the group of P. Kim[16,17] sparked off the exponentially growing graphene research worldwide. The interest in graphene was also fueled by the extensive amount of studies on carbon nanomaterials, such as fullerenes[18] or CNTs,[19] and this had laid a solid foundation for the rapid development of graphene and graphene-based composite materials. Until now, many excellent reviews that focused on graphene have been included[20–26] and, at the same time, novel applications and properties are being investigated. This chapter summarizes the recent progresses in the research of synthesis, structures, and properties of graphene and graphene-related materials and their applications in various fields.

1.2 SYNTHESIS OF GRAPHENE

Until now, the fabrication of defect-free high-quality graphene in large areas, which is critical for nanoelectronic devices, is still a big challenge. Surely, the establishment of controllable approaches for specific applications is the precondition toward real application. Various methods have been developed and more novel fabrication methods are still emerging. In this section, the most important preparation methods that have been used so far will be discussed in detail.

1.2.1 Mechanical and Ultrasonication Exfoliation

1.2.1.1 Mechanical Exfoliation

Mechanical exfoliation is a convenient method to extract thin carbon layers from bulk graphite through cleavage/peeling. We preferentially refer to it as a mechanochemical method rather than a physical method because the peeling along the crystallographic orientation of graphite produces a significantly different material from the parent starting materials in terms of the resulting properties. Although this method appears to be impractical, its contribution to the development and basic understanding of high-quality graphene is outstanding. The development of mechanical exfoliation can be classified into several stages, which are outlined below.

1.2.1.1.1 Nanomanipulation

Nanomanipulation is possible by using atomic force microscope (AFM) and scanning tunneling microscope (STM). Several groups have investigated nanomanipulation of graphitic sheets on the surface of highly oriented pyrolytic graphite (HOPG), mostly by using AFM and/or STM tips as the operating tools. 3D "Origami" graphene was proposed by Ebbesen and Hiura in 1995 as they used scanning probe microscopy to create folds in graphene layers on the surface of HOPG.[27] AFM and STM tips were used to manipulate, fold, and tear graphitic carbon layers while trying to control the properties of graphitic sheets by tailoring their curvature or geometry. Interestingly, these folds and tears occurred preferentially along the symmetry axes of graphite.[28] Ruoff's group developed this technique greatly and made it more reliable and controllable by the pattern pillar approach (Figure 1.1a and 1.1b).[29] To determine the mechanical strength of graphite in the basal plane, HOPG was patterned lithographically by oxygen plasma etching. Periodic arrays of islands, or holes of several microns on an edge, thin graphene sheets, were obtained on freshly cleaved HOPG surfaces.[29,30] Most importantly, it was demonstrated that such peeled graphite platelets can be transferred to another substrate, such as silica on silicon (SiO$_2$/Si) wafers. On the basis of these studies, it is possible to obtain very thin graphite layer from the surface of HOPG. For example, AFM cantilevers were used to rub the lithographically

(a) 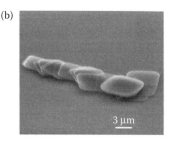 (b) (c)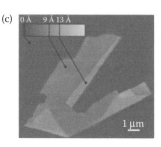

FIGURE 1.1 Mechanical exfoliation of graphite to graphene. (a) SEM of individual HOPG island; (b) stacked thin platelets of graphite; (c) AFM image of graphene from Scotch tape. (From Lu, X. et al., *Nanotechnology*, 10, 269, 1999; Novoselov, K. S. et al., *Proc. Nat. Acad. Sci.*, 102, 10451, 2005. With permission.)

patterned pillar to make thin graphite layers and to study their electron transport properties.[31,32] This way, in earlier studies, graphitic layers with tens and hundreds of graphene layers were produced by the mechanical method.

1.2.1.1.2 Scotch® Tape Method

Scaling up a real yield, Geim's group developed nanofabrication into "macrofabrication" using Scotch tape instead of the sophisticated AFM/STM tips as the manipulation tool (Figure 1.1c).[1,15] Similar to nanomanipulation, HOPG was treated by oxygen plasma etching to create 5 μm deep mesas and these were pressed subsequently into a layer of photoresist, which was baked and the HOPG was cleaved from the resist. Scotch tape was used to repeatedly peel off graphite flakes from the mesas. These thin flakes were then released from the tape by treatment in acetone and transferred to the surface of an SiO_2/Si wafer. Using an SiO_2/Si wafer with a silica layer thickness of 300 nm, sufficient contrast was obtained for visible inspection, which made graphene clearly observable under the optical microscope, and finally identified by AFM. In addition to graphene, a variety of stable 2D crystals with macroscopic areal continuity under ambient conditions, such as single layers of boron nitride (BN), dichalcogenides (MoS_2, $NbSe_2$), and complex oxides ($Bi_2Sr_2CaCu_2Ox$), have been prepared by using micromechanical cleaving.[15]

Large flakes of high-purity and high-quality graphene are crucial for the systematic study of electron transport measurements because it is virtually free from functional surface groups. Up to now, mechanical exfoliation remains an important method for producing single-layer or bilayer graphene with excellent quality for fundamental study. However, this method is not suitable for large-scale production (i.e., several cm²) of single- or few-layer graphene for practical application because of low yield and small size (i.e., several micrometers).

1.2.1.1.3 Ball Milling

Recently, a new method of mechanochemical preparation was reported, which yielded edge-functionalized graphene from graphite.[33] Using this method, a high yield of edge-selectively carboxylated graphite (ECG) was reported by ball milling of pristine graphite in the presence of dry ice. The electrical conductivity of a thermally decarboxylated ECG film was found to be as high as 1214 S/cm. Chen et al.[34] reported that by using a three-roll mill machine with a polymer adhesive, single- and few-layer graphene sheets can be obtained from graphite through continuous mechanical exfoliation. In addition, bulk quantities of graphene nanosheets (only two to five layers thick) and nanodots (diameters in the range of 9–29 nm and heights in the range of 1–16 nm) have been selectively fabricated by mechanical grinding exfoliation of natural graphite in a small quantity of ionic liquids.[35] The resulting graphene sheets and dots are solvent free with low levels of naturally absorbed oxygen, inherited from the starting graphite. The ball milling method can provide simple approaches to low-cost production of graphene nanosheets in large scale for practical applications.

1.2.1.2 Ultrasonication Exfoliation

In contrast to the dry exfoliation of graphite, exfoliation with the assistance of ultrasonication in a liquid phase has also been investigated using various solutions, including organic, aqueous, as well as ionic liquids.

Blake et al. and Hernandez et al. have demonstrated that graphite can be exfoliated in *N*-methyl-pyrrolidone to produce defect-free monolayer graphene, in which the similar surface energy of *N*-methyl-pyrrolidone and graphene facilitates exfoliation.[36,37] Sodium dodecylbenzene sulfonate (SDBS) has been used as surfactant to exfoliate graphite in water.[38] The graphene monolayers are stabilized against reaggregation by a relatively large potential barrier caused by the Coulomb repulsion between surfactant-coated sheets. Similarly, Green and Hersam have used sodium cholate as a surfactant to exfoliate graphite and moved further to isolate the resultant graphene sheets with controlled thickness using density-gradient ultracentrifugation (DGU). The DGU process yielded graphene sheets with thickness that increases as a function of their buoyant density.[20,39]

Ionic liquids can also act as an efficient media for ultrasonic exfoliation of graphite to graphene. In this method, natural graphite flakes were used as starting materials and [Bmim][Tf$_2$N] as the dispersion liquid.[40] Such mixtures were subjected to tip ultrasonication for a total of 60 min using a set of individual cycles of 5–10 min. This method uses a stable suspension of graphene sheets with high concentrations, up to 0.95 mg mL^{-1}, and this dispersion contains sheets with micron-sized edges and stacks of more than 5 graphene layers. The exfoliation of graphite in an ionic liquid in the presence of water and with the assistance of electrochemistry (ionic-liquid electrochemical method) yielded ionic-liquid-functionalized graphite sheets.[41]

1.2.2 Epitaxial Graphene from Silicon Carbide

Graphene has also been synthesized epitaxially on silicon carbide by means of selective extraction of silicon from the carbide crystal lattice.[42] Graphene as a special form of carbide-derived carbon (CDC) was first reported in 1975 by Van Bommel et al., who demonstrated that a carbon layer can be grown on hexagonal silicon carbide in ultrahigh vacuum (UHV) at temperatures above 800°C.[43] However, the electronic properties and the applications potential of carbide-derived graphene had not been considered until 2001.[44] Transport properties, including magnetoresistance measurement, weak electric field effect, and electronic device application of thin carbon layers derived from SiC, were systematically investigated (Figure 1.2) and in the following years, the invention of graphene-based electronics was patented (2003) and published (2004).[45]

At high temperatures (1100–1600°C) and high vacuum (10^{-5}–10^{-10} Torr), silicon will sublimate from the surface of SiC, which allows the Si-depleted surface to rearrange into an epitaxial graphene layer.[46] Generally, the graphene growth rate depends on termination of the SiC crystal. Compared to the C-terminated 0001 face, the growth of a carbon layer on the Si-terminated (0001) face is much slower.

The application of graphene for electronic devices requires very high quality, virtually defect-free graphene in well-defined patterned form. It has been demonstrated that epitaxial graphene films grown on SiC by sublimation in UHV can be patterned using standard microelectronics methods and that the films had 2D electron gas properties.[44] Furthermore, graphene produced on the surface of SiC through UHV annealing requires no transfer before processing devices, which is very important for semiconductor industry. Therefore, the epitaxial graphene on SiC provides the possibility to directly obtain devices for electronic applications. In contrast, graphene formed through conventional methods was often too rich in defects and yielded low mobilities.[46] It should be noted that SiC wafers, because of the high costs and a higher defect concentration at the SiO$_2$/SiC interface, are less attractive for electronic applications and device manufacturing compared to well-established silicon wafers. Controllable sublimation of Si is critical for the quality of graphene. By encapsulating SiC crystals in graphite, de Heer et al.[44] increased the vapor pressure of silicon in the reaction system and consequently decreased the evaporation rate of silicon to make the sublimation process take place

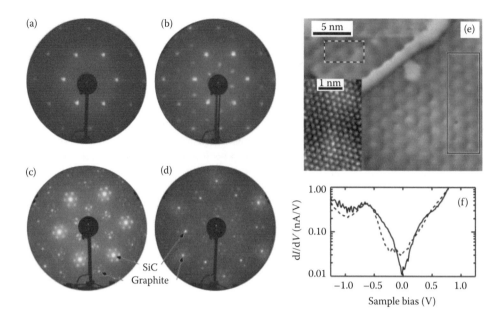

FIGURE 1.2 (a–d) Low-energy electron diffraction (LEED) patterns from graphene on SiC(0001) obtained at different temperatures. (a) 1050°C for 10 min, (b) 1100°C, 3 min, (c) 1250°C, 20 min, (d) 1400°C, 8 min, (e) STM image of a surface region of the sample described in Figure 1d, (f) dI/dV spectra (log scale) acquired from the regions marked with corresponding line types in the image at the top. (With permission from Berger, C. et al., *J. Phys. Chem. B*, 108, 19912, Copyright 2004, American Chemical Society.)

close to equilibrium. This method substantially improves the quality of graphene on both polar faces of the SiC crystals by increasing the growth temperature and decreasing the Si sublimation rate.

The growth of epitaxial graphene with different layers on single-crystal silicon carbide substrates has also been accomplished. This was accomplished on a single crystal that was treated by oxidation or H$_2$ etching to improve surface quality. After removal of the oxide layer by electron bombardment in UHV at 1000°C, the samples were heated to 1250–1450°C, resulting in the formation of thin graphitic layers. This approach offers the advantage that high-quality layers can be grown on large area substrates.[44]

1.2.3 CHEMICAL VAPOR DEPOSITION

CVD growth of graphite layers on the surface of Pt (100), (111), and (110) by thermal decomposition of C$_2$H$_2$ and C$_2$H$_4$ was demonstrated by John May as early as in 1969,[7] which most probably is the first report about CVD growth of multilayer graphene. Later, Jack Blakely et al. systematically studied the surface growth of monolayer or multilayer graphite on the crystalline faces of different transition metals, such as Ni (111), Co (0001), Pd (100), and Pd (111).[8] However, the transfer of graphene from the grown metal substrate and transport measurement has not been investigated during these initial studies.

More recently, hydrocarbons, such as methane, ethylene, acetylene, and benzene, have been used as carbon source for the growth of graphene on various transition metal substrates. Generally, the growth of graphene monolayers is achieved on single crystals of transition metals, such as Co, Pt, Ir, Ru, and Ni, under low pressure or UHV conditions. These requirements, however, strongly limit the use of this method for large-scale applications. To overcome these limitations, Kong et al. reported a low-cost and scalable technique for the synthesis of single- to few-layer graphene films on polycrystalline Ni film at ambient pressure and transferred the graphene produced to alternative substrates (Figure 1.3).[47] The typical fabrication procedure starts with the evaporation of a

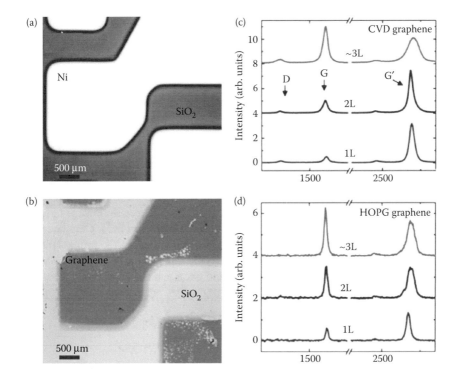

FIGURE 1.3 (a) Optical image of a prepatterned Ni film on SiO$_2$/Si. (b) Optical image of the grown graphene transferred from the Ni surface in panel (a) to another SiO$_2$/Si substrate. (c) Raman spectra of 1, 2, and ≈3 graphene layers from a CVD graphene film on SiO$_2$/Si. (d) Raman spectra of 1, 2, and 3 graphene layers derived by microcleaving of HOPG for comparison. (With permission from Reina, A. et al., *Nano Lett.*, 9, 30, Copyright 2009, American Chemical Society.)

polycrystalline Ni film on an SiO$_2$/Si substrate, subsequently high-temperature annealing of the substrate generates an Ni-film microstructure with single-crystalline grains of sizes between 1 and 20 μm. The Ni-coated substrates are then subjected to CVD growth at 900–1000°C in flowing and highly dilute hydrocarbon (CH$_4$/H$_2$). The graphene film formed on the substrate was released by protecting the graphene film with poly(methyl methacrylate) (PMMA) and etching the underlying Ni with HCl aqueous solution (≈3% vol.). Thereafter, the film was transferred to a substrate for analysis, characterization, and application. The films are continuous over the entire area and can be patterned lithographically or by prepatterning the underlying catalytic Ni film. The optical transmittance is approximately 90% in the 500–1000 nm wavelength regime for a film having 3 nm average thickness, and the resistances of the films are 770–1000 Ω/sq. The high intrinsic quality of CVD graphene films makes them excellent candidates for both optoelectronic and electronic applications.

Similarly, Hong's group reported the growth of graphene patterns on thin nickel layers via CVD and subsequent transferring to arbitrary substrates.[48] Hong's group used an electron beam evaporation method to deposit thin layers of nickel with less than 300 nm thickness on SiO$_2$/Si substrates. After reaction with a flowing gas mixture (CH$_4$, Ar, and H$_2$) at 1000°C, the samples were rapidly cooled to room temperature (25°C) in argon. The monolayers transferred to silicon dioxide substrates showed high electron mobility, greater than 3700 cm^2/V s, and exhibited the half-integer quantum Hall effect. They also demonstrated the macroscopic use of these graphene films as highly conducting and transparent electrodes in flexible, stretchable, and foldable electronics. Commonly, nickel can be etched by strong acid, which often produces hydrogen bubbles and damages graphene.

In this work, Fe^{3+} was used as the etchant to release graphene grown on the surface of substrates via the following reactions:

$$2Fe^{3+}(aq) + Ni(s) \rightarrow 2Fe^{2+}(aq) + Ni^{2+}(aq)$$

The large amount of carbon absorbed on the reactive nickel foils readily leads to thick graphite crystals rather than monoatomically thin graphene films. By varying the thickness of the metal and the temperature-dependent solubility of carbon atoms dissolved in the metal, the thickness of the graphite film can be controlled.[49] However, graphene grown on Ni seems to be limited by its small grain size, the presence of multilayers at the grain boundaries, and high solubility of carbon (≈ 0.1 at.% at 900°C). By taking advantage of low carbon solubility in copper (≈ 0.001 at.%, at 900°C), Ruoff's group developed a large-scale graphene CVD growth process on copper foils (Figure 1.4).[50] The low solubility of carbon in copper appears to help make this growth process self-limiting, and in agreement, the two- and three-layer flakes do not grow larger with time. Compared to graphene growth on Ni, the films grow directly on the surface by a surface-catalyzed process rather than precipitation process. As a result, single-layer graphene, with only a small coverage of few layers, was formed predominantly across copper surface steps and grain boundaries. The graphene film formed on copper was released by iron etching of the copper substrate and transferred to another substrate:

$$2Fe^{3+}(aq) + Cu(s) \rightarrow 2Fe^{2+}(aq) + Cu^{2+}(aq)$$

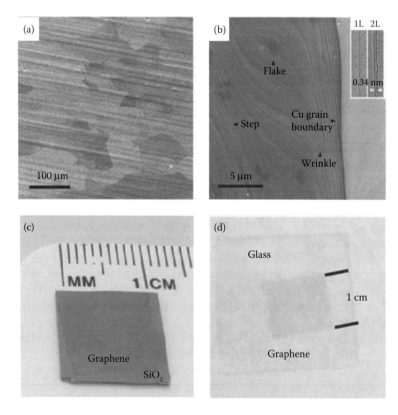

FIGURE 1.4 (a) SEM image of graphene on a copper foil with a growth time of 30 min. (b) High-resolution SEM image showing a Cu grain boundary and steps, two- and three-layer graphene flakes, and graphene wrinkles. (c and d) Graphene films transferred on to an SiO_2/Si substrate and a glass plate, respectively. (From Li, X. et al., *Science*, 324, 1312, 2009. With permission of AAAS.)

Using this method, dual-gated field-effect transistors were fabricated on SiO_2/Si substrates and an electron mobility as high as 4050 cm²/V s with the residual carrier concentration at Dirac point of $n_0 = 3.2 \times 10^{11}$ cm⁻² at room temperature was achieved.

On the basis of the work of the Ruoff group, graphene layers with a diagonal length of up to 30 in. were produced and transferred to transparent flexible substrates with a time- and cost-effective roll-to-roll process (Figure 1.5).[51] In this process, graphene is first grown by CVD on a roll of copper foil. A thin thermal release tape polymer film support is then attached to graphene by passing between two rollers. Thereafter, graphene is released by chemical etching of the copper substrate and transferred to the thin polymer film. Finally, graphene mounted on thermal release tape (polymer film) and the target substrate is inserted between rolls, and, under heat treatment, the detaching graphene film is attached to a target substrate. By repeating these steps on the same substrate, multilayered graphene films can be prepared to obtain enhanced electrical and optical properties. By using the layer-by-layer stacking process, a doped four-layer film was fabricated which gave a sheet resistance as low as ≈30 Ω/sq at ≈90% transparency. The low resistance and high optical transmittance of the graphene films make them promising candidates to replace the commercial transparent electrode indium tin oxide (ITO), which are currently used in flat panel displays and touch screens.

The limitations in a real sample size and material quality are the main bottlenecks in building graphene-based electronic devices currently, and strategies to overcome these limitations are extensively being studied. For example, the fabrication of single-crystalline graphene monolayer as large as a few millimeters with good continuity and perfect crystallinity was reported via thermal

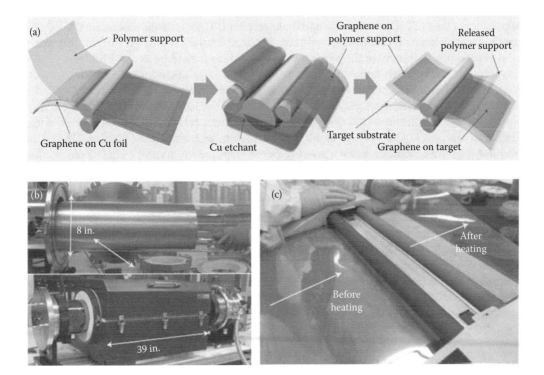

FIGURE 1.5 (a) Schematic illustration of the roll-based production of graphene films grown on a copper foil. The process includes adhesion of polymer supports, copper etching (rinsing), and dry transfer-printing on a target substrate. Wet-chemical doping can be carried out using a set-up similar to that used for etching. (b) Copper foil wrapping around a 7.5-in. quartz tube to be inserted into an 8-in. quartz reactor. The lower image shows the stage in which the copper foil reacts with CH_4 and H_2 gases at high temperatures. (c) Roll-to-roll transfer of graphene films from a thermal release tape to a PET film at 120°C. (With permission from Macmillan Publishers Ltd., *Nat. Nanotechnol.*, Bae, S. et al., 5, 574, Copyright 2010.)

annealing of a ruthenium single crystal (Ru (0001)) containing carbon. Analysis of Moire patterns supported by first-principles calculations shows that the graphene layer is incommensurate with the underlying Ru (0001) surface, forming an $N \times N$ superlattice with an average lattice strain of +0.81%.[53] Also, the growth of millimeter-sized hexagonal single-crystal graphene films was reported on Pt by ambient pressure CVD at low temperature.[54] It was demonstrated that the size of hexagonal single crystalline graphene grain grown on the Pt substrate was strongly dependent on the concentration of CH_4 in CH_4/H_2: lower the CH_4 concentration, larger the hexagonal graphene grain size. After the CVD process, PMMA-coated graphene/Pt was used as a cathode with a Pt foil as the anode in a water electrolysis process. Owing to the formation of H_2 bubbles, graphene was readily detached from the Pt surface. This nondestructive transfer enables the repeated use of the Pt substrate, and graphene obtained from this method shows a high crystalline quality with a carrier mobility greater than 7100 cm^2/V s.

In addition to flat metal surfaces, graphene can also grow on the curved surface of metals. This enables novel ways to design and nanoengineer more complex shapes of graphene materials. Cheng et al. reported the direct synthesis of 3D graphene foams by template-directed CVD.[55] Using commercial Ni foam as the substrate and methane as carbon source, a 3D, flexible, electrically conductive, and highly interconnected graphene network was produced on the Ni foam. Graphene foam/poly(dimethyl siloxane) composites show a very high electrical conductivity of 10 S/cm with graphene foam loading as low as 0.5 wt%. This illustrated the great potential of such composites for flexible, foldable, and stretchable conductors.

PECVD offers graphene synthesis at a lower temperature compared to thermal CVD.[56] The advantages of plasma deposition include very short deposition time (<5 min) and a low growth temperature of only 650°C compared to approximately 1000°C for conventional thermal CVD. The growth mechanism involved a balance between graphene deposition through surface diffusion of carbon-containing species from the precursor gas and etching caused by atomic hydrogen.

It was demonstrated that large area, high-quality graphene with controllable thickness, can be grown from different solid carbon sources, such as polymer films or small molecules deposited on a metal catalyst substrate at temperatures as low as 800°C.[57] The solid carbon source used was a spin-coated PMMA thin film (100 nm), and the metal catalyst substrate was a Cu film. At a temperature as low as 800°C or as high as 1000°C for 10 min, with a reductive gas flow (H_2/Ar) and under low-pressure conditions, a uniform monolayer of graphene was formed on the substrate. Much less expensive carbon sources, such as biomaterials (food, insects, and waste), can be used without purification to grow high-quality monolayer graphene directly on the backside of Cu foils under a flowing H_2/Ar atmosphere.[58] Furthermore, Eduardo Ruiz-Hitzky et al. have prepared graphene-like materials supported on porous solids using natural resources,[59] such as sucrose disaccharide (table sugar) and gelatin, and the synthesis was carried out in the absence of oxygen at relatively low temperatures (<800°C). After carbon deposition, the inorganic host was eliminated using an acid treatment, the remaining well-stacked carbon layers showed good electrical conductivity.

In summary, compared with other methods, CVD can produce graphene with high quality, in large scale, and cost-effectively. By using different catalysts, carbon sources, and integration of novel techniques to production, the number of layers, quality, and areas of graphene can be controlled. Graphene from CVD process shows great potential in the applications of electronics and solar cells.

1.2.4 CHEMICALLY DERIVED GRAPHENE

Chemically derived graphene (CDG) from graphite oxide (GO) has become one of the most promising ways for the production of graphene on a large scale.[60] GO was generally produced by the Hummers' method, in which concentrated sulfuric acid, sodium nitrate, and potassium permanganate were used as oxidants to oxidize graphite.[61] The obtained GO distributed hydroxyl and epoxy groups on the basal plane (Figure 1.6),[62] whereas carbonyl and carboxyl groups were present at the sheet edges. The approaches for removal of these carbon-containing groups to transform GO

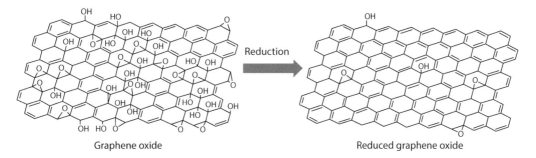

Graphene oxide   Reduced graphene oxide

FIGURE 1.6 Reduction of graphene oxide (Lerf–Klinowski model). (From *Chem. Phys. Lett.*, 287, He, H. et al., 53, Copyright 1998, with permission from Elsevier.)

to graphene have been investigated intensively. The interlayer distance varies and increases from 0.335 nm of pristine graphite to 0.4–0.7 nm in GO. Owing to the abundance of oxygen-containing groups, GO is hydrophilic colloid dispersion in water and is readily exfoliated to form graphene oxide by mild ultrasonication. This is because of the strength of interactions between water and the oxygen-containing functional groups introduced into the basal plane during oxidation. The formation of stable graphene oxide in water was attributed to not only its hydrophilicity but also electrostatic repulsion.[63] However, graphene oxide can easily restack to GO once the water is eliminated. Removal of the oxygen-containing functional groups from graphene oxide surfaces leads to the formation of graphene. Various approaches have been used for GO deoxygenation, such as thermal reduction, chemical reduction, and so on.

1.2.4.1 Thermal Expansions

By rapidly heating stacked GO to high temperature, oxygen functional groups were extruded as carbon dioxide.[64,65] Exfoliation took place when the decomposition rate of the oxygen-containing groups exceeded the diffusion rate of the formed gases; therefore, the yielding pressure will overcome the van der Waals forces holding the graphene sheets together. It was estimated that 2.5 MPa was required to separate GO sheets. Thermal reduction can efficiently remove the oxygen groups and produce approximately 80% single-layer CDG according to the AFM studies, whereas approximately 30% mass loss, abundant vacancies, and structural defects were left behind in the formed CDG, which may affect the mechanical and electrical properties of graphene. The bulk conductivities of the CDG formed were greatly increased to 1000–2300 S/m, indicating the effective reduction and restoration of electronic structures from GO.

The reduction degree of GO is significantly affected by heating temperature and heating rate. It has been estimated that 550°C is the minimum temperature for exfoliation of GO, however, it must be more than 1000°C for full exfoliation of GO to single- or few-layered graphene with a high conductivity. Li et al. have investigated the effects of annealing temperature on chemical structure variation of GO with XPS spectrum, which reveals that high temperature is necessary to obtain high-quality graphene.[66] Using the ultra-high temperatures (>2000°C) yielded during arc discharge, Cheng's group developed a hydrogen arc discharge exfoliation method for the synthesis of graphene sheets with excellent electrical conductivity and good thermal stability from GO.[67] The graphene obtained by hydrogen arc discharge exfoliation exhibits a high electrical conductivity of 2×10^3 S/cm and high thermal stability. Elemental analysis revealed that the exfoliated graphene sheets had a C/O ratio of 15–18.

The annealing atmosphere is one of the most important factors for the reduction of GO. The reduction of GO can be realized at a relatively low temperature when reducing gases, such as H_2 and NH_3, is present in the atmosphere.[68] Wu et al. reported that GO can be well reduced at 450°C for 2 h in an Ar/H_2 (1:1) mixture with a resulting C:O ratio of ~15:1 and conductivity of about 1×10^3 S/cm.

When GO is annealed in the presence of NH_3, simultaneously, reduction and N-doping can take place.[65] Electrical measurements of individual GO sheet devices demonstrate that GO annealed in NH_3 exhibits higher conductivity than those annealed in H_2, suggesting more effective reduction of GO by annealing in NH_3 than in H_2.

Moreover, annealing pressure can also affect thermal expansion significantly. Expansion and exfoliation of GO may occur at low temperature, but the graphitic layers are not completely exfoliated. Yang et al. employed a low-temperature (as low as 200°C) exfoliation approach to produce graphene in a high vacuum environment (below 1 Pa). The graphene materials obtained by this method have a moderately high specific surface area (400 m^2/g) and a high electrochemical capacitance of up to 264 F/g (for the tenth cycle in 6 M KOH electrolyte) without any posttreatments.[69]

1.2.4.2 Microwave Irradiation-Assisted Exfoliation

Microwave irradiation (MWI) has been widely used in the field of organic synthesis,[70] environmental remediation,[71] and preparation of catalysts and activated carbon[72] because of its advantages of heating substances uniformly and rapidly. Recently, microwave techniques were used for the preparation of graphene in different solvents, and they were also used to produce solvent-free graphene sheets by treating the GO precursor in a microwave oven for less than 1 min.[73] MWI-assisted exfoliation has provided a straightforward method to generate exfoliated GO and the reduced GO has a moderately high specific surface area and is electrically conductive. In detail, the role of MWI during GO deoxygenation process was studied by Hu et al. By manipulating the oxygen content in GO and/or graphene-based materials, it is demonstrated that the microwave absorption capacity of carbon materials is highly dependent on their chemical composition and structure. The increase of oxygen in GO decreases its microwave absorption capacity significantly because of the size decrease of the π–π conjugated structure in these materials, and vice versa. It was revealed that graphene is an excellent microwave absorbent, whereas GO has poor microwave absorption capacity; the not-oxidized graphitic region "impurities" in GO act as microwave absorbents to initiate microwave-induced deoxygenation. The addition of a small amount of graphene to GO leads to avalanche-like deoxygenation reaction of GO under MWI and graphene formation, which was used for electrode materials in supercapacitors. The interaction between microwaves and graphene or graphene-based materials may be beneficial for the fabrication of a variety of graphene-based nanocomposites with exceptional properties and several practical applications.

1.2.4.3 Plasma-Assisted Exfoliation

Controllable production of graphene by simultaneous exfoliation and reduction of GO under dielectric barrier discharge (DBD) plasma at atmospheric pressure with various working gases, including H_2 (reducing), Ar (inert), and CO_2 (oxidizing), has been investigated.[75] Using the plasma method, the majority of oxygen-containing functional groups on the surface of GO are removed and oxygen content of the products are manipulated by selecting different working gas and plasma treatment times. Sankaran et al.[76] reported a remote high-pressure plasma process for GO reduction. In this process, hydrogen was continuously dissociated in a microplasma to produce atomic hydrogen and carried by gas flow to react with and remove oxygen functional groups from GO films at low temperatures (<150°C) via radical-assisted chemistry. The low temperature allows GO to be directly reduced on polymeric substrates, such as polyethylene terephthalate (PET) to obtain highly conductive films. The unique combination of low temperature, atmospheric pressure, and high purity suggests that this nonequilibrium chemical approach can be used for roll-to-roll processing of GO films for large-scale, flexible conductor applications.

1.2.4.4 Photothermal Reduction

Flash reduction of free-standing GO films can be achieved with a single, close-up (<1 cm) flash from a xenon lamp. The photo energy emitted by the flash lamp at a close distance (<2 mm: 1 J/cm^2) can provide nine times the thermal energy needed for heating GO (thickness 1 μm) to more than 100°C.[77] The GO films typically expand tens of times after flash reduction because of rapid degassing, and the electrical conductivity of the expanded film is around 10 S/cm using its maximum expanded thickness in the calculation. The advantage of this method compared with others is that reduced GO (rGO) patterns can be easily fabricated with photomasks, which facilitates the direct fabrication of electronic devices based on rGO films. Scott Gilje et al.[78] reported that the photothermally initiated deflagration of GO can take place even in an oxygen-deficient environment without the presence of catalysts. They proposed GO may act as additive to rocket fuels to attain distributed fuel ignition.

1.2.4.5 Chemical Reduction of GO

Chemical reduction of graphene oxide sheets has been performed with different reducing agents. Hydrazine hydrate was widely used to produce very thin and fine graphite-like sheets. In 2007, Stankovich et al. demonstrated a solution-based process for producing single-layer graphene with hydrazine hydrate as a reductant (Figure 1.7).[79] It can be understood that the removal of oxygen groups makes the reduced sheets less hydrophilic and quickly aggregate in solution. Dan Li et al. reported the preparation of water-soluble graphene by raising the pH during reduction which leads to charge-stabilized graphene colloidal dispersions.[80] Sodium borohydride ($NaBH_4$) is another effective chemical reducing agent for the reduction of GO in aqueous solution.[81] The $NaBH_4$ treatment can eliminate all the parent oxygen-containing groups and the resultant solid becomes infrared inactive like pure graphite. Such reduction produced graphene with sheet resistances as low as 59 kΩ/sq (compared to 780 kΩ/sq for a hydrazine reduced sample), and C:O ratios were as high as 13.4:1 (compared to 6.2:1 for hydrazine). High conductivity and high flexibility are necessary for potential application of graphene in transformation conductors. In another study, an effective hydrohalic acid reducing method was applied to make the assembled graphene-based conductive film without destroying their integrity and flexibility at low temperature based on the nucleophilic substitution reaction.[82,83] The reduction maintains good integrity and flexibility, and even improves strength and ductility, of the original GO films. Based on this reducing method, a flexible graphene-based transparent conductive film with a sheet resistance of 1.6 kΩ/sq and 85% transparency was obtained.

1.2.4.6 Green Reduction Agents

Generally, the reductants used in the chemical reduction of GO are highly toxic and cause environmental concerns. As a consequence, using a nontoxic, environmentally friendly, and effective reductant for chemically producing graphene has been studied. Loh et al.[84] presented a hydrothermal dehydration route to convert GO into stable graphene solution. Compared to chemical reduction processes using hydrazine, the "water-only" route has the combined advantage of removing oxygen functional groups from GO and repairing the aromatic structures. Zhang et al. reported the reduction of GO with vitamin C as the reductant and amino acid as the stabilizer.[85] To avoid the

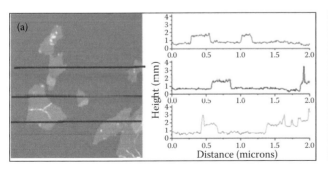
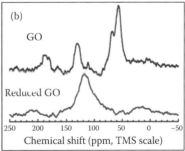

FIGURE 1.7 (a) A noncontact mode AFM image of exfoliated GO sheets with three height profiles acquired in different locations. (b) Solid-state ^{13}C MAS NMR spectra of GO (top) and reduced exfoliated GO (bottom). (From *Carbon*, 45, Stankovich, S. et al., 1558, Copyright 2007, with permission from Elsevier.)

agglomeration and precipitation of the resulting graphene sheets, L-tryptophan was selected as the stabilizer. Ruoff et al.[86] reported a similar method to exfoliate GO into graphene using aqueous vitamin C as a mild and green reducing agent with a small amount of polyethylene glycol. The resulting film has electrical conductivity properties ($\sigma \approx 15$ S/cm) and has fewer defects compared to rGO films obtained by using hydrazine reduction.

Various metals (mainly aluminum, iron, and zinc powder) have also been employed as reductants to reduce GO. Fan et al.[87,88] reported a new green route for the reduction of GO by aluminum and iron powder in the presence of HCl (35%) medium. The bulk conductivity of the prepared graphene sheets is 2.1×10^3 S/m, which is only one order of magnitude lower than that of pristine graphite ($\approx 3.2 \times 10^4$ S/m). Wang et al.[89] employed zinc powder in mild alkaline conditions to synthesize graphene from GO. This method can effectively remove a significant fraction of the oxygen-containing functional groups and yield a C/O ratio as high as 8.09 and 8.59 reduced for 10 min and 60 min under ultrasonication, respectively.

As a typical and traditional material synthesis method, solvothermal and hydrothermal methods have also been used for the fabrication of graphene and graphene-based composites. Hou et al.[90] used expanded graphite as the starting material to prepare monolayer and bilayer graphene sheets in a highly polar acetonitrile solvent by solvothermal-assisted exfoliation. The resulting graphene sheets are of high quality without any significant structure defects demonstrated by electron diffraction and Raman spectroscopy. It was shown that graphene sheets can also be made directly from natural graphite flakes, intercalated by oleum and tetrabutylammonium cations, and reduced in N,N-dimethylformamide by solvothermal method. This method can decrease the number of defects and oxygen content in graphene with a low resistivity.[91] Stride et al.[92] developed a gram-scale production of graphene-like carbon film by a bottom-up approach based on the common laboratory reagents ethanol and sodium. The ability to produce bulk graphene samples from nongraphitic precursors with a scalable, low-cost approach allows real application of graphene.

Exfoliation of GO is the most promising approach for low-cost and scalable production of graphene. Both GO and rGO can be solution processable which facilitates fabrication. However, GO-derived graphene has abundant defects on its surface resulting from the harsh oxidation process, which degrades its physical properties. Generally, the mechanical and electrical properties of GO-derived graphene cannot compete with those of pristine graphene. To obtain graphene with high quality, deep oxidation of graphite to GO should be avoided, and mild oxidation combined with other techniques for exfoliation of graphite should be used.

1.2.5 GRAPHENE NANORIBBONS

Thinning and slicing graphene into narrow strips results in graphene nanoribbons (GNRs). 2D graphene generally exhibits semimetallic behavior; however, GNRs can generally be either metallic or semiconducting depending on the width as well as the topology of their edges.[93] For example, if the width of nanoribbons is less than 10 nm, they will show semiconductor behavior regardless of their edge patterns.[94] Therefore, graphene nanoribbons (width <10 nm) are promising candidates for the fabrication of nanoscale electronic devices because they exhibit semimetallic behavior, quantum confinement, and edge effects. For larger widths, properties of the nanoribbon strongly depend on the shape of the edge. Zigzag-shaped edges are metallic, whereas armchair-shaped edges could be semiconductive or semimetallic depending on their width. In addition, under the effect of an electric field or with the adsorption of polar molecules at the edges, zigzag nanoribbons become semimetallic, thus the chemistry at nanoribbon edges is very interesting and could be exploited in the fabrication of highly sensitive sensors.[95,96] Various routes to the graphene nanoribbons have been established, such as cutting graphene sheets into ribbons, sonochemical, lithographic methods, and through unzipping of CNTs; however, reliable production of graphene nanoribbons with atomic-scale precision remains a big challenge.

1.2.5.1 Direct CVD Method

Edges of GNRs have highly reactive sites; therefore, modulation of electrical, chemical, and magnetic properties of GNRs could be achieved by edge chemistry. The growth of ribbon-like carbon materials (10 μm long, 0.1–0.7 μm wide, 10–200 nm thick) with CVD was first reported by Murayama and Maeda in 1990, with Fe(CO)$_5$ and CO as catalyst precursor and carbon source, respectively.[97] It was found that the annealing treatment of ribbon-like filaments at 2800°C resulted in the formation of loop structures between open edges. The nanoribbons are highly ordered materials in which the graphite layers are oriented perpendicular to their large surface. In 2008, Torrones et al. reported the large-scale production of highly crystalline graphitic nanoribbons (several micrometers long, 20–300 nm wide, and 15 nm thick) with ferrocene, ethanol, and thiophene as the starting materials under CVD conditions.[98] However, the catalytic formation mechanism of the nanoribbons was not clarified. Similarly, Subramanyam[99] reported the fabrication of crystalline carbon nanoribbons by pyrolysis of ferrocene and tetrahydrofuran (THF). Using ZnS nanoribbons as template, Liu's group fabricated carbon nanoribbons (≈3–4 nm thick, 0.5–5 μm wide) by CVD.[100]

1.2.5.2 Nanocutting

Contrary to the CVD process, nanocutting of graphene is a versatile and efficient method for the fabrication of GNRs. Graphene nanoribbon devices have been fabricated and tested by combining electron-beam lithography with plasma etching.[101] By means of catalytic hydrogenation, catalytic nanocutting of graphene has been reported using nickel or iron nanoparticles in the presence of hydrogen at high temperatures. Datta et al. demonstrated that few-layer graphene samples can be etched along crystallographic axes by thermally activated Fe nanoparticles, and crystallographic edges of more than 1 μm were produced by catalytic etching of graphene by hydrogen.[102] Using the single-particle etching technique, GNR with width of 15 nm and lengths on the order of micrometers was fabricated. Similarly, Kampos et al. used Ni nanoparticles to etch single-layer graphene through catalytic hydrogenation of carbon at high temperature, where carbon atoms from exposed graphene edges dissociate into the Ni nanoparticle, and then react with H$_2$ at the Ni surface (C + 2H$_2$ → CH$_4$).[103] Ci et al. demonstrated a multistage cutting technique that is able to produce specific well-defined shapes of graphene on the surface of HOPG.[104]

A simple chemical route to produce GNRs with width below 10 nm employs commercially expandable graphite, which is first exfoliated by thermal treatment at 1000°C for 60 s, and subsequently the resulting exfoliated graphite was dispersed in a 1,2-dichloroethane (DCE) solution of poly(m-phenylenevinylene-co-2,5-dioctoxy-p-phenylenevinylene) (PmPV) by assistance of sonication. During this process, noncovalent polymer functionalization renders the GNRs stably suspended in solvents to form a homogeneous suspension.[105] After centrifugation, GNRs with ultra smooth edges and possibly well-defined zigzag or armchair-edge structures were obtained in the supernatant. They systematically studied the sub-10-nm-wide graphene nanoribbon field-effect transistors (GNRFETs) and showed the possibility of producing all-semiconducting devices with the sub-10 nm GNRFETs. As predicted by theory, electrical transport experiments demonstrated that all the GNRs with width less than 10 nm afforded semiconducting FETs without exception, with I_{on}/I_{off} ratio up to 10^6 and on-state current density as high as ≈2000 A/m. It was estimated that carrier mobility is about 200 cm^2/V s and scattering mean free path about 10 nm in sub-10 nm GNRs.

1.2.5.3 Unzipping of CNTs

The preparation of GNRs from unzipping of CNTs has been investigated intensively by several groups and various routes have been developed (Figure 1.8).[106]

a. *Wet chemical method.* In the case of wet chemical method, CNTs were suspended in concentrated sulfuric acid and treated with KMnO$_4$.[107] The CNTs were then cut longitudinally by oxidation, similar to the case of GOs, and the resulting nanoribbons possessed oxygen-containing groups, such as carbonyls, carboxyls, and hydroxyls, at the edges and on the

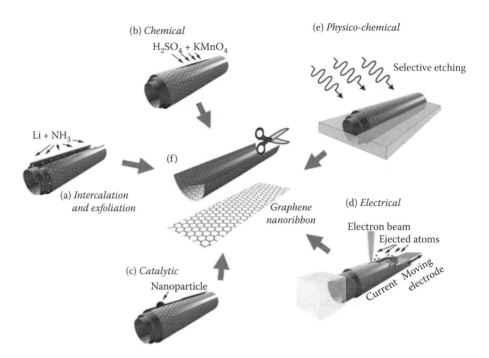

FIGURE 1.8 Different ways to unzip CNTs for production of graphene nanoribbons (GNRs): (a) Intercalation exfoliation; (b) chemical oxidation; (c) catalytic approach; (d) the electrical method; and (f) selective etching by Ar plasma. (From *Nano Today*, 5, Terrones, M. et al., 351, Copyright 2010, with permission from Elsevier.)

surface. The nanoribbons obtained have oxidized edges, making them highly soluble in polar solvents; however, the *p*-conjugated network was disrupted by surface oxidation which made the nanoribbons poorly conductive. Chemical reduction of the nanoribbons restored the conjugation and part of the conductivity. Electrical conductivity of the reduced nanoribbons is not so high as that of the original CNTs because of incomplete reduction and the existence of graphene islands interspersed with regions of tetrahedral sp^3-hybridized carbon atoms. It has been known that carboxyl groups at the edges cannot be eliminated by N_2H_4, and that the remaining carboxyl groups can disrupt the π–π network. With the decrease of oxygen-containing groups in the reduction process, the aggregation of the nanoribbons takes place due to the increase of π–π stacking. To prevent the aggregation of reduced nanoribbons, ammonia was used along with N_2H_4 during reduction process. Using a similar procedure, single-walled carbon nanotubes (SWCNTs) were also used to produce narrow single-layer nanoribbons; however, it was shown that this procedure only provides entangled materials. It was also found that it is possible to obtain highly crystalline graphene nanoribbons exhibiting extremely high electrical conductivities. The method involves efficient chemical oxidation at 60°C, in the presence of a second acid ($C_2HF_3O_2$ or H_3PO_4) besides the H_2SO_4–$KMnO_4$ mixture. Tour et al. demonstrated that GNRs produced by chemical unzipping of CNTs could be conveniently used from solution to hand-paint unidirectional arrays of GNRs atop silicon oxide. Numerous GNR-based devices, including field effect transistors, sensors, and memory storage devices, can be easily fabricated on a single chip. Similarly, a chemical method for unzipping CNTs by oxidation of the CNTs in air was demonstrated.[108] The oxidized CNTs were then subjected to sonication in a DCE-PmPV solution. High yield of unzipped tubes was obtained; furthermore, the edges of the resulting nanoribbons appear to be smooth and produced GNRs that exhibit high conductance.

b. *Selective etching methods.* The resulting graphitic nanoribbons from chemical oxidation are relatively wide and do not exhibit atomically smooth edges because severe acid
treatments required in these techniques damage the edges owing to heavy chemical functionalization. Jiao et al.[109] demonstrated selective etching technique to produce narrow
nanoribbons by assistance of plasma (Figure 1.9). In this method, arc-grown MWCNTs are
partially embedded in a polymer film and are etched selectively with Ar plasma. The film
is then removed using solvent vapor followed by heat treatment of the resulting nanoribbons to remove any residue. This technique can create very narrow graphene nanoribbons
(<10 nm) which exhibit semiconductor behavior. Addition of argon, hydrogen, and ammonia to oxygen gas can also promote isotropic etching of graphene predominantly from the
edges without damaging the basal plane.[110]

c. *Intercalation exfoliation.* CNTs were intercalated by lithium in liquid NH_3 after removal
of tube caps with oxidation and finally exfoliation with heat treatment.[111] In this method,
after dispersion of CNTs in dry tetrahydrofuran by sonication, the dispersion was added
into liquid NH_3, and Li was then added into the liquid in a 10:1 Li:C weight ratio. The
intercalation was carried out during a period of a few hours. After evaporation of NH_3,
a 10% HCl solution was stirred into the flask containing intercalated MWCNTs and then
recovered with a microfiltration membrane. This process can produce small amounts of
exfoliated MWCNTs (≈0–5%), however, mostly resulting in partially exfoliated or structurally damaged nanotubes.

d. *Nanoparticles as nanoscalpels.* Metal nanoparticles (Co, Ni) can act as catalysts for hydroge-
nation/oxidation of carbon to etch the nanotube longitudinally like nanoscalpels and Co or Ni
nanoparticles can be deposited on the surface of CNTs via a solution process or magnetron
sputtering. The catalytic hydrogenation of carbon was performed at 850°C for 30 min, in
which carbon atoms react with H_2 to form methane (CH_4) and leave the CNTs. The nanoribbons produced by this method are typically 15–40 nm wide and 100–500 nm long. Datta
explained how a droplet can etch graphene with a simple model and two factors possibly
contribute to this process: a difference between the equilibrium wettability of graphene and
the substrate that supports it, or the large surface energy associated with the graphene edge.[112]

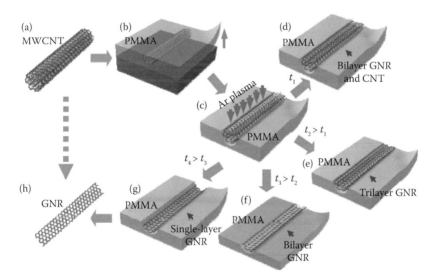

FIGURE 1.9 (a) Pristine MWCNT, (b) MWCNT coated with a PMMA film, (c) the PMMA–MWCNT film
was peeled from the substrate, turned over, and exposed to an Ar plasma etching, (d–g) possible products
generated by controlling the etching time. (With permission from Macmillan Publishers Ltd., *Nature*, Jiao,
L. Y. et al., 458, 877, Copyright 2009.)

e. *Electrical method.* By passing high electrical current, Zettl et al.[113] found that MWCNTs could be easily unwrapped to form GNRs. By using movable electrodes and controlling the voltage bias, the outer wall of MWCNT is severed to form GNR attached to the remaining inner core. GNR formed in this manner is then removed from the MWCNT by sliding between the GNR and the MWCNT inner core. The GNR can be completely removed from the MWCNT or partly removed by controlling the sliding process to achieve a preselected length of GNR fully suspended in vacuum with each end electrically and mechanically attached to a conducting electrode.

At first sight, the unzipping of CNTs seems to be an efficient and viable method to produce graphene nanoribbons in large scale because CNTs can already be produced in quantities of tons per year. However, the harsh oxidation conditions used in the process strongly degrade the quality of the resulting nanoribbons and consequently limit their applications because even minor deviations from the ideal shape of the edges significantly deteriorate the material properties. In fact, there is a close fundamental connection between the electrical performance and their structures, such as the width, the edge periphery, and the occurrence of the defects. Therefore, reliable production of graphene nanoribbons with atomic precision has remained a big challenge up to now. Bottom-up organic synthetic protocols seem to be viable routes to precise production of GNRs for electronics application.

1.2.5.4 Total Organic Synthesis

As the pioneers in organic synthesis of graphene, the Müllen Group made a great breakthrough in solution synthesis of GNRs with a width of about 1 nm and a length of up to 10 nm. On the basis of the design of polyphenylene precursors, various corresponding polycyclic aromatic hydrocarbons (PAHs) have been obtained by a Scholl reaction (oxidative cyclodehydrogenation) (Figure 1.10).[114,115] To increase the solubility of the reactants, polyphenylenes with a range of aliphatic chain substitution are chosen as the starting materials. However, it is difficult to synthesize GNRs with reasonable size for electronic applications because the solubility of the nanoribbons rapidly decreases with increased molecular weight. In addition, structurally perfect GNRs cannot be produced with this method because of abundant defects resulting from incomplete cyclodehydrogenation and side reactions that occurred during the final reaction step; thus, there had been a tremendous need for further optimization of the reaction procedure.[116,117]

Cai et al.[118] reported a bottom-up approach to the atomically precise production of GNRs of different topologies and widths. In this method, they used surface-assisted coupling of molecular precursors into linear polyphenylenes and their subsequent cyclodehydrogenation. The topology, width, and edge periphery of the graphene nanoribbons are defined by the structure of the precursor monomers which can be designed to give access to a wide range of different graphene nanoribbons. Therefore, it appears to provide a promising route to GNRs with engineered chemical and electronic properties.

With the high ratio of non-three coordinated atoms at the edges to three coordinated atoms in the plane network, CNRs exhibit fascinating property, such as semiconductor behavior, which makes CNRs promising in the application of electronics. At this stage, the currently existing methods for fabrication of CNRs cannot meet the requirements both in quality and quantity. Further improvements of CNT unzipping and the entire organic approach will hopefully lead to the controllable fabrication of CNRs with high precision. Future work should focus on the controllable fabrication of CNRs with well-defined edge structures as well as elucidating the correlation of their atomic edge structures with performances in electronic devices. Chemical functionalization on the edges can be used to efficiently tailor the properties of the GNRs. On appropriate chemical functionalization, graphene nanoribbons are rendered semiconducting with a wide range of band gaps, metallic, ferromagnetic, antiferromagnetic, or half-metallic and the transport properties near the armchair edge are expected to differ significantly from that near the zigzag edge. Clearly, a better understanding of the "edge" chemistry and controlling the defects in CNRs will enable them to ultimately be incorporated into the practical and high-performance electronic devices.[106]

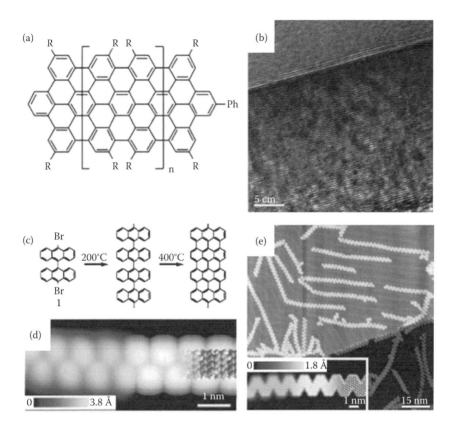

FIGURE 1.10 Polycyclic aromatic hydrocarbons (PAHs) may offer a ground-up synthesis of graphene. (a) Chemical structure of PAHs and (b) TEM image of PAHs in (a). (With permission from Yang, X. et al., *J. Am. Chem. Soc.*, 130, 4216, Copyright 2008, American Chemical Society.) (c) Reaction scheme from precursor 1 to straight $N = 7$ GNRs. (d) STM image taken after surface-assisted C–C coupling at 200°C but before the final cyclodehydrogenation step, showing a polyanthrylene chain (left, temperature $T = 5$ K, voltage $U = 1.9$ V, current $I = 0.08$ nA), and DFT-based simulation of the STM image (right) with partially overlaid model of the polymer (gray, carbon; white, hydrogen). (e) Overview STM image of chevron-type GNRs fabricated on a Au(111) surface ($T = 35$ K, $U = -2$ V, $I = 0.02$ nA). The inset shows a high-resolution STM image ($T = 77$ K, $U = -2$ V, $I = 0.5$ nA) and a DFT-based simulation of the STM image (grayscale) with partly overlaid molecular model of the ribbon. (With permission from Macmillan Publishers Ltd., *Nature*, 466, Cai, J. M. et al., 470, Copyright 2010.)

1.3 STRUCTURES AND PHYSICAL PROPERTIES OF GRAPHENE

1.3.1 STRUCTURES

Graphene, a monoatomic carbon sheet with atoms tightly packed into a honeycomb lattice, has been recognized as the first truly isolated 2D crystalline material.[119] The carbon atoms are bonded via strong in-plane σ-bonds producing the ultra strong sheet and the remaining π-orbits perpendicular to the plane constitute the delocalized network of electrons which makes the structure highly conductive. There are two carbon atoms per unit cell with area of 0.052 nm^2 (Figure 1.11), which gives rise to the specific surface area of approximately 2600–2700 m^2/g. Graphene can be seen as the building block of other carbon materials.[20,120] For example, the sheets can be curved into 0D fullerene, rolled into 1D CNTs, or stacked into 3D graphite.[121] The unique structure endows graphene with marvelous electronic, optical, mechanical, and thermal properties as well as a wealth of potential applications.

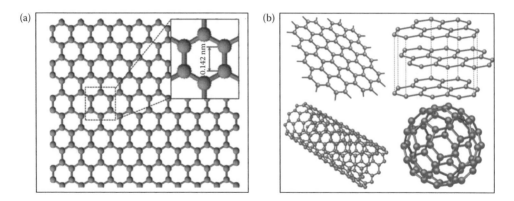

FIGURE 1.11   (a) Schematic illustration of ideal crystal structure as well as the unit cell of graphene. (b) Graphene and the corresponding graphitic carbon built by graphene. (From Neto, A. C. et al., *Phys. World*, 19, 33, 2006. With permission.)

1.3.1.1   Ripples and Wrinkles in Graphene

Before the discovery of graphene, it was commonly believed that 2D materials are unstable thermodynamically. The discovery of graphene seems to disprove the prediction. However, free-standing 2D graphene always vibrates to form lots of ripples on the graphene surface, which indicate strong transformational tendency from 2D to 3D nanostructures. It is predicted that the electrical conduction in a perfectly 2D flat graphene sheet should be ballistic; however, the experimental observation is generally not strictly consistent with this prediction. The reason for this may be attributed to the presence of ripples and/or defects in real graphene sheet (Figure 1.12).[122–126]

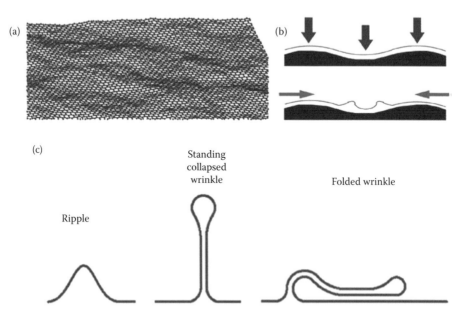

FIGURE 1.12   (a) A typical configuration of graphene at room temperature with intrinsic ripples. (With permission from Macmillan Publishers Ltd., *Nat. Mater.*, 6, Fasolino, A. et al., 858, Copyright 2007.) (b) Generation of wrinkles in graphene on substrate. (From *Solid State Commun.*, 149, Guinea, F. et al., 1140, Copyright 2009, with permission from Elsevier.) (c) Schematics of simple ripple, standing collapsed wrinkle, and folded wrinkle. (With permission from Zhu, W. J. et al., *Nano Lett.*, 12, 3431, Copyright 2012, American Chemical Society.)

Owing to either thermal fluctuations or interaction with a substrate, scaffold, and adsorbents, graphene is readily subject to distortions of its structure.[122] The formation of ripples in graphene sheet may also be induced by their chemical and physical environment. The presence or formation of vacancies and sp^3 bonds can interrupt the conjugation and make the graphene surface buckle over short ranges. As a result, bending of a graphene sheet can cause changes in the distance and relative angle between the carbon atoms, and eventually affect hybridization of the π-orbitals, as in the cases of fullerene and CNTs. Three effects result from bending of the graphene sheet: (i) decrease in the distance between carbon atoms, (ii) rotation of the p_z orbitals, and (iii) rehybridization between π- and σ-orbitals.

It has been demonstrated that Dirac fermions are scattered by ripples of the graphene sheet through a potential that is proportional to the square of the local curvature. The coupling between geometry and electron propagation is unique to graphene, and results in additional scattering and resistivity because of the ripples.[123]

Periodic ripples in a suspended thin film can be induced by different approaches, such as stretching in the axial direction or compression/displacement of the fixed edges in the lateral direction. Furthermore, it is known that graphene deposited on a substrate exhibits corrugations which were also confirmed by numerical simulations. Strong correlations were found between the roughness of the substrate and graphene topography during STM measurements on graphene,[127,128] which indicated that disorder in the substrate translated into disorder in the graphene sheet. Electronic scattering across the section of isolated ripples vanishes at short- and long-electron wavelengths, with a peak for wavelengths comparable to the size of the ripple.

Ripples can develop on the graphene surface when it is suspended or subjected to thermal stress.[124,129] The ripple structure in graphene can be controlled by thermal treatment, which provides a simple and efficient method to tailor the band gap of graphene. It is also known that graphene has a negative thermal expansion coefficient. Using simple thermal manipulation, the orientation, wavelength, and amplitude of these ripples can be controlled. During thermal treatment, different deformations are created between the graphene and its suspended substrates resulting from different thermal expansion coefficients. Because the amplitude of the fluctuations is much smaller than the sample size, the long-range order of graphene is preserved. However, the curvature induced stress in these ripples will generate sites of enhanced reactivity. By controlling the orientation and amplitude of these ripples, it can be envisaged that well-controlled stoichiometric functionalization of graphene can be potentially attained.

The ripples in graphene are intrinsic structures, and different curvatures can be envisioned. With the outlined strong correlation between the graphene sheet structure and the resulting physical and chemical properties, such properties of graphene can be controlled via manipulation of graphene ripples.

1.3.1.2 Structural Defects

Defects are always present in any material. It has been known that graphene with perfect atomic lattices exhibits outstanding electronic and mechanical properties; however, various structural defects unavoidably form during growth or processing of graphene and these defects ultimately deteriorate the properties of graphene. For example, structural defects of the honeycomb lattice induce long-range deformations, modify the electron trajectories, and lead to scattering. Great efforts have been made to restrict the formation of defects in graphene. However, defects can also be used to tailor the local properties of graphene. New functionalities in graphene can be achieved by tuning the density, locality, as well as type of the defects. For example, by reconstruction of its lattice around intrinsic defects or by means of introduction of extrinsic defects including foreign atoms, interesting effects and novel properties of graphene can be created, which enables the design of specific graphene-based devices for potential applications.[130,131]

There are different types of defects in graphene, which can be classified into several groups[106,131–134]:

1. *Structural defects.* Nonhexagonal rings (e.g., pentagons or heptagons) were embedded into graphene lattice. Generally, the presence of nonhexagonal rings significantly distorts the curvature of the hexagonal carbon honeycomb structure.

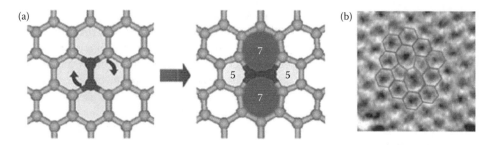

FIGURE 1.13 (a) Molecular model for the Thrower–Stone–Wales (TSW)-type defect (55–77). (With permission from Terrones, H. et al., *Rep. Prog. Phys.*, 75, 062501, Copyright 2012, IOP Science.) (b) Experimental TEM image of TSW defects. (With permission from Meyer, J. C. et al., *Nano Lett.*, 8, 3582, Copyright 2008, American Chemical Society.)

2. *Topological defects.* They are also named as Stone–Thrower–Wales (STW type) defects (Figure 1.13),[131,133] generated by a lattice reconstruction. For example, four hexagons are transformed into two pentagons and two heptagons [SW(55–77) defect] by rotating one of the C–C bonds by 90°. The feature of this type of defect lies in that fact that no atoms are removed or added, and no large curvature distortion is caused in the sheets.

3. *Doping-induced defects.* The replacement of carbon atoms with noncarbon foreign atoms within the hexagonal lattice will create defects.

4. *Non-sp²-hybridized carbon defects.* These defects are caused, for example, by the presence of vacancies, edges, adatoms, interstitials, or carbon chains.[130,134]

Currently, a main problem is high defect concentration in graphene produced from different approaches, which degrade the properties of graphene and significantly limit its application especially in the field of electronics. To improve the physical properties of graphene, restriction of defect formation is crucial. However, the presence of defects can also provide active sites for chemical reactions and facilitate the functionalization of graphene to enrich their properties.

1.3.1.3 Edge States

Graphene edges, especially the edges of GNRs and platelets, strongly influence their optical, magnetic, electrical, and electronic properties. When a graphene sheet is cut into smaller ones, generally, two types of edges referred to as armchair and zigzag are created (Figure 1.14).[131,135–138] For the two basic shapes of graphene edges, armchair and zigzag edges, it has been shown that graphitic networks with zigzag edges have a localized *edge state* at the Fermi level which armchair edges do not have.

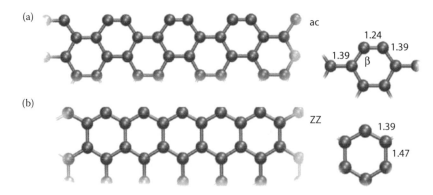

FIGURE 1.14 Armchair (a) and zigzag (b) edges in graphene. (With permission from Koskinen, P. et al., *Phys. Rev. Lett.*, 101, Copyright 2008 by the American Physical Society.)

Theoretical predictions and experimental observations have clarified that the electronic structure in graphene strongly depends on the geometry of edges, especially when the flake size is in a nanometer scale because a large fraction of carbon atoms sit on its edge. The relative importance of the edge state strongly depends on the size of graphene and, for example, the properties of GNRs are indeed dominated by edge effects. However, the effect of edges on large area graphene decreases promptly with increasing of the size of a graphene layer.[136] Depending on width and edge structure, both semiconducting and metallic GNRs exist which may be seen in analogy to CNTs with different chirality.[137] The edges of graphene provide a large number of active sites for the occurrence of chemical reactions. Generally, a chemical reaction readily takes place on the edges rather than in basal plane. Graphene possesses rich edge chemistry and this makes chemical functionalization for tailoring properties of graphene possible for the development of applications in various fields, such as electronics, spintronics, and optoelectronics.

From the energetic point of view, the zigzag edges are more stable than armchair edges.[139] Therefore, the interconversion of zigzag edges to the armchair configuration is difficult because this process requires migration of the edge atoms. However, chemical reactivity is inconsistent with physical stability.[140] Because each carbon atom of the zigzag edge has an unpaired electron, reactants are easily combined on zigzag edge through the interaction with active unpaired electrons, whereas the carbon atoms of the armchair edge side are more stable in chemical reactivity because of a triple covalent bond between the two open edge carbon atoms of each edge hexagonal ring.

Based on the understanding of graphene edges in atomic nanoscale, various types of treatments have been established, such as chemical functionalization to terminate the edge with different molecular or functional groups, chemical etching, as well as structural modification of the edge states by varying the type of molecules adsorbed on graphene.[141] Dangling bonds at the edges of graphene are another significant factor for determining electronic, optical, and magnetic properties of graphene. The covalent bonds are easily formed at graphene edges and/or surface defects with relatively high activity. The higher chemical reactivity of edge carbon atoms relative to the perfectly bonded sp^2 carbon atoms in the basal plane is attributed to the presence of bond disorder and various functional groups at the edges of graphene.

The edge morphology of graphene affects chemical reactivity and edge stability significantly, as well as the band gap, magnetic, and optical behavior of graphene.[142] Although much progress has been made in understanding of edge configuration and their corresponding properties, there are still many challenges because of the limitations of characterization techniques.

1.3.2 PROPERTIES

1.3.2.1 Electronic Properties

On the discovery of graphene, various electronic properties were revealed which make graphene currently the most attractive topic in condensed matter physics.[119] Unlike traditional materials, graphene exhibits totally different band structures.[143] As shown in Figure 1.15,[145] the conductance bands and the value bands of graphene contact each other at two inequivalent points of K and K' in the Fermi surface. Near these points, the energy and momentum show a linear relation [119,143] which can be described as

$$E_{2D} = \hbar v_F \sqrt{k_x^2 + k_y^2}$$

Obviously, the dispersion relation shown above is analogous to the Dirac equation of photons, $E = \hbar c k$ (c equals the velocity of light), only replacing the velocity of light with Fermi velocity (V_F), which is 300 times slower than that of photons.[143] However, the velocity of charge carriers on graphene is much higher than that of conventional materials, making graphene one of the most conductive materials. For example, the conductivity of copper is 10^5 S/cm, whereas the conductivity of graphene is even an order of magnitude higher.[144]

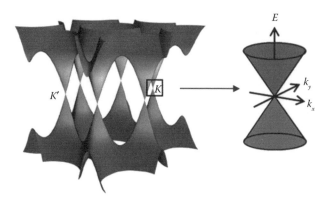

FIGURE 1.15 The band structure of free-standing suspended graphene. (From Rao, C. N. R. et al., *J. Mater. Chem.*, 19, 2457, 2009. With permission of The Royal Society of Chemistry.)

Graphene sheets have been revealed to show a strong ambipolar field effect at room temperature (Figure 1.16a). The conducting channel can be continuously tuned between electrons and holes with charge carriers at a concentration of up to 10^{13} cm^{-2} and mobility as high as 10,000 cm^2/V s when a gate voltage is applied. The positive gate bias can induce electrons impregnating into conduction band and vice versa.[20] Several groups have reported that electron mobility in graphene is independent of the temperature, and that the transport of charge carriers is mainly affected by scattering sites, such as defects, interfacial phonons, and interaction with the underlying substrate.[16,20,144,146,147] Bolotin et al. demonstrated a field effect device with suspended single-layer graphene and achieved carrier mobility as high as 200,000 cm^2/V s, at a concentration of 2×10^{11} cm^{-2}.[146] Recently, researchers revealed that carrier mobility can be further increased to 250,000 cm^2/V s.[148]

More importantly, massless charge carriers in graphene can give rise to a room temperature quantum Hall effect (QHE) that was exclusively observed in two-dimensional electron gas (2DEG) systems earlier. The QHE on graphene is different from conventional cases of semiconductor heterostructures because of the unique band structure.[17] Generally, the Hall resistance as a function of concentration of electrons produces a series of plateaus at $jh/2eB$, which is referred to as integer

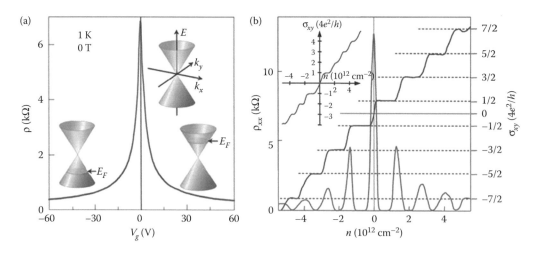

FIGURE 1.16 (a) Ambipolar electric field effect in single-layer graphene. (With permission from Macmillan Publishers Ltd., *Nat. Mater.*, 6, Geim, A. K. and Novoselov, K. S., 183, Copyright 2007.) (b) The half-integer quantum Hall effect in graphene. (With permission from Macmillan Publishers Ltd., *Nature*, 438, Novoselov, K. S. et al., 197, Copyright 2005.)

QHE.[20] However, graphene affords half-integer QHE as both electrons and holes are degenerated at K and K' points in the Fermi surfaces. This phenomenon has been observed[16,17] almost at the same time. As shown in Figure 1.16b, Hall conductivity of the single-layer graphene as a function of carrier concentration measured at $B = 14$ T and $T = 4$ K shows an uninterrupted plateau with distance of $4e^2/h$ which is bigger than that of integer QHE.[16] Another advantage of graphene used as platform for QHE is that this phenomenon is observable even at ambient temperature instead of the boiling point of liquid helium for traditional experiments.[149] Very recently, fractional QHE has been observed on suspended graphene devices.[150,151] In this case, the key is suspending graphene through which the interference from interaction with the substrate can be eliminated.[152] Besides, an insulating phase competing with QHE has been observed. These results pave the way for the research of collective behavior of Dirac fermions.[150]

The unique electronic properties of graphene, especially zero-band gap, have been discovered as fascinating quantum physics phenomena.[152] Unfortunately, the gapless characteristic is not suitable for practical application in electronics. For example, frequent on/off switching required in devices like transistors cannot be achieved on structures with zero-band gap.[20] As a result, the regulating band gap of graphene is crucial for electronic applications. The band structure has close correlation with the number of graphene layers. By developing bi-layer graphene films on silicon carbide, Rotenberg et al.[153,154] demonstrated that controllable band gap could be achieved by modulating the carrier concentration in different layers selectively. Interestingly, the band gap of the bilayer graphene can be tuned continuously from 0.0 to 0.3 eV under external electrical field.[119] As for the trilayer graphene, the Bernal-stacked form is semimetallic, whereas the rhombohedral-stacked form shows a semiconducting property with tunable band gap.[155] In addition to the layers of graphene, the size of a graphene sheet also affects the band structure because of quantum confinement effect. As mentioned earlier, for example, armchair GNRs are either metallic or semiconducting, whereas zigzag GNRs are metallic.[137] Generally, producing GNRs with smooth edges is not easy which restricts the popularization of this method.[156] Recently, Torres et al. proposed a facile approach to tuning the band gap of graphene dynamically by illuminating graphene with laser in the mid-infrared range.[157] Moreover, creating defects on graphene sheets deliberately is an easy and efficient approach to modify the band gap and, thus, reduced graphene oxides (RGOs) decorated with controllable defects are attractive for practical applications.[63,64,79,158] The properties of RGOs prepared via different methods are significantly different from each other; for example, the electrical conductivity may range from less than 1 to 1314 S/cm.[79,159]

1.3.2.2 Optical Properties

The unique one atomic layer structure of graphene causes exceptional optical behaviors.[160,161] It has been revealed that the single-layer graphene absorbs 2.3% of incident white light with negligible reflectance (<0.1%) (Figure 1.17a). The absorption is independent of material parameters and only determined by the fine structure constant $\alpha = e^2/\hbar c \approx 1/137$ (c is the speed of light), where $2.3\% \approx \pi\alpha$.[161] Based on Fermi's golden rule, the calculated absorption of light is consistent with the measured one, which is coincident with the gapless band characteristic.[161] Because the absorption of light is linear with the number of layers, each sheet produces another 2.3% of absorption.[51] The absorption of light remains the same in wide range but shows a drastic absorption peak at 250 nm as incident light with higher energy could result in nonlinear dispersion relation because the linear relation is only valid near the Dirac points.[162] The well-defined relationship of graphene structure with incident light makes absorption spectroscopy one of the most efficient characterization methods for the one-atom-thick sheet.[163] For example, the number of layers can be determined by the transmittance of visible light, whereas the conjugated areas can be reflected from the peaks in the ultraviolet region (Figure 1.17b).[51,80]

In addition, the optical properties of graphene can be tuned via doping, applying external electric fields, and controlling the layer size.[51,164–166] Graphene with defects created intentionally, such as RGO, shows highly tunable optical response (Figure 1.18).[26] The defects on the RGO sheets efficiently enlarge the band gap, thus translating the sheets into semi-metal with finite density of states

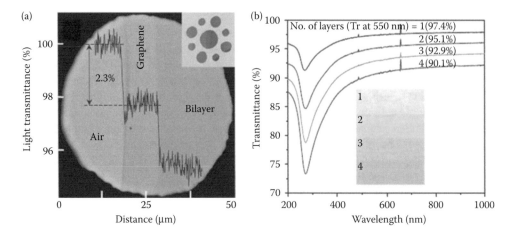

FIGURE 1.17 (a) Photograph of single-layer and bilayer graphene in transmitted light, the inset shows a graphene sheet placed over a metal support with several apertures. (From Nair, R. R. et al., *Science*, 320, 1308, 2008. With permission of AAAS.) (b) UV–vis spectra of graphene with different layers. (With permission from Macmillan Publishers Ltd., *Nat. Nanotechnol.*, 5, Bae, S. et al., 574, Copyright 2010.)

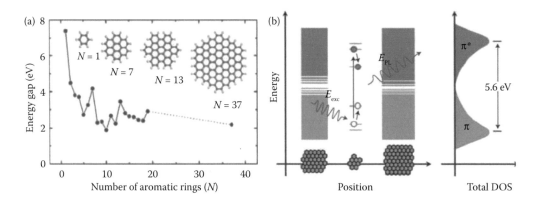

FIGURE 1.18 (a) The energy gap of π–π* transitions as a function of the number of fused aromatic rings. (b) Schematic illustration of electronic band structure of GO. (Eda, G. et al., *Adv. Mater.*, 2010, 22, 505. Copyright Wiley-VCH Verlag GmbH & Co. KGaA. With permission.)

at the Dirac points.[26] The combination of high transparency with good electric conductivity makes RGO a promising substitute for ITO as transparent conductors. The GO/RGO sheets also produce fluorescence spanning from near-infrared to ultraviolet and the wavelength of fluorescence can be controlled by modulating the size of sp^2 clusters on GO/RGO sheets.[167–169]

Furthermore, graphene can be used as nonlinear optical material because of the universal absorption of light over broadband and nonlinear Kerr effect.[26] Both saturable absorption and reverse saturable absorption have been observed on graphene or graphene-based composition. Loh et al.[170] have functionalized graphene with conjugated organic molecules and demonstrated the superior performance as saturable absorber for ultrafast lasers. Also, Liu et al.[171] have produced a graphene-based composite showing optical limiting property with potential application for eye protection.

1.3.2.3 Mechanical Properties

The excellent mechanical properties of graphene are derived from the extremely strong σ-bonds (670 kJ/mol) in the basal plane. In 2002, Bemhoic et al.[172] had predicted that the intrinsic strength of graphene would be superior to any other material by quantum simulation. Lee et al.'s experimental

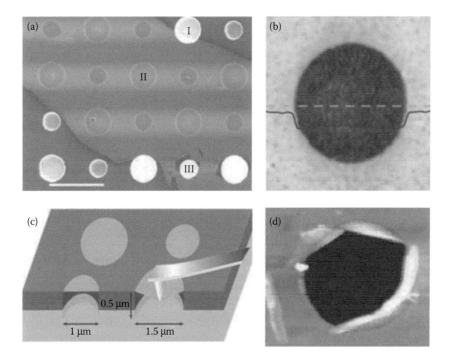

FIGURE 1.19 Measurement of the mechanical properties of suspended graphene. (a) A single-layer graphene on membrane. (b) AFM image of one hole on membrane. (c) Illustration of nanoindentation of graphene. (d) AFM images of fractured graphene. (From Lee, C. et al., *Science*, 321, 385, 2008. With permission of AAAS.)

measurements were carried out on suspended single-layer graphene via nanoindentation technology as illustrated in Figure 1.19. Young's modulus and fractural strength are as high as 1 TPa and 130 GPa, respectively.[173] The in-plane tensile elastic strain can be as high as 25% without plastic deformation and the drastic deformation shows negligible influence on electrical conductivity.[173,174] This highly elastic property combined with stable electrical conductivity under severe strain demonstrates the potential application as flexible conductors.[174]

Although GO/RGO is decorated with an abundance of defects and functional groups, the relatively intact 2D structure inherited from graphene still endows the derivatives with excellent mechanical performances and Young's modulus can be as high as 0.25 TPa which is very close to that of pristine graphene.[175] GO/RGO can also be easily incorporated into the polymer matrix producing composites with enhanced Young's modulus and fracture strength.[176] Furthermore, a series of GO/RGO-based macro-assemblies, such as fibers, films, and monoliths, have been successfully synthesized and the outstanding mechanical properties of graphene have been extended into practical applications.[177–181]

1.3.2.4 Thermal Properties

Efficient heat removal is crucial for further development of electronic industry as the performances of integrated circuits deteriorate dramatically under high temperature; thus, searching for high-performance thermal conductive structure is in urgent demand.[20,182,183] Graphitic carbons regarded as rolled or stacked graphene sheets show high thermal conductivity which can be deduced from the strong sp^2 bonds resulting in dissipation of heat via lattice vibration or phonon scattering.[20,182,183] The thermal conductivity of graphene is demonstrated to be superior to graphite and CNT as well as diamond both theoretically and experimentally.[184,185]

Early in the year 2000, based on molecular dynamics (MD) simulation, the thermal conductivity of suspended single-layer graphene had been predicted to be as high as 6000 W/mK. This value is higher than for any other material.[185] The experimental measurement of thermal conductivity of graphene

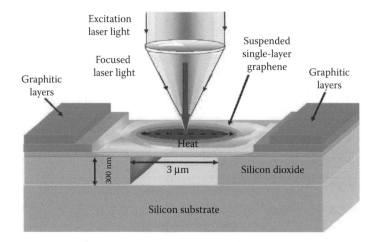

FIGURE 1.20   Schematic illustration of experimental apparatus for measurement of thermal conductivity of suspended graphene. (With permission from Balandin, A. A. et al., *Nano Lett.*, 8, 902, Copyright 2008. American Chemical Society.)

lagged behind because of technological challenges of the current methods.[184] Recently, Balandin et al. determined the thermal conductivity of suspended single-layer graphene from micromechanical exfoliation by an unconventional noncontact method based on confocal micro-Raman spectroscopy (Figure 1.20). As the G-peak of graphene is temperature-dependent, monitoring the shift of the G-peak with variation of temperature is possible which shows high efficiency to determine thermal conductivity. The thermal conductivity of suspended graphene is as high as 5000 W/mK, which is very close to the predicted value.[185] In another spot, the conductivity of graphene derived from CVD was determined to be about 2500 W/mK. This halved value may result from intrinsic thermal resistance because of the defects, wrinkles, or ripples.[186]

In practical applications, the direct contact of thermal conductors with substrates is required, and hence conductivity of supported graphene should also be carefully determined. The thermal conductivity of graphene supported on SiO_2/Si is about 600 W/mK, and the decreased value is attributed to phonons leaking and interface scattering.[187] However, this value is still higher than that of copper and silicon demonstrating the potential application of graphene in thermal management.

Ideal graphene possesses a monoatomically thin, crystalline structure and exceptional optical, mechanical, and thermal properties. However, structural defects, wrinkles, or ripples always exist in real graphene. Graphene edges significantly affect the properties of graphene, especially the properties of GNRs. The discrepancy in properties between graphene sheets from different preparation routes varies remarkably due to their structural discrepancy. More efforts are needed to prepare graphene with near perfect structure by restricting defect formation during growth or subsequent processing of graphene.

1.4   CHEMISTRY OF GRAPHENE

Pristine graphene is a hydrophobic material with a chemically inert surface, which makes solution processing difficult. Surface modification of graphene can improve its solubility and dispersion in solution, which is crucial for practical application in diverse fields, such as polymers, biomaterials, catalytic materials, and so on. Up to now, a series of methodologies, including hydrogenation and fluorination, functionalization by small organic molecules, as well as polymer for the creation of various graphene derivatives with a great many special structures, compositions, and properties, have been developed.[188–192] This section presents a comprehensive outline and state-of-the-art description of the current research status on the fast development of graphene chemistry and functionalization in recent years. The physical and chemical properties of those derivatives have been discussed, along with a forward outlook on their applications in various fields.

1.4.1 Hydrogenation and Fluorination

Graphene consists of a flat monolayer of carbon atoms that are tightly packed into a honeycomb lattice. From the viewpoint of organic chemistry, graphene can be regarded as an analog of a giant aromatic poly-molecule and the structure of graphene can be made up by dehydrogenation of numerous benzene rings. Therefore, the conjugated sp^2 carbon atoms in graphene can react with small elementary molecules such as hydrogen and fluorine.

Previous studies have shown that atomic hydrogen can be chemisorbed on CNTs and atomic hydrogen can create C–H bonds with the carbon atoms in the nanotube walls.[191,192] Recently, the hydrogenation of graphene has also been accomplished by exposing the micromechanical cleavage graphene to H_2/Ar plasma under a low pressure (0.1 mbar). The reaction with hydrogen can transform the highly conductive zero-overlap graphene into an insulator called graphane.[188] This kind of new crystal has the same hexagonal lattice as graphene, but its period is shorter than that of graphene (figure graphane). STM characterization reveals four different hydrogen configurations on the surface of graphene: ortho dimers (structure A in Figure 1.21b), para-dimers (structure B in Figure 1.21b), elongated dimers, as well as monomers (marked in Figure 1.21b, c, and d, respectively). Other studies show that hydrogenation preferentially occurs on the surface of epitaxial graphene grown by heating SiC at high temperature. At low coverage, hydrogen dimers occur preferentially on the protruding areas in the STM topography of the graphene–SiC surface, whereas at higher coverage, random adsorption into larger hydrogen clusters is formed.[189] Hydrogenation of graphene is reversible and hence it is possible to control the electronic properties of graphene. By using an STM tip, researchers have precisely controlled the location of hydrogen atoms on the surface of graphene to form different patterns.[190] Hydrogenated graphene sheets are stable at room temperature, and

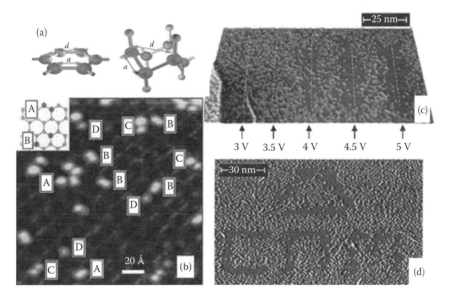

FIGURE 1.21 (a) Schematic pictures of the crystal structure of graphene and theoretically predicted graphane. (From Elias, D. C. et al., *Science*, 323, 610, 2009. With permission of AAAS.) (b) Scanning tunneling microscopy image of hydrogenated graphene and four different types of hydrogen on the surface of graphene. (With permission from Balog, R. et al., *J. Am. Chem. Soc.*, 131, 8744, Copyright 2009. American Chemical Society.) (c) STM image shows the graphene pattern with changes as a function of the positive sample bias. (d) The patterned institutional logo and initials by STM tips. (From Sessi, P. et al., *Nano Lett.*, 9, 4343, 2009. With permission.)

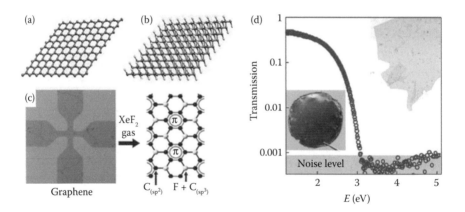

FIGURE 1.22 (a) Structure of graphene. (b) Structure of fluorographene. (With permission from Schrier, J., *ACS Appl. Mat. Interfaces*, 3, 4451, Copyright 2011. American Chemical Society.) (c) Fluorination of graphene by xenon difluoride gas. (d) Photographs of graphene paper before (left) and after (right) fluorination, as well as the optical transparency of a 5 μm fluorographene film as a function of *E*. (Robinson, J. T. et al., *Nano Lett.*, 2010, 10, 3001; Nair, R. R. et al., *Small*, 2010, 6, 2877. Copyright Wiley-VCH Verlag GmbH & Co. KGaA. With permission.)

the electronic properties of the patterned graphene are dependent on the size of the pattern. After removal of saturated hydrogen, the inherent electronic properties of graphene are recovered. This reversible and local mechanism to modify the properties of graphene has significant implications for its application in circuit and nanodevices.

In addition to hydrogen, fluorine is another special gas molecule that can radically react with many allotropes of carbon, such as CNTs[195,196] and carbon nanofibers.[197,198] First principles calculation shows that band splitting of graphene can increase by orders after fluorination,[199] when the fluorine atoms are added to the surface of graphene both (1,2) and (1,4) additions can occur.[200] Fluorographene have been synthesized with various methods, such as plasma-assisted decomposition of CF_4,[201] liquid-phase exfoliation of fluorinated graphite,[202] and treatment of graphene with xenon difluoride gas.[203,204] The physiochemical properties of the prepared fluorographene depend on the synthesis procedure. XeF_2 treatment can fluorinate graphene without etching, when F coverage saturates at 25% (C_4F), the product is optically transparent and more resistive than graphene (more than 6 orders higher).[204] Recent research demonstrates that partially fluorinated graphene produced by XeF_2 treatment can luminesce broadly in the ultraviolet and visible light regions, which has the optical properties of both excitonic and direct optical absorption and emission features.[203] Liquid-phase exfoliation of graphite fluoride with sulfolane can produce stoichiometric graphene fluoride monolayers. It has been verified both theoretically and experimentally that the thermodynamically stable exfoliated fluorographene can transform readily into graphene, with nanodiamonds as a byproduct, by a halide-exchange process.[202] Another stoichiometric derivative of graphene with a fluorine atom chemically bonded to each carbon was reported to be stable up to 400°C even in air, similar to PTFE (polytetrafluoroethylene; Figure 1.22c).[194] This kind of fluorographene is a nonconductive insulator which, however, still shows the mechanical strength of graphene.[204]

1.4.2 COVALENT FUNCTIONALIZATION

Theoretically, graphene is very stable and there are no other functional groups on its hydrophobic surface. It is a great challenge to graft other covalent bonds directly to the inert surface of graphene. Fortunately, chemists have developed a well-established solution-based route to synthesize graphene, which is based on the exfoliation of GO.[205] GO contains large amounts of highly reactive oxygen functional groups, such as carboxylic groups at the edge or epoxy and hydroxyl located on the basal planes. Those negatively charged oxygen species help GO to disperse in water very well, the zeta potential of GO solution can be as negative as −64 mV that through a deoxidation procedure.[206] Compared to zero-gap graphene, GO is an electrically insulating material because of the disrupted sp[2] bonding networks. The most important reaction involving GO is its reduction to produce RGO because the π–π conjugated structure within the graphene nanosheet can recover after the reaction. Therefore, a possible chemical route to functionalized graphene starts from the formation of chemical bonds with GO, and then undergoes a chemical reduction procedure to obtain covalently functionalized graphene derivate.

Fully developed organic chemistry has facilitated the reaction of GO with a variety of organic molecules. A series of functional groups as well as polymers have covalently bonded with graphene. Figure 1.23 summarizes the possible chemical reaction to obtain functionalized graphene via covalent bond.[207] In terms of chemical reactions, esterification[208] and acylation are the most common methods to graft covalent bindings to the surface of graphene. The covalent functionalization of graphene with poly(vinyl alcohol) (PVA) was accomplished by the carbodiimide-activated esterification reaction between the carboxylic acid group on GO and hydroxyl groups on PVA.[209] Such PVA functionalized graphene sheets were readily soluble in hot water and DMSO (dimethyl sulfoxide) to form stable dispersions, which can facilitate solution-phase processing of graphene-based composites without any harmful organic solvent or other foreign substances (Figure 1.23a). The introduction of amines often requires activation of the carboxylic groups by thionyl chloride (SOCl$_2$).[210–212] Porphyrin and fullerene covalently functionalized graphene hybrid has demonstrated large nonlinear optical properties (Figure 1.23b).[211] The formation of covalent bonds is usually characterized by x-ray photoelectron spectroscopy (XPS), Fourier transform infrared spectroscopy (FT-IR), Raman spectroscopy, and NMR spectroscopies. Other organics, such as isocyanates, have the opportunity to react with both the edge carboxyl and surface hydroxyl groups simultaneously. The treatment of GO with different isocyanates has been investigated to result in a series of covalently functionalized graphene sheets (Figure 1.23d).[213]

There are also several reactive epoxy groups that are located on the basal plane of GO, which can react with various amine-terminated organic molecules by nucleophilic ring-opening reaction. Surface-functionalized GO nanosheets are obtained from *in situ* functionalization by octadecylamine (ODA) during the graphite oxidation and exfoliation process. FT-IR analysis confirms that surface grafting occurred via epoxide ring-opening reaction. The resulting ODA-GO plates are readily dispersed in organic solvents to form stable colloidals that are spin coated into thin film with controllable thickness (Figure 1.23c).[215] Graphene sheets covalent functionalized with 3-aminopropyltriethoxysilane (APTS) have also been explored as a reinforcing component of silica monolith via a sol–gel process. The compressive failure strength and toughness of the graphene-reinforced monolith increased by 20% and 92%, respectively, with 0.1 wt% functionalized graphene sheets.[218]

Diazonium chemistry is another very important method to obtain functionalized carbon nanomaterials.[219,220] Many of the sidewall reactions involving CNTs are accomplished throughout diazonium reagent to form a stable covalent aryl bond. The chemical environment of the graphene surface is very similar to that of CNTs. Therefore, covalent functionalization of exfoliated graphene is achieved by organic diazonium salts.[221] Functionalization can successfully prevent reaggregation of the monolayer carbon sheet and provide solubility in organic media as well. The covalent bond between graphene and the aryl group was verified by STM investigation on epitaxial graphene that was functionalized by diazonium salts. A related study shows that the nitrophenyl groups bond with graphene in a perpendicular configuration.[222] Graphene sheets obtained from reduction of GO can also be functionalized with high amounts of varying aryl addends, making these nanosheets disperse readily in organic solvents (Figure 1.23e).[217]

1.4.3 Noncovalent Functionalization

The insolubility of pristine graphene in most solvents makes its solution very difficult, which eventually restricts the practical application of graphene in many fields. As mentioned previously, many

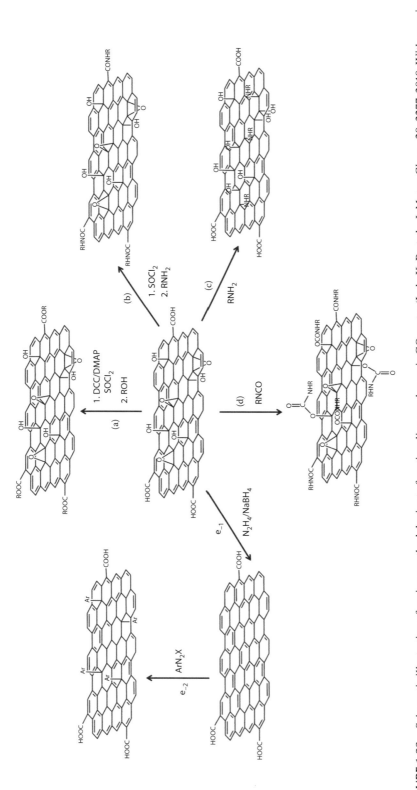

FIGURE 1.23 Schematic illustration of various methodologies to functionalize graphene via GO route. (Loh, K. P. et al., *J. Mater. Chem.*, 20, 2277, 2010. With permission of The Royal Society of Chemistry.) (a) First, activation of COOH groups by SOCl₂ then esterification of GO; (From Salavagione, H. J. et al., *Macromolecules*, 42, 6331, 2009; Veca, L. M. et al., *Chem. Commun.*, 2565, 2009. With permission.) (b) Modification of GO by acylation reaction after SOCl₂ activation; (From Zhang, X. et al., *Carbon*, 47, 334, 2009; Niyogi, S. et al., *J. Am. Chem. Soc.*, 128, 7720, 2006; Xu, Y. et al., *Adv. Mater.*, 21, 1275, 2009. With permission.) (c) Ring open reaction of GO by amine-terminated organic molecules (From Wang, S. et al., *Adv. Mater.*, 20, 3440, 2008; Yang, H. et al., *Chem. Commun.*, 3880, 2009; Touhara, H. and Okino, F., *Carbon*, 38, 241, 2000; Xu, Z. and Gao, C., *ACS Nano*, 5, 2908, 2011. With permission.) (d) Functionalization of GO by isocyanates that can interact with both carboxyl and hydroxyl groups. (From Stankovich, S. et al., *Carbon*, 44, 3342, 2006. With permission.) (e₋₁) Chemical reduction of GO by NaBH₄ or N₂H₄ to obtain RGO, and then covalent functionalization of RGO by diazonium reaction (e₋₂). (From Lomeda, J. R. et al., *J. Am. Chem. Soc.*, 130, 16201, 2008. With permission.)

modification methodologies have been developed to graft functional groups to the surface of graphene with the aim of increasing its solubility and/or endowing new properties. However, the covalent functionalization of graphene can transform carbon hybridization from sp^2 to sp^3, leading to a possible disruption to the conjugated π–π structure in the structure of graphene which may degrade the properties of graphene to some degree. Fortunately, graphene can also form noncovalent interactions with a series of materials, such as small organic molecules, polymers, and so on. The noncovalent bindings include van der Waals forces (such as hydrophobic interactions) and π–π stacking of aromatic molecules. In comparison to covalent functionalization, the noncovalent interaction does not disrupt conjugated π–π structures on graphene. Another advantage is that both GO and graphene can form noncovalent bindings based on different methodology. In fact, the noncovalent functionalization of graphene has been widely used in preparation of graphene.[36,37]

1.4.3.1 Noncovalent Functionalization with Aromatic Molecules

Some aromatic molecules, such as pyrene, porphyrin, and their derivatives, have strong affinity with the basal plan of graphene sheets via π–π interactions. Noncovalent functionalization has been used in the functionalization of CNTs,[223] and as the rise of graphene, it has been used for functionalization of graphene.

Stable aqueous dispersions of graphene sheets have been produced using 1-pyrenebutyrate (PB) as a stabilizer because of the strong interactions between pyrene moiety and the surface of graphene. A vacuum-induced self-assembly method was employed to fabricate flexible graphene film with controllable thickness. As a result of chemical reduction, the conductivity of the fabricated film is 7 orders of magnitude larger than that of the GO.[224] This simple method has provided the opportunity to synthesize various graphene dispersions by using other stabilizers with large planar aromatic rings. Water-soluble graphene can also be produced after noncovalent functionalization by using 7,7,8,8-tetracyanoquinodimethane (TCNQ) anion as a stabilizer and expanded graphite as a starting material.[225]

Large-area transparent graphene films modified with pyrene buanoic acid succidymidyl ester (PBASE) were used as anode for application in organic solar cells.[226] The PBASE molecules can attach to the surface of graphene with a face-to-face orientation via π–π interactions, which improve the surface wettability of graphene without degrading its pristine conductivity. The power conversion efficiency (PCE) of the fabricated photovoltaic devices was 1.71%, which corresponds to ≈55.2% of the PCE of an ITO-based device. These findings have paved the way to apply graphene in plastic electronics and optoelectronics. The formation of graphene film with high conductivity makes application of graphene sheets in the field of energy conversion and storage devices possible.

1.4.3.2 Noncovalent Interaction with Polymers

Graphene-based composites have been fabricated by noncovalent functionalization of reduced graphene (r-G) with sulfonated polyaniline (SPANI). The prepared SPANI/r-G shows good electro activity in both acidic and neutral media.[227] A graphene-based composite was formed by blending conducting polymer poly(3-hexylthiophene) (P3HT) and graphene and served as an active layer in heterojunction (BHJ) polymer photovoltaic cells. The strong electron/energy transfer from P3HT to graphene increases the power conversion efficiency to 1.1% in the presence of 10 wt% of graphene.[228] Few-layered graphene sheets were dispersed in chitosan/acetic solutions, the mixture was then fabricated into biocompatible graphene films by solution-casting method. The elastic modulus of the graphene/chitosan composite films increased significantly with graphene addition at a very low content (0.1–0.3 wt%).

1.4.3.3 Biomolecule-Based Noncovalent Functionalization

Many biomolecules including simple saccharides,[229] bacterias,[230] and DNA[231,232] are widely used for the production of biocompatibility or bioactivity graphene-based materials. Here, we focus on the biosafety of graphene (and GO) as well as their biomolecule-based derivatives.

When graphene is explored for biomedical applications, its biocompatibility must be considered in the first place. Graphene paper was initially explored as a platform for cell culture experiments. Mouse fibroblast cells (L-929) were found to proliferate on the surface of a self-assembled graphene paper made by vacuum-induced filtration method (Figure 1.24a and b).[233] Water-dispersible RGO was synthesized from GO that employs vitamin C as the reductant and amino acid as the stabilizer.[234] This approach used bio-compounds for nontoxic and scalable production of graphene with unique electrical properties that may facilitate the application of graphene for electronic devices as well as biocompatible materials. Other ecofriendly compounds, such as reducing sugars,[229] hydrohalic acids,[235] and hexamethylenetetramine,[236] are also used for green production of graphene that may be used as raw materials for the biomedical application of this novel nanomaterial.

DNA has been explored for noncovalent functionalization of graphene as well. Stable aqueous suspensions of graphene with concentrations as high as 2.5 mg mL^{-1} were produced with the assistance of DNA, and this negatively charged dispersion can be assembled into a graphene-based layered bio-composite that contains intercalated DNA molecules or redox proteins.[231] Multifunctional 3D GO hydrogels can self-assemble by using single-stranded DNA as a binder. It is interesting to find that the GO/DNA self-assembled hydrogel possesses a self-healing property.[232] In other biomedical materials, such as aromatic drugs, insolubility in water has hampered their application for disease treatment.

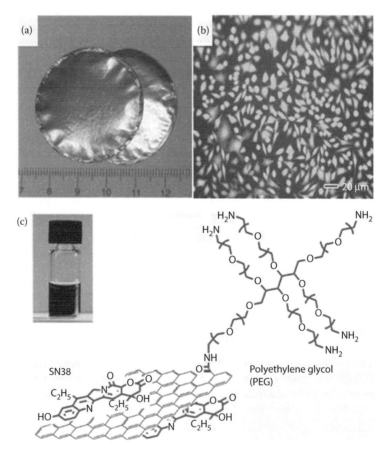

FIGURE 1.24 (a) Photograph of two free-standing graphene papers fabricated by vacuum filtration method; (b) fluorescence microscopy image of calcein-stained L-929 cells growing on graphene paper; (From Chen, H. et al., *Adv. Mater.*, 2008, 20, 3557, Copyright Wiley-VCH Verlag GmbH & Co. KGaA. With permission.) (c) noncovalent functionalized GO for drug delivery. (With permission from Liu, Z. et al., *J. Am. Chem. Soc.*, 130, 10876, Copyright 2008. American Chemical Society.)

Branched polyethylene glycol (PEG) functionalized GO can be dispersed in various biological solutions. This ability can be used to absorb various bioactive hydrophobic aromatic molecules (like camptothecin analog and SN38; the latter is the active metabolite of irinotecan) to its surface via $\pi-\pi$ interactions (Figure 1.24c). The prepared GO/SN38 complex inherited the high cancer-killing potency of SN38 and exhibited high water solubility.[237] These findings have demonstrated that graphene is a promising material for *in vivo* cancer treatment with various aromatic, hydrophobic drugs.

The monoatomically thin film of graphene offers the opportunity to form covalent bonds or $\pi-\pi$ stacking. Chemical processing is a promising alternative for functionalization of graphene toward practical applications. However, the chemistry of graphene has not been clarified to date and much more work is needed. The development of atomically resolved surface characterization technique allows the measurement of functionalization periodicities and density, which enables us to establish an understanding of graphene chemistry. Research of graphene is becoming a truly multidisciplinary subject encompassing scientists from different research fields, including surface science, materials science, physics, chemistry, as well as nanoscience and nanotechnology. The combination of conventional lithographical patterning and chemical functionalization will play a critical role in the generation of novel devices and circuitry.

1.5   APPLICATIONS OF GRAPHENE

Owing to their excellent properties and potential applications in various fields, graphene and graphene-based materials have attracted tremendous interest across the world in recent years. Applications of graphene in almost every field of science and technology have been intensively investigated, such as electronics, energy storage and conversion, sensor, or field effect devices. In this section, we focus on the applications of graphene and its related materials.

1.5.1   ENERGY STORAGE AND CONVERSION

1.5.1.1   Lithium Ion Batteries

Although Li-ion battery has been widely used, high-performance Li-ion battery, with high charge–discharge rates and high capacity, is still a big challenge and in great demand for application in current communication and transportation. Many researchers have explored extensively the possibility of graphene and graphene-based materials in the application of Li-ion battery. Similar to other carbon nanomaterials, graphene can be directly used to act as anode material. It has been demonstrated that the reversible capacity of Li-ion closely depends on the layer-to-layer distance of graphene nanosheets. For example, the *d*-spacing of graphene nanosheets was tuned by inserting CNT and C_{60} and different reversible capacities have been achieved.[238] Obviously, expansion in *d*-spacing of graphene layers could provide additional space and position for accommodation of lithium ions. For the purpose of capacity improvement, the creation of more edges and other defects in graphene nanosheets favor reversible capacities (794–1054 mAh/g).[239] Graphene nanosheets with loose structures show an enhanced theoretical capacity as high as 744 mAh/g, compared with the theoretical graphite capacity of 372 mAh/g.[240] It has been reported that highly hydrogenated graphene (HHG) with H/C = 0.76 gives a capacity of 800 mAh/g after 55 cycles.[241]

Although transition metal oxides possess high theoretical capacities (SnO_2, 790 mAh/g; Co_3O_4, 890 mAh/g; Fe_2O_3, 1006 mAh/g; and Mn_3O_4, 936 mAh/g), poor conductivity, low stability, short cycling life, and inter-particle agglomeration inhibit these materials to be used as practical anode materials. Graphene-transition metal oxide composites seem to be promising for Li-ion applications. Using graphene nanosheets as matrices for *in situ* synthesis and dispersion of the metal oxide nanostructures, most of the problems existing in current anode materials, such as poor conductivity, inter-particle agglomeration, and unbearable volume changes, during the discharge/charge cycles can be overcome. Therefore, active materials and surfaces of the nanoparticles (NPs) can be accessible and participate in Li/electron diffusion. Furthermore, elastic and flexible graphene sheets can accommodate readily volume expansion of the NPs upon Li insertion/extraction. At the same time, the presence of NPs also efficiently reduces the restack of graphene. This structural integrity of the NPs on graphene can be maintained and eventually increase the life span after many cycles of discharge/charge.[25] To further effectively reduce the easy aggregation of nanoparticles with high surface free energy, different graphene/metal oxide composites, such as ordered composites,[242] graphene-wrapped metal oxide,[243] or graphene-encapsulated metal oxide nanoparticles,[244] have been designed which facilitate the formation of an integrated structure with an electron conductive network and shortened ion transport paths. In this manner, eventually enhanced electrochemical performance can be achieved resulting from size effects and interfacial interactions. As a result, the capacity, rate capability, cycling stability, and energy and power densities will be substantially improved. For example, low conductivity and large volume change (200–300%) of SnO_2 during lithiation/delithiation has significantly limited its practical application due to cracking of electrode and loss of electrical contact, resulting from large internal stress. However, SnO_2/graphene composites can deliver a charge capacity of 840 mAh/g after 30 charge/discharge cycles at a current density of 67 mA/g.[245] Similar results have been demonstrated with other graphene-based composites, such as V_2O_5/graphene[246] and Si/graphene.[247,248] In addition, graphene–$LiFePO_4$ composites prepared with different methods were employed as cathode materials, which can remarkably improve electronic conductivity (10^{-9} S/cm^2) and lithium-ion diffusivity (10^{-14}–$10^{-16}cm^2/s$) of $LiFePO_4$.[249–252]

1.5.1.2 Supercapacitors

Graphene can be used as electrode in a supercapacitor. Ruoff et al. reported graphene-based supercapacitor with chemically modified graphene (CMG) as the electrode material. Owing to the high surface area of the CMG agglomerate, both sides of the individual sheets are exposed to the electrolyte, CMG-based supercapacitor exhibits specific capacitances of 135 and 99 F/g in aqueous KOH and organic electrolytes, respectively.[253] Other groups also report graphene materials used in supercapacitor which displays high specific capacitance (205 F/g)[254] and long cycle life.[255]

Composites derived from graphene and electrically conducting polymers (such as polyaniline, PANI) have been demonstrated to improve the performance of supercapacitors due to the interaction of the PANI and graphene materials. First, the addition of graphene sheets can improve electrical conductivity of PANI and, thus, increase power density. Second, two charge storage mechanisms including electric double-layer charging–discharging of graphene sheets and pseudo-capacitive redox reactions of PANI exist in the composite materials resulting in a higher supercapacitor performance.[256] Supercapacitor devices based on composite films of chemically converted graphene (CCG) and polyaniline nanofibers (PANI-NFs) show a large electrochemical capacitance (210 F/g) at a discharge rate of 0.3 A/g.[257] The morphological and electrochemical property changes of PANI-NFs induced by charge/discharge cycling can be greatly reduced.

Composites of graphene–metal oxide or graphene-hydroxide have also been used in supercapacitors. Graphene sheets in the electrode could be separated by nanoparticles, such as RuO_2, MnO_2, and $Ni(OH)_2$. As a result, high specific capacitance, enhanced rate capability, excellent electrochemical stability, and high-energy density can be obtained. The synergistic effect of both components enables a high specific capacitance for $Ni(OH)_2$ nanoplates/graphene sheets[258] and $Co(OH)_2$/graphene composites.[259]

1.5.1.3 Fuel Cells

Opposing the direct combustion of hydrogen and oxygen gases to produce thermal energy, proton exchange membrane fuel cell (PEMFC) transforms the chemical energy liberated during the electrochemical reaction of hydrogen and oxygen into electrical energy at low temperature (50–100°C). PEMFC has been regarded as the next-generation electrochemical device with promising applications in transportation, stationary, as well as portable electronics. Although Pt is the most efficient catalyst used for fuel cells so far, the high price and the poor utilization efficiency of Pt catalyst loading per unit area limit its practical applications.[260] Currently, the research on PEMFC focuses on how to obtain higher catalytic activity than the standard carbon-supported platinum particle catalysts used in current PEMFC, and how to reduce the poisoning of PEM fuel cell catalysts by impurity gases.[261]

Graphene sheets with high conductivity and high surface area have the potential to be used as catalyst support to reduce the aggregation of Pt particles in the PEMFC. High specific surface area (SSA) and good electrochemical durability can be achieved by the Pt/graphene catalyst. Compared with Pt NPs supported on commonly used commercial carbon blacks, Pt/graphene shows not only larger specific surface area and higher oxygen reduction reaction (ORR) activity, but also excellent stability after 5000 cycles.[262] It has been demonstrated that the current density of the 3D graphene catalysts (1.6 mA/cm^2) is higher than that of commercial Pt/C catalysts (1.4 mA/cm^2).[263] In addition, nitrogen defects in graphene could act as anchoring sites for deposition of platinum nanoparticles, the improved performance of fuel cells with N-doping graphene as catalyst support can be attributed to increased electrical conductivity and improved carbon–catalyst binding.[264,265]

Currently, most of the PEMFCs use platinum particles on carbon supports to promote both hydrogen oxidation and oxygen reduction. For the purpose of reducing the cost of PEMFC, nonprecious metal electrocatalysts have caused great attention. As nonprecious metal electrocatalysts, graphene-based composites as catalysts of fuel cells were exploited for replacing Pt electrode. Co_3O_4/graphene was reported as a synergistic catalyst for oxygen reduction reaction. In 1 M and 6 M KOH electrolytes, the Co_3O_4/N-rmGO (N-doped reduced mildly oxidized graphene oxide) ORR catalyst matched the performance of freshly loaded Pt/C catalyst in the same current density, accompanied by a positive shift in the ORR onset potential from 0.1 M KOH. In addition, the hybrid exhibited superior durability to Pt/C catalyst in 0.1–6 M KOH, with little decay in ORR activity over 10,000–25,000 s of continuous operation. In contrast, the Pt/C catalyst exhibited 20–48% decrease in activity in 0.1–6 M KOH, giving lower long-term ORR currents than the stable currents sustained by the Co_3O_4/N-rmGO hybrid catalyst.[266]

To overcome the disadvantages of sulfonated tetrafluoroethylene-based fluoropolymer-copolymer (Nafion) membranes, including high cost, low conductivity at low humidity and/or high temperature, loss of mechanical stability at high temperature, elevated methanol permeability, and restricted operation temperature,[260] functionalized graphene oxide/Nafion nanocomposites membrane[267,268] have been developed and tested in PEMFC. Also, poly ethylene oxide (PEO)/GO membranes without polymer modifications were reported, which had a proton conductivity of 0.09 S/cm at 60°C, and gave a power density of 53 mW/cm^2 in hydrogen PEMFC.[269]

1.5.1.4 Solar Cells

Owing to its high theoretical conductivity and good optical transparency (one-layer graphene transmittance ≈97.7%), graphene has become an attractive candidate for transparent electrode materials.[270] In a pioneering work, solution-processed graphene films were used to serve as transparent conductive anodes for organic photovoltaic cells, and the corresponding values of the transmittance and sheet resistance were 95–85%, and 100–500 kΩ/sq, respectively.[271] To produce ITO-free solar cells, 6–30 nm thick graphene film with an average sheet resistance varying from 1350 to 210 Ω/sq, and an optical transparency from 91% to 72% were employed as transparent and conductive anode to fabricate organic photovoltaic (OPV) devices.[272] Poly(3,4-ethylenedioxythiophene) (PEDOT) was introduced to facilitate hole injection/extraction as a buffer layer to overcome the large hole-injection barrier between graphene film (work function: 4.2 eV) and P3HT (work function: 5.2 eV). C_{60}-grafted graphene nanosheets[273] were used to enhance carrier mobility and suppress charge recombination. Graphene as an acceptor material in an OPV device was first reported by Liu et al.[274] in which organic solution-processed graphene material was used as an acceptor material, and poly (3-hexylthiophene) (P3HT) and poly (3-octylthiophene) (P3OT) as electron donors. The interaction between graphene and P3OT/P3HT makes this composite work well as the active layer. The devices after annealing treatment at 160°C for 20 min had a η value of 1.4%, V_{oc} (open-circuit voltage) of 0.92 V, J_{sc} (short-circuit current density) of 4.2 mA/cm, and FF (filling factor) of 0.37. A graphene oxide film was also incorporated between the photoactive P3HT: PCBM (phenyl-C61-butyric acid methyl ester) layer and the transparent and conducting ITO,[275] and recombination of electrons and holes and leakage currents were decreased. Thus, the efficiencies in the OPV were comparable to devices fabricated with PEDOT:PSS as the hole transport layer. By different functionalization and reduction methods, the energy level of graphene could be tuned. For example, the device adopting only PEDOT: PSS shows J_{sc} of 8.06 mA/cm², V_{oc} of 0.58 V, and PCE of 3.10%. For the device using FLGs and PEDOT:PSS at the ratio of 2:1 (v/v), the J_{sc} and power conversion efficiency (PCE) increased to 10.68 mA/cm² and 3.70%, respectively.[276]

In a dye-sensitized solar cell (DSSC), by incorporating a conductive network into TiO_2 anode, charge transfer can be promoted, and consequently backward recombination can be efficiently reduced. As known, graphene shows extremely high carrier mobility and unique 2D flexible structure.[277] It was reported that the performance of TiO_2/graphene composite is much better than that of pure TiO_2 or TiO_2/CNT, which means that graphene is a good promoter for electron transportation.[278,279] Furthermore, the porous network of the composite enhances light scattering at the photoanode and achieves a power conversion efficiency of about 4–7%; this is much higher than the commercial P25 TiO_2.[277,278] Recently, research on cost-effective alternative catalysts to Pt counter electrode has also attracted considerable attention. It was reported that the composite of TiN-supported on N-doped graphene has comparable electrochemical catalytic performance to Pt for I_3^- reduction. The DSSC with N-doped graphene-supported TiN as the counter electrode gave a power conversion efficiency of about 5.78%, even higher than that with Pt counter electrode (5.03%).[280]

GS films have been used to form Schottky junctions with the combination with n-Si wafer. In that system, GS films both serve as a transparent electrode for light illumination and an active layer for electron–hole separation and hole transport.[281] The design and fabrication of CdSe nanobelt/graphene Schottky junction solar cells were reported by Yu Ye et al.[282] Owing to the work function difference between graphene ($\Phi_G \approx 4.66$ eV) and CdSe ($\Phi_{CdSe} \approx 4.2$ eV), a built-in potential (Vi) forms in the CdSe near the CdSe nanobelts/graphene interface. The photo-generated holes and electrons are driven toward the Schottky electrode (graphene film) and CdSe nanobelts, respectively, by the built-in electric field. The high performance of the cell, with an open-circuit voltage of 0.51 V, a short-circuit current density of 5.75 mA/cm², and an energy conversion efficiency of 1.25% were obtained.[282]

1.5.2 SENSORS

Graphene possesses an exceptionally high surface-to-volume ratio and high conductivity, which makes the electrical resistivity of monolayer graphene exhibit significant changes on exposure to different molecules. Therefore, high sensitivity to molecular disruption on graphene surface can be obtained which enables monolayer graphene to be used as a promising candidate for sensors. Oxygen sensor with monolayer graphene was fabricated.[283] The current of the sensor shows a rapid increase when the O_2 concentration changes to 1.25 vol% in the open cavity with a continuous gas flow. A further increase in the current for the sensor is observed when the O_2 concentration increases to 4.7 vol%. These abrupt current increases are because of the change in charges in graphene upon a shift in O_2 concentration. The sensing mechanism of O_2 on graphene can be reasonably explained as follows: on O_2, molecules are attached on the graphene thin film and they act as p-type dopants. Nanoporous-reduced graphene oxide sheets produced by hydrothermal steaming at 200°C have been used for NO_2 detection, compared with nonporous-reduced graphene oxide annealed at the same temperature, it exhibits nearly two orders of magnitude increase in sensitivity and improved recovery time.[284]

The large electrochemical potential window (~2.5 V in 0.1 mM phosphate buffer saline solution) of graphene enables the detection of molecules with high oxidation or reduction potential. In a typical setup of electrochemical sensing, a graphene thin film is used as the electrode.[285] In addition to the direct use of graphene as working electrode, graphene or graphene derivatives have been used to modify the conventional glassy carbon electrode[286] and Au electrode.[287] Graphene nanoparticle composites have been applied in electrochemical sensing to take advantage of the good catalytic activity of noble metal nanoparticles.[288,289] In a typical FET-based sensor, graphene is used as a conducting channel between drain and source electrodes. A gate potential is applied through back-gate (typical SiO_2 thin layer) or top-gate (electric double layer in electrolyte) for gas and biosensing, respectively. The absorption of analyte molecules or change of local environment leads to change of its electrical conductance. Fluorescence-based detection methods possess the advantages of both high sensitivity and selectivity in the analysis of biomolecules. Induced by the fluorescence resonance energy transfer (FRET), graphene has the capability of fluorescence quenching which can be used in fluorescence-based bio-detection.[290] Various kinds of graphene-based sensors, such as H_2-sensors,[291] glucose bio-sensor,[292] or Cd^{2+}-sensor,[293] have been developed based on different detection mechanisms.

1.5.3 FIELD EFFECT TRANSISTORS

Graphene has significant potential for application in electronics, but cannot be used for effective field-effect transistors operating at room temperature because it is a semimetal with a zero band-gap.[294] As mentioned in Section 1.2.5, processing graphene sheets into GNRs with widths of less than 10 nm can open up a band gap that is large enough for room-temperature transistor operation with on–off ratios of about 10^7 at room temperature.[295] However, the currently made GNR devices often have low driving currents or transconductances.[294] Graphene nanomesh (GNM),[294,296] nanoperforated graphene,[297] or holey reduced graphene oxide (hRGO)[291] have also been developed, which not only open the band gap of graphene but also provide high driving current. It has been demonstrated that FETs fabricated with GNM as the channel material can support current 100 times greater than that of the individual GNR devices. At the same time, by tuning the neck width, the on–off ratio has developed to be comparable with that achieved in individual GNR devices.[282] The chemical functionalization of graphene offers an alternative approach in controlling its electronic properties. The specially designed FETs of bilayer graphene and organic seed layer made from a derivative of polyhydroxystyrene show an on–off current ratio of around 100 at room temperature.[298]

1.5.4 GRAPHENE–POLYMER COMPOSITES

Graphene has shown many unique properties, such as high carrier mobility at room temperature (≈ 1000 cm^2/V s), good optical transparency ($\approx 97.7\%$), high Young's modulus (~ 1 TPa), and excellent thermal conductivity (3000–5000 W/mK). These advantages mean that graphene can be integrated into polymers to make composites that have enhanced properties than pure polymers.

Graphene–polymer composites show enhanced electrical properties only by introducing a small amount of graphene into polymer. In various insulating polymer matrixes, low percolation thresholds were achieved. The first graphene-based composite materials were reported by the Ruoff group;[299] polystyrene–graphene composite with ≈ 1 vol% graphene had a conductivity of ≈ 0.1 S m^{-1}, sufficient for many electrical applications, and the percolation threshold is 0.1 vol%. Thereafter, many other polymer–graphene composite materials with low percolation thresholds were reported, such as poly(vinyl chloride/vinyl acetate) copolymer–graphene composites (with a percolation threshold of 0.15 vol%)[300] and PET–graphene composites (with a conductivity of 2.11 S m^{-1} obtained at a loading of 3.0 vol% graphene).[301] Graphene can also enhance the conductivity of conductive polymers, for example, the conductivity of composite films of chemically converted graphene (CCG) and polyaniline nanofibers (PANI-NFs) containing 44% CCG (5.5×10 S/m) is about 10 times that of PANI-NF films.[257]

The mechanical properties of polymer can be enhanced by addition of graphene. The elastic modulus of PMMA at room temperature was increased by 33% for 0.01 wt% functionalized graphene sheets (FGS)-PMMA.[302] 2D poly(vinyl alcohol)-graphene oxide (PVA-GO) composite films containing 3 wt% GO obtained by vacuum filtration showed a Young's modulus and tensile yield strength of 4.8 GPa and 110 ± 7 MPa, respectively.[303] A significant enhancement in the mechanical properties of pure poly(vinyl chloride) films was obtained with a 2 wt% loading of graphene, such as a 58% increase in Young's modulus and an almost 130% improvement of tensile strength.[304] 3D graphene foam (GF) synthesized by template-directed CVD obviously improved the mechanical properties of the PDMS (poly dimethyl siloxane) matrix; the ultimate strength of the GF/PDMS composites with a GF loading as low as $\approx$0.5 wt% was increased by a factor of $\approx$1.3 compared with the pure PDMS.[55]

Thermal properties generally include thermal stability and thermal conductivity. Compared with pure polymer, graphene–polymer composites show enhanced thermal properties. For example, FGS-PMMA composite yielded an increase of 30°C in T_g (glass transition temperature) with a loading of 0.05 wt% FGS.[302] The inclusion of FGS increased thermal conductivity by up to 6% in the 0.25 wt% FGS-silicone foam composite.[305]

1.6   CONCLUSIONS AND PERSPECTIVES

As a new material with exceptional 2D structures and novel properties, graphene is attracting increasing attention from both the fundamental research and the practical application communities. The emerging graphene-based microelectronics offer promising opportunities for semiconductor industry, from micrometer- to nanometer-scale electronics, as well as for various other fields. However, before the realization of practical applications, there are many challenges that we have to face and overcome.

Most of the existing fabrication methods are limited to producing graphene up to only a few tens of micrometers with still a high density of structural defects. The limitation in sample size and quality is currently the main bottleneck in building graphene-based electronic devices. The exceptional properties can only be observed in defect-free pristine graphene; therefore, the realization of practical application of graphene in the electronic device was ultimately determined by the quality of graphene. The fabrication of large, single-crystalline graphene with few defects by optimizing synthetic strategies and manipulating the structure and properties after synthesis are crucial to electronic devices.

The electrical properties of graphene are closely related to its thickness. Monolayer and bilayer graphene are of most interest to the community. Recently, breakthroughs have been made in the preparation of graphene with relative high quality by CVD and epitaxial growth on silicon carbide. New approaches to the fabrication of defect-free monolayer graphene with large areas are still in high demand.

In addition to the physicists and materials scientists, graphene also brought great opportunities as well as challenges to chemists as chemistry will play a major role in tailoring the electronic structure of graphene for future devices and circuitry.[306] Atom-by-atom covalent functionalization of the graphene surface to introduce predesigned patterning of the conjugated network enables precise manipulation of the energy band gap. Such as construction of surface nanoribbons with well-defined edges or arrays of graphene quantum dots can help to tune the electrical properties of graphene and make graphene available for incorporation into electronic and optical devices. Lithographic fabrication of conducting channels of graphene, P–N junction, and confinement-related energy band gaps can be created in a controlled way by hydrogenation and fluorization, insulating regions with graphane and fluorographene to create patterns. In addition to conventional planar graphene design, the geometry of graphene and 3D graphene may inhibit restacking of planar graphene, from planar graphene to curve or facet structures, producing large 3D structures with graphene walls, building the bridge between 2D and 3D.

The emerging 2D carbon crystalline graphene and graphene-related materials have attracted enormous attention from scientific and technological communities. The rapid development of graphene will enable the real applications of graphene realized in various fields and eventually play an important role in changing our lives in the near future.

ACKNOWLEDGMENTS

This work was supported by the NSFC (grants 51072028, 20876026, 20836002, and 20725619). We thank Dr. Volker Presser, Drexel University, for editing the manuscript, and Dr. Han Hu, Quan

Zhou, Wubo Wan, Yanfeng Dong, Dalian University of Technology, for collecting materials for the manuscript. Collaboration between Drexel University and Dalian University of Technology was supported by the Cheung Kong Scholarship.

REFERENCES

1. Novoselov, K. S. et al., *Science*, 306, 666, 2004.
2. The Royal Swedish Academy of Science, *Scientific Background on the Nobel Prize in Physics 2010, Graphene.*
3. Wallace, P. R., *Phys. Rev.*, 71, 622, 1947.
4. McClure, J. W., *Phys. Rev.*, 104, 666, 1956.
5. Semenoff, G. W., *Phys. Rev. Lett.,* 53, 2449, 1984.
6. Divincenzo, D. P. and Mele, E. J., *Phys. Rev. B*, 29, 1685, 1984.
7. May, J. W., *Surf. Sci.,* 17, 267, 1969.
8. Patil, H. R. and Blakely, J. M., *J. Appl. Phys.,* 45, 3806, 1974.
9. Ebbesen, T. W. and Ajayan, P. M., *Nature*, 358, 220, 1992.
10. Ando, Y. et al., *Carbon*, 35, 153, 1997.
11. Zhao, X. et al., *Carbon*, 35, 775, 1997.
12. Wu, Y. H. et al., *Adv. Mater.,* 14, 64, 2002.
13. Kelly, B. T., *Physics of Graphite*, Springer, London, 1981.
14. Dresselhaus, M. S. and Dresselhaus, G., *Adv. Phys.,* 30, 139, 1981.
15. Novoselov, K. S. et al., *Proc. Nat. Acad. Sci.*, 102, 10451, 2005.
16. Novoselov, K. S. et al., *Nature*, 438, 197, 2005.
17. Zhang, Y. B. et al., *Nature*, 438, 201, 2005.
18. Kroto, H. W. et al., *Nature*, 318, 162, 1985.
19. Iijima, S., *Nature*, 354, 56, 1991.
20. Singh, V. et al., *Prog. Mater Sci.,* 56, 1178, 2011.
21. Rao, C. N. R. et al., *Angew. Chem. Int. Ed.,* 48, 7752, 2009.
22. Pei, S. and Cheng, H., *Carbon*, 50, 3210, 2012.
23. Allen, M. J. et al., *Chem. Rev.,* 110, 132, 2010.
24. Chen, D. et al., *Chem. Rev.*, 112, 6027, 2012.
25. Huang, X. et al., *Chem. Soc. Rev.,* 41, 666, 2012.
26. Loh, K. P. et al., *Nat. Chem.*, 2, 1015, 2010.
27. Ebbesen, T. W. and Hiura, H., *Adv. Mater.,* 7, 582, 1995.
28. Hiura, H. et al., *Nature*, 367, 148, 1994.
29. Lu, X. K. et al., *Nanotechnology*, 10, 269, 1999.
30. Lu, X. K. et al., *Appl. Phys. Lett.,* 75, 193, 1999.
31. Zhang, Y. B. et al., *Appl. Phys. Lett.,* 86, 2005.
32. Liu, S. P. et al., *Acta. Physica. Sinica.*, 54, 4251, 2005
33. Jeon, I. Y. et al., *Proc. Nat. Acad. Sci.*, 109, 5588, 2012.
34. Chen, J. et al., *J. Mater. Chem.,* 22, 19625, 2012.
35. Shang, N. G. et al., *Chem. Commun.,* 48, 1877, 2012.
36. Blake, P. et al., *Nano Lett.,* 8, 1704, 2008.
37. Hernandez, Y. et al., *Nat. Nanotechnol.,* 3, 563, 2008.
38. Lotya, M. et al., *J. Am. Chem. Soc.,* 131, 3611, 2009.
39. Green, A. A. and Hersam, M. C., *Nano Lett.,* 9, 4031, 2009.
40. Wang, X. Q. et al., *Chem. Commun.,* 46, 4487, 2010.
41. Liu, N. et al., *Adv. Funct. Mater.,* 18, 1518, 2008.
42. Fulvio, P. F. et al., *Adv. Funct. Mater.,* 21, 2208, 2011.
43. Van Bommel, A. J. et al., *Surf. Sci.,* 48, 463, 1975.
44. de Heer, W. A. et al., *Proc. Natl. Acad. Sci.*, 108, 16900, 2011.
45. Berger, C. et al., *J. Phys. Chem. B,* 108, 19912, 2004.
46. de Heer, W. A. et al., *Solid State Commun.,* 143, 92, 2007.
47. Reina, A. et al., *Nano Lett.,* 9, 30, 2009.
48. Kim, K. S. et al., *Nature*, 457, 706, 2009.
49. Cai, W. W. et al., *Appl. Phys. Lett.,* 95, 2009.
50. Li, X. S. et al., *Science*, 324, 1312, 2009.

51. Bae, S. et al., *Nat. Nanotechnol.,* 5, 574, 2010.
52. Li, X. et al., *Science,* 324, 1312, 2009.
53. Pan, Y. et al., *Adv. Mater.,* 21, 2739, 2009.
54. Gao, L. B. et al., *Nat. Commun.,* 3, 699, 2012.
55. Chen, Z. P. et al., *Nat. Mater.,* 10, 424, 2011.
56. Wang, J. J. et al., *Carbon,* 42, 2867, 2004.
57. Sun, Z. Z. et al., *Nature,* 468, 549, 2010.
58. Ruan, G. D. et al., *ACS Nano,* 5, 7601, 2011.
59. Ruiz-Hitzky, E. et al., *Adv. Mater.,* 23, 5250, 2011.
60. Pei, S. F. and Cheng, H. M., *Carbon,* 50, 3210, 2012.
61. Hummers, W. S. and Offeman, R. E., *J. Am. Chem. Soc.,* 80, 1339, 1958.
62. He, H. et al., *Chem. Phys. Lett.,* 287, 53, 1998.
63. Hakimi, M. and Alimard, P., *World Appl. Program.,* 6, 377, 2012.
64. Schniepp, H. C. et al., *J. Phys. Chem. B,* 110, 8535, 2006.
65. McAllister, M. J. et al., *Chem. Mater.,* 19, 4396, 2007.
66. Li, X. L. et al., *J. Am. Chem. Soc.,* 131, 15939, 2009.
67. Wu, Z. S. et al., *ACS Nano,* 3, 411, 2009.
68. Panchokarla, L. S. et al., *Adv. Mater.,* 21, 4726, 2009.
69. Lv, W. et al., *ACS Nano,* 3, 3730, 2009.
70. Dallinger, D. and Kappe, C. O., *Chem. Rev.,* 107, 2563, 2007.
71. Liu, X. T. et al., *Appl. Catal., A,* 264, 53, 2004.
72. He, X. J. et al., *Carbon,* 48, 1662, 2010.
73. Zhu, Y. W. et al., *Carbon,* 48, 2118, 2010.
74. Hu, H. et al., *Carbon,* 50, 3267, 2012.
75. Zhou, Q. et al., *J. Mater. Chem.,* 22, 6061, 2012.
76. Lee, S. W. et al., *J. Phys. Chem. Lett.,* 3, 772, 2012.
77. Cote, L. J. et al., *J. Am. Chem. Soc.,* 131, 11027, 2009.
78. Gilje, S. et al., *Adv. Mater.,* 22, 419, 2010.
79. Stankovich, S. et al., *Carbon,* 45, 1558, 2007.
80. Li, D. et al., *Nat. Nanotechnol.,* 3, 101, 2008.
81. Shin, H. J. et al., *Adv. Funct. Mater.,* 19, 1987, 2009.
82. Pei, S. F. et al., *Carbon,* 48, 4466, 2010.
83. Zhao, J. P. et al., *ACS Nano,* 4, 5245, 2010.
84. Zhou, Y. et al., *Chem. Mater.,* 21, 2950, 2009.
85. Gao, J. et al., *Chem. Mater.,* 22, 2213, 2010.
86. Dua, V. et al., *Angew. Chem. Int. Ed.,* 49, 2154, 2010.
87. Fan, Z. J. et al., *Carbon,* 48, 1686, 2010.
88. Fan, Z. J. et al., *ACS Nano,* 5, 191, 2011.
89. Liu, Y. Z. et al., *J. Mater. Chem.,* 21, 15449, 2011.
90. Qian, W. et al., *Nano Res.,* 2, 706, 2009.
91. Wang, H. L. et al., *J. Am. Chem. Soc.,* 131, 9910, 2009.
92. Choucair, M. et al., *Nat. Nanotechnol.,* 4, 30, 2009.
93. Abergel, D. S. L. et al., *Adv. Phys.,* 59, 261, 2010.
94. Yang, L. et al., *Phys. Rev. Lett.,* 99, 186801, 2007.
95. Son, Y. W. et al., *Nature,* 444, 347, 2006.
96. Dalosto, S. D. and Levine, Z. H., *J. Phys. Chem. C,* 112, 8196, 2008.
97. Murayama, H. and Maeda, T., *Nature,* 345, 791, 1990.
98. Campos-Delgado, J. et al., *Nano Lett.,* 8, 2773, 2008.
99. Mahanandia, P. et al., *Mater. Res. Bull.,* 43, 3252, 2008.
100. Wei, D. C. et al., *J. Am. Chem. Soc.,* 131, 11147, 2009.
101. Ryu, S. et al., *Nano Lett.,* 8, 4597, 2008.
102. Datta, S. S. et al., *Nano Lett.,* 8, 1912, 2008.
103. Campos, L. C. et al., *Nano Lett.,* 9, 2600, 2009.
104. Ci, L. et al., *Nano Res.,* 1, 116, 2008.
105. Li, X. L. et al., *Science,* 319, 1229, 2008.
106. Terrones, M. et al., *Nano Today,* 5, 351, 2010.
107. Kosynkin, D. V. et al., *Nature,* 458, 872, 2009.
108. Jiao, L. Y. et al., *Nat. Nanotechnol.,* 5, 321, 2010.

109. Jiao, L. Y. et al., *Nature*, 458, 877, 2009.
110. Wang, X. R. and Dai, H. J., *Nat. Chem.*, 2, 661, 2010.
111. Cano-Marquez, A. G. et al., *Nano Lett.*, 9, 1527, 2009.
112. Datta, S. S., *J. Appl. Phys.*, 108, 024307, 2010.
113. Kim, K. et al., *ACS Nano*, 4, 1362, 2010.
114. Rempala, P. et al., *J. Am. Chem. Soc.*, 126, 15002, 2004.
115. Di Stefano, M. et al., *Chem. Phys.*, 314, 85, 2005.
116. Wu, J. S. et al., *Chem. Rev.*, 107, 718, 2007.
117. Yang, X. et al., *J. Am. Chem. Soc.*, 130, 4216, 2008.
118. Cai, J. M. et al., *Nature*, 466, 470, 2010.
119. Geim, A. K. and Novoselov, K. S., *Nat. Mater.*, 6, 183, 2007.
120. Zhu, Y. et al., *Adv. Mater.*, 22, 3906, 2010.
121. Neto, A. C. et al., *Phys. World*, 19, 33, 2006.
122. Castro Neto, A. H. et al., *Rev. Mod. Phys.*, 81, 109, 2009.
123. Katsnelson, M. I. and Geim, A. K., *Appl. Catal., A*, 366, 195, 2008.
124. Fasolino, A. et al., *Nat. Mater.*, 6, 858, 2007.
125. Guinea, F. et al., *Solid State Commun.*, 149, 1140, 2009.
126. Zhu, W. J. et al., *Nano Lett.*, 12, 3431, 2012.
127. Ishigami, M. et al., *Nano Lett.*, 7, 1643, 2007.
128. Stolyarova, E. et al., *Proc. Natl. Acad. Sci.*, 104, 9209, 2007.
129. Bao, W. et al., *Nat Nano*, 4, 562, 2009.
130. Banhart, F. et al., *ACS Nano*, 5, 26, 2011.
131. Terrones, H. et al., *Rep. Prog. Phys.*, 75, 062501, 2012.
132. Girit, Ç. Ö. et al., *Science*, 323, 1705, 2009.
133. Meyer, J. C. et al., *Nano Lett.*, 8, 3582, 2008.
134. Botello-Mendez, A. R. et al., *Phys. Status Solidi-R.*, 3, 181, 2009.
135. Jia, X. et al., *Nanoscale*, 3, 86, 2011.
136. Nakada, K. et al., *Phys. Rev. B*, 54, 17954, 1996.
137. Son, Y.-W. et al., *Phys. Rev. Lett.*, 97, 216803, 2006.
138. Koskinen, P. et al., *Phys. Rev. Lett.*, 101, 115502, 2008.
139. Wei, D. and Wang, F., *Surf. Sci.*, 606, 485, 2012.
140. Jiang, D. et al., *J. Chem. Phys.*, 126, 134701, 2007.
141. Gorjizadeh, N. and Kawazoe, Y., *J. Nanomater.*, 2010, 513501, 2010.
142. Cruz-Silva, E. et al., *Phys. Rev. Lett.*, 105, 45501, 2010.
143. Avouris, P. et al., *Nat. Nanotechnol.*, 2, 605, 2007.
144. Chen, J. H. et al., *Nat. Nanotechnol.*, 3, 206, 2008.
145. Rao, C. N. R. et al., *J. Mater. Chem.*, 19, 2457, 2009.
146. Bolotin, K. I. et al., *Solid State Commun.*, 146, 351, 2008.
147. Novoselov, K. S. et al., *Science*, 306, 666, 2004.
148. Service, R. F., *Science*, 324, 875, 2009.
149. Novoselov, K. S. et al., *Science*, 315, 1379, 2007.
150. Du, X. et al., *Nature*, 462, 192, 2009.
151. Bolotin, K. I. et al., *Nature*, 475, 7354, 2009.
152. Morpurgo, A. F., *Nature*, 462, 170, 2009.
153. Berger, C. et al., *Science*, 312, 1191, 2006.
154. Ohta, T. et al., *Science*, 313, 951, 2006.
155. Bao, W. et al., *Nat. Phys.*, 7, 948, 2011.
156. Tapaszto, L. et al., *Nat Nano*, 3, 397, 2008.
157. Calvo, H. L. et al., *Appl. Phys. Lett.*, 98, 232103, 2011.
158. Gomez-Navarro, C. et al., *Nano Lett.*, 10, 1144, 2010.
159. Su, Q. et al., *Adv. Mater.*, 21, 3191, 2009.
160. Lim, G.-K. et al., *Nat Photon*, 5, 554, 2011.
161. Nair, R. R. et al., *Science*, 320, 1308, 2008.
162. Kravets, V. G. et al., *Phys. Rev. B*, 81, 155413, 2010.
163. Kim, J. et al., *Mater. Today* 13, 28, 2010.
164. Liu, J. et al., *Appl. Phys. Lett.*, 93, 041106, 2008.
165. Li, Z. Q. et al., *Nat. Phys.*, 4, 532, 2008.
166. Wang, F. et al., *Science*, 320, 206, 2008.

167. Eda, G. et al., *Adv. Mater.,* 22, 505, 2010.
168. Pan, D. et al., *Adv. Mater.,* 22, 734, 2010.
169. Liu, Z. et al., *J. Am. Chem. Soc.,* 130, 10876, 2008.
170. Bao, Q. L. et al., *Adv. Funct. Mater.,* 20, 782, 2010.
171. Liu, Z.-B. et al., *J. Phys. Chem. B,* 113, 9681, 2009.
172. Zhao, Q. et al., *Phys. Rev. B*, 65, 144105, 2002.
173. Lee, C. et al., *Science*, 321, 385, 2008.
174. Huang, M. et al., *Nano Lett.,* 11, 1241, 2011.
175. Gomez-Navarro, C. et al., *Nano Lett.,* 8, 2045, 2008.
176. Bai, H. et al., *Adv. Mater.,* 23, 1089, 2011.
177. Xu, Y. X. et al., *ACS Nano*, 4, 4324, 2010.
178. Chen, C. M. et al., *Adv. Mater.,* 21, 3007, 2009.
179. Chen, H. et al., *Adv. Mater.,* 20, 3557, 2008.
180. Dikin, D. A. et al., *Nature*, 448, 457, 2007.
181. Xu, Z. and Gao, C., *Nat. Commun.*, 2, 571, 2011.
182. Prasher, R., *Science*, 328, 185, 2010.
183. Balandin, A. A., *Nat. Mater.,* 10, 569, 2011.
184. Balandin, A. A. et al., *Nano Lett.,* 8, 902, 2008.
185. Berber, S. et al., *Phys. Rev. Lett.,* 84, 4613, 2000.
186. Cai, W. et al., *Nano Lett.,* 10, 1645, 2010.
187. Seol, J. H. et al., *Science*, 328, 213, 2010.
188. Elias, D. C. et al., *Science*, 323, 610, 2009.
189. Balog, R. et al., *J. Am. Chem. Soc.,* 131, 8744, 2009.
190. Sessi, P. et al., *Nano Lett.,* 9, 4343, 2009.
191. Nikitin, A. et al., *Phys. Rev. Lett.,* 95, 225507, 2005.
192. Khare, B. N. et al., *Nano Lett.,* 2, 73, 2001.
193. Schrier, J., *ACS Appl. Mat. Interfaces,* 3, 4451, 2011.
194. Nair, R. R. et al., *Small*, 6, 2877, 2010.
195. Seifert, G. et al., *Appl. Phys. Lett.*, 77, 1313, 2000.
196. Kelly, K. F. et al., *Chem. Phys. Lett.,* 313, 445, 1999.
197. Hayashi, T. et al., *Nano Lett.,* 2, 491, 2002.
198. Touhara, H. and Okino, F., *Carbon*, 38, 241, 2000.
199. Zhou, J. et al., *Carbon*, 48, 1405, 2010.
200. Osuna, S. L. et al., *J. Phys. Chem. C,* 114, 3340, 2010.
201. Bon, S. B. et al., *Chem. Mater.,* 21, 3433, 2009.
202. Zbořil, R. et al., *Small*, 6, 2885, 2010.
203. Jeon, K.-J. et al., *ACS Nano*, 5, 1042, 2011.
204. Robinson, J. T. et al., *Nano Lett.,* 10, 3001, 2010.
205. Bielawski, C. W. et al., *Chem. Soc. Rev.,* 39, 228, 2010.
206. Xu, Z. and Gao, C., *ACS Nano*, 5, 2908, 2011.
207. Loh, K. P. et al., *J. Mater. Chem.,* 20, 2277, 2010.
208. Salavagione, H. J. et al., *Macromolecules*, 42, 6331, 2009.
209. Veca, L. M. et al., *Chem. Commun.,* 2565, 2009.
210. Zhang, X. et al., *Carbon*, 47, 334, 2009.
211. Liu, Z. B. et al., *J. Phys. Chem. B,* 113, 9681, 2009.
212. Niyogi, S. et al., *J. Am. Chem. Soc.*, 128, 7720, 2006.
213. Stankovich, S. et al., *Carbon*, 44, 3342, 2006.
214. Xu, Y. et al., *Adv. Mater.,* 21, 1275, 2009.
215. Wang, S. et al., *Adv. Mater.,* 20, 3440, 2008.
216. Yang, H. et al., *Chem. Commun.*, 3880, 2009.
217. Lomeda, J. R. et al., *J. Am. Chem. Soc.,* 130, 16201, 2008.
218. Yang, H. et al., *J. Mater. Chem.,* 19, 4632, 2009.
219. Bahr, J. L. et al., *J. Am. Chem. Soc.,* 123, 6536, 2001.
220. Chen, B. et al., *Chem. Mater.,* 17, 4832, 2005.
221. Englert, J. M. et al., *Nat Chem.,* 3, 279, 2011.
222. Zhu, H. et al., *J. Mater. Chem.,* 22, 2063, 2012.
223. Zhao, Y.-L. and Stoddart, J. F., *Acc. Chem. Res.,* 42, 1161, 2009.
224. Xu, Y. et al., *J. Am. Chem. Soc.*, 130, 5856, 2008.

225. Hao, R. et al., *Chem. Commun.,* 6576, 2008.
226. Wang, Y. et al., *Appl. Phys. Lett.*, 95, 063302, 2009.
227. Bai, H. et al., *Chem. Commun.,* 1667, 2009.
228. Liu, Q. et al., *Adv. Funct. Mater.,* 19, 894, 2009.
229. Zhu, C. et al., *ACS Nano*, 4, 2429, 2010.
230. Salas, E. C. et al., *ACS Nano*, 4, 4852, 2010.
231. Patil, A. J. et al., *Adv. Mater.,* 21, 3159, 2009.
232. Xu, Y. et al., *ACS Nano*, 4, 7358, 2010.
233. Chen, H. et al., *Adv. Mater.,* 20, 3557, 2008.
234. Gao, J. et al., *Chem. Mater.,* 22, 2213, 2010.
235. Pei, S. et al., *Carbon*, 48, 4466, 2010.
236. Shen, X. et al., *J. Colloid Interface Sci.,* 354, 493, 2011.
237. Liu, Z. et al., *J. Am. Chem. Soc.*, 130, 10876, 2008.
238. Yoo, E. J. et al., *Nano Lett.,* 8, 2277, 2008.
239. Pan, D. et al., *Chem. Mater.,* 21, 3136, 2009.
240. Dahn, J. R. et al., *Science*, 270, 590, 1995.
241. Chen, W. et al., *Nanoscale*, 4, 2124, 2012.
242. Wang, D. et al., *ACS Nano*, 4, 1587, 2010.
243. Zhou, G. et al., *Chem. Mater.,* 22, 5306, 2010.
244. Yang, S. et al., *Angew. Chem. Int. Ed.,* 49, 8408, 2010.
245. Wang, X. et al., *Carbon*, 49, 133, 2011.
246. Liu, H. and Yang, W., *Energy Environ. Sci.,* 4, 4000, 2011.
247. Chou, S. L. et al., *Electrochem. Commun.,* 12, 303, 2010.
248. Xiang, H. et al., *Carbon*, 49, 1787, 2011.
249. Wang, L. et al., *Solid State Ionics,* 181, 1685, 2010.
250. Zhou, X. et al., *J. Mater. Chem.,* 21, 3353, 2011.
251. Ding, Y. et al., *Electrochem. Commun.,* 12, 10, 2010.
252. Wang, Y. et al., *Mater. Lett.,* 71, 54, 2012.
253. Stoller, M. D. et al., *Nano Lett.,* 8, 3498, 2008.
254. Wang, Y. et al., *J. Phys. Chem. C,* 113, 13103, 2009.
255. Liu, C. et al., *Nano Lett.,* 10, 4863, 2010.
256. Wu, Z. S. et al., *Adv. Funct. Mater.,* 20, 3595, 2010.
257. Wu, Q. ct al., *ACS Nano*, 4, 1963, 2010.
258. Wang, H. L. et al., *J. Am. Chem. Soc.,* 132, 7472, 2010.
259. Chen, S. et al., *J. Phys. Chem. C,* 114, 11829, 2010.
260. Sahoo, N. G. et al., *Adv. Mater.*, 24, 4203, 2012.
261. http://en.wikipedia.org/wiki/Proton_exchange_membrane_fuel_cell
262. Kou, R. et al., *Electrochem. Commun.,* 11, 954, 2009.
263. Maiyalagan, T. et al., *J. Mater. Chem.,* 22, 5286, 2012.
264. Jafri, R. I. et al., *J. Mater. Chem.,* 20, 7114, 2010.
265. Qu, L. et al., *ACS Nano*, 4, 1321, 2010.
266. Liang, Y. et al., *Nat. Mater.,* 10, 780, 2011.
267. Zarrin, H. et al., *J. Phys. Chem. C,* 115, 20774, 2011.
268. Choi, B. G. et al., *ACS Nano*, 5, 5167, 2011.
269. Cao, Y. C. et al., *J. Power Sources,* 196, 8377, 2011.
270. Wan, X. et al., *Adv. Mater.,* 23, 5342, 2011.
271. Wu, J. et al., *Appl. Phys. Lett.,* 92, 263302, 2008.
272. Wang, Y. et al., *Appl. Phys. Lett.,* 95, 063302, 2009.
273. Yu, D. et al., *J. Phys. Chem. Lett.,* 2, 1113, 2011.
274. Liu, Z. et al., *Adv. Mater.,* 20, 3924, 2008.
275. Li, S. S. et al., *ACS Nano*, 4, 3169, 2010.
276. Nguyen, D. et al., *Nanotechnology*, 22, 295606, 2011.
277. Yang, N. et al., *ACS Nano*, 4, 887, 2010.
278. Sun, S. et al., *Appl. Phys. Lett.,* 96, 083113, 2010.
279. Tang, Y. B. et al., *ACS Nano*, 4, 3482, 2010.
280. Wen, Z. et al., *Adv. Mater.,* 23, 5445, 2011.
281. Li, X. et al., *Adv. Mater.,* 22, 2743, 2010.
282. Ye, Y. et al., *Nanoscale*, 3, 1477, 2011.

283. Chen, C. et al., *Appl. Phys. Lett.,* 99, 243502, 2011.
284. Han, T. H. et al., *J. Am. Chem. Soc.,* 133, 15264, 2011.
285. Shang, N. G. et al., *Adv. Funct. Mater.,* 18, 3506, 2008.
286. Wang, Z. et al., *J. Phys. Chem. C,* 113, 14071, 2009.
287. Xie, X. et al., *J. Phys. Chem. C,* 114, 14243, 2010.
288. Zhong, Z. et al., *Biosens. Bioelectron.,* 25, 2379, 2010.
289. Shan, C. et al., *Biosens. Bioelectron.,* 25, 1070, 2010.
290. Huang, X. et al., *Small,* 7, 1876, 2011.
291. Vedala, H. et al., *Nano Lett.,* 11, 2342, 2011.
292. Wang, Y. et al., *ACS Nano,* 4, 1790, 2010.
293. Li, J. et al., *Electrochem. Commun.,* 11, 1085, 2009.
294. Bai, J. et al., *Nat. Nanotechnol.,* 5, 190, 2010.
295. Li, X. et al., *Science,* 319, 1229, 2008.
296. Liang, X. et al., *Nano Lett.,* 10, 2454, 2010.
297. Kim, M. et al., *Nano Lett.,* 10, 1125, 2010.
298. Xia, F. et al., *Nano Lett.,* 10, 715, 2010.
299. Stankovich, S. et al., *Nature,* 442, 282, 2006.
300. Wei, T. et al., *Carbon,* 47, 2296, 2009.
301. Zhang, H. B. et al., *Polymer,* 51, 1191, 2010.
302. Ramanathan, T. et al., *Nat. Nanotechnol.,* 3, 327, 2008.
303. Xu, Y. X. et al., *Carbon,* 47, 3538, 2009.
304. Vadukumpully, S. et al., *Carbon,* 49, 198, 2011.
305. Verdejo, R. et al., *J. Mater. Chem.,* 21, 3301, 2011.
306. Gogotsi, Y., *J. Phys. Chem. Lett.,* 2, 2509, 2011.

2 Fullerene C_{60} Architectures in Materials Science

Francesco Scarel and Aurelio Mateo-Alonso

CONTENTS

2.1 INTRODUCTION

Since the discovery of fullerenes in 1985,[1] different research fields in fundamental science and nanotechnology have investigated this unique carbon nanomaterial. The family of this allotropic form of carbon consists of cage-like molecules with a spheroid shape. The more common cages are C_{60} or C_{70}, whereas bigger ones such as C_{76}, C_{78}, C_{84}, ..., C_{190} are usually only present in trace amounts. Among these aromatic molecules, the highly symmetrical and spherical structure of C_{60} has, in particular, attracted the attention of scientists across the world. The investigation studies of this molecule increased drastically after it became available in bulk quantities.[2] Apart from interesting physical and chemical properties, this molecule shows low toxicity and is almost inert to different environments under mild conditions.

In C_{60}, the 60 chemically equivalent carbon atoms (sp^2) are distributed symmetrically along an icosahedral I_h structure that establish a skeleton comprising 20 hexagonal and 12 pentagonal rings. The subunits composing the buckyball can be thus named as 1,3,5-cyclohexatriene and [5]radialene (Figure 2.1).[3] The crystal structure revealed that the formal double bonds are located on the two hexagons junction (6–6 junction), whereas between a pentagon and a hexagon (5–6 junction), a single bond character is predominant. This is why C_{60} cannot be considered as a superaromatic compound, but rather an electron-deficient polyene.[4,5]

From an electronic point of view, C_{60} is described to have a completely full fivefold degenerated highest occupied molecular orbital (HOMO) located energetically lower than the corresponding lowest unoccupied molecular orbital (LUMO).[6,7] The low energy of the LUMO and the ability of the fullerene to highly delocalize the electron coming from the first reduction on a triply degenerate

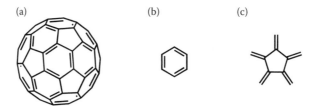

FIGURE 2.1   Representation of (a) C_{60}, (b) 1,3,5-cyclohexatriene, and (c) [5]radialene substructure.

unoccupied molecular orbital confer to C_{60} the property of a good electron acceptor. Indeed, it can reversibly gain up to six electrons on reduction,[8] which are very well delocalized on the spherical surface. However, oxidation is much more difficult to achieve, and only the first three reversible oxidations have been observed.[9]

2.2   ARCHITECTURE OF C_{60}

The molecular and supramolecular chemistry of C_{60} have been explored by scientists during the last few years. One of the problems encountered with fullerenes is their lack of solubility in the most organic solvents, and therefore their difficult processibility.[10] C_{60} presents a good solubility only in aromatic organic solvents, but it is completely insoluble in protic and dipolar aprotic organic solvents. The functionalization of fullerenes allowed solubility in the most common organic solvents and enabled the development of a wide variety of adducts and new materials. Indeed, the special characteristics of the added groups can be combined with the chemical and physical features of the fullerenes (optical, electrochemical, semiconducting,[11] superconducting,[12–14] and magnetic[15–18] properties), which are in most cases maintained or enhanced after functionalization. This relatively easy method by which to obtain new materials increased specific studies in technological applications, such as photovoltaics, light-emitting diodes, electronic devices, and molecular machinery.

2.2.1   COVALENT ARCHITECTURES

Owing to the convex shape of the wall of C_{60}, all the 30 double bonds, located exocyclic to the pentagons, are strained and thus more reactive than formal double bonds in aromatic compounds. Because the electron density is much higher on the [6,6] ring junctions than on the [5,6] junctions, most of the chemical reactions of fullerenes tend to occur across these sites.[19,20] Cycloaddition reactions (widely applied for functionalization of fullerenes), as well as additions of nucleophiles and free radicals, provoke a hybridization change in the carbon atoms involved from a trigonal sp^2 to a tetrahedral sp^3. This release of the double bond strain is the driving force of such reactions.

A wide range of chemical moieties can be easily attached to the C_{60} via covalent bonds, but because of the high number of reactive sites, the reaction mixture can contain both monoadducts and multiadducts. Multiple additions provide a large number of regioisomers. As an explicative example, Figure 2.2 shows the eight possible positions for all bisadducts isomers.

The covalent functionalization of fullerenes, in particular C_{60}, has been reviewed extensively.[21,22] Examples of well-established C_{60} functionalization protocols are represented in Scheme 2.1.

2.2.1.1   Cycloaddition Reactions

The behavior of C_{60} is similar to that of 2π-electron-deficient dienophiles and dipolarophiles, and this allows cycloadditions, such as [1 + 2], [2 + 2], [3 + 2], and [4 + 2] to take place either thermally or photochemically.

The [1 + 2] cycloaddition involves divalent species, such as carbine,[23–27] nitrenes,[28–32] and silylenes,[33] to provide methanofullerenes and heteroanalogs. Sugar-C_{60} type of molecules were achieved through reaction with sugar-derived spirodiazirines (Figure 2.3a).[34] Addition of nitrene

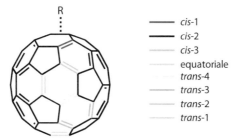

FIGURE 2.2 Possible positions of bisadducts isomers of a general fulleropyrrolidine.

intermediates leads to azafullerenes in the [6,6]-junction (Figure 2.3b). A particularity of this reaction is the high regioselectivity addition of two nitrene units as the second addition occurs in the *cis*-1 position, which is an inaccessible site for other cycloadditions.

Highly reactive free carbene species (i.e., the reaction under pyrolytic conditions of sodium trichloroacetate generates dichlorocarbene) can be considered as reactants for [1 + 2] cycloaddition to

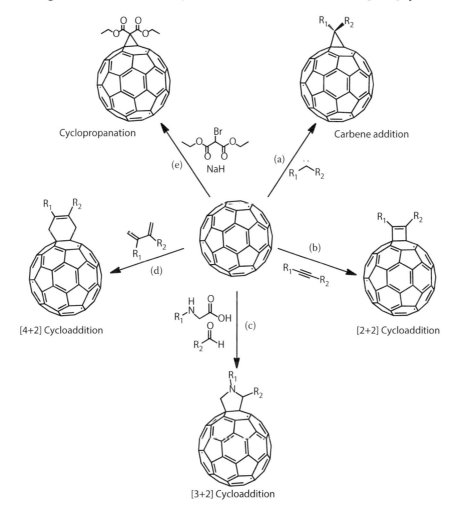

SCHEME 2.1 Examples of C$_{60}$ established functionalization methods. (a) [1 + 2] cycloaddition, (b) [2 + 2] photochemical cycloaddition, (c) [3 + 2] cycloaddition of azomethine ylides, (d) [4 + 2] Diels–Alder cycloaddition, and (e) Bingel–Hirsch reaction.

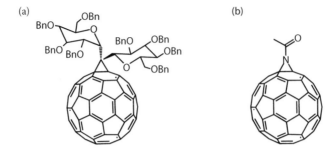

FIGURE 2.3 Examples of (a) fullerene–sugar-type conjugate and (b) azafullerene.

form methanofullerenes.[35] A further development of this species concerns the generation of metha-nofullerene carbenes intermediates to afford all-carbon dimers C_{121} and C_{122} (Scheme 2.2).[36,37]

On [2 + 2] cycloaddition, unsaturated hydrocarbons such as alkenes and alkynes can react chemically and photochemically with C_{60}.[38–41] Reported examples (Figure 2.4a and b) have shown the use of these derivatives as protective groups of the double bonds of C_{60} because of the reversible nature of this reaction. These adducts can easily undergo cycloreversion to C_{60} pristine on irradiation with visible or ultraviolet (UV) light.

Following this method, dimers and polymers[42] of C_{60} have been synthesized and studied. Some experiments were carried out using a dimer of C_{60} following a technique called high-speed vibration milling (HSVM) in the presence of potassium cyanide (Figure 2.4c).[43–45]

[3 + 2] cycloadditions are one of the most successful methods for functionalization of fullerenes in particular. The 1,3-dipolar cycloaddition of azomethine ylides is the most common approach used to yield a heterocyclic five-member ring attached to the fullerene.[46–48] Azomethine ylides are prepared *in situ* via thermal condensation of α-amino acids and aldehydes (otherwise ketones)

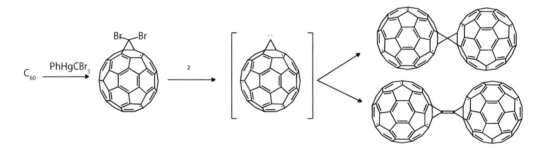

SCHEME 2.2 Generation and reactivity of fullerene carbenes.

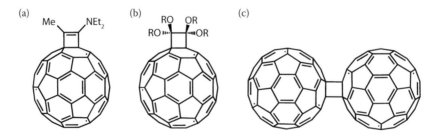

FIGURE 2.4 Examples of [2 + 2] cycloaddition derivatives where C_{60} has been reacted with (a) ynamine, (b) tetraalkyloxyethenes, and (c) a C_{60} dimer.

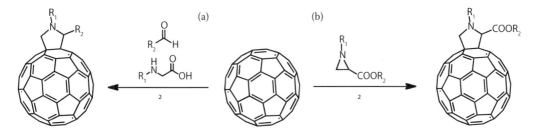

SCHEME 2.3 Two different approaches for synthesis of fulleropyrrolidines. (a) Thermal condensation of α-amino acids and aldehydes, and (b) thermal ring opening of aziridines.

and further decarboxylation of the ammonium salt obtained to provide fulleropyrrolidines. Alternatively, fulleropyrrolidines can be obtained by exploiting the thermal ring opening of aziridines (Scheme 2.3).

A very important characteristic of fulleropyrrolidines is high stability. Retro-cycloaddition reactions, although reported, have to be carried out under harsh conditions, such as high temperatures while trapping the ylide with a metal and a dipolarophile,[49] or by microwave irradiation in ionic liquids.[50,51]

The functionalization via 1,3-dipolar cycloaddition has been used to drastically increase the solubility of C$_{60}$ in the common organic solvents. The addition of hydrophilic building blocks, such as oligoethylene glycol chains bearing ammonium salts to the extremely hydrophobic spheroid of C$_{60}$, led to the formation of amphiphilic derivatives soluble in polar media (Figure 2.5).[52]

Fullerenes can also be involved in [4 + 2] cycloaddition reactions (also called Diels–Alder reaction) with 1,3-dienes added. Here, the fullerene acts as an electron-deficient dienophile and addition takes place on a [6,6]-junction (Scheme 2.1d). Although this reaction allows the control of the degree of functionalization, [4 + 2] adducts are not stable and the reaction is in most cases thermally and photochemically reversible.

An interesting example that takes advantage of this instability involves the synthesis of a water-soluble fullero-dendrimer via Diels–Alder reaction between a dendrimeric anthracene and C$_{60}$. The reversibility character of Diels–Alder reaction was exploited to release C$_{60}$ intracellularly by removing the dendrimer solubilizing group (Figure 2.6a).[53]

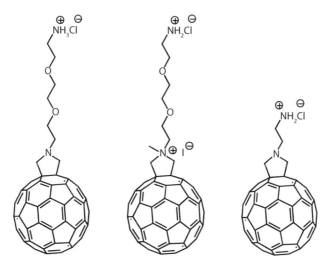

FIGURE 2.5 Amphiphilic water-soluble fulleropyrrolidine salts.

(a)

(b)

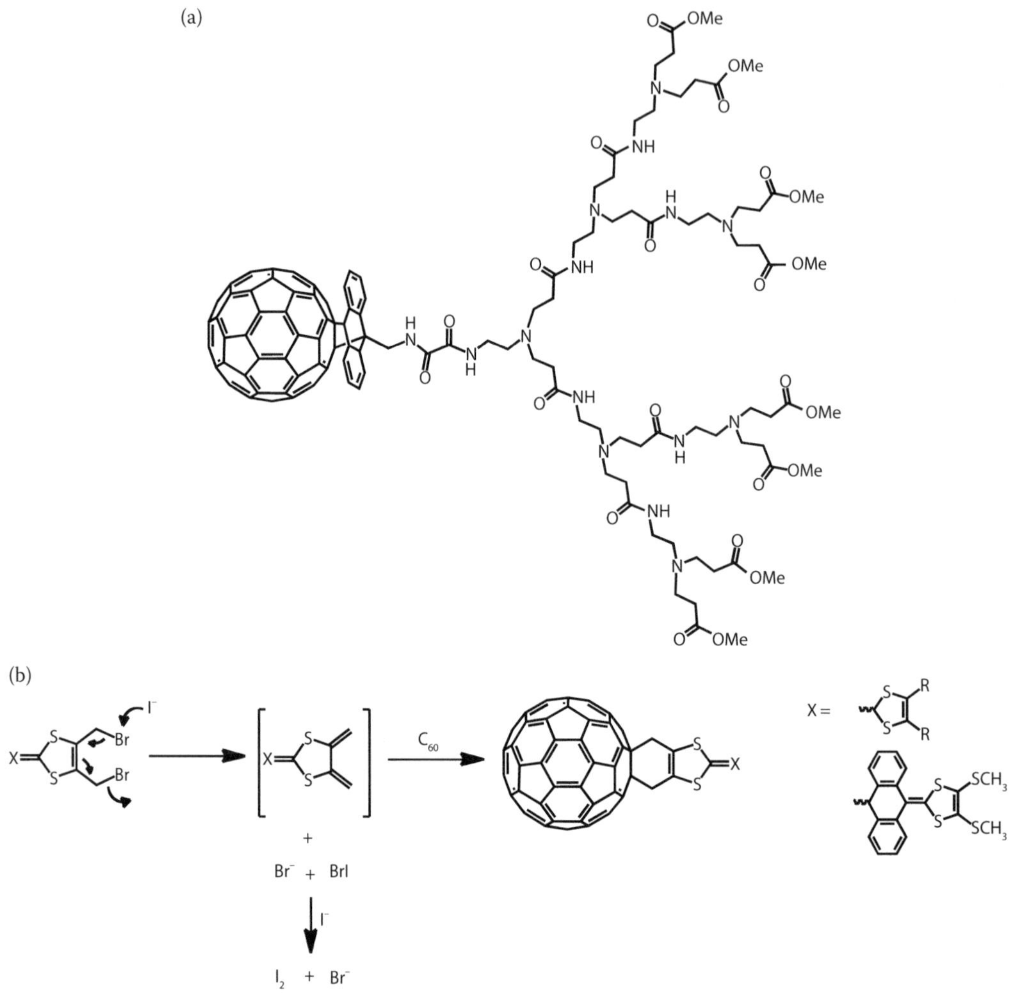

FIGURE 2.6 Examples of different approaches of Diels–Alder reaction: (a) direct formation of a fullerene dendrimer and (b) fullerene–TTF derivatives synthesized by the generation *in situ* of the corresponding diene.

    The dienes used to perform this reaction can also be prepared *in situ*, starting from the appropriate precursors. Reductive elimination of tetrathiafulvalene[54,55] (TTF) dibromo derivatives generates *in situ* dienes that are subsequently trapped by C_{60} (Figure 2.6b).[56]

    The [4 + 2] cycloadditions have been used to synthesize fullerene multiadducts in the past. A six-fold Diels–Alder product[57] has been reported not only by the reaction of an excess of 2,3-dimethyl-1,3-butadiene with C_{60}, but also by different bisadducts connected crown ethers to the fullerene spheroid.[58] The properties of these derivatives have been studied for several potential applications, such as molecular sensors and complexation chemistry.

2.2.1.2 Nucleophilic Addition

The most used nucleophilic synthesis reaction to form fullerenes is cyclopropanation. The original method was developed by Bingel, and it employs the generation of a carbon nucleophile starting from α-halo esters in the presence of a base (such as NaH), and the subsequent addition to C_{60}.[59] After the addition of the anions of α-halo ester, an intramolecular substitution of the halide takes place with the intermediate fullerene anion, giving the corresponding methanofullerene derivative (Scheme 2.1e). Further modification of this method consist of preparing the α-halomalonate *in situ*,

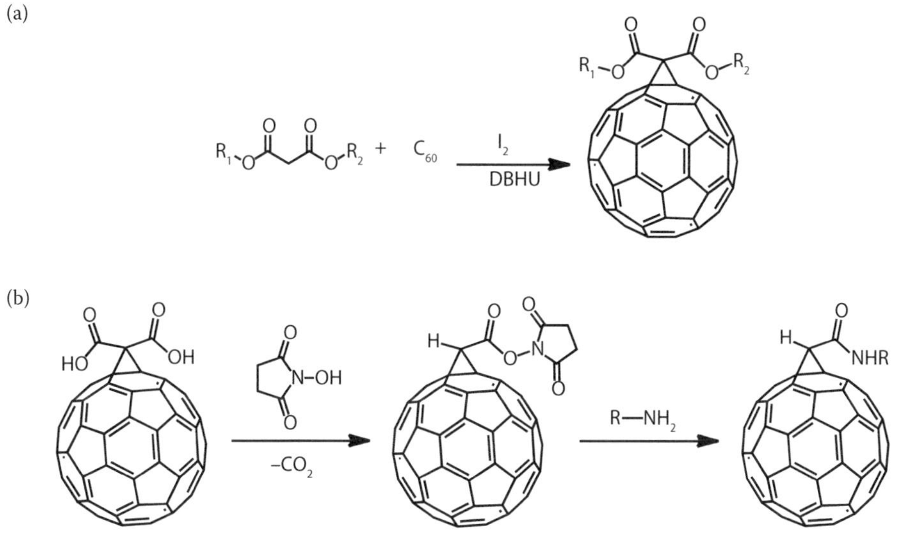

(a)

(b)

SCHEME 2.4 Bingel adducts as promising building blocks.

under mild conditions.[60,61] This introduces the possibility to work with a wide range or functional groups attached to the malonate precursors and to increase the construction of novel methanofullerenes derivatives (Scheme 2.4a). The availability of many starting esters building blocks allows a chemical transformation of the synthesized methanofullerene. As described in the example in Scheme 2.4b, activation of the malonic acid methanofullerene with *N*-hydroxysuccinimide introduced the possibility to obtain a wide range of methanofullerene amides.

In general, the methanofullerenes formed by the Bingel reaction are stable and have been used widely for the synthesis of new materials. However, as a drawback, they show liability under reductive conditions. Electrochemical reduction is sufficient to induce a reverse Bingel reaction in di(alkoxycarbonyl)methanofullerenes, generating the parent fullerene and starting malonate as products.[62,63]

Other approaches that involve nucleophiles as reactants have been studied recently with interesting results. To name a few, organometallic additions (organocopper),[64,65] cyanide anions,[66] Grignard agents,[67] or zwitterions[68] have been recently introduced in the family of nucleophilic reactions on fullerenes.

2.2.2 SUPRAMOLECULAR ARCHITECTURES

The wide range of possible supramolecular architectures and arrays containing C$_{60}$ is driven by means of different noncovalent interactions, such as π–π stacking, hydrogen bonding, and metal-mediated complexation. The attention on these novel structures has increased with time as is illustrated in several reviews.[69–71]

Depending on their nature, these functionalization methodologies can be applied on pristine and derivatized fullerenes. This allows the construction of complex assemblies and architectures.

2.2.2.1 Direct π-Stack Assemblies

Fullerenes have an extended tridimensional π-system with an anisotropic electron distribution over the entire spherical surface, where the π-orbitals are oriented radially. The five-membered ring (pentagons) substructures, responsible for the curvature of the fullerene, are considered electron-deficient areas, and are alternated with electron-rich regions characterized by 1,3,5-cyclohexatriene (hexagons) subunits. Because of this ambivalent electronic nature, C$_{60}$ can easily adapt to different substrates, and associate with other organic architectures with higher propensity compared to the classic bidimensional π-systems.[72]

2.2.2.1.1 Concave–Convex Recognition

The association of two units of γ-cyclodextrin has been described by Wennerström et al.,[73] which allowed the solubilization of C_{60} in water media. In 1994, Raston[74] and Shinkai[75] reported a *p-tert-buthylcalix[8]arene* able to selectively associate and separate C_{60} from soot. Other examples of calixarenes, calixnaphthalenes, cyclotriveratrylenes (CTVs), and cyclodextrines have been studied and extensively reviewed.[76,77]

The flexibility of the structure used to functionalize fullerenes seemed to play an important role. The association constants of some CTV–C_{60} complexes were negatively affected by addition of rigid linkers to connect the CTV units.[78] However, further development of very rigid aromatic nanorings with properly designed size and shape showed the formation of stable complexes with fullerenes, turning the lack of flexibility into an advantage. This concept is clearly remarked by the closed loops formed by conjugated π-orbitals synthesized by Kawase et al. (Figure 2.7).[79–81]

UV/vis titration studies on *p-aryleneacetylene* family of nanorings showed 1:1 supramolecular complexes with C_{60} and C_{70}, and the increased size of the aromatic surface yielded interesting supramolecular nanostructures (such as onion-type complexes).[82]

π-Extended tetrathiafulvalene (ExTTF), a curved TTF derivative, has been widely investigated and used in complex fullerenes because of concave–convex and electronic complementarity (Figure 2.8a and b).[83] Despite the geometrical and electronic properties suitable for the convex surface of fullerenes, exTTF moiety alone was not able to complex C_{60}. With the development of "tweezers" characterized by two exTTF units linked together through an aromatic bridge, the affinity with C_{60} increased remarkably and the association constant was reported as $\log K_a = 3.5$ (Figure 2.8c).

This association constant reached higher values (almost doubled) with another exTTF-based derivative recently synthesized (Figure 2.8d). The covalently bounded bridges linked to the exTTF

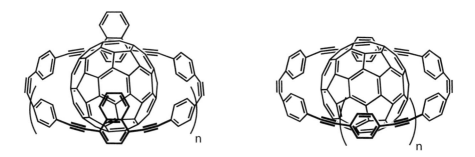

FIGURE 2.7   Examples of carbon nanorings complexed with C_{60}.

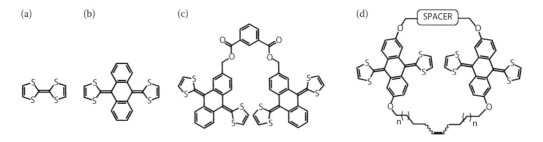

FIGURE 2.8   Structures of (a) TTF, (b) exTTF, (c) exTTF-based tweezers, and (d) exTTF-based receptor.

units increased the preorganization level of this new macrocycle, while setting a determinate distance between the electron-rich units. Modification of the spacer (switching through *p*-phenylene, *m*-phenyl-ene, and 2,6-naphthylene units) and elongation of the alkenyl chain allowed the development of several rings employed in the supramolecular functionalization of C$_{60}$ and C$_{70}$.[84]

Very interesting supramolecular structures are known as carbon peapods, and consist of single-walled carbon nanotubes (SWCNTs) filled with fullerenes. After their first detection via high-resolution transmission electron microscopy (HRTEM) as a side product in the production of carbon nanotubes,[85] different methodologies have been developed to produce such new carbon allotropes. The harsh conditions previously used (such as high temperatures, low pressures, and acidic medias)[86] have been later overcome by mild condition experiments that exploited not completely understood mechanisms of nano-condensation and nanoextraction.[87]

Low temperatures and supercritical fluids (such as CO$_2$) facilitate the penetration of the fullerenes in the nanotubes. Owing to low viscosity, these media could flow inside the nanotubes without remaining trapped, thus leaving free room for other molecules.[88]

It was shown how both endohedral and exohedral fullerenes can be inserted in nanotubes. In the peapods containing endohedral fullerenes (for instance Ce@C$_{82}$), HRTEM images showed interesting rotation and translation motion of the trapped spheroids.[89] Exohedral metallofullerenes, CsC$_{60}$, have been synthesized and successfully encapsulated into SWCNTs via a new chemical reduction of C$_{60}$ molecules into anions.[90] The addition of iodine to already prepared peapods allowed the coalescence of C$_{60}$ directly inside the nanotubes. Indeed, after heating at 550°C, iodine-doped peapods, inside the C$_{60}$ molecules molecules have been transformed in a tubular structure.[91] Khlobystov et al. were able to perform reactions on the inner surface of carbon nanotubes in the presence of catalytically active atoms of rhenium and monitor the whole process via HRTEM.[92]

2.2.2.1.2 *Planar–Convex (Porphyrin–Fullerene) Recognition*

The concept of porphyrins and metalloporphyrins (MPs) that recognize and surround C$_{60}$ in solution opened the way to a new research field, implementing the development of hosts for binding and trapping fullerenes. Transition metals trapped in porphyrin-like structures can also interact directly with the fullerene spheroids. Indeed, the association between these macrocyclic structures and fullerenes takes place even with free base porphyrins (widely used are tetraphenylporphyrins, TPP) and is an association of different interactions such as van der Waal,[93] electrostatic,[94] and charge transfer.[95]

The first structure showing this planar–convex interaction was reported in 1995 for a covalently linked porphyrin, and after this, many other cocrystals have been studied.[96,97] Subsequent to these interesting and remarkable results, the investigation moved to the exploration of a series of metalloporphyrins (MPs) and their interaction with C$_{60}$ and C$_{70}$. The chosen metals (Mn, Co, Ni, Cu, Zn, and Fe) have a direct influence on the length of the complexation contacts that are shorter than ordinary van der Waals ones (2.7–3.0 Å instead of 3.0–3.5 Å).[98–101]

Following these promising studies, a type of bis-porphyrin system, also known as "porphyrine jaw," has been synthesized and detected via matrix-assisted laser desorption/ionization (MALDI) mass spectrometry (Figure 2.9a).[102] Cyclic dimers were realized with two metal biphenyltetrahexylporphyrin moieties linked together through aliphatic spacers (Figure 2.9b).[103]

The inclusion of fullerenes into these cyclic dimers and jaws implies a close proximity of the MP with the spheroid π-system. The effects of this interaction are detectable in the red shifts of the absorption bands (Soret and Q-bands) that usually come together with low extinction coefficients. This is a typical sign of a mutual perturbation of a π-system. Moreover, the emission of the chromophores is also progressively quenched after the addition of fullerene to the solution.

Aida et al. noticed that zinc porphyrin (ZnP) cyclic dimer formed an inclusion complex with C$_{60}$, stable enough to remain associated even under chromatographic conditions.[103] The cyclic dimer model was further extensively studied to improve the association binding constants and thus the selectivity with fullerenes. An exhaustive report of this work has been published recently.[104]

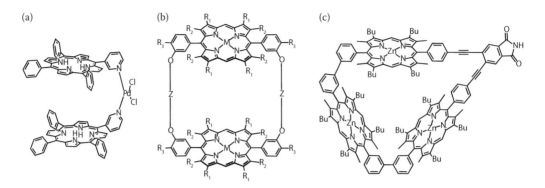

FIGURE 2.9 Different structures able to host fullerenes. (a) An example of porphyrin "jaw," (b) cyclic dimer model with high affinity for C_{60} and C_{70}, and (c) cyclic dimer able to self-assemble in nanotubes.

The modification of the metal and the distance between the two porphyrin rings enabled reaching supramolecular hosts with different cavity sizes. The affinity of ZnP was found comparable to that of free base porphyrins, whereas metal ions such as Co(II) and Rh(III) showed much higher association constants toward fullerenes.[105] In particular, Rh(III) species are known because of their ability to add preferentially on double bonds.[106] The π-electron-rich 6,6 ring junctions of C_{60} and C_{70} are good targets for such hosts bearing Rh(III) porphyrins, and they are the reason for a discriminative behavior of C_{70}. Indeed, this oval fullerene adopts a side-on orientation inside the porphyrin dimer, but it can also adopt an end-on orientation on lowering the temperature until −60°C. The largest association constant among those reported to date has been found between fullerenes and iridium porphyrins. X-ray diffraction and nuclear magnetic resonance (NMR) spectroscopy showed a strong bond-like Ir–C_{60} coordination in an η^2 fashion, forcing an ellipsoidal deformation of C_{60} at low temperatures. Moreover, this strong affinity is able to convert the thermodynamically favored side-on orientation of C_{70} to end-on position.[107]

This field evolved in the development of fashionable cage-like architectures, where several hosts for fullerenes are characterized by the number of interacting porphyrin units and from the way they are connected together. The porphyrin units dedicated to fullerene recognition can be covalently linked through organic bridges.

Recently, Anderson has reported a rigid tri-porphyrin host able to associate with C_{60}, C_{70}, and also C_{86} with extremely high binding constant values such as $\log K_a = 6.2$, 8.2, and >9, respectively (Figure 2.9c). The strong affinity of this trimer to bigger fullerenes cannot be measured by fluorescence titration, and it has been demonstrated with some advanced experiments.[108]

Another example is given by the four Ni–porphyrins linked together with pyridine rings in a supramolecular host for fullerenes called nanobarrels.[109] The concave shape of this ring is complementary with fullerenes, but the association constant ($\log K_a = 5.7$) is not so large as expected from a synergistic effect of the four porphyrin units.

The porphyrin units can also be coordinated in solution by metals to assemble a box for fullerene guests. The case shown in Figure 2.10 is an example of a new class of close-face metallo-supramolecular cubic hosts synthesized, where pyridylimine centers coordinating with Fe(II) constitute the corners and the Ni or Zn porphyrins the faces.[110]

Another approach to trap fullerenes involves the use of C_{60} as a template for the formation of its own porphyrin host. The fullerene is, thus, able to preorganize the MP subunits before the *in situ* cyclization reaction that carries out the formation of the macrocycle host. In the example reported by Langford et al., the porphyrin subunits involved in a metathesis reaction form mainly a dimer without the presence of C_{60} in a dichloromethane solution while they are preferentially forming the host trimer when the fullerene is exploited as the template.[111]

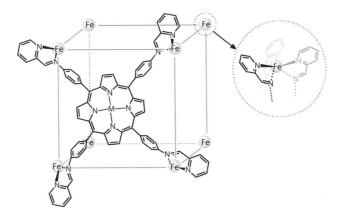

FIGURE 2.10 Self-assembled cubic cage formed through metal coordination.

2.2.2.1.3 Other Receptors

A novel ring-shape derivative was synthesized by means of 1,8-naphthyridine units placed as a connection bridge between two trypticenes (Figure 2.11a). X-ray studies showed two structures of this derived oxacalixarene, with different orientation but with a comparable size of the inner cavity. Moreover, the nitrogen atoms in the 1,8-naphthyridine are all positioned inside the cavity. The high affinity of these rings with C$_{60}$ and C$_{70}$ is probably a consequence of the remarkable rigidity of the trypticenes units, together with the right size of the cavity. The formation of inclusion complexes with a 1:1 stoichiometry was detected via fluorescence spectroscopy.[112]

Pyrrole-based macrocyclic assemblements were also utilized as functionalizing units for fullerenes. Despite the fact that the first example coming to mind is a porphyrine macrocycle, the ability of such π-electron-rich systems to strongly bind fullerenes in a convex–planar supramolecular association will be discussed in Section 2.2.2.2.

The first pyrrole-based moiety able to interact with C$_{60}$ through π–π interactions was described by Sessler et al.[113] and consisted of a calix[4]pyrrole scaffold fused with four units of TTF (Figure 2.11b). This molecule interacted with C$_{60}$ very weakly in its open form, but is able to switch in a cone conformation after the addition of Cl$^-$ ions, forming an electron-rich deep cavity defined by the TTF arms. Indeed, a 2:1 complex was formed after the addition of tetrabutylammonium chloride to a solution of C$_{60}$ and the macrocycle.[114]

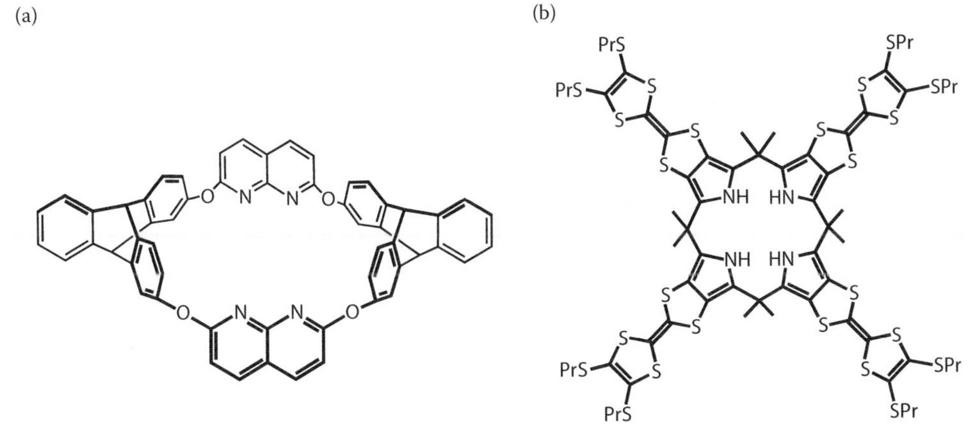

(a)

(b)

FIGURE 2.11 (a) Rigid macrocycle based on triptycene unit and (b) calyx[4]pyrrole scaffold with four TTF units.

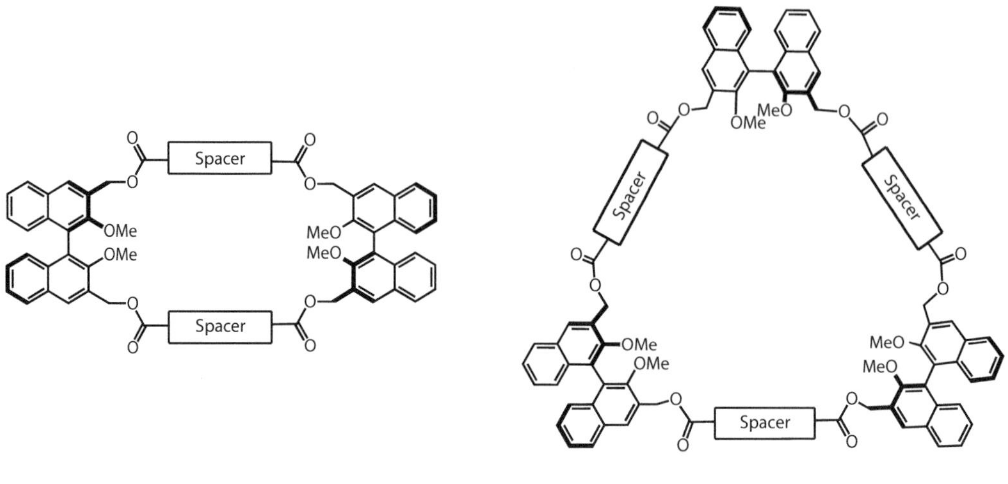

FIGURE 2.12 Chiral macrocycles with recognition properties for C_{60}.

Recently, several aromatic chiral moieties[115,116] showed a good affinity for C_{60}, and this opened some good perspectives for chiral resolution of higher fullerenes (Figure 2.12).

2.2.2.2 Indirect Supramolecular Assemblies

Fullerenes covalently functionalized with appropriate moieties can be arranged in suprastructures by means of hydrogen bonding or metal complexation. Some of the most illustrative examples are described below.

2.2.2.2.1 Hydrogen Bonding Recognition

Hydrogen bonds are widely spread in nature as organizational motifs that are able to define tridimensional structures of living systems, starting from the molecular level. Single hydrogen bonds are very specific and can provide a high degree of directionality, but they suffer from a lack of stability. Nevertheless, this problem can be overcome by the arrangement of multiple hydrogen bonds pattern.

Obviously, pristine fullerenes cannot participate directly in the formation of hydrogen bonds, and hence their covalent functionalization with appropriate building blocks is the first step to be achieved.

A supramolecular C_{60} dimer was built by the recognition between two independent fullerene derivatives, one functionalized with an ammonium salt and the other with a crown ether.[117] The contribution of N^+–$H\cdots O$ and C–$H\cdots O$ hydrogen bonds and ion–dipole interaction is the driving force for this assembled supramolecular system, where the C_{60}-dibenzylammonium adduct was threaded through the cavity of a crown ether C_{60} derivative (Figure 2.13).

Other self-complementary recognition patterns consist of hydrogen donors (D) and hydrogen acceptors (A). The association constants (K_a) for the simplest DA system are observed around 10 M^{-1}, but are increasing exponentially for triple hydrogen bond motifs (10^2–10^3 M^{-1}).[118,119] Even higher values have been calculated for quadruple motifs (10^5 M^{-1}),[120,121] and the unusual binding strength of a DDAA motif with 2-ureido-4-pyrimidones provided a $K_a = 10^7$ M^{-1}.[122]

Quadruple hydrogen bond arrays introduced to functionalized fullerenes have been used in the formation of self-assembled C_{60} dimers (Figure 2.14) through 2-ureido-4-pyrimidones recognition moieties.[123,124]

The assembled fulleropyrrolidine dimers showed fluorescence quench of 50%, and this is an evidence of the strong electronic coupling occurring via hydrogen bonds. Indeed, after the addition of different protic solvents (able to disrupt the formation of the hydrogen bonds), an enhancement of fluorescence was detected along with separation of the dimer.

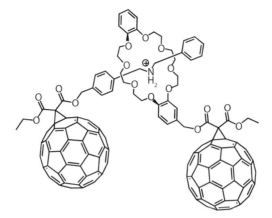

FIGURE 2.13 Pseudorotaxane assembled through hydrogen bond recognition.

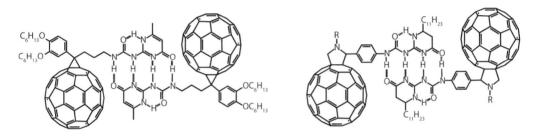

FIGURE 2.14 Examples of supramolecular C$_{60}$ dimers.

2.2.2.2.2 Axial Coordination

The functionalization of fullerenes with nitrogenated groups such as pyridine and imidazole allows such fullerene derivatives to coordinate with metallated porphyrines, phthalocyanines (Pc), and naphthalocyanines (Nc) (Figure 2.15).[125,126] The structure of a noncovalently linked ZnTPP to a C$_{60}$ derivative has been studied with x-ray diffraction, bringing more information regarding conformation and atom distances of the complex.[127]

In other examples reported, the cromophore is connected to an imidazole ring through metal complex coordination, forming a more stable complex in solution. However, fullerenes also functionalized with two pyridine entities have been successfully assembled with TPP and Nc's macrocycles.[128]

(a) M=Zn   (c) M=Zn   (e)
(b) M=Ru(CO) (d) M=Mg

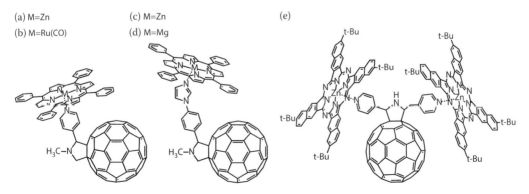

FIGURE 2.15 Structures of different systems exploiting metal coordination with C$_{60}$ derivatives functionalized by pyridine and imidazole heterocycles.

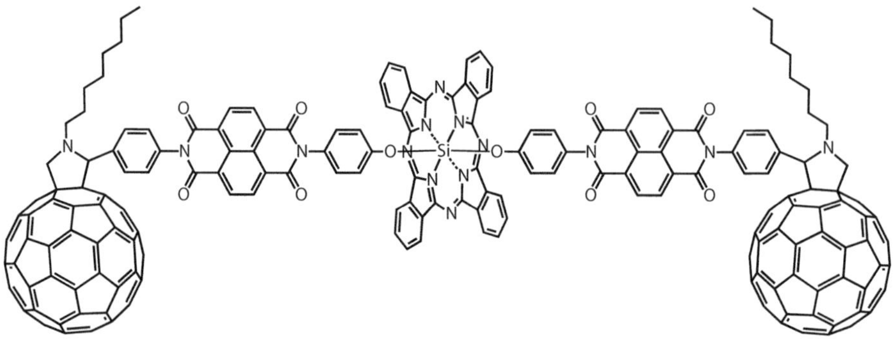

FIGURE 2.16   Silicon phthalocyanine–naphthalenediimide–fullerene connected system.

Metals are also involved in axial coordination systems, as resulted from the employment of Si in a core of phthalocyanines (SiPc). These dyes can indeed coordinate with two ligands at the same time, raising the possibility to build triads through self-assembly methods.[129,130] As an example, a tetra-tert-butyl-$Si^{IV}PcCl_2$ was placed as a bridge between two electron acceptor moieties, namely, 4,5,8-naphthalenenediimide (NDI) and a C_{60} derivative, covalently linked together (Figure 2.16).[131]

2.2.3   MECHANICALLY INTERLOCKED MOLECULAR ARCHITECTURES

Mechanically interlocked molecules consist of two or more submolecular components. The connection between the two molecules to each other cannot be separated without breaking a covalent bond. Molecules built through mechanical bonds are intrinsically multistable as their interlocked components can adopt multiple coconformations. These features have several important implications both in fundamental science and in molecular nanotechnology. Rotaxanes and catenanes are among the most popular mechanically interlocked architectures.[132] In particular, rotaxanes consist of a dumbbell-shaped component (also called thread or axl), which is threaded through a macrocycle (Figure 2.17a). In a rotaxane, thread and macrocycle are interlocked to each other as two stoppers are present at both ends of the thread. This prevents efficient dissociation (unthreading) as the stoppers are larger than the internal diameter of the macrocycle. In a pseudorotaxane, at least one of the stoppers is not present (Figure 2.17a); thus, the macrocycle and ring can dissociate depending on the conditions. Catenanes consist of two or more interlocked macrocycles (Figure 2.17a).

The strategy to synthesize rotaxanes exploits the preorganization of the components through supramolecular recognition, and can be summarized in capping, clipping, and slipping (Figure 2.17b). Capping requires the addition of a second stopper in a previously formed pseudorotaxane to trap the macrocycle around the thread and to avoid dissociation. The clipping strategy uses a preformed thread bearing a partial macrocycle by means of some supramolecular interactions. This half macrocycle is then transformed into a complete ring by a ring-closing reaction around a template motif on the thread to obtain the desired rotaxanes. The third method, slipping, exploits the temperature as a driving force to allow a macrocycle to slip through a preformed thread, bearing appropriate sized stoppers. The slipping process takes place with high temperatures, and then, by cooling the system, it remains kinetically trapped within the thread.

Bistable rotaxanes are particularly interesting to develop practical applications. The position of the macrocycle can vary between two positions along the thread; in this way, the molecule can adopt different coconformations. Bistable rotaxanes are characterized from different positions in the thread (also called stations) where the macrocycle stays preferentially under certain conditions (Figure 2.17c). A change in these conditions promotes submolecular shuttling of the macrocycle from one station to another. These molecular shuttles can be activated through diverse external stimulus such as light irradiation,[133,134] variation of electrochemical potential,[135–138] pH,[139,140] or solvent changes.[141]

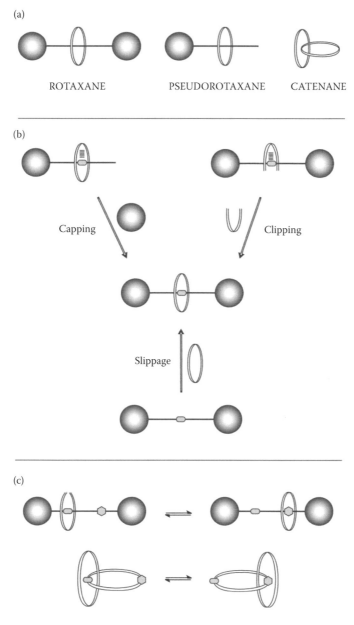

FIGURE 2.17 (a) Schematic representation of rotaxanes, pseudorotaxanes, and catenanes, (b) synthetic strategies, and (c) schematic representation of bistable rotaxanes and catenanes.

Also, in catenanes, one of the rings can be equipped with two different stations; so, the other ring can circumrotate and stop on these programmed positions.[142,143]

Fullerenes, especially C_{60}, have been involved in the architecture of rotaxanes and catenanes to confer novel properties to these supramolecular architectures.

A very interesting rotaxane bearing two C_{60} stoppers was prepared from a well-established method developed by Dietrich-Buchecker and Sauvage,[144] which consists of coupling a macrocycle and a thread, both presenting phenantroline systems and employing copper(I) as a template (Figure 2.18).[145] This system showed relevant differences in the electrochemical, spectroscopic, and photophysical properties when compared to pristine C_{60}.

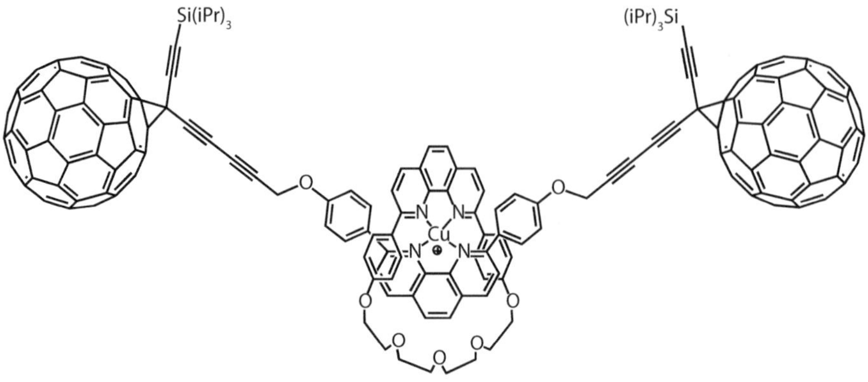

FIGURE 2.18   First rotaxane bearing C_{60} stoppers.

In 1997, a [2]catenane bearing C_{60} was synthesized through Stoddart's methodology,[146] exploiting π–π interactions between aromatic moieties.[147] The double Bingel addition of a [34]crown-10 bismalonate to C_{60} was followed by a ring-closed reaction, where the electron-deficient cyclobis-(paraquat-*p*-phenylene) (CBPQT[4+]) macrocycle recognizes the electron-rich hydroquinone template.

This method has been applied in other studies to obtain a bistable fullerene–rotaxane, where the CBPQT[4+] macrocycle was able to switch from two different electron-rich stations, a 1,5-dihydroxynaphthalene (DNP), and a TTF.[148] The macrocycle closes at the TTF station after the clipping reaction with *p*-xylylene dibromide, because the TTF template unit is more electron-rich than DNP. The further oxidation of TTF to the cation caused an electrostatic repulsion between the station and the positively charged macrocycle that switches to the other electron-rich DNP station (Figure 2.19).

Fullerene–rotaxanes can also be synthesized by using Leigh's methodology.[149] Starting from the precursors at high dilution conditions, the benzylic amide macrocycle closes on a diamidic template by formation of four hydrogen bonds (Figure 2.20). Several fullerene rotaxanes have been synthesized through this method that has the advantage to give rotaxanes in a one-step reaction.[150,151]

Bistable rotaxanes have been designed through this methodology as promising powerful molecular machines, where the fullerene is used to probe the motion of the macrocycle and, more importantly, to induce shuttling.

A fullerene rotaxane reported in 2003 showed an interesting solvent-dependent switching behavior of the benzylic amide macrocycle along a thread composed of a C_{60} stopper, a glycylglycine template station, and a long alkyl chain that separate a phenyl stopper.[152] In the solvent system, this promotes the hydrogen bonds (CH_2Cl_2, $CHCl_3$) and the macrocycle is tied on the glycylglycine station by means of four hydrogen bonds, whereas in solvents such as DMSO (dimethyl sulfoxide), which disturb the formation of such bonds, the macrocycle is free to move along the thread and in this case to reside over the alkyl chain (Figure 2.21a). This shuttling was studied and confirmed through photophysical experiments on the triplet–triplet of the fullerene, whose properties are independent from the solvent in the isolated thread. When the system was studied in CH_2Cl_2, the triplet–triplet spectra and lifetime were affected by the proximity of the macrocycle to the fullerene sphere (the triplet lifetime was found shorter by a factor of 1.7).

A similar rotaxane was further synthesized and experiments showed a reverse shuttling behavior.[153] This reverse performance was explained by means of π–π interactions that in this case are allowed between the C_{60} and the macrocycle. Indeed, in the previous system, there was an additional amide between the fullerene stopper and the template station, and the solvation of all the amides in DMSO hampered the possibility for the macrocycle to get close to the C_{60}. Therefore, this promoted shuttling over the alkyl chain. In the second system, shown in Figure 2.21b, the solvation of the succinamide template released the macrocycle that was attracted by the fullerene via π–π interactions, as demonstrated through cyclic voltammetry experiments.

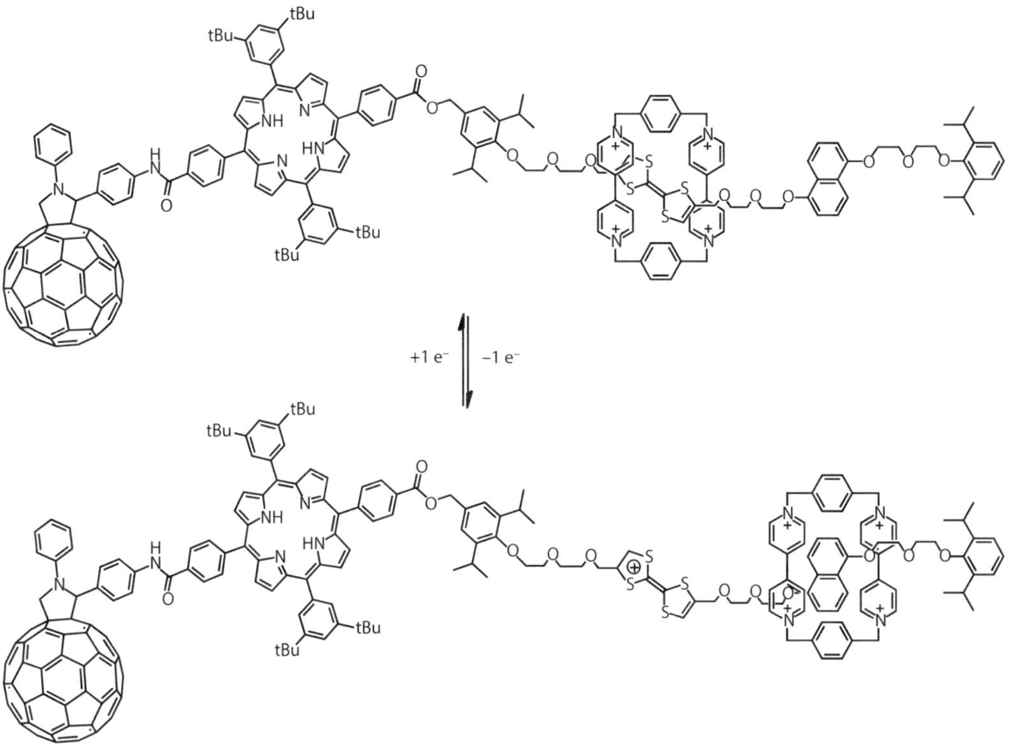

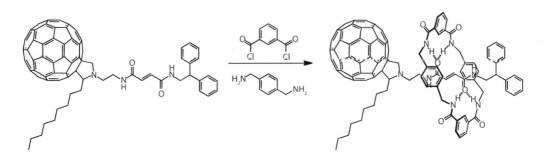

FIGURE 2.19 Bistable rotaxane assembled via π–π interactions. Electron-deficient CBPQT^{4+} macrocycle can switch from two different electron-rich stations.

In addition to the solvent switching behavior, the macrocycle can also be shifted chemically by oxidation of the nitrogen atom of the fulleropyrrolidine ring (Figure 2.21c).[154] The *N*-oxide formed was encapsulated in the macrocycle by means of hydrogen bonding. This resulted in an increased stability of *N*-oxide, which is known to be thermally unstable and difficult to study and characterize.

The synthesis of a new fullerene rotaxane showed that fullerenes can be used as a powerful tool to induce submolecular motion (Figure 2.22).[135] In this case, the template glycylglycine station was placed far away from the fullerene stopper and separated from it by a triethylene glycol spacer. The solvent switching behavior is in agreement with similar previously studied systems. Indeed, when the system was dissolved in solvents, such as THF (tetrahydrofuran) or CHCl$_3$, the macrocycle stayed on the peptide station by means of hydrogen bonds, while in solvents that disturb hydrogen bonds (DMSO or DMF, dimethylformamide), the macrocycle was found to π-stack on the fullerene

FIGURE 2.20 Synthesis of a fullerene-stopped rotaxane via the Leigh protocol.

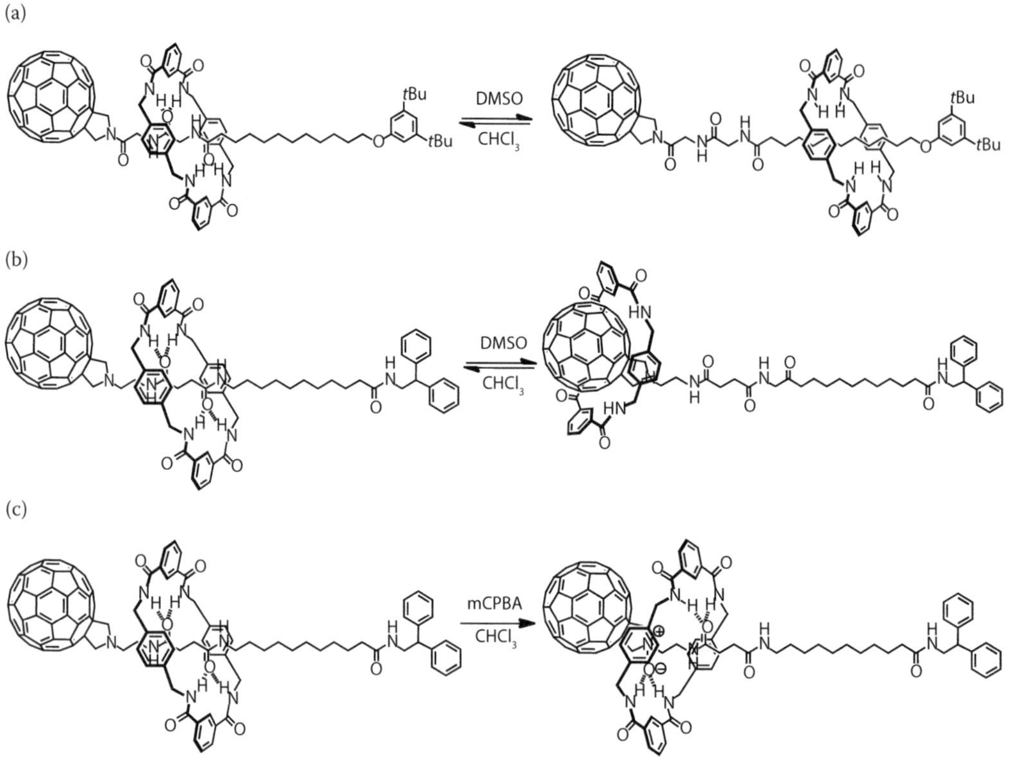

(a)

(b)

(c)

FIGURE 2.21 (a) Shuttling and (b) reverse shuttling of macrocycle, and (c) formation and stabilization of fulleropyrrolidine *N*-oxide.

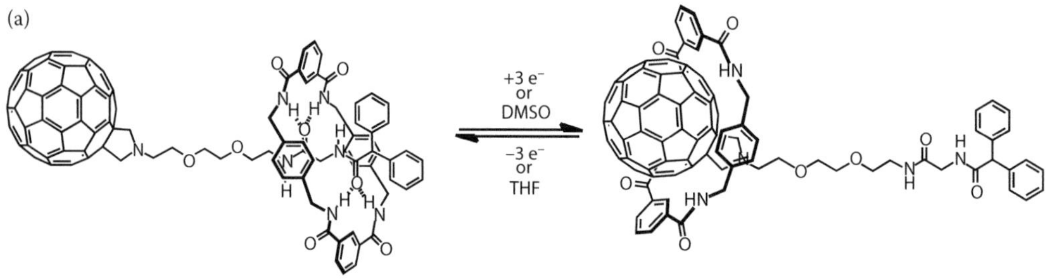

(a)

FIGURE 2.22 Electrochemical and solvent depending shuttling in a fullerene rotaxane.

wall. Remarkably, the electrochemically generated trianion on the C_{60} spheroid in THF induced shuttling of the macrocycle to the fullerene stopper side of the thread, overcoming the hydrogen bonds that, in such a solvent, collocate the macrocycle on the peptide station.

The shuttling can also modulate the nonlinear optical (NLO) response of fullerenes, measured by Z-scan technique.[155] Twofold difference of the NLO response has been observed in rotaxanes because of translocation of the macrocycle.

2.3 EMERGING APPLICATIONS IN MATERIAL SCIENCES

The different methodologies of functionalization of fullerenes, as well as the several types of supramolecular interactions, allowed the preparation of multiple architectures with interesting properties in technological fields. Some of the most promising fields of application of these novel materials include artificial photosynthesis and solar cells.

2.3.1　ENERGY CONVERSION

2.3.1.1　Artificial Photosynthetic Reaction Centers

Research on donor–acceptor systems consists of the development of an artificial model, which can mimic the natural photosynthetic process[156] to transform light into chemical energy. The natural photosynthetic reaction centers are composed of several photoactive units working together in a complicated system where the electron transfer (ET) events take place after light irradiation. This gives a long-distance and long-lived charged separate state (CSS), and these are the necessary conditions to enable the conversion into chemical energy.

To approach an artificial model from the chemical point of view, there is a need to simplify it in a basic donor–acceptor system, formed by one electron donor unit (D) and one electron acceptor unit (A), linked together through covalent or noncovalent interactions (Figure 2.23). This simple system (also called *dyad*) should efficiently give a long-lived CSS after irradiation with light. Donors supply a double function: they not only act as excitation-harvesting antennas but also as primary electron donors that transfer electrons to the acceptor.

To increase the lifetime of these systems, fullerenes, especially C$_{60}$, have been widely used as electron acceptor units. Indeed, C$_{60}$ possesses high electron affinity[8,157,158] and small reorganization energy.[159] The electrons are also highly delocalized in the tridimensional π-system,[160] and all these properties allow the formation of very stable radical pairs.

Another process that affects the lifetime is charge recombination (CR). This is the reverse process that happens after ET, which relaxes the system to the original ground state. Both these electron processes can occur through bond or through space, depending on the nature of the linker (or spacer) between the D and the A. The selection of the spacer is very important because it controls the distance, the orientation, and the electronic coupling between the two units.

2.3.1.1.1　Dyads

Dyads can contain photoactive electron donors, such as porphyrins (P), phthalocyanines (Pc), subphthalocyanines (subPc), and perylene diimides (PDI). The advantage of these chromophores lies in strong absorption in the visible region in addition to minimal structural changes when electrons are released. Porphyrins, composed of four pyrrolic macrocycles connected via methine bridges,

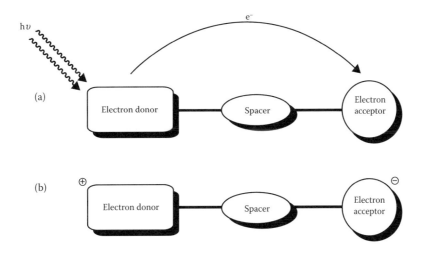

FIGURE 2.23　(a) Photoinduced ET and (b) CSS.

are the most studied chromophores in dyad systems, and are characterized by strong bands at about 420–450 nm (Soret region) and weaker Q-bands in the long wavelength region (650–680 nm). Pc are widely used and similar to macrocycles composed of four indole units conjugated with nitrogen bridges; they are also characterized by strong absorption in the Q-bands region and less intensive bands in the Soret region.

In a D–A system, the donor can also belong to the nonphotoactive class of compounds, such as TTF and exTTF,[54] ferrocene (Fc),[161] ruthenocene,[162] and conjugated oligomers. Because these electron donors do not absorb light, it is here that the fullerene unit (electron acceptor) triggers the ET by getting excited by light irradiation in its singlet state.[163]

The constitutive units of such systems can be linked together through covalent bonds or supramolecular bonds.[164–166]

2.3.1.1.2 Dyads Assembled through Covalent Bonds

Between all the cyclo-reaction protocols, the 1,3-dipolar addition allowed the preparation of several architectures bearing porphyrins in position 2 of fulleropyrrolidines (Figure 2.24).[167,168] The photoexcitation of a zinc imidazoporphyrin–fullerene dyad with a short linkage results in the formation of the CSS with a lifetime of 310 μs in benzonitrile at 278 K. A similar system showed ultra-long-lived CSS, measured to be 230 μs at 25°C and increasing up to 120 s at −150°C.[169]

Similarly, it has been possible to obtain Pc covalently linked to fulleropyrrolidines to generate dyads able to self-aggregate in solution (Figure 2.25). Amphiphilic Pc–C_{60} salts were dispersed in

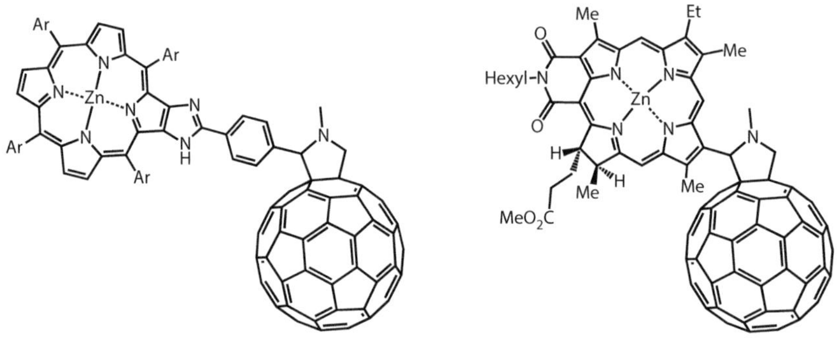

FIGURE 2.24 Covalently linked D–A systems.

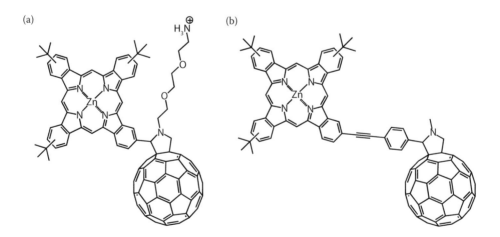

FIGURE 2.25 Pc macrocycles covalently linked to C_{60}.

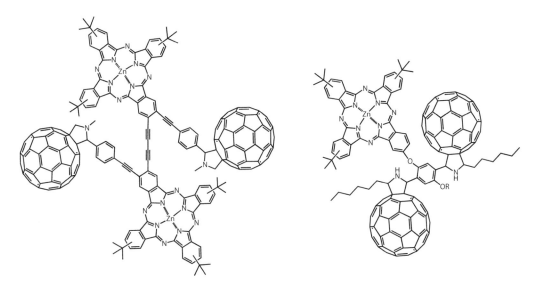

FIGURE 2.26 Molecular architectures containing multiple donor and/or multiple acceptor systems.

water and transmission electron microscopy (TEM) revealed perfectly ordered one-dimensional (1D) nanotubules that were found to be able to generate CSS with lifetimes up to 1 ms.[170] A change in the spacer between the fullerene and the Pc afforded other derivatives illustrated in Figure 2.25b.[171] The use of a [2.2]paracyclophane pseudoconjugated spacer ensures efficient and rapid formation of the radical–ion pair state (with lifetimes up to 458 ns in toluene) and its remarkable stabilization.[172]

Photoinduced ET was studied in fullerene bisadducts synthesized through a Bingel reaction bearing different kinds of macrocycles in a parachute-like shape architecture.[173] The double addition to C$_{60}$ imposes geometrical restrictions that set the macrocycle and the fullerene moieties in spatial proximity and affects the ET as well as the CR processes.

Molecular architectures containing multiple D and/or multiple A systems have been developed to render the light-harvesting process more efficient (Figure 2.26). For the examples reported in the figure, the number of D–A units played an important role in the elongation of radical pair lifetime. In the system formed by two C$_{60}$ and two Pc, linked together through acetylene bridges,[174] only short lifetimes were detected (picoseconds range), probably because of the close proximity of the photoactive units. On the contrary, the linking of two fullerene moieties to one single Pc afforded a radical pair state with 21 ns lifetime.[175]

The reaction of an Si^{IV}Pc with C$_{60}$ functionalized with an alcohol group gave a series of donor–acceptor systems assembled through the formation of silicon–oxygen bonds.[129,176] Dentrimeric structures and pentads have been further developed following such covalent bonds.[177,178]

Nonphotoactive units, such as ex-TTF and Fc, have been covalently linked to C$_{60}$ spheroid by means of different methodologies. When Diels–Alder cycloaddition was used to attach exTTF moieties to the fullerene, efficient electron and energy transfer took place by a through-bond coupling mechanism (Figure 2.27a).[179,180]

Bingel-type adducts[181] have also been studied, as well as fulleropyrrolidines synthesized by means of azomethine ylides with vinyl spacers bearing TTF moieties[182] and with azides bearing TTF and exTTF donors.[183,184]

Fc units were also attached to fullerenes to create D–A systems with either variable-spacing building blocks or rigid linkers (Figure 2.27b and c).[185–187] Liquid-crystalline dyads bearing Fc and a mesomorphic dendrimer, prepared through 1,3-dipolar cycloaddition, showed photoinduced intramolecular electron transfer (Figure 2.27d).[188]

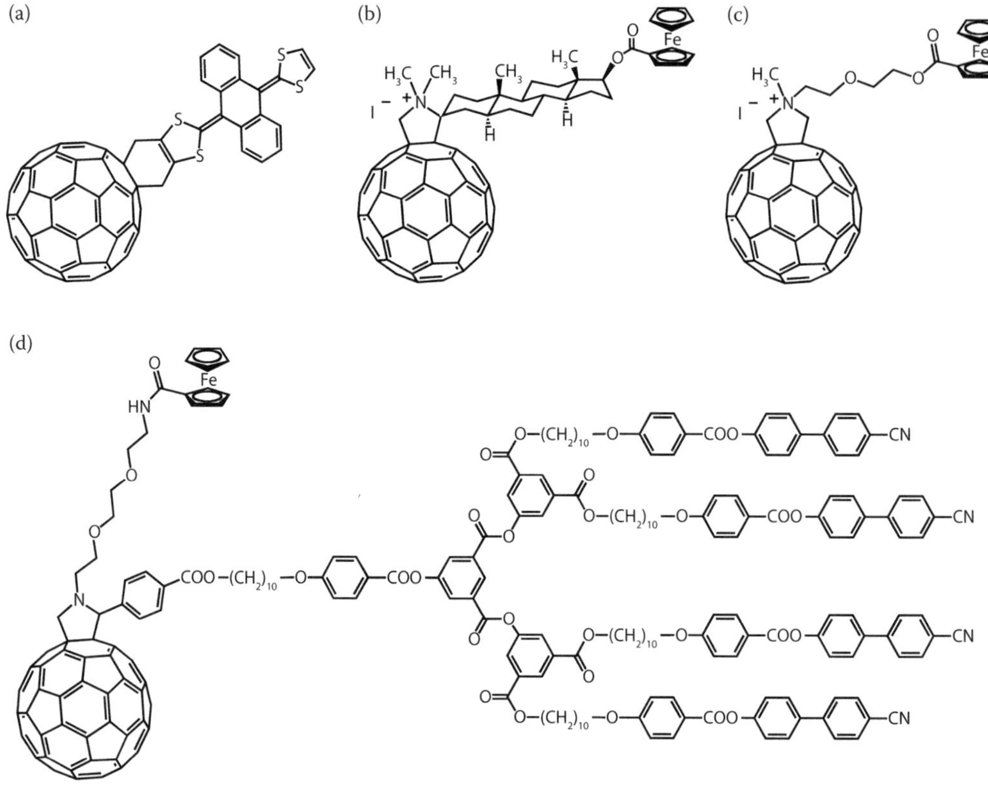

(a)   (b)   (c)

(d)

FIGURE 2.27   Dyads bearing fullerenes covalently linked to nonphotoactive units such as exTTF and Fc.

2.3.1.1.3   Dyads Assembled through Supramolecular Bonds

2.3.1.1.3.1   Axial Coordination   Many D–A systems are constructed through supramolecular bonds. The frequently used method exploited the metal centered inside the macrocycle rings to achieve D–A systems through coordination chemistry.

Fullerene derivatives functionalized with pyridine or imidazole substituents[125,127,189,190] can coordinate with several metal porphyrins, Pc, and Nc in solution by means of self-assembly. Such electron-rich macrocyclic dyes are good electron donors as they absorb light in the visible near-infrared (IR) region and exhibit a usable redox potential.

The light-induced ET occurring in these self-assembled conjugates gave emphasis to the research field, and the first success in this direction was a fulleropyrrolidine coordinated with zinc tetraphenylporphyrin (ZnTPP) (Figure 2.28a).[125] The weak character of the metal–pyridine bond allows an equilibrium between the association and the dissociation of the fullerene acceptor and the ZnTPP donor, facilitating the formation of a CSS with lifetime on the microseconds range. The use of another metal such as ruthenium in the same system[191] revealed a more rapid recombination of the charge-separated species (picoseconds range), and this is probably because of higher stability of the formed complex (Figure 2.28b). Indeed, Ru is known to bond with pyridine with a strong σ-character (compared to Zn). This more stable bond hindered the splitting of the radical pair and induced a faster recombination of the charge.

Zinc is also used as a coordination metal in a recent work that involves a zinc phthalocyanine dyad peripherally substituted with 4,4-difluoro-4-bora-3*a*,4adiaza-*s*-indacene (bodipy). Such panchromatic light-harvesting system (Figure 2.28c) absorbs light over a broad spectral range[192], and the irradiation of this supramolecular ensemble within the visible range leads to the formation of a CSS with a remarkably long lifetime of 39.9 ns in toluene.

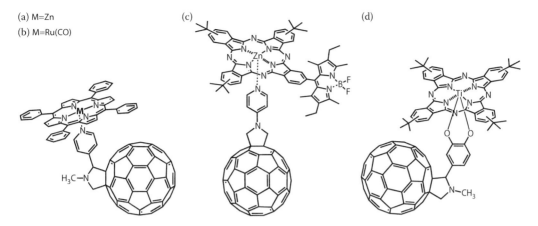

FIGURE 2.28 Supramolecular D–A systems assembled by means of metal coordination.

The groups of Torres and Ito used catechol ligands in fullerene derivatives for the synthesis of supramolecular complexes with Ti(IV), where ET from the photoexcited Ti-Pc to C$_{60}$ has been demonstrated (Figure 2.28d).[193,194]

2.3.1.1.3.2 Crown Ether Ammonium Salt Recognition Supramolecular interlocked D–A systems have been synthesized exploiting the ability of crown ethers to recognize ammonium salts in solution.

The electron donor properties of porphyrins have been studied in rotaxanes bearing C$_{60}$ as the stopper on a thread (Figure 2.29a),[195] giving the formation of CSS with lifetimes of 180 ns.[196]

Similarly, rotaxanes displaying the fullerene spheroid on the macrocycle and triphenylamine[197] or Fc[198,199] as electron donors have also been synthesized. Lifetimes of radical pairs generated from derivatives reported in Figure 2.29c and d have been calculated after photophysical studies as 360 and 13 ns, respectively.

The ET behavior of supramolecular assembled dyads was compared to similar donor–acceptor systems where the donors were covalently bonded to the fullerenes. Surprisingly, in similar covalently linked systems, the lifetime of the radical–ion pairs is lower than those observed in supramolecular assemblies. As an example, the lifetime of the CSS of the fullerene–(zinc phthalocyanine) dyad (Figure 2.29b) assembled through hydrogen bonds was found to be 1.5 µs, when a similar but covalently linked dyad[200] exhibited lifetimes for the radical–ion pair of only 3 ns.

2.3.1.1.3.3 Metal Coordination A different kind of recognition pattern has been developed following the same approach adopted by Sauvage et al.,[108,111] governed by a tetrahedrally coordinated Cu(I) center (Figure 2.30). The position and the amount of the donor and the acceptor resulted in a drastic variation of the lifetimes of the CSS. Relatively low lifetimes of 1.17 µs were measured in the case of the rotaxane presenting two porphyrin stoppers and C$_{60}$ is linked to the macrocycle.[201] In the case when two fullerene stoppers are on the thread and a porphyrine is on the macrocycle, the lifetime increased to 32 µs.[202]

2.3.1.1.3.4 Hydrogen Bonding Directional hydrogen bonding interactions have also been used to assemble D–A systems. Different electron donors have been connected to fullerene derivatives through these multiple hydrogen bond motifs, and the lifetime of the photogenerated radical–ion pairs was improved compared to those reported for related covalently linked dyads. An exhaustive example is given by the dyad represented in Figure 2.31a, where the supramolecular connection between the fulleropyrrolidine acceptor and the porphyrin donor consists of a guanosine–cytidine scaffold. In such a system, the lifetime of CSS was detected to be 2.02 µs.[203]

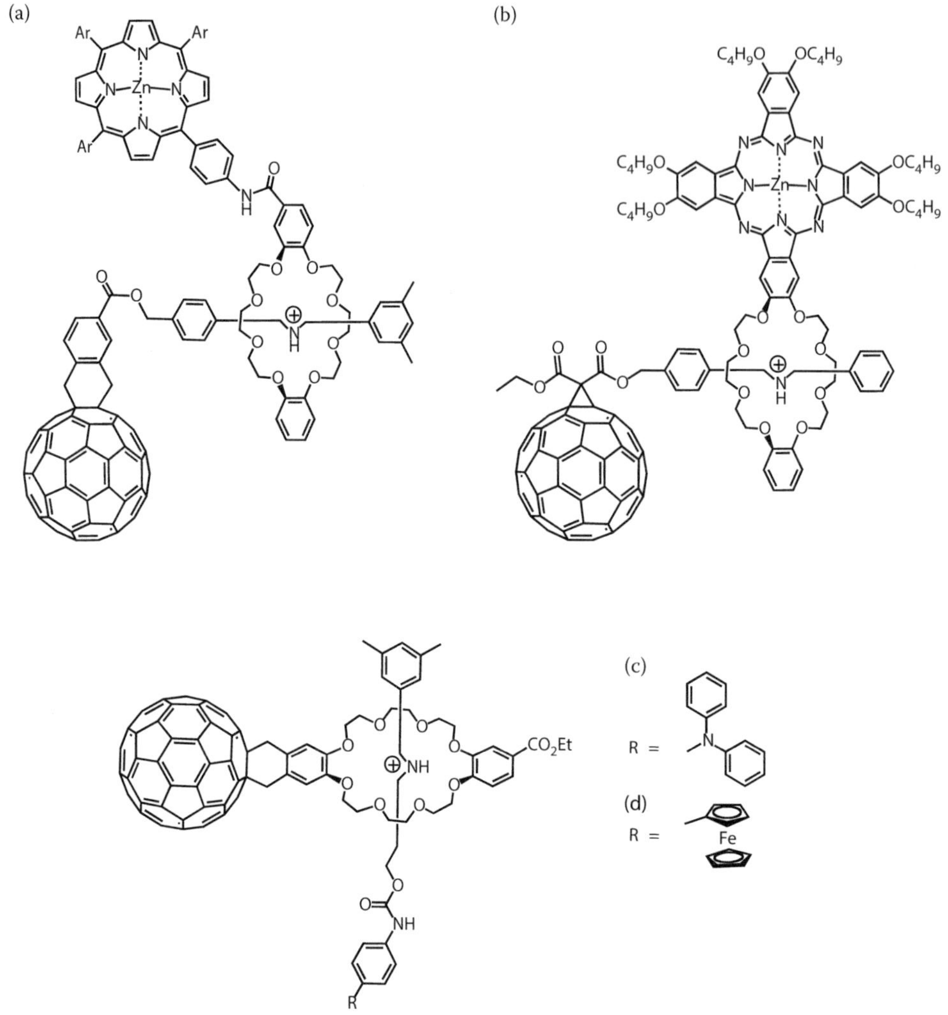

FIGURE 2.29 Electron donors covalently bonded to crown ethers and then mechanically interlocked with C_{60}–ammonium salt derivatives.

Conversely, the self-assembly further developed using cytidine-substitute Pc showed significantly short lifetimes (3 ns), because of the large association constant that is responsible for a strong coupling between donor and acceptor units (Figure 2.31b).[204,205]

Guanidinium and carboxylate ion pairs have been used in a reported series of C_{60}–TTF ensembles with a tunable molecular architecture. Spacers of different lengths (phenyl and biphenyl) and two functional groups (ester and amide) were used to study the ET process (Figure 2.32).[206]

2.3.1.1.3.5 Interlocked Architectures   Different D–A systems including fullerene have been assembled using a molecular interlocked design.

An interesting synthetic technique to link C_{60} covalently to a macrocycle exploited the Vogtle method,[207] where a sulfolene group was transformed at high temperatures to a 1,4-diene, which can react with a double bond of C_{60} (Figure 2.33). This macrocycle, threaded in a two-porphyrin stopper thread, acted as the acceptor for the ET process.[208–210] The length of the thread is directly proportional with the lifetimes of the radical pairs reached after photoexcitation, which are 180, 230, and 625 ns, respectively, for the structures reported in Figure 2.33b.

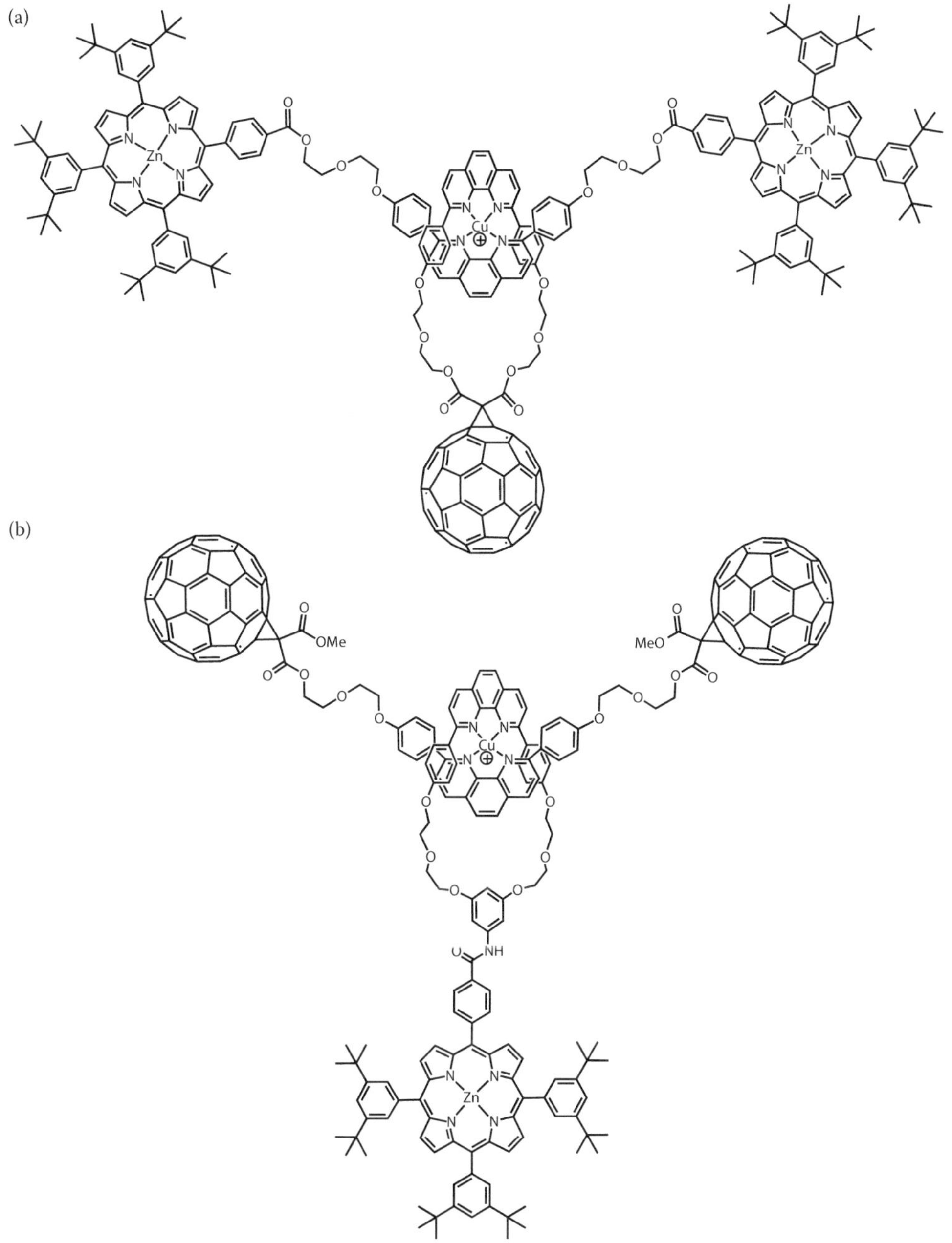

(a)

(b)

FIGURE 2.30 D–A systems assembled adopting Cu(I) as coordination template.

By modification of the Leigh protocol,[149] electron donor Fc derivatives were added to a macrocycle in a fullerene-stoppered rotaxane (Figure 2.34).[211] The distance from the Fc units to the fullerene was important to increase the lifetimes of the radical pair species, which arise from 9 to 26 ns when a triethyleneglycol spacer was added between the 1,4-diamide template and C$_{60}$.[212] Moreover, the addition of a component (hexafluoroisopropanol) able to disturb the hydrogen bonds in the solution decreased the relative distance between the electron donors and the fullerene, together with the lifetimes of the radical pairs, reduced at 13 ns. This, together with other photophysical

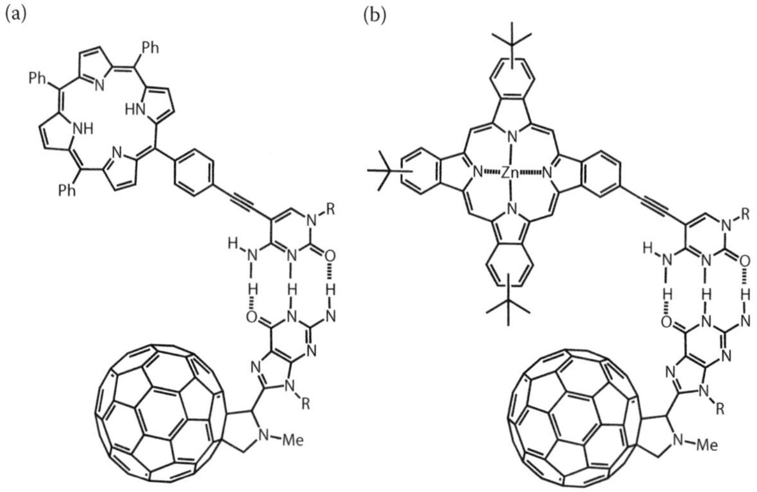

FIGURE 2.31   Donor–acceptor systems assembled by means of directional hydrogen bonding.

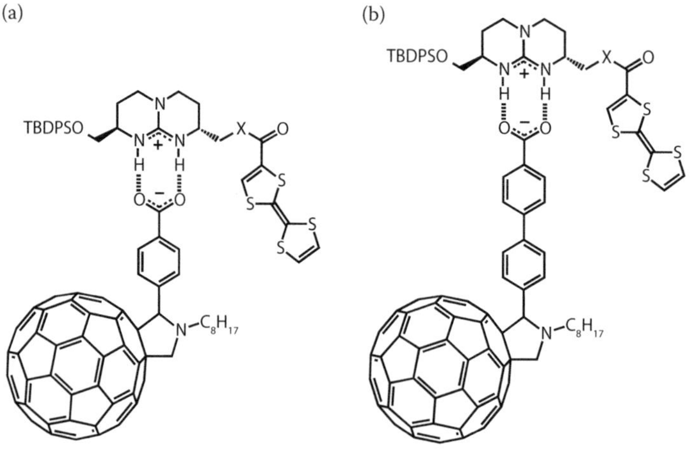

FIGURE 2.32   Examples of D–A systems bearing nonphotoactive TTF units linked to fullerenes via hydrogen pattern recognition.

experiments, demonstrated the contribution of the displacement of the macrocycle to modulation of the ET process.

2.3.1.1.4   Polyads

A polyad consists of a system holding more than two different electroactive units, which can be assembled with different methodologies.[213,214] In these D–A systems called, for example, triads or tetrads (depending on the number of units), a short-distance multistep photoinduced ET is exploited to achieve long-lived CSS, without reducing the efficiency.

Triads and tetrads involving C_{60}, porphyrins, and Fc have been synthesized, investigated, and reviewed by Imahori et al.[215,216] Among these systems representing an artificial reaction center, the first triad based on fullerene–free base porphyrin–Zn porphyrin showed a highly efficient ET and lifetimes of the CSS up to 21 μs.[217] In this system, first an energy transfer occurs from Zn–porphyrin to the free base porphyrin (after photostimulation), and then an ET follows from the excited free base porphyrin to the C_{60} molecule.

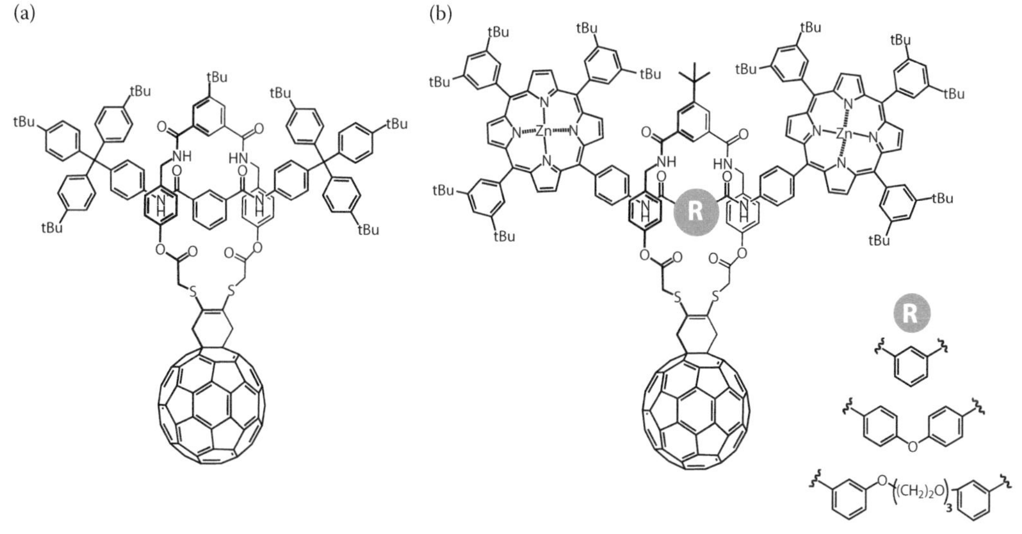

(a)

(b)

FIGURE 2.33   Rotaxanes synthesized via Vogtle method.

The substitution of the terminal metal–porphyrin in the triad with another electron donor such as Fc gave a comparable lifetime of the Fc$^+$–C$_{60}$$^-$ radical pair, with very high quantum yields.[218] In this particular case, after light irradiation, the first ET goes from the porphyrin to the fullerene, and the second ET goes from the Fc to the porphyrin.

A considerable elongation of CSS lifetimes (0.38 s) was reached with the tetrad formed by C$_{60}$–free base porphyrin–Zn porphyrin–Fc, where a multistep ET generated a long-distance radical pair in frozen media and in solution.[219] In a similar triad derivative, a thiol group, added close to the Fc unit, allows the formation of well-packed self-assembled monolayer on gold electrodes, leading to the generation of a photocurrent output (Figure 2.35).[220,221]

ZnP yields the most promising donor chromophores employed in D–A systems. In many examples (some of them reported in Figure 2.36), the ZnP is situated between the electron acceptor (supramolecularly or covalently linked with C$_{60}$) and another additional donor, attached in the opposite side. Among these second electron donors, TTF and exTTF moieties,[222] boron dipyrrin derivatives,[223] subPc,[224] and triphenylamine[225] showed interesting results in terms of lifetimes of CSS generated from the photoinduced ET.

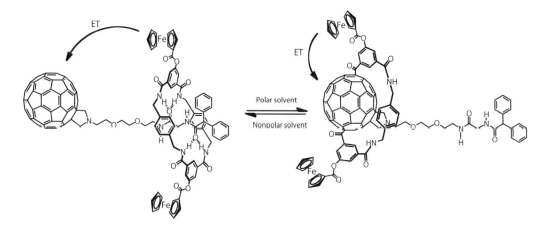

FIGURE 2.34   D–A system with tunable ET.

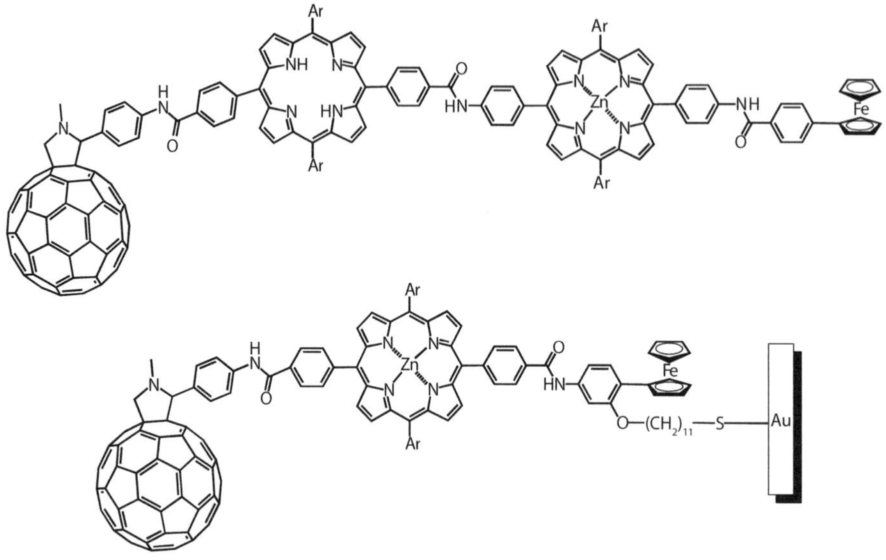

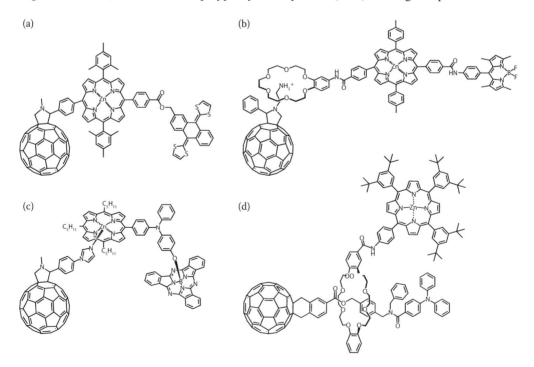

FIGURE 2.35   Molecular structure of fullerene–free base porphyrin–Zn porphyrin–ferrocene tetrad and Au electrode modified with a self-assembled monolayer of fullerene–Zn porphyrin–ferrocene triad.

Specifically, the triad formed by subPc–triphenylamine–ZnP system coordinated to a fulleropyrrolidine (Figure 2.36c) showed CSS with a lifetime of 1.1 ns after light irradiation. Lifetimes of up to 320–420 ns have been recently reported for triads involving ZnP, C_{60}, and triphenylammine as the additional electron donor (Figure 2.36d).

The separation of donors from C_{60} acceptor with conductive oligomers allowed the formation of long-distance CSS, as in the case of polyphenylenevinylene[226] (PPV) and oligothiophene.[227]

FIGURE 2.36   ZnP as electron donors in triads and tetrads, with C_{60} as electron acceptor.

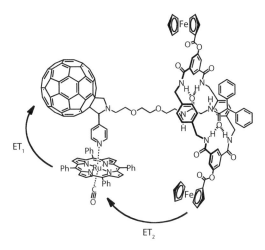

FIGURE 2.37 Triad composed of a fullerene–rotaxane, RuTPP, and ferrocenes units.

Among rotaxanes, a fulleropyrrolidine bearing a pyridyl ring linked to the thread was used to construct a triad with Fc groups bonded to the macrocycle (Figure 2.37). The second electron donor system was a ruthenium tetraphenylporphyrin (RuTPP) coordinated with the pyridine, in a rotaxane fashion.[228] After photoexcitation of the RuTPP, the ET takes place in the C$_{60}$ moiety, followed by a charge shift from the RuTPP to Fc units. As a matter of fact, the charge shift elongates the lifetime of the C$_{60}$ radical pair that remains persistent on the 3.0 ns time scale.

2.3.2 Solar Cells

Organic solar cells are considered a less expensive alternative to the corresponding cells based on inorganic materials. The discovery of the participation of fullerenes in ET process with photoexcitable polymers such as PPV[229] was applied in a photodiode device able to generate current from the CSS.[230] In general, the morphology of the active organic layer is important to increase the efficiency of the solar cells, and functionalized fullerenes are able to mix intimately with the conjugated polymers decreasing the aggregation processes.

The organic solar cells are constructed by deposition of different layers following a *p–i–n* architecture, as shown in Figure 2.38. The transport layers introduced between the organic active layer and the electrodes play an important role, because the active layer normally has poor transport properties. Moreover, there is a need to separate the excitons that form close to the electrodes to avoid quenching processes that would hamper the photocurrent. The corresponding electrodes are usually made of indium tin oxide (ITO) as the anode and some metal, such as Au or Al, as the cathode.

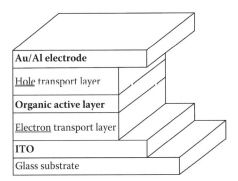

FIGURE 2.38 Schematic representation of a *p–i–n* architecture in a solar cell.

The organic active layer is a blend of a conductive polymer (acting as electron donor) and an electron acceptor characterized by an organic moiety. Indeed, fullerene derivatives showed suitable electronic properties for this role. The most commonly used polymers for solar cells are P3HT (poly(3-hexylthiophene)), but other copolymers bearing thiophene units have also been studied to build plastic solar cells with higher photoconductivity.[231–243]

The most commonly used C_{60} derivative is a methanofullerene called PCBM (1-(3-methoxycarbonyl)-propyl-1-1-(6′6)C_{61}), which proved to be an excellent electron acceptor in multiple bulk heterojunction solar cells (Figure 2.40a).[244–247] The particular ability of PCBM to mix homogeneously with the other components of the organic active layer placed fullerene derivative at the first place in the rank of the electron acceptors suitable for solar cells.[248,249] The C_{70} analog (PC$_{70}$BM) was also developed for these purposes.[250]

The efficiency of these organic solar cells stays around 4–7%, but there are some exceptions where this value has reached even higher levels. In the case of a π-conjugated oligomer–tetrafullerene nanoarray used to build a solar cell with P3HT as electron donor polymer, the efficiency was 15% (Figure 2.39).[251]

Attempts to stabilize the morphology of the polymeric active organic layer achieved different polymerizable C_{60} derivatives, such as in the case reported by Drees et al.[252] In this work, a new synthetic C_{61}–butyric acid glycidol ester (PCBG) was able to change the nanomorphology of the blended layer, increasing the operational stability of these plastic solar cells.

A wide series of fullerene–polymers composited has led to focused attention in the construction of polymer bulk–heterojunctions solar cells.[253–255] Cross-linked polymers, end-capped polymers (where the C_{60} moieties are located in the terminal position of the chains), and star-shaped polymers (known as flagellenes) are some examples. Fullerenes are located as pendants in the skeleton in the side chain. Novel double-cable polymers consist of donor cables (p-type π-conjugated backbones), which are covalently connected with acceptor cables bearing fullerenes (Figure 2.40b).[256] A possible approach to develop materials for solar cells is self-assembling because it allows us to control the morphology of the system and therefore to obtain better devices. 2,6-Diacylaminopyridine–C_{60} derivatives were reported to be able to recognize through hydrogen bonds a uracil moiety covalently linked to polyphenylenevinylenecarbazole (Figure 2.40c).[256]

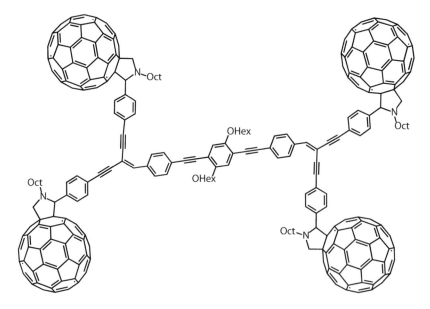

FIGURE 2.39 Efficient π-conjugated oligomer–tetrafullerene derivative employed in solar cells.

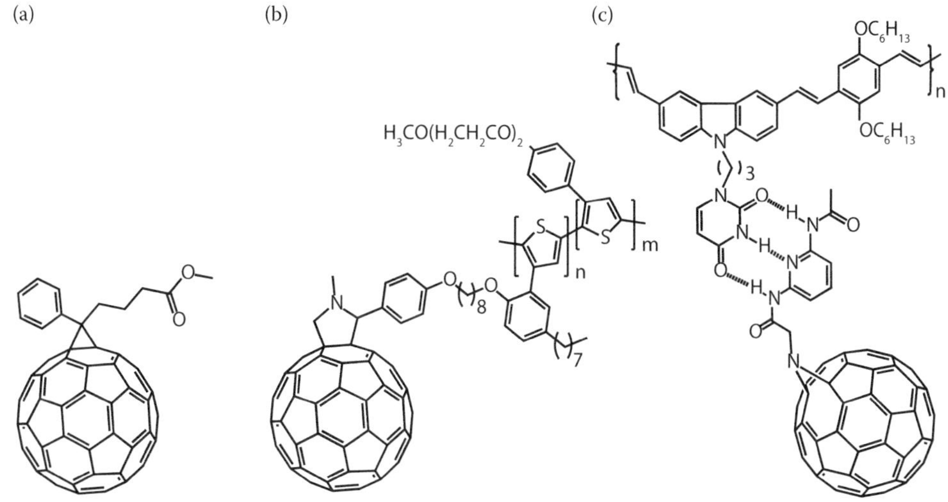

(a) (b) (c)

H$_3$CO(H$_2$CH$_2$CO)$_2$

OC$_6$H$_{13}$

OC$_6$H$_{13}$

FIGURE 2.40 (a) Most used fullerene derivative in solar cells PCBM, (b) example of double-cable polymer, and (c) supramolecular polymer assembled through hydrogen bonding.

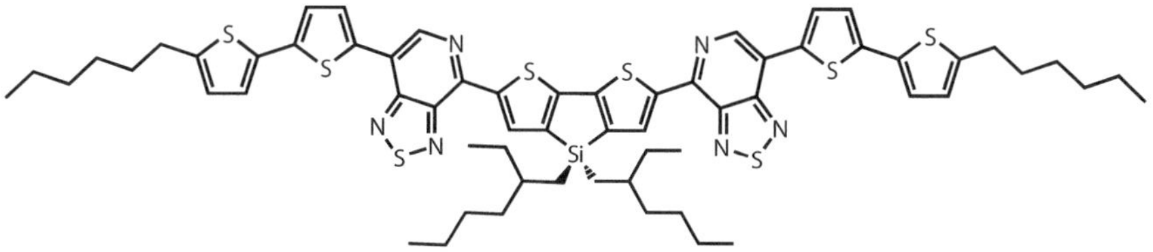

FIGURE 2.41 A novel electron donor DTS(PTTh$_2$)$_2$ used in a high-performance small-molecule solar cell device.

Although polymer-based solar cells remain the most efficient devices constructed, small molecules such as organic semiconductors could represent a promising solar energy technology.[257,258] However, the weak point of these devices has always been the small efficiency in comparison with the levels obtained with polymeric cells.[259]

Recently, a remarkable efficiency of 6.7% has been performed by a solution-processed small-molecule solar cell, composed of a novel electron donor DTS(PTTh$_2$)$_2$ (Figure 2.41) and PC$_{70}$BM as acceptor.[260] Such highest efficiency was allowed by the introduction of a very small amount of 1,8-diiodooctane (DIO) as a solvent additive (0.25 vol.% compared with the 1–3 vol.% usually needed in polymer-based devices).[261]

2.4 CONCLUSIONS

The development of new methods to functionalize fullerenes has opened the door for the design of novel materials with a wide variety of applications. One prominent application is related to energy conversion that has been the subject of this chapter. Such methods are based on the preparation of covalent, supramolecular, and mechanically interlocked architectures that are used to exploit the basic properties of fullerenes. Plenty of possibilities remain available, yet there are still many challenges to overcome, such as control of the supramolecular organization of fullerenes in the solid state that is key to developing practical applications. This is something that cannot be solved from one single perspective and fullerene science still remains as one of the most fascinating and multidisciplinary fields of research with new and unexpected discoveries to come.

REFERENCES

1. Kroto, H. W., Heath, J. R., O'Brien, S. C., Curl, R. F. and Smalley, R. E. C_{60}: Buckminsterfullerene. *Nature* **318**, 162–163, 1985.
2. Haufler, R. E. et al. Efficient production of C_{60} (buckminsterfullerene), $C_{60}H_{36}$, and the solvated buckide ion. *J. Phys. Chem.* **94**, 8634–8636, 1990.
3. Hawkins, J. M., Meyer, A., Lewis, T. A., Loren, S. and Hollander, F. J. Crystal structure of osmylated C_{60}: Confirmation of the soccer ball framework. *Science* **252**, 312–313, 1991.
4. Haddon, R. C. Electronic structure, conductivity and superconductivity of alkali metal doped (C_{60}). *Acc. Chem. Res.* **25**, 127–133, 1992.
5. Haddon, R. C. Chemistry of the fullerenes: The manifestation of strain in a class of continuous aromatic molecules. *Science* **261**, 1545–1550, 1993.
6. Haddon, R. C., Brus, L. E. and Raghavachari, K. Rehybridization and π-orbital alignment: The key to the existence of spheroidal carbon clusters. *Chem. Phys. Lett.* **131**, 165–169, 1986.
7. Yang, S. H., Pettiette, C. L., Conceicao, J., Cheshnovsky, O. and Smalley, R. E. Ups of buckminsterfullerene and other large clusters of carbon. *Chem. Phys. Lett.* **139**, 233–238, 1987.
8. Echegoyen, L. and Echegoyen, L. E. Electrochemistry of fullerenes and their derivatives. *Acc. Chem. Res.* **31**, 593–601, 1998.
9. Bruno, C. et al. Electrochemical generation of C_{60}^{2+} and C_{60}^{3+}. *J. Am. Chem. Soc.* **125**, 15738–15739, 2003.
10. Korobov, M. V. and Allan, L. M. In *Fullerenes: Chemistry, Physics, and Technology*, eds. K. M. Kadish and R. S. Ruoff, Wiley Interscience, New York, NY, 2000, p. 53.
11. Haddon, R. C. et al. C_{60} thin film transistors. *Appl. Phys. Lett.* **67**, 121–123, 1995.
12. Hebard, A. F., Rosseinsky, M. J., Haddon, R. C., Murphy, D. W., Glarum, S. H., Palstra, T.T.M., Ramirez, A. P. and Kortan, A. R. Superconductivity at 18 K in potassium-doped C_{60}. *Nature* **350**, 600–601, 1991.
13. Grant, P. Superconductivity: Up on the C_{60} elevator. *Nature* **413**, 264–265, 2001.
14. Dagotto, E. Superconductivity: Enhanced: The race to beat the cuprates. *Science* **293**, 2410–2411, 2001.
15. Allemand, P.-M. et al. Organic molecular soft ferromagnetism in a fullerene C_{60}. *Science* **253**, 301–302, 1991.
16. Lappas, A. et al. Spontaneous magnetic ordering in the fullerene charge-transfer salt (TDAE)C_{60}. *Science* **267**, 1799–1802, 1995.
17. Narymbetov, B. et al. Origin of ferromagnetic exchange interactions in a fullerene–organic compound. *Nature* **407**, 883–885, 2000.
18. Mizoguchi, K. et al. Pressure effect in TDAE-C—{60} ferromagnet: Mechanism and polymerization. *Phys. Rev. B* **63**, 140417, 2001.
19. Hirsch, A. *The Chemistry of the Fullerenes*. John Wiley & Sons, Weinheim, 2002.
20. Wilson, S. R., Schuster, D. I., Nuber, B., Meier, M. S., Maggini, M., Prato, M. and Taylor, R. In *Fullerenes: Chemistry, Physics, and Technology*, eds. K. M. Kadish and R. S. Ruoff, Wiley Interscience, New York, NY, 2000, p. 91.
21. Prato, M. Fullerene materials. *Top. Curr. Chem.* **199**, 173–187, 1993.
22. López, A. M., Mateo-Alonso, A. and Prato, M. Materials chemistry of fullerene C_{60} derivatives. *J. Mater. Chem.* **21**, 1305–1318, 2011.
23. Prato, M. et al. Energetic preference in 5,6 and 6,6 ring junction adducts of C_{60}: Fulleroids and methanofullerenes. *J. Am. Chem. Soc.* **115**, 8479–8480, 1993.
24. Eiermann, M., Wudl, F., Prato, M. and Maggini, M. Electrochemically induced isomerization of a fulleroid to a methanofullerene. *J. Am. Chem. Soc.* **116**, 8364–8365, 1994.
25. Janssen, R. A. J., Hummelen, J. C. and Wudl, F. Photochemical fulleroid to methanofullerene conversion via the Di–pi–methane (Zimmerman) rearrangement. *J. Am. Chem. Soc.* **117**, 544–545, 1995.
26. Gonzalez, R., Hummelen, J. C. and Wudl, F. The specific acid-catalyzed and photochemical isomerization of a robust fulleroid to a methanofullerene. *J. Org. Chem.* **60**, 2618–2620, 1995.
27. Grösser, T., Prato, M., Lucchini, V., Hirsch, A. and Wudl, F. Ring expansion of the fullerene core by highly regioselective formation of diazafulleroids. *Angew. Chem. Int. Ed. Engl.* **34**, 1343–1345, 1995.
28. Kuwashima, S. et al. Synthesis and structure of nitrene–C_{60} adduct $C_{60}NPhth$ (Phth = phthalimido). *Tetr. Lett.* **35**, 4371–4374, 1994.
29. Nakahodo, T. et al. [2 + 1] Cycloaddition of nitrene onto C_{60} revisited: Interconversion between an aziridinofullerene and an azafulleroid. *Angew. Chem. Int. Ed.* **47**, 1298–1300, 2008.
30. Ishida, T., Tanaka, K. and Nogami, T. Fullerene aziridine. Facile synthesis and spectral characterization of fullerene urethane, $C_{60}NCO_2CH_2CH_3$. *Chem. Lett.* **23**, 561–562, 1994.

31. Prato, M., Li, Q. C., Wudl, F. and Lucchini, V. Addition of azides to fullerene C$_{60}$: Synthesis of azafulleroids. *J. Am. Chem. Soc.* **115**, 1148–1150, 1993.

32. Schick, G., Hirsch, A., Mauser, H. and Clark, T. Opening and closure of the fullerene cage in *cis*-bisimino adducts of C$_{60}$: The influence of the addition pattern and the addend. *Chem. Eur. J.* **2**, 935–943, 1996.

33. Akasaka, T., Ando, W., Kobayashi, K. and Nagase, S. Reaction of C$_{60}$ with silylene, the first fullerene silirane derivative. *J. Am. Chem. Soc.* **115**, 1605–1606, 1993.

34. Vasella, A., Uhlmann, P., Waldraff, C. A. A., Diederich, F. and Thilgen, C. Fullerene sugars: Preparation of enantiomerically pure, spiro-linked C-glycosides of C$_{60}$. *Angew. Chem. Int. Ed.* **31**, 1388–1390, 1992.

35. Tsuda, M., Ishida, T., Nogami, T., Kurono, S. and Ohashi, M. C$_{61}$Cl$_2$. Synthesis and characterization of dichlorocarbene adducts of C$_{60}$. *Tetr. Lett.* **34**, 6911–6912, 1993.

36. Osterodt, J. and Vögtle, F. C$_{61}$Br$_2$: A new synthesis of dibromomethanofullerene and mass spectrometric evidence of the carbon allotropes C$_{121}$ and C$_{122}$. *Chem. Commun.* **4**, 547–548, 1996.

37. Dragoe, N. et al. Carbon allotropes of dumbbell structure: C$_{121}$ and C$_{122}$. *Chem. Commun.* **1**, 85–86, 1999.

38. Liou, K.-F. and Cheng, C.-H. Phosphine-mediated [2 + 2] cycloaddition of internal alk-2-ynoate and alk-2-ynone to [60]fullerene. *J. Chem. Soc., Chem. Commun.* **24**, 2473–2474, 1995.

39. Zhang, X., Fan, A. and Foote, C. S. [2 + 2] Cycloaddition of fullerenes with electron-rich alkenes and alkynes. *J. Org. Chem.* **61**, 5456–5461, 1996.

40. Zhang, X. and Foote, C. S. [2 + 2] Cycloadditions of fullerenes: Synthesis and characterization of C$_{62}$O$_3$ and C$_{72}$O$_3$, the first fullerene anhydrides. *J. Am. Chem. Soc.* **117**, 4271–4275, 1995.

41. Zhang, X., Romero, A. and Foote, C. S. Photochemical [2 + 2] cycloaddition of *N,N*-diethylpropynylamine to C$_{60}$. *J. Am. Chem. Soc.* **115**, 11024–11025, 1993.

42. Rao, A. M., Zhou, P., Wang, K.-A., Hager, G. T., Holden, J. M., Ying Wang, W.-T. L., Bi, X.-X. et al. Photoinduced polymerization of solid C$_{60}$ films. *Science, New Ser.* **259**, 955–957, 1993.

43. Wang, G.-W., Komatsu, K., Murata, Y. and Shiro, M. Synthesis and x-ray structure of dumb-bell-shaped C$_{120}$. *Nature* **387**, 583–586, 1997.

44. Chen, X. and Yamanaka, S. Single-crystal x-ray structural refinement of the "tetragonal" C$_{60}$ polymer. *Chem. Phys. Lett.* **360**, 501–508, 2002.

45. Chen, X., Yamanaka, S., Sako, K., Inoue, Y. and Yasukawa, M. First single-crystal x-ray structural refinement of the rhombohedral C$_{60}$ polymer. *Chem. Phys. Lett.* **356**, 291–297, 2002.

46. Maggini, M., Scorrano, G. and Prato, M. Addition of azomethine ylides to C$_{60}$: Synthesis, characterization, and functionalization of fullerene pyrrolidines. *J. Am. Chem. Soc.* **115**, 9798–9799, 1993.

47. Prato, M. and Maggini, M. Fulleropyrrolidines: A family of full-fledged fullerene derivatives. *Acc. Chem. Res.* **31**, 519–526, 1998.

48. Tagmatarchis, N. and Prato, M. The addition of azomethine ylides to [60]fullerene leading to fulleropyrrolidines. *Synlett* **6**, 768–779, 2003.

49. Martín, N. et al. Retro-cycloaddition reaction of pyrrolidinofullerenes. *Angew. Chem. Int. Ed.* **45**, 110–114, 2006.

50. Guryanov, I. et al. Metal-free, retro-cycloaddition of fulleropyrrolidines in ionic liquids under microwave irradiation. *Chem. Commun.* **26**, 3940–3942, 2009.

51. Guryanov, I. et al. Microwave-assisted functionalization of carbon nanostructures in ionic liquids. *Chem. Eur. J.* **15**, 12837–12845, 2009.

52. Kordatos, K. et al. Novel versatile fullerene synthons. *J. Org. Chem.* **66**, 4915–4920, 2001.

53. Takaguchi, Y. et al. Reversible binding of C$_{60}$ to an anthracene bearing a dendritic poly(amidoamine) substituent to give a water-soluble fullerodendrimer. *Angew. Chem. Int. Ed.* **41**, 817–819, 2002.

54. Segura, J. L. and Martín, N. New concepts in tetrathiafulvalene chemistry. *Angew. Chem. Int. Ed.* **40**, 1372–1409, 2001.

55. Kreher, D. et al. Rigidified tetrathiafulvalene–[60]fullerene assemblies: Towards the control of through-space orientation between both electroactive units. *J. Mater. Chem.* **12**, 2137–2159, 2002.

56. Cava, M. P., Deana, A. A. and Muth, K. Condensed cyclobutane aromatic compounds. VIII. The mechanism of formation of 1,2-dibromobenzocyclobutene; a new Diels–Alder synthesis. *J. Am. Chem. Soc.* **81**, 6458–6460, 1959.

57. Kräutler, B. and Maynollo, J. A. Highly symmetric sixfold cycloaddition product of fullerene C$_{60}$. *Angew. Chem. Int. Ed. Engl.* **34**, 87–88, 1995.

58. Nakamura, Y., Asami, A., Ogawa, T., Inokuma, S. and Nishimura, J. Regioselective synthesis and properties of novel [60]fullerene bisadducts containing a dibenzocrown ether moiety. *J. Am. Chem. Soc.* **124**, 4329–4335, 2002.

59. Bingel K. *Chem. Ber.* **126**, 1957, 1993.

60. Nierengarten, J., Gramlich, V., Cardullo, F. and Diederich, F. Regio- and diastereoselective bisfunction-alization of C_{60} and enantioselective synthesis of a C_{60} derivative with a chiral addition pattern. *Angew. Chem. Int. Ed. Engl.* **35**, 2101–2103, 1996.

61. Camps, X. and Hirsch, A. Efficient cyclopropanation of C_{60} starting from malonates. *J. Chem. Soc., Perkin Trans.* **1**, 1595–1596, 1997.

62. Herranz, M. Á., Cox, C. T. and Echegoyen, L. Retrocyclopropanation reactions of fullerenes: Complete product analyses. *J. Org. Chem.* **68**, 5009–5012, 2003.

63. Ángeles Herranz, M. et al. Chemical retro-cyclopropanation reactions in methanofullerenes: Effect of the 18-crown-6 moiety. *J. Supramol. Chem.* **1**, 299–303, 2001.

64. Sawamura, M. et al. Stacking of conical molecules with a fullerene apex into polar columns in crystals and liquid crystals. *Nature* **419**, 702–705, 2002.

65. Zhong, Y.-W., Matsuo, Y. and Nakamura, E. Lamellar assembly of conical molecules possessing a fullerene apex in crystals and liquid crystals. *J. Am. Chem. Soc.* **129**, 3052–3053, 2007.

66. Keshavarz, K. M., Knight, B., Srdanov, G. and Wudl, F. Cyanodihydrofullerenes and dicyanodihydrofullerene: The first polar solid based on C_{60}. *J. Am. Chem. Soc.* **117**, 11371–11372, 1995.

67. Wudl, A., Hirsch, A., Khemani, K. C., Suzuki, T., Allemand, P.-M., Koch, A., Eckert, H., Srdanov, G. and Webb, H. M. Survey of chemical reactivity of C_{60}, electrophile and dieno-polarophile par excellence. *ACS Symp. Ser.* **481**, 161–175, 1992.

68. Zhang, W. and Swager, T. M. Functionalization of single-walled carbon nanotubes and fullerenes via a dimethyl acetylenedicarboxylate-4-dimethylaminopyridine zwitterion approach. *J. Am. Chem. Soc.* **129**, 7714–7715, 2007.

69. Diederich, F. and Gómez-López, M. Supramolecular fullerene chemistry. *Chem. Soc. Rev.* **28**, 263–277, 1999.

70. Guldi, D. M. and Martín, N. Fullerene architectures made to order; biomimetic motifs—Design and features. *J. Mater. Chem.* **12**, 1978–1992, 2002.

71. Canevet, D., Pérez, E. M. and Martín, N. Wraparound hosts for fullerenes: Tailored macrocycles and cages. *Angew. Chem. Int. Ed.* **50**, 9248–9259, 2011.

72. Mirkin, C. A. and Brett Caldwell, W. Thin film, fullerene-based materials. *Tetrahedron* **52**, 5113–5130, 1996

73. Andersson, T., Nilsson, K., Sundahl, M., Westman, G. and Wennerström, O. C_{60} embedded in γ-cyclodextrin: A water-soluble fullerene. *J. Chem. Soc., Chem. Commun.* 604–606, 1992.

74. Atwood, J. L., Koutsantonis, G. A. and Raston, C. L. Purification of C_{60} and C_{70} by selective complexation with calixarenes. *Nature* **368**, 229–231, 1994.

75. Suzuki, T., Nakashima, K. and Shinkai, S. Very convenient and efficient purification method for fullerene (C_{60}) with 5,11,17,23,29,35,41,47-Octa-tert-butylcalix[8]arene-49,50,51,52,53,54,55,56-octol. *Chem. Lett.* **23**, 699–702, 1994.

76. Delgado, J. L. and Nierengarten, J. F. *Calixarenes in the Nano-World*. Springer, Dordrecht: 2007.

77. Lhoták, P. and Kundrát, O. Fullerene receptors based on calixarene derivatives. *Artif. Receptors Chem. Sens.* **8**, 249–272, 2010.

78. Matsubara, H. et al. Supramolecular inclusion complexes of fullerenes using cyclotriveratrylene derivatives with aromatic pendants. *Chem. Lett.* **27**, 923–924, 1998.

79. Kawase, T., Tanaka, K., Fujiwara, N., Darabi, H. R. and Oda, M. Complexation of a carbon nanoring with fullerenes. *Angew. Chem. Int. Ed.* **42**, 1624–1628, 2003.

80. Kawase, T., Tanaka, K., Seirai, Y., Shiono, N. and Oda, M. Complexation of carbon nanorings with fullerenes: Supramolecular dynamics and structural tuning for a fullerene sensor. *Angew. Chem. Int. Ed.* **42**, 5597–5600, 2003.

81. Kawase, T., Tanaka, K., Shiono, N., Seirai, Y. and Oda, M. Anion-type complexation based on carbon nanorings and a buckminsterfullerene. *Angew. Chem. Int. Ed.* **43**, 1722–1724, 2004.

82. Kawase, T. and Kurata, H. Ball-, bowl-, and belt-shaped conjugated systems and their complexing abilities: Exploration of the concave–convex π–π interaction. *Chem. Rev.* **106**, 5250–5273, 2006.

83. Pérez, E. M. et al. Weighting non-covalent forces in the molecular recognition of C_{60}. Relevance of concave–convex complementarity. *Chem. Commun.* **38**, 4567–4569, 2008.

84. Canevet, D. et al. Macrocyclic hosts for fullerenes: Extreme changes in binding abilities with small structural variations. *J. Am. Chem. Soc.* **133**, 3184–3190, 2011.

85. Smith, B. W., Monthioux, M. and Luzzi, D. E. Encapsulated C_{60} in carbon nanotubes. *Nature* **396**, 323–324, 1998.

86. Luzzi, D. E. and Smith, B. W. Carbon cage structures in single wall carbon nanotubes: A new class of materials. *Carbon* **38**, 1751–1756, 2000.

87. Yudasaka, M. et al. Nano-extraction and nano-condensation for C$_{60}$ incorporation into single-wall carbon nanotubes in liquid phases. *Chem. Phys. Lett.* **380**, 42–46, 2003.

88. Britz, D. A. et al. Selective host–guest interaction of single-walled carbon nanotubes with functionalised fullerenes. *Chem. Commun.* **2**, 176–177, 2004.

89. Khlobystov, A. N. et al. Molecular motion of endohedral fullerenes in single-walled carbon nanotubes. *Angew. Chem. Int. Ed.* **43**, 1386–1389, 2004.

90. Sun, B.-Y. et al. Entrapping of exohedral metallofullerenes in carbon nanotubes: (CsC$_{60}$)n@SWNT nano-peapods. *J. Am. Chem. Soc.* **127**, 17972–17973, 2005.

91. Guan, L. et al. Coalescence of C$_{60}$ molecules assisted by doped iodine inside carbon nanotubes. *J. Am. Chem. Soc.* **129**, 8954–8955, 2007.

92. Chamberlain, T. W. et al. Reactions of the inner surface of carbon nanotubes and nanoprotrusion processes imaged at the atomic scale. *Nat. Chem.* **3**, 732–737, 2011.

93. Schuster, D. I., Jarowski, P. D., Kirschner, A. N. and Wilson, S. R. Molecular modelling of fullerene–porphyrin dyads. *J. Mater. Chem.* **12**, 2041–2047, 2002.

94. Sun, D., Tham, F. S., Reed, C. A. and Boyd, P. D. W. Extending supramolecular fullerene–porphyrin chemistry to pillared metal–organic frameworks. *PNAS* **99**, 5088–5092, 2002.

95. Guldi, D. M. et al. Parallel (face-to-face) versus perpendicular (edge-to-face) alignment of electron donors and acceptors in fullerene porphyrin dyads: The importance of orientation in electron transfer. *J. Am. Chem. Soc.* **123**, 9166–9167, 2001.

96. Drovetskaya, T., Reed, C. A. and Boyd, P. A. Fullerene porphyrin conjugate. *Tetr. Lett.* **36**, 7971–7974, 1995.

97. Sun, Y. et al. Fullerides of pyrrolidine-functionalized C$_{60}$. *J. Org. Chem.* **62**, 3642–3649, 1997.

98. Evans, D. R. et al. π-Arene/cation structure and bonding. Solvation versus ligand binding in iron(III) tetraphenylporphyrin complexes of benzene, toluene, *p*-xylene, and [60]fullerene. *J. Am. Chem. Soc.* **121**, 8466–8474, 1999.

99. Boyd, P. D. W. et al. Selective supramolecular porphyrin/fullerene interactions 1. *J. Am. Chem. Soc.* **121**, 10487–10495, 1999.

100. Olmstead, M. M. et al. Interaction of curved and flat molecular surfaces. The structures of crystalline compounds composed of fullerene (C$_{60}$, C$_{60}$O, C$_{70}$, and C$_{120}$O) and metal octaethylporphyrin units. *J. Am. Chem. Soc.* **121**, 7090–7097, 1999.

101. Konarev, D. V. et al. New molecular complexes of fullerenes C$_{60}$ and C$_{70}$ with tetraphenylporphyrins (M[tpp]), in which M = H$_2$, Mn, Co, Cu, Zn, and FeCl. *Chem. Eur. J.* **7**, 2605–2616, 2001.

102. Sun, D., Tham, F. S., Reed, C. A., Chaker, L. and Boyd, P. D. W. Supramolecular fullerene–porphyrin chemistry. Fullerene complexation by metallated "Jaws Porphyrin" hosts. *J. Am. Chem. Soc.* **124**, 6604–6612, 2002.

103. Tashiro, K. et al. A cyclic dimer of metalloporphyrin forms a highly stable inclusion complex with C$_{60}$. *J. Am. Chem. Soc.* **121**, 9477–9478, 1999.

104. Tashiro, K. and Aida, T. Metalloporphyrin hosts for supramolecular chemistry of fullerenes. *Chem. Soc. Rev.* **36**, 189–197, 2007.

105. Zheng, J. et al. Cyclic dimers of metalloporphyrins as tunable hosts for fullerenes: A remarkable effect of rhodium(III). *Angew. Chem. Int. Ed.* **40**, 1857–1861, 2001.

106. Wayland, B. B., Van Voorhees, S. L. and Del Rossi, K. J. Formation and dehydration of an (.alpha.,.beta.-dihydroxyethyl)rhodium porphyrin complex: Potential relevance to coenzyme B12-substrate complexes. *J. Am. Chem. Soc.* **109**, 6513–6515, 1987.

107. Yanagisawa, M., Tashiro, K., Yamasaki, M. and Aida, T. Hosting fullerenes by dynamic bond formation with an iridium porphyrin cyclic dimer: A "chemical friction" for rotary guest motions. *J. Am. Chem. Soc.* **129**, 11912–11913, 2007.

108. Gil-Ramírez, G. et al. A cyclic porphyrin trimer as a receptor for fullerenes. *Org. Lett.* **12**, 3544–3547, 2010.

109. Song, J., Aratani, N., Shinokubo, H. and Osuka, A. A porphyrin nanobarrel that encapsulates C$_{60}$. *J. Am. Chem. Soc.* **132**, 16356–16357, 2010.

110. Meng, W. et al. A self-assembled M8L6 cubic cage that selectively encapsulates large aromatic guests. *Angew. Chem. Int. Ed.* **50**, 3479–3483, 2011.

111. Mulholland, A. R., Woodward, C. P. and Langford, S. J. Fullerene-templated synthesis of a cyclic porphyrin trimer using olefin metathesis. *Chem. Commun.* **47**, 1494–1496, 2011.

112. Hu, S.-Z. and Chen, C.-F. Triptycene-derived oxacalixarene with expanded cavity: Synthesis, structure and its complexation with fullerenes C$_{60}$ and C$_{70}$. *Chem. Commun.* **46**, 4199–4201, 2010.

113. Nielsen, K. A. et al. Supramolecular receptor design: Anion-triggered binding of C$_{60}$. *Angew. Chem. Int. Ed.* **45**, 6848–6853, 2006.

114. Nielsen, K. A. et al. Binding studies of tetrathiafulvalene–calix[4]pyrroles with electron-deficient guests. *Tetrahedron* **64**, 8449–8463, 2008.

115. Caricato, M., Coluccini, C., Dondi, D., Griend, D. A. V. and Pasini, D. Nesting complexation of C_{60} with large, rigid D2 symmetrical macrocycles. *Org. Biomol. Chem.* **8**, 3272–3280, 2010.

116. Coluccini, C. et al. Structurally-variable, rigid and optically-active D2 and D3 macrocycles possessing recognition properties towards C_{60}. *Org. Biomol. Chem.* **8**, 1640–1649, 2010.

117. Diederich, F., Echegoyen, L., Gómez-López, M., Kessinger, R. and Stoddart, J. F. The self-assembly of fullerene-containing [2]pseudorotaxanes: Formation of a supramolecular C_{60} dimer. *J. Chem. Soc., Perkin Trans.* **2**, 1577–1586, 1999.

118. Whitesides, G. M. et al. Noncovalent synthesis: Using physical–organic chemistry to make aggregates. *Acc. Chem. Res.* **28**, 37–44, 1995.

119. Beijer, F. H. et al. Hydrogen-bonded complexes of diaminopyridines and diaminotriazines: Opposite effect of acylation on complex stabilities. *J. Org. Chem.* **61**, 6371–6380, 1996.

120. Beijer, F. H., Kooijman, H., Spek, A. L., Sijbesma, R. P. and Meijer, E. W. Self-complementarity achieved through quadruple hydrogen bonding. *Angew. Chem. Int. Ed.* **37**, 75–78, 1998.

121. Beijer, F. H., Kooijman, H., Spek, A. L., Sijbesma, R. P. and Meijer, E. W. Durch vier Wasserstoffbrückenbindungen vermittelte selbstkomplementarität. *Angew. Chem.* **110**, 79–82, 1998.

122. Beijer, F. H., Sijbesma, R. P., Kooijman, H., Spek, A. L. and Meijer, E. W. Strong dimerization of ureido-pyrimidones via quadruple hydrogen bonding. *J. Am. Chem. Soc.* **120**, 6761–6769, 1998.

123. Rispens, M. T., Sánchez, L., Knol, J. and Hummelen, J. C. Supramolecular organization of fullerenes by quadruple hydrogen bonding. *Chem. Commun.* **2**, 161–162, 2001.

124. González, J. J. et al. A new approach to supramolecular C_{60}-dimers based on quadruple hydrogen bonding. *Chem. Commun.* **2**, 163–164, 2001.

125. Ros, T. D. et al. A noncovalently linked, dynamic fullerene porphyrin dyad. Efficient formation of long-lived charge separated states through complex dissociation. **7**, *Chem. Commun.* 635–636, 1999.

126. Da Ros, T., Prato, M., Guldi, D. M., Ruzzi, M. and Pasimeni, L. Efficient charge separation in porphyrin–fullerene–ligand complexes. *Chem. Eur. J.* **7**, 816–827, 2001.

127. D'Souza, F., Rath, N. P., Deviprasad, G. R. and Zandler, M. E. Structural studies of a non-covalently linked porphyrin–fullerene dyad. *Chem. Commun.* **3**, 267–268, 2001.

128. D'Souza, F. and Ito, O. Supramolecular donor–acceptor hybrids of porphyrins/phthalocyanines with fullerenes/carbon nanotubes: Electron transfer, sensing, switching, and catalytic applications. *Chem. Commun.* **33**, 4913–4928, 2009.

129. Martín-Gomis, L., Ohkubo, K., Fernández-Lázaro, F., Fukuzumi, S. and Sastre-Santos, Á. Synthesis and photophysical studies of a new nonaggregated C_{60}–silicon phthalocyanine–C_{60} triad. *Org. Lett.* **9**, 3441–3444, 2007.

130. Martín-Gomis, L., Ohkubo, K., Fernández-Lázaro, F., Fukuzumi, S. and Sastre-Santos, A. Adiabatic photoinduced electron transfer and back electron transfer in a series of axially substituted silicon phthalocyanine triads. *J. Phys. Chem. C* **112**, 17694–17701, 2008.

131. El-Khouly, M. E. et al. Synthesis and photoinduced intramolecular processes of light-harvesting silicon phthalocyanine–naphthalenediimide–fullerene connected systems. *Chem. Eur. J.* **15**, 5301–5310, 2009.

132. Diederich, F. and Stang, P. J. *Templated Organic Synthesis*. Wiley-VCH, Weinheim: 1999.

133. Brouwer, A. M. et al. Photoinduction of fast, reversible translational motion in a hydrogen-bonded molecular shuttle. *Science* **291**, 2124–2128, 2001.

134. Balzani, V. et al. Autonomous artificial nanomotor powered by sunlight. *PNAS* **103**, 1178–1183, 2006.

135. Mateo-Alonso, A. et al. An electrochemically driven molecular shuttle controlled and monitored by C_{60}. *Chem. Commun.* **19**, 1945–1947, 2007.

136. Tseng, H., Vignon, S. A. and Stoddart, J. F. Toward chemically controlled nanoscale molecular machinery. *Angew. Chem. Int. Ed.* **42**, 1491–1495, 2003.

137. Altieri, A. et al. Electrochemically switchable hydrogen-bonded molecular shuttles. *J. Am. Chem. Soc.* **125**, 8644–8654, 2003.

138. Durola, F. and Sauvage, J. Fast electrochemically induced translation of the ring in a copper-complexed [2]rotaxane: The biisoquinoline effect. *Angew. Chem. Int. Ed.* **46**, 3537–3540, 2007.

139. Elizarov, A. M., Chiu, S.-H. and Stoddart, J. F. An acid–base switchable [2]rotaxane. *J. Org. Chem.* **67**, 9175–9181, 2002.

140. Keaveney, C. M. and Leigh, D. A. Shuttling through anion recognition. *Angew. Chem. Int. Ed.* **43**, 1222–1224, 2004.

141. Leigh, D. A. et al. Patterning through controlled submolecular motion: Rotaxane-based switches and logic gates that function in solution and polymer films. *Angew. Chem. Int. Ed.* **44**, 3062–3067, 2005.

142. Collier, C. P. et al. A [2]catenane-based solid state electronically reconfigurable switch. *Science* **289**, 1172–1175, 2000.

143. Hernández, J. V., Kay, E. R. and Leigh, D. A. A reversible synthetic rotary molecular motor. *Science* **306**, 1532–1537, 2004.

144. Dietrich-Buchecker, C. O. and Sauvage, J. P. Interlocking of molecular threads: From the statistical approach to the templated synthesis of catenands. *Chem. Rev.* **87**, 795–810, 1987.

145. Diederich, F., Dietrich-Buchecker, C., Nierengarten, J.-F. and Sauvage, J.-P. A copper(I)-complexed rotaxane with two fullerene stoppers. *J. Chem. Soc., Chem. Commun.* **7**, 781–782, 1995.

146. Griffiths, K. E. and Stoddart, J. F. Template-directed synthesis of donor/acceptor [2]catenanes and [2]rotaxanes. *Pure Appl. Chem.* **80**, 485–506, 2008.

147. Ashton, P. R. et al. Self-assembly of the first fullerene-containing [2]catenane. *Angew. Chem. Int. Ed. Engl.* **36**, 1448–1451, 1997.

148. Saha, S. et al. A redox-driven multicomponent molecular shuttle. *J. Am. Chem. Soc.* **129**, 12159–12171, 2007.

149. Kelly, T., Kay, E. and Leigh, D. *Molecular Machines*. **262**, Springer, Berlin: 2005.

150. Mateo-Alonso, A. et al. Photophysical and electrochemical properties of a fullerene-stopped rotaxane. *Photochem. Photobiol. Sci.* **5**, 1173–1176, 2006.

151. Mateo-Alonso, A. and Prato, M. Synthesis of a soluble fullerene–rotaxane incorporating a furamide template. *Tetrahedron* **62**, 2003–2007, 2006.

152. Da Ros, T. et al. Hydrogen bond-assembled fullerene molecular shuttle. *Org. Lett.* **5**, 689–691, 2003.

153. Mateo-Alonso, A. et al. Reverse shuttling in a fullerene-stopped rotaxane. *Org. Lett.* **8**, 5173–5176, 2006.

154. Mateo-Alonso, A., Brough, P. and Prato, M. Stabilization of fulleropyrrolidine *N*-oxides through intrarotaxane hydrogen bonding. *Chem. Commun.* **14**, 1412–1414, 2007.

155. Mateo-Alonso, A., Iliopoulos, K., Couris, S. and Prato, M. Efficient modulation of the third order nonlinear optical properties of fullerene derivatives. *J. Am. Chem. Soc.* **130**, 1534–1535, 2008.

156. Norris, J. R. and Deisenhofer, J. *The Photosynthetic Reaction Center*. Academic Press, San Diego: 1993.

157. Echegoyen, L., Diederich, F. and Echegoyen, L.E. Electrochemistry of fullerenes. *Fullerenes: Chem. Phys. Technol.* 1, 2000.

158. Arias, F., Echegoyen, L., Wilson, S. R., Lu, Q. and Lu, Q. Methanofullerenes and methanofulleroids have different electrochemical behavior at negative potentials. *J. Am. Chem. Soc.* **117**, 1422–1427, 1995.

159. Hiroshi, I. et al. The small reorganization energy of C$_{60}$ in electron transfer. *Chem. Phys. Lett.* **263**, 545–550, 1996.

160. Guldi, D. M. Fullerenes: Three dimensional electron acceptor materials. *Chem. Commun.* **5**, 321–327, 2000.

161. Chuard, T. and Deschenaux, R. First fullerene[60]-containing thermotropic liquid crystal. Preliminary communication. *Helv. Chim. Acta* **79**, 736–741, 1996.

162. Oviedo, J. J., de la Cruz, P., Garín, J., Orduna, J. and Langa, F. Ruthenocene as a new donor fragment in [60]fullerene–donor dyads. *Tetr. Lett.* **46**, 4781–4784, 2005.

163. Guldi, D. M. and Prato, M. Excited-state properties of C$_{60}$ fullerene derivatives. *Acc. Chem. Res.* **33**, 695–703, 2000.

164. Guldi, D. M., Illescas, B. M., Atienza, C. M., Wielopolski, M. and Martín, N. Fullerene for organic electronics. *Chem. Soc. Rev.* **38**, 1587, 2009.

165. Wróbel, D. and Graja, A. Photoinduced electron transfer processes in fullerene–organic chromophore systems. *Coord. Chem. Rev.* **255**, 2555–2577, 2011.

166. Sandanayaka, A. S. D., Sasabe, H., Takata, T. and Ito, O. Photoinduced electron transfer processes of fullerene rotaxanes containing various electron-donors. *J. Photochem. Photobiol. C: Photochem. Rev.* **11**, 73–92, 2010.

167. Kashiwagi, Y. et al. Long-lived charge-separated state produced by photoinduced electron transfer in a zinc imidazoporphyrin-C$_{60}$ dyad. *Org. Lett.* **5**, 2719–2721, 2003.

168. Guldi, D. M. et al. Langmuir–Blodgett and layer-by-layer films of photoactive fullerene–porphyrin dyads. *J. Mater. Chem.* **14**, 303–309, 2004.

169. Ohkubo, K. et al. Production of an ultra-long-lived charge-separated state in a zinc chlorine–C$_{60}$ dyad by one-step photoinduced electron transfer. *Angew. Chem. Int. Ed.* **43**, 853–856, 2004.

170. Guldi, D. M. et al. Nanoscale organization of a phthalocyanine–fullerene system: Remarkable stabilization of charges in photoactive 1-D nanotubules. *J. Am. Chem. Soc.* **127**, 5811–5813, 2005.

171. Bottari, G. et al. Highly conductive supramolecular nanostructures of a covalently linked phthalocyanine–C$_{60}$ fullerene conjugate. *Angew. Chem. Int. Ed.* **47**, 2026–2031, 2008.

172. Kahnt, A., Guldi, D. M., Escosura de la, A., Martínez-Díaz, M. V. and Torres, T. [2.2]Paracyclophane: A pseudoconjugated spacer for long-lived electron transfer in phthalocyanine–C$_{60}$ dyads. *J. Mater. Chem.* **18**, 77–82, 2007.

173. Isosomppi, M. et al. Photoinduced electron transfer of double-bridged phthalocyanine–fullerene dyads. *Chem. Phys. Lett.* **430**, 36–40, 2006.

174. Kahnt, A., Quintiliani, M., Vázquez, P., Guldi, D. M. and Torres, T. A. Bis(C$_{60}$)–bis(phthalocyanine) nanoconjugate: Synthesis and photoinduced charge transfer. *ChemSusChem* **1**, 97–102, 2008.

175. Gouloumis, A. et al. Photoinduced electron transfer in a new bis(C$_{60}$)–phthalocyanine triad. *Org. Lett.* **8**, 5187–5190, 2006.

176. Kim, K. N., Choi, C. S. and Kay, K.-Y. A novel phthalocyanine with two axial fullerene substituents. *Tetr. Lett.* **46**, 6791–6795, 2005.

177. El-Khouly, M. E. et al. Silicon–phthalocyanine-cored fullerene dendrimers: Synthesis and prolonged charge-separated states with dendrimer generations. *Chem. Eur. J.* **13**, 2854–2863, 2007.

178. El-Khouly, M. E. et al. Synthesis and photoinduced intramolecular processes of light-harvesting silicon phthalocyanine–naphthalenediimide–fullerene connected systems. *Chem. Eur. J.* **15**, 5301–5310, 2009.

179. Llacay, J. et al. The first Diels–Alder adduct of [60]fullerene with atetrathiafulvalene. *Chem. Commun.* **7**, 659–660, 1997.

180. Mas-Torrent, M. et al. Isolation and characterization of four isomers of a C$_{60}$ bisadduct with a TTF derivative. Study of their radical ions. *J. Org. Chem.* **67**, 566–575, 2001.

181. González, S., Martín, N. and Guldi, D. M. Synthesis and properties of Bingel-type methanofullerene–π-extended-TTF diads and triads. *J. Org. Chem.* **68**, 779–791, 2002.

182. Martín, N. et al. Photoinduced electron transfer between C$_{60}$ and electroactive units. *Carbon* **38**, 1577–1585, 2000.

183. Guldi, D. M. et al. Efficient charge separation in C$_{60}$-based dyads: Triazolino[4',5':1,2][60]fullerenes. *J. Org. Chem.* **65**, 1978–1983, 2000.

184. González, S., Martín, N., Swartz, A. and Guldi, D. M. Addition reaction of azido-exTTFs to C$_{60}$: Synthesis of fullerotriazoline and azafulleroid electroactive dyads. *Org. Lett.* **5**, 557–560, 2003.

185. Prato, M. et al. Synthesis and electrochemical properties of substituted fulleropyrrolidines. *Tetrahedron* **52**, 5221–5234, 1996.

186. Maggini, M. et al. Ferrocenyl fulleropyrrolidines: A cyclic voltammetry study. *J. Chem. Soc. Chem. Commun.* **5**, 589–590, 1994.

187. Guldi, D. M. et al. Photoactive nanowires in fullerene–ferrocene dyad polyelectrolyte multilayers. *Nano Lett.* **2**, 775–780, 2002.

188. Campidelli, S. et al. Liquid-crystalline fullerene–ferrocene dyads. *J. Mater. Chem.* **14**, 1266–1272, 2004.

189. Armaroli, N. et al. A new pyridyl-substituted methanofullerene derivative. Photophysics, electrochemistry and self-assembly with zinc(II) meso-tetraphenylporphyrin (ZnTPP). *New J. Chem.* **23**, 77–83, 1999.

190. Hauke, F., Swartz, A., Guldi, D. M. and Hirsch, A. Supramolecular assembly of a quasi-linear heterofullerene–porphyrin dyad. *J. Mater. Chem.* **12**, 2088–2094, 2002.

191. Da Ros, T., Prato, M., Guldi, D. M., Ruzzi, M. and Pasimeni, L. Efficient charge separation in porphyrin–fullerene–ligand complexes. *Chem. Eur. J.* **7**, 816–827, 2001.

192. Rio, Y. et al. A panchromatic supramolecular fullerene-based donor–acceptor assembly derived from a peripherally substituted bodipy–zinc phthalocyanine dyad. *Chem. Eur. J.* **16**, 1929–1940, 2010.

193. Chen, Y., EI-Khouly, M. E., Sasaki, M., Araki, Y. and Ito, O. Synthesis of the axially substituted titanium Pc–C$_{60}$ dyad with a convenient method. *Org. Lett.* **7**, 1613–1616, 2005.

194. Ballesteros, B. et al. Synthesis and photophysical characterization of a titanium(IV) phthalocyanine–C$_{60}$ supramolecular dyad. *Tetrahedron* **62**, 2097–2101, 2006.

195. Sasabe, H. et al. Synthesis of [60]fullerene-functionalized rotaxanes. *Tetrahedron* **62**, 1988–1997, 2006.

196. Sasabe, H. et al. Axle charge effects on photoinduced electron transfer processes in rotaxanes containing porphyrin and [60]fullerene. *Phys. Chem. Chem. Phys.* **11**, 10908–10915, 2009.

197. Sandanayaka, A. S. D. et al. Photoinduced electron-transfer processes between [C60]fullerene and triphenylamine moieties tethered by rotaxane structures. Through-space electron transfer via excited triplet states of [60]fullerene. *J. Phys. Chem. A* **108**, 5145–5155, 2004.

198. Rajkumar, G. A. et al. Prolongation of the lifetime of the charge-separated state at low temperatures in a photoinduced electron-transfer system of [60]fullerene and ferrocene moieties tethered by rotaxane structures. *J. Phys. Chem. B* **110**, 6516–6525, 2006.

199. Sandanayaka, A. S. D. et al. Photoinduced electron transfer processes in rotaxanes containing [60]fullerene and ferrocene: Effect of axle charge on light-induced molecular motion. *Aust. J. Chem.* **59**, 186–192, 2006.

200. Guldi, D. M., Gouloumis, A., Vázquez, P. and Torres, T. Charge-transfer states in strongly coupled phthalocyanine fullerene ensembles. *Chem. Commun.* **18**, 2056–2057, 2002.

201. Li, K., Schuster, D. I., Guldi, D. M., Herranz, M. Á. and Echegoyen, L. Convergent synthesis and photophysics of [60]fullerene/porphyrin-based rotaxanes. *J. Am. Chem. Soc.* **126**, 3388–3389, 2004.

202. Li, K. et al. [60]Fullerene-stoppered porphyrinorotaxanes: Pronounced elongation of charge-separated-state lifetimes. *J. Am. Chem. Soc.* **126**, 9156–9157, 2004.

203. Sessler, J. L. et al. Synthesis and photophysics of a porphyrin–fullerene dyad assembled through Watson–Crick hydrogen bonding. *Chem. Commun.* **14**, 1892–1894, 2005.

204. Torres, T. et al. Photophysical characterization of a cytidine–guanosine tethered phthalocyanine–fullerene dyad. *Chem. Commun.* **3**, 292–294, 2007.

205. Sessler, J. L. et al. Guanosine and fullerene derived de-aggregation of a new phthalocyanine-linked cytidine derivative. *Tetrahedron* **62**, 2123–2131, 2006.

206. Segura, M., Sánchez, L., de Mendoza, J., Martín, N. and Guldi, D. M. Hydrogen bonding interfaces in fullerene•TTF ensembles. *J. Am. Chem. Soc.* **125**, 15093–15100, 2003.

207. Vögtle, F., Dünnwald, T. and Schmidt, T. Catenanes and rotaxanes of the amide type. *Acc. Chem. Res.* **29**, 451–460, 1996.

208. Watanabe, N. et al. Photoinduced intrarotaxane electron transfer between zinc porphyrin and [60]fullerene in benzonitrile. *Angew. Chem. Int. Ed.* **42**, 681–683, 2003.

209. Sandanayaka, A. S. D. et al. Synthesis and photoinduced electron transfer processes of rotaxanes bearing [60]fullerene and zinc porphyrin: Effects of interlocked structure and length of axle with porphyrins. *J. Phys. Chem. B* **109**, 2516–2525, 2005.

210. Sandanayaka, A. S. D. et al. Syntheses of [60]fullerene and *N,N*-bis(4-biphenyl)aniline-tethered rotaxane: Photoinduced electron-transfer processes via singlet and triplet states of [60]fullerene. *Bull. Chem. Soc. Jpn.* **78**, 1008–1017, 2005.

211. Mateo-Alonso, A., Guldi, D. M., Paolucci, F. and Prato, M. Fullerenes: Multitask components in molecular machinery. *Angew. Chem. Int. Ed.* **46**, 8120–8126, 2007.

212. Mateo-Alonso, A. et al. Tuning electron transfer through translational motion in molecular shuttles. *Angew. Chem. Int. Ed.* **46**, 3521–3525, 2007.

213. Guldi, D. M. Fullerene–porphyrin architectures; photosynthetic antenna and reaction center models. *Chem. Soc. Rev.* **31**, 22–36, 2002.

214. Imahori, H. Porphyrin–fullerene linked systems as artificial photosynthetic mimics. *Org. Biomol. Chem.* **2**, 1425–1433, 2004.

215. Imahori, H., Mori, Y. and Matano, Y. Nanostructured artificial photosynthesis. *J. Photochem. Photobiol. C: Photochem. Rev.* **4**, 51–83, 2003.

216. Imahori, H. Giant multiporphyrin arrays as artificial light-harvesting antennas. *J. Phys. Chem. B* **108**, 6130–6143, 2004.

217. Luo, C., Guldi, D. M., Imahori, H., Tamaki, K. and Sakata, Y. Sequential energy and electron transfer in an artificial reaction center: Formation of a long-lived charge-separated state. *J. Am. Chem. Soc.* **122**, 6535–6551, 2000.

218. Imahori, H. et al. Modulating charge separation and charge recombination dynamics in porphyrin–fullerene linked dyads and triads: Marcus-normal versus inverted region. *J. Am. Chem. Soc.* **123**, 2607–2617, 2001.

219. Imahori, H. et al. Charge separation in a novel artificial photosynthetic reaction center lives 380 ms. *J. Am. Chem. Soc.* **123**, 6617–6628, 2001.

220. Imahori, H., Yamada, H., Ozawa, S., Sakata, Y. and Ushida, K. Synthesis and photoelectrochemical properties of a self-assembled monolayer of a ferrocene–porphyrin–fullerene triad on a gold electrode. *Chem. Commun.* **13**, 1165–1166, 1999.

221. Imahori, H., Yamada, H., Nishimura, Y., Yamazaki, I. and Sakata, Y. Vectorial multistep electron transfer at the gold electrodes modified with self-assembled monolayers of ferrocene–porphyrin–fullerene triads. *J. Phys. Chem. B* **104**, 2099–2108, 2000.

222. Kodis, G. et al. Photoinduced electron transfer in π-extended tetrathiafulvalene–porphyrin–fullerene triad molecules. *J. Mater. Chem.* **12**, 2100–2108, 2002.

223. Maligaspe, E. et al. Photosynthetic antenna-reaction center mimicry: Sequential energy- and electron transfer in a self-assembled supramolecular triad composed of boron dipyrrin, zinc porphyrin and fullerene. *J. Phys. Chem. A* **113**, 8478–8489, 2009.

224. El-Khouly, M. E., Ju, D. K., Kay, K., D'Souza, F. and Fukuzumi, S. Supramolecular tetrad of subphthalocyanine–triphenylamine–zinc porphyrin coordinated to fullerene as an "antenna-reaction-center" mimic: Formation of a long-lived charge-separated state in nonpolar solvent. *Chem. Eur. J.* **16**, 6193–6202, 2010.

225. Sandanayaka, A. S. D. et al. Axle length effect on photoinduced electron transfer in triad rotaxane with porphyrin, [60]fullerene, and triphenylamine. *J. Phys. Chem. A* **114**, 5242–5250, 2010.

226. Giacalone, F., Segura, J. L., Martín, N. and Guldi, D. M. Exceptionally small attenuation factors in molecular wires. *J. Am. Chem. Soc.* **126**, 5340–5341, 2004.

227. Nakamura, T. et al. Control of photoinduced energy- and electron-transfer steps in zinc porphyrin–oligothiophene–fullerene linked triads with solvent polarity. *J. Phys. Chem. B* **109**, 14365–14374, 2005.

228. Mateo-Alonso, A., Ehli, C., Guldi, D. M. and Prato, M. Charge transfer reactions along a supramolecular redox gradient. *J. Am. Chem. Soc.* **130**, 14938–14939, 2008.

229. Sariciftci, N. S., Smilowitz, L., Heeger, A. J. and Wudl, F. Photoinduced electron transfer from a conducting polymer to buckminsterfullerene. *Science* **258**, 1474–1476, 1992.

230. Sariciftci, N. S. et al. Semiconducting polymer–buckminsterfullerene heterojunctions: Diodes, photodiodes, and photovoltaic cells. *Appl. Phys. Lett.* **62**, 585–587, 1993.

231. Günes, S., Neugebauer, H. and Sariciftci, N. S. Conjugated polymer-based organic solar cells. *Chem. Rev.* **107**, 1324–1338, 2007.

232. Mozer, A. J. et al. Charge transport and recombination in bulk heterojunction solar cells studied by the photoinduced charge extraction in linearly increasing voltage technique. *Appl. Phys. Lett.* **86**, 112104–1121043, 2005.

233. Mihailetchi, V. D., Xie, H. X., de Boer, B., Koster, L. J. A. and Blom, P. W. M. Charge transport and photocurrent generation in poly(3-hexylthiophene): Methanofullerene bulk-heterojunction solar cells. *Adv. Funct. Mater.* **16**, 699–708, 2006.

234. Mozer, A. J. and Sariciftci, N. S. Negative electric field dependence of charge carrier drift mobility in conjugated, semiconducting polymers. *Chem. Phys. Lett.* **389**, 438–442, 2004.

235. Hoppe, H. et al. Nanoscale morphology of conjugated polymer/fullerene-based bulk-heterojunction solar cells. *Adv. Funct. Mater.* **14**, 1005–1011, 2004.

236. Li, G. et al. High-efficiency solution processable polymer photovoltaic cells by self-organization of polymer blends. *Nat. Mater.* **4**, 864–868, 2005.

237. Youngkyoo, K. et al. A strong regioregularity effect in self-organizing conjugated polymer films and high-efficiency polythiophene:fullerene solar cells. *Nat. Mater.* **5**, 197–203, 2006.

238. Kim, J. Y. et al. Efficient tandem polymer solar cells fabricated by all-solution processing. *Science* **317**, 222–225, 2007.

239. Lenes, M. et al. Fullerene bisadducts for enhanced open-circuit voltages and efficiencies in polymer solar cells. *Adv. Mater.* **20**, 2116–2119, 2008.

240. Park, S. H. et al. Bulk heterojunction solar cells with internal quantum efficiency approaching 100l[percent]l. *Nat. Photon.* **3**, 297–302, 2009.

241. Lee, C. H. et al. Sensitization of the photoconductivity of conducting polymers by C_{60}: Photoinduced electron transfer. *Phys. Rev. B* **48**, 15425–15433, 1993.

242. Cravino, A. and Sariciftci, N. S. Double-cable polymers for fullerene based organic optoelectronic applications. *J. Mater. Chem.* **12**, 1931–1943, 2002.

243. Janssen, R. A. J. et al. Photoinduced electron transfer from π-conjugated polymers onto buckminsterfullerene, fulleroids, and methanofullerenes. *J. Chem. Phys.* **103**, 788–793, 1995.

244. Yu, G., Gao, J., Hummelen, J. C., Wudl, F. and Heeger, A. J. Polymer photovoltaic cells: Enhanced efficiencies via a network of internal donor–acceptor heterojunctions. *Science* **270**, 1789–1791, 1995.

245. Shaheen, S. E. et al. 2.5% efficient organic plastic solar cells. *Appl. Phys. Lett.* **78**, 841–843, 2001.

246. Brady, M. A., Su, G. M. and Chabinyc, M. L. Recent progress in the morphology of bulk heterojunction photovoltaics. *Soft Matter* **7**, 11065, 2011.

247. Dennler, G., Scharber, M. C. and Brabec, C. J. Polymer–fullerene bulk-heterojunction solar cells. *Adv. Mater.* **21**, 1323–1338, 2009.

248. Chen, H., Hegde, R., Browning, J. and Dadmun, M. D. The miscibility and depth profile of PCBM in P3HT: Thermodynamic information to improve organic photovoltaics. *Phys. Chem. Chem. Phys.* **14**, 5635, 2012.

249. Moujoud, A., Oh, S. H., Hye, J. J. and Kim, H. J. Improvement in stability of poly(3-hexylthiophene-2,5-diyl)/[6,6]-phenyl-C_{61}-butyric acid methyl ester bulk heterojunction solar cell by using UV light irradiation. *Sol. Energy Mater. Sol. Cells* **95**, 1037–1041, 2011.

250. Brunetti, F. G., Kumar, R. and Wudl, F. Organic electronics from perylene to organic photovoltaics: Painting a brief history with a broad brush. *J. Mater. Chem.* **20**, 2934–2948, 2010.

251. Atienza, C. M. et al. Light harvesting tetrafullerene nanoarray for organic solar cells. *Chem. Commun.* **5**, 514–516, 2006.

252. Drees, M. et al. Stabilization of the nanomorphology of polymer–fullerene "bulk heterojunction" blends using a novel polymerizable fullerene derivative. *J. Mater. Chem.* **15**, 5158–5163, 2005.
253. Cheng, Y.-J., Yang, S.-H. and Hsu, C.-S. Synthesis of conjugated polymers for organic solar cell applications. *Chem. Rev.* **109**, 5868–5923, 2009.
254. Brabec, C. J. et al. Polymer–fullerene bulk-heterojunction solar cells, polymer–fullerene bulk-heterojunction solar cells. *Adv. Mater.* **22**, 3839–3856, 2010.
255. Boudreault, P.-L. T., Najari, A. and Leclerc, M. Processable low-bandgap polymers for photovoltaic applications. *Chem. Mater.* **23**, 456–469, 2010.
256. Giacalone, F. and Martín, N. Fullerene polymers: Synthesis and properties. *Chem. Rev.* **106**, 5136–5190, 2006.
257. Mishra, A., Bäuerle, P., Mishra, A. and Bäuerle, P. Small molecule organic semiconductors on the move: Promises for future solar energy technology. *Angew. Chem. Int. Ed.* **51**, 2020–2067, 2012.
258. Walker, B., Kim, C. and Nguyen, T.-Q. Small molecule solution-processed bulk heterojunction solar cells. *Chem. Mater.* **23**, 470–482, 2010.
259. Roncali, J. Molecular bulk heterojunctions: An emerging approach to organic solar cells. *Acc. Chem. Res.* **42**, 1719–1730, 2009.
260. Sun, Y. et al. Solution-processed small-molecule solar cells with 6.7% efficiency. *Nat. Mater.* **11**, 44–48, 2012.
261. Peet, J. et al. Efficiency enhancement in low-bandgap polymer solar cells by processing with alkane dithiols. *Nat. Mater.* **6**, 497–500, 2007.

3 Graphite Whiskers, Cones, and Polyhedral Crystals

Svetlana Dimovski and Yury Gogotsi

CONTENTS

3.1 INTRODUCTION

Both planar graphite[1] and carbon nanotubes[2] have been extensively studied, and their structure and properties are well documented in the literature. This section is a review of the current understanding of some less-common nonplanar graphitic materials, such as graphite whiskers, cones, and polygonized carbon nanotubes (graphite polyhedral crystals [GPCs]). Although nonplanar graphitic microstructures in the shape of cones were reported as early as 1957,[3,4] it is only recently[5–8] that attention has been paid to these exotic classes of graphitic materials. There is no doubt that this growing interest has been triggered by the discovery of fullerenes[9] and nanotubes,[2,10] which has stimulated intensive research on carbonaceous nanomaterials during the past 25–30 years. Although fullerenes and nanotubes have been discussed in several books during the past decade, carbon cones, whiskers, and other similar structures have received much less attention. Here, our intention is to provide an overview of the current understanding of their structure, synthesis methods, properties, and potential applications.

Some of the engineering disciplines that could benefit from the emergence of these forms of carbon are

- *Materials engineering*: Graphitic cones and polyhedral crystals enable the development of new functional nanomaterials and fillers for nanocomposites.
- *Chemistry and biomedicine*: In the development of new chemical sensors, cellular probes, and micro-/nanoelectrodes.
- *Analytical tools and instrumentation development*: Cones and polyhedral crystals can potentially act as probes for atomic force and scanning tunneling microscopes.
- *Energy, transportation, and electronic devices*: As materials for energy storage, field emitters, and components for nanoelectromechanical systems.

The common features of carbon whiskers, cones, scrolls, and GPCs, besides their chemistry and the graphitic nature of their bonds, are their morphology and the high length-to-diameter aspect ratio, which places them between graphite and carbon nanotube materials. In the following sections, we examine the various types of these materials, show the effect of structural conformation, as well as describe its properties and potential applications.

3.2 GRAPHITE WHISKERS AND CONES

Graphite whiskers, cones, and polyhedral crystals are needle-like structures—meaning that their length is considerably larger than their width or diameter. The major difference between the graphitic cones and polyhedral crystals, besides their shape, is in their texture, that is, in the orientation of the atomic planes within the structure. Although GPCs comprise (0 0 0 1) planes parallel to their main axis, carbon cones and whiskers may have various textures. Therefore, we will consider GPCs separately from cones and whiskers. The orientation and the stacking arrangement of the planes are closely related to the nucleation mechanism and the growth conditions of the cones. The following section explains the currently available methods for synthesis of carbon whiskers and cones. Also, two different types of cones have also been observed in natural deposits of carbon. These are briefly described in a separate section. A detailed explanation of their structure and properties in relation to their potential applications follows.

3.2.1 SYNTHETIC WHISKERS AND CONES

3.2.1.1 Whiskers

Graphite whiskers are the first known nonplanar graphitic structures that were obtained through a controlled preparation. Bacon[11] succeeded in growing high-strength graphite whiskers on carbon electrodes using a direct current (DC) arc under an argon pressure of 92 atm. The temperatures developed in the arc were sufficiently close to the sublimation point of graphite (>3600°C), which enabled carbon to vaporize from the tip of the positively charged electrode and form cylindrical deposits embedded with whiskers of up to 3 cm in length and a few microns in diameter.[12] Carbon deposition under extreme conditions, such as a "flash CVD" process, also resulted in the growth of very peculiar micron-sized tree-like carbon structures.[13]

Whiskers and filaments of graphite have also been observed to form during pyrolytic deposition of various hydrocarbon materials. Hillert and Lange[14] studied the thermal decomposition of *n*-heptane and reported the formation of filamentous graphite on iron surfaces at elevated temperatures. Pyrolysis of methane[15] and carbon monoxide[16] on iron surfaces or heated carbon filaments[17] also resulted in the formation of similar structures as well as the thermal decomposition of acetylene on Nichrome alloy wires below 700°C.[18]

Whisker growth during pyrolytic-deposition process is generally considered as being catalyzed by metals.[19,20] However, Haanstra et al.[21] observed noncatalytic columnar growth of carbon on β-SiC crystals by pyrolysis of carbon monoxide at 1 atm pressure above 1800°C (Figure 3.1). The experiments showed that the growth of carbon whiskers in that case was defined by rotation twinning and stacking faults on {1 1 1} habit plane of the β-SiC substrate. The cylindrical carbon columns formed by this mechanism were observed to consist of parallel conical graphitic layers stacked along the column axis. Most specimens of a run had the diameter between 3 and 6 μm, and they were several tens of microns long. The apex angle of the conical mantle was measured to be about 141°. The conical nuclei of these columns were produced on defects in twinned β-SiC, such as the dislocation with a screw component perpendicular to the surface.

Very similar "needle"- or "spine"-like graphitic materials were also reported by Knox et al.[22] In an attempt to synthesize porous graphitic carbon material that would be capable of withstanding considerable shear forces, such as those seen in high-performance liquid chromatography, they produced porous glassy carbon spheres, which in most cases contained graphitic needles. Knox et al. impregnated the high-porosity silica gel spheres with a melt of phenol and hexamethylenetetramine (hexamine) in a 6:1 weight ratio. Impregnated material was first heated gradually at 150°C to form phenol–formaldehyde resin within the pores of the silica gel and then carbonized slowly at 900°C in a stream of oxygen-free nitrogen. The silica was then dissolved with hot aqueous potassium hydroxide (at least 99% complete), and the remaining porous glassy carbon was consequently heated up to 2500°C in oxygen-free argon.

Besides the expected glassy carbon structure, the resulting product often contained considerable amounts of needle-like material that was determined to have a three-dimensional graphitic structure. The graphitic whiskers resulting from the experiments were usually a few microns long and about 1 μm thick. Electron diffraction and transmission electron microscopy (TEM) revealed the twinned structure of the whiskers. The angle between the layers was measured to be about 135°.

Because this material was a side product of their experiment, Knox et al. neither provided further details of its structure nor explained the nucleation mechanism. However, it is highly probable that their graphitic needles were nucleated and grown approximately the same manner as Haanstra's whiskers as shown in Figure 3.1. Incomplete dissolution of the silica matrix could have caused formation of twinned β-SiC phase during the glassy carbon pyrolysis between 1000°C and 2500°C, which further induced growth of columnar graphite from disproportionated CO within the porous glassy carbon spheres. The 135° whiskers have also been synthesized recently at 2100°C from gaseous CO and ball-milled natural graphite[23,24] contaminated with zirconia particles during milling. When heated above 1900°C, the zirconia particles react with the carbon to form ZrC.[25] The growth of the graphitic whiskers was probably initiated by screw dislocations on the surfaces of ZrC particles.

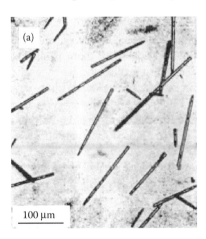
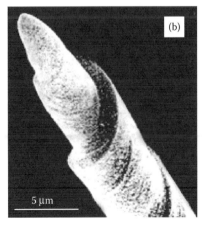

FIGURE 3.1 Electron micrographs of (a) pencil-like carbon whiskers; (b) columnar carbon whisker with screw-like markings on side. (Adapted from H. B. Haanstra, W. F. Knippenberg, and G. Verspui, *J. Cryst. Growth*, 16, 71, 1972.)

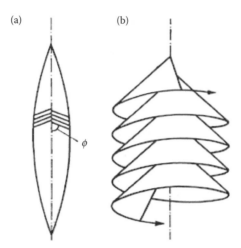

FIGURE 3.2 Model illustrating the formation of cigar-like graphite: (a) longitudinal cross section of the cigars showing their texture; (b) cone–helix structure of graphitic filaments. (Adapted from J. Gillot, W. Bollman, and B. Lux, *Carbon*, 6, 381, 1968.)

Similarly, Gillot et al.[26] studied the heat treatment of products of martensite electrolytic dissolution and observed the formation of "cigar"-shaped crystals of graphite at 2800°C. The model of the texture they obtained is shown in Figure 3.2a. The length of the crystals ranged from a few microns to 250 μm with a length-to-diameter ratio of about 10. It was suggested that the growth mechanism of the "cigars" involved mass transfer through the gas phase. The graphite layers in the whisker had the shape of an obtuse cone, the axis of which was coincident with the axis of the whisker, and had basically the same structure noted but not fully described by Knox et al. Such whiskers were assumed to be formed by a single graphene sheet coiled around the axis in a helix, each turn of the helix having the shape of a cone (Figure 3.2b). The angular shift θ of the (h k 0) crystallographic directions from one whorl to the next one in the helix was measured to be $\theta \approx 60°$. All graphitic layers were found to have the same stacking arrangement as of a perfect graphite crystal. Later, Double and Hellawell[27] proposed the cone–helix growth mechanism of such structures, which relies on the formation of a negative wedge disclination within a graphene sheet. This model will be explained in detail in the following section.

3.2.1.2 Cones

Ge and Sattler,[6] Sattler,[7] and Krishnan et al.[8] were among the first to observe and study fullerene nanocones, that is, seamless conical structures formed when one or more pentagonal rings are incorporated into a graphene network. Incorporation of pentagonal and heptagonal defects into graphene sheets and nanotubes had at that time been already discussed by Iijima et al.,[28] Ajayan et al.,[29] Ajayan and Ijima,[30] Ebbesen,[31] Ebbesen and Takada,[32] and others,[33–35] to explain conical morphologies of carbon nanotube tips observed by high-resolution transmission electron microscopy (HRTEM). The importance of the pentagonal defects in the formation of three-dimensional conical graphitic structures, however, was not fully recognized until the thorough investigation of their electron-diffraction patterns by Amelinckx et al.,[5,36,37] who studied helically wound conical graphite whiskers (Figure 3.3a) by electron microscopy and electron diffraction. Whiskers gave rise to unusual diffraction effects consisting of periodically interrupted circular ring patterns (Figure 3.3b). Very similar diffraction patterns had been previously obtained from whiskers described in Ref. 21. Amelinckx et al. proposed a growth mechanism whereby the initial graphite layer adopts a slitted dome-shaped configuration (Figures 3.4a and b) by removing a sector β and introducing a fivefold carbon ring in the sixfold carbon network (Figure 3.4c). Successive graphene sheets were then rotated with respect to the previous graphene sheet over a constant angle, thus realizing a

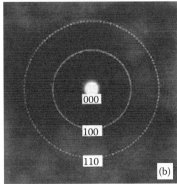

FIGURE 3.3 Conical graphite whiskers. (a) SEM micrograph of a cleavage fragment of conically wound graphite whisker. Note the 140° apex angle of the conical cleavage plane. (b) Electron diffraction pattern with the incident electron beam along the normal to the cleavage "plane" of the conically wound whiskers. Note the 126-fold rotation symmetry of the pattern. (Adapted from W. Luyten et al., *Ultramicroscopy*, 49, 123, 1993.)

helical cone around a "disclination," with a fivefold carbon ring core. The model explains the morphological features and the particular diffraction effects observed on these reproducibly prepared columnar graphite crystals and it also builds on the other cone models.[21,27]

The first true multishell fullerene graphitic cones consisting of seamless axially stacked conical surfaces (Figure 3.5) were observed in the products of chlorination of silicon carbide at temperatures above 1000°C in 1972[38] and then reported by Millward and Jefferson[39] in 1978. Because these structures were rather singular observations in the products of the reaction, they were not recognized as a new material until much later.[38] Similar structures in large quantities were for the first time successfully produced by Ge and Sattler.[6] Up to 24 nm in length and 8 nm in base diameter, these nanometer-sized structures were generated by vapor condensation of carbon atoms on a highly oriented pyrolytic graphite (HOPG) substrate. All the cones had the same apex angle of ~19°, which is the smallest among five possible opening angles for perfect graphitic cones (Figure 3.6a). The growth of these nanostructures is thought to be initiated exclusively by fullerene-type nucleation seeds with a different number of pentagons. Fullerene cones of other apex angles corresponding to 1–4 pentagons were produced and reported 3 years later by Krishnan et al.[8] (Figures 3.6c–f). They also reproduced the ~19° cone (Figure 3.6b).

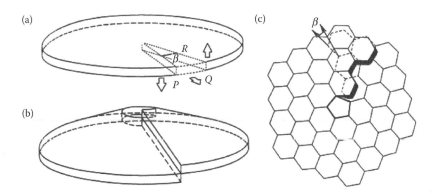

FIGURE 3.4 Model illustrating the formation of a conical helix. (a) Sector β is removed from a disc. (b) The angular gap is closed and a cone is formed. (c) Twisted nucleus of the conical helix containing one pentagonal ring in the graphene network. Conical helix is formed through rotation of successive graphene sheets over a constant angle. (Adapted from W. Luyten et al., *Ultramicroscopy*, 49, 123, 1993.)

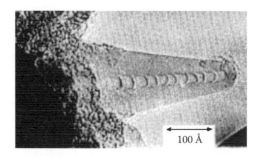

FIGURE 3.5 Carbon cone showing separation of layers in fullerenic end cups. (Adapted from H. P. Boehm, *Carbon*, 35, 581, 1997.)

Graphite conical crystals of very small apex angles (from ~3 to 20°) and perfectly smooth surfaces (Figure 3.7) have been reported to form in the pores of glassy carbon at high temperatures,[40,41] in addition to other various axial graphitic nano- and microcrystals.[42] Graphitic structures from the glassy carbon pores were produced from carbon-containing gas formed during decomposition of phenol–formaldehyde. The size of these graphite conical crystals ranged from about 100 to 300 nm in the cone base diameter, and their lengths ranged from about 500 nm to several micrometers. Similarly, few other conical structures of graphite were produced by thermal decomposition of hydrocarbons[43] with or without the aid of a catalyst, or by employing various thermochemical routes.[44] The structure of the majority of catalyst-free cones observed is consistent with the cone–helix growth model; however, some of the small apex angle cones (~2.7°), as seen in Figure 3.7c,[40,41] do not conform to this rule. These are most likely carbon scroll structures.[45,46] Orientation of layers in catalytically produced cones is closely related to and resembles the shape of the catalyst particle.[47–49] Catalytically produced cones can adopt open-,[49–51] helical,[49,51] or close-shell structures.[49]

Several other types of cones have been reported that are actually composed of cylindrical graphite sheets.[52,53] The so-called tubular graphite cones (TGCs) (Figure 3.8) have been synthesized on

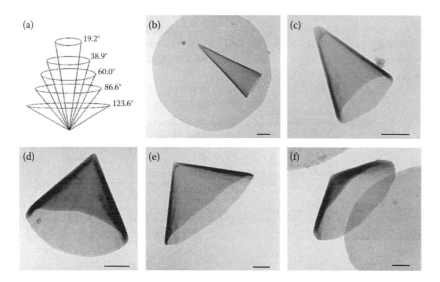

FIGURE 3.6 Fullerene cones. (a) The five possible seamless graphitic cones, with cone angles of 19.2°, 38.9°, 60°, 86.6°, and 123.6°. (Adapted from M. Ge and K. Sattler, *Chem. Phys. Lett.*, 220, 192, 1994; K. Sattler, *Carbon*, 33, 915, 1995.) Electron micrographs of the corresponding five types of cones (scale bars in b–f, 200 nm). Apex angles: (b) 19.2°, (c) 38.9°, (d) 60.0°, (e) 84.6°, and (f) 112.9°. (Adapted from A. Krishnan et al., *Nature*, 388, 451, 1997.)

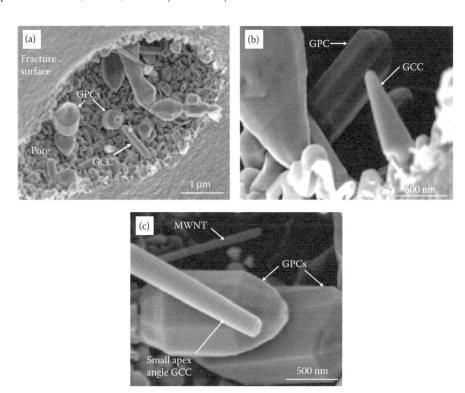

FIGURE 3.7 SEM micrographs of carbon nano- and microcrystals found in pores of glassy carbon. (a) Fracture surface, showing crystals in a pore. (b) Graphite polyhedral crystals (GPCs) and graphite conical crystals (GCCs). (c) A small apex angle GCC growing along with GPCs, and a stylus-like multiwall carbon nanotube (MWNT). (Adapted from Y. Gogotsi, S. Dimovski, and J. A. Libera, *Carbon*, 40, 2263, 2002.)

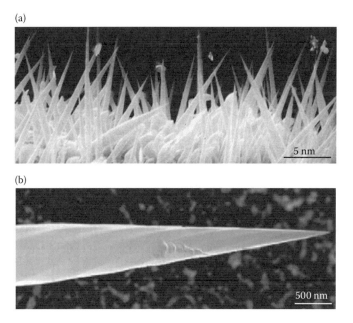

FIGURE 3.8 Tubular graphite cones. (a) Aligned TGCs grown on an iron needle surface. (b) A high-resolution view of one TGC shows the faceted and helical appearance. (Adapted from G. Y. Zhang, X. Jiang, and E. G. Wang, *Science*, 300, 472, 2003.)

an iron needle using a microwave–plasma-assisted chemical vapor deposition (MWCVD) method[52] in a CH_4/N_2 gaseous environment. Corn-shaped carbon nanofibers with metal-free tips have also been synthesized by an MWCVD method using CH_4 and H_2 gases.[53] Graphitic coils wound around a tapered carbon nanotube core have also been produced by the same technique using different substrate material.[54] What makes these and similar structures cone shaped is not purely an inclination of their graphitic layers with respect to the cone axis, but rather the continuous shortening of graphitic wall layers from the interior to the exterior of the structure,[55] or a combination of both mechanisms, as in the case of carbon nanopipettes.[54] Although their morphology resembles a cone, intrinsically, their microstructure is that of the multiwall carbon nanotubes. This implies different mechanical and electronic properties. Tailoring carbon nanotubes to cone shapes can now be done routinely.[56–59]

3.2.2 OCCURRENCE OF GRAPHITE WHISKERS AND CONES IN NATURE

Graphite whiskers and cones have also been observed growing on natural Ticonderoga graphite crystals,[60] Gooderham carbon aggregates,[61] and friable, radially aligned fibers of Kola graphite.[62] In their brief communication, Patel and Deshapande[60] reported the growth of 65- to 125-μm-thick graphite whiskers in ⟨0 0 0 1⟩ direction, the (0 0 0 1) planes of graphite being perpendicular to the whisker axis. The growth of the whiskers was presumed to be the result of a screw dislocation mechanism during the growth of graphite, but no details indicating the relationship between the structure and the geological origin of the sample were given. Several other exotic forms of graphite have been observed recently from two different geological environments: arrays of graphite cones in calcite from highly sheared metamorphic rocks in Eastern Ontario (Gooderham graphite, Figure 3.9a),[61] cones, and scrolls of tubular graphite in syenitic igneous rock from the Kola Peninsula of Russia (Figures 3.9b–d).[62]

In a few geological occurrences, graphite forms compact spherical aggregates with radial internal textures,[62–66] similar to those observed in graphite spheres in cast iron.[27] One prominent natural

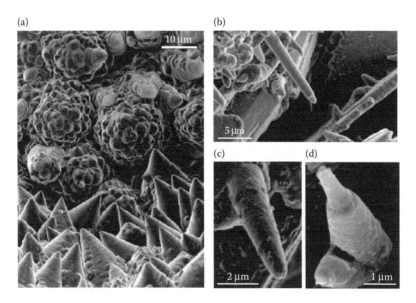

FIGURE 3.9 FESEM images of graphite cones from (a) Gooderham, Ontario, Canada. (Adapted from J. A. Jaszczak et al., *Carbon*, 41, 2085, 2003.) (b–d) Graphite cones, scrolls, and tubes from Hackman Valley, Kola Peninsula, Russia. A scroll-type structure is suggested in (b). Some of the Kola cones appear to be hollow, as indicated by a fractured structure (d). (Adapted from S. Dimovski et al., *Biennial Conference on Carbon*, American Carbon Society, RI, 2004.)

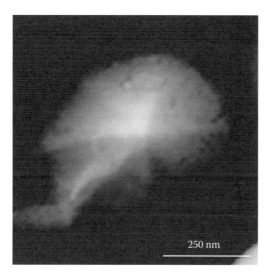

FIGURE 3.10 FESEM micrograph of a graphite "protocone" having a faceted tip. (Adapted from S. Dimovski et al., *Biennial Conference on Carbon*, American Carbon Society, RI, 2004.)

occurrence is in metasedimentary rocks exposed at a road cut, south of Gooderham, Ontario, Canada.[67,68] In this region, graphite crystallizes in calcite in various forms of tabular flakes, spherical, spheroidal, and triskelial polycrystalline aggregates,[62,69] some of which were found to contain large arrays of graphitic cones dominating the surfaces of the samples.[61] Cone heights ranged from less than a micron to 40 μm, and unlike most laboratory-produced cones, they showed a wide distribution of apex angles. The apex angles were found to vary from 38° to ~140°, with 60° being the most common. The cone structure can be well described by the Double and Hellawell[27] disclination model. Other than full and solid cones, some Gooderham samples also revealed partly conical hollow structures composed of curved graphite shells ("protocones").[62] These indicate a possible earlier growth stage for the cones as reported in Ref. 61. Unlike large solid cones, many of these graphitic structures have partly faceted surfaces (Figure 3.10). The tips of the polygonal cones typically have six facets, and these facets only extend part way down the surfaces of the cones, which maintain a circular base.[62] The faceted cones are reminiscent of the polyhedral graphite crystals from glassy carbon pores.[42] The morphology, the surface topography of the cones, and petrologic relations of the samples suggest that the cones formed from metamorphic fluids.

Numerous scroll-type graphite whiskers, up to 15 μm in length and up to 1 μm in diameter (Figures 3.9b–d), were discovered to cover the inner and outer surfaces of channels comprising tabular graphite crystals. They have been found in samples of alkaline syenitic pegmatite of Kola Peninsula, Russia.[70] The surfaces of cavities in the host rock were coated with a fine-grained graphite layer composed solely of such whiskers. Some of the Kola natural graphite whiskers are cigar-like (Figure 3.9b), whereas others exhibit true conical (Figure 3.9c) morphologies with dome-shaped tips. The conical whiskers appear to be significantly larger and more abundant than the tube-like whiskers. Many Kola cones show distinct spiral growth steps at the surfaces of their tips, suggesting that they have a scroll-type structure, as seen previously in other synthetic whiskers.[11,17,21,26] Scanning electron microscopy (SEM) images of some broken cones reveal that they are hollow (Figure 3.9d).

3.2.3 Structure: Geometrical Considerations

In the previous section, we have seen that, on the basis of their structure, a distinction can be made between the two major classes of graphitic cones. One type has a "scroll-helix" structure, whereas the second type comprises seamless conical graphene layers stacked over each other along their axis (therefore called "fullerene cones"). This classification may be considered as an equivalent to differentiating between "scroll" and "Russian-doll" type of multiwall carbon nanotubes.[71]

Pure "scroll-helix" cones are made up of a single graphene sheet that coils around an axis, each layer having a cone shape. The nucleation of this kind of structure is generally controlled by a line defect (dislocation), although we will see later that in addition, it always involves a screw dislocation and some kind of point defect at the terminated side of the dislocation line, as indicated in Figure 3.4. However, an ideal "fullerene" cone contains only point defects in the form of pentagonal, heptagonal, or lower/higher order carbon rings and their various combinations. It is also possible that some of the actual graphitic cones are neither purely helical nor purely fullerene structures, but rather a combination of the two.

Euler's theorem[72] has been found to be particularly useful in explaining geometrical aspects and generation of fullerenes and fullerene cones. Suppose that a polyhedral object is formed by enclosing a space with polygons. The number of polygons is therefore equal to the number of faces (F) of such object. If V is the number of vertexes, E the number of edges, and g a genus of the structure, then the four parameters correlate as follows:

$$V - E + F = 2(1 - g) \tag{3.1}$$

For bulk three-dimensional solids, g is equivalent to the number of cuts required to transform a solid structure into a structure topologically equivalent to a sphere (for instance, $g = 0$ for a polygonal sphere such as C_{60} or C_{70}, and $g = 1$ for a torus). Suppose, further, that the object is formed of polygons having different (i) number of sides. The total number of faces (F) is then

$$F = \sum N_i \tag{3.2}$$

where N_i is the number of polygons with i sides. Each edge, by definition, is shared between two adjacent faces and each vertex is shared between three adjacent faces, which is represented as

$$E = \frac{1}{2} \sum i N_i \tag{3.3}$$

and

$$V = \frac{2}{3} E \tag{3.4}$$

By substituting Equations 3.2 through 3.4 into Equation 3.1, Euler's postulate for $i \geq 3$ is represented as

$$3N_3 + 2N_4 + N_5 - N_7 - 2N_8 - 3N_9 - \cdots - = 12(1 - g) \tag{3.5}$$

$$\sum (6 - i)N_i = 12(1 - g) \tag{3.6}$$

It can be observed from Equation 3.5 that the number of hexagons does not play a role, and a balance between the number of pentagons and higher order polygons ($i \geq 7$) is required to form an enclosed structure. If each vertex is considered as an atomic site containing an sp^2-hybridized C atom and each edge is assigned to one C–C bond, then according to Equations 3.5 and 3.6, only 12

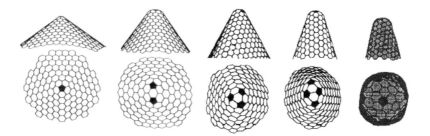

FIGURE 3.11 Distribution of pentagonal defects within the cone tip. The apex angle changes with the number of pentagons. (Adapted from M. Endo et al., *Carbon*, 33, 873, 1995.)

pentagons are needed to form a fullerene or a nanotube. If one heptagon is present, then 13 pentagons will close the structure.

The total disclination in a completely closed structure, such as a sphere, is 720° (i.e., 4π). Each of 12 pentagons contributes a positive disclination of 720°/12 = 60°, and a heptagon, similarly, creates a negative 60° disclination. Incorporation of a heptagon in a graphene sheet will, therefore, produce a saddle-like deformation,[73] whereas adding pentagons will result in conical structures. Exactly five different cones (Figure 3.6) are generated by having, respectively, 1–5 pentagonal rings in their structure, as experimentally observed[6,7] and mentioned in the previous section. Careful examination of such cones suggests that pentagons are isolated from each other by hexagonal rings, as in fullerene molecules and fullerene nanotube caps. The apex angles for these cones can be calculated from the following relation:[74]

$$\sin(\theta/2) = 1 - (N_5/6) \tag{3.7}$$

where N_5 is the number of pentagons in the cone structure.

Topo-combinatoric conformations of i-polygonal carbon rings (where $i = 1, 2, 3, 4, 5, 7, 9, \ldots$) within a hexagonal carbon network had been studied in detail even before the discovery of fullerenes, carbon nanotubes, and graphitic cones.[75,76] Growing interest in this topic resulted in the number of publications[73,77–80] that revealed the fine structure of the cone tip, such as the reconfiguration of carbon atoms and distribution of defects in the near vicinity of the tip. It had also been shown that the pentagons separated by hexagons (Figure 3.11)[74] make the most stable conformation of the cone tip structure, as observed experimentally. Establishing valid theoretical models of structure later helped in calculating the electronic properties of cones and curved carbon surfaces.[81–84]

Apart from seamless cones, there are conical structures that are formed by introducing a wedge disclination (Figure 3.12a) and a screw dislocation (Figure 3.12b) in a graphite sheet, as observed

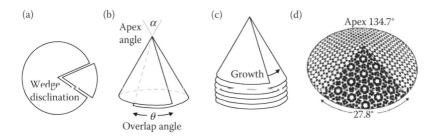

FIGURE 3.12 Formation of helical cones. (a) Positive wedge disclination is created after a sector is removed from a graphene sheet and (b) the cone is formed by an overlap through a screw dislocation. (c) Model illustrating growth of columnar carbon, and (d) one of several energetically preferred stacking arrangements. (Adapted from D. D. Double and A. Hellawell, *Acta Metall.*, 22, 481, 1974.)

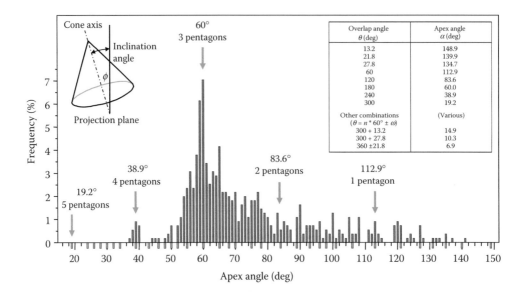

FIGURE 3.13 Frequency of occurrence of various apex angles for natural graphite cones. The maximum is observed at 60°. Apex angles that correspond to "goodness" of fit are listed in the table. When measuring apex angles of cones, the inclination of cones to the projection plane of the microscope has been taken into account (inset, left). (Adapted from J. A. Jaszczak et al., *Carbon*, 41, 2085, 2003.)

experimentally by various groups.[4,5,21,22,26,36] The cone–helix model[27] is based on growth around a positive disclination with a screw dislocation component (Figure 3.12c). As a graphene sheet wraps around the disclination, adjacent overlapping layers are rotated with respect to one another by an angle equal to the disclination angle. Among a practically unlimited number of disclination angles, some of them should be energetically more favorable (Figure 3.12c and inset table in Figure 3.13). Their value can be calculated from the following equation:

$$\alpha = n \times 60°, \quad \text{or } \alpha = n \times 60° \pm \omega \tag{3.8}$$

where $n = 0, 1, 2, \ldots, 6$, and $\omega = 13.2, 21.8, 27.8°, \ldots$ are expected low-energy (0 0 1) twist grain-boundary angles based on lattice coincides, which are a measure of "goodness of fit," but do not account for atomistic interactions and the curvature of the sheets. Disclinations with overlap angles equal to integer multiples of 60° should be energetically the most favorable because they preserve the graphite crystal structure without stacking faults, provided the screw component of the disclination has a Burgers vector corresponding to an even multiple of the graphite's c-axis interplanar spacing. Values of corresponding apex angles are calculated from the following relation:

$$\theta = 2\sin^{-1}(1 - \alpha/360°) \tag{3.9}$$

and they range from 6° to 149°. Graphitic cones having such apex angles should predominate over others.[61] The apex-angle distribution in a sample of natural cones is shown in Figure 3.13. Among all possible apex angles, the 60° angle is found to be the most frequent. Cones with smaller apex angles may be disfavored because of the higher elastic energy due to bending required to form the corresponding disclinations.

The dislocation line usually terminates with a point defect that includes bond recombination within the hexagonal network to form a pentagon or some other kind of polygon, as shown in Figure 3.4c.

3.2.4 Properties and Applications

3.2.4.1 Electronic Properties of Synthetic Whiskers and Cones

Carbon nanotubes are known to be either metallic or semiconducting, depending on their diameter and chirality.[85–89] The role of pentagon, heptagon, or pentagon–heptagon pair topological defects in structural and electronic properties of nanotubes has also been studied theoretically[81,84,90] and experimentally by means of scanning tunneling microscopy (STM) and scanning tunneling spectroscopy (STS).[84] Special attention has been paid to curved surfaces of capped carbon nanotube tips because these can be considered as regions of high density of defects. As the density of defect states increases at the tube ends, it can be expected that the electronic band structure of the end significantly differs from that elsewhere on the tube.[81–84] This has been successfully demonstrated by means of spatially resolved STM/STS carried out on a conically shaped tube end (Figures 3.14a and b),[84] which in fact is a fullerene-type carbon nanocone structure.

An STM image of one such conical tip is shown in Figure 3.14a. The apex of the cone has a diameter of 2.0 nm. The tunneling spectra were acquired at four different positions along the tube (marked with white letters in Figure 3.14a). Local densities of states (LDOS), derived from

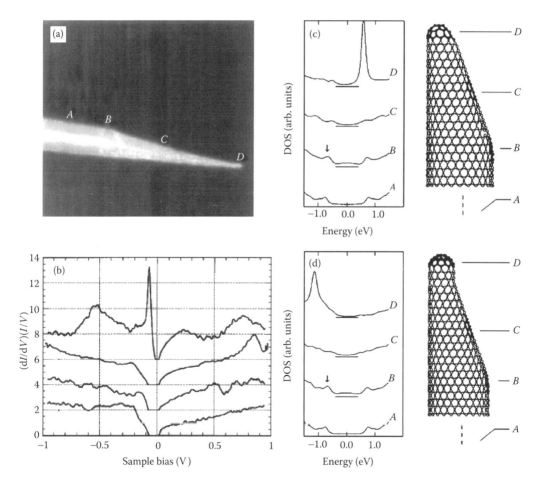

FIGURE 3.14 Electronic structure and localized states at carbon nanotube tips: (a) STM image of a fullerene carbon cone; (b) local densities of states for a cone tip derived from scanning tunneling spectra at four (A–D) points along the tip; and (c–d) tight-binding calculations for two different configurations of cone tips. (Adapted from D. L. Carroll et al., *Phys. Rev. Lett.*, 78, 2811, 1997.)

the scanning tunneling spectra, are represented in Figure 3.14b. In addition, the tight-binding calculations performed on two different tip morphologies are given in Figures 3.14c and d.

As we move along the tube from position A to position D, the density of topological defects increases because the topological defects are concentrated in a smaller volume. As a result, the effect of confinement on the electronic structure becomes increasingly pronounced. This does not seem to be very striking in the case of the conduction band, where only a slight and broad enhancement has been noted in the LDOS at the cone apex. However, the valence band is found to alter considerably, exhibiting sharp resonant states at the cone tip (Figure 3.14b, curve D). The strength and position of these resonant states with respect to the Fermi level are, in addition, very sensitive to the distribution and position of defects within the cone. This is illustrated with two models of cones having different morphologies obtained by altering the position of pentagons within the tip structure (Figures 3.14c and d). In the two examples, the (A), (B), and (C) LDOS calculated by the tight-binding method are very similar. Strong and sharp peaks in (D) LDOS have different shape and position in the case of models I and II. The values calculated for model II show better fit to the experimental values given in Figure 3.14b. The distribution of the defects and their effect on electronic properties of the cones have been studied in detail elsewhere.[91] LDOS of helix-type carbon cones are obtained by establishing the tight-binding model of a screw dislocation in graphite.[81]

Localized resonant states are very important in predicting the electronic behavior of carbon cones. They can also strongly influence the field emission properties of cones.

3.2.4.2 Raman Spectra

Owing to its sensitivity to changes in the atomic structure of carbons, Raman spectroscopy has proven to be a useful tool in understanding the vibrational properties and the microstructure of graphitic crystals and various disordered carbon materials.[92–97] The relationship between the spectra and the structure has been extensively discussed in the literature, and the studies cover a wide range of carbon materials, such as pyrolytic graphite (PG)[94,95] and HOPG,[95,98,99] microcrystalline graphite, amorphous carbon and glassy carbon, fullerenes, carbon onions, nanotubes, and so on. Little work has been carried out on the Raman scattering from graphite whiskers,[100–102] which usually consist of carbon layers oriented parallel to the growth axes. For such structures, it is expected that their Raman spectra will be similar to those of disordered graphite crystals and carbon fibers.

Figure 3.15 shows the Raman spectra of an individual graphite whisker and turbostratically stacked particles, using 632.8 nm excitation wavelength. Whiskers were synthesized in a graphitization furnace using a high-temperature heat-treatment method.[23] Carbon layers in these whiskers are almost perpendicular to their growth axes. Most of the first- and second-order Raman modes in whiskers, such as the D, G, and D' modes at ~1333, 1582, and 1618 cm^{-1}, respectively, can be assigned to the corresponding modes in HOPG and PG.

In contrast to other carbon materials, the Raman spectra of whiskers exhibit several distinct characteristics. For example, the intensity of the two-dimensional (2D) overtone is found to be 13 times stronger than that of the first-order G mode in whiskers. The strong enhancement of the D and 2D modes is also found in the Raman spectra of whiskers with 488.0 and 514.5 nm laser excitations.[100] Second, there are two additional low-frequency sharp peaks located around 228 and 355 cm^{-1}, and two additional strong modes (around 1833 and 1951 cm^{-1}) observed in the second-order frequency region. The line widths of the D, G, D', 2D, and $2D'$ modes in whiskers are 17, 18, 10, 20, and 14 cm^{-1}, respectively. Because the frequencies of the L_1 and L_2 modes are in the frequency region of acoustic modes, these two modes are supposed to be the resonantly excited acoustic modes in the transverse- and longitudinal-acoustic phonon branches. The two high-frequency modes at 1833 and 1951 cm^{-1} are designated as $L_1 + D'$ and $L_2 + D'$ modes, respectively. The observed excitation-energy dependence (140 cm^{-1} eV^{-1}) of the 1833 cm^{-1} mode is in excellent agreement with the theoretical value of 139 cm^{-1} eV^{-1} of the $L_1 + D'$ mode.[100]

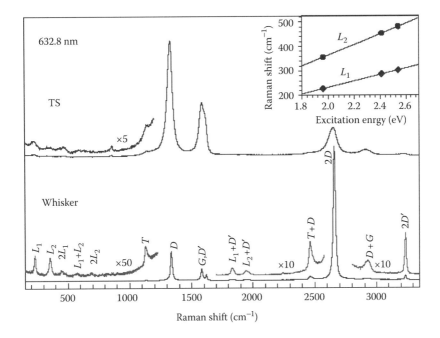

FIGURE 3.15 Raman spectra of turbostratically stacked (TS) particles and an individual graphite whisker excited with 632.8 nm laser excitation. The inset gives the energy dependence of the frequencies of the L_1 and L_2 modes. (Adapted from P. H. Tan, S. Dimovski, and Y. Gogotsi, *Phil. Trans. R. Soc. Lond.* A, 362, 2289, 2004.)

The intensity enhancement of the dispersive modes indicates that double-resonance Raman scattering may be responsible for this phenomenon.[103] Such enhancement of the 2D mode is also observed in GPCs (Figure 3.17) that have a similar loop-edge structure in brim regions.[42]

Raman spectra from tubular, helix-type, and naturally occurring carbon cones are available elsewhere in the literature.[61,102]

3.3 GPCs: POLYGONAL MULTIWALL TUBES

3.3.1 SYNTHESIS

The structure of single- and multiwall carbon nanotubes, and single-wall carbon nanotube ropes have been widely studied during the past 10 years.[2,31,71,104–108]

Although the ability of carbon to form multiwall tubular nanostructures is well known and these tubes have been studied extensively, much less information is available about carbon nanotube structures having polygonal cross sections. An occurrence of polygonal vapor-grown carbon fibers with a core carbon nanotube protrusion was noted by Speck et al.[109] as early as 1989, but no details were given about core fiber structure and its polygonization.

Zhang et al.[110] have studied the structure of an arc-discharge-produced carbon soot by the HRTEM, and they were the first to indicate the possibility of polygonal multiwall carbon nano tubes, assuming that the tubes consisted of closed coaxial concentric layers. The first evidence for the occurrence of polygonized carbon nanotubes came from Liu and Cowley,[104,105,108] who used nanodiffraction in conjunction with HRTEM and selected area electron diffraction to investigate the structures of carbon nanotubes having diameters of a few nanometers. Nanodiffraction is a form of convergent beam electron diffraction, which allows one to obtain a diffraction pattern from regions of the specimen about 1 nm or less in diameter. The tubes used in this study were produced by a variant of Kratschmer–Huffman arc-discharge method[111] in helium gas at a pressure of 550 Torr. The DC voltage applied to electrodes was 26–28 V and the corresponding current was 70 A. The carbon nanotubes obtained at the given experimental conditions consisted of 3–30 carbon sheets and had a length of up to 1 μm. The inner diameters of these tubes ranged from 2.2 to 6 nm, and the outer diameters ranged from 5 to 26 nm. In addition to nanotubes of circular cylindrical cross section, with zero, one, or several helix angles, there were many tubes having polygonal cross sections, made up of flat regions joined by regions of high and uniform curvature.[105] An HRTEM image of one such structure is given in Figure 3.16a. Polygonization of the cross section is observed indirectly through formation of uneven patterns of lattice fringes on the two sides of the tube, with spacings varying from 0.34 nm from the circular cylinder tubes, to 0.45 nm from the regions of high curvature (Figure 3.16b).

In their study of the intershell spacing of multiwall carbon nanotubes prepared by the same Kratschmer–Huffman arc-discharge method, Kiang et al.,[112] similarly, found that the intershell spacing in carbon nanotubes ranged from 0.34 to 0.39 nm among different nanotubes, decreasing with the increase in the tube diameter (Figure 3.16c). Other reports have also shown variation of the values from 0.344 nm (obtained by the electron and powdered x-ray diffraction measurements)[113] to 0.375 nm (based on the HRTEM images).[114]

Faceted multiwall carbon nanotubes with larger diameters, called GPCs, have been reported to grow at high temperatures in the pores of a glassy carbon material (Figures 3.17a and b).[42] The glassy carbon containing polyhedral tubes was made from a thermoset phenolic resin by carbonization at 2000°C in N_2 atmosphere at ~10 Torr. The density of glassy carbon was 1.48 g/cm^3 with an open porosity of ~1%; its microstructure and properties are typical of other glassy carbons. After the structure of the matrix was set and some closed pores were formed, polyhedral nanotubes grew from C–H–O (N_2) gas trapped within these pores during the resin carbonization phase.

GPCs have a very complex morphology. Their size ranges from 100 to 1000 nm in diameter and up to few micrometers in length. The number of facets can vary from 5 to 14 and more, and they may possess a helical habit or may be axially true. Many of the crystals terminate with a thin

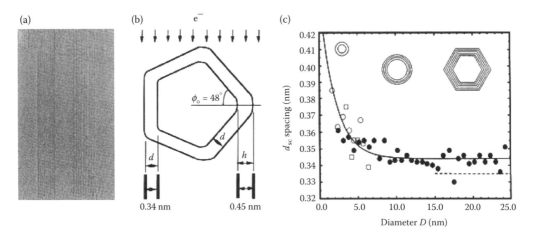

FIGURE 3.16 (a) HRTEM image of a nine-sheet nonsymmetric tube. A d spacing of 0.34 nm is found on the left side and a d spacing of 0.45 nm is seen on the right side. (Adapted from M. Liu and J. M. Cowley, *Carbon*, 32, 393, 1994.) (b) Model illustrating the formation of the nonsymmetric fringes from a tube (a) with polygonal cross section. (Adapted from M. Liu and J. M. Cowley, *Carbon*, 32, 393, 1994.) (c) The graphitic interplanar spacing decreases as the tube diameter increases, and approaches 0.344 nm at roughly $D = 10$ nm. The data were measured from three different nanotubes indicated by different symbols. Hollow circles: from a seven-shell tube with innermost diameter $D_{min} = 1.7$ nm. For large D, graphitization may occur resulting in a polygonal cross section. The broken line indicates the expected decrease in interplanar spacing owing to local graphitic stacking. (Adapted from C.-H. Kiang et al., *Phys. Rev. Lett.*, 81, 1869, 1998.)

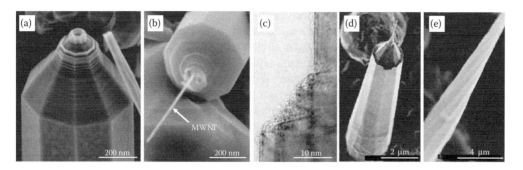

FIGURE 3.17 Graphite polyhedral crystals (GPCs). (a) SEM micrograph of a faceted GPC. (Adapted from P. H. Tan, S. Dimovski, and Y. Gogotsi, *Phil. Trans. R. Soc. Lond.* A, 362, 2289, 2004.) (b) A carbon nanotube stylus is connected to a microsize body. (c) TEM image of a GPC's lattice fringes indicates that GPCs are highly graphitic and that the basal planes are terminated by a closed-loop structure. (Adapted from P. H. Tan, S. Dimovski, and Y. Gogotsi, *Phil. Trans. R. Soc. Lond.* A, 362, 2289, 2004.) (d–e) GPCs produced by using the flame combustion method. (Adapted from H. Okuno et al., *Carbon*, 43, 692, 2005.)

protruding needle that appears to be a multiwall nanotube (Figure 3.17b), typically with a core diameter of about 5–20 nm and a conical, dome-capped, or semitoroidal tip. There is no evidence about catalytic nucleation of the GPCs. Formation of highly ordered structures is promoted with the high temperature of treatment, supersaturation of the environment with carbon atoms, slow reaction kinetics, and the presence of active species such as hydrogen and oxygen atoms that balance the crystal growth rate with the surface-etching rate. This explains the surprisingly large number of ordered carbon layers (up to 1500) growing on the core nanotube, resulting in complex axis-symmetric structures. GPCs of somewhat less-perfect structures (Figures 3.17c and d) have been successfully produced recently by using the flame-combustion method.[115]

Annealing of carbon nanotubes with a circular cross section at high temperatures causes polygonization of their walls. An HRTEM image of a CVD carbon nanotube sample before and after annealing is shown in Figure 3.18. The tubes were annealed for 3 h in a 10^{-6} Torr vacuum at 2000°C. High-temperature annealing of carbon nanotubes in a vacuum or an inert environment allows for the transformation of circular tubes into polygonal ones. However, polygonization will not be uniform along the tube, nor will the cross section take the shape of a regular polygon.

To the best of our knowledge, natural counterparts of polygonal carbon multiwall nanotubes have not been observed so far, but it would not come as a surprise if they are discovered in the near future.

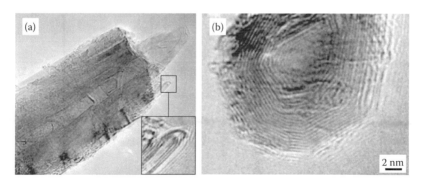

FIGURE 3.18 Transformation of multiwall carbon nanotubes into polygonal GPC-like structures by annealing in vacuum at 2000°C. (a) TEM image of a tube with a core nanotube in the form of stylus extension. Arched semitoroidal structures, similar to that of GPCs, have been also formed (inset) through elimination of dangling bonds at high temperature. (b) Polygonized cross section and graphitic walls of an annealed hollow tube. (TEM micrographs courtesy of H. Ye.)

Very short needle-like polygonal multiwall carbon nanotubes have also been synthesized from a supercritical C–H–O fluid by hydrothermal treatment of various carbon precursor materials[116,117] with and without the aid of a metal catalyst. Hollow carbon nanotubes, with multiwall structures, comprising well-ordered concentric graphitic layers, have been produced by treating amorphous carbon in pure water at 800°C and 100 MPa.[117] HRTEM analysis of the reaction products indicates the presence of carbon nanotubes with polygonal cross sections (varying contrast and lattice spacing along the tube diameter) within these samples. The experimental conditions for hydrothermal synthesis of nanotubes resemble to a great extent the conditions of geological metamorphic fluids, and it is possible that some polygonal tubes are present in the Earth's crust along with the natural graphitic cones and tubules but have not been found yet.

3.3.2 STRUCTURE OF POLYGONAL TUBES

One of the earliest works dealing with the polygonization of the cross section of carbon nanotube is a report by Zhang et al.[110] On the basis of experimental observations, it was suggested that the fine structure of carbon nanotubes is determined by two competing accommodation mechanisms (Figure 3.19a). As a result of one of the mechanisms, the successive tube can adopt different helical shapes to accommodate the change in circumference, thereby keeping an orientationally disordered (turbostratic) stacking. As a result of the second mechanism, the rows of hexagons in successive tubes remain parallel and adopt a graphitic stacking, thus inducing some regions of stacking faults because of the deformation of hexagons. The regions of stacking fault are assumed to be evenly distributed along the tube circumference[110] and are separated from the graphitic structure by interfacial dislocations (bold lines in Figure 3.19a). Therefore, polygonization of carbon nanotubes may appear as a result of the necessity to allow graphitic stacking of the layers, as often seen in carbon onions (Figure 3.19b).[110,118,119] It is easy to envision a trade-off between the energy associated with turbostratic stacking versus the strain energy associated with shape changes and stacking faults. The mechanism prevailing strongly depends on the tube size. The relative strain required to maintain graphitic order is smaller with increasing tube size; therefore, thicker tubes are expected to have graphitic ordering and polygonal cross sections with discrete regions of stacking faults, whereas thinner tubes are more likely to retain turbostratic concentric structures with cylindrical cross section and varying helicity between the individual shells.

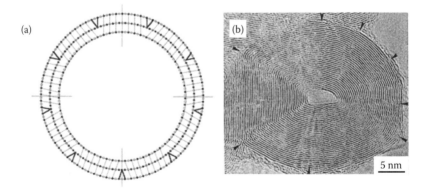

FIGURE 3.19   (a) Schematic model of a nanotube cross section. "Interfacial dislocations" (bold lines) are introduced to accommodate the strains on the tube surfaces. The graphite stacking is maintained in the tube walls (as indicated by lines). The full circles represent the atoms in the paper plane and the open circles are projected positions of the atoms of the paper plane. (b) Defect regions in the HRTEM image of a carbon "onion." The defect regions are characterized by their abnormal image contrast. (Adapted from X. F. Zhang et al., *J. Cryst. Growth*, 130, 368, 1993.)

In the near-planar regions of the polygonized tubes, an ordering of the stacking sequence of the carbon layers gives rise to hexagonal and possibly rhombohedral graphite structure.[108] The near-planar regions are connected to seamless shells through the regions of high curvature (Figure 3.16b). A small value for the radius of curvature is preferred in regions of bending of the carbon sheets between the extended near-planar regions because of the nature of local perturbations of the carbon-bonding arrangement.[104,105,108]

A tube-structure model has been proposed to explain the variation of intershell spacing as a function of tube diameter (Figure 3.20a). Individual intershell spacings as a function of tube diameters were measured in real space from HRTEM images of various nanotubes.[112] The empirical equation for the best fit to these data is given as

$$d_{002}^i = 0.344 + 0.1e^{-D/2} \tag{3.10}$$

The function is plotted in Figure 3.20b. The large full circles show experimental values.

The increase in the intershell spacing with decreased nanotube diameter is attributed to the high shell curvature of small diameter tubes, and it has also been suggested that polygonization of the tube cross section will occur for inner tube diameters larger than ~12 nm (see Figure 3.16c). This observation is in agreement with the model suggested by Zhang et al.[110]

Furthermore, it has also been proposed that multiwall nanotubes most likely consist of circular core shells and polygonal outer layers.[120] In their pioneering work on nanodiffraction from carbon nanotubes, Liu and Cowley[104,105] noted a possibility that there might be some nanotubes with neither entirely polygonal nor fully cylindrical cross sections. Such tubes could be considered as a mixture of the two possible morphologies, with the structure varying along the tube length and the shell diameter. A schematic illustrating this model, taking into account variations of intershell spacing, is given in Figure 3.20.

To obtain direct evidence of tube microstructure, there have been several cross-sectional TEM studies.[121–124] These studies were conducted on both carbon nanotubes produced by arc-discharge method and the tubes produced by a chemical vapor deposition. A large number of defects in CVD

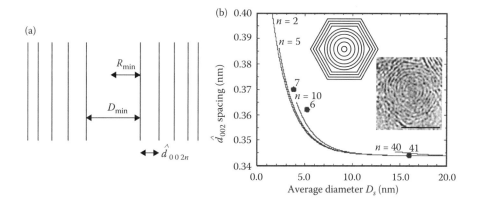

FIGURE 3.20 Effect of tube diameter on interplanar spacing. (a) Model for a nanotube crystal with a varying intershell spacing. (b) The d spacing is given an exponential function of tube diameter (Equation 3.10). The intershell spacing d is plotted as a function of the average tube diameter (D_a), where n is the number of shells in a nanotube. The curves are calculated for $n = 2, 5, 10$, and 40, using the above model and Equation 3.10. The three data points shown by the large full circles were obtained based on experimental observations. (Adapted from C.-H. Kiang et al., *Phys. Rev. Lett.*, 81, 1869, 1998.) Insets: Model illustrating change of interplanar spacing d and polygonization of tube cross section with increase of tube diameter, as observed in some TEM micrographs. (TEM image: courtesy of S. Welz; scale bar, 5 nm.)

tubes is very common, and it is an intrinsic property of the CVD process and, therefore, will not be discussed further here. HRTEM images of the cross sections of tubes reveal their nested structure, but they do not confirm models proposed by Liu et al.,[105] Zhang et al.,[110] and others.[112,120] Instead, rather random dislocation lines extending radially have been recorded.[121]

Besides their polygonal cross section, GPCs[42,115] possess another important feature that may affect their electrical, chemical, and mechanical properties to a great extent. This is the several-nanometer-thick loop-like layer (Figure 3.17c and inset in Figure 3.18a) formed by zipping of the adjacent graphitic shells at their terminations.[125,126] This phenomenon is also observed on edge planes of certain high-temperature planar graphites[125,127] and on the surfaces of cup-like multiwalled carbon nanotubes annealed in argon atmosphere above 900°C.[128] In the case of planar graphite, zipping of graphitic layers (also known as "lip–lip" interactions) forms nanotube-like sleeves, whereas in the case of multiwall nanotubes, the resulting structure resembles concentric polygonal hemitoroidal structures.[126] "Lip–lip" interactions are especially pronounced when samples are annealed at temperatures above 1600°C.[42,51,125,128–130] The reactive edge sites transform into stable multiloops through the elimination of dangling bonds because of enhanced carbon mobility at higher temperatures. Multiloops are typically built by 2–6 adjacent graphitic layers. Typically, single-loop structures are formed between 900°C and 1200°C, whereas 1500°C is considered as the threshold for the formation of multilayer loops.[128] The radius of curvature of the outer layer is similar to the average radius of double-walled nanotubes.[131]

3.3.3 PROPERTIES AND APPLICATIONS

3.3.3.1 Electronic Band Structure

Electronic properties of cylindrical single- and multiwall carbon nanotubes have been widely studied both theoretically and experimentally during the past 15 years, and findings have been summarized in several books about carbon nanotubes[71,132] as well as in Chapter 2 of this book.

Electronic structure of polygonal single-wall carbon nanotubes has been investigated theoretically within tight-binding and *ab initio* frameworks,[133,134] and it has been found that polygonization changes the electronic band structure qualitatively and quantitatively. An example of a zigzag nanotube is given. Considered is the (10, 0) tube with a circular (a) and pentagonal (b) cross section (Figure 3.21). The (10, 0) carbon nanotube with a circular cross section is a semiconductor,

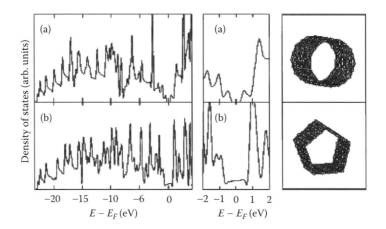

FIGURE 3.21 Tight-binding densities of states (states/eV/cell) for the (10, 0) cylindrical (a), and the (10, 0)^5 pentagonal (b) cross-section nanotubes. The Fermi level is positioned at zero energy. Both nanotubes are also represented in the inset on the right of their respective DOS. (Adapted from J.-C. Charlier, P. Lambin, and T. W. Ebbesen, *Phys. Rev.* B, 54, R8377, 1996.)

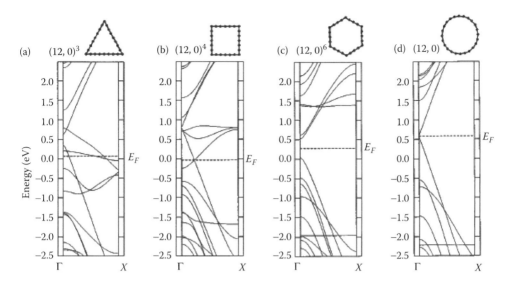

FIGURE 3.22   Tight-binding band structures of the metallic (12, 0) nanotube, illustrating the effect of the degree of polygonization of the cross section on electronic behavior. Given here are examples of (a) the triangle $(12, 0)^3$, (b) square $(12, 0)^4$, and (c) hexagonal $(12, 0)^6$ geometries that are compared to the pure cylinder case tube (d). (Adapted from J.-C. Charlier, P. Lambin, and T. W. Ebbesen, *Phys. Rev. B*, 54, R8377, 1996.)

with a band gap of 0.8 eV. In calculating the band structure of a polygonal tube, it is reasonable to assume that the zones of strong curvatures near the edges of the polygonal tube will introduce a σ^*–π^* hybridization of carbon bonds. This local variation of bonding with strong sp^3 character in the folds creates a sort of defect line in the sp^2 carbon network.[135] In addition to the effect of bond hybridization, polygonization of the cross section lowers the symmetry from a ten- to a fivefold axis. Furthermore, out-of-plane bending of the hexagonal carbon rings along the polygonal edges brings new pairs of atoms closer than the second-neighbor distance in graphite. All this leads to the modification of the electronic band structure, and as a consequence, the semiconducting band gap of the (10, 0) polygonal tube is almost completely closed (Figure 3.21b). The *ab initio* calculations[134] confirm these tight-binding results and predict a gap of 0.08 eV for the pentagonal cross section. Electronic behavior of metallic armchair nanotubes is not so strongly altered with polygonization because the σ^*–π^* hybridization is not possible in the case of armchair configurations.

Theoretical studies also suggest that the perturbation of electronic properties of carbon nanotubes will be different for various degrees of polygonization (i.e., various numbers of facets).[134] An example is given for a (12, 0) nanotube. Zigzag (12, 0) nanotube of circular cross section is metallic. When different polygonal cross sections (triangle, square, and hexagon) are considered, the results indicate that all kinds of electronic properties arise (Figure 3.22). The first two cases are metallic, whereas the third is a 0.5 eV band gap semiconductor. It is important to remember here that these calculations are given for a carbon nanotube comprising a single shell.

3.3.3.2   Raman Spectra

Vibrational properties of GPCs have been studied by means of Raman spectroscopy,[42,102,115] and it has been confirmed that they are highly graphitized structures with the distinct disorder-induced (*D*) band and the strong graphitic (*G*) band of about the same full width at half-maximum (FWHM = 14 cm⁻¹) as in crystals of natural graphite.[136] The selective micro-Raman spectra from the crystal's side face and tip are shown in Figure 3.23. Spectra from the crystal faces correspond to perfect graphite with a narrow *G* band at 1580 cm⁻¹. In addition to 1580 cm⁻¹ peak, the spectra from the tips feature a weak *D* band at 1352 cm⁻¹, and an unusually strong 2D (2706 cm⁻¹) overtone that exceeds the intensity of

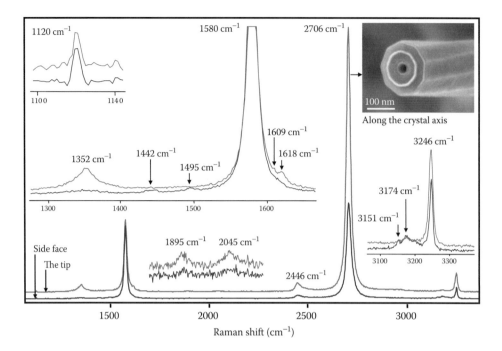

FIGURE 3.23 Fundamental modes, combination modes, and overtones in Raman spectra taken from the side face and the tip of an individual graphite polyhedral crystal (514.5 nm excitation). Inset: an SEM micrograph of a different crystal having structure similar to the one used to record Raman spectra. (Adapted from P. H. Tan, S. Dimovski, and Y. Gogotsi, *Phil. Trans. R. Soc. Lond.* A, 362, 2289, 2004.)

the *G* band and is known to originate from free-standing graphene, probably at the edges, where separation of layers was observed. A similar effect was observed on graphite scrolls (Figure 3.15). Raman spectra of GPCs contain two additional bands in the second-order frequency range at ca. 1895 and 2045 cm^{-1} (Figure 3.23). A number of weak low-frequency bands including a doublet at 184/192 cm^{-1} have been observed in some samples.[42] These low-frequency bands, typical for single-wall nanotubes,[137] may come from the innermost carbon nanotube shells protruding sometimes from the GPCs (see Figure 3.17b). Low-intensity bands are also present in the range of 1440–1500 cm^{-1} (Figure 3.24), the origin of which has not been unambiguously determined yet.

3.3.3.3 Chemical, Thermal, and Mechanical Stability

Functionalization and chemical activity of carbon nanotubes are the subject of intensive study, and many breakthroughs have been made in this field during the past 5 years. Side walls of single-wall carbon nanotubes have been successfully functionalized with fluorine,[138] carboxylic acid groups,[139] isocyanate groups,[140] dichlorocarbene,[141] polymethyl methacrylate,[142] and polystyrene.[143] Various functional groups have been attached to tube edge sites. This has made it possible to use carbon nanotubes and other tubular carbon materials in the fabrication of sorbents,[144] catalyst supports,[145,146] gas-storage materials,[147] and polymer matrix composites.[148,149] The effect of polygonization on chemical behavior of carbon nanotubes has not been thoroughly investigated yet; however, it is expected that polygonal tubes have chemical properties similar to those of circular multiwall carbon nanotubes and graphite, and that they may be more reactive (less stable) along the polygonal edges than in the extended near-planar regions.[120] Another interesting property of GPCs is polygonized hemitoroidal edge plane terminations.[150] Transformation of active sites into loops promotes the GPC into a more stable (or chemically inert) structure. Moreover, because of their specific spatial conformation, they allow for easier intercalation with foreign atoms such as lithium and others.[127,151]

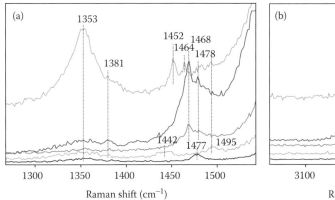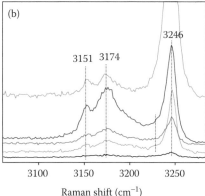

FIGURE 3.24   Raman spectra of graphite polyhedral crystals showing additional weak bands in the range: (a) 1400 to 1500 cm⁻¹ and (b) 3100 to 3200 cm⁻¹ (for 514.5 nm excitation). (Adapted from P. H. Tan, S. Dimovski, and Y. Gogotsi, *Phil. Trans. R. Soc. Lond.* A, 362, 2289, 2004.)

Several methods have been used to probe the chemical activity of polygonal tubes. In one of them, GPCs were intercalated with 50:50 H_2SO_4/HNO_3 for 1 h, washed in deionized water, dried for 24 h, and then exfoliated by rapid heating at about 980°C for about 15 s or until maximum volume expansion is reached. GPCs survived very severe intercalation and exfoliation conditions, most of them retaining their original shape of faceted axial whiskers, although damage in the form of cracks along the axis and striations on the hemitoroidal surfaces was observed on most of the crystals. Corrosion studies of graphitic polyhedral crystals showed that exposure of GPCs to overheated steam at normal pressure for 1 h will cause their complete oxidation at temperatures between 600°C and 700°C. Similarly, 700°C was determined as an onset temperature for oxidation in air. However, they were more stable than disordered glassy carbon in supercritical water.[42]

Polygonization of the cross section of a tube is not expected to significantly affect its tensile strength, nor should it drastically affect the bending modulus of the tube. However, for a helical polygonal multiwall nanotube, the pullout strength is expected to be higher than in the case of its cylindrical counterpart because of the more favorable stress distribution upon loading.

3.4   CONCLUSIONS

Extended growth of nanotubes or fullerene cones leads to formation of a myriad of nano- and microstructures, which are larger relatives of carbon nanotubes. Graphite cones and whiskers have much in common with carbon nanotubes and nanofibers. They inherit the conical structure of scrolled tubes (also known as "herring-bone" structure in carbon nanotubes). Seamless graphite nano- and microcones can have five different apex angles with "magic" numbers of 19.2°, 38.9°, 60.0°, 83.6°, and 112.9°, which correspond to 60°, 120°, 180°, 240°, and 300° disclinations in graphite. Scrolled conical structures may virtually have any apex angle and very small (2–3°) angles have been observed. A few other specific values also occur more frequently as they are energetically preferred because they allow the registry between graphene sheets in the cone.

Polygonization of nanotubes accompanied by growth in the radial direction leads to the formation of GPCs. They have nanotube cores and graphite crystal faces. Unusual symmetries have been observed in GPCs. Most of them are built of coaxial graphite shells.

Cones and large nanotubes have been produced synthetically in the laboratory, and several methods for their synthesis are known. GPCs have only been discovered in synthetic carbon materials and can be grown by CVD or hydrothermally. However, since many natural graphites have been formed from hydrothermal deposits in nature, it would not be surprising to learn that GPCs exist in nature as well. Large carbon nanotubes have already been observed in natural deposits along with carbon cones.

Cones, whiskers, and GPC can bridge the nano- and microworlds and may have numerous applications, where sizes between nanotubes and carbon fibers are required. They may also have interesting electronic and mechanical properties determined by their geometry. However, although their structure has been well understood, very little is known about their properties. Properties need to be studied before their wide-scale applications can be explored.

ACKNOWLEDGMENT

This work was supported by the U.S. Department of Energy grant DE-FJ02-01ER45932.

REFERENCES

1. B. T. Kelly, *Physics of Graphite*, Applied Science Publishers, London, 1981.
2. S. Iijima, *Nature*, 354, 56, 1991.
3. T. Tsuzuku, *Proceedings of the 3rd Conference on Carbon*, Pergamon Press, University of Buffalo, New York, 1957, p. 433.
4. T. Tsuzuku, *J. Phys. Soc. Jpn.*, 12, 778, 1957.
5. S. Amelinckx, W. Luyten, T. Krekels et al., *J. Cryst. Growth*, 121, 543, 1992.
6. M. Ge and K. Sattler, *Chem. Phys. Lett.*, 220, 192, 1994.
7. K. Sattler, *Carbon*, 33, 915, 1995.
8. A. Krishnan, E. Dujardin, M. M. J. Treacy et al., *Nature*, 388, 451, 1997.
9. H. W. Kroto, J. R. Heath, S. C. O'Brien et al., *Nature*, 318, 162, 1985.
10. S. Iijima, *MRS Bull.*, 19, 43, 1994.
11. R. Bacon, *J. Appl. Phys.*, 31, 283, 1960.
12. D. W. McKee, *Annu. Rev. Mater. Sci.*, 03, 195, 1973.
13. P. M. Ajayan, J. M. Nugent, R. W. Siegel et al., *Nature*, 404, 243, 2000.
14. M. Hillert and N. Lange, Z. *Kristallogr. Kristallgeometrie Krystallphys. Kristallchem.*, 111, 24, 1958.
15. S. D. Robertson, *Carbon*, 8, 365, 1970.
16. P. L. J. Walker, J. F. Rakszawski, and G. R. Imperial, *J. Phys. Chem.* 63, 133, 1959.
17. M. L. Lieberman, C. R. Hills, and C. J. Miglionico, *Carbon*, 9, 633, 1971.
18. P. A. Tesner, E. Y. Robinovich, I. S. Rafalkes et al., *Carbon*, 8, 435, 1970.
19. W. R. Davis, R. J. Slawson, and G. R. Rigby, *Trans. Br. Ceram. Soc.*, 56, 67, 1957.
20. A. Fonseca, K. Hernadi, P. Piedigrosso et al., *Appl. Phys. A: Mater. Sci. Process*, 67, 11, 1998.
21. H. B. Haanstra, W. F. Knippenberg, and G. Verspui, *J. Cryst. Growth*, 16, 71, 1972.
22. J. H. Knox, B. Kaur, and G. R. Millward, *J. Chromatogr.*, 352, 3, 1986.
23. J. Dong, W. Shen, B. Zhang et al., *Carbon*, 39, 2325, 2001.
24. J. Dong, W. Shen, F. Kang et al., *J. Cryst. Growth*, 245, 77, 2002.
25. P. T. B. Shaffer, *Plenum Press Handbooks of High-Temperature Materials No. 1 Materials Index*, Plenum Press, New York, 1964.
26. J. Gillot, W. Bollman, and B. Lux, *Carbon*, 6, 381, 1968.
27. D. D. Double and A. Hellawell, *Acta Metall.*, 22, 481, 1974.
28. S. Iijima, T. Ichihashi, and Y. Ando, *Nature*, 356, 776, 1992.
29. P. M. Ajayan, T. Ichihashi, and S. Iijima, *Chem. Phys. Lett.*, 202, 384, 1993.
30. P. M. Ajayan and S. Iijima, *Nature*, 358, 23, 1992.
31. T. W. Ebbesen, *Annu. Rev. Mater. Sci.*, 24, 235, 1994.
32. T. W. Ebbesen and T. Takada, *Carbon*, 33, 973, 1995.
33. A. L. Mackay and H. Terrones, *Nature*, 352, 762, 1991.
34. H. Terrones and A. L. Mackay, *Carbon*, 30, 1251, 1992.
35. B. I. Dunlap, *Phys. Rev. B*, 46, 1933, 1992.
36. W. Luyten, T. Krekels, S. Amelinckx et al., *Ultramicroscopy*, 49, 123, 1993.
37. S. Amelinckx, A. Lucas, and P. Lambin, *Rep. Prog. Phys.*, 62, 1471, 1999.
38. H. P. Boehm, *Carbon*, 35, 581, 1997.
39. G. R. Millward and D. A. Jefferson, in *Chemistry and Physics of Carbon*, vol. 14, P. L. J. Walker and P. A. Thrower, eds., Dekker, New York, 1978, p. 1.
40. S. Dimovski, J. Libera, and Y. Gogotsi, *Mat. Res. Soc. Symp. Proc.*, 706, Z6.27.1, 2002.

41. Y. Gogotsi, S. Dimovski, and J. A. Libera, *Carbon*, 40, 2263, 2002.
42. Y. Gogotsi, J. A. Libera, N. Kalashinkov et al., *Science*, 290, 317, 2000.
43. N. Muradov and A. Schwitter, *Nano Lett.*, 2, 673, 2002.
44. J. Liu, W. Lin, X. Chen et al., *Carbon*, 42, 669, 2004.
45. S. F. Braga, V. R. Coluci, S. B. Legoas et al., *Nano Lett.*, 4, 881, 2004.
46. L. M. Viculis, J. J. Mack, and R. B. Kaner, *Science*, 299, 1361, 2003.
47. N. M. Rodriguez, A. Chambers, and R. T. K. Baker, *Langmuir*, 11, 3862, 1995.
48. V. I. Merkulov, A. V. Melechko, M. A. Guillorn et al., *Chem. Phys. Lett.*, 350, 381, 2001.
49. V. I. Merkulov, M. A. Guillorn, D. H. Lowndes et al., *Appl., Phys., Lett.*, 79, 1178, 2001.
50. H. Terrones, T. Hayashi, M. Muños-Navia et al., *Chem. Phys. Lett.*, 343, 241, 2001.
51. M. Endo, Y. A. Kim, T. Hayashi et al., *Appl. Phys. Lett.*, 80, 1267, 2002.
52. G. Y. Zhang, X. Jiang, and E. G. Wang, *Science*, 300, 472, 2003.
53. Y. Hayashi, T. Tokunaga, T. Soga et al., *Appl. Phys. Lett.*, 84, 2886, 2004.
54. R. C. Mani, X. Li, M. K. Sunkara et al., *Nano Lett.*, 3, 671, 2003.
55. P. Liu, Y. W. Zhang, and C. Lu, *Appl. Phys. Lett.*, 85, 1778, 2004.
56. Z. F. Ren, Z. P. Huang, D. Z. Wang et al., *Appl. Phys. Lett.*, 75, 1086, 1999.
57. Q. Yang, C. Xiao, W. Chen et al., *Diam. Relat. Mater.*, 13, 433, 2004.
58. H. Lim, H. Jung, and S.-K. Joo, *Microelectron. Eng.*, 69, 81, 2003.
59. Y. K. Yap, J. Menda, L. K. Vanga et al., *Mat. Res. Soc. Symp. Proc.*, 821, P3.7.1, 2004.
60. A. R. Patel and S. V. Deshapande, *Carbon*, 8, 242, 1970.
61. J. A. Jaszczak, G. W. Robinson, S. Dimovski et al., *Carbon*, 41, 2085, 2003.
62. S. Dimovski, J. A. Jaszczak, G. W. Robinson et al., Extended abstracts of carbon, *Biennial Conference on Carbon*, American Carbon Society, RI, 2004.
63. J. A. Jaszczak, in *Mesomolecules: From Molecules to Materials*, G. D. Mendenhall, A. Greenberg, and J. F. Liebman, eds., Chapman & Hall, New York, 1995, p. 161.
64. V. N. Kvasnitsa and V. G. Yatsenko, *Mineralogicheskii Zh.*, 13, 95, 1991.
65. V. G. Kvasnitsa, V. N. Yatsenko, and V. M. Zagnitko, *Mineralogicheskii Zh.*, 20, 34, 1998.
66. C. Lemanski, *Picking Table*, 32, 1, 1991.
67. B. A. Van der Pluijm and K. A. Carlson, *Geology*, 17, 161, 1989.
68. K. A. Carlson, B. A. Van der Pluijm, and S. Hanmer, *Geol. Soc. Am. Bull.*, 102, 174, 1990.
69. J. A. Jaszczak and G. W. Robinson, *Rocks Miner.*, 75, 172, 2000.
70. S. N. Britvin, G. U. Ivanyuk, and V. N. Yakovenchuk, *World Stones*, 5/6, 26, 1995.
71. P. J. F. Harris, *Carbon Nanotubes and Related Structures*, Cambridge University Press, Cambridge, 1999
72. E. A. Lord and C. B. Wilson, *The Mathematical Description of Shape and Form*, Halsted Press, New York, 1984.
73. S. Ihara, S. Itoh, K. Akagi et al., *Phys. Rev. B*, 54, 14713, 1996.
74. M. Endo, K. Takeuchi, K. Kobori et al., *Carbon*, 33, 873, 1995.
75. J. R. Dias, *Carbon*, 22, 107, 1984.
76. A. T. Balaban, D. J. Klein, and X. Liu, *Carbon*, 32, 357, 1994.
77. D. J. Klein, *Phys. Chem. Chem. Phys.*, 4, 2099, 2002.
78. D. J. Klein, *Intl. J. Quantum Chem.*, 95, 600, 2003.
79. K. Kobayashi, *Phys. Rev. B*, 61, 8496, 2000.
80. H. A. Mizes and J. S. Foster, *Science*, 244, 559, 1989.
81. R. Tamura, K. Akagi, M. Tsukada et al., *Phys. Rev. B*, 56, 1404, 1997.
82. R. Tamura and M. Tsukada, *Phys. Rev. B*, 52, 6015, 1995.
83. R. Tamura and M. Tsukada, *Phys. Rev. B*, 49, 7697, 1994.
84. D. L. Carroll, P. Redlich, P. M. Ajayan et al., *Phys. Rev. Lett.*, 78, 2811, 1997.
85. B. I. Dunlap, *Phys. Rev. B*, 49, 5643, 1994.
86. J. W. Mintmire, B. I. Dunlap, and C. T. White, *Phys. Rev. Lett.*, 68, 631, 1992.
87. R. Saito, M. Fujita, G. Dresselhaus et al., *Phys. Rev. B*, 46, 1804, 1992.
88. J.-C. Charlier and J.-P. Issi, *Appl. Phys. A: Mater. Sci. Process*, 67, 79, 1998.
89. J. W. G. Wilder, L. C. Venema, A. G. Rinzler et al., *Nature*, 391, 59, 1998.
90. J. C. Charlier, T. W. Ebbesen, and P. Lambin, *Phys. Rev. B*, 53, 11108, 1996.
91. S. Berber, Y.-K. Kwon, and D. Tománek, *Phys. Rev. B*, 62, 2291, 2000.
92. R. Vidano and D. B. Fischbach, *J. Am. Cer. Soc.*, 61, 13, 1978.
93. P. Lespade, R. Al-Jishi, and M. S. Dresselhaus, *Carbon*, 20, 427, 1982.
94. G. Katagiri, H. Ishida, and A. Ishitani, *Carbon*, 26, 565, 1988.
95. Y. Kawashima and G. Katagiri, *Phys. Rev. B*, 52, 10053, 1995.

96. G. G. Samsonidze, R. Saito, A. Jorio et al., *Phys. Rev. Lett.*, 90, 027403, 2003.
97. F. Tuinstra and J. L. Koenig, *J. Chem. Phys.*, 53, 1126, 1970.
98. R. J. Nemanich and S. A. Solin, *Phys. Rev. B*, 20, 392, 1979.
99. P. H. Tan, Y. M. Deng, and Q. Zhao, *Phys. Rev. B*, 58, 5435, 1998.
100. P. H. Tan, C. Y. Hu, J. Dong et al., *Phys. Rev. B*, 6421, 214301, 2001.
101. J. Dong, W. C. Shen, and B. Tatarchuk, *Appl. Phys. Lett.*, 80, 3733, 2002.
102. P. H. Tan, S. Dimovski, and Y. Gogotsi, *Phil. Trans. R. Soc. Lond. A*, 362, 2289, 2004.
103. C. Thomsen and S. Reich, *Phys. Rev. Lett.*, 85, 5214, 2000.
104. M. Q. Liu and J. M. Cowley, *Mater. Sci. Eng. A*, 185, 131, 1994.
105. M. Liu and J. M. Cowley, *Ultramicroscopy*, 53, 333, 1994.
106. S. Amelinckx, D. Bernaerts, X. B. Zhang et al., *Science*, 267, 1334, 1995.
107. N. G. Chopra, L. X. Benedict, V. H. Crespi et al., *Nature*, 377, 135, 1995.
108. M. Liu and J. M. Cowley, *Carbon*, 32, 393, 1994.
109. J. S. Speck, M. Endo, and M. S. Dresselhaus, *J. Cryst. Growth*, 94, 834, 1989.
110. X. F. Zhang, X. B. Zhang, G. Van Tendeloo et al., *J. Cryst. Growth*, 130, 368, 1993.
111. W. Kratschmer, L. D. Lamb, K. Fostriopoulos et al., *Nature*, 347, 354, 1990.
112. C.-H. Kiang, M. Endo, P. M. Ajayan et al., *Phys. Rev. Lett.*, 81, 1869, 1998.
113. Y. Saito, T. Yoshikawa, S. Bandow et al., *Phys. Rev. B*, 48, 1907, 1993.
114. M. Bretz, B. G. Demczyk, and L. Zhang, *J. Cryst. Growth*, 141, 304, 1994.
115. H. Okuno, A. Palnichenko, J.-F. Despres et al., *Carbon*, 43, 692, 2005.
116. J. M. Calderon-Moreno and M. Yoshimura, *Mater. Trans.*, 42, 1681, 2001.
117. J. M. Calderon Moreno and M. Yoshimura, *J. Am. Chem. Soc.*, 123, 741, 2001.
118. S. Iijima, *J. Cryst. Growth*, 50, 675, 1980.
119. D. Ugarte, *Nature*, 359, 707, 1992.
120. Y. Maniwa, R. Fujiwara, H. Kira et al., *Phys. Rev. B*, 64, 073105, 2001.
121. L. A. Bursill, P. Ju-Lin, and F. Xu-Dong, *Philos. Mag. A (Phys. Condens. Matter, Defects Mech. Prop.)*, 71, 1161, 1995.
122. S. Q. Feng, D. P. Yu, G. Hub et al., *J. Phys. Chem. Solids*, 58, 1887, 1997.
123. G. Hu, X. F. Zhang, D. P. Yu et al., *Solid State Commun.*, 98, 547, 1996.
124. J.-B. Park, Y.-S. Cho, S.-Y. Hong et al., *Thin Solid Films*, 415, 78, 2002.
125. S. V. Rotkin and Y. Gogotsi, *Mater. Res. Innov.*, 5, 191, 2002.
126. A. Sarkar, H. W. Kroto, and M. Endo, *Carbon*, 33, 51, 1995.
127. K. Moriguchi, Y. Itoh, S. Munetoh et al., *Phys. B: Condens. Matter*, 323, 127, 2002.
128. M. Endo, B. J. Lee, Y. A. Kim et al., *New J. Phys.*, 5, 121.1, 2003.
129. H. Murayama and T. Maeda, *Nature*, 345, 791, 1990.
130. M. Endo, Y. A. Kim, T. Hayashi et al., *Carbon*, 41, 1941, 2003.
131. Z. Zhou, L. Ci, X. Chen et al., *Carbon*, 41, 337, 2003.
132. M. S. Dresselhaus, G. Dresselhaus, and P. C. Eklund, *Science of Fullerenes and Carbon Nanotubes*, Academic Press, New York, 1996.
133. P. Lambin, A. A. Lucas, and J.-C. Charlier, *J. Phys. Chem. Solids*, 58, 1833, 1997.
134. J.-C. Charlier, P. Lambin, and T. W. Ebbesen, *Phys. Rev. B*, 54, R8377, 1996.
135. H. Hiura, T. W. Ebbesen, J. Fujita et al., *Nature*, 367, 148, 1994.
136. K. Ray and R. L. McCreery, *Analyt. Chem.*, 69, 4680, 1997.
137. A. M. Rao, E. Richter, S. Bandow et al., *Science*, 275, 187, 1997.
138. E. T. Mickelson, I. W. Chiang, J. L. Zimmerman et al., *J. Phys. Chem. B*, 103, 4318, 1999.
139. J. Chen, M. A. Hamon, H. Hu et al., *Science*, 282, 95, 1998.
140. C. Zhao, L. Ji, H. Liu et al., *J. Solid State Chem.*, 177, 4394, 2004.
141. H. Hu, B. Zhao, M. A. Hamon et al., *J. Am. Chem. Soc.*, 125, 14893, 2003.
142. H. Kong, C. Gao, and D. Yan, *J. Am. Chem. Soc.*, 126, 412, 2004.
143. G. Viswanathan, N. Chakrapani, H. Yang et al., *J. Am. Chem. Soc.*, 125, 9258, 2003.
144. Y. Xu, J. W. Zondlo, H. O. Finklea et al., *Fuel Process Technol.*, 68, 189, 2000.
145. J. Ma, C. Park, N. M. Rodriguez et al., *J. Phys. Chem. B*, 105, 11994, 2001.
146. M. Endo, Y. A. Kim, M. Ezaka et al., *Nano Lett.*, 3, 723, 2003.
147. A. Chambers, C. Park, R. T. K. Baker et al., *J. Phys. Chem. B*, 102, 4253, 1998.
148. R. Andrews and M. C. Weisenberger, *Curr. Opin. Solid State Mater. Sci.*, 8, 31, 2004.
149. V. Datsyuk, C. Guerret-Piecourt, S. Dagreou et al., *Carbon*, 43, 873, 2005.
150. J. Han, *Chem. Phys. Lett.*, 282, 187, 1998.
151. K. Moriguchi, S. Munetoh, M. Abe et al., *J. Appl. Phys.*, 88, 6369, 2000.

4 Epitaxial Graphene and Carbon Nanotubes on Silicon Carbide

Goknur C. Büke

CONTENTS

4.1   EPITAXIAL GRAPHENE AND CNTs ON SiC

Graphene (i.e., a single layer of graphite) and carbon nanotubes (CNTs; i.e., graphene rolled into a cylinder) are excellent candidate materials for advanced applications because of their unique electrical, optical, and mechanical properties combined with a high surface area. The successful development of graphene-/CNT-based technology depends on large-scale availability of the high-quality, reproducible, and uniformly ordered material. One of the most versatile methods to produce vertically, self-aligned CNTs and epitaxial graphene is the vacuum annealing of silicon carbide single crystals [1,2]. This is a very versatile method because carbon is supplied from the carbide lattice as known from the synthesis of carbide-derived carbons (CDCs, see Figure 4.1) and, as no catalysts or secondary phases are utilized; the produced graphene and CNTs exhibit extremely high purity. However, to increase the grain/domain size and quality of these carbon nanostructures, further control of the process is needed.

Understanding the effects of experimental parameters of the processes is the basis for tailoring the resulting material so as to enable applications. Therefore, this chapter describes the factors (as summarized in Figure 4.2) influencing the vacuum-annealing process of SiC and the reaction mechanisms.

4.1.1   HISTORY ON THERMAL DECOMPOSITION OF SiC

Initially, studying thermal decomposition of solids was for the purpose of examining thermal stability of materials. With the discovery of epitaxial growth of graphene and self-aligned CNTs on the crystal surfaces of silicon carbide, thermal decomposition has developed into a facile method of producing catalyst-free, high-purity, and highly homogeneous carbon.

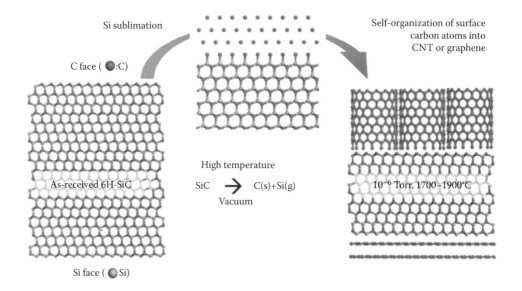

FIGURE 4.1　Schematic drawing of the formation of SiC-derived carbon resulting in layers of CNT and graphene.

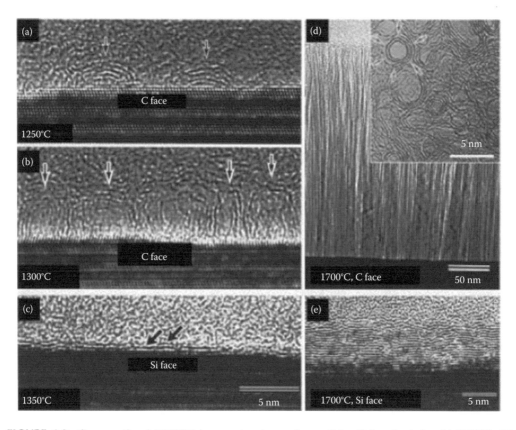

FIGURE 4.2　Cross-sectional HRTEM images showing surfaces of the C face heated at (a) 1250°C, (b) 1300°C, (c) the Si face heated at 1350°C for half an hour, (d) C face, and (e) Si face heated at 1700°C for half an hour. Inset in (d) shows the CNT film observed along a CNT axis plane–view direction. (Adapted and modified from Kusunoki, M. et al., *Philosophical Magazine Letters*, 1999. **79**(4): 153–161; Kusunoki, M. et al., *Applied Physics Letters*, 2000. **77**(4): 531–533.)

The first documented observation of carbon formation by vacuum annealing of SiC was published in 1962 (employing temperatures up to 2150°C) [3]. A decade later, Bommel et al. [4] found monolayers of graphite on SiC after annealing at 800°C and enhanced graphitization around 1500°C. Later, in the 1980s and 1990s, the work of Bommel was confirmed by Muehlhoff et al. [5], Forbeaux et al. [6], and Charrier et al. [7] who reported similar results. However, graphitic structures were not the only structures reported. A new interest in the thermal decomposition of SiC arose in 1997 when Kusunoki et al. [8] discovered the growth of self-organized carbon nanotube films by heating SiC wafer at 1700°C for half an hour in a vacuum furnace ($P = 1 \times 10^{-4}$ Torr) [9]. Later, a model was proposed for the formation mechanism of the CNT film on SiC that consisted of three stages. At temperatures above 1000°C (stage I), several graphene sheets were formed parallel to the (0001) SiC plane. At around ≈1300°C (stage II), carbon nanocaps grow. At stage III, the graphite sheets align upward and CNTs grow toward the interior of the SiC crystal.

Further investigations were carried out on both the Si- and C-terminated basal plane of hexagonal silicon carbide faces by using cross-sectional high-resolution transmission electron microscopy (HRTEM) at lower temperatures [10]. A remarkable difference between the carbon structures formed on (0001)Si and ($000\bar{1}$) C-faces of a 6H-SiC single crystals was shown. Although CNT growth was observed under various experimental conditions on the C-face, only flat graphite sheets parallel to the surface were found to be stable on the Si-face (Figure 4.2).

To study the initial growth process of CNTs at lower temperatures (≈1300°C), Watanabe et al. [11] observed the *in situ* formation of CNTs on β-SiC (111) single crystal inside a TEM and proposed that the caps of the CNTs were formed by a lift of a part of the graphene through the generation of pentagons and heptagons. It was confirmed that around 1360°C amorphous carbon with many holes and defects forms because of the desorption of Si atoms. Afterward, these carbon layers crystallize into graphitic layers. Successive heating at the appropriate temperature results in two types of CNTs: *single-walled* carbon nanotubes (SWCNTs) with a diameter of 0.8 and 1.5 nm and *double-walled* carbon nanotubes (DWCNTs; Figure 4.3). It was confirmed by scanning tunneling microscopy (STM) measurements that the ends of the CNTs were closed (capped).

However, these studies did not explain why different carbon structures occur on Si and C faces and why the results provided by different research groups differed so greatly. In 2004, the parallel publication of the electrical response of graphene by Novoselov et al. [12] and Berger et al. [13] provided new motivation to optimize the growth conditions of graphene on SiC that required a deep understanding of the processes involved at the atomic level during decomposition.

4.1.2 FACTORS AFFECTING THE CARBON GROWTH ON SiC

The reactions that take place when SiC is heated can be summarized as follows:

1. SiC surface reconstruction and surface-oxide removal
2. SiC decomposition: breaking of Si–C bonds (SiC → C(s) + Si(g))
3. Si desorption
4. C diffusion on the surface (self-organization of carbon atoms into carbon nanostructures) that results in carbon islands/graphene nucleation
5. Growth of the nucleus

The nucleation and growth of graphene/CNT are highly dependent on SiC surface morphology and the general growth conditions [14] (as summarized in Figure 4.4), which are outlined in more detail in the following sections.

4.1.2.1 SiC Polytypism

The morphology of epitaxial graphene/CNT is highly influenced by the underlying SiC structure. The basic building block of a silicon carbide is a tetrahedron with four carbon atoms and a single

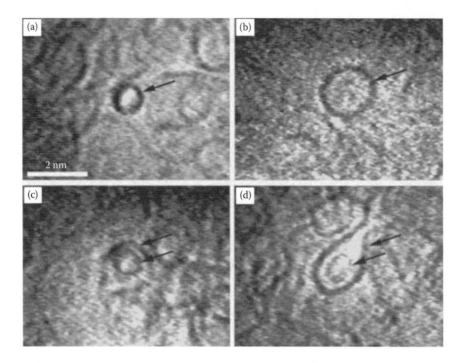

FIGURE 4.3 Examples of *in situ* HRTEM images of CNTs at 1360°C. (a) SWNT with a diameter of 0.8 nm, near the size of a fullerene C_{60}, 0.7 nm, (b) SWNT with a diameter of 1.5 nm, (c) DWNT, and (d) DWNT shaped like a water drop. An inner ring in concentric circles has a weak intensity compared to the outer ring. It is suggested that the inner ring forms after the formation of the outer ring. (From Watanabe, H. et al., *Journal of Microscopy,* 2001. **203**(1): 40–46.)

silicon atom in the center (Figure 4.5a). SiC crystals are constructed with these units that structurally share one joint corner. Silicon carbide exists in a variety of more than 250 different crystal structures called polytypes that are cubic, hexagonal, or rhombohedral. Different polytypes of SiC are composed of different stacking sequences of Si–C bilayers where each single Si–C bilayer can be viewed as a planar sheet of silicon atoms coupled with a planar sheet of carbon atoms (Figure 4.5b).

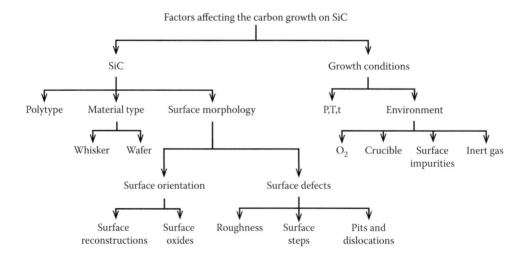

FIGURE 4.4 Summary of the factors that are covered in the chapter.

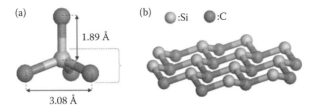

FIGURE 4.5   (a) Si–C$_4$ tetrahedron and (b) Si–C bilayer.

The plane formed by a bilayer sheet of silicon and carbon atoms is known as the basal plane, and the stacking direction [0001] (direction for hexagonal structure), and [111] (direction for cubic structure) are defined as normal to Si–C bilayer plane. In this direction, all bilayers are stacked on top of each other with possible positions as *A*, *B*, and *C* by arranging the sheets in a specific repetitive order.

The different polytypes are constructed by permutations of various stacking of these three positions. The only cubic polytype in SiC is called 3C-SiC, following the Ramsdell notation, which has the stacking sequence ABC. The simplest hexagonal structure is 2H-SiC, which has a stacking sequence AB and the two commercially and application-related important polytypes, 6H-SiC and 4H-SiC, have stacking sequences ABCACB and ABCB, respectively. The number in the Ramsdell notation first determines the number of layers before the sequence repeats itself along the stacking direction and the latter determines the resulting structure of the crystal: C for cubic, H for hexagonal, and R for rhombohedral. Cubic SiC is also referred to as β-SiC that has a zinc blende structure. The noncubic polytypes of SiC are also usually summarized and referred to as α-SiC, which shows a characteristic Wurtzite-type structure [15]. Studies on the epitaxial growth of graphene on SiC have, so far, focused on hexagonal polytypes. The unit cells of the two commonly used SiC hexagonal polytypes 6H-SiC and 4H-SiC are given along with the data for 3C-SiC in Figure 4.6.

When considering the carbon density in SiC, to produce a single graphene layer, the decomposition of approximately three bilayers of SiC is needed. Interestingly, it was shown that the growth of graphene appears to be independent of the polytype (whether 4H- or 6H-SiC) [17].

4.1.2.2   SiC Material Type

4.1.2.2.1   SiC Whiskers
Figure 4.7 shows the ordered graphitic CDC structures obtained by annealing of SiC whiskers for 30 min in vacuum at 1700°C. The small whisker diameter enables direct observation in the transmission electron microscope (TEM). The vacuum-annealed whiskers kept their original shape as shown in scanning electron microscopy (SEM) micrographs. On increasing the vacuum-treatment temperature to 2000°C, the entire whisker transforms into well-ordered graphitic carbon. The same cross-sectional shape as found in the SiC whisker was confirmed by SEM and TEM after complete transformation. Raman spectra (Figure 4.7) obtained from vacuum-treated whiskers at 1700°C and 2000°C verified the ordered graphite structure of the CDC by demonstrating a high-intensity graphitic *G* band. Increasing the vacuum-treatment temperature from 1700°C to 2000°C resulted in a further increase in graphite ordering.

Some large multiwalled CNT-like structures were also observed on the basal plane of the whiskers with a broken tip that has sufficiently large flat area (Figure 4.8). Owing to the irregular shape of the whisker and its relatively small diameter, the shape of the nanotubes was highly distorted and the average wall thickness of the formed structures noticeably exceeded that of CNTs obtained by Kusunoki et al. [19].

4.1.2.2.2   SiC Wafers
Most of the vacuum-decomposition experiments were performed on SiC hexagonal single-crystalline wafers; however, there has been a controversy in structures formed on the Si- and C-terminated

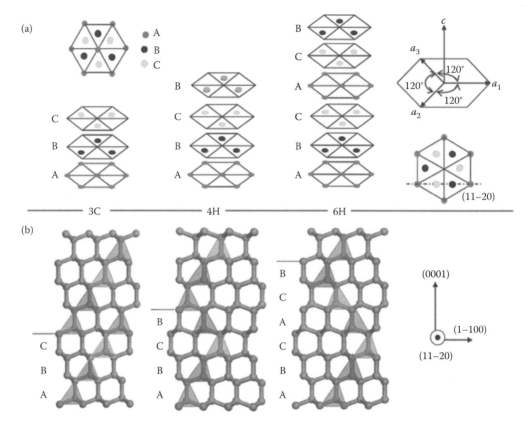

FIGURE 4.6 SiC crystal structure showing (a) the stacking sequence of bilayers along *c*-axis (0001) and (b) the plane of the three most common SiC polytypes. (Cambaz, G.Z., *Formation of Carbide Derived Carbon Coatings on SiC*, in *Materials Science and Engineering*. 2007: Philadelphia, PA: Drexel University.)

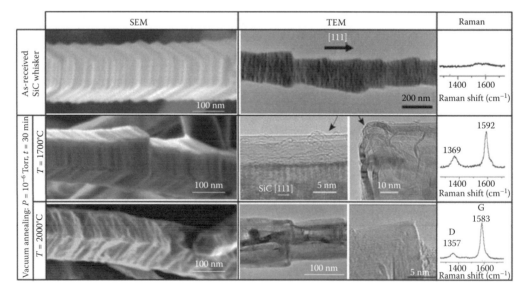

FIGURE 4.7   SEM, TEM, and Raman studies on as-received and vacuum-annealed SiC whiskers.

(a) (b)

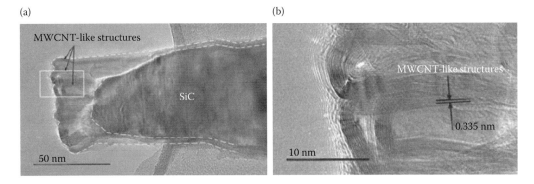

FIGURE 4.8 Tip of a broken SiC whisker annealed in vacuum at 1700°C for 30 min: (a) TEM and (b) HRTEM of the framed section of the whisker demonstrating growing MWCNT-like structures.

face (i.e., the (0001) plane and (000$\bar{1}$) plane, respectively). This crystallographic aspect is discussed in more detail below. Indeed, some results from certain studies could not be reproduced by other groups although the presumably same synthesis conditions had been applied. Cambaz et al. [14] pointed out that, in fact, both graphene and CNT formation are possible on either Si- or C-face depending on the surface morphology and experimental parameters.

4.1.2.3 SiC Substrate Surface Morphology

4.1.2.3.1 Surface Orientation

As mentioned previously, a big difference was observed in the behavior of (0001) and (000$\bar{1}$) lanes and usually, CNTs are observed on the (000$\bar{1}$) plane, whereas graphene is observed on the (0001) plane. Moreover, Kusunoki [20] also studied the decomposition of the (000$\bar{1}$) ($\bar{1}$100) and (11$\bar{2}$0) planes. It was shown that although CNTs formed on the (11$\bar{2}$0) plane, they were disordered; on the ($\bar{1}$100) and (000$\bar{1}$) planes, well-aligned CNTs were formed.

SiC is a polar semiconductor along the c-axis, in that one surface normal to the c-axis is terminated with silicon atoms, whereas the opposite normal c-axis surface is terminated with carbon atoms; these surfaces are referred to as the Si-face (0001) and C-face (000$\bar{1}$), respectively (Figure 4.1) [21]. Although both faces exhibit the same processes (carbon formation) during heating, both the growth of graphene and its structure are intrinsically different for the C-face compared to the Si-face [17,22]. Although for the Si face, the graphene growth proceeds in a nearly layer-by-layer manner; on the C-face, it proceeds in a more three-dimensional (3D) manner [5,23,24].

There are also differences in the carbon structure formed on the two faces. On the Si-face, graphene grows layer by layer epitaxially with an orientational phase rotated 30° relative to SiC, whereas C-face films can have multiple orientational phases [4,25]. Moreover, on the Si-face, small graphene domains (30–100 nm in diameter) are formed, whereas on the C-face, larger domains (≈200 nm) of multilayered graphene are observed [26]. The different decomposition of SiC on the Si- and C-face is related to differences in their surface reconstruction and surface energies, but possibly also to the presence of surface oxides and nanofacetting.

4.1.2.3.1.1 Surface Reconstruction

Before transformation into carbon, the SiC surfaces undergo a number of surface reconstructions that are dependent on the temperature and the crystal polarity (i.e., different processes occur on the Si- or C-terminated face). In the case of the Si-face, heating more than 1050°C results in a well-defined $\sqrt{3} \times \sqrt{3}$ (R phase; further heating more than 1100°C causes a mixture of $6\sqrt{3} \times 6\sqrt{3}$) R and $\sqrt{3} \times \sqrt{3}$ phases to develop. Continued heating more than 1200°C results in a $6\sqrt{3} \times 6\sqrt{3}$ (R pattern with the epitaxial graphene layers forming a well-ordered $6\sqrt{3} \times 6\sqrt{3}$) superstructure [6].

On the C-face, heating SiC results in several different reconstructions: heating at 1050°C for 15 min removes the native oxide layer and gives a (3 × 3) reconstruction [27,28]. Continued heating at 1075°C produces a (2 × 2) C phase in coexistence with the (3 × 3) phase [29]. Graphene grown on the C-face is much more randomly oriented that is indicative of a much weaker substrate influence [26].

4.1.2.3.1.2 Surface Oxides The oxide layers on C- and Si-terminated faces of SiC differ in composition and structure [27,30]. The oxide layer is directly connected to the topmost SiC bilayer by a Si–C bond on $(000\overline{1})$ whereas a linear Si–O–Si bridge makes the contact on (0001). The latter structure leads to a stronger bonding and a slower decomposition rate of the Si-face. Zinovev et al. [31] observed a thicker oxide layer on the Si-face compared to the C-face. X-ray photoelectron spectroscopy (XPS) analysis of SiC wafers detected SiO_xC_y on the C-face and Si-face; however, the oxygen content was higher on the C-face. Therefore, the thermal stability of the surface oxides at elevated temperatures (>1300°C) was reported to be higher on the Si-face compared to the C-face.

4.1.2.3.2 Surface Defects

4.1.2.3.2.1 Roughness Although very high-quality SiC wafers are already state of the art in industrial-scale production, some surface scratches from polishing remain in the highest-grade SiC wafers even after mechanochemical polishing. Cambaz et al. [14] introduced scratches on the polished Si-face of SiC, wafer and it was shown that scratching and related structural deterioration leads to accelerated transformation of SiC on the Si-face to carbon [14] (Figure 4.9).

An effective method to remove surface oxides and residual polishing damage/scratches on SiC substrate is H_2-etching at high temperatures [32–34] at 1500°C for 10–30 min. However, such treatments also change the surface morphology and cause typically significant step bunching [35–37]. This type of microstep formation was explained in terms of etching kinetics and surface energies by Nakamura et al. [35]. It is important to clarify the most stable surface termination on the vicinal surface because the surface energy affects the kinetics (adsorption, desorption, and diffusion) [38].

4.1.2.3.2.2 Surface Steps SiC wafers are generally grown, cut, and polished such that the *c*-axis direction is perpendicular to the surface of the substrate. However, in some cases, for example, to control the defect density, off-axis cuts can be preferable. Miscut tolerances can be as high as ±0.5° off-axis, leaving terraces on the surface that can measure multiple unit cells height. The atomic structure of the step depends on the crystal structure, vicinal angle, and the direction.

The formation of step bunches and/or facets on hydrogen-etched Si- and C-faces of 6H-SiC has been studied by Nie et al. [38], using both nominally on-axis and intentionally miscut (i.e., vicinal) substrates. For nominally on-axis substrates, H_2-etching produced uniformly distributed step-terrace arrays on both Si- and C-faces, as shown in Figure 4.10. The step arrays form because of the miscut of the surface (the average miscut values for the substrates of Figure 4.10a and b are

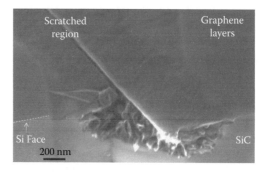

FIGURE 4.9 Accelerated transformation of Si face to carbon at 1700°C induced by a scratch on Si face. (From Cambaz, Z.G. et al., *Carbon*, 2008. **46**(6): 841–849.)

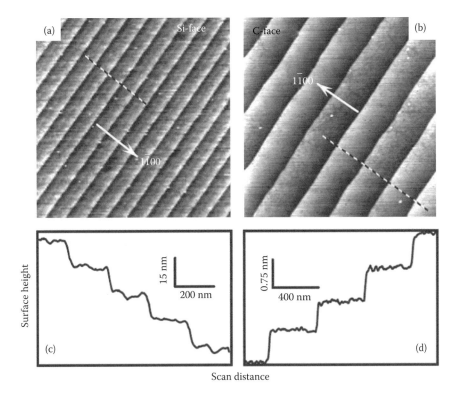

FIGURE 4.10 Atomic force microscopy (AFM) images of nominally on-axis 6H-SiC surfaces: (a) Si-face and (b) C-face. Step edges are normal to $\langle \bar{1}100 \rangle$ directions. Dashed lines indicate the paths of the line scans displayed in (c) and (d), respectively. The images are displayed with gray-scale ranges (height difference between white and black gray levels) of (a) 1.4 nm and (b) 0.9 nm. A linear background subtraction has been applied to the data of (c) and (d) compared to that of (a) and (b). (From Nie, S. et al., *Surface Science*, 2008. 602(17):2936–2942.)

0.28° and 0.18°, respectively [38]). Step edges perpendicular to the [$\bar{1}$100] directions were observed for both faces. Other studies show that these step directions have the lowest surface energy on the (0001) basal plane; in other words, the H$_2$-etching is slowest in these directions [35,39,40]. The observed step height from Figure 4.10 is 1.5 nm (i.e., six bilayers; a full unit-cell height) for the 6H-SiC Si-face whereas it is 0.75 nm (three bilayers; a half unit-cell height) for the 6H-SiC C-face.

TEM studies by Norimatsu et al. [41] and Robinson et al. [42] showed the importance of nano-facets because they are the initial points for graphene formation; this is related to the low number and weaker bonding with silicon atoms [41,42]. Therefore, understanding the dynamic behavior of steps is important for the control of graphene formation on SiC. On the vicinal surfaces, the width of the terraces and the morphology of steps depend on miscut direction and angle. It was shown that depending on whether the miscut angle is toward the [$\bar{1}$100] or [11$\bar{2}$0] direction, the nanofacet makes an angle of ≈25 or 13–14° from the basal plane [38]. Much less facetting is observed on miscut ($000\bar{1}$) surfaces (Figure 4.11).

Many researchers have interpreted these steps as the in-preferred defects. However, by defect engineering, Sprinkle et al. [43] used the facets as templates for bottom-up growth of graphene ribbons for top-gated transistors (Figure 4.12).

4.1.2.3.2.3 Pits and Dislocations   Another localized region, where the step density is higher, is related to the pits that form during heating. Hannon et al. [44] imaged the evolution of surface morphology during annealing above 1200°C using low-energy electron microscopy (LEEM) on the

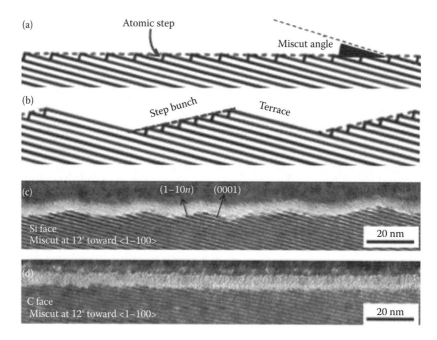

FIGURE 4.11 (a) Stepped vicinal surface consisting of equidistant steps of monolayer height. (b) An alternating array of singular facets (terraces) separated by vicinal facets. (c and d) Cross-sectional TEM images obtained from (c) Si and (d) C-face of 6H-SiC miscut at 12° toward $\langle 1\bar{1}00 \rangle$. In (c), facet orientations of (0001) and $(1\bar{1}0n)$ with $n \approx 12$ are seen on the surface, whereas in C-face, only an average orientation 12° from $(000\bar{1})$ is observed. Both images are acquired with the electron beam directed along $\langle 11\bar{2}0 \rangle$ (c and d are from Nie, S. et al., *Surface Science*, 2008. **602**(17): 2936–2942.)

Si-face. Although the starting surface consisted of a well-ordered array of straight steps, the state of the surface when graphene forms was found to be rough. The steps were no longer straight and deep pits were observed.

In another study, Hite et al. [45] also showed that at the initial stages graphene was formed in localized areas on C-face 6H-SiC substrates. Graphene areas were determined to lie below the level of the surrounding substrate and showed different morphologies based on size. Employing electron-channeling contrast imaging, the presence of threading screw dislocations was indicated near the centers of each of these areas. After graphene was removed, these dislocations were revealed to lie

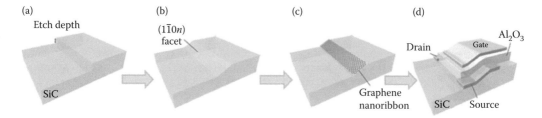

FIGURE 4.12 Process for tailoring of the SiC crystal for selective graphene growth and device fabrication. (a) A nanometer-scale step is etched into SiC crystal by fluorine-based RIE. (b) The crystal is heated at 1200–1300°C (at low vacuum), inducing step flow and relaxation to the $(1\bar{1}0n)$ facet. (c) On further heating at 1450°C, self-organized graphene nanoribbon forms on the facet. (d) Complete device with source and drain contacts, graphene nanoribbon channel, Al_2O_3 gate dielectric, and metal top gate. (From Sprinkle, M. et al., *Nature Nanotechnology*, 2010. **5**(10): 727–731.)

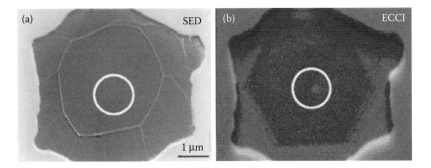

FIGURE 4.13 Secondary electron image (a) and electron-channeling contrast image (b) showing that screw dislocations were near the centers of graphene areas. (From Hite, J.K. et al., *Nanoletters*, 2011. **11**(3): 1190–1194.)

within the SiC substrate. Such hallow-core screw dislocation (often called micropipes) is indeed a common structural defect in SiC wafers. In the light of observations, they suggested that screw dislocations act as preferred nucleation sites for graphene growth on C-face SiC (Figure 4.13).

4.1.2.4　Growth Conditions

4.1.2.4.1　Temperature, Time, and Pressure

The minimum temperature for the formation of graphitic bonds, as determined by *K*-resolved inverse photoelectron spectroscopy (KRIPES) spectra, is approximately the same for graphene grown in ultrahigh vacuum (UHV) on the Si-face or C-face and ranges between 1080°C and 1100°C measured by an infrared pyrometer using an assumed emissivity of 0.9 [25]. However, the data in Figure 4.14 show that graphene starts to form at a significantly lower temperature on the C-face (≈1100°C) compared to the Si-face (≈1250°C) [5,17,24]. XPS, atomic emission spectroscopy (AES), and electron energy loss spectroscopy (EELS) indicated that on the Si-face, the UHV-grown graphene film thickness increases slowly, whereas on the C-face the films grow much faster. Thus, for a given temperature, a thicker graphene film forms on the C-face than on the Si-face [17,24]. It can be seen from Figure 4.14 that about nine monolayers are formed on the C-face at about 1250°C, whereas only one monolayer is formed on the Si-face [24]. Moreover, Hass et al. [17] showed that the number of graphene layers grown on Si-face in UHV is more insensitive to the growth time and depends more strongly on the growth temperature.

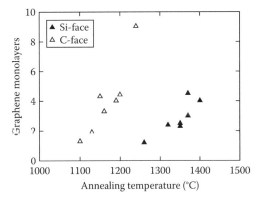

FIGURE 4.14 Graphene thickness as a function of annealing temperature for 6H-SiC {0001}-*r* faces, showing results for C-face (annealing time 20 min) and Si-face (annealing time 40 min). (From Luxmi et al., *Physical Review B*, 2010. **82**(23): 235406.)

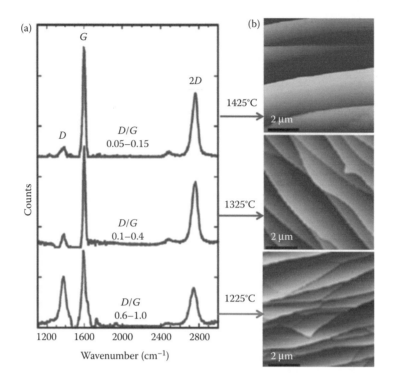

FIGURE 4.15 (a) Raman spectra and (b) AFM of SiC substrates in vacuum annealed at various temperatures. Raman indicates that the *D/G* band ratio increases as growth temperature is decreased suggesting that crystallite size is proportional to growth temperature. (Adapted from Robinson, J. et al., *ACS Nano*, 2010. **4**(1): 153–158.)

Kusunoki et al. [2] also studied the effect of temperature, time, and pressure. SiC wafers were heated at several temperatures (1300°C, 1500°C, and 1700°C) for 0.5 h in a vacuum of 1×10^{-4} Torr, then were observed using TEM. At 1500°C, the length of the CNTs formed was 50 and 180 nm for 0.5 and 6 h, respectively. Moreover, when a wafer was heated at 1500°C for 0.5 h in a better vacuum (1×10^{-6} Torr), the length of the CNTs was only 25 nm; thus, decreasing the pressure results in CNTs with a shorter length.

Similarly, to investigate the graphene formation on SiC, Robinson et al. [42] studied the graphene formation on SiC at different temperatures (1225°C, 1325°C, and 1425°C; pressure $= 1 \times 10^{-6}$ Torr). Under these conditions, graphene exclusively formed at terrace-step edges and other topological defects. A variation in surface coverage as a function of growth temperature was also demonstrated with a lower coverage at high temperatures [42]. Moreover, as the temperature was decreased, the disorder in graphene increased (Figure 4.15).

4.1.2.4.2 Environment

4.1.2.4.2.1 Effect of Oxygen Kusunoki et al. [2] mentioned that the segregation of oxygen was observed at the interface between the CNT film and SiC and this led to the suggestion that residual oxygen in the vacuum chamber might play an important role in the formation of CNTs. Perhaps, that was the reason why CNTs could not be reproduced by other research groups for a long time.

To investigate the effect of oxygen on carbon structures formed on SiC, Cambaz et al. [14] introduced a small amount of oxygen into a vacuum chamber during the thermal treatment of SiC. On the Si-face, a thicker graphite film was formed with wider wrinkles (up to 50 nm) (Figure 4.16a). On the C-face, the transformation was not as conformal as in high vacuum and surface features were not conserved. Moreover, SEM studies of fractured surfaces of the carbon coating

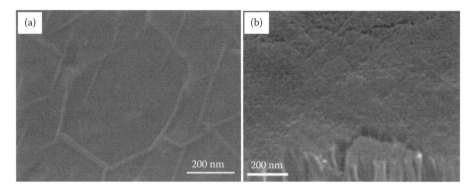

FIGURE 4.16   SEM images of (a) the graphite coating formed on Si face. (b) Dense CNT forest formed on the C-face by vacuum annealing for 4 h at 1900°C in a low vacuum. (From Cambaz, G.Z., *Formation of Carbide Derived Carbon Coatings on SiC*. 2007: Philadelphia, PA: Drexel University.)

showed that the carbon was primarily nanotubular with graphite walls growing normal to the surface (Figure 4.16b). In that study, the crucible and the heating elements were all graphite; therefore, this raised the question of carbon transport via oxidation from the walls of a reaction vessel or crucible.

It is known that there are two modes of oxidation of the SiC surface: (1) passive oxidation, in which SiO_2 layer is formed on SiC surface passivating the surface and (2) active oxidation, in which the silicon-containing products are gaseous, such as Si or SiO [46–50]. Figure 4.17 shows a phase diagram for the interaction of O_2 with SiC surface [50].

Si removal from SiC may follow two possible paths [50,51]: SiC decomposition with carbon as a solid product (if there is no oxidizing species present) and active oxidation, as shown in the following reactions:

$$SiC(s) \rightarrow Si(g) + C(s) \qquad\qquad (R.4.1)$$

$$SiC(s) + \frac{1}{2}O_2(g) \rightarrow SiO(g) + C(s) \qquad\qquad (R.4.2)$$

(with g denoting a gaseous phase and s denoting a solid phase).

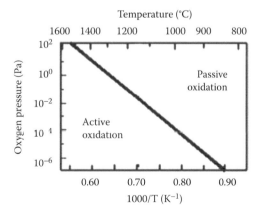

FIGURE 4.17   Phase diagram for the interaction of O_2 with $(000\bar{1})$ H–SiC showing the active–passive boundary. (From Song, Y. and F.W. Smith. *Applied Physics Letters*, 2002. **81**(16): 3061–3063.)

The rate of SiC Reaction 4.1 is faster than that of Reaction 4.2 [50]. Additionally, taking into account that all lining and heating elements inside the vacuum furnace are made of carbon; carbon transport may occur in the presence of oxygen by Reaction 4.3 (Bouduard reaction).

$$2C(s) + O_2(g) \leftrightarrow 2CO(g) \qquad (R.4.3)$$

These results suggest that at higher oxygen pressures, the carbon-formation rate is increased due to the parallel processes of Reactions 4.1 through 4.3 including gas-phase carbon transport.

4.1.2.4.2.2 Effect of Crucible   To investigate the hypothesis that carbon transport occurs from carbon lining, Cambaz [16] placed SiC wafers with the Si face and C face down onto graphite plates, assuming that a short distance (compared to the transport from the crucible or remote furnace heaters) would assist and accelerate the transport.

This experiment produced a much thicker CNT coating on C face (≈9 μm) and the surface features were lost. Moreover, this time, nanotube carpets were formed on the Si-face as well (Figure 4.18c); however, these nanotubes were shorter yet more ordered.

4.1.2.4.2.3 Effect of CO_2   To further corroborate the observed effect of carbon transport through the gas phase, a small amount of CO_2 into the vacuum chamber was introduced through a leak valve [14]. This led to a CNT coating on both C and Si faces (Figure 4.19), where single nanotubes were observed protruding from the surface (Figure 4.19b), suggesting growth from an external carbon source. Thus, carbon deposition by gas-phase transport at a low partial pressure of oxygen contributes to the growth of nanotubes upon decomposition of SiC. Similarly, the addition of CO also leads to a stabilized CNT growth.

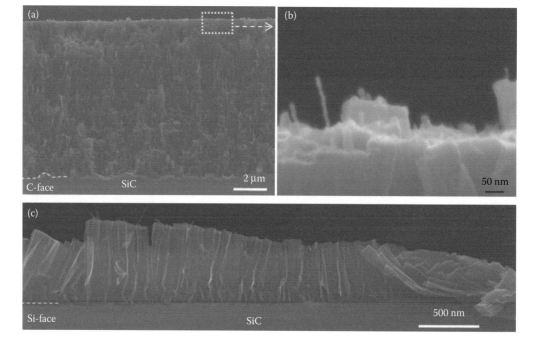

FIGURE 4.18   SEM micrographs of CNTs formed after 4 h at low vacuum (10^{-4} Torr) at 1900°C on the (a,b) C-face and (c) Si-face. Synchronous tilting of nanotubes after scratching the surface shows that their sides are not interconnected, and they are free to slide against each other.

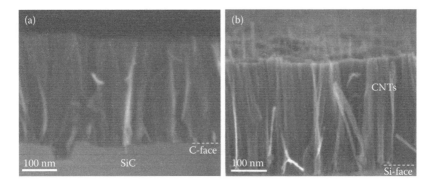

FIGURE 4.19   SEM micrographs of CNTs formed (a) on the C-face with a thickness of 400 nm and (b) on the Si-face with a thickness of 310 nm by annealing for 4 h at 1700°C in CO_2. (From Cambaz, Z.G. et al., *Carbon*, 2008. **46**(6): 841–849.)

4.1.2.4.2.4   Impurities on the Surface   To study the effect of impurities, a SiC wafer was patterned (Figure 4.20a) using organic materials and permanent markers that contain carbon. Surface painting led to an accelerated transformation of SiC on both the Si- and C-face to carbon (Figure 4.20b,c). Moreover, CNT growth was observed on both faces in the region covered by paint. These CNTs are embedded in the substrate with the tubes not rising above the wafer surface and, therefore, may offer significant advantages in device fabrication and printable electronics.

4.1.2.4.2.5   Presence of Inert Gas   Surface-limited reactions are sensitive to the ease of removal of volatile products that may be hampered by the presence of an inert gas. Emtsev et al. [52] showed that the growth of epitaxial graphene on SiC-(0001) in an Ar atmosphere close to atmospheric pressure provides morphologically superior graphene layers in comparison with vacuum graphitization. The key factor in achieving an improved growth is the significantly higher annealing temperature of 1650°C that is attainable for graphene formation under argon at a pressure of 900 mbar as compared with 1280°C in UHV.

For a given temperature, the presence of a high pressure of Ar leads to a reduced Si evaporation rate because the silicon atoms desorbing from the surface have a finite probability of being reflected back to the surface by collision with Ar atoms. In an Ar atmosphere, no sublimation of Si from the surface is observed at temperatures up to 1500°C, whereas Si desorption commences at ≈1100°C in vacuum. The significantly higher growth temperature thus attained results in an enhancement of surface diffusion such that the restructuring of the surface is completed before graphene is formed. Ultimately, this leads to a significantly improved surface morphology.

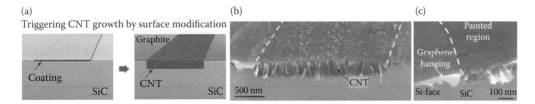

FIGURE 4.20   (a) Coating with a fluxing agent leads to localized formation of CNTs on the (b) C-face and (c) the Si-face of SiC. SEM micrograph showing that CNTs grow on the patterned areas, whereas graphite covers the remaining surface. Wafers were treated at 1700°C. (From Cambaz, Z.G. et al., *Carbon*, 2008. **46**(6): 841–849.)

4.1.3 CNT and Graphene-Formation Mechanism on SiC

Explanation of all the details is not easily verified because *in situ* investigation of the aforementioned processes is hard and limited at high temperatures. However, imaging the produced features on the surface with atomic-scale resolution (using AFM, SPM, and HRTEM) provided considerable insight into the growth mechanism of carbon on SiC. Studies showed that the resultant carbon structure strongly depends on the nucleation rate of carbon at the initial stages. If the nucleation rate is high, nonuniform nucleation takes place as islands on the terraces on the (0001)-surface. This mostly results in disordered thick graphitic structures. However, if there is an extra carbon supply present, whether by CO, CO_2, or carbon transport from the crucible walls through oxidation, these islands become curved forming distinct caps that turn into CNTs as the reaction proceeds.

To form uniform graphene, the nucleation rate should be small and this can be achieved by pretreatment. For the H_2-etched wafers, Kusunoki and coworkers [41] and Robinson et al. [42] proposed a very similar graphene growth mechanism (Figure 4.21) based on HRTEM results. According to the observations, the initial stages of the process can be summarized with the following subsequent steps:

a. Silicon atoms at the step are preferentially sublimed.
b. Multilayer graphene nucleates at the step edge.
c. The outermost graphene layer grows over the terrace region making the formation of single-layer graphene possible.
d. Graphene grows on other terraces.
e. Coalescence in the vicinity of the step.

The quality of graphene can be improved by annealing at higher temperature. However, in that case, to delay the reaction, the purge of inert gases such as Ar is needed.

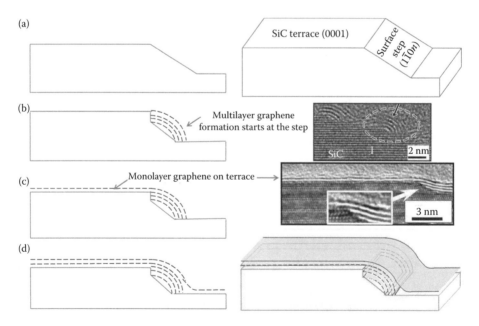

FIGURE 4.21 Schematic illustration of the early stages of graphene formation. (a) Silicon atoms at the step are preferentially sublimed, (b) multilayer graphene nucleates at the edge step, (c) the outermost graphene layer grows over the terrace region making the formation of single-layer graphene possible, and (d) coalescence in the vicinity of step. (HRTEM images are adapted from Norimatsu, W. and M. Kusunoki, *Physica E-Low-Dimensional Systems and Nanostructures*, 2010. **42**(4): 691–694; Robinson, J. et al., *ACS Nano*, 2010. **4**(1): 153–158.)

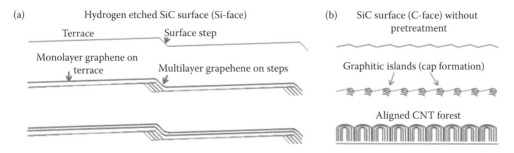

FIGURE 4.22 (a) Graphene and (b) CNT formation on SiC depending on the pretreatment.

4.2 SUMMARY

Annealing SiC single crystals in a vacuum furnace leads to the sublimation of silicon atoms from the surface and the subsequent formation of CNT or graphene layers by the remaining carbon atoms on the surface. For the controlled growth of the carbon structures (Figure 4.22) with high quality, factors affecting the process and the details of the formation process were covered. The nucleation and growth of monolayer graphene on SiC are highly dependent on the SiC surface morphology (surface orientation, defects) and growth conditions (temperature, time, pressure, and environmental gases). Moreover, pretreatments such as hydrogen etching or annealing at high temperature result in the transformation of the surface into a 3D surface with terraces exhibiting (0001) and ($1\bar{1}0n$) planes. Although the (0001) surface exhibits uniform bonding of all silicon surface atoms, many of the Si atoms along the ($1\bar{1}0n$) plane exhibit dangling bonds [35]. The formation of graphene is thought to begin at these step edges [28,42,52–55]. Having fewer steps along well-defined crystallographic directions reduces the nucleation density of multilayer graphene and, thus, single-layer graphene with a high quality may be obtained by few stepped, large-terraced SiC under Ar flow at very high temperature. Moreover, the graphene nanoribbons with controlled thickness can be produced by controlling the step width.

ACKNOWLEDGMENT

The author would like to thank Professor Tarık Oğurtanı and Assistant Professor Ersin Emre Ören for the valuable discussions.

REFERENCES

1. Berger, C. et al., Electronic confinement and coherence in patterned epitaxial graphene. *Science*, 2006. **312**(5777): 1191–1196.
2. Kusunoki, M. et al., Formation of self-aligned carbon nanotube films by surface decomposition of silicon carbide. *Philosophical Magazine Letters*, 1999. **79**(4): 153–161.
3. Badami, D.V., Graphitization of α-silicon carbide. *Nature*, 1962. **193**: 569.
4. Van Bommel, A.J., J. E. Crombeen, and A.V. Tooren, LEED and Auger electron observations of the SiC(0001) surface. *Surface Science*, 1975. **48**: 463–472.
5. Muehlhoff, L. et al., Comparative electron spectroscopic studies of surface segregation on SiC (0001) and SiC (0001). *Journal of Applied Physics*, 1986. **60**: 2842.
6. Forbeaux, I., J.M. Themlin, and J.M. Debever, Heteroepitaxial graphite on 6H-SiC(0001): Interface formation through conduction-band electronic structure. *Physical Review B*, 1998. **58**(24): 16396–16406.
7. Charrier, A. et al., Solid-state decomposition of silicon carbide for growing ultra-thin heteroepitaxial graphite films. *Journal of Applied Physics*, 2002. **92**(5): 2479–2484.
8. Kusunoki, M., M. Rokkaku, and T. Suzuki, Epitaxial carbon nanotube film self-organized by sublimation decomposition of silicon carbide. *Applied Physics Letters*, 1997. **71**(18): 2620–2622.
9. Kusunoki, M. et al., Aligned carbon nanotube film self-organized on a SiC wafer. *Japanese Journal of Applied Physics Part 2-Letters and Express Letters*, 1998. **37**(5B): L605–L606.

10. Kusunoki, M. et al., A formation mechanism of carbon nanotube films on SiC(0001). *Applied Physics Letters*, 2000. **77**(4): 531–533.

11. Watanabe, H. et al., *In situ* observation of the initial growth process of carbon nanotubes by time resolved high resolution transmission electron microscopy. *Journal of Microscopy*, 2001. **203**(1): 40–46.

12. Novoselov, K.S. et al., Electric field effect in atomically thin carbon films. *Science*, 2004. **306**(5696): 666–669.

13. Berger, C. et al., Ultrathin epitaxial graphite: 2D electron gas properties and a route toward graphene-based nanoelectronics. *Journal of Physical Chemistry B*, 2004. **108**(52): 19912–19916.

14. Cambaz, Z.G. et al., Noncatalytic synthesis of carbon nanotubes, graphene and graphite on SiC. *Carbon*, 2008. **46**(6): 841–849.

15. Saddow, S.E. and A. Agarwal, *Advances in Silicon Carbide Processing and Applications*. 2004: Norwood: Artech House Publishers.

16. Cambaz, G.Z., *Formation of Carbide Derived Carbon Coatings on SiC*, in *Materials Science and Engineering*. 2007: Philadelphia, PA: Drexel University.

17. Hass, J., W.A. de Heer, and E.H. Conrad, The growth and morphology of epitaxial multilayer graphene. *Journal of Physics-Condensed Matter*, 2008. **20**(32): 323202.

18. Cambaz, Z.G. et al., Formation of carbide-derived carbon on beta-silicon carbide whiskers. *Journal of the American Ceramic Society*, 2006. **89**(2): 509–514.

19. Kusunoki, M. et al., Selective synthesis of zigzag-type aligned carbon nanotubes on SiC (000-1) wafers. *Chemical Physics Letters*, 2002. **366**(5–6): 458–462.

20. Kusunoki, M., High-density and well-aligned carbon nanotube films on silicon-carbide wafers. *Journal of the Ceramic Society of Japan*, 2005. **113**(1322): 637–641.

21. Powell, J.A., P. Pirouz, and W.J. Choyke, Growth and characterization of silicon carbide polytypes for electronic applications, in *Semiconductor Interfaces, Microstructures, and Devices: Properties and Applications*, Z.C. Feng, ed. 1993: UK: Institute of Physics Publishing: p. 257.

22. Srivastava, N. et al., Graphene formed on SiC under various environments: Comparison of Si-face and C-face. *Journal of Physics D-Applied Physics*, 2012. **45**(15): 154001.

23. Luxmi et al., Morphology of graphene on SiC(000$\bar{1}$) surfaces. *Applied Physics Letters*, 2009. **95**(7): 3207757.

24. Luxmi et al., Comparison of graphene formation on C-face and Si-face SiC {0001} surfaces. *Physical Review B*, 2010. **82**(23): 235406.

25. Forbeaux, I. et al., Solid-state graphitization mechanisms of silicon carbide 6H-SiC polar faces. *Applied Surface Science*, 2000. **162**: 406–412.

26. Hass, J. et al., Highly ordered graphene for two dimensional electronics. *Applied Physics Letters*, 2006. **89**(14): 143106.

27. Bernhardt, J. et al., Epitaxially ideal oxide–semiconductor interfaces: Silicate adlayers on hexagonal (0001) and (000$\bar{1}$)over-bar) SiC surfaces. *Applied Physics Letters*, 1999. **74**(8): 1084–1086.

28. Johansson, L.I., P.A. Glans, and N. Hellgren, A core level and valence band photoemission study of 6H-SiC(000(1)over-bar). *Surface Science*, 1998. **405**(2–3): 288–297.

29. Bernhardt, J. et al., Stable surface reconstructions on 6H-SiC(0001). *Materials Science and Engineering B-Solid State Materials for Advanced Technology*, 1999. **61–2**: 207–211.

30. Nagano, T., Y. Ishikawa, and N. Shibata, Effects of surface oxides of SiC on carbon nanotube formation by surface decomposition. *Japanese Journal of Applied Physics Part 1-Regular Papers Short Notes and Review Papers*, 2003. **42**(3): 1380–1385.

31. Zinovev, A.V. et al., Etching of hexagonal SiC surfaces in chlorine-containing gas media at ambient pressure. *Surface Science*, 2006. **600**(11): 2242–2251.

32. Berger, C. et al., Electronic confinement and coherence in patterned epitaxial graphene. *Science*, 2006. **312**(5777): 1191–1196.

33. de Heer, W.A. et al., Epitaxial graphene. *Solid State Communications*, 2007. **143**(1–2): 92–100.

34. Emtsev, K.V. et al., Towards wafer-size graphene layers by atmospheric pressure graphitization of silicon carbide. *Nature Materials*, 2009. **8**(3): 203–207.

35. Nakamura, S. et al., Formation of periodic steps with a unit-cell height on 6H-SiC (0001) surface by HCl etching. *Applied Physics Letters*, 2000. **76**(23): 3412–3414.

36. Fujii, M. and S. Tanaka, Ordering distance of surface nanofacets on vicinal 4H-SiC(0001). *Physical Review Letters*, 2007. **99**(1): 016102.

37. Hayashi, K. et al., Stable surface termination on vicinal 6H-SiC(0001) surfaces. *Surface Science*, 2009. **603**(3): 566–570.

38. Nie, S. et al., Step formation on hydrogen-etched 6H-SiC{0001} surfaces. *Surface Science*, 2008. **602**(17): 2936–2942.

39. Wulfhekel, W. et al., Regular step formation on concave-shaped surfaces on 6H-SiC(0001). *Surface Science*, 2004. **550**(1–3): 8–14.

40. Nakajima, A. et al., Step control of vicinal 6H-SiC(0001) surface by H$_2$ etching. *Journal of Applied Physics*, 2005. **97**(10): 104919.

41. Norimatsu, W. and M. Kusunoki, Formation process of graphene on SiC (0001). *Physica E-Low-Dimensional Systems and Nanostructures*, 2010. **42**(4): 691–694.

42. Robinson, J. et al., Nucleation of epitaxial graphene on SiC(0001). *ACS Nano*, 2010. **4**(1): 153–158.

43. Sprinkle, M. et al., Scalable templated growth of graphene nanoribbons on SiC. *Nature Nanotechnology*, 2010. **5**(10): 727–731.

44. Hannon, J.B. and R.M. Tromp, Pit formation during graphene synthesis on SiC(0001): *In situ* electron microscopy. *Physical Review B*, 2008. **77**(24): 241404.

45. Hite, J.K. et al., Epitaxial graphene nucleation on C-face silicon carbide. *Nano Letters*, 2011. **11**(3): 1190–1194.

46. Luthra, K.L., Some new perspectives on oxidation of silicon carbide and silicon nitride. *Journal of the American Ceramic Society*, 1991. **74**(5): 1095–1103.

47. Balat, M. et al., Active to passive transition in the oxidation of silicon carbide at high temperature and low pressure in molecular and atomic oxygen. *Journal of Materials Science*, 1992. **27**(3): 697–703.

48. Schneider, B. et al., A theoretical and experimental approach to the active-to-passive transition in the oxidation of silicon carbide—Experiments at high temperatures and low total pressures. *Journal of Materials Science*, 1998. **33**(2): 535–547.

49. Goto, T. and H. Homma, High-temperature active/passive oxidation and bubble formation of CVD SiC in O$_2$ and CO$_2$ atmospheres. *Journal of the European Ceramic Society*, 2002. **22**(14–15): 2749–2756.

50. Song, Y. and F.W. Smith, Phase diagram for the interaction of oxygen with SiC. *Applied Physics Letters*, 2002. **81**(16): 3061–3063.

51. Maruyama, T. et al., STM and XPS studies of early stages of carbon nanotube growth by surface decomposition of 6H-SiC(000-1) under various oxygen pressures. *Diamond and Related Materials*, 2007. **16**(4–7): 1078–1081.

52. Emtsev, K.V. et al., Towards wafer-size graphene layers by atmospheric pressure graphitization of silicon carbide. *Nature Materials*, 2009. **8**(3): 203–207.

53. Hupalo, M., E.H. Conrad, and M.C. Tringides, Growth mechanism for epitaxial graphene on vicinal 6H-SiC(0001) surfaces: A scanning tunneling microscopy study. *Physical Review B*, 2009. **80**(4): 041401.

54. Poon, S.W. et al., Probing epitaxial growth of graphene on silicon carbide by metal decoration. *Applied Physics Letters*, 2008. **92**(10): 104102.

55. Virojanadara, C. et al., Substrate orientation: A way towards higher quality monolayer graphene growth on 6H-SiC(0001). *Surface Science*, 2009. **603**(15): L87–L90.

5 Cooperative Interaction, Crystallization, and Properties of Polymer–Carbon Nanotube Nanocomposites

Eric D. Laird, Matthew A. Hood, and Christopher Y. Li

CONTENTS

5.1 INTRODUCTION

Polymer nanocomposites have been an area of active research for the past 20 years. The field is relatively young, essentially begun during the late 1980s, with the discovery by Toyota Research and Development Labs [1,2] that some important properties of polymers could be improved even as their total material cost was reduced by introducing clays as additives. As such, understanding of the interplay between nanoparticle (NP) and matrix has advanced rapidly within a short period of time. The consequences of this have been threefold: (1) nanocomposites have made considerable advances beyond laboratory benchtops to be adopted in industry for industrial and consumer goods; (2) there has been a substantial push to improve specific properties of bulk materials, sometimes at the expense of real structure/property understanding; and (3) the imperative to gain a competitive edge in this field has allowed basic science to regain the upper hand. Understanding the role of filler materials in improving the properties of matrix materials is now crucial to advancing the frontier in nanocomposite technology. This can be done primarily through studying the chemistry of the interface and the unique properties found at the interphase region of the polymer matrix.

Of all potential fillers for polymer nanocomposites, carbon nanotubes (CNTs) would seem to have natural advantages over all other alternative materials (see Figure 5.1). They have extremely high-aspect ratios (with length-to-diameter ratio ranging from single digits up to tens or even hundreds of millions) [3,4], low density (≈ 1.0–1.7 g/cm³) [5], and a relatively high specific surface area (≈ 100–1300 m²/g) [6]. Their high-aspect ratio combined with their theoretical Young's modulus on the order of several TPa [7–10] and ultimate tensile strength around 100 GPa [11,12] has earned them a reputation as "the most perfect fiber that has ever been fabricated" [13]. It is without question that they have a huge range of possible properties: CNTs can be open-ended or capped, rigid or semiflexible, single- or multiwalled, almost defect free, or only slightly more structured than soot. They come with or without impurities, and with or without functional groups. They can be either hydrophilic or hydrophobic, metallic or semiconducting. With this variety of properties that can be obtained, CNTs are suitable for an incredibly wide range of applications. However, the wide range of properties found in CNTs also complicates the problem of characterization and experimental repeatability. In this chapter, the problems inherent in CNTs as nanofillers will be addressed, with particular attention devoted to the question of polymer structures and interfaces.

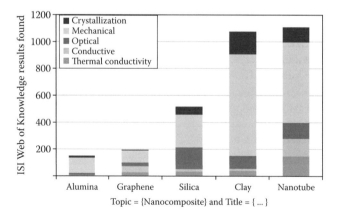

FIGURE 5.1 Number of listed results as of June 2012 for ISI Web of Knowledge searches containing the topic strings shown. "Nanotube" title search string ranks either first or second for all applications listed (clay shows the largest number of listed results for mechanical nanocomposites and crystallization; silica ranks first for optical nanocomposites).

5.2 CNTs AND CNT–POLYMER HYBRIDS

5.2.1 DIMENSIONALITY IN CARBON NANOMATERIALS

As fibers, CNTs can be considered one-dimensional molecules. Their radius virtually reduces to zero, whereas their lengths are typically on the order of micrometers. A useful simplification describes CNTs as hybrids of two other low-dimensional molecules: graphene sheets (2D) are rolled up into tubes and capped at either end with fullerene (0D) hemispheres. It is worth considering some of the properties of these materials in detail to better understand certain properties of CNTs.

The so-called "zero-dimensional" carbon allotropes include the family of spherical fullerenes as well as nanodiamonds [14]. Spherical fullerenes with one layer are commonly called buckyballs: cage-like Archimedean solids comprised purely of sp^2 carbon atoms [15]. When multiple concentric layers are present, the material is, depending on structural parameters, termed nano onion, carbon onion, or onion-like carbon [16,17]. The most common form of fullerenes is the C_{60} molecule, although both lower and higher order fullerenes exist as well. Geometric constraints require that to form a regular shape any fullerene must contain 12 pentagonal rings (a hemisphere such as a CNT cap contains six) [18]. Carbons in a C_{60} molecule form the corners of a truncated icosahedron [19,20]. They are as close to perfect spheres as it is possible to achieve at the molecular level, so in forming a minimal surface in low-pressure unreactive atmosphere [21] this is the favored structure [22]. The angle strain associated with the surface curvature, the fully conjugated structure, and the participation in five-membered rings contribute to remarkable electronic properties: C_{60} can support up to six reversible solution reduction levels [23], and superconductivity has been achieved by doping C_{60} with alkali metals [24]. Another consequence of high angle strain and conjugated double bonding is that simple chemical reactions can form functionalized versions of C_{60}, like the [6,6]-phenyl-C_{61}-butyric acid methyl ester molecule frequently studied for bulk heterojunction solar cell applications [25]. C_{20} [26] and C_{36} molecules can also be produced, but high reactivities limit practical applications [27]. On the higher end, C_{70} is the second most common fullerene in typical synthesis techniques.

Acting as the CNT sidewall material, graphene [28,29] is considered 2D, despite its high degree of waviness and frequent curvature and folds. Along with graphene oxide (GO) [30–32] and reduced graphene oxide (RGO) [33–37], this family of exfoliated graphite sheets are considered as 2D carbons. Both mechanical and chemical exfoliation have been demonstrated ("top-down"), as well as direct synthesis of graphene ("bottom-up").

Combining these zero- and two-dimensional allotropes, one returns to the one-dimensional structure of the carbon nanotube (CNT) which will be the main focus of this chapter. As mentioned previously, CNTs can be visualized as rolled-up sheets of graphene capped at either end with a buckyball hemisphere. Nanotubes with single-layer graphene shells are called single-walled nanotubes (SWCNTs), whereas multiple layers nested concentrically like a matryoshka doll are called multiwalled nanotubes (MWCNTs). A separate classification is sometimes made for double-walled CNTs (DWCNTs). The discovery of MWCNTs is commonly misattributed to Iijima et al. in 1991 [41]. However, there is much evidence to suggest that several other groups made the discovery independently before this dating back at least to the 1950s (see Figure 5.2a–c) [38,39,42]. SWCNTs have an electronic structure that can be metallic if the chiral vector (chirality) is (n,n). If the chiral vector is $(n,n + 3j)$ for integer values of j, the nanotube becomes a small-band gap (E_g) semiconductor and shows a dependency of the diameter, D, with $E_g \propto 1/D^2$. For all other cases, the nanotube behaves as a semiconductor with band gap of $E_g \propto 1/D$ [43,44].

5.2.2 OVERVIEW OF PRISTINE NANOTUBES

CNTs contain many types of defects and impurities [45]. These may include (but are not limited to) NP catalyst impurities, other carbon allotrope impurities, oxygen-containing functional groups, carbon vacancies, Stone–Wales defects, and other kinds of odd-numbered rings in sidewalls causing

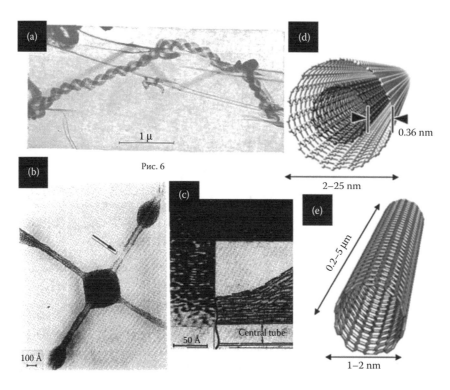

FIGURE 5.2 Classical observations of MWCNTs and schematic representations of MWCNT and SWCNT. (a) First known TEM observation of MWCNTs published in 1952 in the Soviet *Journal of Physical Chemistry* [38]. (b) Filamentous CNT grown by decomposition of benzene and published in the *Journal of Crystal Growth* in 1976 [39]. Electron diffraction confirmed the structures as MWCNTs. (c) High-resolution TEM image of some of the fringes of a branching MWCNT from the same article as in (b). Inset shows a hand-drawn depiction of the proposed structure. (d) Schematic representation of a simplified MWCNT with important dimensions noted [40] (From S. Iijima, *Physica B: Condensed Matter*, 323, 2002, 1–5.). (e) Schematic representation of an SWCNT (From A. Hirsch, *Angew. Chem. Int. Ed.*, 41, 2002, 1853–1859.).

kinking, and coiling/deflection of the tube axis. For a detailed review on the various synthesis methods, see Refs. [43,45–47]. It should be noted that different synthesis methods contribute different types of impurities and defects which affect overall CNT nanocomposite properties. As one example, CNTs can be produced by laser ablation of a graphite target in a horizontal flow chamber under inert atmosphere [48]. This production method results mostly in SWCNTs with a fairly high density of point defects and catalyst impurities. Arc discharge has been used between conducting spheres to produce high-quality nanotubes [41] (both SWCNT and MWCNT can be produced by this method). The main drawback to this technique is low yield: only small amounts of CNTs can be produced for each pair of charge accumulators. Non-nanotube carbon allotropes are also commonly found as a by-product in arc-discharge production. Chemical vapor deposition (CVD) can also produce high-quality CNT and with a significantly higher yield than arc discharge [49], but NP impurities are common by-products. Acid treatment is a suitable method to remove NPs, but not all NPs can be removed in this manner. In particular, some NP impurities can be coated with a graphitic shell which presents an added challenge for purification. One variation on this technique is the high-pressure carbon monoxide (HiPCO) technique [50], which is popular for SWCNT production, and cobalt–molybdenum catalyst (CoMoCat) [51], another SWCNT production method that produces rigid tubes. One appealing extension to the traditional CVD growth technique for CNTs is the water-assisted supergrowth method developed at the Nanotube Research Center in Japan [52–54]. A pilot program for industry-scale production at this facility is able to rapidly produce 0.5 m² SWCNT and DWCNT mats with controllable tube density.

The high surface free energy of CNTs is often a problem as it leads to the formation of tight bundles of CNTs [55–58]. Owing to their extraordinary length, the relatively weak van der Waals interaction between CNTs becomes an important problem, because it results in a cumulative binding energy that can be large because of the extent of contact that is formed between neighboring CNTs. This energy is estimated to be on the order of 180–480 meV/Å depending on the diameter of the tubes [57,59]. To minimize surface free energy, CNTs pack into tight twine-like bundles [60]. In addition to van der Waals interactions, π–π stacking also contributes to robust binding in CNT bundles. This issue is even more pronounced in the stacking of graphene layers.

5.2.3 Carbon Nanotube Processing

5.2.3.1 Purification of Carbon Nanotubes

Purification methods most commonly involve acid treatments and/or high-temperature annealing treatments. A typical acid functionalization might involve sonication in 1:3 HNO_3:H_2SO_4 concentrated acid solution for 1 h, followed by filtration and multiple rinsing steps using deionized water [61]. Acid sonication is fairly destructive, and reflux may be used as a gentler alternative [62]. The purpose of acid functionalization treatments is to remove NP catalysts, but additional functionality can be imparted by acid treatments as well. The aforementioned acid mixture produces nanotubes with mostly carboxylic acid functional groups. In general, acid treatments tend to preferentially functionalize end caps at the five-membered rings and sidewall defects of CNTs, which are sensitive to attack by acids because of bond angle strain. Long acid treatment results in open-ended nanotubes and can potentially disrupt the sidewall structure or shorten CNTs which actually is sometimes desirable. For instance, peapod-like structures can be fabricated using buckyballs or NPs loaded into tubes, and it is known that small molecule impurities and polymers will explore the interiors of CNTs [63].

Adjusting the purification recipe can produce CNTs with different surface functional groups. Acid treatments can make CNTs soluble in water, allow nucleic acid or NP attachment, or prepare initiator or catalyst sites for later polymerization reactions. Both grafting-to and grafting-from approaches have been used to fabricate CNTs with covalently attached polymer chains [64]. Grafting-from techniques, in particular, can be effective for the preparation of isotropic nanocomposites. If open-ended tubes are not desired, or if defects dominate the CNT lattice, high-temperature annealing in an inert atmosphere can be used to improve the structure. This so-called graphitization step is effective because CNTs have an extremely high degradation temperature. CNTs can survive up to 900°C in air, and as high as 3000°C in nitrogen [65]. Thus, high temperatures can provide the activation energy needed to improve the aromaticity of the sidewalls and end caps. High temperatures will also degrade organic impurities and evaporate small molecules. When combined with acid treatment, graphitization can result in a low impurity, low defect concentration CNTs that follow theoretical models for specific diameters and chiralities as well as for bundles and crosslinks [66].

5.2.3.2 Dissolution of CNTs

CNTs are almost completely insoluble in organic solvents. *N*-methyl-pyrrolidone has been shown to be one of the few solvents capable of spontaneously exfoliating nanotubes, and even then only at low concentrations [67]. Other nonhydrogen bonding Lewis bases have also been shown to be fair dispersants for CNTs, with dimethylformamide and a few other amide solvents in particular having shown good dispersion at low CNT concentrations [68]. Most CNT dispersants require vigorous and sometimes destructive techniques to produce metastable sols. These techniques include mechanical stirring and sonication. Both techniques can introduce defects and can shorten the length of CNTs. Sonication can also ablate layers from MWCNTs and damage CNT sidewalls. Other solvents have also been shown to be able to produce mostly bundles with some individualized CNTs, such as xylenes and halogenated benzenes, and *ortho*-dichlorobenzene (DCB) in particular has a high solubility limit for CNTs. Cheng et al. [69] showed that DCB was able to support the highest concentration of CNTs in solution of any of the solvents they tested. They found that the majority of DCB solution CNTs participated in bundles, whereas other solvents may have lower solubilities but most CNTs in these solutions were exfoliated. Therefore, centrifugation is generally required for DCB to remove bundles that remain after sonication [70]. In DCB, free radicals formed during sonication lead to ring-opening sonopolymerization of halogenated benzenes. DCB sonopolymer is frequently an unexpected by-product which can be suppressed using small amounts of alcohol [71]. As these solvents do not form stable suspensions on their own, individualized CNTs produced by sonication must be readied for processing quickly after.

Ultracentrifugation in the presence of surfactants can also produce metastable dispersions and even separate CNTs by size or species [70,72], but long centrifugation times at very high rotation rates are required. The yield from this method is also very low. Three-phase systems, such as water and single-stranded DNA (ssDNA) or polypeptides, can also individualize CNTs [73]. Surfactants such as Triton X-100 or dodecylsulfate salt are also able to adjust hydrophilicity of CNTs for solution in water and such dispersions have reasonably good stability [74]. In the absence of surfactants, evaporation of the solvent phase must be discouraged if reaggregation is to be prevented. Solvent annealing of CNTs results in bundles that are extremely robust [53]. In some cases, this can be beneficial; however, for polymer nanocomposites it is seldom acceptable to have a large percentage of the CNTs in the sample participate in bundles, as bundled CNTs do not contribute to efficient mechanical or electrical percolation.

5.2.3.3 Physical Functionalization of CNTs

"Physical functionalization" of CNTs refers to noncovalent adhesion of an intermediary molecule to the CNT sidewall, hindering aggregation. This intermediary can be a small molecule such as an anionic surfactant [72,75], a cholesterol-derived cation [76], chiral porphyrin derivatives [77], an oligomeric surfactant [78], or a zwitterion [79]. Surfactants are amphiphilic molecules: they have both hydrophobic and hydrophilic regions, allowing them to stabilize nonpolar CNTs in polar media. Surfactants are able to compatibilize CNTs with water, making processing simpler and safer [80]. One major downside to stabilizing CNTs with surfactants is that the latter can be difficult to rinse off after they have exhausted their usefulness. Surfactants affect bulk conductivity and typically weaken the interaction between CNTs and the polymer matrix [81]. They can also lead to a phase separation and crystallize to form impurity domains in a nanocomposite system. Polymer wrapping has also been used to functionalize CNTs [73,82,83]. In general, this is not as simple as functionalization of CNTs via an additional surfactant phase, but it has the advantage of not requiring removal of such surfactants. Several techniques have been demonstrated. Melt spinning [84], bulk casting [85], electrospinning [86–89], and interfacial polymerization [90] are some of the techniques that have been used to produce polymer composites based on noncovalent CNT functionalization.

A unique case of physical functionalization is ssDNA wrapping of CNTs with chemical functionalization consequences. Although the ssDNA wrapping of CNT has the essential features of physical functionalization, ssDNA introduces an unpaired electron every 0.7 nm. The helical structure of the ssDNA chain combined with the electron-rich sequence in close proximity to the CNT results in either broken symmetry [91] or a higher order repeat unit which is a unique and periodic superstructure composed from the substituent materials [92]. The CNT-ssDNA superstructure has different chemistry and allowed phonon modes. The result is that ssDNA-wrapped CNTs do not match the description of physical functionalization in terms of leaving the conjugated structure unperturbed, as would be expected for a physical wrapping of the CNT by polymers and biomacromolecules that do not form such complexes. However, this technique does have major advantages. CNT-ssDNA complex allowed CNTs to be dispersed into aqueous solution. It was further demonstrated that the $d(GT)_n$: $n = 10$–45 sequence self-assembled around individual CNTs such that the electrostatic properties of the hybrid were amplified based on the tube type, enabling CNT species fractionation by anion-exchange chromatography [73].

Biopolymers have also been used to wrap CNTs in efforts to functionalize them for aqueous dispersion. Chitosan, poly(L-lactide) [93], and cellulose [94] are some of the polymers that are commonly employed. Chitosan in particular has been the subject of considerable research because of the ease with which it is functionalized. For instance, enzyme attachment has been used to produce robust chemical sensors for glucose [95,96]. Layered arrangement of chitosan/CNT composite produced a low-power requirement electrochemical actuator with superior mechanical properties and repeatability [97].

5.2.3.4 Chemical Functionalization of CNTs

Although physical functionalization methods are typically thought of as employing steric hindrance and chemical functionalization methods employ more active repulsion, in fact both physical and chemical functionalization techniques are able to keep nanotubes separate using both physical barriers and electrostatic interactions. Covalent functionalization techniques have advantages as well as disadvantages. On the one hand, covalent attachment allows excellent incorporation and load transfer into the matrix polymer, and, therefore, can produce nanocomposites with an exceptionally high strength relative to most nanocomposites that make use of physically functionalized or kinetically trapped CNTs. Chemical modification is a possible way to obtain composites with a high volume fraction of nanotubes due to CNT shortening and increased solubility. Some chemical modification techniques can also be suitable as techniques for species differentiation: etching in piranha solution (a 4:1 solution of 96% H_2SO_4/30% H_2O_2 was used) resulted in cut tubes, and selectively dissolved narrow diameter tubes [98]. On the other hand, covalent functionalization of CNTs causes diminishment of properties for the CNTs. Modified CNTs generally suffer from some combination of reduced mechanical strength, reduced length relative to pristine CNTs, and loss of electrical conductivity [99]. As defect sites disrupt the aromaticity and scatter phonons, structural instability, and lowered electrical and thermal conductivity often result from covalent attachments [100]. Thorough reviews on the subject of chemical functionalization can be found [101–105], but a few examples will be included here owing to the importance and the broad ranging consequences of this set of techniques.

Among covalent approaches to CNT modification, processing by oxidation in acidic solutions is the most common [64], but many other methods have been used, including peroxide etching [106], electrochemical oxidation [107] and reduction [108], microwave radiation [109], sol–gel chemistry [110–112], plasma treatments [113–117], carbene insertion [118], and nucleophilic addition [119]. For example, through a standard sulfuric acid/nitric acid reflux process, mostly carboxyls and some carbonyl groups form, preferentially at defect sites and end caps of SWCNTs. These functionalized CNTs could be treated at elevated temperatures with thionyl chloride to generate acyl chloride groups as a functional intermediate. Amino-decarboxylation reactions of these acyl chloride groups can be used to form open-ended SWCNTs terminated with primary amines [120]. This process is quite standard in literature, but the downside is that, though not overly complicated, it consumes a lot of time, energy, and labor. At elevated temperatures, fluorine gas was used to directly reduce CNT sidewalls to produce sidewall-fluorinated CNTs (see Figure 5.3). As cathode materials in an electrochemical cell, fluorinated CNTs had a higher cell potential than commercial fluorinated graphite [121]. Sidewall fluorine atoms could be exchanged for terminal diamines to again produce primary amine-functionalized CNTs, and in this case amine functionality was preferentially located at sidewalls [122]. Radio frequency (RF) plasma treatments at 13.56 MHz [115] and plasma-enhanced chemical vapor deposition [123] have also been used to directly functionalize CNTs with primary amines, although RF plasma functionalization introduces amide, imide, and imine functional groups as well. Amine functionality is tremendously useful because of the relative ease with which they can be reacted further with polymers and biomolecules [124]. An obvious example of how such materials could be well dispersed and covalently incorporated into nanocomposites is through the ring opening of epoxides for high-strength adhesives [125,126]. Reactions of functionalized CNTs have been used to attach covalently to moieties of a wide range of different polymers

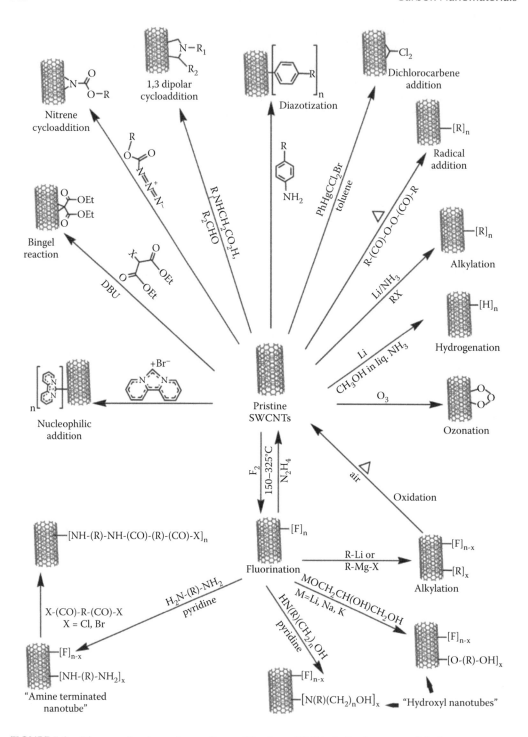

FIGURE 5.3 Diagram showing pathways for modification of CNTs, indicating some of the important routes to chemical functionalization preparation for use in nanocomposites. Descriptions of most of the pathways shown can be found in Ref. [98].

and nanomaterials for targeted applications. Silanes [127], dendrimers [128], porphorins [129–131], polymers [132–134], proteins [135,136], antibodies [137], DNA [138–140], and other nucleic acids [141,142] have all been covalently attached to CNT sidewalls, and covalently attached ligands on CNT surfaces to bond metal NPs have also been demonstrated [143–145].

5.3 CNT-INDUCED CRYSTALLIZATION, POLYMER SINGLE CRYSTAL CASE

5.3.1 POLYMER–CNT INTERFACES AND INTERPHASES

Polymer composites take advantage of the appealing properties of the matrix polymer (i.e., good processability, toughness, and damping properties, high resistance to dielectric breakdown, optical transmittance) while improving or imparting specific properties using nanofillers [146]. For a few applications, it is reasonable to assume that the composite material will follow the rule of a simple mixture for some desired property [147]. This is seldom the case for nanocomposites because of the complexity of the region of matrix polymers in the vicinity of the filler [148]. Three regions of the polymer part in the nanocomposite must be considered. These include the bulk, the interphase [149,150], and the interface [151]. The bulk phase can be considered indistinguishable from neat polymer and the polymer is sufficiently far away from any interface that the chain dynamics and physical properties in the bulk region are not affected by the filler material. The interphase is in the vicinity of the NP, without being precisely in the area of closest approach. There is evidence to suggest that the interphase region can extend a great distance from the nanofiller and actually dominate the properties of the nanocomposite matrix [152]. Here, properties, such as glass transition and diffusion length, are influenced by the presence of the filler particles as the chain dynamics are disrupted indirectly. The interface region of the polymer refers to chains that are in direct contact with the filler particle. Polymers in this region often behave in a way that can be very different from the bulk polymer. Electrostatic interactions and π–π stacking between the filler and the polymer as well as Leonard Jones potential-well formation can cause disruptions to the chain behavior that might dramatically change properties such as melting temperature, crystal nucleation rate, chemical potential, dielectric constant, glass transition, and diffusion length.

Local changes to the polymer matrix notwithstanding, CNT nanocomposites can be expected to have chemistry similar to the neat polymer except at high loading contents [153], and the polymer can be selected to suit a certain application. One general guideline is that there should be some level of affinity between the matrix polymer and the CNT if functionalization repulsion is not used to disperse the CNTs [154]. Thus, probing the chemical properties of CNTs has been an important research field during the last decade. The results from different test methods are not consistent in their reporting of surface free energy of CNTs. Using contact angle measurements of polymer beads on the MWCNT surfaces with electron microscopy, Nuriel, Barber, and Wagner estimated the nanotube surface energy as approximately 45 mJ/m^2 [155]. Ma et al. used sessile drops to estimate the surface energy of nonfunctionalized MWCNTs as 50 mJ/m^2 [156]. Zhang et al., using inverse gas chromatography, determined the surface energy of MWCNT formed by chemical vapor deposition (CVD) to be $\approx$120 mJ/m^2 [157]. From a practical standpoint, the surface energy as estimated by dispersion in organic solvents might be most convincing, as this method has been used to accurately predict the types of solvents that should be effective media for CNT exfoliation. Bergin et al. reported [67] a surface free energy for SWCNTs of approximately 65–70 mJ/m^2 based on the dispersibility of CNT at various concentrations dispersed in organic solvents with a range of multicomponent solubility parameters. In any case, there is general consensus that CNTs have high surface energy. This makes it fairly easy to wet the surfaces of CNTs or CNT-bundles with a wide range of polymers. CNT-based hybrids and composites have been formed using polymers of an enormous range of different solubility parameters. Table 5.1 lists a selection of some of the most important polymers tested for polymer/CNT nanocomposites.

TABLE 5.1
Selection of Polymers Used in CNT Nanocomposites and Some of Their Proposed Applications

Polymer	Structure Repeat Unit	Major Potential Applications	References
Polystyrene (PSt)		Controllable phase separation for molecular channels	[158,159]
Poly(methyl methacrylate) (PMMA)		Targeted applications based on copolymer phase separation or side chain functionalization	[160–162]
Polycarbonate (PC)		Conductive and optically transparent parts	[163–166]
		Environmentally responsive sensors	[167]
Epoxy polymers		Adhesives	[168–175]
Isotactic polypropylene (iPP)		Structural and conductive components produced using traditional melt processing techniques	[176–179]
		Supramolecular structures	[177,180–182]
Polyethylene (PE)		Structural and conductive components	[183–187]
		Flame resistant and heat-deflecting parts	[188]
		Nanohybrid shish kebab	[189–196]
Poly(ethylene terephthalate) (PET)		Fibers	[197]
Poly(vinylidene fluoride) (PVDF)		Piezoelectric sensor	[198,199]
		Binder material	[200,201]
		Carbon fiber precursor	[202]
Polyamide (PA)		Continuously spun fibers	[90]

Polymer	Structure	Applications	References
Poly(ethylene oxide) (PEO)		Biocompatible surfaces	[203]
		Lithium ion carrier	[204]
		Controlled phase separation	[205]
Polyurethane (PU)		Elastomers	[206]
		Adhesives	[207]
		Shape-memory	[208,209]
Polyaniline (shown in partially oxidized state, "emeraldine")		Sensors	[210]
		Charge storage	[109,211–213]
Poly([3,4]-ethylene dioxythiophene)		Sensors	[214]
		Biocompatible surfaces	[215]
		Biological interfacing	[216,217]
Polythiophenes		Photovoltaics	[218]
		Charge storage	[219]
Chitosan		Glucose sensors via enzyme attachment	[95,96]
Deoxyribonucleic acid (DNA)		Hybrid structures	[220,221]
		Engineered surfaces	[222]
		NP and biomolecule attachment	[223]
		Aqueous dispersion of CNTs	[73,139]
		Molecular recognition	[128,137,210,224]

5.3.2 CNT-Induced Transcrystallinity

The CNT–polymer interface is of great importance to the physical properties of the resultant nano-composites. For crystalline polymers in bulk state, transcrystallization at the interface often takes place. Transcrystallinity induced by fibers in semicrystalline polymer matrices has been recently reviewed in Ref. [225]. In transcrystalline composites, the interphase layer is composed of aligned columnar crystallites that are heterogeneously nucleated from a suitable fiber, that is, the c-axis of the polymer chain runs parallel to the filler axis, and lamellae are forced into a columnar, perpendicular orientation. In some cases, secondary nucleation can produce off-axis crystals but the majority of the crystals in these interphase regions are highly aligned. The high density of crystal nuclei forces impingement of the crystals, and thus restricts them to radially splayed, oriented growth in the interphase rather than spherulitic growth. In addition to confining the PSCs, fibers serve as impurities that can lower the activation energy necessary for nucleus formation. Thus, crystallization can take place more rapidly and at higher temperatures than for the neat polymer. Surface chemistry plays the most important role in determining whether transcrystallinity will occur. Carbon fiber with the surface dominated by the graphite basal plane induced transcrystallinity in iPP, whereas carbon fiber with plane end exposure did not [226,227]. Other properties of fillers that have been suggested to assist in promotion of transcrystalline phases include high surface energy [228], high modulus [229], and high thermal conductivity inducing a steeper thermal gradient between filler and matrix. In line with these observations, it has also been shown that CNTs themselves can also initiate crystallinity (see Figure 5.4, [182]).

The authors differ in their analysis of the effects of transcrystallinity on the properties of fiber-reinforced composites, with some showing improvements in mechanical properties due to interfacial shear, and others finding no improvement in mechanical interfacial strength. Of the latter, Wang and Hwang performed single fiber pull-out tests of polytetrafluoroethylene (PTFE) in a polypropylene matrix at a range of crystallization temperatures to show that the transcrystalline interphase did not affect the interfacial adhesion [230]. Conversely, Chen and Hsiao used single fiber pull-out tests of polyacrylonitrile-based carbon fiber to show that adhesion was considerably higher for strands in transcrystalline phases than noncrystalline regions. This effect was still observed although it was shown to be diminished at higher filler loadings [227]. Similarly, uniaxially aligned Kevlar-reinforced Nylon 6,6 showed dramatic improvements in both Young's modulus and ultimate tensile strength [231]. The relatively high surface energy and hydrogen bonding between Kevlar and polyamide likely play a role in promoting the interfacial stress transfer. It was found that a strong correlation exists between surface energy of the filler and interfacial adhesion in transcrystalline composites [228]. Some of the important polymers in which CNTs serve as nuclei for transcrystalline behavior include PE, iPP, PA, and poly(vinyl alcohol) (PVA).

Rapid nucleation of crystalline polymers has an additional advantage from a scientific standpoint: heterogeneous nucleation serves to study the crystallization of polymers and oligomers at CNT surfaces when the polymer is present in lower concentrations. This in turn makes it possible to use PSCs to functionalize CNTs.

5.3.3 Nano Hybrid Shish Kebabs

5.3.3.1 Nano Hybrid Shish Kebabs Formation on CNT in Dilute Solution

Diluents/solvents affect the crystallization of polymers in several ways. One is that they lower the concentration of the polymer, limiting the rate of nucleation and growth. The other is that they lower the equilibrium dissolution temperature. Diluents/solvents can, therefore, be used to get controlled conditions for PSC with very regular crystal structures. Just as with composites cooled from a melt, CNTs can be used to nucleate PSCs in dilute and semi-dilute solutions as well. The precise control offered by dilution allows CNTs to crystallize polymer at temperatures above the homogeneous nucleation temperature. In some cases, this results in the transcrystallinity discussed in the last section, and in other

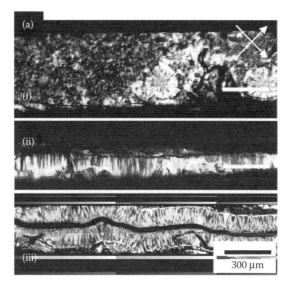

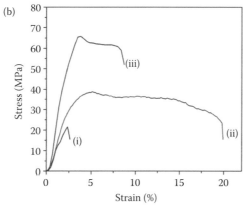

FIGURE 5.4 Transcrystallinity induced by a CNT fiber. (a) Tensile test specimens. (i) Nontranscrystalline spherulitic iPP. (ii) Transcrystalline iPP interphase cleaved from the CNT fiber. (iii) Transcrystalline polymer with embedded CNT fiber. (b) Stress–strain curves of fibers shown in (a). (From S. Zhang et al., *Polymer*, 49, 2008, 1356–1364.)

cases with lower polymer concentrations, single crystals are grown from the CNT sidewall (Figure 5.5). This is different from transcrystallinity insofar as the extent and direction of crystal growth is not limited by another impinging polymer crystal growth front. In such cases, a unique morphology is produced with a periodic pattern of PSC-decorated CNTs (see Figure 5.5). There are several techniques that can be used to produce this effect, including isothermal and nonisothermal solution crystallization [194] and introduction of supercritical CO_2 [193]. The result in either case is termed a nanohybrid shish kebab (NHSK) structure, and was first demonstrated using PE and Nylon 6,6 PSC decorations on MWCNTs [189]. The NHSK term comes from the classical shish kebab structures described in the literature in the 1960s by Pennings [232]. In Pennings' experiment, ultra-high molecular weight PE (UHMWPE) was homogeneously crystallized during extensional flow and as a result, elongated UHMWPE fibers were formed. Flow instabilities at the fiber surface caused chain ends and lower molecular weight PE to crystallize in lamellae in direct epitaxy to the fiber surface [233].

However, no extensional flow is needed to produce the NHSK structure, although there is a striking visual similarity. It is believed that the NHSK is formed as a result of affinity between the

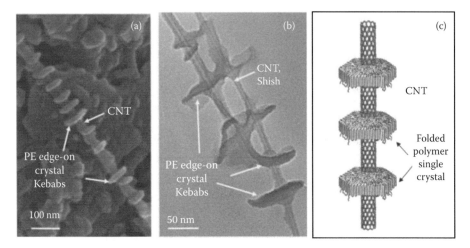

FIGURE 5.5   First published description of the nanohybrid shish kebab structure showing (a) SEM and (b) TEM image of MWCNT used to template PE single crystals. (c) Early schematic depiction of PE single crystals grown on a SWCNT. (Adapted from C.Y. Li et al., *Adv. Mater.*, 17, 2005, 1198–1202.)

polymer chain and the CNT sidewall. As the polymer continues to chainfold, the crystal orientation is directed by the geometrical confinement of the narrow-diameter CNT axis. This causes the lamella to grow perpendicularly. As the local concentration of polymer is exhausted by incorporation into the lamella, it prevents crystal nucleation from areas of the CNT between the existing kebabs. Thus, a semi-periodic structure is achieved [190,234].

Both isothermal and nonisothermal NHSK crystallization begin in the same manner: a polymer solution is prepared at a sufficiently high temperature so as to ensure homogeneity in the mixture. Separately, CNTs are dispersed in the same solvent and heated to the same temperature. The two solutions are then mixed. At this point, in nonisothermal solution crystallization, the sample is then allowed to cool slowly to a certain temperature. This allows the sample to scan the range of conditions between dissolution of the polymer in liquid down to precipitation from solution. In isothermal crystallization, the mixed sample is stepped down from the temperature of complete dissolution to slightly above the temperature of homogeneous crystallization. This prevents homogeneous nucleation but the system is still in a supercooled state for CNT-induced heterogeneous nucleation. At the end of isothermal crystallization, the sample can be rinsed with temperature-equilibrated clean solvent to remove any polymer that has not undergone crystallization. Nonisothermally crystallized NHSK are marginally easier to produce, but the kebab size distribution can be much broader than for isothermally crystallized NHSK (although this depends strongly on the cooling rate) [235]. A comparison of the two structures formed using PE crystallized from dichlorobenzene is shown in Figure 5.6. Supercritical CO_2 has been used to destabilize solution mixtures of polymer and CNTs to form shish kebab crystal structures. By this method, the solvent solubility is lowered controllably. This method has been used to produce well-formed nanohybrid shish kebabs from a number of functionally important polymers, including PE [236] and poly(vinyl alcohol) [83]. This technique is well suited to shish kebab formation, and like similar nonsolvent titration, is an effective method for producing nanohybrid shish kebabs with high precision [193,237].

5.3.3.2 Nano Hybrid Shish Kebabs Formed by Solvent Evaporation and Physical Vapor Deposition

Several major applications require flat, in-plane architectures, such as conductive coatings, solar panels, and semiconductor electronics. Microfabrication and thin-film composites require 2D/quasi-2D architectures as well as specialized patterned surfaces. Dilute solution crystallization, while useful for producing 3D NHSK, is not particularly well suited to such applications.

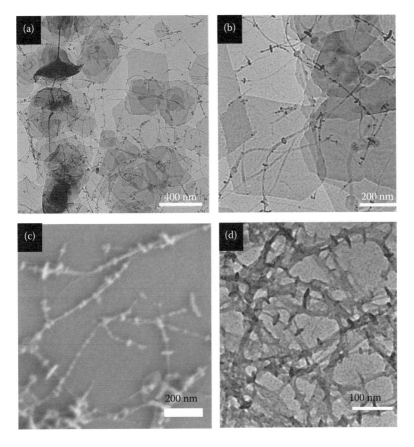

FIGURE 5.6 NHSK formed via solution crystallization. (a,b) TEM micrographs of NHSK formed via non-isothermal crystallization. (Adapted from H. Uehara et al., *J. Phys. Chem. C*, 111, 2007, 18950–18957.) (c) SEM and (d) TEM image of isothermally crystallized NHSK.

To that end, centipede-like polymer-decorated CNTs have been produced by several mechanisms [191,192,205,238]. To produce quasi-2D NHSK, nanotubes were spin-coated over glass, mica, or silicon substrates. The CNT-coated substrates were then either drop cast with a dilute PE solution or placed in a vacuum chamber for physical vapor deposition (PVD) of thermally cleaved PE oligomers. Centipede-like NHSK produced using these two methods are shown in Figure 5.7a–c. Crystallization of the solution-deposited PE and PVD-grown PE oligomers clearly was responsible for producing the quasi-2D NHSK structures. Annealing of the structures was performed to gain an insight into the mechanisms by which PE chains recrystallized. Figure 5.7b, e, and f shows the results of centipede-like NHSK subjected to annealing treatments. When the solution-grown 2D NHSKs were annealed, the chains melted and then recrystallized into the original 2D NHSK structure. This suggested that the PE chains had found their preferential conformation during the initial solution deposition step. Although thermal annealing melted the PE single crystals, on cooling they regrew in similar conformations with a better defined structure. Conversely, when PE oligomers grown via PVD were melted and allowed to recrystallize, the resulting structure was quite different, more closely resembling a homogeneous PSC. Comparing these two processes, one apparent difference is the PE concentration. In the PVD process, PE oligomers sequentially deposit on the substrate so that during the nucleation process, PE concentration is relatively low, hence heterogeneous nucleation occurred. In the recrystallization process, all the PE oligomers are on the substrate and the local concentration of PE is high. This significantly enhances homogenous nucleation, which leads to flat-on PE lamellae.

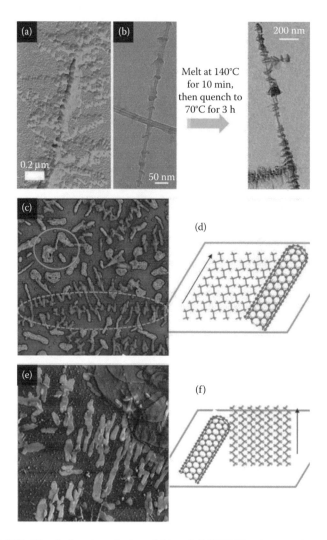

FIGURE 5.7 (a) 0.05% PE solution deposited on 0.01 wt% MWCNT spin coated on carbon-coated glass substrate. (b) TEM image of solution-deposited 2D NHSK before and after recrystallization. (c) PE oligomers deposited from the vapor phase onto SWCNT. Islands of homogeneous PE oligomers can be observed, as well as rod-like templated NHSKs. (d) Proposed arrangement of rod-like PE oligomers at the vicinity of the CNT. (e) Homogeneous PE oligomer morphology developed after recrystallization. (f) Proposed recrystallized structure of morphology observed in (e). (Adapted from L.Y. Li et al., *Polymer*, 52, 2011, 3633–3638.)

Although less energetically stable than polymeric NHSK, a variety of other oligomeric NHSK have also been demonstrated. Oligomers are excited into a vapor phase via thermal degradation of polymeric precursors using a PVD process. Some of these PVD–NHSKs have been formed using polymeric precursors that, for a variety of different reasons, have proven to be difficult to solution crystallize on CNT. Examples of 2D NHSK that have been deposited in this manner include poly(L-lysine) and PVDF [238]. As a polypeptide, two-dimensionally patterned poly(L-lysine) NHSK have opportunities for biosensors and cell culturing substrates. NHSK formed from PVDF, a piezoelectric polymer, could find applications in pressure sensing.

5.3.3.3 Size-Dependent Soft Epitaxy

There are two possible factors that control NHSK growth: the epitaxial growth of PE on CNT and geometric confinement. Epitaxial growth of PE on the surface of highly ordered pyrolytic graphite (HOPG) dictates the PE chain direction or $\langle 001 \rangle$ of the PE crystal to be parallel with $\langle 2\,\overline{1}\,\overline{1}\,0 \rangle$ of the underneath graphite. However, because of their small diameter, CNTs can be considered as rigid macromolecules and polymer chains prefer to align along the tube axis regardless of the lattice matching between the polymer chain and the graphitic sheet, rendering a geometric confinement on polymer chains. This mechanism can be attributed to soft epitaxy and strict lattice matching is not required while a cooperative orientation of the polymer chains and the CNT axes is needed [239]. The growth mechanism for CNT-induced PE crystallization involves both these aspects and it is size dependent. On the surface of the fiber/tube with a diameter much larger than the polymer size, the polymer behaves as if it was on a flat surface and epitaxy becomes the main growth mechanism. As the fiber/tube diameters decreased to the order of the polymer size, the polymer starts to crystallize on the surface and geometric confinement is the major factor; also, the polymer chains are exclusively parallel to the CNT axis, disregarding the CNT chirality. As a consequence, the PE crystal lamellae arrange perpendicular to the CNT axis and orthogonal orientation is obtained which was confirmed by selective area electron diffraction (SAED). RGO and graphene represent the extreme limit of a radius equal to infinity: such two-dimensional surfaces nucleate PE crystals to grow in virtually any orientation [240]. It should be noted that any compatible (CNT)–(crystalline polymer)–(weak solvent) system is a candidate for NHSK formation, but many polymers crystallize in a helical conformation, limiting their contact with the CNT sidewall. Thus, chains may simply wrap the CNT surface or form poor crystals if the epitaxial relationship is unfavorable.

It should be noted that there is also an alternative theory to explain the NHSK structure formation [235]. Zhang et al. suggested that NHSK form the following three steps: First, PE chains near the MWCNTs are adsorbed on the nanotube surface as soon as they are mixed in the solution, to reduce the high surface free energy of MWCNTs. Second, with the increasing of wrapping chains, they began to slide along tubes and change into extended chain conformation, forming a homogeneous coating around MWCNTs with few subglobules. Third, with decrease in temperature, PE chains epitaxially grow from subglobules of the homogeneous coating and formed the crystal lamellae. As further evidence, they offered molecular dynamics simulations by Yang et al. [241]. The formation of an adsorbed polymer layer depends on polymer molecular weight, annealing temperature, and time. Simulation by Yang et al. also suggested folded chain instead of extended chain conformation for the crystallites. Epitaxial growth from a subglobule is rather confusing as the subglobule should be considered as the nucleus of the kebab crystal. Moreover, the chain rotation suggested in the PE coating layer is speculative and it is not clear whether a homogeneous PE coating layer is needed for subsequent kebab crystal growth.

5.4 STRUCTURE OF CNT-BASED NANOCOMPOSITES

5.4.1 PREPARATION OF CNTS BY POLYMERIZATION *IN SITU*

Various approaches have been used to fabricate CNT nanocomposite, including melt blending, solution blending, and direct polymerization. Among these methods, direct polymerization of the solvent phase is perhaps the best technique that can be employed and thermosets are excellent starting materials for this purpose. Epoxy matrices have been considered a model system for direct polymerization amid CNT dispersion with various degrees of alignment and filler loadings and epoxy-based composites have been studied at least since 1997 [242]. Such studies have been used to study mechanical percolation [243], rheology [244], electrical percolation behavior [245,246], and thermal conductivity [247]. Epoxy composites are still an area of intense research; recently, Chapartegui et al. [248] showed that in a melt-mixed resin, the presence of MWCNT accelerates epoxy curing, even below rheological percolation. Three hypotheses were proposed to explain this surprising result: (1) chemical reactions between the CNT and the epoxy matrix, (2) acceleration of the ring-opening reactions because of the presence of catalyst NPs, and/or (3) efficient thermal conductivity of MWCNTs improving the transmission of thermal energy to the polymer matrix. Citing several reports on the incorporation of functionalized polymers into similar epoxy resins [249,250], they did not consider MWCNT–resin crosslinking reactions to be likely, and attributed the observed behavior to thermal conductivity enhancements although acceleration because of NP catalyst was not ruled out.

In situ polymerization of thermoplastics is a common technique. Tang and Xu used an *in situ* polymerization technique to wrap poly(phenylacetylene) (PPA) around MWCNTs, and used TEM, XRD, and NMR spectroscopy, as well as a control experiment of prepolymerized PPA in the presence of CNTs to show that polymer wrapping had occurred [82]. Despite this, it was not clear whether the polymer wrapping was noncovalent. Jia et al. used azobisisobutylnitrile as an initiator to polymerize methyl methacrylate *in situ*, but they found that the polymerized monomer had chemically reacted with MWCNT [251]. Park et al. were able to use a sonication technique to form a nanocomposite with isotropically dispersed SWCNTs within a polyimide matrix using a sonication procedure, but again, no evidence that CNT were not chemically modified was presented [252]. Later, it was confirmed that nanocomposites could be trapped in a polyimide matrix during polymerization [253]. Raman spectroscopy of CNT composites was first reported in detail in 1993 [254], but was not used as a standard characterization technique until later [14,255,256]. Regardless of whether CNTs are distributed within an *in situ* polymerized matrix by means of chemical functionalization or kinetic trapping, this technique offers the ability to form composites with a wide range of filler loading contents which would be difficult using standard processing techniques. Processing techniques performed in the melt or in an uncured syrup generally experience a catastrophic increase in viscosity at high loading contents [257]. In general, isotropic distribution of CNTs is difficult to achieve for heavily filled nanocomposites without the majority of the processing being done in dilute solution.

An interesting example of how processing technology can take full advantage of the physical properties of CNTs was demonstrated using electrophoretic alignment. Bubke et al. were the first to show that electrophoresis can be used to impose anisotropy on dispersed CNTs [258]. Later, Prasse showed that carbon nanofibers (CNFs) can self-assemble reversibly into wires when suspended in a viscous medium under an applied electric field [259]. This technique was demonstrated for uncured epoxy; when the epoxy was cured at elevated temperature, the CNFs were effectively frozen into the matrix in a preformed conductive wire. Martin et al. later extended this work to MWCNTs [172], and Wang [260] showed that this technique not only improved the electrical conductivity 10-fold along the transient wire formation direction, but improved mechanical properties as well. Park et al. showed that this type of network formation is also possible using SWCNTs in a photopolymerizable blend of urethane dimethacrylate and 1,6-hexanediol dimethacrylate [261]. In this system, a light source was used to freeze the nanotubes in place within the photopolymerizable monomer.

5.4.2 CHARACTERIZATION OF DISPERSION, INTERCALATION, AND ALIGNMENT

Characterizing the mixture of CNTs in a polymer matrix for a given processing technique is important in nanocomposites, as will be explained in more detail throughout this section. Polymer/CNT systems have three CNT self-association aspects that are of concern, namely, dispersion, exfoliation, and alignment, which may be defined as follows. Dispersion can be considered as the homogeneity of the mixture of CNTs into the matrix. SEM at high accelerating voltage is a useful technique in identifying areas of high CNT concentration which appear as bright spots of accumulated charge. Polarized light microscopy (PLM) is another excellent method to survey the dispersion of CNTs in a polymer matrix, as shown in Figure 5.8. If agglomerates are known to be present in a two-phase mixture, small-angle x-ray scattering (SAXS) can be a useful technique to determine size distributions. Exfoliation can be considered the closeness/association of nanofiller particles. For example, a small bundle of SWCNTs might be well separated from other bundles in the polymer matrix, but

(a) (b) (c) (d)

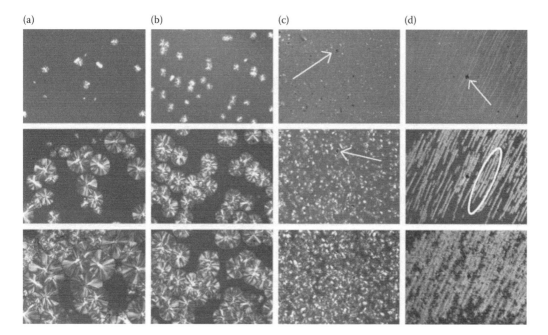

FIGURE 5.8 Crystallization growth of melt extruded iPP and iPP/MWCNT at different drawing length/die diameter ratios studied by PLM. (a) Neat PP, draw ratio of 1. (b) Neat PP, draw ratio of 7. (c) iPP/0.3 wt% MWCNT, draw ratio of 1. (d) iPP/0.3 wt% MWCNT, draw ratio of 7. Arrows indicate agglomerates, circle indicates MWCNT orientation. (Adapted from Z. Hou et al., *Polymer*, 49, 2008, 3582–3589.)

each might contain as little as two or as many as hundreds of tightly entwined tubes. Thus, the composite would have good dispersion but poor exfoliation. Techniques which may be used to quantify exfoliation include conventional x-ray diffraction (XRD) [60,262,263], Raman spectroscopy [72], atomic force microscopy (AFM), and TEM [74].

Alignment of CNTs has important consequences for applications, and can be investigated using a range of techniques, but perhaps the best are XRD and SAXS. PLM and polarized Raman spectroscopy are useful as well. With XRD, a quantitative parameter known as Herman's orientation factor, f, can be determined, which varies from 1 for perfectly oriented crystals to 0 for random orientation, and –1/2 for opposite orientation (i.e., if the orientation along the c-axis is 1, the orientation along the a- and b-axes would be –1/2). A single arc corresponding to a given planar reflection can be used to determine the plane's degree of alignment, as in Equation 5.1 [264].

$$f(\cos\theta) = \frac{3\langle\cos^2\theta\rangle - 1}{2},$$

$$\langle\cos^2\theta\rangle \equiv \frac{\displaystyle\int_0^{\pi/2} I(\theta)\cos^2\theta\sin\theta\,d\theta}{\displaystyle\int_0^{\pi/2} I(\theta)\sin\theta\,d\theta} \qquad (5.1)$$

where θ is the angle formed between an arc and a given reference direction of interest, and I is the scattering intensity. Pujari et al. showed that using SAXS, the second moment tensor of a diffraction pattern can be used to find the orientation of CNTs even in the absence of PSCs. They used this method to quantify the orientation of MWCNTs aligned by shear flow in pure epoxy resin [265]. This technique was based on a method described in 1985 by Salem and Fuller using light microscopy [266]. If x-, y-, and z-directions have reciprocal space projections along ξ, η, and ζ, respectively, then the second moment tensors for scattering patterns obtained in x–y and x–z reciprocal space are given as

$$\langle \mathbf{qq} \rangle_{\xi\eta} = \begin{pmatrix} \langle q_\xi q_\xi \rangle & \langle q_\xi q_\eta \rangle \\ \langle q_\xi q_\eta \rangle & \langle q_\eta q_\eta \rangle \end{pmatrix} \equiv \frac{\iint \mathbf{q} \cdot \mathbf{q} I(\mathbf{q}) \, d\mathbf{q}_\xi \, d\mathbf{q}_\eta}{\iint q^2 I(\mathbf{q}) \, d\mathbf{q}_\xi \, d\mathbf{q}_\eta} \tag{5.2}$$

$$\langle \mathbf{qq} \rangle_{\xi\zeta} = \begin{pmatrix} \langle q_\xi q_\xi \rangle & \langle q_\xi q_\zeta \rangle \\ \langle q_\xi q_\zeta \rangle & \langle q_\zeta q_\zeta \rangle \end{pmatrix} \equiv \frac{\iint \mathbf{q} \cdot \mathbf{q} I(\mathbf{q}) \, d\mathbf{q}_\xi \, d\mathbf{q}_\zeta}{\iint q^2 I(\mathbf{q}) \, d\mathbf{q}_\xi \, d\mathbf{q}_\zeta} \tag{5.3}$$

for all possible scattering vectors $\mathbf{q}$. An anisotropy factor, AF, for each pattern can be defined, which has a maximum of 1 for perfect orientation along any direction and becomes 0 for a completely random orientation.

$$\mathrm{AF} = \sqrt{\left(\langle q_\xi q_\xi \rangle - \langle q_k q_k \rangle \right)^2 + 4 \langle q_\xi q_k \rangle^2} \tag{5.4}$$

where $k = \eta$ or ζ. Of course, this technique works for off-axis projections as well and is valid for any reciprocal space scattering pattern of interest. Using any of these techniques, the anisotropy of a nanocomposite system can be quantified.

5.4.3 CRYSTALLIZATION OF POLYMER/CNT NANOCOMPOSITES

5.4.3.1 Quiescent Growth

The presence of CNT inclusions in semicrystalline polymers complicates the crystallization process. Polymers can take advantage of the CNT as a preformed nucleus; in doing so they sacrifice spherulitic growth. Until now, it has not been shown that the presence of CNTs distorts the polymer lattice, but they do alter the appearances of lamellae and spherulites. The one-dimensional PSC growth at CNT surfaces is not a natural conformation for most polymers. Polymers, therefore, attempt to reestablish more natural crystalline morphologies in the interphase. Many crystalline polymer matrices decrease in crystallinity (χ_c) and increase in crystallization temperature (T_c) when CNTs are added, but this is by no means a universal trend. Whether T_c increases or decreases depends on the affinity of the polymer chains to the CNT surface, that is, the effectiveness of the CNT as a preformed PSC nucleus while the increase or decrease of χ_c is determined by the structure of the transcrystalline region. Large changes may also suggest a different crystal morphology. For polymers with polymorphic crystal structures, CNTs promote specific forms at the expense of others. For instance, several authors have determined that pristine or fluorinated CNTs encourage the formation of the technically important β-phase of PVDF at the expense of the more common α- and γ-phases [199,267–269]. In such cases, the composite crystallization kinetics may not be directly comparable with those of homopolymer systems: homopolymer PVDF ordinarily shows almost no β-phase morphology under quiescent conditions. Direct comparison is possible with many other nanocomposite systems as shown in Table 5.2, and in such cases, the Avrami equation is useful in studying the associated changes in crystallization kinetics. The Avrami equation for isothermal crystallization describes the rate of crystal growth for a material that is stepped to a supercooled temperature [270–272]. The relative crystallinity at time t, $\phi(t)$, is given by Equation 5.5.

$$\phi(t) = 1 - \exp\left[-k_A t^N \right] \tag{5.5}$$

TABLE 5.2
Selected Crystallization Studies That Used CNTs as Polymer Nanocomposite Fillers

Polymer	Polymer Characteristics	Nanotube Type	Nanotube Preparation Steps	Compositing Technique	Filler Loading (wt%)	N_{neat}/N_{comp}	$\log(k_{A,neat})/$ $\log(k_{A,comp})$ (k_A in s^{-1})	ΔT_c (°C)	ΔT_m (°C)	$\left(\dfrac{t_{1/2,comp}}{t_{1/2,neat}}\right)$	χ_c (%)	References
iPP	MFI = 3, M_W = 399,000, PDI = 4.6	MWCNT	Chengdu Institute, as-received	CNT/ anhydrous ethanol dispersion, mixed into iPP/xylene at 140°C. Solution was evaporated, dried, melt-pressed	0.1	4.00/4.00 + 1.82 1.82/2.52[a]	—	—	—	0.138 0.125[a]	—	[180]
iPP	M_W = 340,000, PDI = 3.51	MWCNT-COOC18H37	(see Ref. [273])	CNT/xylene suspension added to iPP/ xylene solution at 120°C, cast into methanol	0.2 ≈ 9.1	2.4 ≈ 3.0/ 2.5 ≈ 3.9	−5.2/−4.2 ≈ −3.3	—	—	0.098 ≈ 0.139	44.5 ≈ 51.0/ 38.4 ≈ 48.2	[273]
iPP	MFI = 4	MWCNT-COOH	Chengdu Institute, as-received	Compounded w/HAAKE rheometer	1 ≈ 7	4.3/2.8 ≈ 3.4	—	12.6	−1.3	0.517 ≈ 0.792	≈ 40	[179]
iPP	Polyolefin, Amoco powder, MFI = 17	Purified HiPCO SWCNT	Rice University, as-received	Compounded in HAAKE mixer at 240°C	1	2.8 ≈ 3.4/ 3.4 ≈ 3.5	−7.74 ≈ −7.70/ −5.57 ≈ −4.40	11.3	—	0.048	—	[274]

continued

TABLE 5.2 (continued)
Selected Crystallization Studies That Used CNTs as Polymer Nanocomposite Fillers

Polymer	Polymer Characteristics	Nanotube Type	Nanotube Preparation Steps	Compositing Technique	Filler Loading (wt%)	N_{neat}/N_{comp}	$\log(k_{A,neat})/\log(k_{A,comp})$ (k_A in s^{-1})	ΔT_c (°C)	ΔT_m (°C)	$\left(\dfrac{t_{1/2,comp}}{t_{1/2,neat}}\right)$	χ_c (%)	References
iPP	MFI = 2.9	SWCNT	CarboLex, as-received	Melt-blended	5 ≈ 20	2.6/2.54 ≈ 2.46	−3.3/0.42 ≈ 0.69	18.3	−2.8	0.118 ≈ 0.137	—	[275]
iPP	MFI = 1.8, Honam Petrochem Co.	CVD MWCNT	Iljin Nanotech Co., acid reflux	Melt-blended, 190°C 10 min; hot-pressed	1 ≈ 5	2.6 ≈ 3.2/ 3.1 ≈ 3.3	−7.64 ≈ −7.60/ −5.52 ≈ −4.41	12	−2	0.037 ≈ 0.075	—	[276]
iPP	MFI = 88, Fina Dypro	CoMoCat SWCNT	Acid refluxed, amide functionalized	CNT decalin solution added to iPP solution; solvent evaporated; melted on hot plate	0.6, 1.8	2.05 ≈ 2.40/ 1.84 ≈ 2.86	−6.93 ≈ −5.11/ −7.79 ≈ −3.68	5.3	−2.1	0.433 ≈ 0.474	50.5 ≈ 60.6/ 59.1 ≈ 72.6	[277]
iPP/P(E-PP-diene) blend	iPP, MFI = 2.9; Rubber, Mooney viscosity = 55	SWCNT	CarboLex, as-received	Melt-blended	0.25 ≈ 1.0	2.46/ 2.00 ≈ 2.46	−0.72/ −0.09 ≈ 0.79	4.1	2.5	0.516 ≈ 0.659	40.1 ≈ 44.9	[278]
Nylon 6	Polyamide, used as-received from Formosa Chemical Co.	MWCNT-COOH	Ethylene CVD, acid purification	Mechanical mixing in formic acid, cast on glass slide	0.25 ≈ 3	3.08 ≈ 3.18/ 2.02 ≈ 2.45	−0.72 ≈ 2.1/ −0.11 ≈ 3.2	—	−4.1	0.272 ≈ 0.348	—	[279]

Nylon 6,6	$M_N = 10,000$ g/mol	MWCNT	Aldrich, acid purified	Preformed NHSK mixed with additional polymer solution in glycerol, 240°C	0.1 ≈ 2	2.2/1.7	$-1 \approx 0.59/$ $-1 \approx 0.77$	—	—	0.519 ≈ 3.16	—	[280]
PE	$M_W = 53,600$, PDI = 2.35	SWCNT	Carbon Nanotechnologies Inc., as-received	Dispersed in DCB, added to hot PE solution in DCB, cast into methanol	0.02 ≈ 0.5	2.8/2.9 ≈ 3.1	—	5	—	0.075	—	[85]
PE	$[\eta] = 0.955$ g cm³, Ziegler-Natta-MWCNT; $M_W = 230,000$, PDI = 18.4	MWCNT	(see Ref. [281])	Mixed in screw extruder	0.52	3.3 ≈ 3.6/ 2.6 ≈ 2.9	$-13.3 \approx -5.24/$ $-6.39 \approx 1.30$	1.1	-2	0.618	59/62	[281]
PE	$M = 50,000$ g/mol, 78% crystallinity	SWCNT	NASA Johnson Space Center, baked at 300°C	CNT sonicated in DCB, PE solution added, coagulated and dried	1 ≈ 30	2.7/1.6	$-8.5 \approx -4/$ $-5.2 \approx -2$	2	—	0.036	—	[84]
PEO	$M_W = 20,000$ g/mol	MWCNT MWCNT-OH MWCNT-COOH	Not reported, although standard techniques exist. See Section 5.2.3.4	CNTs dispersed in water, mixed into 5% PEO solution (aq.), cast on glass plate	0.1, 0.5 1 0.1, 0.5	2.2 ≈ 2.5/ 2.1 ≈ 2.5 2.2 ≈ 2.5/ < 2 2.2 ≈ 2.5/2.1 ≈ 2.5	—	-28.4 -33.1 -28.4	-23.9 -22.9 -17.5	1.272 1.182 1.364	86.7/14.1 ≈ 61.4 86.7/14.1 ≈ 58.7 86.7/35.4 ≈ 60.7	[282]

continued

TABLE 5.2　(continued)
Selected Crystallization Studies That Used CNTs as Polymer Nanocomposite Fillers

Polymer	Polymer Characteristics	Nanotube Type	Nanotube Preparation Steps	Compositing Technique	Filler Loading (wt%)	N_{neat}/N_{comp}	$\log(k_{A,neat})/\log(k_{A,comp})$ (k_A in s^{-1})	ΔT_c (°C)	ΔT_m (°C)	$\left(\dfrac{t_{1/2,comp}}{t_{1/2,neat}}\right)$	χ_c (%)	References
PLLA	Biomass-derived linear aliphatic polyester, MFI = 8	MWCNT-COOH	Chengdu Institute, as-received	Melt-blended	2	2.78 ≈ 2.96/ 2.43 ≈ 2.65	0.75 ≈ 0.88/ 1.3 ≈ 2.2	6.7	b	0.633 ≈ 0.668 (MC)/ 1.26 ≈ 1.85 (CC)	—	[93]
Poly(1-butene)	Polyolefin, M_w = 570,000 g/mol	MWCNT	Aldrich, as-received	Compounded w/HAAKE mixer, 150°C, 7 min	3, 5, 7	4.2 ≈ 4.6/ 3.6 ≈ 4.8	−2.3 ≈ 0.54/ −1.86 ≈ 0.96	16.4	—	0.867	48.2/ 54.3 ≈ 55.1	[283]
Poly(butylene succinate)	Biodegradable polyester, MFI = 10	MWCNT-COOH	Chengdu Institute, as-received	Compounded w/Braebender mixer, 150°C, 10 min	2	2.6 ≈ 3.1/ 2.8 ≈ 3.1	−9.7 ≈ 0.1/ −5.3 ≈ 1.3	9.5	(see Ref. [284])	0.251 ≈ 0.269	—	[284]
PEN	Polyester, PET analog, [η] = 0.97 dL/g	CVD MWCNT	Iljin Nanotech Co., dried, as-received	Compounded w/HAAKE rheometer	0.1 ≈ 2	3.9 ≈ 5.8/ 4.6 ≈ 6.5	—	−13.8	−0.8 ≈ 0.3	0.808 ≈ 1.068	20.8/22.3 ≈ 31.1	[285]

Polymer	Description	CNT	Treatment	Processing	Loading						Ref.
PTMT	Custom-prepared polyester, $[\eta] = 0.795$ dL/g	MWCNT-terbutyl titanate	Refluxed in HNO₃, rinsed, functionalized	Sonicated w/ required amount of polymer in 1:1 TCE: phenol, cast into aluminum mold	0.05	$3.2 \approx 3.7/$ $2.6 \approx 3.0$	$-1.9 \approx 3.2/0.90$ ≈ 3.47	16.6	$-0.9 \approx 1.7$	—	[286]
PCL	Aliphatic, biodegradable polyester	MWCNT-COOH	Ethylene CVD, acid purification	Mixed in THF, cast, and melt-pressed	$0.25 \approx 1$	$2.58 \approx 2.86/$ $2.23 \approx 2.86$	$-3.4 \approx -0.40/$ $-0.59 \approx 3.3$	—	7.9	$19.7 \approx 38.7$	—	[287]
PCL	Polyester	SWCNT-PCL	(undisclosed)	Polymerized *in situ*	0.35, 1.8, 4.6	$2.3 \approx 2.6/$ $2.1 \approx 2.4$	$1.9 \approx 3.0/$ $2.0 \approx 3.1$	16.2	3.8	0.833	$48 \approx 58/$ $48 \approx 60$	[79,288]

Acronyms and variables: ccld crystallization (CC); cobalt molybdenum catalyst (CoMoCAT); chemical vapor deposition (CVD); dichlorobenzene (DCB); high-pressure carbon monoxide (HiPCO); isotactic polypropylene (iPP); melt crystallization (MC); melt flow index (MFI); poly(ε-caprolactone) (PCL); polydispersity index (PDI); polyethylene (PE); polyethylene oxide (PEO); poly(ethylene-propylene-diene) [P(E-PP-diene)]; poly(ethylene 2,6-naphthalate) (PEN); poly(ethylene terephthalate) (PET); poly(L-lactic acid) (PLLA); poly(trimethylene terephthalate) (PTMT); poly(vinyl alcohol) (PVA); 1,1,2,2-tetrachloroethane (TCE); tetrahydrofuran (THF); weight-average molecular weight $\equiv M_W$; number-average molecular weight $\equiv M_N$; Avrami exponent for neat polymer $\equiv N_{neat}$; Avrami exponent for CNT composite $\equiv N_{comp}$; Avrami rate constant for neat polymer $\equiv k_{A,neat}$; Avrami exponent for composite $\equiv k_{A,comp}$; neat polymer crystallization temperature subtracted from composite crystallization temperature $\equiv \Delta T_c$; neat polymer melting temperature subtracted from composite melting temperature $\equiv \Delta T_m$; time to half of complete relative crystallinity for neat polymer $\equiv t_{1/2,neat}$; time to half of complete relative crystallinity for composite $\equiv t_{1/2,comp}$; percent crystallinity $\equiv \chi_c$; intrinsic viscosity $\equiv [\eta]$.

[a] First set of values refers to crystallization under quiescent conditions. CNTs nucleated secondary crystallites (mother–daughter lamellae). Second set of values show Avrami exponents for crystallization during a step shear of 20 s⁻¹ for 5 s.

[b] Multiple melting events due to cold crystallization.

where k_A is a kinetic rate constant and N is the Avrami exponent. The Avrami exponent has important physical meaning. Exponents of 3 or 4 indicate instantaneous or sporadic nucleation of 3D spherulites, respectively, while an exponent of 2 indicates rather a disk-like growth, and an Avrami exponent of 1 is indicative of instantaneous nucleation of rod-shaped crystals. Fractional values are common, and are generally because of secondary nucleation or crystallization within spherulites. Haggenmueller et al. showed that the addition of SWCNTs changes the Avrami exponent in PE from ≈2.5–3, to a value of ≈1.6 for PE/CNT nanocomposite [84]. Thus, CNT within the PE matrix discourages spherulite formation in favor of plate-like lamellae, suggesting that the presence of nanotubes alters the growth habit. Table 5.3 shows some of the results that have been obtained for CNT nanocomposites formed from various crystalline polymers. It should be noted that Avrami crystallization kinetics can merely serve as a guide in nanocomposite systems. Nucleation at the CNT surface differs dramatically from that of neat polymer, and interphases as well as bulk regions have Avrami exponents and growth rates that differ from those associated with polymer at the interface.

As Table 5.2 shows, the ability for CNTs to nucleate crystallinity in semicrystalline polymers varies widely between different polymer composite systems. Nevertheless, several general conclusions can be drawn. First, it is evident that CNTs increase the nucleation rate of crystallizable polymer systems. In some cases, this rate increase was dramatic, and in certain other cases only slight, but as a general trend we see an overall increase in the crystallization rate on introduction of CNTs in all studies. Second, the Avrami constant decreases more often than it increases for polymer/nanotube composites. However, still, both increased and decreased Avrami constants are observed, which suggests that the purity and quality of CNTs affects polymer crystallization in ways that goes beyond simple dimensional rearrangements. In some systems, such as Mitchell's SWCNT/PCL nanocomposite [79,288], the change in the Avrami exponent would seem to change the geometry of the polymer crystal structure, but incongruously and the presence of CNTs barely affects the crystallization rate. In the case of MWCNT in poly(1-butene), even high CNT loadings had only a marginal impact on the Avrami constant and crystallization rate [283] which is indicative of extremely weak interactions. In other studies, such as the PE/CNT and the MWCNT/Nylon 6 couples, changes were much more pronounced. It is reasonable to assume stronger interactions between nanotubes and the crystalline matrix as the Avrami kinetics likely correlate with the degree of interaction between CNT and polymer.

Ozawa extended the Avrami model to quantify polymer crystallization kinetics using nonisothermal data [289]. It was reasoned that nonisothermal crystallization amounted to infinitesimal short crystallization times at isothermal conditions, given a crystallization temperature T [290]. This analysis led to the following equation:

$$X(T) = 1 - \exp\left[\frac{-K(T)}{\varphi^m}\right] \tag{5.6}$$

$K(T)$ is the Ozawa rate parameter, m the Ozawa exponent, and φ the heating rate. It is important to note that the Ozawa exponent and rate parameter do not have direct physical meaning that can correspond to the nonisothermal case which is related to constant temperature changes influencing the nucleation and growth of crystallites. The importance of this analysis lies in the relative ease with which the experiment can be performed. This is especially important for technically complicated systems wherein isothermal conditions are difficult to maintain. Such a study was conducted by Probst et al. [290] concerning surfactant-stabilized SWCNTs in a water-soluble polymer, PVA. PVA is hygroscopic and water vapor-accelerated degradation can occur below the melting point of PVA. To study crystallization in this system, PVA/SWCNT aqueous dispersions were initially dried using a rotavapor, followed by spin coating, and drying in a vacuum oven. Temperature scans were performed to determine crystallization kinetics through Ozawa analysis, and it was found that

TABLE 5.3
Electrical Percolation Data for Some CNT/Polymer Composite Systems Reported Since 2008

Matrix	CNT Type	Loading (wt% Unless Specified)	State (b/nb, a, l)	CNT Purification	D/L_N (nm)	Compositing Method	Φ_c (wt%, Unless Specified)	a	σ_{max} (S/m)	References
EBBA	CVD-MWCNT	0.01–1	b.	—	20/5000–10,000	Sonication of the mixed material	0.1	—	2.5	[348]
EPON 862 vinyl ester-styrene resin	CVD-MWCNT	0.1–0.75	—.	—	—	CNTs calendared in vinyl ester monomer; styrene mixed; initiator added, cured	0.1	—	0.13	[349]
Epoxy	MWCNT	0.05–0.5	b.	—	30–50/10,000–20,000	CNTs added to epoxy resin, sonicated, degassed; DMPS added, poured in mold	0.51	2.12	$1 \cdot 10^{-6}$	[350]
	MWCNT-NH$_2$	0.2–2	nb.	Ultrasonicated in EtOH; 3,6-diamino-1,2,4,5-tetrazine added, refluxed, filtered, dried			0.13	2.04	$6 \cdot 10^{-6}$	
Epoxy	MWCNT	14.73–36.05	l.	—	30–50/10,000–20,000	Repeated drop-casting of CNT solution/resin solution in turns	—	—	12	[351]
Gellan gum	CVD-MWCNT	0.002–0.091	—	—	—	Tip-sonication in polymer solution, drop casting	0.010–0.048	—	110	[352]
HDPE	HiPCO-SWCNT	13–70	nb.	—	0.8–1.2/1000	Sonicated, decorated with PSCs, vacuum filtered	≈1	2.0	193,000	[195]
HDPE	MWCNT	0.17–3.38	b.	—	20–40/—	Dispersed in alcohol, mixed; alcohol evaporated, compression molded	0.142	1.67	1	[353]
HDPE	CVD-MWCNT	0.5–7	nb.	—	40–50/≤$1 \cdot 10^5$	Twin-screw extrusion	1–2.5	—	0.7	[354]
LDPE			b.				≈2.5	—	0.04	

continued

TABLE 5.3   (continued)

Electrical Percolation Data for Some CNT/Polymer Composite Systems Reported Since 2008

Matrix	CNT Type	Loading (wt% Unless Specified)	State (b/nb, a, l) a, l	CNT Purification	D/L_N (nm)	Compositing Method	Φ_c (wt%, Unless Specified)	a	σ_{max} (S/m)	References
LDPE	CVD-MWCNT, research grade	0.03125–4 (mg/mL)	1.	—	9.5/1500	Tip sonicated in THF to promote dispersion, sonicated w/PE film, dried	4 (vol%)	1.8	$4.15 \cdot 10^4$	[355]
LDPE	MWCNT	0.5–7 (v)	—	—	10/10,000–30,000	MWCNT added to polymer solution, stirred, sonicated, drop-cast	$1.01–4.36^a$ (vol%)	$1.12–1.84^a$	0.0079	[356]
iPP	CVD-MWCNT	1–6	a.	—	9.5/1500	Twin-screw extrusion; vibration injection-molded	≈ 1	—	1.2	[357]
iPP-g-maleic anhydride	CVD-MWCNT HiPCO-SWCNT	0.1–2 0.02–2	—	—	—	Surfactant-stabilized, stirred into polymer emulsion, freeze-dried, compression molded	0.11 0.04	— —	68 7	[358]
PSt	CVD-MWCNT HiPCO-SWCNT	0.5–1.7 0.5–1.8	—	—	—		0.81 0.55	— —	46 9	
P(St-b-B.-b-MMA)	CVD-MWCNT	1–6	—	—	—	Sonication in acetone with SBM terpolymer, cast and dried either under ambient or sat. vapor conditions	$< \approx 1\%$	—	—	[359]
P3HT	SWCNT	0–30	b.	—	0.7–1.2/7100–7800	Sonicated and shaken in 5 mg/ml polymer; settled, decanted, spin-coated	2	1.5	0.00013	[360]

Polymer	Nanotube	Loading		Treatment	Dimensions	Processing				Ref.
PDMS	CVD-MWCNT	0.45–3	nb.	—	50–100/ 5000–15,000	Sonication in petroleum ether, shear mixed for 1 h, evaporation, cast, cured	≈ 0.5%, thin; ≈2%, bulk	≈1	0.0003, fresh; 0.00009, aged 4 weeks	[361]
						PDMS poured over MWCNT, left to soak with petroleum ether, shear mixed, cast, cured	< ≈0.3%, thin; ≈2%, bulk	≈1.6	2.4×10^{-5}, fresh; 2.8×10^{-5}, aged 4 weeks	
PEEK	CVD-MWCNT	0.25–17	—	—	9.5/1500	Twin-screw extrusion, compression molded	≈1	—	1	[362]
PFA	CVD-MWCNT	0.2–0.58 0.02–0.18	nb.	—	$50/1 \cdot 10^5$	Dispersed in SDBS (aq.); mixed with aq. PFA colloid, sonicated, sprayed onto heated substrate	0.43 (vol%); 0.17 (vol%)	1.67 2	0.006 0.0037	[363]
PET	HiPCO-SWCNT	0.5–3	a.	—	—	Dispersed in HF2P; PET flakes added and mixed, poured, doctor-bladed	0.5	—	$4 \cdot 10^{-6}$, $1.3 \cdot 10^{-5}$ @ 9.4T magnetic field	[364]
Poly(BPA-carbonate)	CVD-MWCNT	0–10	—	HNO_3-H_2O-treated	—	Stirred in CH_2Cl_2 for 2 days, drop-cast	0.3 0.19	1.29 1.02	$4.35 \cdot 10^4$ $1.60 \cdot 10^5$	[365]
PA 12	CVD-MWCNT	5, 7	a.	—	10–15/ ≈ 2000	Solution-mixed, then single screw extrusion	0.19–0.21 (vol%)^c	—	0.36	[366]
PA 6,6	MWCNT	0.25–10	—	HF followed by HCl	7.3/— 10.1/— 12.1/— 15.5/—	Mixed in a twin-screw microcompounder	1.48 1.49 2.25 —	3.8 4.3 6.7 —	16 79 5 $3.2 \cdot 10^{-13}$	[367]

continued

TABLE 5.3   (continued)
Electrical Percolation Data for Some CNT/Polymer Composite Systems Reported Since 2008

Matrix	CNT Type	Loading (wt% Unless Specified)	State (b/nb, a, l)	CNT Purification	D/L_N (nm)	Compositing Method	Φ_c (wt%, Unless Specified)	a	σ_{max} (S/m)	References
PEI	CVD-MWCNT	0.01–3	—	Nitric acid sonicated; separated and decanted, filtered, rinsed	$8/5\cdot10^5$	Polymer dissolved in H$_2$O, CNTs added gradually during 3 h sonication	0.02	—	4400	[368]
PSt	FeNP-MWCNT	0.1–7	nb.	—	$50\text{–}80/2\text{–}3\cdot10^5$	Briefly sonicated in To.; mixed, sonicated, dried under vacuum	0.21	1.95 ± 0.08	0.3	[369]
PSt	HiPCO-SWCNT	22–82	b.	—	Bundle diameter ≈13	Tip-sonicated in PSt/NMP solution for 3 min, bath sonicated 4 h, filtered, rinsed, dried		2.2 ± 0.2	100 $-1.5\cdot10^4$	[370]
PSt	CVD-MWCNT	0.5–7	b.	—	$10\text{–}100/500\text{–}40,000$	4 h sonication in DMF, mechanical mixing, electrospinning	3.5^b	0.795^b	0.0085	[371]
PSt, with St-B.-St copolymer	CVD-MWCNT	1–5		—			4^b		0.005	
PSt	CVD-MWCNT, industrial grade	0.08–80	—	—	9.5/1500	Sonication in styrene, polymerized *in situ*; PSt beads added subsequently; melt pressed	0.045	2.11	16	[372]
s-PSt	CVD-MWCNT	1–10	nb.	—	40/–	Shear-mixed CNT in NMP, sonicated, added to s-PSt/NMP solution and bulk-cast into water	2–3	—	0.135	[373]
PU foam	CVD-MWCNT	1–2	nb.	—	$20\text{–}40/10,000\text{–}30,000$	CNTs dispersed in EtOH, mixed with polyether polyol, sonicated; surfactant, catalyst, water added; isocyanate added	1.2	—	0.0005	[374]

TPU	CVD-MWCNT	—	0.2–5	—	9.5/1500	Twin-screw extrusion; hot-pressed	0.13	4.6	1.2	[375]
TPU	MWCNT MWCNT-HFU	—	0.5, 1	One-pct polycondensation to graft HPU to CNTs	40–60/–	CNTs added to P(EO-THF), mixed with aliphatic polyisocyanate, dibutyltin dilaurate, degassed, cast into molds	≈ 0.5 >1	—	$1.1 \cdot 10^{-12}$ $1.2 \cdot 10^{-7}$	[376]
PVDF	CVD-MWCNT	—	0.05–8	—	10–50/4000–10,000	PVDF solution in DMAc probe sonicated with CNT, stirred, poured; ambient solvent evaporation	0.07	—	63	[377]
Unsat. Polyester	CVD-MWCNT	—	0.05–0.3	—	9.5/1500	Mixed using styrene plasticizer at 1000 rpm for 7 h, cured in an oven	0.026	2.55	0.13	[378]
Vulc. PP/P(E-P-diene)	CVD-MWCNT	nb.	1–16 (v)	—	10–30/5000–15,000	HAAKE mixer, 200°C, 100 rpm, 20 min; hot pressed	0.07 (vol%)	4.5	2	[379]

Abbreviations: aligned (a.); butadiene (B.); bundled (b.); block (*b*); bisphenol A (BPA); chemical vapor deposition (CVD); dimethyl acetamide (DMAc); dimethyl formamide (DMF); 4,4-diamino-diphenylsulfone (DMPS); 4-ethoxybenzylidene-4-*n*-butylaniline (EBBA); ethanol (EtOH); iron nanoparticle (FeNP); high-density polyethylene (HDPE); 1,1,1,3,3,3-hexafluoro-2-propanol (HF2P); high-pressure carbon monoxide (HiPCO); hyperbranched poly(urea-urethane) (HPU); graft (*g*); low-density polyethylene (LDPE); methyl methacrylate (MMA); not bundled (nb.); *N*-methyl pyrrolidone (NMP); polyamide (PA); poly(ethylene oxide-tetrahydrofuran) [P(EO-THF)]; perfluoroalkoxy (PFA); polyethyleneimine (PEI); poly(ethylene terephthalate) (PET); polystyrene (PSt); polyurethane (PU); poly(vinylidene fluoride) (PVDF); syndiotactic (*s*); styrene (St); tetrahydrofuran (THF); toluene (To.); thermoplastic polyurethane (TPU); unsaturated (Unsat.); volume percent (vol%); vulcanized (Vulc.); weight percent (wt%)

Variables: critical exponent (*a*); diameter (*D*); length (L_N); maximum conductivity (σ_{max}); critical loading percolation threshold (Φ_c).

a Varied with film thickness.

b As reported.

c Extrapolated from shift factor calculated through uniaxial extension.

in the neat polymer, m varied between 0.56 and 3.31, but was dramatically lower for the samples incorporating CNTs, varying from 0.24 to 1.42 which indicates a change in the crystal morphology.

5.4.3.2 Crystallization of Polymer/CNT Composites under Shear

In addition to the *in situ* polymerization method, low concentrations of CNTs can simultaneously be dispersed and incorporated into a polymer matrix using melt processing techniques. This can be as straightforward as pouring CNT and polymer pellets into a hopper and mixing them in a screw extruder. Single-screw extruders have a very simple processing geometry resulting in a local shear alignment [291]. In general, samples produced in this manner have a high degree of anisotropy. Twin-screw extruders improve the isotropic dispersion of CNTs in the polymer matrix [292,293]. It is important to note that one effect of incorporating CNTs into a melted polymer is a dramatic decrease in melt flow index, especially at high CNT content. This can lead to reduced processing throughput and shortened equipment lifetime. The tradeoff is that nanotubes dispersed in highly viscous media take considerably longer to reaggregate. This widens the processing window and enables additional techniques for alignment and the facile introduction of additives.

Observations of the nanostructure of CNT composites produced by melt spinning [294], injection molding [187,295,296], and recrystallization of polymer thin films [297] reflect the structure of the NHSK previously described. Using the injection molding process, bars of HDPE/MWCNT composite can be produced. As is typical of injection molded parts, orientation varied throughout the samples, showing almost no orientation at the core region but a highly aligned shear layer (Herman's orientation parameter of 0.84 at a depth of 800 µm; see Figure 5.9). Fast cooling occurred at the skin layer and, thus, no specific change of the crystallization was observed in a polymer that rested against a cooling plate. However, the high degree of alignment of the polymer chains coupled with slower cooling produced folded chain structures at both the shear layer and core layer for dynamic packing injection-molded bars.

5.4.4 NHSK as Polymer Nanocomposite Filler Materials

PE and Nylon 6,6 have been used to produce polymer nanocomposites via solution mixing as well as via blending techniques. In the presence of CNTs, it was shown that the triclinic α-form of Nylon 6,6 is the preferred crystal structure [280]. Depending on the solvent used to produce Nylon 6,6 NHSK, different morphological features have been observed. NHSK formed using glycerol at 180°C showed either a more traditional banded NHSK structure or a wrapped coating. When NHSKs were formed in dimethyl sulfoxide at 144°C, overgrown crystals were produced with possible transcrystallinity of the densely nucleated crystallites.

After an NHSK crystallization experiment during which the solution was still at a predetermined crystallization temperature (185°C in glycerol), additional Nylon 6,6/glycerol solution was added to the mixture, then allowed additional time to crystallize, and finally cooled to room temperature. Negative spherulites were formed by this process. Figure 5.10c shows a diagram of the process used to produce PA spherulites with fully incorporated MWCNT.

Curiously, although individual Nylon lamellae were clearly visible, MWCNTs could no longer be discerned within the material, as the kebab structures were overgrown by crystallizing polymer and unable to prevent CNT assimilation into the growing spherulite. To demonstrate that the NHSK had not been phase separated out of the spherulites, nitric acid etching was used to open hole areas into these structures. The acid etching exposed CNTs within the spherulite, indicating that the nanotubes had been fully incorporated into the nylon matrix. Melt pressing the material resulted in a nanocomposite with fine crystalline grains as observed by PLM. Darker regions showed where nanoconfined polymer was interspersed within the matrix. Thus, NHSK served as the material for incorporation into an affine polymer matrix enabled by physical functionalization. A similar experiment has been reported for PE NHSK used as filler materials in a PE matrix [234]. In this experiment, SWCNT or MWCNT formed in *p*-xylene solution was mixed at elevated temperatures with

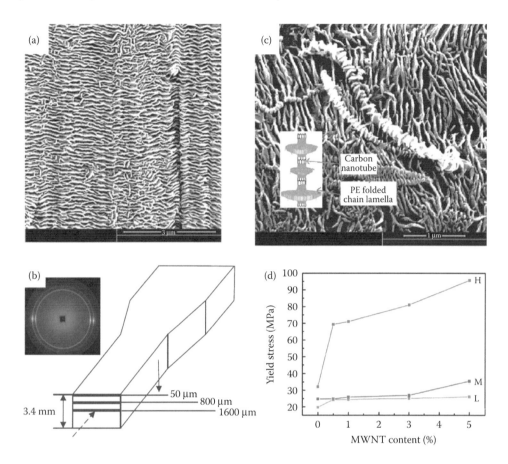

FIGURE 5.9 Folded-chain PE/MWCNT structures produced by injection molding (a,b) and melt spinning (c,d). (a) SEM micrograph of composite layer observed at a depth of 800 μm (Reused from J. Yang et al., *Polymer*, 51, 2010, 774–782.). (b) Depiction of the depths of injection molded samples studied. Inset: 2-D WAXS pattern of 800 μm depth sample shown in (a) (Reused from J. Yang et al., *Polymer*, 51, 2010, 774–782.). (c) SEM micrograph of melt-spun composite showing hybrid superstructure (Taken from F. Mai et al., *J. Phys. Chem. B*, 114, 2010, 10693–10702.). (d) Yield stress versus CNT content for high, medium, and low drawing ratios (denoted as H, M, and L, respectively). (Adapted from L. Wang et al., *Colloid Polym. Sci.*, 289, 2011, 1661–1671.)

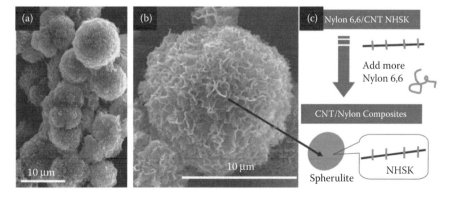

FIGURE 5.10 (a) and (b) Spherulites formed by the addition of Nylon 6,6 to Nylon 6,6 shish kebabs showing negative spherulites. (c) Schematic representation of the production method. (From L. Li et al., *Polymer*, 48, 2007, 3452–3460.)

dissolved HDPE, also in *p*-xylene. SWCNT fillers at low loading contents were assimilated into PE microspheres, whereas MWCNT were still visible after crystallization, and showed an overgrowth PE crystal structure with still visible lamellae.

5.4.5 NHSK PAPER

Perhaps, the most intriguing aspect of the NHSK functionalization technique is its ability to form a well-dispersed nanocomposite paper without any matrix. With controlled structure, the relatively simple system of PE kebabs on SWCNTs offers control over several factors simply by adjusting the relative ratio of polymer to CNT in the feeding mixture [195]:

- Surface roughness of the nanostructure is tuned by the diameter of the PE kebab.
- As graphite and CNTs are slightly hydrophilic, while PE single crystals are hydrophobic, the wettability of the structure can be tuned by varying the kebab size.
- Electrical conductivity varies with the square of the kebab diameter, as the PSC spacer effect controls the number of contacts that form between neighboring tubes.
- Films formed using NHSK materials have variable pore sizes from 10 to 100 nm, also controlled by the relative distances between neighboring tubes.
- As NHSK materials are functionalized and dispersed at controlled distances from one another, they are ideally suited to produce novel nanocomposites which can be tuned both through dispersion and steric hindrance [234].

PE single crystals enabled controllable spacing between individual nanotubes and an expanded form of buckypaper could be produced. To produce free-standing NHSK films, solutions of NHSK were vacuum-deposited over PTFE membranes, rinsed with methanol, and finally allowed to dry. SWCNT loading contents, including 13, 20, 25, 52, and 70 wt% films were produced, corresponding to approximately 8.3, 13, 17, 40, and 59 vol% when the pore volume fraction of the film is ignored. These samples were compared to classical SWCNT buckypaper. "NHSK buckypaper" had a thickness of 5–15 μm, and thicker films were found to delaminate easily without careful control of the drying conditions. In contrast to that, thinner films were difficult to obtain in a free-standing fashion. It was found that these films had weak orientation in the plane of the film. Waviness of the orientation of NHSK deposited from solution was attributed to entanglements of the quasi-3D structure which formed cloud-like aggregates in solution. Figure 5.11 shows a comparison of NHSK films and SWCNT buckypaper. With NHSK films, it was noted that there was near-complete exfoliation of individual CNTs, whereas the SWCNT buckypaper surface was covered mostly by CNT bundles with an average size of approximately 13 nm. A noteworthy aspect of the NHSK paper was its high apparent pore volume. BET surface area analysis confirmed that pores averaging ≈30 nm pervaded the 25 CNT wt% film, with porosity accounting for ≈40 vol% of the material. Electrical conductivity was affected in an unusual way by the presence of the NHSK spacers. It was noted that, despite the fact that films should theoretically have been far above the percolation threshold of the CNT loading fraction for an SWCNT/PE composite, conductivity scaled with a power law exponent of 2 instead of 1. In typical composites far above the percolation threshold, it is assumed that polymer is replaced by filler particles isotropically and this effect is described in detail in Section 5.5.2. This unusual behavior was believed to be because of the way in which the polymer was replaced by conductive filler in the composite film. In NHSK composites, this replacement is not isotropic and when a conductive filler is added at the expense of flat, nearly circular edge-on lamellae decrease the average distance between conductive materials by a power of 1/2 as the weight fraction increases.

NHSK paper has a unique structure: PSCs are oriented with their lamellae directed outward from the film surface. The surface also has furrows, mimicking the disorganized CNT alignment within

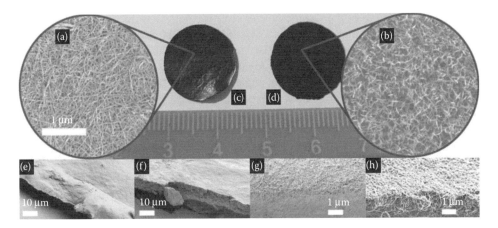

FIGURE 5.11 Comparison of SWCNT buckypaper and NHSK paper. (a) Top surface of SWCNT buckypaper. Bundles could be observed on the surface. (b) NHSK paper surface. Photographs of (c) SWCNT buckypaper and (d) NHSK paper show the differences in reflectivity of the surfaces due to efficient light scattering by the NHSK paper. (e) Cross-sectional view of SWCNT buckypaper. (f) Freeze-fractured cross-section of NHSK paper. (g) Close-up of SWCNT paper. (h) Close-up of high PE content NHSK paper showing relaxation of the PE/CNT. (Adapted from E.D. Laird et al., *ACS Nano*, 6, 2012, 1204–1213.)

the film. Therefore, a hierarchical roughness system is created: at the submicrometer scale, inter-NHSK pores create an undulating pattern on the surface. Within this pattern, intra-NHSK distances have a high-aspect ratio nanoscale roughness due to the perpendicularly oriented kebabs. Because of this unusual roughness pattern (Figure 5.12), an interesting wetting behavior was observed which differs from the one of bare HOPG or buckypaper surfaces.

The sessile drop in Figure 5.12 shows that SWCNT–NHSK with a 25 wt% loading of CNTs demonstrate superhydrophobicity at a static contact angle of 152.3°, which, however, was lost after annealing. Throughout the range of different SWCNT loadings, contact angles first increase then decrease. At lower PE contents, the nanoscale surface roughness decreases. Consequently, the

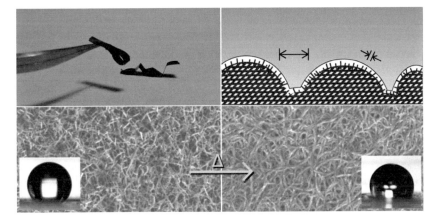

FIGURE 5.12 Characteristics of NHSK paper, composed of 25 wt% SWCNT. The materials are flexible and workable, advantages over the relatively brittle SWCNT buckypaper. Superhydrophobicity is achieved by a mechanism similar to the "petal effect" as explained in the text. Annealing of the NHSK films results in a relatively smooth composite where PE has completely wetted the surface.

droplet is able to wet more of the submicrometer scale topology and the contact angle decreases toward the intrinsic contact angle of the bare SWCNT buckypaper, measured to be 82°. The contact angle of SWCNT paper reaches a maximum at 25 wt% SWCNT content and at higher loadings, the PE single crystals begin to collapse, the roughness decreases, and the droplet contact angle begins to depress toward the intrinsic contact angle of PE single crystals (≈94°) [298]. Ordinarily, water droplets roll off superhydrophobic surfaces very easily; however, it is important to note that albeit the very high contact angle, this was found not to be the case for NHSK. Droplets as large as 5 μL could be tilted to any angle without roll-off and even suspended upside-down from superhydrophobic NHSK films. Several surfaces have been shown to exhibit this type of statically superhydrophobic but dynamically nonsliding behavior, including red rose petals [299]. The cause of this curious wetting property has been shown to be the dual length scale roughness [300–304]. Droplets deposited on dual-roughness surfaces have hydrophobic wetting characteristics that are dependent on the aspect ratio of the asperity over which the water droplet sits. There are two modes of wetting: the Cassie–Baxter and the Wenzel wetting behavior. High-aspect ratio asperities might have Cassie–Baxter wetting behavior, that is, the droplet will remain suspended above a certain feature without actually wetting it. Hydrophobic surfaces that have rough surface features with shallower profiles might be able to be wetted by droplets (Wenzel wetting behavior). Now, it has been shown that NHSK films exhibit Cassie–Baxter wetting at the nanoscale and Wenzel wetting at the submicrometer scale. Therefore, a droplet could deform to fill the submicrometer inter-NHSK morphology, but would still be suspended above the NHSK film by virtue of its inability to wet the nanoscale intra-NHSK features. Extending the functionality of NHSK films via an additional coating step using a low surface energy material [305,306] has been found to further enhance this repulsive behavior.

5.5 PROPERTIES OF POLYMER/CNT COMPOSITES

5.5.1 Mechanical Properties of CNT Composites

At present, it is not yet clear whether CNTs will become cost-effective fillers for the enhancement of mechanical properties alone, without also being used for uniquely specialized applications where there are requirements for exceptional chemical stability, elevated or cryogenic temperatures [168,169,307], or in microfabricated parts [102]. The cost and manufacturing challenges inherent to CNTs are relatively high compared with micrometer scale fillers, such as fiberglass, carbon fibers, silica [308], or clays [309–311]. The main problem for CNT-based mechanical reinforcement, though, is the unevenness of the quality and predictability of reinforcement. Processing issues of CNT nanocomposites make the final properties difficult to engineer, and anisotropy, changes in thermal history, storage time, and residual solvents can greatly influence the final material. Although improvements in toughness, wear resistance, and tensile strength can be achieved using CNTs instead of fiberglass or carbon fiber, cost, and variability make this an uncertain direction. Part of the problem boils down to the fundamental characteristics of CNTs which have extraordinary tensile strength, toughness, and strain-to-break, but are weak against shear. Off-axis loading of the CNT can cause dramatic reduction in the mechanical properties of the nanocomposite in that local area which can significantly contribute to the ultimate failure of a device or component. CNTs do contribute to enhancement of mechanical properties though, in a wide range of polymer [168,186,312–319] as well as inorganic matrices [320,321], and it is likely that this will contribute to their adoption in widespread industrial and consumer products.

Although there is enormous variability in mechanical properties because of local concentration and isotropy variation, the Halpin–Tsai model has been shown to provide reasonably good approximations of the elastic modulus of randomly distributed CNT-based nanocomposites [322,323]. The Halpin–Tsai model (Equations 5.7, 5.8) is specifically for glassy matrix polymers, but it has also shown to be fairly effective for semicrystalline polymers as well [324].

$$E = E_M \left[\frac{3}{8} \frac{1 + 2(L_N/D)\eta_L v}{1 - \eta_L v} + \frac{5}{8} \frac{1 + 2\eta_T v}{1 - \eta_T v} \right] \quad (5.7)$$

Here, E_M is the elastic modulus of the polymer matrix, L_N the length of the nanotubes (ideally all of the same length), D the diameter, and v the volume fraction. The dimensionless parameters η_T and η_L can be predicted from the elastic modulus of the nanotube* (E_N) and the modulus of the matrix material assuming the length/diameter ratio is known:

$$\eta_L = \frac{(E_N/E_M) - 1}{(E_N/E_M) + 2(L_N/D)}$$

$$\eta_T = \frac{(E_N/E_M) - 1}{(E_N/E_M) + 2} \quad (5.8)$$

The Halpin–Tsai model shows two instabilities, one at $\eta_L \cdot v = 1$ and the other at $\eta_T \cdot v = 1$ (as η_L and η_T are fixed, these values increase with increased filler loading). Neither of these have actually a real physical meaning but at these points the effective overall modulus drops as the term begins to work against the overall modulus. However, this is not a real effect: in reality, the stiffness increases generally with increased CNT loading content [316]. The Halpin–Tsai model underestimates nanocomposite stiffness at higher volume fractions for simple glassy polymer systems. A few systems are slightly more complicated. For matrices such as epoxies, where crosslinks develop that span through the glassy matrix during curing, the storage modulus can decrease with increased volume fraction [325]. Alignment has been shown to improve tensile modulus for components which is helpful if the stress distribution is known *a priori* [291,326–328]. Also, the yield strength has been shown to either decrease [317,329,330] or increase [185,327,331] with CNT loading. Overall, it is difficult to generalize or predict the mechanical properties of CNT/polymer nanocomposites [332]. CNTs have enormous variability in their own mechanical properties, and load transfer between polymer matrix and CNT which is directly related to pull-out stress can sometimes be poor—even when chemical functionalization is employed. Reasons for poor interfacial adhesion could include bundled nanotubes, sliding between shells of multiwalled tubes, chemical functionalities incompatible with matrix polymer, or uneven functionalization. In general, however, chemical crosslinks are known to improve load transfer from matrix to filler [333,334]. Physical functionalization is generally not as well suited to producing strong mechanical property enhancements, but PSC decorations do promote strong interfacial adhesion between CNT and polymer matrix [335].

Although there have been few studies directly focusing on this aspect, the actual processing history has often a major impact on the final properties of nanocomposites. Residual stresses build up especially around dense regions of CNTs, causing diminishing returns in terms of work-to-failure, ultimate tensile strength, and so on, which diminishes the loss of polymer chain entropy due to stretching and entanglement. Polymer/CNT nanocomposites researchers often use a postprocessing annealing step to remove any residual solvents and to even out the processing history. One of the few studies devoted to the impact of the process history on nanocomposite properties was the study by Koerner et al. on MWCNT-filled thermoplastic polyurethane

---

* The elastic moduli of numerous species of CNT have been measured directly and predicted from experimental models. Direct measurement of CNTs is difficult. Tensile tests between adjacent AFM tips have been performed with both MWCNT [408] and SWCNT bundles [409]. AFM-based tensile testing has also been performed by deflecting a bundle of SWCNT that was suspended across a channel [330]. The influence of shear stress on these materials was probably significant. Although indirect, perhaps the clearest technique for measurement of Young's modulus for CNTs is the observation of a standing wave in a cantilevered tube by room-temperature thermal vibration. For MWCNT, Young's modulus values range from 0.4 to 4.1 TPa [410] and for SWCNT the modulus was 1.3 TPa [411].

(TPU) [208]. They observed that based on the solvent removal rate, initial orientation of the MWCNT, thermal history, and length of time of sample storage, modulus differences of up to 100% were shown even for similar samples. The impact of processing history was found to be more pronounced for soft segment linear diol regions than the cyclic hard segments. In these soft segment regions, MWCNTs were found to increase the crystalline content which could improve the tensile modulus. However, this would be to the detriment of chain entropy, particularly after long periods of sample storage during which soft segment regions were able to develop their crystalline domains.

CNT-based nanocomposite production benefits from the long history of research on carbon fiber and fiberglass-based composites. One promising avenue of nanocomposite development is represented by crafting sandwich structures and functionally graded plates [336–338]. For composites with the same mass fraction of CNT, the bending moment was calculated to increase for functionally graded plates as opposed to uniform composites, whereas the center deflection was predicted to be almost unchanged. Much work remains to be done to understand and predict CNT reinforcement, yet it is worth to attempt inventing engineered structures such as these or others [339] as there is considerable room for extending the properties of polymer-based nanocomposites.

5.5.2 CONDUCTIVITY OF CNT COMPOSITES

CNTs have a tremendous advantage over other composite fillers in terms of electrical conductivity. Small volume fractions of MWCNT and even smaller amounts of SWCNT are needed to make an electrically conductive nanocomposite, as will be explained below. At the low loading fractions needed for electrical conductivity, composites can often maintain transparency which makes them attractive as next-generation conductive coatings and may allow them to replace indium tin oxide glass [340–343]. The reason for the low loading fraction needed for conductivity is the high-aspect ratio of these nanofillers and the odds of points of contact forming between two randomly oriented filler particles quickly become very high (see Figure 5.13b,c). Pike and Seager were the first to consider this aspect [344]. An extreme case is the so-called "slender rod limit" [345], an assumption that can be used to calculate the critical percolation volume fraction for very high aspect ratio nanofillers (~1:500 or even larger) of uniform length, provided they are all conducting and rigid [346]. In the simplest model for electrical percolation in a two-component mixture, one phase is a perfect insulator and the other phase is a good conductor with a conductivity of σ_C. It also follows that if the entire volume of the insulating phase would be replaced by conductive filler, the conductivity of the entire composite will also be σ_C. If not sufficient conducting particles are isotropically blended into the insulating matrix, as in Figure 5.13a, the conductivity should be zero. Between these two extremes, at some loading fraction, sufficient conducting particles that represent the percolation threshold have accumulated in the matrix to form a conducting network (Figure 5.13b, conducting network highlighted). Above this threshold, the conducting network quickly begins to accumulate the previously isolated conductive particles, and the probability of a particle in the matrix not participating in a conductive pathway is β. During this transition region, islands of previously discontinuous and unconnected conducting material are rapidly brought into the conducting network. This transition is characterized by a power law, scaling as $\log(\sigma) \propto a \log(v - v_C)$. For high-aspect ratio filler particles, β quickly approaches zero, and every incremental replacement of insulating matrix with conducting filler contributes monotonically to the conductivity of the system. Similarly, a characteristic sigmoid shape is predicted for the electrical percolation transition. Within the transition region, the so-called "critical exponent" a is predicted to have a value of 1 for a 1D system, 1.33 for a 2D system, and 2 for a 3D system. Far above the percolation threshold, this value is expected to drop to 1 as β drops to 0 and the system is reduced to isotropic replacement. In practice, exponents much higher than 2 have been observed because of nonideal conditions. Winey et al. have done some critical work both in experimental and theoretical percolation in nanocomposites, including predicting and experimentally showing some of the effects alignment has on network formation. It

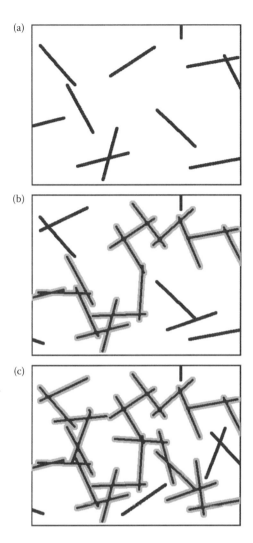

FIGURE 5.13 Schematic representation of the onset of percolation in CNT composites. (a) Below percolation, CNTs do not form a conducting pathway. (b) At percolation, CNTs in contact with one another can stretch across the sample. (c) At higher loadings, more CNTs are accumulated into the conducting network.

was shown that orientation of CNTs within a polymer matrix has the effect of ultimately increasing the conductivity, but also increasing the volume fraction needed for percolation [347]. Percolation behavior of numerous CNT/composite systems has now been experimentally verified [81]. Bauhofer and Kovacs reviewed CNT nanocomposite conductivity in 2008; a collection of CNT composite experiments reporting percolation behavior since then is presented in Table 5.3.

5.5.3 THERMAL CONDUCTIVITY OF COMPOSITES

CNTs have a high thermal conductivity,[*] therefore it seems quite natural that they should be attractive candidates for thermally conductive composites. Perhaps, surprisingly, this is only

---

[*] Thermal conductivity for CNTs is expected to be on the order of 2000–6600 W/m K for isolated CNTs at room temperature [388]. These values are fairly optimistic for defect-prone CNTs, but experiments have found thermal conductivity for an isolated MWCNT to be 3000 W/m K. Chemical functional groups dramatically reduce the theoretical and experimental thermal conductivity.

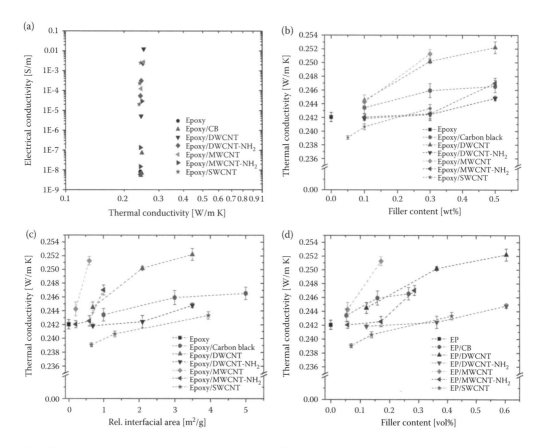

FIGURE 5.14 Thermal conductivity as a function of filler property. (a) Invariance of thermal conductivity with electrical conductivity. (b–d) Thermal conductivity plotted as a function of filler content (b), interfacial area (c), and composite volume (d), respectively. (Adapted from F.H. Gojny et al., *Polymer*, 47, 2006, 2036–2045.)

occasionally true [183,380–382], and CNT/polymer composites do not approach the rule of mixtures for this property (see Figure 5.14). Gojny et al. [383] found that thermal conductivity was nearly invariable with addition of various carbon fillers including carbon SWCNT, DWCNT, and MWCNT. This was in agreement with previous work done on a similar system [384] although Biercuk et al. found the opposite effect [385]. Poor thermal performance in such materials is thought to be because of disruption of phonons within the CNT due to interfacial scattering and damping of the CNT's vibrational modes. Observing that MWCNT and DWCNT composites showed the best thermal conductivity, they ascribed the enhancements for these nanocomposites to three factors: (1) relatively low interfacial area, (2) weak interfacial adhesion between nanotube and polymer, and (3) shielded internal layers. Thermal conductivity, therefore, apparently has requirements for nanocomposites which are directly incompatible with those needed for mechanical property enhancements.

Certain techniques can be used to improve the thermal conductivity of CNT composites. For instance, relatively high enhancements of the thermal conductivity can be identified for nanocomposites with aligned CNTs [386]. In particular, SWCNT nanocomposites were prepared by *in situ* injection molding using a sacrificial adhesive coating on the cooling platens to form composites with exposed CNT ends [387]. Using this method, thermal conductivity was enhanced by 280% over the neat polymer, as opposed to 5% for dispersed CNTs. However, it was noted that the improvement observed through this experiment was still an order of magnitude lower than expected values.

5.5.4 THERMAL STABILITY AND FLAMMABILITY OF POLYMER/CNT NANOCOMPOSITES

Although the thermal conductivity of CNT nanocomposites is still relatively low, heat dissipation into a CNT composite has empirically been widely studied throughout the literature. This heat dissipation within a polymer matrix has important consequences. In many cases, thermal degradation follows Arrhenius behavior as in Equation 5.9. In this equation, α is the fraction of remaining polymer, the temperature and universal gas constant are T and R, respectively, $f_T(\alpha)$ is a defining function of the degradation, k_T is the heating rate, A is an exponential prefactor (a material propery-dependent constant), and E_a is the activation energy for thermal degradation.

$$\frac{d\alpha}{dT} = \frac{f_T(\alpha)}{k_T} A \exp\left[\frac{E_a}{RT}\right] \tag{5.9}$$

Ordinarily, the degradation function $f_T(\alpha)$ is continuous but arbitrary, making it difficult to predict. However, it can be directly calibrated against certain material life properties. CNTs frequently delay the onset of thermal degradation in composites because of the improved bulk heat dissipation imparted by CNTs and the ability of CNTs to adsorb free radicals [388].

An important related application of CNT nanocomposites is flame retardancy. It was found that during cone calorimetry experiments CNTs drove degradation toward char formation instead of auto-accelerated burning, resulting in a much lower heat release profile [389]. CNT composites remain largely intact after burning with a thin nanotube network at the surface. Initial flame retardancy experiments used MWCNTs produced by acetylene decomposition mixed with ethylene-vinyl acetate copolymer. More recently, SWCNTs have been shown to be even better for this purpose with a strong dependency on the quality of dispersion [390]. Carbon black is not useful for flame retardancy, and the loading fraction needed for CNF to be of use is prohibitively high. Numerous matrix polymers as disparate as iPP [391–396], PA [397–399], unsaturated polyester [400,401], PMMA [388], poly(L-lactide) [402], and chloroprene rubber [403] among others have shown improved flame retarding properties, suggesting that the chemistry of the polymer matrix is not the main contributing factor. Evidence suggests that the physical mechanism behind the flame retarding properties of CNT nanocomposites is the formation of a heat-shielding surface layer. Kashiwagi et al. [390] showed that improvement in flame retardancy* could be directly correlated with the formation of a mechanically percolated filler network. During the flammability test (see Figure 5.15), a neat PMMA sample bubbled vigorously and burned quickly leaving no residue at the bottom of the pan. A PMMA/carbon black (CBP) sample bubbled and left a small amount of island-like black soot at the bottom of the container. However, a sample that had 0.5 wt% well dispersed and exfoliated SWCNT in the PMMA matrix bubbled gently for a short time, then stopped. After the mass loss rate profile dropped to zero, there remained a porous structure with a well-developed solid network only slightly thinner than the original composite. The differences between these materials could be traced to their viscoelastic behavior: the frequency-dependent storage modulus of the SWCNT sample plateaued at low frequency, approaching Hookean behavior, whereas samples with a more intense heat release profile (e.g., neat polymer, carbon black, low-loading carbon fiber) remained viscoelastic even at low frequencies.

5.6 CONCLUSIONS AND FUTURE OUTLOOK

In this chapter, structure and properties of CNT-based polymer nanocomposites have been discussed. In particular, the ability of CNTs to serve as heterogeneous nuclei for PSCs has been considered, along with the particulars of the interplay between polymer and nanotube that favor such interactions.

---

* This experiment used high-temperature nitrogen gasification which involves no flaming but simulates fire conditions.

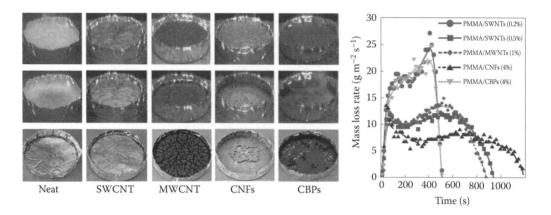

FIGURE 5.15 Results of cone calorimetry experiments for various PMMA/carbon allotrope systems. Heavily filled (4 wt%) CNF composites and lightly filled (0.5 wt%) SWCNT composites performed the best due to robust network formation. A quantity of 4 wt% CBP and 0.2 wt% SWCNT samples did not show noticeable improvement in the flammability of the solid due to incomplete network formation. (From T. Kashiwagi et al., *Nat Mater.*, 4, 2005, 928–933.)

Numerous studies have indicated that CNTs can nucleate PSCs, and it was suggested that both isothermal and nonisothermal crystallization behavior can be used as a bellwether of CNT/polymer "soft epitaxial" matching for templated crystallinity. With the addition of CNT fillers into polymer matrices, some new properties (electrical conductivity, flame resistance) can be imparted, while others (mechanical strength, thermal conductivity) can be enhanced. Again, these properties are dependent on the interplay between CNT and polymer matrix, as well as the organization of the two.

It is well known that CNTs and CNFs can be used as filler materials for strengthening of structural composites. Carbon fiber has been known since the 1950s, and both CNF and CNT nanocomposites seek to displace it as the next-generation fibrous filler material. The difficulty here lies in consistently producing parts with an evenly distributed load, which requires control over the local processing stresses in the polymer, interfacial load transfer, and controlled placement of the CNTs. These are not trivial challenges and they are the key to optimizing such systems for mechanical property enhancements. This challenge has important implications, because even in a short time CNT-based nanocomposites have been studied, samples have been produced which have mechanical strength as high as the very best carbon fiber composites.

Electrical properties of CNT-based nanocomposites can have a much more immediate impact, as neither carbon fiber nor CNF composites can reasonably compete in terms of performance. Composites can be produced that have conductivity as high as metals, and semiconductor-level conductivity can be managed easily. Electromechanical actuation has been demonstrated in several reports, and CNT nanocomposites may one day be used in artificial muscles. CNTs can be grown from catalyst substrates using preexisting semiconductor manufacturing equipment. This makes them excellent candidates for conductive shunts between functional layers in semiconductor devices. As an important inroad for miniaturization and portable devices, electromagnetic shielding or antistatic materials can be produced at small fractions of a percent of CNT loading.

In future research, surface chemistry and surface geometry should take on added importance. There can be little doubt that as CNT-based composites begin to serve as monitors as well as electronic or structural materials, the orientation and chemical affinity of such surfaces will dominate research focus. Perhaps, the most underdeveloped area of research for these materials is in engineered textiles. Great challenges lie ahead in terms of producing woven or crosshatched CNT mats for specially designed anisotropy in composites [404–407]. NHSK will also have an important place in such materials: the galleries between PSCs can be tuned for mechanical properties or used to serve as microreactors for ion and analyte accumulation. They can perform this function without sacrificing conductivity, dispersion, or exfoliation of the material.

Cooperative interaction between polymer and CNT has one other major advantage. As Figure 5.1 implies, CNT-based nanocomposites have great potential for a range of different individual properties, but perhaps the most attractive feature of them is that they are *multi*functional. Multifunctionality will become a major advantage for CNT nanocomposites in the future, as engineered smart materials become increasingly important. Polymer/CNT nanocomposites will begin to take on dual-use functions—for instance, acting as structural, wear-resistant, and chemical-resistant materials, while handling sophisticated functions like pressure sensing and chemical detection. Materials with such properties as optically transparent shape memory are not beyond the scope of possibility. As we look to the future of these materials, three directions will take paramount importance. First, material behavior at the interface and in the interphase regions is a complex case and much research is needed to advance our understanding. Second, polymer/CNT nanocomposites will require processing optimization to be able to carefully control their alignment, dispersion, and exfoliation, while minimizing cost and maximizing throughput. Although this scaling problem is not particularly prestigious or novel, it will nevertheless be critical for widespread commercial adoption. Third, there is an overarching need to continue to push the limit of innovation: to find novel uses for CNT nanocomposites, and to continue to ignite the imaginations of the public, who will one day be both the vanguards and the beneficiaries of this important technology.

REFERENCES

1. A. Okada, M. Kawasumi, A. Usuki, Y. Kojima, T. Kurauchi, O. Kamigaito, *MRS Proc.*, 171, 1989, 45.
2. Y. Kojima, A. Usuki, M. Kawasumi, A. Okada, Y. Fukushima, T. Kurauchi, O. Kamigaito, *J. Mater. Res.*, 8, 1993, 1185–1189.
3. X. Wang, Q. Li, J. Xie, Z. Jin, J. Wang, Y. Li, K. Jiang, S. Fan, *Nano Lett.*, 9, 2009, 3137–3141.
4. Q. Wen, R. Zhang, W. Qian, Y. Wang, P. Tan, J. Nie, F. Wei, *Chem. Mat.*, 22, 2010, 1294–1296.
5. M.J. Green, et al., *Polymer*, 50, 2009, 4979–4997.
6. A. Peigney, C. Laurent, E. Flahaut, R.R. Bacsa, A. Rousset, *Carbon*, 39, 2001, 507–514.
7. G. Overney, W. Zhong, D. Tomanek, *Z. Phys. D: Atoms Mol. Clusters*, 27, 1993, 93–96.
8. B.I. Yakobson, C.J. Brabec, J. Bernholc, *Phys. Rev. Lett.*, 76, 1996, 2511–2514.
9. J.P. Lu, *Phys. Rev. Lett.*, 79, 1997, 1297–1300.
10. R.S. Ruoff, D. Qian, W.K. Liu, *Comptes Rendus Physique*, 4, 2003, 993–1008.
11. B.I. Yakobson, M.P. Campbell, C.J. Brabec, J. Bernholc, *Comput. Mater. Sci.*, 8, 1997, 341–348.
12. T. Belytschko, S.P. Xiao, G.C. Schatz, R.S. Ruoff, *Phys. Rev. B*, 65, 2002, 8.
13. P.M. Ajayan, O.Z. Zhou, Applications of carbon nanotubes, in: M.S. Dresselhaus, G. Dresselhaus, P. Avouris, Eds., *Carbon Nanotubes: Synthesis, Structure, Properties, and Applications*, Springer-Verlag: Berlin, 2001.
14. S. Osswald, G. Yushin, V. Mochalin, S.O. Kucheyev, Y. Gogotsi, *J. Am. Chem. Soc.*, 128, 2006, 11635–11642.
15. W. Kratschmer, L.D. Lamb, K. Fostiropoulos, D.R. Huffman, *Nature*, 347, 1990, 354–358.
16. D. Pech, M. Brunet, H. Durou, P. Huang, V. Mochalin, Y. Gogotsi, P.-L. Taberna, P. Simon, *Nat. Nanotechnol.*, 5, 2010, 651–654.
17. G.N. Yushin, S. Osswald, V.I. Padalko, G.P. Bogatyreva, Y. Gogotsi, *Diam. Relat. Mat.*, 14, 2005, 1721–1729.
18. Y. Kasahara, R. Tamura, M. Tsukada, *Phys. Rev. B*, 67, 2003, 115419.
19. B. Kostant, *Proc. Natl. Acad. Sci.*, 91, 1994, 11714–11717.
20. R.C. Haddon, *J. Am. Chem. Soc.*, 119, 1997, 1797–1798.
21. O. Berné, A.G.G.M. Tielens, *Proc. Natl. Acad. Sci.*, 109, 2012, 401–406.
22. G. Zheng, S. Irle, K. Morokuma, *Chem. Phys. Lett.*, 412, 2005, 210–216.
23. Q.S. Xie, E. Perezcordero, L. Echegoyen, *J. Am. Chem. Soc.*, 114, 1992, 3978–3980.
24. T.T.M. Palstra, R.C. Haddon, *Solid State Commun.*, 92, 1994, 71–81.
25. G. Li, V. Shrotriya, J. Huang, Y. Yao, T. Moriarty, K. Emery, Y. Yang, *Nat. Mater.*, 4, 2005, 864–868.
26. H. Prinzbach, A. Weller, P. Landenberger, F. Wahl, J. Worth, L.T. Scott, M. Gelmont, D. Olevano, B. von Issendorff, *Nature*, 407, 2000, 60–63.

27. C. Piskoti, J. Yarger, A. Zettl, *Nature*, 393, 1998, 771–774.
28. K.S. Novoselov, A.K. Geim, S.V. Morozov, D. Jiang, Y. Zhang, S.V. Dubonos, I.V. Grigorieva, A.A. Firsov, *Science*, 306, 2004, 666–669.
29. A.H. Castro Neto, F. Guinea, N.M.R. Peres, K.S. Novoselov, A.K. Geim, *Rev. Mod. Phys.*, 81, 2009, 109–162.
30. H.C. Schniepp, et al., *J. Phys. Chem. B*, 110, 2006, 8535–8539.
31. K.N. Kudin, B. Ozbas, H.C. Schniepp, R.K. Prud'homme, I.A. Aksay, R. Car, *Nano Lett.*, 8, 2008, 36–41.
32. D.R. Dreyer, S. Park, C.W. Bielawski, R.S. Ruoff, *Chem. Soc. Rev.*, 39, 2010, 228–240.
33. S. Stankovich, D.A. Dikin, R.D. Piner, K.A. Kohlhaas, A. Kleinhammes, Y. Jia, Y. Wu, S.T. Nguyen, R.S. Ruoff, *Carbon*, 45, 2007, 1558–1565.
34. C. Gomez-Navarro, R.T. Weitz, A.M. Bittner, M. Scolari, A. Mews, M. Burghard, K. Kern, *Nano Lett.*, 7, 2007, 3499–3503.
35. H.A. Becerril, J. Mao, Z. Liu, R.M. Stoltenberg, Z. Bao, Y. Chen, *ACS Nano*, 2, 2008, 463–470.
36. G. Eda, G. Fanchini, M. Chhowalla, *Nat. Nanotechnol.*, 3, 2008, 270–274.
37. K.S. Kim, Y. Zhao, H. Jang, S.Y. Lee, J.M. Kim, J.H. Ahn, P. Kim, J.Y. Choi, B.H. Hong, *Nature*, 457, 2009, 706–710.
38. L.V. Radushkevich, V.M. Lukyanovich, *Russ. J. Phys. Chem.*, 26, 1952, 88–95.
39. A. Oberlin, M. Endo, T. Koyama, *J. Cryst. Growth*, 32, 1976, 335–349.
40. R.M. Reilly, *J. Nucl. Med.*, 48, 2007, 1039–1042.
41. S. Iijima, *Nature*, 354, 1991, 56–58.
42. M. Monthioux, V.L. Kuznetsov, *Carbon*, 44, 2006, 1621–1623.
43. M.S. Dresselhaus, G. Dresselhaus, P. Avouris, *Carbon Nanotubes: Synthesis, Structure, Properties, and Applications*, Springer, Berlin; New York, 2001.
44. V.H. Crespi, M.L. Cohen, A. Rubio, *Phys. Rev. Lett.*, 79, 1997, 2093–2096.
45. R.C. Haddon, J. Sippel, A.G. Rinzler, F. Papadimitrakopoulos, *MRS Bull.*, 29, 2004, 252–259.
46. T.W. Ebbesen, P.M. Ajayan, *Nature*, 358, 1992, 220–222.
47. Y.N. Xia, P.D. Yang, Y.G. Sun, Y.Y. Wu, B. Mayers, B. Gates, Y.D. Yin, F. Kim, Y.Q. Yan, *Adv. Mater.*, 15, 2003, 353–389.
48. T. Guo, P. Nikolaev, A. Thess, D.T. Colbert, R.E. Smalley, *Chem. Phys. Lett.*, 243, 1995, 49–54.
49. O.A. Louchev, Y. Sato, H. Kanda, *Appl. Phys. Lett.*, 80, 2002, 2752–2754.
50. M. Cinke, J. Li, B. Chen, A. Cassell, L. Delzeit, J. Han, M. Meyyappan, *Chem. Phys. Lett.*, 365, 2002, 69–74.
51. D.E. Resasco, W.E. Alvarez, F. Pompeo, L. Balzano, J.E. Herrera, B. Kitiyanan, A. Borgna, *J. Nanopart. Res.*, 4, 2002, 131–136.
52. D.N. Futaba, K. Hata, T. Yamada, K. Mizuno, M. Yumura, S. Iijima, *Phys. Rev. Lett.*, 95, 2005, 056104.
53. D.N. Futaba, K. Hata, T. Yamada, T. Hiraoka, Y. Hayamizu, Y. Kakudate, O. Tanaike, H. Hatori, M. Yumura, S. Iijima, *Nat. Mater.*, 5, 2006, 987–994.
54. B. Zhao, D.N. Futaba, S. Yasuda, M. Akoshima, T. Yamada, K. Hata, *ACS Nano*, 3, 2009, 108–114.
55. J. Chen, M.A. Hamon, H. Hu, Y.S. Chen, A.M. Rao, P.C. Eklund, R.C. Haddon, *Science*, 282, 1998, 95–98.
56. M.A. Hamon, J. Chen, H. Hu, Y.S. Chen, M.E. Itkis, A.M. Rao, P.C. Eklund, R.C. Haddon, *Adv. Mater.*, 11, 1999, 834–840.
57. D. Cao, W. Wang, *Chem. Eng. Sci.*, 62, 2007, 6879–6884.
58. S.B. Hutchens, L.J. Hall, J.R. Greer, *Adv. Funct. Mater.*, 20, 2010, 2338–2346.
59. L.A. Girifalco, M. Hodak, R.S. Lee, *Phys. Rev. B*, 62, 2000, 13104–13110.
60. Y. Miyata, K. Yanagi, Y. Maniwa, T. Tanaka, H. Kataura, *J. Phys. Chem. C*, 112, 2008, 15997–16001.
61. Y. Xing, L. Li, C.C. Chusuei, R.V. Hull, *Langmuir*, 21, 2005, 4185–4190.
62. M.N. Tchoul, W.T. Ford, G. Lolli, D.E. Resasco, S. Arepalli, *Chem. Mat.*, 19, 2007, 5765–5772.
63. H.J. Gao, Y. Kong, D.X. Cui, C.S. Ozkan, *Nano Lett.*, 3, 2003, 471–473.
64. D. Tasis, N. Tagmatarchis, A. Bianco, M. Prato, *Chem. Rev.*, 106, 2006, 1105–1136.
65. R. Andrews, D. Jacques, D. Qian, E.C. Dickey, *Carbon*, 39, 2001, 1681–1687.
66. P.E. Lyons, S. De, F. Blighe, V. Nicolosi, L.F.C. Pereira, M.S. Ferreira, J.N. Coleman, *J. Appl. Phys.*, 104, 2008, 044302–044308.
67. S.D. Bergin, et al., *ACS Nano*, 3, 2009, 2340–2350.
68. S. Giordani, S. D. Bergin, V. Nicolosi, S. Lebedkin, M. M. Kappes, W. J. Blau, J. N. Coleman *J. Phys. Chem. B*, 110, 2006, 15708–15718.
69. Q. Cheng, S. Debnath, E. Gregan, H.J. Byrne, *J. Phys. Chem. C*, 112, 2008, 20154–20158.
70. A.P. Yu, E. Bekyarova, M.E. Itkis, D. Fakhrutdinov, R. Webster, R.C. Haddon, *J. Am. Chem. Soc.*, 128, 2006, 9902–9908.

71. S. Niyogi, M.A. Hamon, D.E. Perea, C.B. Kang, B. Zhao, S.K. Pal, A.E. Wyant, M.E. Itkis, R.C. Haddon, *J. Phys. Chem. B*, 107, 2003, 8799–8804.
72. J.F. Cardenas, A. Gromov, *Nanotechnology*, 20, 2009, 8.
73. M. Zheng, A. Jagota, E.D. Semke, B.A. Diner, R.S. McLean, S.R. Lustig, R.E. Richardson, N.G. Tassi, *Nat. Mater.*, 2, 2003, 338–342.
74. K. Shen, S. Curran, H.F. Xu, S. Rogelj, Y.B. Jiang, J. Dewald, T. Pietrass, *J. Phys. Chem. B*, 109, 2005, 4455–4463.
75. M. Kaempgen, J. Ma, G. Gruner, G. Wee, S.G. Mhaisalkar, *Appl. Phys. Lett.*, 90, 2007, 264104–264102.
76. H.J. Barraza, F. Pompeo, E.A. O'Rea, D.E. Resasco, *Nano Lett.*, 2, 2002, 797–802.
77. X. Peng, N. Komatsu, S. Bhattacharya, T. Shimawaki, S. Aonuma, T. Kimura, A. Osuka, *Nat. Nanotechnol.*, 2, 2007, 361–365.
78. R. Rastogi, R. Kaushal, S.K. Tripathi, A.L. Sharma, I. Kaur, L.M. Bharadwaj, *J. Colloid Interface Sci.*, 328, 2008, 421–428.
79. C.A. Mitchell, R. Krishnamoorti, *Macromolecules*, 40, 2007, 1538–1545.
80. R. Haggenmueller, et al., *Langmuir*, 24, 2008, 5070–5078.
81. M. Moniruzzaman, K.I. Winey, *Macromolecules*, 39, 2006, 5194–5205.
82. B.Z. Tang, H. Xu, *Macromolecules*, 32, 1999, 2569–2576.
83. Y. Li, C. Chen, S. Zhang, Y. Ni, J. Huang, *Appl. Surf. Sci.*, 254, 2008, 5766–5771.
84. R. Haggenmueller, J.E. Fischer, K.I. Winey, *Macromolecules*, 39, 2006, 2964–2971.
85. K. Jeon, et al., *Polymer*, 48, 2007, 4751–4764.
86. R. Sen, B. Zhao, D. Perea, M.E. Itkis, H. Hu, J. Love, E. Bekyarova, R.C. Haddon, *Nano Lett.*, 4, 2004, 459–464.
87. L.Y. Yeo, J.R. Friend, *J. Exp. Nanosci.*, 1, 2006, 177–209.
88. Z. Qinghua, C. Zhenjun, Z. Meifang, M. Xiumei, C. Dajun, *Nanotechnology*, 18, 2007, 115611.
89. K. Saeed, S.-Y. Park, H.-J. Lee, J.-B. Baek, W.-S. Huh, *Polymer*, 47, 2006, 8019–8025.
90. J. Gao, M.E. Itkis, A. Yu, E. Bekyarova, B. Zhao, R.C. Haddon, *J. Am. Chem. Soc.*, 127, 2005, 3847–3854.
91. V.I. Puller, S.V. Rotkin, *Europhys. Lett.*, 77, 2007, 5.
92. S. Snyder, S. Rotkin, *JETP Lett.*, 84, 2006, 348–351.
93. D. Wu, L. Wu, W. Zhou, M. Zhang, T. Yang, *Polym. Eng. Sci.*, 50, 2010, 1721–1733.
94. L.A. Berglund, T. Peijs, *MRS Bull.*, 35, 2010, 201–207.
95. X.H. Kang, Z.B. Mai, X.Y. Zou, P.X. Cai, J.Y. Mo, *Anal. Biochem.*, 369, 2007, 71–79.
96. M.G. Zhang, A. Smith, W. Gorski, *Anal. Chem.*, 76, 2004, 5045–5050.
97. J.Z. Li, W.J. Ma, L. Song, Z.G. Niu, L. Cai, Q.S. Zeng, X.X. Zhang, H.B. Dong, D. Zhao, W.Y. Zhou, S.S. Xie, *Nano Lett.*, 11, 2011, 4636–4641.
98. J.Z. Kirk, G. Zhenning, S. Jonah, C. Zheyi, L.F. Erica, J.S. Daniel, C. Candace, H.H. Robert, E.S. Richard, *Nanotechnology*, 16, 2005, S539.
99. X. Blase, L.X. Benedict, E.L. Shirley, S.G. Louie, *Phys. Rev. Lett.*, 72, 1994, 1878–1881.
100. J.C. Charlier, *Acc. Chem. Res.*, 35, 2002, 1063–1069.
101. S. Banerjee, T. Hemraj-Benny, S.S. Wong, *Adv. Mater.*, 17, 2005, 17–29.
102. Z. Spitalsky, D. Tasis, K. Papagelis, C. Galiotis, *Prog. Polym. Sci.*, 35, 2010, 357–401.
103. K. Balasubramanian, M. Burghard, *Small*, 1, 2005, 180–192.
104. A. Hirsch, O. Vostrowsky, Functionalization of carbon nanotubes, in: A.D. Schluter, Ed., *Functional Molecular Nanostructures*, Springer-Verlag, Berlin, 2005, pp. 193–237.
105. S. Bose, R.A. Khare, P. Moldenaers, *Polymer*, 51(5), 2010, 975–993.
106. Y. Miyata, Y. Maniwa, H. Kataura, *J. Phys. Chem. B*, 110, 2006, 25–29.
107. G.A. Snook, G.Z. Chen, D.J. Fray, M. Hughes, M. Shaffer, *J. Electroanal. Chem.*, 568, 2004, 135–142.
108. J.L. Bahr, J. Yang, D.V. Kosynkin, M.J. Bronikowski, R.E. Smalley, J.M. Tour, *J. Am. Chem. Soc.*, 123, 2001, 6536–6542.
109. B.A. Kakade, V.K. Pillai, *Appl. Surf. Sci.*, 254, 2008, 4936–4943.
110. Y.Y. Liu, J. Tang, X.Q. Chen, R.H. Wang, G.K.H. Pang, Y.H. Zhang, J.H. Xin, *Carbon*, 44, 2006, 165–167.
111. Z. Zhou, S.F. Wang, S. Lu, Y.X. Zhang, Y. Zhang, *Compos. Sci. Technol.*, 68, 2008, 1727–1733.
112. Q.L. Liao, Y.X. Wang, Y. Gao, H.M. Li, *Curr. Nanosci.*, 6, 2010, 243–248.
113. R. Ionescu, E.H. Espinosa, E. Sotter, E. Llobet, X. Vilanova, X. Correig, A. Felten, C. Bittencourt, G. Van Lier, J.C. Charlier, J.J. Pireaux, *Sens. Actuators*, B, 113, 2006, 36–46.
114. Y.H. Yan, J. Cui, M.B. Chan-Park, X. Wang, Q.Y. Wu, *Nanotechnology*, 18, 2007, Article number: 115712.
115. A. Felten, C. Bittencourt, J.J. Pireaux, G. Van Lier, J.C. Charlier, *J. Appl. Phys.*, 98, 2005, 074308–074309.
116. D. Kolacyak, J. Ihde, U. Lommatzsch, *Surf. Coat. Technol.*, 205, 2011, S605–S608.

117. A.A. Dameron, et al., *Appl. Surf. Sci.*, 258, 2012, 5212–5221.
118. P. Luksirikul, B. Ballesteros, G. Tobias, M.G. Moloney, M.L.H. Green, *J. Mater. Chem.*, 21, 2011, 19080–19085.
119. R. Graupner, J. Abraham, D. Wunderlich, A. Vencelova, P. Lauffer, J. Rohrl, M. Hundhausen, L. Ley, A. Hirsch, *J. Am. Chem. Soc.*, 128, 2006, 6683–6689.
120. A. Gromov, S. Dittmer, J. Svensson, O.A. Nerushev, S.A. Perez-Garcia, L. Licea-Jimenez, R. Rychwalski, E.E.B. Campbell, *J. Mater. Chem.*, 15, 2005, 3334–3339.
121. K.F. Kelly, I.W. Chiang, E.T. Mickelson, R.H. Hauge, J.L. Margrave, X. Wang, G.E. Scuseria, C. Radloff, N.J. Halas, *Chem. Phys. Lett.*, 313, 1999, 445–450.
122. J.L. Stevens, A.Y. Huang, H. Peng, I.W. Chiang, V.N. Khabashesku, J.L. Margrave, *Nano Lett.*, 3, 2003, 331–336.
123. J.Y. Yook, J. Jun, S. Kwak, *Appl. Surf. Sci.*, 256, 2010, 6941–6944.
124. L. Wang, S.A. Feng, J.H. Zhao, J.F. Zheng, Z.J. Wang, L. Li, Z.P. Zhu, *Appl. Surf. Sci.*, 256, 2010, 6060–6064.
125. J. Zhu, H. Peng, F. Rodriguez-Macias, J.L. Margrave, V.N. Khabashesku, A.M. Imam, K. Lozano, E.V. Barrera, *Adv. Funct. Mater.*, 14, 2004, 643–648.
126. F.L. Jin, C.J. Ma, S.J. Park, *Mater. Sci. Eng. A-Struct. Mater. Prop. Microstruct. Process.*, 528, 2011, 8517–8522.
127. P.C. Ma, J.K. Kim, B.Z. Tang, *Carbon*, 44, 2006, 3232–3238.
128. K.Q. Yang, W.L. Qin, H. Tang, L. Tan, Q.J. Xie, M. Ma, Y.Y. Zhang, S.Z. Yao, *J. Biomed. Mater. Res. Part A*, 99A, 2011, 231–239.
129. Z. Guo, F. Du, D. Ren, Y. Chen, J. Zheng, Z. Liu, J. Tian, *J. Mater. Chem.*, 16, 2006, 3021–3030.
130. M. Alvaro, P. Atienzar, P. de la Cruz, J.L. Delgado, V. Troiani, H. Garcia, F. Langa, A. Palkar, L. Echegoyen, *J. Am. Chem. Soc.*, 128, 2006, 6626–6635.
131. H. Li, R.B. Martin, B.A. Harruff, R.A. Carino, L.F. Allard, Y.P. Sun, *Adv. Mater.*, 16, 2004, 896–900.
132. R. Blake, Y.K. Gun'ko, J. Coleman, M. Cadek, A. Fonseca, J.B. Nagy, W.J. Blau, *J. Am. Chem. Soc.*, 126, 2004, 10226–10227.
133. B. Philip, J.N. Xie, J.K. Abraham, V.K. Varadan, *Polym. Bull.*, 53, 2005, 127–138.
134. B. Ruelle, S. Peeterbroeck, C. Bittencourt, G. Gorrasi, G. Patimo, M. Hecq, R. Snyders, S.D. Pasquale, P. Dubois, *React. Funct. Polym.* 72, 2012, 383–392.
135. K. Jiang, L.S. Schadler, R.W. Siegel, X. Zhang, H. Zhang, M. Terrones, *J. Mater. Chem.*, 14, 2004, 37–39.
136. C. Salvador-Morales, E.V. Basiuk, V.A. Basiuk, M.L.H. Green, R.B. Sim, *J. Nanosci. Nanotechnol.*, 8, 2008, 2347–2356.
137. K. Maehashi, T. Katsura, K. Kerman, Y. Takamura, K. Matsumoto, E. Tamiya, *Anal. Chem.*, 79, 2007, 782–787.
138. S.E. Baker, W. Cai, T.L. Lasseter, K.P. Weidkamp, R.J. Hamers, *Nano Lett.*, 2, 2002, 1413–1417.
139. D. Chris, G. Martin, F. Michael, W. Sean, S. Richard, E. Dorothy, *Nanotechnology*, 13, 2002, 601.
140. M.J. Moghaddam, S. Taylor, M. Gao, S. Huang, L. Dai, M.J. McCall, *Nano Lett.*, 4, 2003, 89–93.
141. K.V. Singh, R.R. Pandey, X. Wang, R. Lake, C.S. Ozkan, K. Wang, M. Ozkan, *Carbon*, 44, 2006, 1730–1739.
142. I. Palchetti, M. Mascini, *Anal. Bioanal. Chem.*, 402, 2012, 3103–3114.
143. R. Zanella, E.V. Basiuk, P. Santiago, V.A. Basiuk, E. Mireles, I. Puente-Lee, J.M. Saniger, *J. Phys. Chem. B*, 109, 2005, 16290–16295.
144. Y.-Y. Ou, M.H. Huang, *J. Phys. Chem. B*, 110, 2006, 2031–2036.
145. W. Zhao, H. Wang, X. Qin, X. Wang, Z. Zhao, Z. Miao, L. Chen, M. Shan, Y. Fang, Q. Chen, *Talanta*, 80, 2009, 1029–1033.
146. B.M. Novak, *Adv. Mater.*, 5, 1993, 422–433.
147. M. Griebel, J. Hamaekers, *Comput. Meth. Appl. Mech. Eng.*, 193, 2004, 1773–1788.
148. C.J. Brinker, Y. Lu, A. Sellinger, H. Fan, *Adv. Mater.*, 11, 1999, 579–585.
149. H. Zhang, Z. Zhang, K. Friedrich, C. Eger, *Acta Mater.*, 54, 2006, 1833–1842.
150. E.T. Thostenson, W.Z. Li, D.Z. Wang, Z.F. Ren, T.W. Chou, *J. Appl. Phys.*, 91, 2002, 6034–6037.
151. A.E. Frise, G. Pages, M. Shtein, I.P. Bar, O. Regev, I. Furo, *J. Phys. Chem. B*, 116, 2012, 2635–2642.
152. D. Ciprari, K. Jacob, R. Tannenbaum, *Macromolecules*, 39, 2006, 6565–6573.
153. X.D. Cao, H. Dong, C.M. Li, L.A. Lucia, *J. Appl. Polym. Sci.*, 113, 2009, 466–472.
154. M.Q. Tran, J.T. Cabral, M.S.P. Shaffer, A. Bismarck, *Nano Lett.*, 8, 2008, 2744–2750.
155. S. Nuriel, L. Liu, A.H. Barber, H.D. Wagner, *Chem. Phys. Lett.*, 404, 2005, 263–266.
156. P.-C. Ma, N.A. Siddiqui, E. Mäder, J.-K. Kim, *Compos. Sci. Technol.*, 71, 2011, 1644–1651.

157. X.L. Zhang, D. Yang, P. Xu, C.C. Wang, Q.G. Du, *J. Mater. Sci.*, 42, 2007, 7069–7075.
158. D.B. Zax, D.-K. Yang, R.A. Santos, H. Hegemann, E.P. Giannelis, E. Manias, *J. Chem. Phys.*, 112, 2000, 2945–2951.
159. A. Karim, T.M. Slawecki, S.K. Kumar, J.F. Douglas, S.K. Satija, C.C. Han, T.P. Russell, Y. Liu, R. Overney, O. Sokolov, M.H. Rafailovich, *Macromolecules*, 31, 1998, 857–862.
160. K.K.S. Lau, K.K. Gleason, *Macromol. Biosci.*, 7, 2007, 429–434.
161. L. Biasci, M. Aglietto, G. Ruggeri, F. Ciardelli, *Polymer*, 35, 1994, 3296–3304.
162. H. Tas, L.J. Mathias, *J. Polym. Sci., Part A: Polym. Chem.*, 48, 2010, 2302–2310.
163. M.R.S. Castro, N. Al-Dahoudi, P.W. Oliveira, H.K. Schmidt, *J. Nanopart. Res.*, 11, 2009, 801–806.
164. D.S. Hecht, K.A. Sierros, R.S. Lee, C. Ladous, C.M. Niu, D.A. Banerjee, D.R. Cairns, *J. Soc. Inf. Disp.*, 19, 2011, 157–162.
165. D.S. Hecht, D. Thomas, L.B. Hu, C. Ladous, T. Lam, Y. Park, G. Irvin, P. Drzaic, *J. Soc. Inf. Disp.*, 17, 2009, 941–946.
166. K. Peng, L.Q. Liu, Y. Gao, M.Z. Qu, Z. Zhang, *J. Nanosci. Nanotechnol.*, 10, 2010, 7386–7389.
167. T. Villmow, A. John, P. Pötschke, G. Heinrich, *Polymer*, 53, 2012, 2908–2918.
168. M.L. Auad, M.A. Mosiewicki, C. Uzunpinar, R.J.J. Williams, *Comp. Sci. Technol.*, 69, 2009, 1088–1092.
169. Z.K. Chen, J.P. Yang, Q.Q. Ni, S.Y. Fu, Y.G. Huang, *Polymer*, 50, 2009, 4753–4759.
170. X.C. Gui, H.B. Li, L.H. Zhang, Y. Jia, L. Liu, Z. Li, J.Q. Wei, K.L. Wang, H.W. Zhu, Z.K. Tang, D.H. Wu, A.Y. Cao, *ACS Nano*, 5, 2011, 4276–4283.
171. N. Li, Y. Huang, F. Du, X.B. He, X. Lin, H.J. Gao, Y.F. Ma, F.F. Li, Y.S. Chen, P.C. Eklund, *Nano Lett.*, 6, 2006, 1141–1145.
172. C.A. Martin, J.K.W. Sandler, A.H. Windle, M.K. Schwarz, W. Bauhofer, K. Schulte, M.S.P. Shaffer, *Polymer*, 46, 2005, 877–886.
173. P.-C. Ma, S.-Y. Mo, B.-Z. Tang, J.-K. Kim, *Carbon*, 48, 2010, 1824–1834.
174. C.E. Pizzutto, J. Suave, J. Bertholdi, S.H. Pezzin, L.A.F. Coelho, S.C. Amico, *Mater. Res.-Ibero-Am. J. Mater.*, 14, 2011, 256–263.
175. Z. Spitalsky, G. Tsoukleri, D. Tasis, C. Krontiras, S.N. Georga, C. Galiotis, *Nanotechnology*, 20, 2009, 405702.
176. W.E. Dondero, R.E. Gorga, *J. Polym. Sci., Part B: Polym. Phys.*, 44, 2006, 864–878.
177. Z. Hou, K. Wang, P. Zhao, Q. Zhang, C. Yang, D. Chen, R. Du, Q. Fu, *Polymer*, 49, 2008, 3582–3589.
178. W. Leelapornpisit, M.T. Ton-That, F. Perrin-Sarazin, K.C. Cole, J. Denault, B. Simard, *J. Polym. Sci., Part B: Polym. Phys.*, 43, 2005, 2445–2453.
179. D. Wu, Y. Sun, L. Wu, M. Zhang, *J. Appl. Polym. Sci.*, 108, 2008, 1506–1513.
180. Y.H. Chen, G.J. Zhong, J. Lei, Z.M. Li, B.S. Hsiao, *Macromolecules*, 44, 2011, 8080–8092.
181. K. Lu, N. Grossiord, C.E. Koning, H.E. Miltner, B.v. Mele, J. Loos, *Macromolecules*, 41, 2008, 8081–8085.
182. S. Zhang, M.L. Minus, L.B. Zhu, C.P. Wong, S. Kumar, *Polymer*, 49, 2008, 1356–1364.
183. R. Haggenmueller, C. Guthy, J.R. Lukes, J.E. Fischer, K.I. Winey, *Macromolecules*, 40, 2007, 2417–2421.
184. F. Mai, K. Wang, M. Yao, H. Deng, F. Chen, Q. Fu, *J. Phys. Chem. B*, 114, 2010, 10693–10702.
185. R.D. Maksimov, J. Bitenieks, E. Plume, J. Zicans, R.M. Meri, *Mech. Compos. Mater.*, 48, 2012, 47–56.
186. M.A. Samad, S.K. Sinha, *Tribol. Int.*, 44, 2011, 1932–1941.
187. J. Yang, C. Wang, K. Wang, Q. Zhang, F. Chen, R. Du, Q. Fu, *Macromolecules*, 42, 2009, 7016–7023.
188. S. Peeterbroeck, F. Laoutid, J.M. Taulemesse, T. Monteverde, J.M. Lopez-Cuesta, J.B. Nagy, M. Alexandre, P. Dubois, *Adv. Funct. Mater.*, 17, 2007, 2787–2791.
189. C.Y. Li, L.Y. Li, W.W. Cai, S.L. Kodjie, K.K. Tenneti, *Adv. Mater.*, 17, 2005, 1198–1202.
190. L. Li, B. Li, M.A. Hood, C.Y. Li, *Polymer*, 50, 2009, 953–965.
191. L. Li, Y. Yang, G. Yang, X. Chen, B.S. Hsiao, B. Chu, J.E. Spanier, C.Y. Li, *Nano Lett.*, 6, 2006, 1007–1012.
192. L.Y. Li, W.D. Wang, E.D. Laird, C.Y. Li, M. Defaux, D.A. Ivanov, *Polymer*, 52, 2011, 3633–3638.
193. X. Zheng, Q. Xu, Z. Li, *Sci. China Chem.*, 53, 2010, 1525–1533.
194. H. Uehara, K. Kato, M. Kakiage, T. Yamanobe, T. Komoto, *J. Phys. Chem. C*, 111, 2007, 18950–18957.
195. E.D. Laird, W. Wang, S. Cheng, B. Li, V. Presser, B. Dyatkin, Y. Gogotsi, C.Y. Li, *ACS Nano*, 6, 2012, 1204–1213.
196. W. Wang, E.D. Laird, B. Li, L. Li, C. Li, *Sci. China Chem.*, 55, 2012, 802–807.
197. K.A. Anand, T.S. Jose, U.S. Agarwal, T.V. Sreekumar, B. Banwari, R. Joseph, *Int. J. Polym. Mater.*, 59, 2010, 438–449.
198. C.Z. Geng, J.C. Wang, Q. Zhang, Q. Fu, *Polym. Int.*, 61, 2012, 934–938.
199. G.H. Kim, S.M. Hong, Y. Seo, *Phys. Chem. Chem. Phys.*, 11, 2009, 10506–10512.

200. E. Markevich, G. Salitra, D. Aurbach, *Electrochem. Commun.*, 7, 2005, 1298–1304.

201. S.H. Ng, J. Wang, Z.P. Guo, J. Chen, G.X. Wang, H.K. Liu, *Electrochim. Acta*, 51, 2005, 23–28.

202. Y. Yang, A. Centrone, L. Chen, F. Simeon, T.A. Hatton, G.C. Rutledge, *Carbon*, 49, 2011, 3395–3403.

203. E.M. Di Meo, A. Di Crescenzo, D. Velluto, C.P. O'Neil, D. Demurtas, J.A. Hubbell, A. Fontana, *Macromolecules*, 43, 2010, 3429–3437.

204. D.M. Smith, B. Dong, R.W. Marron, M.J. Birnkrant, Y.A. Elabd, L.V. Natarajan, V.P. Tondiglia, T.J. Bunning, C.Y. Li, *Nano Lett.*, 12, 2012, 310–314.

205. B. Li, L. Li, , B. Wang, C. Y. Li, *Nat. Nanotechnol.*, 4, 2009, 358–362.

206. M.A. Hood, B. Wang, J.M. Sands, J.J. La Scala, F.L. Beyer, C.Y. Li, *Polymer*, 51, 2010, 2191–2198.

207. M. Fernández, M. Landa, M.E. Muñoz, A. Santamaría, *Int. J. Adhes. Adhes.*, 30, 2010, 609–614.

208. H. Koerner, W. Liu, M. Alexander, P. Mirau, H. Dowty, R.A. Vaia, *Polymer*, 46, 2005, 4405–4420.

209. J.W. Cho, J.W. Kim, Y.C. Jung, N.S. Goo, *Macromol. Rapid Commun.*, 26, 2005, 412–416.

210. S.R. Ali, Y. Ma, R.R. Parajuli, Y. Balogun, W.Y.C. Lai, H. He, *Anal. Chem.*, 79, 2007, 2583–2587.

211. Q.A. Li, J.H. Liu, J.H. Zou, A. Chunder, Y.Q. Chen, L. Zhai, *J. Power Sources*, 196, 2011, 565–572.

212. C.Z. Meng, C.H. Liu, S.S. Fan, *Electrochem. Commun.*, 11, 2009, 186–189.

213. R.K. Srivastava, A. Srivastava, V.N. Singh, B.R. Mehta, R. Prakash, *J. Nanosci. Nanotechnol.*, 9, 2009, 5382–5388.

214. O.S. Kwon, E. Park, O.Y. Kweon, S.J. Park, J. Jang, *Talanta*, 82, 2010, 1338–1343.

215. M.R. Abidian, D.C. Martin, *Biomaterials*, 29, 2008, 1273–1283.

216. S.M. Richardson-Burns, J.L. Hendricks, B. Foster, L.K. Povlich, D.-H. Kim, D.C. Martin, *Biomaterials*, 28, 2007, 1539–1552.

217. S.M. Richardson-Burns, J.L. Hendricks, D.C. Martin, *J. Neural Eng.*, 4, 2007, L6–L13.

218. G. Li, V. Shrotriya, J.S. Huang, Y. Yao, T. Moriarty, K. Emery, Y. Yang, *Nat. Mater.*, 4, 2005, 864–868.

219. G.A. Snook, P. Kao, A.S. Best, *J. Power Sources*, 196, 2011, 1–12.

220. X. Han, Y. Li, Z. Deng, *Adv. Mater.*, 19, 2007, 1518–1522.

221. M.R. Jones, R.J. Macfarlane, B. Lee, J. Zhang, K.L. Young, A.J. Senesi, C.A. Mirkin, *Nat. Mater.*, 9, 2010, 913–917.

222. J. Zhang, Y. Liu, Y. Ke, H. Yan, *Nano Lett.*, 6, 2006, 248–251.

223. X. Dong, C.M. Lau, A. Lohani, S.G. Mhaisalkar, J. Kasim, Z. Shen, X. Ho, J.A. Rogers, L.-J. Li, *Adv. Mater.*, 20, 2008, 2389–2393.

224. C. Staii, A.T. Johnson, *Nano Lett.*, 5, 2005, 1774–1778.

225. H. Quan, Z.-M. Li, M.-B. Yang, R. Huang, *Compos. Sci. Technol.*, 65, 2005, 999–1021.

226. S.Y. Hobbs, *Nature-Phys. Sci.*, 234, 1971, 12–13.

227. E.J.H. Chen, B.S. Hsiao, *Poly. Eng. Sci.*, 32, 1992, 280–286.

228. K. Cho, D. Kim, S. Yoon, *Macromolecules*, 36, 2003, 7652–7660.

229. J. Thomason, A. Van Rooyen, *J. Mater. Sci.*, 27, 1992, 889–896.

230. C. Wang, L.M. Hwang, *J. Polym. Sci., Part B: Polym. Phys.*, 34, 1996, 1435–1442.

231. H. Nuriel, N. Klein, G. Marom, *Compos. Sci. Technol.*, 59, 1999, 1685–1690.

232. A.M. Kiel, A.J. Pennings, *Kolloid-Z.*, 205, 1965, 160.

233. J. Smook, A.J. Pennings, *J. Mater. Sci.*, 19, 1984, 31–43.

234. S.L. Kodjie, L.Y. Li, B. Li, W.W. Cai, C.Y. Li, M. Keating, *J. Macromol. Sci., Part B: Phys.*, 45, 2006, 231–245.

235. T. Tao, L. Zhang, C. Li, *Polymer*, 50, 2009, 3835–3840.

236. Z. Zhang, Q. Xu, Z. Chen, J. Yue, *Macromolecules*, 41, 2008, 2868–2873.

237. S.E. Bozbag, D. Sanli, C. Erkey, *J. Mater. Sci.*, 47, 2012, 3469–3492.

238. L.Y. Li, B. Li, G.L. Yang, C.Y. Li, *Langmuir*, 23, 2007, 8522–8525.

239. L.Y. Li, C.Y. Li, C.Y. Ni, *J. Am. Chem. Soc.*, 128, 2006, 1692–1699.

240. S. Cheng, X. Chen, Y.G. Hsuan, C.Y. Li, *Macromolecules*, 45, 2012, 993–1000.

241. H. Yang, Y. Chen, Y. Liu, W.S. Cai, Z.S. Li, *J. Chem. Phys.*, 127, 2007, 6.

242. Q.H. Wang, T.D. Corrigan, J.Y. Dai, R.P.H. Chang, A.R. Krauss, *Appl. Phys. Lett.*, 70, 1997, 3308–3310.

243. Y. Chen, R.C. Haddon, S. Fang, A.M. Rao, W.H. Lee, E.C. Dickey, E.A. Grulke, J.C. Pendergrass, A. Chavan, B.E. Haley, R.E. Smalley, *J. Mater. Res.*, 13, 1998, 2423–2431.

244. Y.S. Song, J.R. Youn, *Carbon*, 43, 2005, 1378–1385.

245. J. Sandler, M.S.P. Shaffer, T. Prasse, W. Bauhofer, K. Schulte, A.H. Windle, *Polymer*, 40, 1999, 5967–5971.

246. M.B. Bryning, M.F. Islam, J.M. Kikkawa, A.G. Yodh, *Adv. Mater.*, 17, 2005, 1186–1191.

247. M.B. Bryning, D.E. Milkie, M.F. Islam, J.M. Kikkawa, A.G. Yodh, *Appl. Phys. Lett.*, 87, 2005, 3.

248. M. Chapartegui, N. Markaide, S. Florez, C. Elizetxea, M. Fernandez, A. Santamaria, *Polym. Eng. Sci.*, 52, 2012, 663–670.
249. M. Abdalla, D. Dean, P. Robinson, E. Nyairo, *Polymer*, 49, 2008, 3310–3317.
250. D. Dean, A.M. Obore, S. Richmond, E. Nyairo, *Compos. Sci. Technol.*, 66, 2006, 2135–2142.
251. Z. Jia, Z. Wang, C. Xu, J. Liang, B. Wei, D. Wu, S. Zhu, *Mater. Sci. Eng., A*, 271, 1999, 395–400.
252. C. Park, Z. Ounaies, K.A. Watson, R.E. Crooks, J. Smith Jr, S.E. Lowther, J.W. Connell, E.J. Siochi, J.S. Harrison, T.L.S. Clair, *Chem. Phys. Lett.*, 364, 2002, 303–308.
253. M.J. Arlen, D. Wang, J.D. Jacobs, R. Justice, A. Trionfi, J.W.P. Hsu, D. Schaffer, L.-S. Tan, R.A. Vaia, *Macromolecules*, 41, 2008, 8053–8062.
254. H. Hiura, T.W. Ebbesen, K. Tanigaki, H. Takahashi, *Chem. Phys. Lett.*, 202, 1993, 509–512.
255. S. Osswald, M. Havel, Y. Gogotsi, *J. Raman Spectrosc.*, 38, 2007, 728–736.
256. C.A. Cooper, R.J. Young, M. Halsall, *Composites Part A*, 32, 2001, 401–411.
257. Q.H. Zhang, S. Rastogi, D.J. Chen, D. Lippits, P.J. Lemstra, *Carbon*, 44, 2006, 778–785.
258. K. Bubke, H. Gnewuch, M. Hempstead, J. Hammer, M.L.H. Green, *Appl. Phys. Lett.*, 71, 1997, 1906–1908.
259. T. Prasse, J.-Y. Cavaillé, W. Bauhofer, *Compos. Sci. Technol.*, 63, 2003, 1835–1841.
260. M.W. Wang, *Jap. J. Appl. Phys.*, 48, 2009, Article number: 035002.
261. C. Park, J. Wilkinson, S. Banda, Z. Ounaies, K.E. Wise, G. Sauti, P.T. Lillehei, J.S. Harrison, *J. Polym. Sci., Part B: Polym. Phys.*, 44, 2006, 1751–1762.
262. A. Thess, et al., *Science*, 273, 1996, 483–487.
263. Y. Maniwa, et al., *Phys. Rev. B*, 64, 2001, 3.
264. M. Kakudo, N. Kasai, *X-ray Diffraction by Polymers, Kodansha*; Elsevier, Tokyo, 1972.
265. S. Pujari, S.S. Rahatekar, J.W. Gilman, K.K. Koziol, A.H. Windle, W.R. Burghardt, *J. Chem. Phys.*, 130, 2009, 9.
266. A.J. Salem, G.G. Fuller, *J. Colloid Interface Sci.*, 108, 1985, 149–157.
267. D.J. Kang, K. Pal, D.S. Bang, J.K. Kim, *J. Appl. Polym. Sci.*, 121, 2011, 226–233.
268. N. Levi, R. Czerw, S. Xing, P. Iyer, D.L. Carroll, *Nano Lett.*, 4, 2004, 1267–1271.
269. X. Huang, P. Jiang, C. Kim, F. Liu, Y. Yin, *Eur. Polym. J.*, 45, 2009, 377–386.
270. M. Avrami, *J Chem Phys.*, 7, 1939, 1103–1112.
271. M. Avrami, *J. Chem. Phys.*, 8, 1940, 212–224.
272. M. Avrami, *J. Chem. Phys.*, 9, 1941, 177–184.
273. D.H. Xu, Z.G. Wang, *Polymer*, 49, 2008, 330–338.
274. A.R. Bhattacharyya, T.V. Sreekumar, T. Liu, S. Kumar, L.M. Ericson, R.H. Hauge, R.E. Smalley, *Polymer*, 44, 2003, 2373–2377.
275. L. Valentini, J. Biagiotti, M.A. Lopez-Manchado, S. Santucci, J.M. Kenny, *Polym. Eng. Sci.*, 44, 2004, 303–311.
276. M.K. Seo, J.R. Lee, S.J. Park, *Mater. Sci. Eng., A*, 404, 2005, 79–84.
277. B.P. Grady, F. Pompeo, R.L. Shambaugh, D.E. Resasco, *J. Phys. Chem. B*, 106, 2002, 5852–5858.
278. L. Valentini, J. Biagiotti, J.M. Kenny, M.A.L. Manchado, *J. Appl. Polym. Sci.*, 89, 2003, 2657–2663.
279. E.-C. Chen, T.-M. Wu, *J. Polym. Sci., Part B: Polym. Phys.*, 46, 2008, 158–169.
280. L. Li, C.Y. Li, C. Ni, L. Rong, B. Hsiao, *Polymer*, 48, 2007, 3452–3460.
281. J.F. Vega, J. Martinez-Salazar, M. Trujillo, M.L. Arnal, A.J. Muller, S. Bredeau, P. Dubois, *Macromolecules*, 42, 2009, 4719–4727.
282. J. Jin, M. Song, F. Pan, *Thermochim. Acta*, 456, 2007, 25–31.
283. S.D. Wanjale, J.P. Jog, *Polymer*, 47, 2006, 6414–6421.
284. L. Song, Z. Qiu, *Polym. Degrad. Stabil.*, 94, 2009, 632–637.
285. J.Y. Kim, H.S. Park, S.H. Kim, *Polymer*, 47, 2006, 1379–1389.
286. Y. Xu, H.-B. Jia, J.-N. Piao, S.-R. Ye, J. Huang, *J. Mater. Sci.*, 43, 2008, 417–421.
287. T.M. Wu, E.C. Chen, *J. Polym. Sci., Part. B: Polym. Phys.*, 44, 2006, 598–606.
288. C.A. Mitchell, R. Krishnamoorti, *Polymer*, 46, 2005, 8796–8804.
289. T. Ozawa, *Polymer*, 12, 1971, 150–158.
290. O. Probst, E.M. Moore, D.E. Resasco, B.P. Grady, *Polymer*, 45, 2004, 4437–4443.
291. C.A. Cooper, D. Ravich, D. Lips, J. Mayer, H.D. Wagner, *Compos. Sci. Technol.*, 62, 2002, 1105–1112.
292. T. Villmow, P. Potschke, S. Pegel, L. Haussler, B. Kretzschmar, *Polymer*, 49, 2008, 3500–3509.
293. F. Thiebaud, J.C. Gelin, *Compos. Sci. Technol.*, 70, 2010, 647–656.
294. F. Mai, D.D. Pan, X. Gao, M.J. Yao, H. Deng, K. Wang, F. Chen, Q. Fu, *Polym. Int.*, 60, 2011, 1646–1654.
295. N. Sun, B. Yang, L. Wang, J.-M. Feng, B. Yin, K. Zhang, M.-B. Yang, *Polym. Int.*, 61, 2012, 622–630.

296. L. Wang, B. Yang, W. Yang, N. Sun, B. Yin, J.M. Feng, M.B. Yang, *Colloid Polym. Sci.*, 289, 2011, 1661–1671.
297. J.J. Hernández, M.-C. García-Gutiérrez, D.R. Rueda, T.A. Ezquerra, R.J. Davies, *Compos. Sci. Technol.*, 72, 2012, 421–427.
298. H. Schonhorn, F.W. Ryan, *J. Phys. Chem.*, 70, 1966, 3811–3815.
299. L. Feng, Y.A. Zhang, J.M. Xi, Y. Zhu, N. Wang, F. Xia, L. Jiang, *Langmuir*, 24, 2008, 4114–4119.
300. H. Caps, D. Vandormael, J. Loicq, S. Dorbolo, N. Vandewalle, *Europhys. Lett.*, 88, 2009, 4.
301. Z.G. Guo, W.M. Liu, *Appl. Phys. Lett.*, 90, 2007, 3.
302. H.E. Jeong, M.K. Kwak, C.I. Park, K.Y. Suh, *J. Colloid Interface Sci.*, 339, 2009, 202–207.
303. H.E. Jeong, S.H. Lee, J.K. Kim, K.Y. Suh, *Langmuir*, 22, 2006, 1640–1645.
304. N. Zhao, Q. Xie, X. Kuang, S. Wang, Y. Li, X. Lu, S. Tan, J. Shen, X. Zhang, Y. Zhang, J. Xu, C. Han, *Adv. Funct. Mater.*, 17, 2007, 2739–2745.
305. X. Deng, L. Mammen, H.J. Butt, D. Vollmer, *Science*, 335, 2012, 67–70.
306. J. Yang, Z.Z. Zhang, X.H. Men, X.H. Xu, X.T. Zhu, *J. Colloid Interface Sci.*, 346, 2010, 241–247.
307. J.-P. Yang, Z.-K. Chen, Q.-P. Feng, Y.-H. Deng, Y. Liu, Q.-Q. Ni, S.-Y. Fu, *Composites Part B*, 43, 2012, 22–26.
308. Z.C. Pu, J.E. Mark, J.M. Jethmalani, W.T. Ford, *Polym. Bull.*, 37, 1996, 545–551.
309. P. Reichert, H. Nitz, S. Klinke, R. Brandsch, R. Thomann, R. Mulhaupt, *Macromol. Mater. Eng.*, 275, 2000, 8–17.
310. M. Alexandre, P. Dubois, *Mater. Sci. Eng.*, R, 28, 2000, 1–63.
311. T.D. Fornes, D.L. Hunter, D.R. Paul, *Macromolecules*, 37, 2004, 1793–1798.
312. P. Miaudet, S. Badaire, M. Maugey, A. Derre, V. Pichot, P. Launois, P. Poulin, C. Zakri, *Nano Lett.*, 5, 2005, 2212–2215.
313. N.H. Tai, M.K. Yeh, H.H. Liu, *Carbon*, 42, 2004, 2774–2777.
314. N.H. Tai, M.K. Yeh, T.H. Peng, *Composites Part B*, 39, 2008, 926–932.
315. M.K. Yeh, N.H. Tai, Y.J. Lin, *Composites Part A*, 39, 2008, 677–684.
316. J.N. Coleman, U. Khan, Y.K. Gun'ko, *Adv. Mater.*, 18, 2006, 689–706.
317. M.T. Byrne, W.P. McNamee, Y.K. Gun'ko, *Nanotechnology*, 19, 2008, 8.
318. B.X. Yang, K.P. Pramoda, G.Q. Xu, S.H. Goh, *Adv. Funct. Mater.*, 17, 2007, 2062–2069.
319. A. Wall, J.N. Coleman, M.S. Ferreira, *Phys. Rev. B*, 71, 2005, 5.
320. K. Kang, G. Bae, B. Kim, C. Lee, *Mater. Chem. Phys.*, 133, 2012, 495–499.
321. K. Chiba, M. Tada, *Thin Solid Films*, 520, 2012, 1993–1996.
322. P.K. Mallick, in: M. Dekker, Ed., *Fiber-Reinforced Composites: Materials, Manufacturing, and Design*, 2nd Edition, CRC Press: New York, 1993.
323. D. Qian, E.C. Dickey, R. Andrews, T. Rantell, *Appl. Phys. Lett.*, 76, 2000, 2868–2870.
324. L.Q. Liu, A.H. Barber, S. Nuriel, H.D. Wagner, *Adv. Funct. Mater.*, 15, 2005, 975–980.
325. A.L. Gerson, H.A. Bruck, A.R. Hopkins, K.N. Segal, *Composites Part A*, 41, 2010, 729–736.
326. Y. Zhang, E. Suhir, Y. Xu, *J. Mater. Res.*, 21, 2006, 2948–2954.
327. J.K.W. Sandler, S. Pegel, M. Cadek, F. Gojny, M. van Es, J. Lohmar, W.J. Blau, K. Schulte, A.H. Windle, M.S.P. Shaffer, *Polymer*, 45, 2004, 2001–2015.
328. R.E. Gorga, R.E. Cohen, *J. Polym. Sci., Part B: Polym. Phys.*, 42, 2004, 2690–2702.
329. R. Andrews, D. Jacques, M. Minot, T. Rantell, *Macromol. Mater. Eng.*, 287, 2002, 395–403.
330. B.M. Amoli, S.A.A. Ramazani, H. Izadi, *J. Appl. Polym. Sci.*, 125, 2012, E453–E461.
331. T. Takeda, Y. Shindo, F. Narita, Y. Mito, *Mater. Trans., JIM*, 50, 2009, 436–445.
332. F.M. Blighe, P.E. Lyons, S. De, W.J. Blau, J.N. Coleman, *Carbon*, 46, 2008, 41–47.
333. S.J.V. Frankland, A. Caglar, D.W. Brenner, M. Griebel, *J. Phys. Chem. B*, 106, 2002, 3046–3048.
334. N.G. Sahoo, H.K.F. Cheng, H. Bao, Y. Pan, L. Li, S.H. Chan, *Soft Matter*, 7, 2011, 9505–9514.
335. J.N. Coleman, M. Cadek, R. Blake, V. Nicolosi, K.P. Ryan, C. Belton, A. Fonseca, J.B. Nagy, Y.K. Gun'ko, W.J. Blau, *Adv. Funct. Mater.*, 14, 2004, 791–798.
336. B. Li, M.G. Hahm, Y.L. Kim, H.Y. Jung, S. Kar, Y.J. Jung, *ACS Nano*, 5, 2011, 4826–4834.
337. H.S. Shen, *Compos. Struct.*, 91, 2009, 9–19.
338. Z.X. Wang, H.S. Shen, *Composites Part B*, 43, 2012, 411–421.
339. J.F. Gao, D.X. Yan, B. Yuan, H.D. Huang, Z.M. Li, *Compos. Sci. Technol.*, 70, 2010, 1973–1979.
340. L.B. Hu, G. Gruner, J. Gong, C.J. Kim, B. Hornbostel, *Appl. Phys. Lett.*, 90, 2007, 3.
341. M.K. Massey, C. Pearson, D.A. Zeze, B.G. Mendis, M.C. Petty, *Carbon*, 49, 2011, 2424–2430.
342. W.R. Small, M.I.H. Panhuis, *Small*, 3, 2007, 1500–1503.
343. Z.C. Wu, Z.H. Chen, X. Du, J.M. Logan, J. Sippel, M. Nikolou, K. Kamaras, J.R. Reynolds, D.B. Tanner, A.F. Hebard, A.G. Rinzler, *Science*, 305, 2004, 1273–1276.

344. G.E. Pike, C.H. Seager, *Phys. Rev. B*, 10, 1974, 1421–1434.
345. A.L.R. Bug, S.A. Safran, I. Webman, *Phys. Rev. Lett.*, 54, 1985, 1412–1415.
346. S.I. White, R.M. Mutiso, P.M. Vora, D. Jahnke, S. Hsu, J.M. Kikkawa, J. Li, J.E. Fischer, K.I. Winey, *Adv. Funct. Mater.*, 20, 2010, 2709–2716.
347. F.M. Du, J.E. Fischer, K.I. Winey, *Phys. Rev. B*, 72, 2005, 4.
348. A.I. Goncharuk, N.I. Lebovka, L.N. Lisetski, S.S. Minenko, *J. Phys. D: Appl. Phys.*, 42, 2009, 165411.
349. E.T. Thostenson, S. Ziaee, T.-W. Chou, *Compos. Sci. Technol.*, 69, 2009, 801–804.
350. A. Zhang, J. Luan, Y. Zheng, L. Sun, M. Tang, *Appl. Surf. Sci.*, 22, 2012, 8492–8497.
351. Q.-P. Feng, J.-P. Yang, S.-Y. Fu, Y.-W. Mai, *Carbon*, 48, 2010, 2057–2062.
352. C.J. Ferris, M. in het Panhuis, *Soft Matter*, 5, 2009, 1466–1473.
353. J. Du, L. Zhao, Y. Zeng, L. Zhang, F. Li, P. Liu, C. Liu, *Carbon*, 49, 2011, 1094–1100.
354. O. Valentino, M. Sarno, N.G. Rainone, M.R. Nobile, P. Ciambelli, H.C. Neitzert, G.P. Simon, *Physica E*, 40, 2008, 2440–2445.
355. I. O'Connor, S. De, J.N. Coleman, Y.K. Gun'ko, *Carbon*, 47, 2009, 1983–1988.
356. M. Fu, Y. Yu, J.J. Xie, L.P. Wang, M.Y. Fan, S.L. Jiang, Y.K. Zeng, *Appl. Phys. Lett.*, 94, 2009, 012904–012903.
357. L. Yang, F. Liu, H. Xia, X. Qian, K. Shen, J. Zhang, *Carbon*, 49, 2011, 3274–3283.
358. N. Grossiord, M.E.L. Wouters, H.E. Miltner, K. Lu, J. Loos, B.V. Mele, C.E. Koning, *Eur. Polym. J.*, 46, 2010, 1833–1843.
359. T. Périé, A.-C. Brosse, S. Tencé-Girault, L. Leibler, *Carbon*, 50, 2012, 2918–2928.
360. I. Singh, P.K. Bhatnagar, P.C. Mathur, I. Kaur, L.M. Bharadwaj, R. Pandey, *Carbon*, 46, 2008, 1141–1144.
361. Y.Y. Huang, J.E. Marshall, C. Gonzalez-Lopez, E.M. Terentjev, *Mater. Express*, 1, 2011, 315–328.
362. D.S. Bangarusampath, H. Ruckdäschel, V. Altstädt, J.K.W. Sandler, D. Garray, M.S.P. Shaffer, *Chem. Phys. Lett.*, 482, 2009, 105–109.
363. X. Zhao, A.A. Koos, B.T.T. Chu, C. Johnston, N. Grobert, P.S. Grant, *Carbon*, 47, 2009, 561–569.
364. B.W. Steinert, D.R. Dean, *Polymer*, 50, 2009, 898–904.
365. S.A. Curran, J. Talla, S. Dias, D. Zhang, D. Carroll, D. Birx, *J. Appl. Phys.*, 105, 2009, 073711–073715.
366. F. Grillard, C. Jaillet, C. Zakri, P. Miaudet, A. Derré, A. Korzhenko, P. Gaillard, P. Poulin, *Polymer*, 53, 2012, 183–187.
367. C. Caamaño, B. Grady, D.E. Resasco, *Carbon*, 50, 2012, 3694–3707.
368. G. Zou, M. Jain, H. Yang, Y. Zhang, D. Williams, Q. Jia, *Nanoscale*, 2, 2010, 418–422.
369. R. Bhatia, V. Prasad, R. Menon, *Mater. Sci. Eng.*, B, 175, 2010, 189–194.
370. F.M. Blighe, Y.R. Hernandez, W.J. Blau, J.N. Coleman, *Adv. Mater.*, 19, 2007, 4443–4447.
371. S. Mazinani, A. Ajji, C. Dubois, *Polymer*, 50, 2009, 3329–3342.
372. N.K. Shrivastava, B.B. Khatua, *Carbon*, 49, 2011, 4571–4579
373. G. Sun, G. Chen, Z. Liu, M. Chen, *Carbon*, 48, 2010, 1434–1440.
374. D.-X. Yan, K. Dai, Z.-D. Xiang, Z.-M. Li, X. Ji, W.-Q. Zhang, *J. Appl. Polym. Sci.*, 120, 2011, 3014–3019.
375. R. Zhang, A. Dowden, H. Deng, M. Baxendale, T. Peijs, *Compos. Sci. Tech.*, 69, 2009, 1499–1504.
376. J.-C. Zhao, F.-P. Du, X.-P. Zhou, W. Cui, X.-M. Wang, H. Zhu, X.-L. Xie, Y.-W. Mai, *Composites Part B*, 42, 2011, 2111–2116.
377. Z. Zhao, W. Zheng, W. Yu, B. Long, *Carbon*, 47, 2009, 2118–2120.
378. A. Battisti, A.A. Skordos, I.K. Partridge, *Compos. Sci. Technol.*, 70, 2010, 633–637.
379. Z.-M. Dang, K. Shehzad, J.-W. Zha, A. Mujahid, T. Hussain, J. Nie, C.-Y. Shi, *Compos. Sci. Technol.*, 72, 2011, 28–35.
380. J.L. Wang, G.P. Xiong, M. Gu, X. Zhang, J. Liang, Acta Phys. *Sinica*, 58, 2009, 4536–4541.
381. Z.D. Han, A. Fina, *Prog. Polym. Sci.*, 36, 2011, 914–944.
382. H.M. Tu, L. Ye, *J. Appl. Polym. Sci.*, 116, 2010, 2336–2342.
383. F.H. Gojny, M.H.G. Wichmann, B. Fiedler, I.A. Kinloch, W. Bauhofer, A.H. Windle, K. Schulte, *Polymer*, 47, 2006, 2036–2045.
384. A. Moisala, Q. Li, I.A. Kinloch, A.H. Windle, *Compos. Sci. Technol.*, 66, 2006, 1285–1288.
385. M.J. Biercuk, M.C. Llaguno, M. Radosavljevic, J.K. Hyun, A.T. Johnson, J.E. Fischer, *Appl. Phys. Lett.*, 80, 2002, 2767–2769.
386. P. Gonnet, Z. Liang, E.S. Choi, R.S. Kadambala, C. Zhang, J.S. Brooks, B. Wang, L. Kramer, *Curr. Appl. Phys.*, 6, 2006, 119–122.
387. H. Huang, C.H. Liu, Y. Wu, S. Fan, *Adv. Mater.*, 17, 2005, 1652–1656.
388. T. Kashiwagi, F. Du, K.I. Winey, K.M. Groth, J.R. Shields, S.P. Bellayer, H. Kim, J.F. Douglas, *Polymer*, 46, 2005, 471–481.
389. G. Beyer, *Fire Mater.*, 26, 2002, 291–293.

390. T. Kashiwagi, F. Du, J.F. Douglas, K.I. Winey, R.H. Harris, J.R. Shields, *Nat Mater.*, 4, 2005, 928–933.

391. I. Aranberri, L. German, L. Matellanes, M.J. Suarez, E. Abascal, M. Iturrondobeitia, J. Ballestero, *Plast. Rubber Compos.*, 40, 2011, 133–138.

392. B. Du, Z. Fang, *Nanotechnology*, 21, 2010, Article number: 315603.

393. A. Fereidoon, M. Hemmati, N. Kordani, M. Kameli, M.G. Ahangari, N. Sharifi, *J. Macromol. Sci., Part B: Phys.*, 50, 2011, 665–678.

394. T. Kashiwagi, E. Grulke, J. Hilding, K. Groth, R. Harris, K. Butler, J. Shields, S. Kharchenko, J. Douglas, *Polymer*, 45, 2004, 4227–4239.

395. P. Song, L. Zhao, Z. Cao, Z. Fang, *J. Mater. Chem.*, 21, 2011, 7782–7788.

396. P.a. Song, L. Xu, Z. Guo, Y. Zhang, Z. Fang, *J. Mater. Chem.*, 18, 2008, 5083–5091.

397. G. Cai, A. Dasari, Z.-Z. Yu, X. Du, S. Dai, Y.-W. Mai, J. Wang, *Polym. Degrad. Stabil.*, 95, 2010, 845–851.

398. S.C. Lao, C. Wu, T.J. Moon, J.H. Koo, A. Morgan, L. Pilato, G. Wissler, *J. Compos. Mater.*, 43, 2009, 1803–1818.

399. B. Schartel, P. Potschke, U. Knoll, M. Abdel-Goad, *Eur. Polym. J.*, 41, 2005, 1061–1070.

400. T.D. Hapuarachchi, E. Bilotti, C.T. Reynolds, T. Peijs, *Fire Mater.*, 35, 2011, 157–169.

401. B. Kaffashi, F.M. Honarvar, *J. Appl. Polym. Sci.*, 124, 2012, 1154–1159.

402. T.D. Hapuarachchi, T. Peijs, *Composites Part A*, 41, 2010, 954–963.

403. K. Subramaniam, A. Das, L. Haeussler, C. Harnisch, K.W. Stoeckelhuber, G. Heinrich, *Polym. Degrad. Stabil.*, 97, 2012, 776–785.

404. C.P. Deck, J. Flowers, G.S.B. McKee, K. Vecchio, *J. Appl. Phys.*, 101, 2007, 023512–023519.

405. L.J. Hall, V.R. Coluci, D.S. Galvão, M.E. Kozlov, M. Zhang, S.O. Dantas, R.H. Baughman, *Science*, 320, 2008, 504–507.

406. M. De Volder, S.H. Tawfick, S.J. Park, D. Copic, Z. Zhao, W. Lu, A.J. Hart, *Adv. Mater.*, 22, 2010, 4384–4389.

407. X.W. Zhang, *Adv. Mater.*, 20, 2008, 4140–4144.

408. M.F. Yu, O. Lourie, M.J. Dyer, K. Moloni, T.F. Kelly, R.S. Ruoff, *Science*, 287, 2000, 637–640.

409. M.F. Yu, B.S. Files, S. Arepalli, and R.S. Ruoff, *Phys. Rev. Lett.*, 84, 2000, 5552–5555.

410. M.M.J. Treacy, T.W. Ebbesen, J.M. Gibson, *Nature*, 381, 1996, 678–680.

411. A. Krishnan, E. Dujardin, T.W. Ebbesen, P.N. Yianilos, M.M.J. Treacy, *Phys. Rev. B*, 58, 1998, 14013–14019.

412. S. Berber, Y.K. Kwon, D. Tomanek, *Phys. Rev. Lett.*, 84, 2000, 4613–4616.

413. S. Iijima, *Physica B: Condensed Matter*, 323, 2002, 1–5.

414. A. Hirsch, *Angew. Chem. Int. Ed.*, 41, 2002, 1853–1859.

415. J. Yang, K. Wang, H. Deng, F. Chen, and Q. Fu., *Polymer*, 51, 2010, 774–782.

6 Carbon Nanotube Biosensors

Mei Zhang, Pingang He, and Liming Dai

CONTENTS

6.1 INTRODUCTION

Carbon nanotubes (CNTs) have attracted considerable attention because of the unique combination of their one-dimensional (1D) molecular symmetries and physicochemical properties.[1] Among the many potential applications,[1,2] CNTs have become promising functional materials for the development of advanced biosensors. It has been demonstrated that CNTs promote electron transfer with various redox-active proteins, ranging from glucose oxidase[3,4] with a deeply embedded redox center to cytochrome c[5,6] and horseradish peroxidase[7,8] with surface redox centers. For using CNTs in biosensing applications, however, it is essential to immobilize biomolecules on the CNT structure without diminishing their bioactivity. Therefore, various intriguing approaches have been devised for functioning CNTs with various biomolecules,[9–15] such as deoxyribonucleic acid (DNA), proteins, and enzymes, either on their sidewalls or at the end caps.[16–20]

Although the CNT bioconjugates provide functional materials for the development of advanced biosensors, device design and fabrication also play an important role in regulating the biosensing performance. For many electrochemical biosensing applications, randomly entangled CNTs have been coated onto conventional electrodes.[17,21–24] The use of vertically aligned CNTs,[25] coupled with well-defined chemical functionalization, may offer additional advantages for facilitating the development of advanced biosensors with a high sensitivity and good selectivity.[20,26–28]

In this chapter, we present an overview of the CNT functionalization and electrode fabrication for biosensing applications by spotlighting certain selected examples from the research and development carried out in our own and some other research groups. As the main aim of this chapter is to demonstrate important concepts and rational methods for the development of CNT biosensors with no intention for a comprehensive literature survey of the subject, there will be no doubt that the examples to be presented in this chapter do not cover the entire range reported in literature.

6.2   STRUCTURE AND CHEMICAL REACTIVITY OF CNTs

As can be seen in Figure 6.1A and B, CNTs may be viewed as a graphite sheet that is rolled up into a nanoscale tube to form single-walled CNTs (SWCNTs) or with additional graphene tubes around the core of an SWCNT to generate multiwalled CNTs (MWCNTs).[1,29,30] These elongated nanotubes consist of carbon hexagons arranged in a concentric manner, with both ends of the tubes normally capped by fullerene-like structures containing pentagons (Figure 6.1C). Because the graphene sheet can be rolled up with varying degrees of twist along its length, CNTs may exhibit a variety of chiral structures.[1,30] Depending on their diameter and helicity of the arrangement of graphitic rings on the walls, CNTs can exhibit a large variety of interesting electronic properties attractive for potential applications in areas as diverse as sensors, actuators, molecular transistors, electron emitters, conductive fillers for polymers, metal-free catalysts, and even biomimetic dry adhesives.[1,2,31–33]

To meet specific requirements for particular applications (e.g., biocompatibility for nanotube biosensors and interfacial strength for blending with polymers), chemical modification of CNTs is essential. The pentagon-containing tips of CNTs are more reactive than sidewalls so that a variety

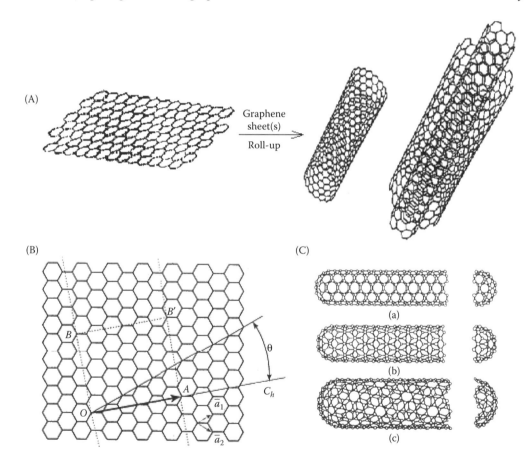

FIGURE 6.1   Schematic representation of (A) single-/multiwalled CNT formation by rolling up graphene sheet(s). (B) CNT formation based on a 2D graphene sheet of lattice vectors a_1 and a_2, the roll-up chiral vector $C_h = na_1 + ma_2$, and the chiral angle θ between C_h and a_1. When the graphene sheet is rolled up to form the cylindrical part of the nanotube, the chiral vector forms the circumference of nanotube's circular cross section with its ends meeting each other. The chiral vector (n, m) defines the tube helicity. (C) Schematic representation of SWCNTs. (a) (5,5) Armchair nanotube, (b) (9,0) zigzag nanotube, (c) (10,5) chiral nanotube. (From L. Dai et al., *Chem. Phys. Chem.* **2003**, 4, 1150. Copyright Wiley-VCH Verlag GmbH & Co. KGaA. With permission.)

of chemical reagents have been attached to the nanotube tips. Many interesting reactions have also been devised for chemical modification of both the inner and outer nanotube walls although the seamless arrangement of hexagon rings has rendered the sidewalls relatively unreactive. Judicious application of these site-selective reactions to nonaligned and aligned CNTs has opened up a rich field of CNT chemistry, which has been reviewed in several review articles (see Refs. 9–15). Below, we summarize some important work on the functionalization of CNTs with biomolecules, particularly those relevant to the development of CNT biosensors.

6.3   FUNCTIONALIZATION OF CNTs

6.3.1   Noncovalent Functionalization

Many biological species, such as carbohydrates,[34] nucleic acids,[35,36] peptides,[37,38] and proteins,[39] can be noncovalently adsorbed on the CNT surfaces through hydrophobic, π–π stacking, and electrostatic interactions or can even be trapped inside the nanotube hollow cavity through supramolecular inclusion/self-assembling.[16,39,40] For instance, SWCNTs have been dissolved in an aqueous solution of starch and iodine by forming starch-wrapped nanotube supramolecular complex.[31] Similarly, SWCNTs were also found to be soluble in an amylase-encapsulated form in an aqueous solution of amylase.[32] More interestingly, Dodziuk et al.[34] have recently reported the solubility of SWCNTs in an aqueous solution of η-cyclodextrins (η-CDs), which has a 12-membered ring structure with an inner cavity of ≈1.8 nm in diameter (Figure 6.2a), by encapsulating individual nanotubes within the η-CD ring structure, as schematically shown in Figure 6.2b. However, it is unlikely to solubilize SWCNTs in an aqueous solution of γ-CD because of its smaller ring.[33]

In addition to the aforementioned noncovalent surface modification of CNTs by macromolecular wrapping or ring inclusion, physical adsorption of small molecular moieties has also been exploited as an alternative approach for noncovalent functionalization of the nanotube sidewalls. In this context, Chen et al.[39] reported a simple approach to noncovalent functionalization of the inherently hydrophobic surface of SWCNTs by irreversibly adsorbing a bifunctional molecule, 1-pyrenebutanoic acid succinimidyl ester, from dimethylformamide (DMF) or methanol solvent. The π–π stacking interaction between the pyrenyl group of 1-pyrenebutanoic acid and the graphitic nanotube sidewall serves as the driving force for the irreversible adsorption. The succinimidyl ester groups anchored to SWCNTs are highly reactive for nucleophilic substitution by primary and secondary amines that exist in abundance on the surface of many proteins (Figure 6.3). This study, therefore, opened up an avenue for immobilizing a wide range of biomolecules, including ferritin, streptavidin, and biotinyl-3,6-dioxaoctanediamine (biotin–PEO–amine), on CNTs.[39] Indeed, Besteman et al.[41] have successfully adopted this concept for immobilizing glucose oxidase (GOx) on as-grown SWCNTs supported by a silicon wafer. The GOx-decorated SWCNTs were then used for sensing enzyme activities.

Similarly, a large number of proteins have been demonstrated to strongly bind to the outer surface of MWCNTs via nonspecific adsorption. For instance, Balvlavoine et al.[42] observed that proteins, such as streptavidin and hydrogen uptake protein regulator (HupR), could be spontaneously adsorbed onto MWCNTs to form a close-packed helical structure. On the basis of the same principle, Dieckmann et al.[37] have created a peptide structure that folds into an amphiphilic R helix to coat CNTs. Considering that the hydrophobic face of the helix can interact noncovalently with the aromatic surface of CNTs while the hydrophilic face can promote self-assembly through the charged peptide–peptide interactions, the amphiphilic peptides could be used not only to solubilize SWCNTs but also to assemble the peptide-coated nanotubes into supramolecular structures with a controllable size and geometry. It is also possible to assemble the peptide-wrapped nanotubes onto substrate surfaces prepatterned with antibodies for creating architectures useful in electrical circuits for molecular-sensing applications.

In addition to the aforementioned noncovalent functionalization of CNTs, a photochemical method has also been developed to functionalize CNTs with photoreactive reagents (e.g., aziridothymidine, AZT), followed by the coupling of single-strand DNA (ssDNA) chains onto the CNTs

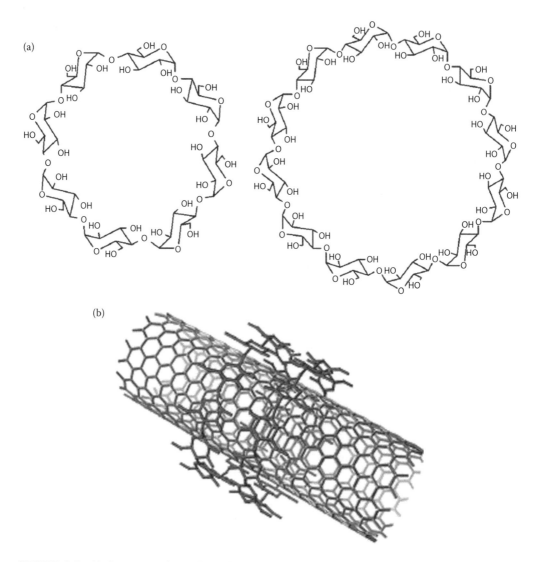

FIGURE 6.2 (a) Structures of γ, η-CDs. (b) Two η-CDs in the head-to-head arrangement threaded on SWCNT. (Adapted from H. Dodziuk et al., *Chem. Commun.* **2003**, 8, 986. With permission of The Royal Society of Chemistry.)

through the photoadduct and coating the nanotube sidewalls with complementary deoxyribonucleic acid (cDNA)-modified gold nanoparticles via self-assembling through DNA hybridization.[43] As can be seen in Figure 6.4A(a), the ultraviolet (UV) irradiation caused the formation of very active nitrene groups in the vicinity of CNTs (Figure 6.4A(a), step (i)). These nitrene groups were then coupled to the nanotube structure via a cycloaddition reaction to form aziridine photoadducts (Figure 6.4A(a), step (i)). After thoroughly rinsing the nanotube photoadducts with pure water, the nanotube sample is placed in an empty reaction chamber of an automated DNA synthesizer and a DNA oligonucleotide of defined base sequence was built *in situ* by the sequential addition of protected nucleotides using standard phosphoramidite chemistry (Figure 6.4A(a), steps (ii)–(iv)). Here, the free hydroxyl group at the 5′-position of the deoxyribose moiety in each AZT group was used as the site of attachment from which the DNA molecule was built.

After the DNA synthesis, the blocking groups on the DNA strand were removed by heating in an ammonia solution (Figure 6.4A(a), step (v)). The resultant CNT sample attached with ssDNA

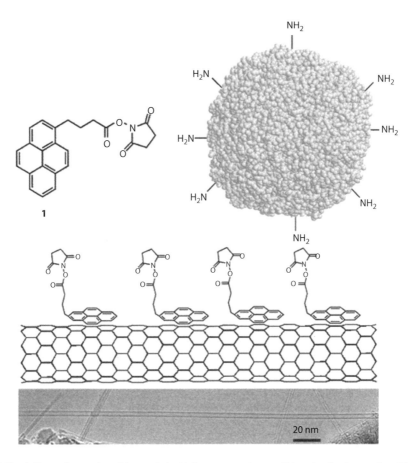

FIGURE 6.3 1-Pyrenebutanoic acid, succinimidyl ester was irreversibly adsorbed onto the sidewall of an SWCNT via π–π stacking. Protein immobilization was then achieved by reacting amino groups on the protein with the succinimidyl ester to form amide bonds. Lower panel shows a TEM image of an as-grown SWCNT on a gold TEM grid. (Adapted with permission from R.J. Chen et al., *J. Am. Chem. Soc.* 123, 3838. Copyright 2001, American Chemical Society.)

chains was washed with water until the supernatant was neutral and then stored in water at 4°C for further use. The locations of the DNA molecules attached onto the CNTs can be determined by a visual assay using the self-assembled cDNA-modified gold nanoparticles and transmission electron microscopy (TEM). Figure 6.4A(b) shows gold nanoparticles positioned in close proximity to the surfaces of the nanotubes. The corresponding sample was prepared by self-assembling cDNA-attached gold nanoparticles with a 16 nm diameter to the ssDNA chains grafted on the nanotubes, followed by a further hybridizing of the remaining unbound cDNA on the 16 nm gold nanoparticles to their complementary DNA (cDNA) chains attached to gold nanoparticles 38 nm in diameter.

More recently, it has also been demonstrated that a wide range of multicomponent structures of CNTs can be constructed by DNA-directed self-assembling of CNTs and gold nanoparticles (Figure 6.4B).[44] Figure 6.4B(a) shows the reaction steps for the DNA-directed self-assembling of multiple CNTs using the gold nanoparticles as a linkage, which involves the acid (HNO_3) oxidation of CNTs to introduce carboxylic end groups for ssDNA grafting. The ssDNA-attached CNTs were then subjected to hybridization with cDNA chains grafted on gold nanoparticles through the highly specific thiol–gold interaction.[45] Figure 6.4B(b) and B(c) shows the formation of various self-assembled multiple CNT structures, including one SWCNT connected to one MWCNT through a nanoparticle (Figure 6.4B(a)) and MWCTNs connected to a single nanoparticle core (Figure 6.4B(b)), depending on the reaction and

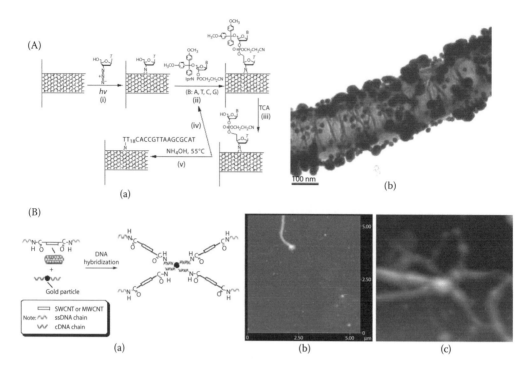

FIGURE 6.4 (A) (a) *In situ* DNA synthesis from sidewalls of CNTs photoetched with azidothymidine. Aligned MWCNTs on a solid support are coated with a solution of azidothymidine (AZT) and UV irradiation to produce photoadducts each with a hydroxyl group. (i) The hydroxyl group reacts with a phosphoramidite mononucleotide to initiate synthesis of the DNA molecule. (ii) Trichloroacetic acid deprotects the hydroxyl group (iii) for reaction with the next nucleotide (iv) and the cycle is repeated until the molecule with desired base sequence is made. Finally, the supported nanotubes are heated in ammonia solution to remove blocking groups from the nucleotides (v) to produce DNA-coated nanotubes. (b) DNA-directed modification of the surface of CNTs with gold nanoparticles of different diameters. The sample represented by this TEM image was prepared by binding the remaining unbound cDNA (on 16 nm gold nanoparticles precoated onto the nanotube) to cDNA attached to gold nanoparticles 38 nm in diameter. Scale bar is 100 nm. (Adapted with permission from M.J. Moghaddam et al., *Nano Lett.* 4, 89. Copyright 2004, American Chemical Society.) (B) (a) Schematic representation of procedures for DNA-directed self-assembling of multiple CNTs and nanoparticles, typical atomic force microscopy (AFM) images of (b) the self-assembly of an ssDNA–SWNT and ssDNA–MWCNT through a cDNA–Au nanoparticle (scanning area: 5.60 × 5.60 µm). (c) Multiple ssDNA–SWCNTs and ssDNA–MWNTs connected to a cDNA–Au nanoparticle core (scanning area: 1.10 × 1.10 µm). (Adapted with permission from S. Li et al., *J. Am. Chem. Soc.* 127, 14. Copyright 2005, American Chemical Society.)

self-assembling conditions. In view of the availability of various CNTs and DNA chains of different base sequences, these results represent an important advance in constructing many multiple CNT self-assembled structures for various applications (e.g., sensor chips, electronic circuits).

6.3.2 CHEMICALLY COVALENT MODIFICATION

Compared with the noncovalent surface modifications discussed above, covalent chemical modification can provide not only a stronger linkage between the functional groups and nanotubes but also additional advantages for controlling the attachment site and accessibility to the attached biomolecules, leading to an improved selectivity and stability for sensing applications. Among many CNT chemistries already reported,[9–15] covalent functionalization of CNTs with biomolecules has been achieved either through chemical derivatization of the carboxylic acid groups on oxidized nanotubes or by direct chemical modification of the pristine CNTs via addition reactions.

6.3.2.1 Chemical Functionalization via Carboxylic Acid Groups

It is well known that a variety of oxygenated groups, including carboxylic acid moieties, can be introduced at the nanotube ends and/or the structural defect sites along the tube walls through strong acid oxidization (e.g., sonication of SWCNTs in mixtures of sulfuric/nitric acids or sulfuric acid/hydrogen peroxide[46,47]) and ozone treatment[48,49] of SWCNTs, respectively. The oxidized CNTs can be further functionalized with thionyl chloride,[50,51] diimide-activated amidation with EDAC (1-ethyl-3-(3-dimethylaminopropyl)carbodiimide), or DCC (*N,N'*-dicyclohexylcarbodiimide) under ambient conditions (Figure 6.5).

Sun and coworkers[52,53] have chemically attached poly(propionylethylenimine-*co*-ethylenimine), which is a widely used aminopolymer in peptide synthesis, to the acid-oxidation-induced carboxylic acid groups on CNTs via diimide-activated amidation with EDAC under ambient conditions. Various biomolecules, including DNA,[54–56] PNA (peptide nucleic acid),[57] and proteins,[58] have also been successfully coupled to the acid-oxidized CNTs through the carboxylic acid functional linkers.

Figure 6.6A shows a typical procedure for attaching amine-containing proteins onto CNTs. As can be seen, CNTs were first acid oxidized to introduce surface carboxylic acid groups. The carboxylic acid groups were then activated by EDAC to form a highly reactive *O*-acylisourea active intermediate, which can be converted into a more stable active ester (succinimidyl intermediate) in the presence of *N*-hydroxysuccinimide (NHS). Finally, the active ester underwent nucleophilic substitution reaction with the amine groups of biomolecules (e.g., proteins). With this approach, Jiang et al.[58] have successfully attached proteins of ferritin or bovine serum albumin (BSA) onto the acid-oxidized CNTs.

Using the carboxylic acid group as a functional linker, Baker et al.[59] have also developed a multistep route toward the covalent grafting of DNA oligonucleotides onto SWCNTs. As shown in Figure 6.6B, the modification procedure involved the activation of the carboxylic acid groups of oxidized SWCNTs at the nanotube ends and sidewalls by thionyl chloride, followed by the reaction with ethylenediamine to produce amine-terminated sites. The amines were then reacted with the heterobifunctional cross-linker succinimidyl 4-(*N*-maleimidomethyl)-cyclohexane-1-carboxylate (SMCC), terminating the surface with maleimide groups. Finally, thiol-terminated DNA chains were reacted with the maleimide groups, resulting in the formation of DNA-grafted SWCNTs.

More recently, a simple but effective photochemical approach has been developed to directly graft different chemical reagents onto the opposite tube ends of individual CNTs by sequentially floating the substrate-free vertically aligned carbon nanotube (VA-CNT) film on two different photoreactive solutions with only one side of the VA-CNT film being contacted with a photoreactive solution and exposed to UV light each time, as schematically shown in Figure 6.7A.[60]

The methodology mentioned above can be used to asymmetrically end-functionalize CNTs with ssDNA and cDNA chains onto their opposite tube ends, which can be very useful for site-selective self-assembling of CNTs into many novel functional structures for various potential applications, including nanotube sensors and sensor chips. In a similar but independent study, asymmetric end

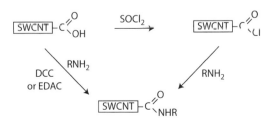

FIGURE 6.5   A typical functionalization protocol for COOH-containing CNTs.

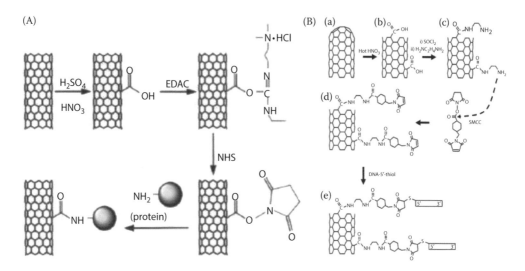

FIGURE 6.6 (A) Schematic view of the attachment of proteins to CNTs via a two-step process of diimide-activated amidation. (Adapted from K. Jiang et al., *J. Mater. Chem.* **2004**, 14, 37. With permission of The Royal Society of Chemistry.) (B) Overall scheme for the preparation of covalently linked DNA nanotubes. (Adapted with permission from S.E. Baker et al., *Nano Lett.* 2, 1413. Copyright 2002, American Chemical Society.)

functionalization has also been achieved through sidewall protection by impregnating a VA-MWCNT array with polystyrene (PS),[61] as schematically shown in Figure 6.7B. The resultant asymmetrically end-functionalized aligned CNT arrays have been demonstrated to be promising membranes for bioseparation and purification.[62]

6.3.2.2 Chemical Functionalization via Direct Addition Reactions

In view of the presence of unsaturated π-electrons in CNTs, Smalley and coworkers[63–65] and Holzinger et al.[66] have functionalized the sidewall of SWCNTs directly by fluorination and azide–thermolysis, respectively. Direct addition of functionalities to the CNTs was enabled by nitrene cycloaddition, nucleophilic carbine addition, and radical addition.

As shown in Figure 6.8a, the addition of nitrenes could be carried out by heating SWCNTs up to 160°C in the presence of a 200-fold excess of alkyl azidoformate as the nitrene precursor. After thermally induced N_2 extrusion, the nitrene addition led to the formation of alkoxycarbonylaziridino-SWCNTs. However, nucleophilic carbene addition onto the sidewall of SWCNTs has been achieved by reacting the nanotube electrophilic π-system with dipyridyl imidazolidene (Figure 6.8b). To demonstrate radical addition, Holzinger et al.[66] have used the photoinduced addition of perfluorinated alkyl radicals onto SWCNTs by illuminating the nanotube suspension with a medium-pressure mercury lamp (150 W) for 4 h in the presence of a 200-fold excess of heptadecafluorooctyl iodide (Figure 6.8c). Subsequent derivatization of the fluorine-containing side groups allowed the introduction of various sidewall functionalities into the nanotube structure.[12] Moghaddam et al.[43] have also used the azide–photochemistry to functionalize CNTs by attaching ssDNA chains onto the nanotube sidewalls and tips.

Besides, Prato and coworkers[67,68] have reported that CNTs underwent 1,3-dipolar cycloaddition when heated in DMF in the presence of α-amino acid and aldehyde. This reaction provides a versatile and powerful approach to attach different functionalities to CNTs and these functional groups can be further coupled with amino acids and bioactive peptides. It has been demonstrated that some of the peptide-functionalized CNTs can enhance the immunization to virus-specific neutralizing antibody responses.[69–71]

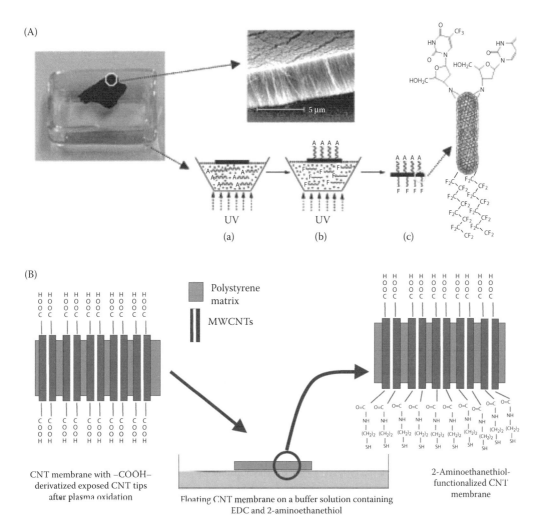

FIGURE 6.7 (A) A free-standing film of aligned CNTs floating on the top surface of (a) a 3′-azido-3′-deoxythymidine solution in ethanol for UV irradiation at one side of the nanotube film for 1 h and (b) a perfluorooctyl iodide solution in 1,1,2,2-tetrachloroethane for UV irradiation at the opposite side of the nanotube film. (Adapted with permission from K.M. Lee, L.C. Li, L. Dai, *J. Am. Chem. Soc.* 127, 4122. Copyright 2005, American Chemical Society.) (B) Plasma-oxidized VA–MWNT membrane (cross-sectional view) with carboxylic acid-derivatized MWNTs when floated on a buffer solution containing EDC and 2-aminoethanethiol, resulted in bifunctional CNTs in the membrane structure. (Adapted from N. Chopra, M. Majumder, B.J. Hinds, *Adv. Funct. Mater.* 2005, 15, 858. Copyright Wiley-VCH Verlag GmbH & Co. KGaA. With permission.)

6.4 FABRICATION OF CNT ELECTRODES

6.4.1 Nonaligned CNT Electrodes

As electrode materials, CNTs possess many advantages, including their excellent electrical property, high chemical stability, strong mechanical strength, and relatively large surface area. As a result, CNT-based electrodes have recently been used in electrochemical applications and as biosensors.[5,21,23] In particular, Britto et al.[21] have used unmodified CNTs as electrodes to study the oxidation of dopamine (DA) and observed a high-oxidation reversibility with the pristine CNT

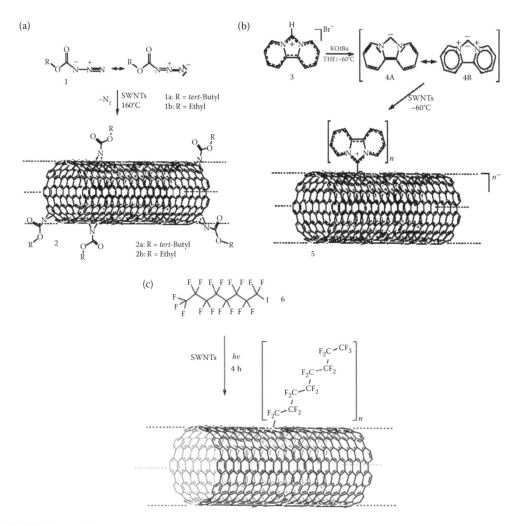

FIGURE 6.8 Schematic presentations of (a) nitrene cycloaddition, (b) nucleophilic carbine addition, and (c) radical addition. (Adapted from M. Holzinger et al., *Angew. Chem. Int. Ed.* **2001**, 40, 4002. Copyright Wiley-VCH Verlag GmbH & Co. KGaA. With permission.)

electrodes. Alternatively, Luo et al.[22] prepared an SWCNT-modified glassy carbon (GC) electrode by spin casting a thin layer of the nitric acid oxidized SWCNTs (see Section 6.3.2.1) on a GC electrode. The GC electrode modified with the oxidized SWCNTs was found to show favorable electrocatalytic behavior toward the oxidation of biomolecules, such as DA, epinephrine, and ascorbic acid (AA), with a very stable electrochemical reactivity.

Furthermore, Davis et al.[17] reported a significantly enhanced faradic response from redox-active sulfonated anthraquinone using an SWCNT-modified GC electrode. As can be seen in Figure 6.9, the nanotube-modified electrode shows a much stronger voltammetric response to anthraquinone-2,6-disulfonate compared to a simple GC electrode.

Using a CNT-coated graphite electrode, Wang et al.[24] have demonstrated a good voltammetric resolution for DA and AA with an anodic potential difference of 270 mV in a pH 5.0 phosphate buffer, which indicates a high selectivity for DA over AA. The observed difference in electrocatalysis toward DA and AA was attributed to a three-dimensional (3D) porous interfacial layer associated with the CNT-modified electrode. The CNT-modified electrodes were also found to exhibit a strong and stable electrocatalytic response to NADH (reduced nicotinamide adenine dinucleotide) with a

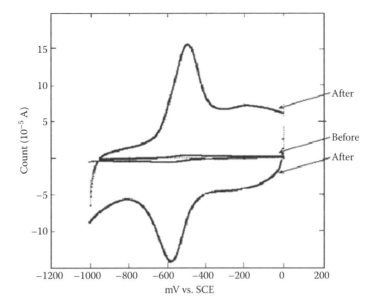

FIGURE 6.9 Comparative faradic responses (100 mV/s) of a GC macroelectrode to 2 mM 1,5-AQDS (anthraqui-none-2,6-disulfonate) before and after modification with the acid-oxidized SWCNTs. (Adapted from J.J. Davis et al., *Chem. Eur. J.* **2003**, 9, 3732. Copyright Wiley-VCH Verlag GmbH & Co. KGaA. With permission.)

low redox potential,[72] and to be useful as electrochemical detectors on a capillary electrophoresis microchip[73] for *in situ* determination of oxidizable amino acids in ion chromatography.[74]

Along with the development of CNT electrodes through modification of conventional electrodes by CNTs with and without chemical modification, CNT composite electrodes have been investigated for biosensing applications. For instance, Hrapovic et al.[75] reported that electrochemical sensors made of SWCNTs mixed with 2–3 nm small platinum nanoparticles displayed a remarkably improved sensitivity toward hydrogen peroxide. In this study, Nafion, a perfluorosulfonated polymer, was used to solubilize SWCNTs and interact with Pt nanoparticles to form a network that connected Pt nanoparticles to the nanotube electrode surface (Figure 6.10A).

By supporting CNTs with a metal substrate of a redox potential lower than that of the metal ions to be reduced into nanoparticles, a facile yet versatile and effective substrate-enhanced electroless deposition (SEED) method for decorating CNTs with various metal nanoparticles has been developed. This includes those nanoparticles otherwise impossible by more conventional electroless deposition methods, in the absence of any additional reducing agent.[76,77] In particular, the SEED method can be employed to deposit Cu, Ag, Au, Pt, and Pd nanoparticles on SWCNTs and MWCNTs, as exemplified in Figure 6.10B.

Asymmetric sidewall modification by self-assembling of differently shaped metal nanoparticles onto the inner wall and outer wall of CNTs was also achieved by the SEED method.[77] As shown in Figure 6.10C, VA-CNTs were first deposited in a commercially available anodic alumina membrane directly. Pt nanospheres were then deposited onto the inner wall of the template-synthesized VA-CNTs supported by a Cu foil (Figure 6.10C(b–e), *left panel in e*), in which the template acts as a protective layer for the nanotube outer wall. Upon completion of the modification of the inner wall, these nanotubes were released by dissolving the alumina template in aqueous HF (Figure 6.10C (f), *left panel in f*). As expected, Figure 6.10C (a and b, *right panel*) shows a very smooth outer wall for the inner-wall-modified CNTs. The presence of nanospheres within the nanotube inner wall was clearly seen in Figure 6.10C (b, *right panel*). To deposit Pt nanocubes onto the outer wall, the inner-wall-modified CNTs were supported by side (Figure 6.10C (*left panel*, g–j)) on a Cu foil and exposed to an aqueous K_2PtCl_4 solution containing $CuCl_2$ for 1 min. While Figure 6.10C (c, *right panel*) shows the presence

(A)

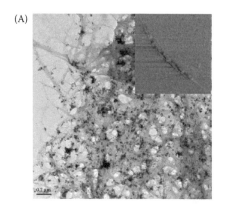

(B)

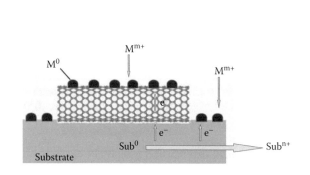

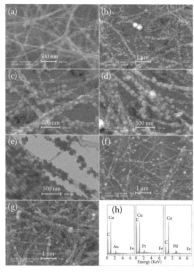

(C)

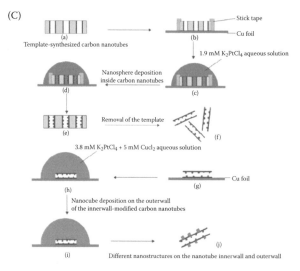

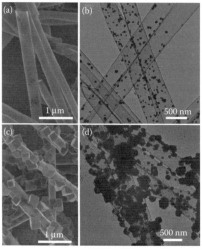

FIGURE 6.10

of nanocubes on the nanotube outer wall, Figure 6.10C (d, *right panel*) reveals the asymmetrically modified CNTs with nanocubes and nanospheres self-assembled onto the nanotube outer wall and inner wall, respectively. The resultant CNTs, either in an aligned or nonaligned form, decorated with metal nanoparticles (e.g., Pt), are promising for biosensing applications as many biosensors are based on the electrochemical detection of hydrogen peroxide.[2,78]

Using CNT and PTFE (polytetrafluoroethylene) composite materials without metal nanoparticles, Wang et al.[78] developed a simple method for preparing effective CNT electrochemical biosensors. These authors first produced the CNT–PTFE composite material by hand mixing desired amounts of CNTs with granular PTFE in dry state. They then added a desired amount of enzyme (GOx or ADH) and NAD$^+$ cofactor into the CNT/PTFE composite (30/70 wt%). Like its graphite-based counterpart, the bulk of the resulting carbon composite electrode serves as a reservoir for the enzyme. However, unlike its graphite-based counterpart, the composite electrode combines the major advantages of CNTs with those of bulk composite electrode, displaying a remarkably high electrocatalytic activity for hydrogen peroxide and NADH.

Nonfluorine-containing polymers have also been used for preparing composite electrodes with CNTs. Rege et al.[79] prepared an enzyme-containing polymer and SWCNT composite biosensor from a suspension of SWCNT, α-chymotrypsin, and poly(methyl methacrylate) in toluene. Although Gavalas et al.[80,81] used CNT sol–gel composite as electrochemical sensing materials and enzyme-friendly platforms for the development of stable biosensors, Wohlstadter et al.[82] compounded CNTs with poly(ethylene vinylacetate), EVA, to produce a CNT and EVA composite sheets. These CNT composite electrodes showed the properties of nanoscopic materials with the advantages of macroscopic systems.

Interestingly, the SWCNT-modified electrode was found to possess capabilities to promote direct electron transfer with certain redox-active enzymes. To demonstrate the direct electron transfer capability, Campbell et al.[83] designed a microelectrode consisting of a CNT attached to a sharpened Pt wire (Figure 6.11A). Although the high-resolution TEM image given in Figure 6.11A(c) shows an opened nanotube end, Figure 6.11A(d) shows the sidewall of the nanotube coated by a thin layer of electrically insulating polyphenol. The nanotube electrode thus prepared shows a sigmoidal voltammetric response to Ru(NH$_3$)$_6^{3+}$ solution, characteristic of direct electron transfer with a long cylindrical ultra-microelectrode. The recent availability of super long ($\approx$5 mm) vertically aligned carbon nanotubes (SLVA-CNTs)[84] enabled the preparation of long CNT electrodes useful for biosensing applications, as shown in Figure 6.11B.[85] By region-selective masking, the nanotube with a nonconducting polymer coating (e.g., PS) for the electrolyte to access only the nanotube sidewall

---

FIGURE 6.10 (continued) (A) TEM micrograph of SWCNT in the presence of Pt nanoparticles. Inset: AFM tapping mode phase image (size, 1 μm × 1 μm; data scale, 20 nm) of one SWCNT in the presence of Pt nanoparticles. (Adapted with permission from L.T. Qu, L. Dai, *J. Am. Chem. Soc.* 127, 10806. Copyright 2005, American Chemical Society.) (B) (Left panel) Schematic illustration of metal nanoparticle deposition on CNTs via the SEED process. (Right panel) SEM images of MWCNTs supported by a copper foil after being immersed into an aqueous solution of HAuCl$_4$ (3.8 mM) for different periods of time: (a) 0 s, (b) 10 s, (c) same as for (b) under a higher magnification, (d) 30 s. (e) A TEM image of Au nanoparticle-coated MWCNTs, (f) the Cu-supported MWCNTs after being immersed into an aqueous solution of K$_2$PtCl$_4$ (4.8 mM) for 10 s, (g) the Cu-supported MWCNTs after being immersed into an aqueous solution of (NH$_4$)$_2$PdCl$_4$ (7.0 mM) for 10 s. (h) EDX spectra (EDX: energy-dispersive x-ray spectroscopy) for the Au, Pt, and Pd nanoparticle-coated MWNTs on Cu foils. (Adapted with permission from M. Musameh et al., *Electrochem. Commun.* 4, 743. Copyright 2002, American Chemical Society.) (C) (Left panel) Procedures for the CNT inner wall modification and the asymmetric modification of the nanotube inner wall with Pt nanospheres and the outer wall with nanocubes. (Right panel) (a) SEM and (b) TEM images of CNTs having their inner wall modified with Pt nanospheres by immersing the template-synthesized (c) SEM and (d) TEM images of the inner-wall-modified CNTs followed by outer-wall modification in an aqueous solution of K$_2$PtCl$_4$ containing CuCl$_2$ with the inner-wall-modified CNTs being supported by a Cu foil in a sidewall-on configuration. (Adapted with permission from L.T. Qu, L. Dai, E. Osawa, *J. Am. Chem. Soc.* 128, 5523. Copyright 2006, American Chemical Society.)

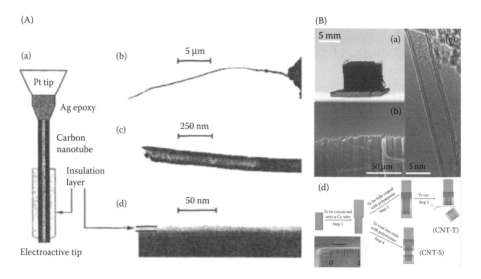

FIGURE 6.11 (A) (a) Schematic representation of a partially insulated CNT electrode. TEM images of mounted nanotubular electrodes showing (b) a 30 μm long electrode, (c) the tip of a ≈100 nm diameter uninsulated nanoelectrode, and (d) ≈10-nm-thick insulation layer of polyphenol on a ≈220 nm diameter nanotube. (Adapted with permission from J.K. Campbell et al., *J. Am. Chem. Soc.* 121, 3779. Copyright 1999, American Chemical Society.) (B) (a) A digital photograph, (b) SEM, and (c) TEM image of the *as-synthesized* aligned super long CNTs. (d) A schematic representation of the procedure for preparing the CNT electrodes with only the nanotube tip (CNT-T) or sidewall (CNT-S) accessible to electrolyte. Inset in (d) shows a digital photograph of a nanotube electrode thus prepared with an aligned super long CNT bundle connected to a copper wire. (Adapted from K. Gong, S. Chakrabarti, L. Dai, *Angew. Chem. Int. Ed.* **2008**, 47, 5446. Copyright Wiley-VCH Verlag GmbH & Co. KGaA. With permission.)

or tip, we have performed the nanotube tip-/sidewall-specific electrochemistry. Depending on the electrochemical species, it was found that both the nanotube tip and sidewall could play a dominating role in the electrochemistry of the CNT electrode.

6.4.2 Aligned CNT Electrodes

The use of ordered (aligned/patterned) CNTs as biosensing electrodes may provide further advantages in addition to the aforementioned electrodes based on the nonaligned, randomly entangled CNTs. Many research groups have used porous membranes (e.g., mesoporous silica, alumina nanoholes) as the template for preparing well-aligned CNTs with uniform diameters and lengths,[25] as exemplified by Figure 6.12.[86]

Without using a template, it is also possible to prepare large-scale MWCNTs aligned perpendicular to quartz substrates by pyrolysis of iron (II) phthalocyanine, and FePc, under Ar/H₂ at 800–1100°C.[87] As can be seen in Figure 6.13a, the constituent CNTs have a fairly uniform length and diameter. The same group has also developed microfabrication methods for patterning the aligned CNTs with a sub-micrometer resolution (Figure 6.13b).[25,87,88]

To construct aligned CNT electrodes, several techniques have been developed for transferring the aligned CNT arrays, in either patterned or nonpatterned fashion, to various other substrates of particular interest (e.g., polymer films for organic optoelectronic devices or metal substrates for electrochemistry).[87,89] Just like their nonaligned counterpart, the use of aligned CNTs for biosensing will inevitably require modification of their surface characteristics to meet the specific requirements for this particular application (e.g., biocompatibility). Given that the alignment is an additional advantage for the use of CNTs in many devices, including biosensors, a particularly attractive option is the surface modification of CNTs while largely retaining their aligned structure.

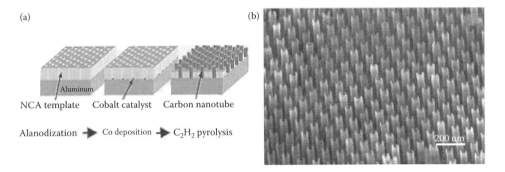

FIGURE 6.12 (a) Schematic of fabrication process. (b) SEM image of the resulting hexagonally ordered array of CNTs. (Adapted with permission from J. Li, C. Papadopoulos, J.M. Xu, *Appl. Phys. Lett.* 75, 367. Copyright 1999, American Institute of Physics.)

To prevent the aligned nanotubes from collapsing during subsequent chemical modification, Meyyappan and coworkers[55,90] have filled the gaps between the aligned nanotubes with a spin-on glass (SOG) and then modified the aligned nanotube tips through, for example, the acid-oxidative reaction for further grafting appropriate biomolecules to the carboxylic acid groups on the acid-oxidized aligned CNTs for biosensing applications.[91]

In addition to the activation of aligned CNTs by acid oxidation, we have developed a simple but effective approach for activating the aligned CNTs with appropriate plasma treatment, followed by further chemical modification under benign conditions through reactive characteristic of the plasma-generated functional groups.[92] By doing so, these authors have successfully immobilized aminodextran chains on acetaldehyde–plasma-treated aligned CNTs through the formation of a Schiff-base linkage, which was further stabilized by reduction with sodium cyanoborohydride (Scheme 6.1). The polysaccharide-grafted nanotubes are very hydrophilic; they are potentially useful for various biological-related applications (e.g., biosensing).

Subsequently, it has also been demonstrated that H_2O–plasma etching can be used to open the aligned CNTs and hence allow the chemical modification of the inner, outer, or both surfaces of the aligned nanotubes.[93]

Apart from the plasma surface modification, aligned CNT electrodes have also been modified by electrochemically depositing a thin layer of the appropriate conducting polymer (e.g., polyaniline, polypyrrole) onto individual nanotubes to form the so-called conducting polymer–carbon nanotube (CP–CNT) aligned coaxial nanowires.[27,89] The scanning electron microscopy (SEM) image for the CP–CNT coaxial nanowires given in Figure 6.14b shows the same features as the pristine aligned

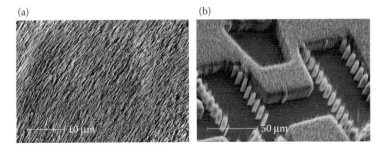

FIGURE 6.13 (a) A typical SEM image of the aligned CNT film prepared by pyrolysis of FePc. The misalignment seen for some of the nanotubes at the edge was caused by the peeling action used in the SEM sample preparation. (Adapted with permission from S. Huang, L. Dai, A.W.H. Mau, *J. Phys. Chem. B* 103, 4223. Copyright 1999, American Chemical Society.) (b) Aligned CNT micropattern. (Adapted with permission from Y. Yang et al., *J. Am. Chem. Soc.* 121, 10832. Copyright 1999, American Chemical Society.)

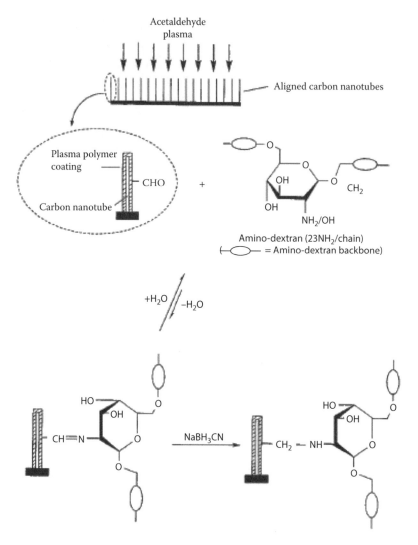

SCHEME 6.1 Covalent immobilization of amino-dextran chains onto acetaldehyde–plasma-activated CNTs. For reasons of clarity, only one of the many plasma-induced aldehyde surface groups is shown for an individual nanotube.

FIGURE 6.14 SEM images of (a) pure aligned CNT array before PPy deposition and (b) aligned PPy–CNT coaxial nanowires; inset shows clear image of single tube coated with PPy. Potentiostatic oxidation (1 V) of 0.10 M pyrrole was carried out in an electrolyte solution containing 0.1 M NaClO$_4$ in pH 7.45 buffer for 1 min. (Adapted from M. Gao, L. Dai, G.G. Wallace, *Electroanalysis* **2003**, 15, 1089. Copyright Wiley-VCH Verlag GmbH & Co. KGaA. Reproduced with permission.)

nanotube array of Figure 6.14a, but with a larger diameter due to the presence of the electropoly-merized polypyrrole coating in this particular case.[27] The uniform conducting polymer coating was also evidenced by TEM imaging.[89] The resultant CP–CNT coaxial nanowires with enzyme (e.g., glucose oxidizer) trapped into the conducting polymer layer have been demonstrated to be attractive for biosensing (e.g., glucose detection) applications.[27]

6.5 CNT BIOSENSORS

6.5.1 PROTEIN AND ENZYME BIOSENSORS

To further support the observation that CNTs can promote electron transfer with certain proteins and enzymes,[6,8,83] Wang et al.[6] have investigated the electrochemical behavior of cytochrome c on an SWCNT film electrode (Figure 6.15). As can be seen in Figure 6.15, cytochrome c gave no obvi-ous electrochemical response at the bare GC electrode (Figure 6.15a), whereas it showed irrevers-ible behavior with a difference between the anodic peak potential and the cathodic peak potential ($\Delta E = 265$ mV) at the unactivated SWCNT-modified electrode (Figure 6.15b). However, a pair of well-defined redox peaks with $\Delta E = 73.7$ mV were observed for cytochrome c at the same electrode after modification with an activated SWCNT film (Figure 6.15c). These results indicate a signifi-cantly improved electron-transfer process for the CNT-modified electrode.

Wang et al.[94] have also demonstrated that Nafion-coated CNT electrodes can enhance the redox activity of hydrogen peroxide so dramatically that a remarkable decrease in the over-oxidation potential for hydrogen peroxide was observed. The accelerated electron transfer from hydrogen peroxide allowed for glucose measurements at very low potentials (−50 mV), where interfering reac-tions of acetaminophen, uric acid, and AA were minimized. Similarly, Luong et al.[95] used 3-amino-propyltriethoxysilane (APTES) as both a solubilizing agent for CNTs and an immobilization matrix for GOx to construct a mediator-less, highly efficient glucose sensor. In this study, a well-defined redox response from glucose was observed at −0.45 V (vs. Ag/AgCl), whereas no response was seen

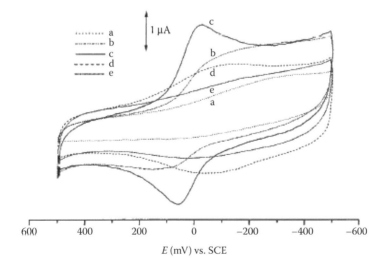

FIGURE 6.15 Cyclic voltammograms at a bare GC electrode. (a) An unactivated SWCNT-modified GC elec-trode (b, e) and an activated SWCNT-modified GC electrode (c, d) in the absence of cytochrome c (d, e) and in the presence of $5.0 \cdot 10^{-4}$ M cytochrome c (a–c) in 0.1 M phosphate buffer solution (pH = 6.24). The potential scan rate is 0.02 V/s. The activation was achieved by scanning the SWCNT-modified electrode in a fresh 0.1 M phosphate buffer solution (pH = 6.24) over +1.5 to −1.0 V at a scan rate of 1 V/s for 1.5 min. (Adapted with per-mission from J.X. Wang et al., *Anal. Chem.* 74, 1993. Copyright 2002, American Chemical Society.)

for three common interfering species, namely uric acid, AA, and acetaminophen, at a concentration corresponding to the physiological level of 0.1 mM.

Unlike the above glucose biosensors based on nonaligned CNTs, Ren and coworkers[96–98] developed a glucose biosensor by covalently immobilizing GOx onto an aligned carbon nanotube nanoelectrode ensemble (CNT-NEE) through the amide formation between the GOx and carboxylic acid groups (Section 6.3.2.1) on the aligned CNT tips (Figure 6.16).

Owing to the high surface packing density of the individual nanoelectrodes within the CNT-NEE assembly, the prepared aligned CNT glucose biosensor was found to be ultrasensitive with a high selectivity for electrochemical analysis of glucose over common interferents (e.g., acetaminophen, or uric, and AAs) (Figure 6.17).

Yu et al.[99] have covalently linked myoglobin and horseradish peroxidase to an aligned SWCNT electrode through the amide linkage. Quasi-reversible Fe^{3+}/Fe^{2+} voltammetry was observed for the iron heme enzymes, myoglobin, and horseradish peroxidase. The results demonstrated that the "trees" in the nanotube forest are potent current collectors that conduct electrons from the external circuit to the redox sites of the enzymes.

Starting with nonaligned CNTs, Gooding et al.[18] have demonstrated that a self-assembly of the oxidation-shortened SWCNTs aligned normal to an electrode surface can act as aligned molecular wires to allow electrical communication between the underlying electrode and redox proteins covalently attached onto the top ends of the SWCNTs. As schematically shown in Figure 6.18, a small redox protein, microperoxidase MP-1, obtained by proteolytic digestion of horse heart cytochrome c, was attached to the free ends of the aligned nanotubes.

In this particular case, electron transfer could occur easily as the redox-active center (i.e., an iron protoporphyrin IX) in MP-11 was not shielded by a polypeptide, and it was demonstrated that the rate of electron transfer was not affected by the tube length. The rate constant for electron transfer

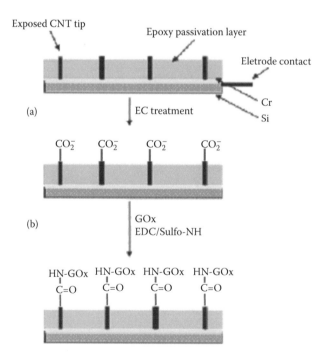

FIGURE 6.16 Fabrication of a glucose biosensor based on CNT nanoelectrode ensembles: (a) Electrochemical treatment of the CNT-NEE for functionalization. (b) Coupling of the enzyme (GOx) to the functionalized CNT-NEE. (Adapted with permission from Y. Lin et al., *Nano Lett.* 4, 191. Copyright 2004, American Chemical Society.)

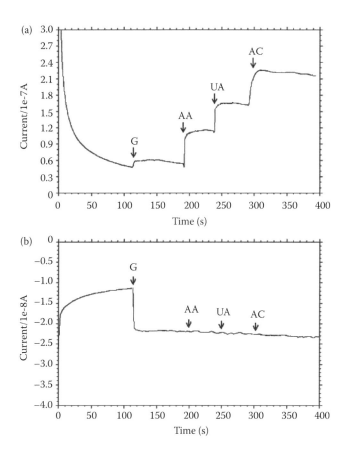

FIGURE 6.17 Amperometric responses of the CNT-NEE glucose biosensor to glucose (G), AA, uric acid (UA), and acetaminophen (AC) at potentials of +0.4 V (a) and −0.2 V (b). (Adapted with permission from Y. Lin et al., *Nano Lett.* 4, 191. Copyright 2004, American Chemical Society.)

for the nanotube-modified electrodes was found to be similar to that for MP-11 attached directly to a cysteamine-modified gold electrode, though the redox peak intensity is much higher for the former because of its large surface area (Figure 6.19).

Patolsky et al.[4] have investigated the long-range electron transfer from redox enzymes chemically bound to the aligned SWCNT structure.

Figure 6.20 shows the procedure used by these authors to assemble the aligned SWCNTs onto a gold electrode. As can be seen, the oxidized SWCNTs were first covalently coupled with 2-thioethanol/cystamine mixed monolayer (3:1 ratio) assembled on an Au electrode. The incorporation of 2-thioethanol in the mixed monolayer was anticipated to prevent nonspecific adsorption of the surfactant-protected SWCNTs onto the electrode surface. The amino-derivative of flavin adenine dinucleotide (FAD) cofactor was then coupled to the carboxylic acid groups at the free top ends of the aligned SWCNTs. Finally, GOx was reconstituted on the FAD units linked to the ends of the aligned SWCNTs for glucose sensing.

Figures 6.21A and B show that the bioelectrocatalytic oxidation of glucose on the Au electrode modified with GOx-attached CNTs (average nanotube length of 25 nm) occurred at $E > 0.18$ V, and the electrocatalytic anodic current became higher as the concentration of glucose increased. Figure 6.21C reproduces the calibration curves corresponding to the anodic currents generated by the GOx-reconstituted SWCNT electrodes of different SWCNT lengths in the presence of variable glucose concentrations, whereas Figure 6.21D shows a strong linear dependence between

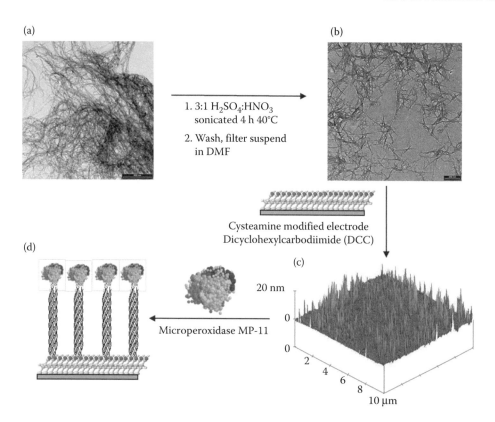

FIGURE 6.18 A schematic representation showing the steps involved in the fabrication of aligned SWCNT arrays (a, b, and c) for direct electron transfer with enzymes such as microperoxidase MP-11 (d). (Adapted with permission from J.J. Gooding et al., *J. Am. Chem. Soc.* 125, 9006. Copyright 2003, American Chemical Society.)

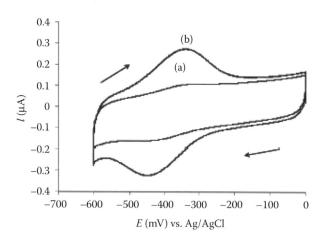

FIGURE 6.19 Cyclic voltammograms of (a) Au/cysteamine after being immersed in DMF and MP-11 solution and (b) Au/cysteamine/SWCNTs/MP-11 in 0.05 M phosphate buffer solution pH 7.0 containing 0.05 M KCl under argon gas at a scan rate of 100 mV/s versus Ag/AgCl. (Adapted with permission from J.J. Gooding et al., *J. Am. Chem. Soc.* 125, 9006. Copyright 2003, American Chemical Society.)

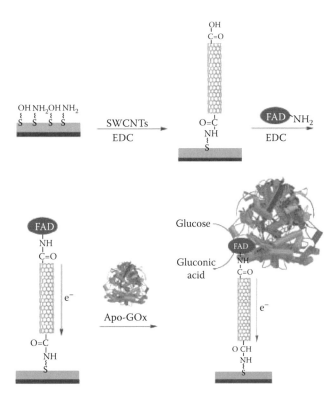

FIGURE 6.20 Assembly of the SWCNT electrode with chemically bound GOx. (Adapted from F. Patolsky, Y. Weizmann, I. Willner, *Angew. Chem. Int. Ed.* **2004**, 43, 2113. Copyright Wiley-VCH Verlag GmbH & Co. KGaA. With permission.)

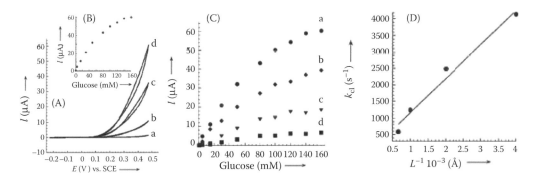

FIGURE 6.21 (A) Cyclic voltammograms corresponding to the electrocatalyzed oxidation of different concentrations of glucose by the GOx reconstituted on the 25 nm long FAD-functionalized CNTs assembly: (a) 0 mm glucose, (b) 20 mm glucose, (c) 60 mm glucose, and (d) 160 mm glucose. Data are recorded in phosphate buffer, 0.1 M, pH = 7.4, and scan rate 5 mV/s. (B) Calibration curve corresponding to the amperometric responses of the reconstituted GOx/CNTs (25 nm) electrode (at E – 0.45 V) in the presence of different concentrations of glucose. (C) Calibration curves corresponding to the amperometric responses (at E = 0.45 V) of reconstituted GOx–CNTs electrodes in the presence of variable concentrations of glucose and different CNT lengths as electrical connector units: (a) about 25 nm SWCNTs, (b) about 50 nm SWCNTs, (c) about 100 nm SWCNTs, and (d) about 150 nm SWCNTs. (D) Dependence of the electron-transfer turnover rate between the GOx redox center and the electrode on the lengths of the SWCNTs comprising the enzyme electrodes. (Adapted from F. Patolsky, Y. Weizmann, I. Willner, *Angew. Chem. Int. Ed.* **2004**, 43, 2113. Copyright Wiley-VCH Verlag GmbH & Co. KGaA. With permission.)

the turnover rate of electron transfer and L^{-1} (L is the nanotube length). The observed length-controlled electron transfer was further supported by examining the interfacial electron transfer to the FAD sites at the ends of the aligned SWCNTs, although the mechanism of the tube length-dependent electron transfer has not been fully understood. Given that no tube length dependence was observed in Gooding's study,[18] the length-dependent charge transfer observed in Figure 6.21 most likely results from defect sites introduced along the nanotube wall during the oxidative-shortening process. However, the detailed mechanism of the tube-length-dependent charge transfer deserves further investigation.

To avoid the tedious processes for preparing aligned SWCNTs, we have used the FePc-generated aligned MWCNTs (Section 6.4.2) as a novel electrode platform for biosensing.[27,100] In a typical experiment, the aligned MWCNTs are coated with conducting polymers by electrochemical deposition of a thin layer of polypyrrole or polyaniline onto individual CNTs along their tubular length (Figure 6.14) in the presence of glucose oxidase. It was found that the detection of H_2O_2 at these aligned MWCNT electrodes coated with GOx-containing conducting polymers can also be achieved at low anodic potentials, leading to glucose sensors with a very high sensitivity and selectivity.[101,102]

6.5.2 DNA Sensors

A major feature of the Watson–Crick model of DNA is that it provides a vision of how a base sequence of one strand of the double helix can precisely determine the base sequence of the partner strand for passing the genetic information in all living species. The principle learned from this has now been applied to the development of biosensors for DNA analysis and diagnosis through the very specific DNA-pairing interaction.[103,104] Owing to their high sensitivity, facial data processing, great simplicity, and good compatibility with electronic detection technologies, DNA electrochemical biosensors have gained considerable interest. The unique electronic properties and relatively large surface area have made CNTs a promising electrode material for constructing advanced biosensors. This, coupled with various chemical reactions reported for the attachment of DNA chains onto CNTs (Sections 6.3.2.1 and 6.3.2.2), has significantly facilitated the development of CNT DNA sensors.

Using a glass carbon electrode (GCE) modified with MWCNTs, Cai et al.[105] observed an enhanced sensitivity for electrochemical DNA biosensor based on CNTs. Figure 6.22 schematically shows the steps for constructing the nanotube–DNA biosensor. To start, carboxylic acid functionalized MWCNTs (COOH–MWCNTs) were dropped on a GCE electrode, single-strand DNA oligonucleotides (ss-DNAs) were then covalently bonded onto the COOH–MWCNTs via amide formation. The hybridization reaction on the electrode was monitored by differential pulse voltammetry (DPV) using an electroactive daunomycin intercalator as the indicator.

Compared to conventional DNA sensors with oligonucleotides being directly incorporated onto carbon electrodes, CNT-based DNA sensors show a dramatically increased DNA attachment density and cDNA detection sensitivity because of the relatively large surface area and good charge-transport characteristics of CNTs. These promising results have further prompted the development of novel biosensors for direct electrochemical detection of DNA hybridization by alternating current (AC) impedance measurements,[106,107] electrodepositing polypyrrole, and oligonucleotide probe onto an MWCNT-coated GCE electrode. The surface area of CNTs allowed a large volume of polypyrrole/oligonucleotide to be deposited on the electrode, whereas the polypyrrole coating on individual CNTs remained sufficiently thin. Owing to their high electrical conductivity, CNTs can reflect any resistance change in the polypyrrole/oligonucleotide thin film caused by hybridization reaction (Figure 6.23).

In a somewhat related but separate study, Guo et al.[108] used electrochemical impedance spectroscopy to investigate the process of the electrostatic assembly of calf thymus DNA on MWCNTs-modified gold electrodes in the presence of a cationic polyelectrolyte PDDA (polydiallyldimethylammonium chloride). The gold electrode modified with carboxylic acid functionalized MWCNTs was

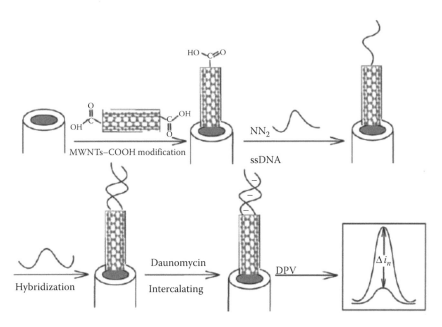

FIGURE 6.22 Schematic representation of the enhanced electrochemical detection of DNA hybridization by DNA biosensor based on the COOH–MWCNTs. (Adapted with kind permission from Springer Science+Business Media: *Anal. Bioanal. Chem.* 375, **2003**, 287. H. Cai et al.)

dipped alternately in aqueous solutions of PDDA and calf thymus DNA to form a multilayer film. Electrochemical impedance spectroscopy was then used to characterize the interfacial properties for the modified electrode.

A typical impedance spectrum presented in the form of the Nyquist plot includes a semicircle portion at higher frequencies corresponding to the electron-transfer-limited process and a linear part at a lower-frequency range associated with the diffusion-limited process. The semicircle diameter in the impedance spectrum equals the electron-transfer resistance, R_{et}, which is related to the electron-transfer kinetics of the redox probe at the electrode surface. As can be seen in Figure

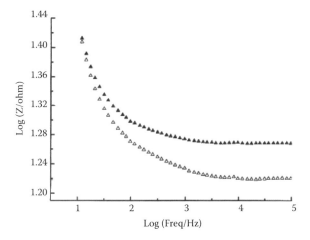

FIGURE 6.23 Impedance curves before (solid triangle) and after (hollow triangle) hybridization in 0.3 M PBS solution on an open-circuit voltage. Frequency range: $10–10^5$ Hz, amplitude: 5 mV versus saturated calomel electrode, SCE. (Adapted from H. Cai et al., *Electroanalysis* **2003**, 15, 23. Copyright Wiley-VCH Verlag GmbH & Co. KGaA. With permission.)

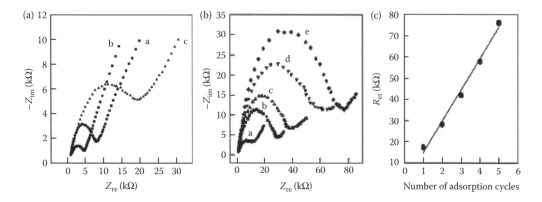

FIGURE 6.24 (A) Nyquist plots (Z_{im} vs. Z_{re}) of: (a) bare gold electrode, (b) MWCNTs-modified gold electrode, and (c) MWCNTs-modified gold electrode after the assembly of PDDA and DNA in 10 mM $K_4Fe(CN)_6$ + 10 mM $K_3Fe(CN)_6$ + 0.1 M KCl solutions. (B) Nyquist plots (Z_{im} vs. Z_{re}) in 10 mM $K_4Fe(CN)_6$ + 10 mM $K_3Fe(CN)_6$ + 0.1 M KCl solutions at MWCNT-modified gold electrode after different numbers of PDDA/DNA adsorption cycles: (a) 1, (b) 2, (c) 3, (d) 4, and (e) 5. (C) Relationship between the electron-transfer resistance (R_{et}) and the number of PDDA/DNA adsorption cycles. (Adapted from *Electrochim. Acta* 49, M. Guo et al., 2637, Copyright 2004, with permission from Elsevier.)

6.24A, R_{et} decreased from 8.01 to 6.75 kΩ on surface modification of the electrode with MWCNTs, attributable to the good electrical conductivity and large surface area of MWCNTs. However, the semicircle diameter increased significantly to 26.72 kΩ after the adsorption of PDDA/DNA layer because of its nonconducting nature.

The impedance spectra for the MWCNT-modified microelectrode after adsorption of different layer numbers of PDDA/DNA are given in Figure 6.24B, which shows an increase in the semicircle diameter with increasing layer number for the adsorbed multilayer film. The linear relationship between R_{et} and the cycle number shown in Figure 6.24C suggests the formation of a uniform PDDA/DNA multilayer film on the MWCNT-modified gold electrode.

To amplify electrical detection of DNA hybridization, Wang et al.[109–111] developed a CNT-based dual amplification route, in which CNTs play a dual amplification role (namely carrying numerous enzyme tags and accumulating the product of the enzymatic reaction) for amplifying enzyme tag numbers in sensing of DNAs and proteins (Figure 6.25). The process shown in Figure 6.25 involved the sandwich DNA hybridization (a) or antigen–antibody binding (b) along with magnetic separation

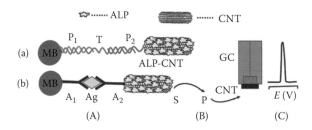

FIGURE 6.25 Schematic representation of the analytical protocol: (A) Capture of the ALP-loaded CNT tags to the streptavidin-modified magnetic beads by a sandwich DNA hybridization (a) or A_1–Ag–A_2 interaction (b). (B) Enzymatic reaction. (C) Electrochemical detection of the product of the enzymatic reaction at the CNT-modified GC electrode; MB, Magnetic beads; P_1, DNA probe 1; T, DNA target; P_2, DNA probe 2; A_1, first antibody; Ag, antigen; A_2, secondary antibody; S and P, substrate and product, respectively, of the enzymatic reaction; GC, glassy carbon electrode; CNT, carbon nanotube layer. (Adapted with permission from J. Wang, G. Liu, M.R. Jan, *J. Am. Chem. Soc.* 126, 3010. Copyright 2004, American Chemical Society.)

of the analyte-linked magnetic-bead/CNT assembly (A), followed by enzymatic amplification (B), and chronopotentiometric stripping detection of the product at the CNT-modified electrode (C).

Figure 6.26 shows the dramatic signal enhancement associated with the CNT-based dual amplification route for DNA hybridization (a) and Ag–Ab (b) bioassays. As can be seen in Figure 6.26, the conventional protocols, based on the single-enzyme tag and a GCE, were not responding to either 10 pg/mL DNA target (A, a) or 80 pg/mL IgG (B,a). The first amplification step based on the alkaline phosphatase (ALP)-loaded CNTs (b) offered convenient detection of these low analyte concentrations. Further enhancements of the DNA and protein signals were observed in the second amplification path, employing the CNT-modified electrode (c). The latter reflects the strong adsorptive accumulation of the liberated α-naphthol on the CNT layer. Two series of six repetitive measurements of 1 pg/mL DNA target or 0.8 ng/mL IgG yielded reproducible signals with relative standard deviations of 5.6% and 8.9%, respectively.

Using aligned CNT electrodes, we have recently developed novel nanotube–DNA sensors of a high sensitivity and selectivity by grafting ssDNA chains onto aligned CNTs generated from FePc (Figure 6.27).[27] In this study, aligned CNTs supported by gold substrate were first treated with acetic acid–plasma to introduce the surface carboxylic acid groups for grafting ssDNA chains with an amino group at the 5′-phosphate end (i.e., [AmC6]TTGACACCAGACCAACTGGT-3′, I). cDNA chains labeled with ferrocenecarboxaldehyde, FCA, (i.e., [FCA-C6] ACCAGTTGGTCTGGTGTCAA-3′, II) were then used for hybridizing with the surface-immobilized oligonucleotides to form the double-strand DNA (dsDNA) helices on the aligned CNT electrodes (Figure 6.27).

The performance of the aligned CNT–DNA sensors for sequence-specific DNA diagnoses was demonstrated in Figure 6.28. The strong oxidation peak seen at 0.29 V in curve *a* of Figure 6.28 is attributed to ferrocene and indicates the occurrence of hybridization of FCA-labeled cDNA (II) chains with the nanotube-supported ssDNA (I) chains, leading to a long-range electron transfer from the FCA probe to the nanotube electrode through the DNA duplex. In contrast, the addition of FCA-labeled

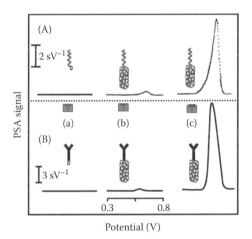

FIGURE 6.26 Chronopotentiometric signals for 10 pg/mL target oligonucleotide (A) and 80 pg/mL IgG (B) using the GC transducer and (a) a single ALP tag and (b) CNT loaded with multiple ALP tags; (c) same as (b) but using the CNT-modified GC electrode. Amount of magnetic beads, 50 μg; sandwich assay with 20 and 30 min for each hybridization event and Ag/Ab association, respectively; sample volume, 50 μL. Detection, addition of 50 μL R-naphthyl phosphate (50 mM) solution with a 20-min enzymatic reaction. Measurements of the R-naphthol product were performed at the bare or modified GC electrodes, using a 2-min accumulation at +0.2 V in a stirred phosphate buffer solution (0.05 M, pH = 7.4; 1 mL), followed by a 10-s rest period (without stirring) and application of an anodic current of +5.0 μA. See supporting information for the concentrations of the oligonucleotide probes, antibody, and sequence of oligonucleotide probes, levels, and preparation of the ALP–DNA–CNT and ALP–streptavidin–CNT conjugates. (Adapted with permission from J. Wang, G. Liu, M.R. Jan, *J. Am. Chem. Soc.* 126, 3010. Copyright 2004, American Chemical Society.)

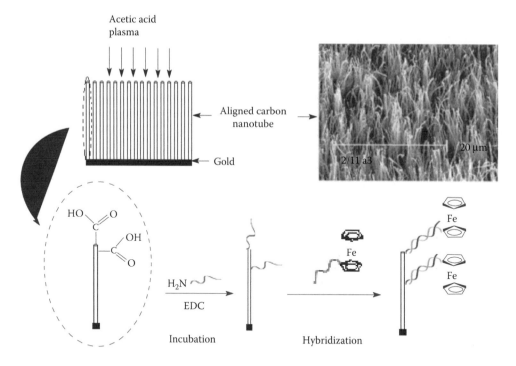

FIGURE 6.27 Schematic illustration of the aligned nanotube–DNA electrochemical sensor. The upper right SEM image shows the aligned CNTs after having been transferred onto a gold foil. For reasons of clarity, only one of the many carboxyl groups is shown at the nanotube tip and wall, respectively. (Adapted from P. He, L. Dai, *Chem. Commun.* **2004**, 3, 348. With permission of The Royal Society of Chemistry.)

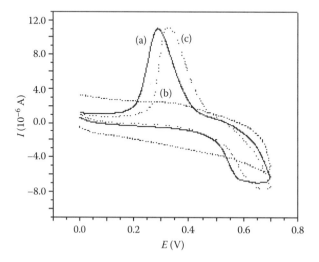

FIGURE 6.28 Cyclic voltammograms of the ssDNA (I)-immobilized aligned CNT electrode after hybridization with FCA-labeled cDNA (II) chains (a) in the presence of FCA-labeled non-cDNA (III) chains (b) and after hybridization with target DNA (IV) chains in the presence of the FCA-labeled non-cDNA (III) chains (c). All the cyclic voltammograms were recorded in 0.1 M H_2SO_4 solution with a scan rate of 0.1 V/s. The concentration of the FCA-labeled DNA probes is 0.05 mg/mL. (Adapted from P. He, L. Dai, *Chem. Commun.* **2004**, 3, 348. With permission of The Royal Society of Chemistry.)

non-cDNA chains (i.e., [FCA-C6]CTCCAGGAGTCGTCGCCACC- 3′, III) under the same conditions did not show any redox response of FCA (curve *b* in Figure 6.28). Subsequent addition of target DNA chains (i.e., 5′-GAGGTCCTCAGCAGCGGTGGACCAGTTGGTCTGGTGTCAA-3′, IV) into the above solution, however, led to a strong redox response from the FCA-labeled DNA (III) chains (curve *c* in Figure 6.28) because the target DNA (IV) contains complementary sequences for both DNA (I) and DNA (III) chains.

Meyyappan and coworkers[28,91] have developed a more advanced micropatterned ultrasensitive DNA biosensor based on aligned CNTs. As shown in Figure 6.29a, aligned MWCNTs were directly grown on individual metal microcontacts, followed by encapsulating the MWCNT arrays and the substrate surface with an SOG layer to expose only the nanotube ends at the surface. Each of the SOG-encapsulated individual MWCNTs function like a nanoelectrode, and there are about 100 or more MWCNT nanoelectrodes on each microcontact. Abundant carboxylic acid groups were produced at the end of MWCNTs by electrochemical etching and then functionalized with a specific oligonucleotide probe through the amide formation. Figure 6.29b and c schematically show the mechanism of the MWCNT nanoelectrode array for the DNA detection. A specific probe [Cy3]5′-CTIIATTTCICAIITCCT-3′-[AmC7-Q] was used in that study, which contains the sequence of the normal allele of the BRCA1 gene associated with the occurrence of several cancers. Electroactive guanine groups in the probe are replaced with nonelectroactive inosines. Genomic DNA from a healthy donor was amplified by polymerase chain reaction (PCR) and functioned as control.

Figure 6.29d shows the electrophoresis results of a DNA molecular weight standard (ΦX174R FDNA-HaeIII digest) and the two PCR amplicons, respectively. The nonspecific binding is removed through stringent washing in three steps using $2 \times$ SSC/0.1% SDS, $1 \times$ SSC, and 0.1% SSC,

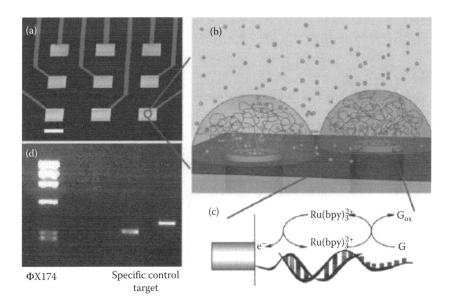

FIGURE 6.29 (a) SEM image of an individually addressable 3×3 microcontact array with an MWCNT nanoelectrode array on each site. The scale bar is 200 μm. (b) Schematic of the mechanism to detect DNA hybridization using an MWCNT nanoelectrode array. The long ssDNA PCR amplicons are hybridized to the short oligonucleotide probes that are functionalized at the very end of the MWCNTs. Ru(bpy)$_3^{2+}$ mediators are used to transfer electrons from the guanine groups to the MWCNT nanoelectrode for all target molecules within the hemispherical diffusion layer of the nanoelectrodes. (c) The schematic mechanism for the guanine oxidation amplified with Ru(bpy)$_3^{2+}$ mediators. (d) The gel electrophoresis. The lanes from left to right are DNA molecular weight standard (ΦX174RFDNA-HaeIII digest), a specific PCR amplicon target with ≈300 bases, and a control sample with an unrelated PCR amplicon with ≈400 bases, respectively. (Adapted from J. Koehne et al., *Nanotechnology* **2003**, *14*, 1239.)

respectively, by shaking the sample solutions at 40°C for 15 min. $Ru(bpy)_3^{2+}$ mediators were used to efficiently transport electrons from the guanine bases to the MWCNT nanoelectrode and to provide an amplified guanine oxidation signal as long as target DNA molecules are within the 3D diffusion layer.

6.6 CONCLUSIONS

CNTs possess superior electronic, chemical, thermal, and mechanical properties with a large surface area and are attractive for a wide range of potential applications. The discovery of the sensing capability of CNTs is significantly intriguing. The high surface area and good electronic property provided by CNTs are attractive feature in the advancement of chemical biosensors. However, the use of CNTs for biosensing inevitably requires modification of their surface characteristics to meet the specific requirements for this particular application. The CNT synthesis and surface modification methods, together with the electrode fabrication techniques, highlighted in this chapter have allowed the development of various CNT biosensors with desirable characteristics. We have given a brief summary of the important research development in this exciting field. Even this brief account has revealed the versatility of CNTs for making biosensors with high sensitivity and selectivity for probing DNA chains of specific sequences as well as proteins and enzymes with different redox properties. Continued research and development in this field will lead to the development of cost-effective, highly sensitive, and selective biosensors to detect diseases at early stage, which will certainly benefit many aspects of our human life.

ACKNOWLEDGMENTS

The authors are very grateful for the financial support from AFOSR, DOD-AFOSR-MURI, AFRL/DAGSI, U.S. AFOSR-Korea NBIT, NSF, and NSF-NSFC WMN.

REFERENCES

1. P.J.F. Harris, *Carbon Nanotubes and Related Structures—New Materials for the Twenty-First Century*, Cambridge University Press: Cambridge, 2001.
2. L. Dai, *Carbon Nanotechnology: Recent Developments in Chemistry, Physics, Materials Science and Device Applications*, Elsevier Science: London, 2006.
3. K. Yamamoto, G. Shi, T.S. Zhou, F. Xu, J.M. Xu, T. Kato, J.Y. Jin, L. Jin, *Analyst* **2003**, 128, 249.
4. F. Patolsky, Y. Weizmann, I. Willner, *Angew. Chem. Int. Ed.* **2004**, 43, 2113.
5. J.J. Davis, R.J. Coles, H.A.O. Hill, *J. Electroanal. Chem.* **1997**, 440, 279.
6. J.X. Wang, M.X. Li, Z.J. Shi, N.Q. Li, Z.N. Gu, *Anal. Chem.* **2002**, 74, 1993.
7. G. Wang, J.J. Xu, H.Y. Chen, *Electrochem. Commun.* **2002**, 4, 506.
8. Y.D. Zhao, W.D. Zhang, H. Chen, Q.M. Luo, S.F.Y. Li, *Sens. Actuators B* **2002**, 87, 168.
9. A. Hirsch, *Angew. Chem. Int. Ed.* **2002**, 41, 1853.
10. D. Tasis, N. Tagmatarchis, A. Bianco, M. Prato, *Chem. Rev.* **2006**, 106, 1105.
11. L. Dai, T. Lin, T. Ji, V. Bajpai, *Aust. J. Chem.* **2003**, 56, 635.
12. V.N. Khabashesku, W.E. Billups, J.L. Margrave, *Acc. Chem. Res.* **2002**, 35, 1087.
13. Y.-P. Sun, K. Fu, Y. Lin, W. Huang, *Acc. Chem. Res.* **2002**, 35, 1096.
14. S. Niyogi, M.A. Hamon, H. Hu, B. Zhao, P. Bhowmik, R. Sen, M. Itkis, R.C. Haddon, *Acc. Chem. Res.* **2002**, 35, 1105.
15. L. Qu, L. Dai, Sidewall functionalization of carbon nanotubes, in *Chemistry of Carbon Nanotubes*, V.A. Basiuk, E.V. Basiuk (eds.), American Scientific Publisher: California, 2007.
16. S.C. Tsang, J.J. Davis, M.L.H. Green, H.A.O. Hill, Y.C. Leung, P.J. Sadler, *Chem. Commun.* **1995**, 17, 1803.
17. J.J. Davis, K.S. Coleman, B.R. Azamian, C.B. Bagshaw, M.L.H. Green, *Chem. Eur. J.* **2003**, 9, 3732.
18. J.J. Gooding, R. Wibowo, J. Liu, W. Yang, D. Losic, S. Orbons, F.J. Mearns, J.G. Shapter, D.B. Hibbert, *J. Am. Chem. Soc.* **2003**, 125, 9006.
19. A. Callegari, S. Cosnier, M. Marcaccio, D. Paolucci, F. Paolucci, V. Georgakilas, N. Tagmatarchis, E. Vázquez, M. Prato, *J. Mater. Chem.* **2004**, 14, 807.

20. P. He, L. Dai, *Chem. Commun.* **2004**, 3, 348.
21. P.J. Britto, K.S.V. Santhanam, P.M. Ajayan, *Bioelectrochem. Bioenerg.* **1996**, 41, 121.
22. H. Luo, Z. Shi, N. Li, Z. Gu, Q. Zhuang, *Anal. Chem.* **2001**, 73, 915.
23. Q. Zhao, Z. Gan, Q. Zhuang, *Electroanalysis* **2002**, 14, 1609.
24. Z. Wang, J. Liu, Q. Liang, Y. Wang, G. Luo, *Analyst* **2002**, 127, 653.
25. L. Dai, A. Patil, X. Gong, Z. Guo, L. Liu, Y. Liu, D. Zhu, *Chem. Phys. Chem.* **2003**, 4, 1150; and references cited therein.
26. P. Diao, Z. Liu, B. Wu, X. Nan, J. Zhang, Z. Wei, *Chem. Phys. Chem.* **2002**, 10, 898.
27. M. Gao, L. Dai, G.G. Wallace, *Electroanalysis* **2003**, 15, 1089.
28. J. Koehne, H. Chen, J. Li, A.M. Cassell, Q. Ye, H.T. Ng, J. Han, M. Meyyappan, *Nanotechnology* **2003**, 14, 1239.
29. P.M. Ajayan, *Chem. Rev.* **1999**, 99, 1787.
30. R. Saito, G. Deesselhaus, M.S. Dresselhaus, *Physical Properties of Carbon Nanotubes*, Imperial College Press: London, 1998.
31. K. Gong, F. Du, Z. Xia, M. Durstock, L. Dai, *Science* **2009**, 323, 760.
32. D. Yu, L. Dai, *J. Phys. Chem. Lett.* **2010**, 1, 467.
33. L. Qu, L. Dai, M. Stone, Z. Xia, Z.L. Wang, *Science* **2008**, 322, 238.
34. H. Dodziuk, A. Ejchart, W. Anczewski, H. Ueda, E. Krinichnaya, W. Dolgonos, W. Kutner, *Chem. Commun.* **2003**, 8, 986.
35. M. Zheng, A. Jagota, E.D. Semke, B.A. Diner, R.S. McLean, S.R. Lustig, R.E. Richardson, N.G. Tassi, *Nat. Mater.* **2003**, 2, 338.
36. N. Nakashima, S. Okuzono, H. Murakami, T. Nakai, K. Yoshikawa, *Chem. Lett.* **2003**, 32, 456.
37. G.R. Dieckmann, A.B. Dalton, P.A. Johnson, J. Razal, J. Chen, G.M. Giordano, E. Munoz, I.H. Musselman, R.H. Baughman, R.K. Draper, *J. Am. Chem. Soc.* **2003**, 125, 1770.
38. S. Wang, E.S. Humphreys, S.-Y. Chung, D.F. Delduco, S.R. Lustig, H. Wang, K.N. Parker et al., *Nat. Mater.* **2003**, 2, 196.
39. R.J. Chen, Y. Zhang, D. Wang, H. Dai, *J. Am. Chem. Soc.* **2001**, 123, 3838.
40. L. Dai, Self-assembling of carbon nanotubes, in *Self-Organized Organic Semiconductors: From Materials to Device Applications,* Q. Li (ed.), John Wiley & Sons, Inc.: New York, 2010.
41. K. Besteman, J.-O. Lee, F.G.M. Wiertz, H.A. Heering, C. Dekker, *Nano Lett.* **2003**, 3, 727.
42. F. Balvlavoine, P. Schultz, C. Richard, V. Mallouh, T.W. Ebbeson, C. Mioskowski, *Angew. Chem. Int. Ed. Eng.* **1999**, 38, 1912.
43. M.J. Moghaddam, S. Taylor, M. Gao, S. Huang, L. Dai, M.J. McCall, *Nano Lett.* **2004**, 4, 89.
44. S. Li, P. He, J. Dong, Z. Gao, L. Dai, *J. Am. Chem. Soc.* **2005**, 127, 14.
45. A. Ulman, *Chem. Rev.* **1996**, 96, 1533 and references cited therein.
46. J. Chen, M.A. Hamon, H. Hu, Y. Chen, A.M. Rao, P.C. Eklund, R.C. Haddon, *Science* **1998**, 282, 95.
47. J. Liu, A.G. Rinzler, H. Dai, J.H. Hafner, R.K. Bradley, P.J. Boul, A. Lu et al., *Science* **1998**, 280, 1253.
48. D.B. Mawhinney, V. Naumenko, A. Kuznetsova, J.T. Yates, Jr., *J. Am. Chem. Soc.* **2000**, 122, 2383.
49. D.B. Mawhinney, V. Naumenko, A. Kuznetsova, J.T. Yates, Jr., J. Liu, R.E. Smalley, *Chem. Phys. Lett.* **2000**, 324, 213.
50. M.A. Hamon, J. Chen, H. Hu, Y. Chen, M.E. Itkis, A.M. Rao, P.C. Eklund, R.C. Haddon, *Adv. Mater.* **1999**, 11, 834.
51. J. Chen, A.M. Rao, S. Lyuksyutov, M.E. Itkis, M.A. Hamon, H. Hu, R.W. Cohn et al., *J. Phys. Chem. B* **2001**, 105, 2525.
52. W. Huang, Y. Lin, S. Taylor, J. Gaillard, A.M. Rao, Y.-P. Sun, *Nano Lett.* **2002**, 2, 231.
53. J.E. Riggs, Z. Guo, D.L. Carroll, Y.-P. Sun, *J. Am. Chem. Soc.* **2000**, 122, 5879.
54. K.A. Williams, P.T.M. Veenhuizen, B.G. de la Torre, R. Eritja, C. Dekker, *Nature* **2002**, 420, 761.
55. C.V. Nguyen, L. Delzeit, A.M. Cassell, J. Li, J. Han, M. Meyyappan, *Nano Lett.* **2002**, 2, 1079.
56. C. Dwyer, M. Guthold, M. Falvo, S. Washburn, R. Superfine, D. Erie, *Nanotechnology* **2002**, 13, 601.
57. M. Hazani, R. Naaman, F. Hennrich, M.M. Kappes, *Nano Lett.* **2003**, 3, 153.
58. K. Jiang, L.S. Schadler, R.W. Siegel, X. Zhang, H. Zhangc, M. Terronesd, *J. Mater. Chem.* **2004**, 14, 37.
59. S.E. Baker, W. Cai, T.L. Lasseter, K.P. Weidkamp, R.J. Hamers, *Nano Lett.* **2002**, 2, 1413.
60. K.M. Lee, L.C. Li, L. Dai, *J. Am. Chem. Soc.* **2005**, 127, 4122.
61. N. Chopra, M. Majumder, B.J. Hinds, *Adv. Funct. Mater.* **2005**, 15, 858.
62. M.A. Shannon, P.W. Bohn, M. Elimelech, J.G. Georgiadis, B.J. Mariñas, A.M. Mayes, *Nature* **2007**, 452, 301.
63. E.T. Michelson, C.B. Huffman, A.G. Rinzler, R.E. Smalley, R.H. Hauge, J.L. Margrave, *Chem. Phys. Lett.* **1998**, 296, 188.

64. E.T. Michelson, I.W. Chiang, J.L. Zimmerman, P.J. Boul, J. Lozano, J. Liu, R.E. Smalley, R.H. Hauge, J.L. Margrave, *J. Phys. Chem. B* **1999**, 103, 4318.
65. P.J. Boul, J. Liu, E.T. Mickelson, C.B. Huffman, L.M. Ericson, I.W. Chiang, K.A. Smith et al., *Chem. Phys. Lett.* **1999**, 310, 367.
66. M. Holzinger, O. Vostrowsky, A. Hirsch, F. Hennrich, M. Kappes, R. Weiss, F. Jellen, *Angew. Chem. Int. Ed.* **2001**, 40, 4002.
67. D. Tasis, N. Tagmatarchis, V. Georgakilas, M. Prato, *Chem. Eur. J.* **2003**, 9, 4000.
68. V. Georgakilas, K. Kordatos, M. Prato, D.M. Guldi, M. Holzinger, A. Hirsch, *J. Am. Chem. Soc.* **2002**, 124, 760.
69. D. Pantarotto, J. Hoebeke, R. Graff, C.D. Partidos, J.-P. Briand, M. Prato, A. Bianco, *J. Am. Chem. Soc.* **2003**, 125, 6160.
70. D. Pantarotto, C.D. Partidos, J. Hoebeke, F. Brown, E. Kramer, J.-P. Briand, S. Muller, M. Prato, A. Bianco, *Chem. Biol.* **2003**, 10, 961.
71. A. Bianco, M. Prato, *Adv. Mater.* **2003**, 15, 1765.
72. M. Musameh, J. Wang, A. Merkoci, Y. Lin, *Electrochem. Commun.* **2002**, 4, 743.
73. J. Wang, G. Chen, M.P. Chatrathi, M. Musameh, *Anal. Chem.* **2004**, 76, 298.
74. J. Xu, Y. Wang, Y. Xian, L. Jin, K. Tanaka, *Talanta* **2003**, 60, 1123.
75. S. Hrapovic, Y. Liu, K.B. Male, J.H.T. Luong, *Anal. Chem.* **2004**, 76, 1083.
76. L.T. Qu, L. Dai, *J. Am. Chem. Soc.* **2005**, 127, 10806.
77. L.T. Qu, L. Dai, E. Osawa, *J. Am. Chem. Soc.* **2006**, 128, 5523.
78. J. Wang, M. Musameh, *Anal. Chem.* **2003**, 75, 2075.
79. K. Rege, N.R. Raravikar, D.-Y. Kim, L.S. Schadler, P.M. Ajayan, J.S. Dordick, *Nano Lett.* **2003**, 3, 829.
80. V.G. Gavalas, R. Andrews, D. Bhattacharyya, L.G. Bachas, *Nano Lett.* **2001**, 1, 719.
81. V.G. Gavalas, S.A. Law, J.C. Ball, R. Andrews, L.G. Bachasa, *Anal. Biochem.* **2004**, 329, 247.
82. J.N. Wohlstadter, J.L. Wilbur, G.B. Sigal, H.A. Biebuyck, M.A. Billadeau, L. Dong, A.B. Fischer et al., *Adv. Mater.* **2003**, 15, 1184.
83. J.K. Campbell, L.I. Sun, R.M. Crooks, *J. Am. Chem. Soc.* **1999**, 121, 3779.
84. H. Chen, A. Roy, J.-B. Baek, L. Zhu, J. Qu, L. Dai, *Mater. Sci. Eng. Rep.* **2010**, 70, 63.
85. K. Gong, S. Chakrabarti, L. Dai, *Angew. Chem. Int. Ed.* **2008**, 47, 5446.
86. J. Li, C. Papadopoulos, J.M. Xu, *Appl. Phys. Lett.* **1999**, 75, 367.
87. S. Huang, L. Dai, A.W.H. Mau, *J. Phys. Chem. B* **1999**, 103, 4223.
88. Y. Yang, S. Huang, H. He, A.W.H. Mau, L. Dai, *J. Am. Chem. Soc.* **1999**, 121, 10832.
89. M. Gao, S. Huang, L. Dai, G. Wallace, R. Gao, Z. Wang, *Angew. Chem. Int. Ed.* **2000**, 39, 3664.
90. J. Li, A. Cassell, L. Delzeit, J. Han, M. Meyyappan, *J. Phys. Chem. B* **2002**, 106, 9299.
91. J. Li, H.T. Ng, A. Cassell, W. Fan, H. Chen, Q. Ye, J. Koehne, J. Han, M. Meyyappan, *Nano Lett.* **2003**, 3, 597.
92. Q. Chen, L. Dai, M. Gao, S. Huang, A.W.H. Mau, *J. Phys. Chem. B* **2001**, 105, 618.
93. S. Huang, L. Dai, A.W.H. Mau, *J. Phys. Chem. B* **2002**, 3543, 106.
94. J. Wang, M. Musameh, Y. Lin, *J. Am. Chem. Soc.* **2003**, 125, 2408.
95. J.H.T. Luong, S. Hrapovic, D. Wang, F. Bensebaa, B. Simard, *Electroanalysis* **2004**, 16, 132.
96. Y. Lin, F. Lu, Y. Tu, Z. Ren, *Nano Lett.* **2004**, 4, 191.
97. Y. Tu, Y. Lin, Z.F. Ren, *Nano Lett.* **2003**, 3, 107.
98. Y. Tu, Z.P. Huang, D.Z. Wang, J.G. Wen, Z.F. Ren, *Appl. Phys. Lett.* **2002**, 80, 4018.
99. X. Yu, D. Chattopadhyay, I. Galeska, F. Papadimitrakopoulos, J.F. Rusling, *Electrochem. Commun.* **2003**, 5, 408.
100. M. Gao, L. Dai, G.G. Wallace, *Synth. Met.* **2003**, 137, 1393.
101. S.G. Wang, Q. Zhang, R. Wang, S.F. Yoona, *Biochem. Biophys. Res. Commun.* **2003**, 311, 572.
102. S. Sotiropoulou, N.A. Chaniotakis, *Anal. Bioanal. Chem.* **2003**, 375, 103.
103. W.C.I. Homs, *Anal. Lett.* **2002**, 35, 1875.
104. J.J. Gooding, *Electroanalysis* **2002**, 14, 1149.
105. H. Cai, X. Cao, Y. Jiang, P. He, Y. Fang, *Anal. Bioanal. Chem.* **2003**, 375, 287.
106. H. Cai, Y. Xu, P. He, Y. Fang, *Electroanalysis* **2003**, 15, 23.
107. Y. Xu, Y. Jiang, H. Cai, P. He, Y. Fang, *Anal. Chim. Acta* **2004**, 516, 19.
108. M. Guo, J. Chen, L. Nie, S. Yao, *Electrochim. Acta* **2004**, 49, 2637.
109. J. Wang, G. Liu, M.R. Jan, *J. Am. Chem. Soc.* **2004**, 126, 3010.
110. J. Wang, G. Liu, M.R. Jan, Q. Zhu, *Electrochem. Commun.* **2003**, 5, 1000.
111. J. Wang, A.-N. Kawde, M. Musameh, *Analyst* **2003**, 128, 912.

7 Carbon Nanostructures in Biomedical Applications

Masoud Golshadi and Michael G. Schrlau

CONTENTS

7.1   INTRODUCTION

Since the 1990s, fibrous carbon nanostructures, such as carbon nanotubes (CNTs) and carbon nano-fibers (CNFs), have proven to be one of the most versatile materials used in the field of engineering, science, and medicine. Carbon nanostructures have made broad and significant societal impacts, from enhancing the properties of material composites and miniaturizing electronics to providing more efficient means of storing energy and facilitating the early detection and treatment of diseases. It is this diversity and versatility of carbon nanostructures that make them attractive and versatile for various applications: they are nanoscopic in size, have high length-to-width and surface-to-volume ratio, can have a hollow core, are chemically inert to many reagents but can be easily modified to possess different surface chemistry, are mechanically robust, and have conducting or semiconducting and unique optical properties. This diversity enables intelligent design and precise tailoring of the carbon structure and properties to suit a certain application.

One of the most exciting applications for carbon nanostructures, and nanotechnology in general, is in the field of biomedicine, where their versatility has the potential to improve the detection of biological threats, efficiently and compactly monitor environmental conditions and screen the health of patients, and detect the early onset of disease. Because of their unique properties, carbon nanostructures can be used for various biomedical applications; from *in vivo* targeted drug-delivery and regenerative tissue scaffolds to single-cell probes and implanted sensors. In this chapter, we focus on the development of several CNT-based sensing platforms and recount the current utilization and application trends in biomedicine.

7.2 1-D CARBON NANOSTRUCTURES

7.2.1 BRIEF PERSPECTIVE

During the past century, carbon fiber research has progressed from thick carbon fibers and filaments to CNFs and nanotubes.[1] Research has also progressed from hollow graphitic nanofibers to the higher ordered structure of CNTs. In the decade after the discovery of multiwalled CNTs in 1991,[2] the majority of research focused on the fabrication, characterization, and determination of properties for CNTs and related structures. So far, in the twenty-first century, one of the main focuses has been on applying their unique qualities as sensors. In this section, we will briefly introduce CNTs and related carbon-based nanostructures and, in the context of biomedical applications, describe how these nanostructures are fabricated and explain their distinct properties.

To simplify our discussion of a broad field, we adopt the earlier proposed nomenclature to describe all hollow tube-like carbon structures with at least one dimension of 100 nm or less as CNTs.[3] Likewise, we describe all solid fiber-like carbon structures with at least one dimension of 100 nm or less as CNFs. Albeit, the reader should note where indicated that distinct differences in CNT structure give distinguishing properties.

7.2.2 FABRICATION

CNTs of various diameters, lengths, and structure can be fabricated using different methods, including electric arc discharge,[4–7] laser ablation,[8,9] catalytic chemical vapor deposition (C-CVD),[10] plasma-enhanced C-CVD (C-PECVD),[11] and template-based CVD.[12–14] Electric arc discharge and laser ablation are very efficient methods for producing high-quality single-wall CNTs (SWCNTs) and multiwall CNTs (MWCNTs) in large quantities. Nanotubes are obtained using these methods, by removing unwanted carbon particles and other materials present in the synthesis yield through purification and filtering processes. CNTs made in this fashion are free to be suspended and used in random dispersions.

CNTs are also produced in large quantities by various other CVD methods. In general, CVD involves the decomposition of hydrocarbon gases to synthesize controllable CNTs and CNFs. In C-CVD, the size of catalyst particles controls the uniform diameter of the carbon products. Catalyst particles can be prefabricated into patterned arrays to produce well-aligned CNT assemblies and even complex patterns are possible.[15,16] The location, alignment, geometry, and structure of CNTs and CNFs can be controlled during synthesis with C-PECVD.

As an extension of CVD processing, CNTs and CNFs can be produced by depositing carbon inside porous substrates, such as glass capillaries and anodic aluminum oxide (AAO) membranes (Figure 7.1).[12] After the carbon has been deposited, the template is removed to obtain aligned arrays,[14,17] individual nanotubes, or single integrated nanostructures.[18–20] Pioneered by the Martin group, template-synthesis processing allows CNTs to be produced in different dimensions, geometries, and shapes, as well as produce customized nanotubes with wall-embedded nanoparticles[21] or nanoparticle-filled inner bores.[22]

7.2.3 PROPERTIES

The unique mechanical, electrical, chemical, thermal, and optical properties of CNTs and CNFs have been well reviewed,[11,14,23] and a detailed discussion is beyond the scope of this chapter. In addition, CNTs and CNFs have extremely large length-to-diameter aspect ratios and high surface-to-volume ratios. Their characteristics primarily depend on their structure and can vary significantly between the type and origin of the carbon material.

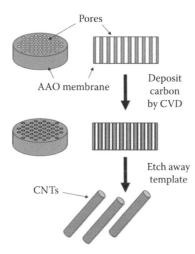

FIGURE 7.1 Schematic depiction of CNT fabrication by non-catalytic template-assisted CVD. Carbon is deposited on AAO membranes by CVD. The AAO is chemically etched away to release the CNTs formed in the AAO pores.

CNTs exhibit high tensile strength, stiffness, and ductility. For instance, the tensile strength of SWCNTs is 100 times that of steel, making them the strongest known material.[24] CNTs also behave elastically: when pushed against a hard surface, CNTs bend and buckle without fracture and subsequently return to their original shape when the force is removed.[19,25] CNTs and related structures exhibit a wide range of electrical characteristics. For example, SWCNTs can be metallic and semiconducting depending on their helicity, whereas MWCNTs exhibit a range of electronic behavior (metallic, semiconducting, and semimetallic).[23] CNTs with defect sites, such as those grown using template-based synthesis methods, are good electrical conductors; yet, they can be annealed to further improve conductivity.[26] Additionally, the reactivity of carbon surfaces varies greatly with surface microstructure, cleanliness, and functional groups.[27] Thus, carbon nanostructures provide a diverse platform to attach biomolecules for a variety of drug-delivery, biological interfacing, and sensing applications.

7.3 MANUFACTURED DEVICES

Carbon nanostructured devices come in different configurations depending on the desired application. The configurations include scaffolds with nanotubes, randomly dispersed nanotubes on their surfaces, substrates or electrodes consisting of aligned nanotube arrays or forests, probes tipped with individual or bundled nanotubes, transistors made up of a single nanotube or networks of nanotubes, or even three-dimensional nanotube-based electromechanical devices.[28] This section will highlight the different CNT-based device platforms.

7.3.1 Devices with Randomly Oriented CNTs

Several sensor configurations are manufactured from randomly oriented CNTs, conductive CNT films, and CNT-coated electrode surfaces to transistor-based sensors. CNT films and composites are well suited for macro- and microscale sensing platforms, such as smart wearable fabrics[29] and lab-on-chip devices,[30] in which concentrations of analytes are above several micromolars. Smaller electrodes, with tips ranging from hundreds of nanometers to hundreds of micrometers, have been coated with CNTs to enhance their sensitivity.

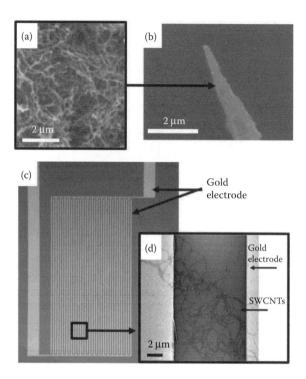

FIGURE 7.2 Devices with randomly oriented CNTs. (a,b) CNT-coated carbon fiber nanoelectrode. (Adapted from Chen, R.J. et al., *P Natl Acad Sci USA*, 100, 4984–4989, 2003. With permission.) (c,d) CNT-based FET. (Adapted from Li, J. et al., *Nano Lett*, 3, 929–933, 2003. With permission.)

For example, SWCNTs (Figure 7.2a) were coated on flame-etched carbon fiber nanoelectrodes and used as electrochemical sensors (Figure 7.2b).[31] The CNT-modified electrode, with tips ranging from 100 to 300 nm, exhibited overall detection capabilities in the nanomolar range, which is an order of magnitude lower than the conventional carbon fiber electrodes of similar geometry. These small electrodes would be well suited for electrochemical applications in ultra-low fluid volumes or biological systems such as tissue or cells.

The size and high sensitivity of CNT-based field-effect transistor (FET) sensors make them well suited to detect trace analytes in restricted spaces. FETs, first developed in 1998,[32,33] have been made either by depositing or growing SWCNTs on a flat SiO$_2$/Si wafer substrate and then patterning connective electrodes, or, in reverse order, patterning electrodes and then depositing SWCNTs on the top. The former is the simpler and more common method and has been used extensively by few research groups. In this method, semiconducting SWCNTs are grown on SiO$_2$/Si substrates by CVD from catalyst particles or islands. The result produces randomly oriented nanotubes across the substrate. Electrical contacts are then patterned over the nanotube dispersion using shadow mask evaporation or electron beam- or photolithographic processes. The resulting sensor may consist of either single[34,35] or connected, networked[36] CNT-based FET sensors. The latter method has been used to create larger interdigitated electrodes (Figure 7.2c) with reproducible performance, where, as shown in Figure 7.2d, networks of randomly dispersed nanotubes lay on the electrodes to bridge the fingers.

7.3.2 DEVICES FROM VERTICALLY ALIGNED CNTS AND CNFS

Sensing devices and cell-culture scaffolds have been made from vertically aligned (normal to the substrate) CNT and CNF forests and patterned arrays. The multiple detection sites of vertically aligned CNT arrays improve the signal-to-noise ratio and temporal response by several orders of magnitude better than the conventional flat electrodes of similar size and material.[37] When grown in patterns, vertically aligned arrays enable a deterministic spatial resolution for a variety of sensing applications, for example, plated living cells.

Even when different substrates and growth methods are used, all vertically aligned CNT or CNF devices are manufactured using a similar strategy, as shown in Figure 7.3. The device occasionally utilizes nonconducting substrates, such as SiO_2/Si, but typically conducting substrates are employed, in the form of a metal wire, sheet, or plate or a substrate with a deposited metal film (Figure 7.3a). The catalyst, such as Fe or Ni films or nanoparticles, is then deposited on the substrate (Figure 7.3b). For deterministic arrays, the catalyst films are patterned. C-PECVD is used to grow the CNTs or CNFs from the catalyst and vertically orient the tubes on the substrate (Figure 7.3c). These devices can be used as they are so that the substrate and nanotubes act as the sensor (Figure 7.3d) or are embedded in an insulating film to expose only the ends of the nanotubes. In the latter, the insulating layer can be applied such that the nanotube tips protrude (Figure 7.3e) or are polished so that the tips are flushed with the surface (Figure 7.3f).

Different methods were developed to grow vertically aligned CNTs from ion nanoparticles embedded in the pores of mesoporous silica[15] and CNFs from Ni films sputtered on glass.[38] These methods produced large-scale, well-aligned vertical carbon nanostructures that were isolated from one another.

Furthermore, various methods were developed to grow patterned, vertically aligned CNT and CNF arrays at predetermined locations by patterning the catalyst on substrates (Figure 7.4a).[39,40] These methods have since been used to produce an assortment of vertically aligned carbon nanostructures using C-PECVD for applications ranging from nanoelectrode ensembles for biosensing to needle arrays for the parallel delivery of reagents to multiple single cells.

As shown in Figure 7.4b, arrays of conically shaped CNFs (200 nm tip diameter, 6–20 μm long, deterministic nanofiber separation) can be produced from Ni catalyst dots patterned on Si substrates.[41] The planar arrangement, vertical alignment, and small dimensions of the CNFs make them well suited to penetrate multiple cells simultaneously and deliver surface-adsorbed deoxyribonucleic acid (DNA) into the cells.[42] Later, a similar technique was used to produce a nanoelec-

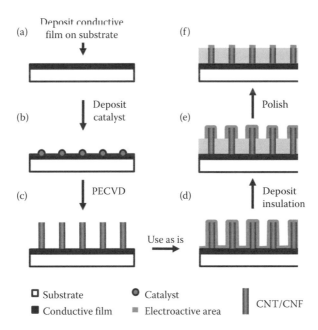

FIGURE 7.3 Fabrication of vertically aligned CNT arrays and device configurations.

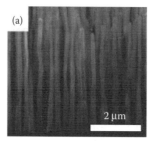 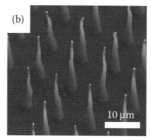

FIGURE 7.4 Vertically aligned arrays of carbon nanostructures. (a) CNT array. (Adapted from Ren, Z.F. et al., *Appl Phys Lett*, 75, 1086–1088, 1999. With permission.) (b) Patterned CNF array. (Adapted from Melechko, A.V. et al., *Nanotechnology*, 14, 1029–1035, 2003. With permission.)

trode array of CNFs (30–160 nm diameter, ≈5 μm long, deterministic nanofiber separation) from patterned Ni catalyst dots on Cr-coated Si wafer.[37]

CNT electrode ensembles consisting of a dense array of CNTs (15–80 nm diameter, 30–100 μm long, with a 100–200 nm nanotube separation) were produced from an ion-sputtered Fe catalyst film on an Al-coated Si wafer.[43] As depicted in Figure 7.3d, the CNTs and the conducting substrate act as the electrode and increase the electroactive surface area. Nanoelectrodes were made from carbon nanopipettes (CNPs) (10–15 nm tip diameter, several micrometers long) on Pt substrates[44] and low-density arrays of vertically aligned CNTs (50–80 nm diameter, 10–12 μm long, with a nanotube separation of more than 5 μm) from electrodeposited Ni nanoparticles on a Cr-coated Si substrate.[45] To limit the electroactive area to the tips of the CNTs or CNFs, these arrays can be coated with an epoxy to insulate the conductive substrate to expose needle-like tips or surface-polished tips so as to expose only the ends of the nanotubes.

7.3.3 CARBON NANOSTRUCTURE-TIPPED DEVICES

In contrast to CNT arrays, devices tipped with a single CNT or bundle of CNTs can interrogate a single feature on a substrate or region of interest, such as a single cell or intracellular location, with high spatial resolution. Carbon nanostructure-tipped devices have been used in several applications, including scanning probe microscopy (SPM), nanoscale electrochemistry, and cellular and intracellular studies. Their utilization differs from other CNT-based devices in that they provide remote manipulation with nanoscale resolution and precise positioning of the sensor.

MWCNTs have been attached to the tips of atomic force microscope (AFM) cantilevers to take advantage of the precise manipulation capabilities of the AFM (Figure 7.5a). By bringing the tip in contact with an adhesive and then a bundle of tubes dispersed on a surface, CNTs can be adhered onto the surface of the tip.[25] These devices, tipped with CNTs ranging from 5 to 200 nm diameters, were used for SPM of surfaces[46–50] to probe and deliver substances to cells.[51,52] CNTs have also been grown directly on the tips of AFM cantilevers.[53] Although CNT-tipped AFM probes are relatively easy to use and have high spatial resolution because of the AFM, they are limited by fabrication, equipment, sensing, and delivery capabilities.

Instead of AFM tips, CNTs are attached to the tips of thin wires using similar contact-assembly methods. One of the earliest examples of such a probe was made by bringing the tip of a micron-sized Pt wire, coated with Ag-conducting epoxy, in contact with aligned MWCNTs (Figure 7.5b).[54] The conductive surfaces were insulated with polyphenol. Thereafter, a −1 V potential was applied in the electrolyte to expose only the CNT tip of the probe. It was demonstrated that these MWCNT bundle-tipped probes can be used as electrochemical nanoelectrodes.

Alternating current (AC) electric fields were used to capture bundles of CNTs at the tips of tungsten microelectrodes to form nanoprobes with a macroscopic handle.[55] Two tungsten microelectrodes

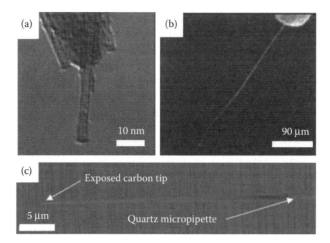

FIGURE 7.5 Probes tipped with carbon nanostructures. (a) CNT-tipped AFM probe. (Adapted from Hafner, J.H. et al., *J Phys Chem B*, 105, 743–746, 2001. With permission.) (b) CNT-tipped electrode. (Adapted from Kaempgen, M. and Roth S., *Synthetic Met*, 152, 353–356, 2005. With permission.) (c) Carbon nanopipette. (Adapted from Schrlau, M.G. et al., *Nanotechnology*, 19, 015101 (4pp), 2008. With permission.)

were submerged in aqueous solutions, containing either SWCNTs or MWCNTs, such that their tips opposed each other. By applying an AC field across the two electrodes, CNTs were attracted to the regions with high-field intensity to form narrow bundles at the electrode tip. Instead of AC fields, magnetic forces have also been employed for patterning by positioning ferromagnetic CNTs at the tips of glass micropipettes.[56] Magnetic CNTs placed inside a glass micropipette were driven to its tip by magnetic fields, where the protruding CNT was fixed in place by epoxy. Magnetic positioning can be alternatively replaced with an evaporative fluid-flow technique to reduce manufacturing complexity.[57]

Glass micropipettes have also been used as templates for forming CNPs. Using CVD, a carbon film was deposited on the surface of the glass micropipettes[18] or preferentially on the inner surface with the catalyst[19] and CVD process controls.[20] After the glass was selectively removed from the pipette tip, an integrated carbon nanotube-like structure remained at the tip of the larger glass micropipette (Figure 7.5c). Diameters as small as 10 nm have been achieved using this technique.

7.3.4 ENHANCEMENT AND SELECTIVITY

To enhance the properties of carbon nanostructures, several strategies have been developed. For example, carbon surfaces are functionalized, modified, and customized to selectively detect molecules, chemicals, and biological compounds in liquid or gas phases. Noncovalent attachment can be utilized to preserve the structure of CNTs by adsorbing the material onto their surface. However, covalent attachment needs the surface of CNTs to have defect sites, often requiring the surface to be chemically activated to bind molecules to their surface. Alternatively, CNTs can be embedded or filled with material, as we will discuss later in this section.

As mentioned earlier, one simple way to modify surfaces of CNTs is by utilizing noncovalent attachment. A simple method is to deposit the material onto CNTs through incubation or drying. For example, CNT-coated glassy carbon electrodes were incubated in a combination of proteins and surfactants to enhance interfacial electron transfer,[58] whereas single-stranded DNA was deposited and left to dry on CNT-based FETs to detect vapors.[59]

A more elegant method is to use the adsorption of aromatic compounds. In general, this strategy involves attaching linker molecules to hydrophobic surfaces of CNTs via π–π stacking. Subsequently, nanoparticles, molecules, or proteins can be selectively attached to the linker molecules. This method was used to selectively bind streptavidin.[60] The surfactant polyethylene glycol (PEG) was used to bind to hydrophobic CNTs to prevent the nonspecific binding of proteins to their surface. However, when CNTs were coated with diamino-PEG, amine-reactive protein reagents (in their case amine-reactive biotin) could be bound to the CNTs. This gave the CNT selectivity by permitting only selective binding to high-affinity molecules, in this case streptavidin. These methods were used with different linker molecules and selective reagents (proteins, receptors, and even chemicals) to enhance the selectivity of FETs and chemical sensors.[30,35,61]

Covalent functionalization is accomplished by chemically attaching molecules to the CNT surface.[62] In general, CNTs that lack defect sites need to undergo processes, such as sonication, acid treatment, electrochemical oxidation, or plasma treatment to actually activate their surface and provide hydrophilic surface functional groups for covalent attachment. CNTs and CNFs that have defect sites, such as those grown from template-based synthesis methods, will have some binding sites but will often require a degree of surface activation.

A common approach is to use two linker molecules to covalently attach biomolecules to the surface of CNTs. CNTs are first activated by acid, electrochemical treatment, or other means to produce carboxylic functional groups on their surface. CNTs are then incubated in standard coupling agents, such as 1-ethyl-3-(3-dimethylaminopropyl) carbodiimide (EDC) and derivatives of N-hydroxysuccinimide (NHS) or 2[N-morpholina]ethane sulfonic acid (MES). The combination of the linker molecules allows biomolecules to bind to the CNT surface by forming linkages between biomolecule amine groups and CNT carboxyl groups. This functionalization method has been successful in many applications, for instance, attaching peptide nucleic acid to CNTs to preferentially bind to DNA sequences,[63] attaching DNA to the surface of magnetic CNTs for intracellular delivery,[64] and binding glucose oxidase to CNTs for the electrochemical detection of glucose.[65]

Several methods have been developed to make CNT–nanoparticle hybrids. The tips of CNT-based nanoelectrodes were coated with Au, Fe, and Ag nanoparticles using short-pulse, nonthermal corona discharge in liquids containing metal–salt solutions.[66] Metal nanoparticles can also be attached to CNT surfaces through electrostatic interactions. A positively charged polyelectrolyte was used to attach negatively charged Au nanoparticles to the negatively charged surface of acid-treated nitrogen-doped MWCNTs.[67] Electrostatic functionalization was also used to attach gold nanoparticles to the surfaces of CNT-tipped probes to enable surface-enhanced Raman spectroscopy (SERS).[57,68]

Besides attachment to CNT surfaces, nanoparticles can be embedded into CNTs or filled inside CNTs. For instance, CNTs grown by C-PECVD contain ferromagnetic nanoparticles enclosed in their tips. It was demonstrated that vertically aligned CNTs grown from Ni can be manipulated by external magnetic forces and be directed to actually spear cells.[64] Similar capabilities were demonstrated by embedding iron oxide nanoparticles into the walls of CNTs[21] or filling them with magnetic nanoparticles[22] or smaller carbon nanostructures[69] during template-based CNT synthesis processes.

7.4 BIOMEDICAL APPLICATIONS

The unique properties of CNTs and other carbon nanostructures make them well suited for a broad range of sensing applications. For example, the mechanical properties of CNTs enable them to be used, among other things, as strain gauges,[70,71] pressure transducers,[72–74] torsional sensors,[75,76] textile-based sensors,[77] atomic mass sensors,[78] fluid flow sensors,[79] and displacement sensors.[80] The thermal responses of CNTs can be used as thermometers,[81,82] environmental monitors,[83,84] and infrared sensors,[85] whereas their optical properties have found uses as strain gauges,[86] pressure gauges,[87] and photodetectors.[88] In this section, we will focus on the biological sensing applications of CNTs and other carbon-based nanostructures. The reviews of additional CNT-based biological sensors can be found in literature.[89–91]

7.4.1 ELECTROCHEMICAL SENSORS

Whether randomly dispersed or grown on conductive surfaces, the electrical properties and modifiable surfaces of CNTs enable them to be used as electrochemical sensors. Compared to conventional electrodes, CNT-based electrochemical sensors offer higher ratio of surface area to volume that significantly increases the signal-to-noise ratio for sensitive detection. CNT-based electrodes can be further modified by surface activation to enhance interfacial electron transfer, selectively detect analytes, or attach reagents that prevent nonspecific binding but allow the specific binding of analytes.[92]

Randomly dispersed CNT films can be used as electrochemical sensors. For instance, ferrocene was noncovalently attached to acid-treated SWCNT films to detect glutamate.[30] The combination of SWCNTs and ferrocene provided enhanced surface area, direct electron transfer, and catalytic effect. Electrodes can also be coated with CNTs to enhance performance. Indeed, CNT-coated carbon nanoelectrodes significantly outperform the conventional carbon fiber electrodes of similar geometry.[31]

CNT coating with CNTs can also enhance other electrodes, such as glassy carbon and Pt microelectrodes. It was found that CNT-coated glassy carbon electrodes offered superior performance to noncoated electrodes.[93] It was also shown that the electrochemical treatment of microperoxidase immobilized on the surface of CNT-coated Pt microelectrodes catalyzed the reduction of O_2 and H_2O_2.[94]

CNT arrays, as depicted in Figure 7.3e and f, can be used directly as electrochemical sensors. The Ren group developed an electrochemical sensor (similar to Figure 7.3f) from an array of CNTs (50–80 nm diameter, 10–12 μm long) grown from electrodeposited Ni nanoparticle catalyst on a Cr-coated Si substrate.[45] The low-density array exhibited sigmoidal behavior during cyclic voltammetry and provided a high signal-to-noise ratio. The electrochemical sensor was used to detect glucose (as well as ascorbic acid, uric acid, and acetaminophen) by attaching glucose oxidase to the CNT surface through the EDC/sulfo-NHS linker moiety.[65] The sensor was later used to detect lead in low-electrolyte conditions[95] and trace cadmium (II) and lead (II) at sub-ppb level with a total detection limit of 0.04 μg/L.[96]

Electrochemical sensors have also been developed from arrays of CNTs that protrude an insulating surface, as depicted in Figure 7.3e. Starting with CNPs grown on a Pt-wire substrate,[97] a nanoelectrode array with protruding conical carbon tips was produced that exhibited sigmoidal electrochemical behavior.[44] By oxidizing the protruding nanoscopic tips, the sensor was able to distinguish electrochemically between ascorbic acid (oxidation-shifted peak) and dopamine. Electrochemical sensors of a similar configuration using C-PECVD-grown CNT arrays have also been reported.[98]

7.4.2 DETECTION OF GASES AND VAPORS

The conductance of SWCNT- and MWCNT-based FETs is significantly affected by changes on or near their surface and as such provides attractive sensors for detecting gases and vapors with high sensitivity. Although they can be manufactured in various ways, CNT-based FETs, in general, consist of a nanotube or network of nanotubes that bridge the gap between two electrodes (Figure 7.6). Semiconducting CNTs are particularly attractive for FET applications because their conductance is highly sensitive to the environment and varies significantly with changes in electrostatic charges and surface adsorption of molecules.

The resistance of semiconducting SWCNT-based FETs operating at room temperature dramatically changes when they are exposed to gaseous molecules such as nitrogen dioxide, ammonia, and oxygen.[34] Owing to their extremely high sensitivity, SWCNT-based FETs had an order of magnitude faster response than solid-state sensors. Similar SWCNT-based FETs were later used to detect butylamine and 3′-(aminopropyl)-triethoxysilane in the vapor phase.[99] Later, this work was

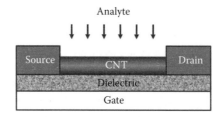

FIGURE 7.6 Configuration of a typical CNT-based FET.

extended to detect low levels of NO_2 (44 ppb) and nitrotoluene (262 ppb) using a network of semi-conducting SWCNTs connecting interdigitated electrodes (Figure 7.2c and d).[36]

CNT-based FETs were used as "electronic noses" to detect odors.[59] Metal electrodes were patterned over CVD-grown CNTs on an SiO_2/Si substrate with electron beam lithography to form the FET. To enhance the detection and provide selectivity, single-stranded DNA was deposited and dried on the device (Figure 7.7a). The sensor was used to detect and distinguish between methanol and propionic acid, as shown in Figure 7.7b, as well as other odors such as trimethylamine, dinitro-toluene, and dimethyl methylphosphonate.

Gas ionization sensors were developed from vertically aligned MWCNT arrays to overcome several limitations of CNT-based FETs, such as poor diffusion kinetics, inability to identify gases

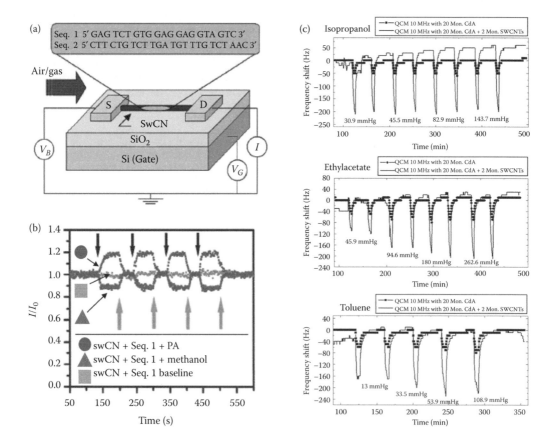

FIGURE 7.7 CNT-based gas and vapor sensing. (a,b) Odor sensing with FET. (Adapted from Staii, C. et al., *Nano Lett*, 5, 1774–1778, 2005. With permission.) (c) Gas sensing by microbalance. (Adapted from Penza, M. et al., *Nanotechnology*, 16, 2536–2547, 2005. With permission.)

with low adsorption energies, low capacity to distinguish between gases and gas mixtures, and high susceptibility to changes in moisture, temperature, and the gas-flow velocity.[100] These sensors utilized differences in breakdown voltage to distinguish between micromolar concentrations of distinct gases, measurements that were unaffected by moisture, temperature, and gas flow.

Acoustic and optical gas sensors, respectively, were developed by coating CNTs onto the chips of a quartz crystal microbalances (QCM) and silica optical fibers (SOFs).[101] As shown in Figure 7.7c, CNT-coated QCM crystals were roughly 1–2 orders of magnitude more sensitive than uncoated QCM crystals. Also, it was demonstrated that CNT-coated SOFs had similar low limits of detection and enhanced sensitivity.

7.4.3 DETECTION OF BIOMOLECULES

CNT-based sensors are attractive for detecting biomolecules commonly found in aqueous environments. For instance, researchers at the NASA Ames Research Center used vertically aligned CNT arrays[43] to bind DNA to the surface of CNTs,[102] detect ribonucleic acid (RNA), DNA, and DNA polymerase chain reaction (PCR) amplicons in the solution,[37,103,104] and even deliver DNA plasmids into cells.[42] For detection applications, CNT-based devices similar to those depicted in Figure 7.3f were manufactured and DNA probes were covalently attached to the ends of CNTs using the EDC/sulfo-NHS method. When combined with the Ru(bpy)32+-mediated guanine oxidation method, the CNT-based sensor was capable of detecting less than a few attomoles of oligonucleotide targets, showing orders of magnitude improvement in sensitivity over other electrochemical detection of DNA immobilization,[104] as well as detecting a low number of PCR amplicon targets comparable to fluorescence-based DNA microarray techniques.[37,103] The electronic platform facilitates DNA and RNA detection and makes it possible to be integrated into hand-held microfluidic lab-on-chip devices.

Although their sensing capabilities were initially demonstrated with the detection of gases and vapors, CNT-based FETs were soon used in liquid phases[105] to detect chemicals and biomolecules in aqueous environments. The development of such SWCNT-based FETs with the selective binding and recognition of target proteins was reported.[35] FETs were manufactured by dispersing a network of polyethylene oxide (PEO)-functionalized SWCNTs between two electrodes. The PEO coating, noncovalently bound to the CNT surface by π–π stacking, prevented strongly nonspecific binding. However, when its hydroxyl termini were activated, the receptors of specific targets could be conjugated to the PEO-coated CNTs to enable the selective binding of target proteins. CNT-based FETs conjugated with staphylococcal protein A (SpA) were extremely sensitive to the addition of a protein that has a high affinity to SpA (such as the immunoglobulin G, IgG, protein) but showed negligible response to others. Additional sensors were developed for streptavidin–biotin and U1A-mAbs 10E3 to demonstrate their usefulness in disease detection and proteomics.

The Strano group developed SWCNT-based near-infrared sensors to detect biomolecules *in situ*. These optical sensors detect the presence of specific targets by monitoring the fluorescence of functionalized SWCNTs at wavelengths that are transparent to the biological tissue (900–1600 nm). In one application, a porous vessel containing glucose oxidase-coated SWCNTs was implanted beneath a human epidermal tissue sample.[106] The implanted sensor was capable of distinguishing between slight variations in local glucose concentration within the range of blood glucose. In contrast to flux-measuring electrochemical glucose sensors, concentration-measuring optical sensors were less prone to biofouling and significantly more stable during longer periods of time.[107] The group later reported detecting DNA hybridization and conformational polymorphism in cells with similar sensors.[108,109]

7.4.4 SINGLE-CELL PROBES AND SENSORS

CNT-tipped probes offer the ability to probe small aqueous environments, such as ultra-low volumes of liquids or single living cells. In contrast to CNT arrays or planar configurations, CNT-tipped probes are needle-like configurations, such as those shown in Figures 7.2b and 7.5, which can be easily maneuvered to interrogate regions of interest in such things as fluid droplets, single cells, or living tissue.

As discussed earlier, CNT-tipped AFM cantilevers were some of the first CNT-tipped probes to be developed.[25] With high spatial resolution and precise manipulation, CNT-tipped AFM cantilevers were used in SPM applications[46–50] and were able to probe and deliver surface-bound compounds to individual cells.[51,52] Despite their relatively easy use and high spatial resolution, these CNT-tipped probes were limited by laborious fabrication and inadequately suited for intracellular fluid delivery and electrical sensing. In contrast, the development of CNT-tipped pencil-shaped probes and their use in biological sensing have progressed more rapidly. This is mostly because of their ability to readily fit the standard cell physiology equipment, such as optical microscopes, micromanipulators, cell-injection systems, and electrophysiology amplifiers.

One of the earliest CNT-tipped biological sensor reported was an SWCNT-coated carbon fiber nanoelectrode (100–300 nm tip diameter, as shown in Figure 7.2b).[31] Using cyclic voltammetry, the CNT-tipped electrode could detect dopamine, epinephrine, and norepinephrine at concentrations on an order of magnitude lower than noncoated probes. The demonstration was significant because the dimensions of the CNT-tipped probe would make it possible to study the functions of living cells and tissue with a minimal level of intrusion. Additionally, the pencil-like shape of the probe readily fits the standard cell physiology equipment and facilitated its use.

Later, pencil-shaped probes tipped with a single or bundle of CNTs were developed, and it was reported that a probe tipped with a bundle of MWCNTs could be used as an electrochemical sensor.[54] This demonstration helped to spark the further development of CNT-based probes to measure intracellular signals.[110] Referred to as CNPs earlier, these CNT-tipped probes were used to penetrate the membrane of single living cells and electrically measure changes in their transmembrane potential upon extracellular chemical and pharmacological stimulation. It was later shown that a CNT-tipped probe was capable of electrically stimulating and measuring intracellular and extracellular responses more efficiently than the electrolyte-filled glass micropipettes typically used in cell electrophysiology.[111] It was found that similar CNT probes had comparable performance to conventional Ag/AgCl (silver/silver chloride) electrodes when employed to monitor both extracellular and intracellular neural activity.[112] Recently, CNT-tipped endoscopes have been developed to provide minimally invasive, multifunctional single-cell analysis, such as intracellular injection, electrochemistry, and SERS.[57,68,113]

7.4.5 Implantable Sensors and Biological Interfaces

The ability to track implanted cells and monitor the progress of tissue formation *in vivo* is important especially in large biological organisms and tissue engineering.[114] Labeling implanted cells not only help to evaluate the viability of the tissue but also understand the distribution and migration pathways of transplanted cells. Traditional methods such as intravital microscopy or flow cytometry are extremely time consuming, and a much more attractive alternative is *in vivo* labeling and contrast imaging. However, there remains a need for contrasting agents that are biocompatible, stable during long periods of time, and provide high image contrast between labeled and unlabeled specimens. Preliminary work has suggested that CNTs are attractive imaging contrast agents for optical imaging,[115] magnetic resonance imaging (MRI),[116] and radiotracking.[117]

Carbon nanoparticle contrasting agents have also been employed to monitor physiological events of single cells, such as ion transport, enzyme/cofactor interactions, protein secretion, and matrix adhesion. Tissue and cellular responses have been observed using other nanoparticle contrast agents, such as inflammation of pancreatic islets,[118] apoptosis (a genetically determined process of cell self-destruction),[119–121] and angiogenesis (a physiological process involving the growth of new blood vessels from preexisting vessels).[122,123] The ability to monitor apoptosis and angiogenesis with high spatial resolution is not only advantageous to cell and tissue culturing but also to disease progress and therapy response. To complement intracellular nanocarbon-based contrasting agents, nanosensors could be employed to provide continuous monitoring of the performance of the cultured tissues.

One method of monitoring tissues *in vivo* would be to use implantable sensors capable of relaying information outside the body. Such a sensor would provide real-time data related to the physiologically relevant parameters, such as pH[124], pO_2, and glucose levels.[65,107] There are several noticeable advantages to using nanosensors for evaluating tissues *in vivo*. First, because the sensing element is nanometer sized and the overall probe miniaturized, implanting within the tissue would not adversely disturb the system. Second, because of the large surface-to-volume ratio, a relatively large active area is available for immobilizing numerous biological and chemical compounds, including DNA[125] and proteins,[126] for improved detection sensitivity.

MWCNT electrodes have been developed to monitor the electrochemical oxidation of insulin, a pancreas-produced hormone that plays a key role in the regulation of carbohydrates and fat metabolism in the body.[127] This provides a possible method to evaluate the quality of pancreatic islets before their transplantation. By coating similar MWCNT electrodes with platinum microparticles, thiols containing amino acids can be detected in rat striatal that is the subcortical part of the forebrain.[128] CNT-based sensors can be incorporated into flexible biocompatible substrates to facilitate *in vivo* sensing. For instance, free cholesterol in blood can be measured using MWCNT electrodes placed on a biocompatible substrate[129] whereas flexible pH sensors can be formed from polyaniline (a conductive polymer) and nanotube composites.[130]

Nanostructured carbon surfaces and scaffolds have been shown to not hinder but significantly promote cell growth. For example, neural cells[131] and mouse fibroblast cells[132] were successfully cultured on CNT scaffolds. The findings encourage the application of nanostructured carbon devices as implantable sensors and tissue scaffolds. For instance, ectopic formation of bone tissue was observed after MWCNT scaffolds were implanted in muscle tissue, suggesting that the nanostructured carbon substrates could encourage cells to grow within the body.[133]

Recent advances in micro- and nanofabrication techniques have provided promising new device platforms and implantable interfaces for studying neural communication. For instance, MWCNT-coated stainless-steel electrodes were implanted and used for wide-frequency stimulation in deep brain structures, showing superior performance to uncoated electrodes.[134] An emerging but promising platform, however, are arrays of carbon nanostructures patterned on rigid or flexible substrates. For example, arrays of addressable CNFs have been used for highly resolved neural stimulation and recording, providing multiple channels of electrical, chemical, and mechanical information.[42,135,136] In a related work, vertically aligned CNFs were employed for extracellular stimulation and recording of neuroelectrical activity. Here, it was reported that, despite the small size of these fibers, CNFs could inject sufficient charge to stimulate organotypic hippocampal tissue.[137,138]

7.5   OUTLOOK

As illustrated in this chapter, the diversity in size, shape, structure, and configuration of carbon nanostructures enables a broad range of capabilities for biomedicine and biological applications. During the past two decades, research related to CNTs and carbon nanostructures has grown from discovery, fabrication, and characterization to utilization and application. However, the latter is still in its infancy. During the next few decades, the applications of CNTs and carbon nanostructures, especially drug-delivery and sensing applications, are expected to show significant growth and will develop into a multitude of industrial applications, clinical usage, and consumer products.

ACKNOWLEDGMENTS

Part of the work described in this chapter was supported by a grant from the W.M. Keck Foundation to establish the Keck Institute for Attofluidic Nanotube-Based Probes at Drexel University and by the Texas Instrument/Douglass Harvey Faculty Development Award administered through the Rochester Institute of Technology.

REFERENCES

1. Saito, R., Dresselhaus, G. and Dresselhaus, M.S., *Physical Properties of Carbon Nanotubes*, London World Scientific Publishing Co., London, UK, 1998.
2. Iijima, S., *Nature*, 354, 56–58, 1991.
3. Martin, C.R. and Kohli, P., *Nat Rev Drug Discov*, 2, 29–37, 2003.
4. Bethune, D.S. et al., *Nature*, 363, 605–607, 1993.
5. Journet, C. et al., *Nature*, 388, 756–758, 1997.
6. Ebbesen, T.W. and Ajayan, P.M., *Nature*, 358, 220–222, 1992.
7. Colbert, D.T. et al., *Science*, 266, 1218–1222, 1994.
8. Thess, A. et al., *Science*, 273, 483–487, 1996.
9. Guo, T. et al., *J Phys Chem-USA*, 99, 10694–10697, 1995.
10. Amelinckx, S. et al., *Science*, 265, 635–639, 1994.
11. Melechko, A.V. et al., *J Appl Phys*, 97, 041301 (39pp), 2005.
12. Martin, C.R., *Science*, 266, 1961–1966, 1994.
13. Kyotani, T., Tsai, L.F. and Tomita, A., *Chem Mater*, 7, 1427–1428, 1995.
14. Che, G. et al., *Chem Mater*, 10, 260–267, 1998.
15. Li, W.Z. et al., *Science*, 274, 1701–1703, 1996.
16. Terrones, M. et al., *Nature*, 388, 52–55, 1997.
17. Che, G.L. et al., *Nature*, 393, 346–349, 1998.
18. Kim, B.M., Murray, T. and Bau, H.H., *Nanotechnology*, 16, 1317–1320, 2005.
19. Schrlau, M.G. et al., *Nanotechnology*, 19, 015101 (4pp), 2008.
20. Singhal, R. et al., *Nanotechnology*, 21, 015304 (9pp), 2010.
21. Mattia, D. et al., *Nanotechnology*, 18, 155305 (7pp), 2007.
22. Korneva, G. et al., *Nano Lett*, 5, 879–884, 2005.
23. Ajayan, P.M., *Chem Rev*, 99, 1787–1799, 1999.
24. Ruoff, R.S., Qian, D. and Liu, W.K., *Comptes Rendus Phys*, 4, 993–1008, 2003.
25. Dai, H.J. et al., *Nature*, 384, 147–150, 1996.
26. Mattia, D. et al., *J Phys Chem B*, 110, 9850–9855, 2006.
27. Chen, P.H. and McCreery, R.L., *Anal Chem*, 68, 3958–3965, 1996.
28. Hayamizu, Y. et al., *Nat Nanotechnol*, 3, 289–294, 2008.
29. Shim, B.S. et al., *Nano Lett*, 8, 4151–4157, 2008.
30. Huang, X.J. et al., *J Phys Chem C*, 111, 1200–1206, 2007.
31. Chen, R.S. et al., *Anal Chem*, 75, 6341–6345, 2003.
32. Tans, S.J., Verschueren, A.R.M. and Dekker, C., *Nature*, 393, 49–52, 1998.
33. Martel, R. et al., *Appl Phys Lett*, 73, 2447–2449, 1998.
34. Kong, J. et al., *Science*, 287, 622–625, 2000.
35. Chen, R.J. et al., *P Natl Acad Sci USA*, 100, 4984–4989, 2003.
36. Li, J. et al., *Nano Lett*, 3, 929–933, 2003.
37. Koehne, J. et al., *J Mater Chem*, 14, 676–684, 2004.
38. Ren, Z.F. et al., *Science*, 282, 1105–1107, 1998.
39. Ren, Z.F. et al., *Appl Phys Lett*, 75, 1086–1088, 1999.
40. Merkulov, V.I. et al., *Appl Phys Lett*, 76, 3555–3557, 2000.
41. Melechko, A.V. et al., *Nanotechnology*, 14, 1029–1035, 2003.
42. McKnight, T.E. et al., *Nanotechnology*, 14, 551–556, 2003.
43. Li, J. et al., *J Phys Chem B*, 106, 9299–9305, 2002.
44. Lowe, R.D. et al., *Electrochem Solid State*, 9, H43–H47, 2006.
45. Tu, Y., Lin, Y.H. and Ren, Z.F., *Nano Lett*, 3, 107–109, 2003.
46. Hafner, J.H., Cheung, C.L. and Lieber, C.M., *J Am Chem Soc*, 121, 9750–9751, 1999.
47. Hafner, J.H. et al., *J Phys Chem B*, 105, 743–746, 2001.
48. Akita, S. et al., *J Phys D Appl Phys*, 32, 1044–1048, 1999.
49. Yenilmez, E. et al., *Appl Phys Lett*, 80, 2225–2227, 2002.
50. Patil, A. et al., *Nano Lett*, 4, 303–308, 2004.
51. Vakarelski, I.U. et al., *Langmuir*, 23, 10893–10896, 2007.
52. Chen, X. et al., *P Natl Acad Sci USA*, 104, 8218–8222, 2007.
53. Hafner, J.H., Cheung, C.L. and Lieber, C.M., *Nature*, 398, 761–762, 1999.
54. Kaempgen, M. and Roth, S., *Synthetic Met*, 152, 353–356, 2005.
55. Kouklin, N.A. et al., *Appl Phys Lett*, 87, 173901 (3pp), 2005.

56. Freedman, J.R. et al., *Appl Phys Lett*, 90, 103108 (3pp), 2007.
57. Singhal, R. et al., *Nat Nanotechnol*, 6, 57–64, 2011.
58. Yan, Y.M. et al., *Langmuir*, 21, 6560–6566, 2005.
59. Staii, C. et al., *Nano Lett*, 5, 1774–1778, 2005.
60. Shim, M. et al., *Nano Lett*, 2, 285–288, 2002.
61. Chen, R.J. et al., *J Am Chem Soc*, 123, 3838–3839, 2001.
62. Bahr, J.L. and Tour, J.M., *J Mater Chem*, 12, 1952–1958, 2002.
63. Williams, K.A. et al., *Nature*, 420, 761–761, 2002.
64. Cai, D. et al., *Nat Methods*, 2, 449–454, 2005.
65. Lin, Y.H. et al., *Nano Lett*, 4, 191–195, 2004.
66. Bhattacharyya, S. et al., *Adv Mater*, 21, 4039, 2009.
67. Jiang, K.Y. et al., *Nano Lett*, 3, 275–277, 2003.
68. Niu, J.J. et al., *Small*, 7, 540–545, 2011.
69. Singhal, R. et al., *Sci Rep*, 2, 510, 2012.
70. Li, Z.L. et al., *Adv Mater*, 16, 640, 2004.
71. Grow, R.J. et al., *Appl Phys Lett*, 86, 093104 (3pp), 2005.
72. Wood, J.R. et al., *J Phys Chem B*, 103, 10388–10392, 1999.
73. Stampfer, C. et al., *Nano Lett*, 6, 233–237, 2006.
74. Hierold, C. et al., *Sensor Actuat a-Phys*, 136, 51–61, 2007.
75. Meyer, J.C., Paillet, M. and Roth, S., *Science*, 309, 1539–1541, 2005.
76. Cohen-Karni, T. et al., *Nat Nanotechnol*, 1, 36–41, 2006.
77. Laxminarayana, K. and Jalili, N., *Text Res J*, 75, 670–680, 2005.
78. Jensen, K., Kim, K. and Zettl, A., *Nat Nanotechnol*, 3, 533–537, 2008.
79. Ghosh, S., Sood, A.K. and Kumar, N., *Science*, 299, 1042–1044, 2003.
80. Hall, L.J. et al., *Science*, 320, 504–507, 2008.
81. Wood, J.R. et al., *Phys Rev B*, 62, 7571–7575, 2000.
82. Shi, L., Yu, C.H. and Zhou, J.H., *J Phys Chem B*, 109, 22102–22111, 2005.
83. Kawano, T. et al., *Nano Lett*, 7, 3686–3690, 2007.
84. Liu, L.T. et al., *Sensors-Basel*, 9, 1714–1721, 2009.
85. Pradhan, B. et al., *Carbon*, 47, 1686–1692, 2009.
86. Frogley, M.D., Zhao, Q. and Wagner, H.D., *Phys Rev B*, 65, 113413 (4pp), 2002.
87. Amer, M.S., El-Ashry, M.M. and Maguire, J.F., *J Chem Phys*, 121, 2752–2757, 2004.
88. Khairoutdinov, R.F. et al., *J Phys Chem B*, 108, 19976–19981, 2004.
89. Merkoci, A. et al., *Trac-Trend Anal Chem*, 24, 826–838, 2005.
90. Kim, S.N., Rusling, J.F. and Papadimitrakopoulos, F., *Adv Mater*, 19, 3214–3228, 2007.
91. Sinha, N., Ma, J.Z. and Yeow, J.T.W., *J Nanosci Nanotechnol*, 6, 573–590, 2006.
92. Huang, X.J. and Choi, Y.K., *Sensor Actuat B-Chem*, 122, 659–671, 2007.
93. Wang, J.X. et al., *Anal Chem*, 74, 1993–1997, 2002.
94. Wang, M.K. et al., *Biosens Bioelectron*, 21, 159–166, 2005.
95. Tu, Y. et al., *Electroanal*, 17, 79–84, 2005.
96. Liu, G.D. et al., *Analyst*, 130, 1098–1101, 2005.
97. Mani, R.C. et al., *Nano Lett*, 3, 671–673, 2003.
98. Miserendino, S. et al., *Nanotechnology*, 17, S23–S28, 2006.
99. Kong, J. and Dai, H.J., *J Phys Chem B*, 105, 2890–2893, 2001.
100. Modi, A. et al., *Nature*, 424, 171–174, 2003.
101. Penza, M. et al., *Nanotechnology*, 16, 2536–2547, 2005.
102. Nguyen, C.V. et al., *Nano Lett*, 2, 1079–1081, 2002.
103. Koehne, J. et al., *Nanotechnology*, 14, 1239–1245, 2003.
104. Li, J. et al., *Nano Lett*, 3, 597–602, 2003.
105. Kruger, M. et al., *Appl Phys Lett*, 78, 1291–1293, 2001.
106. Barone, P.W. et al., *Nat Mater*, 4, U86–U16, 2005.
107. Barone, P.W., Parker, R.S. and Strano, M.S., *Anal Chem*, 77, 7556–7562, 2005.
108. Heller, D.A. et al., *Science*, 311, 508–511, 2006.
109. Jeng, E.S. et al., *Nano Lett*, 6, 371–375, 2006.
110. Schrlau, M.G., Dun, N.J. and Bau, H.H., *Acs Nano*, 3, 563–568, 2009.
111. de Asis, E.D. et al., *Appl Phys Lett*, 95, 153701 (3pp), 2009.
112. Yeh, S.R. et al., *Langmuir*, 25, 7718–7724, 2009.
113. Orynbayeva, Z. et al., *Nanomedicine*, 8, 590–598, 2012.

114. Harrison, B.S. and Atala, A., *Biomaterials*, 28, 344–353, 2007.
115. Shi, D.L. et al., *Adv Mater*, 18, 189, 2006.
116. Gupta, A.K. and Gupta, M., *Biomaterials*, 26, 3995–4021, 2005.
117. Singh, R. et al., *P Natl Acad Sci USA*, 103, 3357–3362, 2006.
118. Denis, M.C. et al., *P Natl Acad Sci USA*, 101, 12634–12639, 2004.
119. Jung, H.I. et al., *Bioconjug Chem*, 15, 983–987, 2004.
120. Zhao, M. et al., *Nat Med*, 7, 1241–1244, 2001.
121. Schellenberger, E.A. et al., *Chembiochem*, 5, 275–279, 2004.
122. Winter, P.M. et al., *Circulation*, 108, 2270–2274, 2003.
123. Schmieder, A.H. et al., *Magnet Reson Med*, 53, 621–627, 2005.
124. Xu, Z. et al., *Biosens Bioelectron*, 20, 579–584, 2004.
125. Singh, R. et al., *J Am Chem Soc*, 127, 4388–4396, 2005.
126. Chen, R.J. et al., *J Am Chem Soc*, 126, 1563–1568, 2004.
127. Wang, J. and Musameh, M., *Anal Chim Acta*, 511, 33–36, 2004.
128. Xian, Y.Z. et al., *J Chromatogr B*, 817, 239–246, 2005.
129. Tan, X.C. et al., *Anal Biochem*, 337, 111–120, 2005.
130. Kaempgen, M. and Roth, S., *J Electroanal Chem*, 586, 72–76, 2006.
131. Lovat, V. et al., *Nano Lett*, 5, 1107–1110, 2005.
132. Correa-Duarte, M.A. et al., *Nano Lett*, 4, 2233–2236, 2004.
133. Abarrategi, A. et al., *Biomaterials*, 29, 94–102, 2008.
134. Minnikanti, S. et al., *J Neural Eng*, 7, 16002, 2010.
135. Nguyen-Vu, T.D. et al., *Small*, 2, 89–94, 2006.
136. Nguyen-Vu, T.D.B. et al., *IEEE T Bio-Med Eng*, 54, 1121–1128, 2007.
137. Haque, S. et al., *Drug Dev Ind Pharm*, 38, 387–411, 2012.
138. Melechko, A.V. et al., *J Phys D Appl Phys*, 42, 193001 (28pp), 2009.

8 Field Emission from Carbon Nanotubes

Peng-Xiang Hou, Chang Liu, and Hui-Ming Cheng

CONTENTS

8.1 FUNDAMENTAL PRINCIPLES OF FIELD EMISSION

Field emission (FE) (also known as field electron emission or electron field emission) is the emission of electrons from a solid surface into vacuum induced by an electrostatic field. FE was first explained by quantum tunneling of electrons in the late 1920s [1], and the theory of FE from bulk metals was proposed by Fowler and Nordheim [2]. A family of approximate equations, called Fowler–Nordheim equations (F–N equations), are named in their honor and have been shown in terms of experimentally measured quantities as

$$\ln\left(\frac{I}{V^2}\right) = \frac{1}{V}\left(\frac{\alpha\varphi^{3/2}d}{\beta}\right) + \text{offset} \tag{8.1}$$

where φ is the work function (WF) of the emitter, d is the interelectrode distance, β is the field amplification factor (field enhancement factor) determined by the geometry of the emission region, and α is a constant [3]. The hypothesis of this model includes

1. Electrons are fermions, and their distribution meets Fermi–Dirac statistics.
2. The surface of the metal is a smooth plane, and its atomic-scale irregularities can be ignored.
3. The influence of the classic mirror-image force is taken into consideration.
4. The distribution of work functions is uniform.

Strictly speaking, F–N equations are applicable only to FE from bulk metals and with suitable modifications from bulk crystalline solids. However, they are often used as a rough approximation to describe FE from other materials and nanomaterials too.

In one-dimensional (1D) materials, electrons are confined in the radial and circumferential direction and propagate only axially. As a consequence, the quantum confinement effect of the electrons is significant, which is very different from the case of macroscopic emitters. The inherent narrow diameter of nanomaterials helps to develop a stronger and more concentrated electric field at their tips. Such a quantum confinement effect influences not only the electron transport characteristics [4–6] but also the FE process and properties [7–10] of 1D nanosystems, such as carbon nanotubes (CNTs). Indeed, electron emission is closely linked to electron transport for any emitter.

An ideal field emitter should be capable of emitting a high current at a low applied electric field and should show good FE uniformity and long-term emission stability. The FE properties of a material can be evaluated mainly by its turn-on field, threshold field, FE current density, field enhancement factor, and FE current stability. These parameters will be reviewed in more detail below:

1. The turn-on electric field and threshold electric field are the most important parameters that indicate a degree of difficulty for electrons to escape from a specific material. They are usually defined as the electric fields required to produce current densities of 10 $\mu A/cm^2$ (electric field) and 10 mA/cm^2 (threshold electric field), respectively [11], which would meet the requirement of applications in flat panel displays.
2. The emission current density reflects the capability of emitting electrons in a specific area.
3. The field enhancement factor is defined as the ratio between the local electric field at the tip of an emitter and the applied macroscopic electric field, reflecting the field enhancement ability of the emitter, and can be calculated from the slope of the F–N plot using the F–N law [12].
4. The emission current stability of a material is important for practical applications. It demonstrates the fluctuation and degradation of the emission current density during prolonged FE.

8.2   FIELD EMISSION PROPERTIES OF CARBON NANOTUBES

In the F–N equation, the parameters that affect the effective tunneling barrier at the interface between a metal surface and vacuum are the WS and field enhancement factor. Because of the unique band structure in which the vacuum level is located below the bottom of the conduction band, the electrons in the conduction band of a diamond film are able to leave the surface easily without any external energy to overcome the surface potential barrier. CNTs are another type of carbon emitter, such as diamond or diamond-like carbon (DLC) [13]. Their high electrical and thermal conductivity and desirable mechanical strength make CNTs promising for use as a field emitter. In addition, CNTs are usually one to several nanometers in diameter, and tens of microns in length. This very high aspect ratio leads to a high field enhancement factor that greatly reduces the threshold electric field for emission [13].

In 1995, FE from a multiwalled CNT (MWCNT) film was reported by De Heer et al. [14], and the FE property of an isolated MWCNT was reported by Rinzler et al. [15]. FE of electrons from individually mounted CNTs was found to be dramatically increased when the nanotube tips were opened by laser evaporation or oxidative etching. Emission currents of 0.1–1 μA were readily obtained at room temperature with bias voltages of less than 80 V. Dai et al. [16] attached individual nanotubes with a length of several micrometers to the silicon cantilever of a conventional atomic force microscope (AFM) using a soft acrylic adhesive. Because of their flexibility, the tips are resistant to damage from tip crashes, whereas their slenderness permits the imaging of sharp recesses in surface topography.

In the following years, several experimental and theoretical studies on the FE properties of CNTs were performed [17–21]. These were mainly conducted using isolated [16,22], fiber-like [23,24], arrayed [25–30], and film-like CNTs [31–37]. In comparison with traditional metallic emitters, the merits of CNTs as a field emitter can be summarized as (1) low turn-on field (E_{to}) or threshold field (E_{thr}), (2) enhanced FE current, (3) high field enhancement factor, and (4) good FE stability.

8.2.1 Field Enhancement Factor (β) of CNTs

The field enhancement factor measured from an ensemble of CNTs on a substrate is about 1000–3000 for MWCNTs [8,13,38,39]. As for single-walled CNTs (SWCNTs), their amplification factor estimated from the F–N plot is in the range of 2500–30,000 [13,24,31,40–42]. The higher field enhancement factor of SWCNTs can be attributed to their smaller diameter and higher aspect ratio. On the contrary, Bonard et al. [43] reported that the efficiency of field amplification can change significantly regardless of tube diameter and the source of the discrepancy was studied using a manipulator inside a scanning electron microscopy (SEM). In general, the field enhancement factor is roughly linearly proportional to its macroscopic aspect ratio when the microstructure of the emission tip is kept constant [24]. Chen et al. synthesized centimeter-long strands of well-aligned SWCNTs by a hydrogen–argon arc discharge and investigated their FE properties. It was found that the enhancement factors of SWCNT strands with a length of 1.9, 1.2, and 0.4 mm were 27,129, 19,869, and 7273, respectively (Figure 8.1) [24]. A two-grade enhancement model for the SWCNT strands was proposed where the enhancement factor is related to both the macrogeometry of the strand and the microstructure of its emission tip [24].

In addition, the enhancement factor is also closely related to the distance between CNT emitters. CNT emitters that are too closely packed would cause a screening effect that causes a large decrease in the enhancement factor. The field enhancement factor of many CNT emitters measured experimentally was found to be smaller than that expected from the ratio of height to diameter of the tubes [43]. Jeong et al. [44] investigated the variation of local field on the surface of vertically grown and patterned MWCNTs. It was found that the electric field is stronger at the edge of each pattern than in the central region. This is because the closely packed CNTs at the center are not effectively amplifying an applied field because of the screening effect caused by adjacent CNTs. Therefore, the high enhancement factors reported for CNT emitters are usually based on extraordinarily long and protruding individual emitters, which are well isolated from their surroundings.

Wang et al. [45,46] calculated the emission current density from CNTs with the help of floating sphere mirror-image model and F–N equation. Their result showed that the FE from a CNT array is at maximum when the intertube distance is equal to one-tenth of the tube height. When the intertube distance becomes larger, the field enhancement factor increases, but the emission current

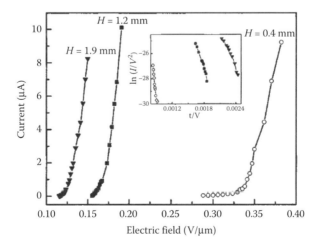

FIGURE 8.1 Characteristic plots of the electron emission current as a function of the applied electric field for an SWCNT strand with different heights (H). The inset is the corresponding F–N ($1/V$–$\ln(I/V^2)$)) plots. (Reprinted with permission from Liu, C. et al., Field emission properties of macroscopic single-walled carbon nanotube strands. *Applied Physics Letters*, 2005, 86(22): 223114. Copyright (2005), American Institute of Physics.)

density decreases rapidly. For arrays with an even higher CNT density, the field enhancement factor decreases greatly because of the screening of the electric field.

8.2.2 FIELD EMISSION CURRENT SUPPRESSION OF CNTS

In the F–N plot, a linear region with a moderate slope is characteristic of a tunneling process assisted by an external field. However, CNTs under an electric field show a deviation from the linear slope as the applied field becomes higher, that is, a decrease of the slope in the F–N curve because of the suppression of the current [28,29,47]. It has now been well established that adsorption states are responsible for this observed deviation at a large electric field or at high temperatures [48–53].

Dean et al. [54] classified three distinct behavioral states in FE for SWCNTs between temperatures of 300 and 1800 K, and confirmed that the adsorbate, water (H_2O), enhances the emission current. Other experiments showed that the emission current of CNTs is significantly increased and easily reaches saturation because of the physisorption of H_2O on the tube surface [49,55]. Lim et al. [48] verified that gas adsorbates play a crucial role in the suppression of current at high fields. A pronounced hysteresis loop, during the rise and fall sweeps, originates from the presence of gas adsorbates. Oxygen and nitrogen gases exhibit different behavior compared to hydrogen gas with respect to changes of the emission properties (Figure 8.2). The changes in the slopes, turn-on voltages, and saturation currents at high field are strongly correlated to the electronegativity of the individual species and nature of the adsorbate. They suggested that oxygen gas dominates the FE properties on adsorption and even

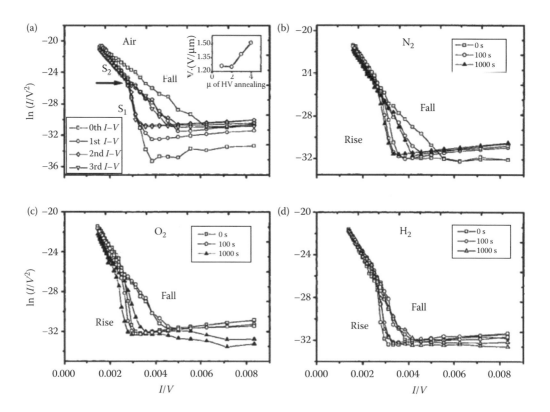

FIGURE 8.2 (a) Characteristic F–N curves of air-exposed CNTs measured during rise and fall sweeps. This measurement was repeated three times. Similar I–V curves for oxygen, nitrogen, and hydrogen are shown in (b–d), respectively, for different gas exposure times. (Lim, S.C. et al., Effect of gas exposure on field emission properties of carbon nanotube arrays. *Advanced Materials*. 2001. 13(20). 1563. Copyright Wiley–VcH Verlag GmbH & Co. KGaA. Reproduced with permission.)

degrades the surface morphologies possibly by an oxidative-etching process, whereas hydrogen gas affects the FE properties least, and simply cleans the surface of the CNT-FE array. It has been reported that resistive heating by a large emission current is effective to get rid of gas adsorbates [55]. As the applied bias increases, the emission current grows and increases the temperature at the end of the CNTs. As a consequence, a large emission current cleans the surface of CNTs.

The effects of various gas species on the electronic structures and resultant FE properties have also been studied both experimentally and theoretically [48,50,53,56–66]. Generally, the adsorption of high electronegative gases, such as O_2, N_2, and NH_3, on CNTs degrade the FE properties with increasing threshold voltages and decreasing emission currents [3]. Inert gases, such as He and Ar, barely affect the emission current and emission stability [48,62–65]. CO and CO_2 gas exposures decrease the emission current of CNTs, which is dependent on the specific partial pressure and exposure time. On the contrary, CH_4 and C_2H_4 exposure can increase the emission current of CNTs but the current stability deteriorates [66]. Furthermore, the arrangement of methane molecules on CNTs also strongly influences the emission current [60].

8.2.3 High Emission Current Density of CNTs

CNTs have very good electrical conductivity and high current durability. Dai et al. [67] measured electrical transport in individual multishell and well-ordered concentric graphitic CNTs with diameters between 7 and 20 nm by depositing a drop of a nanotube suspension on a flat insulating surface and covering it with a uniform layer of Au. The current flowing through the tubes is in the order of 1 μA, corresponding to a current density of $\approx 10^6$ A/cm^2, which is much higher than that of copper. Besides a high electrical conductivity, the thermal conductivity of a CNT is comparable to that of graphite [68]. In transport studies, multiple carbon layers can participate in conducting electrons, resulting in a large current flow. In FE, although each carbon layer participates in electron emission, electrons from the innermost wall should eventually conduct to the outermost wall. Therefore, the threshold current in FE is presumably dominated only by the outermost wall [13]. The highest emission current from a single CNT tip is only a few microamperes, corresponding to 10–4000 mA/cm^2 [11,16,69–71]. Yet, this value is still much higher than for conventional field emitters. The threshold current increases in proportion to the tube diameter, number of walls, as well as the CNT quality. For example, high-quality SWCNT strands synthesized by arc discharge showed a high emission current density of 14 A/cm^2 [24].

As the emission current continues to grow, another emission mechanism, thermionic emission, becomes important. The portion of thermally emitted electrons in the total emission current depends on field strength, work function, and tip temperature [13,72,73]. Lim et al. suggested that the CNT-FE mechanism should be categorized into three regions: FE, thermal FE, and thermionic emission (Figure 8.3). As the field increases, field-driven emission is first initiated. Subsequently, resistive heating gradually increases the tip temperature and at high emission current, thermionic emission dominates. Between field and thermionic emission, there exists a narrow transition zone, which is called thermal FE [13,72]. Thus, the FE current of CNTs increased greatly at high temperature, compared to that of room temperature. For example, the emission current at 720 K is more than five times that at room temperature [72].

8.2.4 Degradation and Failure of CNT Field Emitters

The structure of a carbon nanotube is built by covalent bonds, in which each carbon atom is bound to three other carbon atoms by sp$^2(\sigma)$-bonds. The threshold energy for the removal of one carbon atom is 17 eV [74], whereas the activation energy for surface migration of a tungsten atom is 3.2 eV. Moreover, the melting temperature of a carbon nanotube ($\approx$4800 K) is much higher than that of tungsten (3695 K) (See Ref. 33 in [75]). For these reasons, "it is expected that the structure of a nanotube will hardly change when it is subject to a large electric field, as long as the field does not exceed a

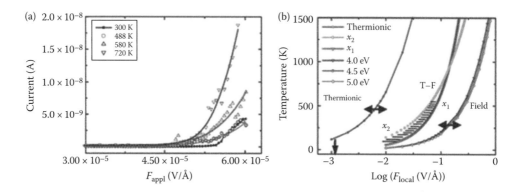

FIGURE 8.3 (a) I–V curves and their fits for emission current obtained at 300, 400, 580, and 720 K. (b) Thermionic, thermal-field, and FE zone boundary for an emitter with the WS of 4.5 eV. In addition, FE zones with WSs of 4.0 and 5.0 eV are presented. The down arrow on the x axis indicates the field at which the I–T curve was measured. (Reprinted from *Carbon*, 43(13), Lim, S.C. et al., Extracting independently the work function and field enhancement factor from thermal-field emission of multi-walled carbon nanotube tips, 2801–2807, copyright (2005), with permission from Elsevier.)

certain limit" [76]. In fact, CNT emitters have good long-term FE stability. So far, the longest stability was reported by Saito et al. [17] when a CNT cathode had a life of more than 625 days at an emission current of 210 μA. However, degradation and failure are unavoidable, especially at large currents.

The degradation of FE mainly results from tip structure change or destruction owing to the chemical and physical modifications under FE conditions. In chemical modifications, for example, chemical adsorption of gas species on the surface of an emitter may change the electronic structure of the emitter surface. Therefore, current saturation by gas adsorption will change, and consequently, the emission properties will change too. As for physical modification, the shape of the tip changes because of resistive heating, ion bombardment, and electrostatic stress [77–79]. The geometric area changes cover not only the near-cap region but also the side walls of CNTs. The observed physical changes of CNTs include cap opening, cap closing, fracturing, splitting of bundles (debundling), sharpening of caps, removing impurities on the side wall, and voiding the inner space. Furthermore, interactions between adjacent carbon atoms [80,81] and peeling-off of the outmost layer [75,82] have also been reported as processes that eventually lead to degradation and failure.

In addition, the FE stability of CNTs is directly influenced by their structural integrity and the applied initial current. We studied the FE stability of SWCNT strands synthesized by the hydrogen arc discharge method (Figure 8.4). An initial current of 20 μA was applied corresponding to a current density of 1.4 A/cm² under a constant applied voltage. There was no decrease of the emission current for 100 h, and the current fluctuation was less than 1.5%. As the initial current was increased to 200 μA (corresponding to a current density of 14 A/cm²), the emission current was decreased by 10% in the first 20 h and then gradually stabilized over a 175-h longer emission period. The good emission persistence and stability of the SWCNT strands even at a very high emission current stability can be ascribed to their good structural stability shown by the relatively high oxidation onset temperature in air (≈600°C) [24].

8.3 APPLICATIONS OF CNTs AS FIELD EMITTERS

The excellent FE properties of CNTs, such as low-voltage operation, good stability, stable and high emission current, and large field enhancement factor, have opened new application fields in CNT-based devices. The use of CNTs as a source of electrons under an applied electric field is one of the most promising applications. As shown in Figure 8.5, it is easy to control the FE pattern of CNTs to

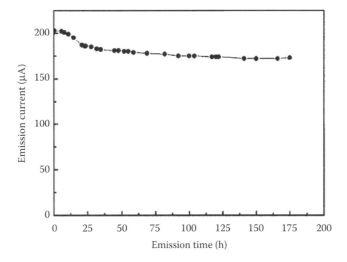

FIGURE 8.4 Emission stability of an SWCNT strand under a vacuum of 5×10^{-7} Pa at room temperature with an initial emission current of 200 μA (corresponding to a current density of 14 A/cm^2) in 175 h. (Reprinted with permission from Liu, C. et al., Field emission properties of macroscopic single-walled carbon nanotube strands. *Applied Physics Letters*, 2005, 86(22): 223114. Copyright (2005), American Institute of Physics.)

a desired area and shape. Therefore, many applications have been demonstrated, including flat panel displays, lighting elements, high-brightness electron microscopy sources, radio-frequency amplifiers, portable x-ray systems, and ionization vacuum gauges [84].

The first CNT-FE display (FED) was proposed by Wang et al. [85] and was fabricated using a CNT–epoxy composite as the electron emission source. Choi et al. [86] demonstrated a fully sealed 4.5-in. CNT-FED diagonal with a high brightness of 1800 cd/cm^2. The fully sealed diode-type display (200 μm glass spacer) turned on at less than 1 V/μm and emitted 1.5 mA at 3 V/μm [87]. The largest colored CNT-FED, 35 in. in size, was demonstrated by Samsung SDI Co. Zeng et al. [88] developed a technique to fabricate a fully printed CNT-FED in which all inner cells were fabricated by a screen-printing process. On the basis of this technique, fully printed matrix-addressable diode

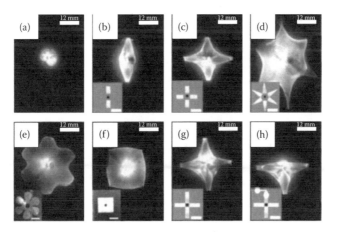

FIGURE 8.5 Photograph of emission patterns (a) in the absence of dielectric materials, (b) through (d) in the presence of glass flakes, (e) and (f) in the presence of copper plates on glass, and (g) and (h) in the presence of ITO (indium tin oxide) glass flakes. One of the ITO glass flakes was grounded in (h). (Reprinted with permission from Tong, Y. et al., Controlling field-emission patterns of isolated single-walled carbon nanotube rope. *Applied Physics Letters*, 2005, 87(4): 043114. Copyright (2005), American Institute of Physics.)

CNT-FEDs, which could display moving images and be driven by the integrated drive circuits of a commercial plasma display panel (PDP), were subsequently fabricated. A very high brightness of 1×10^4 cd/m^2 was achieved at 220 V.

An exemplary application of electron multipliers was demonstrated by Yi et al. [89]. In their experiment, vertically aligned chemical vapor deposition (CVD)-grown MWCNTs were coated with a layer of MgO as a second electron emission (SEE) material. When the surface of MgO was bombarded by the primary electrons, a huge electron emission was detected. The SEE yield was measured to be up to 22,000 and was very sensitive to the voltage applied to the MWCNTs [90]. The large field enhancement factor of CNTs highly exaggerates the electric field at the tip and triggers the secondary electron emission. SEE yields from various CNT carpets coated with different materials, such as MgO, CsI, and ZnO, under different conditions have been reported [89–92]. Alam et al. [93] systematically investigated the SEE behavior of bare CNTs [93–95] and found that CNT forests do not have the highest pure SEE yield among all the materials used. By using a simple experimental procedure under an optical microscope, they made suspended CNTs that are free from interaction with the substrate during electron yield measurements (Figure 8.6a through c). It was found that the secondary electron yield from isolated suspended CNTs was less uniform and decreased as a function of primary electron energy (Figure 8.6d) [95].

Luminescent lighting elements using CNTs have also been demonstrated [96–102]. For instance, Satio et al. [97] developed a flashlight structured with triode electrodes. This device can operate in a high vacuum and still sustain stable emission currents for more than 4000 h. Bonard et al. [98] first demonstrated a cylindrical lighting element. The brightness from this 50 mm long and 21 mm wide tube was reported to be 10,000 cd/cm^2, which is comparable to the current fluorescent tubes but the power consumption is higher because of low efficiency. Fully sealed fluorescent lamps have also been fabricated by Huang et al. using a wire-type CNT cold cathode. Luminescence efficiencies of 37 and 21 lm/W were obtained for a fully sealed lamp with green and white phosphor screens [100]. Yang et al. [118] constructed a luminescent tube using a CNT wire as the cathode and a transparent CNT film as the anode. These cold cathodes may replace conventional hot cathodes that are widely used in vacuum electronic devices. The novel two-layer structure can be applied to any substrate, and is simple and inexpensive to fabricate.

A CNT-based x-ray tube has also been demonstrated [103–108]. Yue et al. deposited CNTs on a substrate with an iron adhesive layer using dielectrophoretic deposition. The metal layer was found to increase not only the emission current but also the emission stability [109]. In this study, the

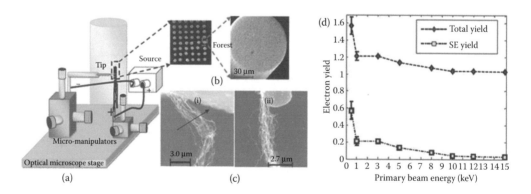

FIGURE 8.6 (a) Schematic of the experimental setup for extracting CNTs from the forests, (b) micrograph of the CNT forests, (c) extracted CNTs on tungsten tips, and (d) electron yield measured from the suspended CNTs (total yield, diamond marks, SE yield, and square marks). The fluctuations in the beam current or specimen current are included in the error bars. (Reprinted with permission from Alam, M.K. et al., Secondary electron yield of multiwalled carbon nanotubes. *Applied Physics Letters*, 2010, 97(26): 261902. Copyright (2010), American Institute of Physics.)

cathode easily yielded a current density of 30 mA/cm^2 with negligible fluctuations during an operation time of 18 h. An array-type x-ray tube was reported in Ref. [103] and the key component of the device is a gated CNT-FE cathode with an array of electron-emitting pixels that are individually addressable by a metal-oxide semiconductor field effect transistor-based electronic circuit (Figure 8.7). This device can potentially lead to a fast data-acquisition rate for laminography and tomosynthesis with a simplified experimental setup. An x-ray gun has shown a high spatial resolution of 30 μm, comparable to the current state-of-the-art thermionic-type guns in the market [110].

The emission from an individual CNT can constitute as a point emitter [111–114]. CNTs exhibit significantly better emission stability than current tungsten-based field emitter, and this stability can be achieved at a moderate temperature rather than at room temperature. A CNT-based point electron emitter was fabricated using cavity-confined dielectrophoresis by Lim et al. [114] for which the emission current of an individual MWCNT was stable up to 10 μA and reached ≈2 mA (1.7×10^8 A/cm^2). At a low electric field, the current fluctuated in a stepwise manner. At higher current emission, the stability was interfered with plateau-like fluctuations that are associated with field-induced unraveling and Joule heating of the cap structures of MWCNTs.

Another application of CNTs to scientific instrumentation is presented by ionization gauges [115–123]. In an ionization gauge, CNTs facilitate the decomposition of residual gases and the ionization current shows good linearity with pressures ranging from 10^{-6} to 10^{-10} Torr [124]. Yang et al. [118] fabricated a triode ionization gauge with low sensitivity using screen-printed CNT cathodes that showed good linearity in He, N$_2$, Ar, and air, and a wide measurement range from 10^{-7} to 1.0 Torr (Figure 8.8). The cathode's emission current degradation in low vacuum was reduced by sputtering a 20 nm thick HfC polycrystalline film on the CNT cathode. This gauge has the virtues of compact size, extremely low power consumption, and a cold cathode with no heating effect, and is promising for use in low-vacuum applications, such as CVD or sputtering.

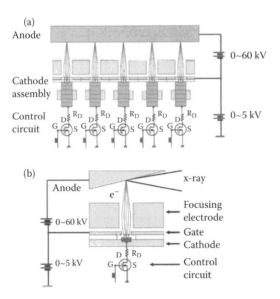

FIGURE 8.7 (a) Schematic of a multibeam x-ray source with an FE cathode that contains five emitting pixels and a molybdenum target at 10^{-7} Torr pressure. The pixels are evenly spaced with a center-to-center spacing of 1.27 cm. Each pixel is capable of emitting 1 mA current. The anode voltage is set at 40 kV. The gate voltage varies depending on the flux required. Switching the x-ray beam from pixel to pixel is controlled by sweeping a 0–5 V direct current (dc) pulse through the MOSFET. (b) Each emitting pixel comprises a 1.5 mm diameter CNT film coated on a metal disk, a 150 mm thick dielectric spacer, extraction gate, and a focusing electrode. (Reprinted with permission from Zhang, J. et al., Stationary scanning x-ray source based on carbon nanotube field emitters. *Applied Physics Letters*, 2005, 86(18): 184104. Copyright (2005), American Institute of Physics.)

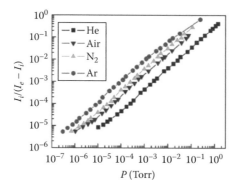

FIGURE 8.8 Normalized ion current versus chamber pressure for helium, air, nitrogen, and argon. (Reprinted with permission from Yang, Y. C. et al., A low-vacuum ionization gauge with HfC-modified carbon nanotube field emitters. *Applied Physics Letters*, 2008, 92(15): 325707. Copyright (2008), American Institute of Physics.)

8.4 SUMMARY AND OUTLOOK

The large aspect ratio, high mechanical strength and electrical conductivity, good thermal stability, and chemical inertness endow CNTs with desirable FE properties, such as a high field enhancement factor, low threshold field, high emission current density, excellent emission uniformity, and long-term durability. Therefore, the applications of CNT-based field emitters, for example, in flat panel displays, lighting elements, high-brightness electron microscope sources, radio-frequency amplifiers, portable x-ray devices, and ionization vacuum gauges, have been explored. However, there are still challenging issues to be solved before CNT field emitters can be practically used. These issues are mainly related to the stability, uniformity, reproducibility, and the cost of CNT-based devices. To solve these issues, it is important to achieve controlled synthesis of CNTs with desired structures, to fully understand the FE and decay mechanisms of CNTs and to optimize the structure of CNT-based devices.

1. *Controlled synthesis of CNTs with desired structures*   It is well known that the properties of CNTs are closely related to their structures; thus, the reported FE properties based on different CNTs are quite diverse. For example, high-quality and high-purity few-walled CNTs show notably better electron FE properties than the conventional commercially available nanotubes [125]. At the same time, the length, carbon layer crystallinity, diameter, tip geometry, and orientation of CNTs are also important factors that influence their FE performance [17,27,126–129]. In addition, β- or *N*-doped CNTs are reported to exhibit improved FE properties [130,131]. Therefore, reproducible synthesis of high-quality, high-purity CNTs with a uniform and desired structure (including diameter, length, number of carbon layers, orientation, and conductivity) is urgently required.

   The electronic WS is an important parameter determining the emission properties of a material. SWCNTs can be either metallic or semiconducting depending on the wrapping angle and diameter of the graphene sheet. It is reasonable to predict that SWCNTs with different chiralities possess different FE properties. Therefore, the selective synthesis of SWCNTs with controllable chirality may further improve their FE performance. In addition, the requirements of CNTs for applications in flat panel displays, lighting elements, high-brightness electron microscope sources, radio-frequency amplifiers, portable x-ray systems, or ionization vacuum gauges may be different. After all, it is important to achieve the controlled fabrication of CNTs.

2. *Understanding on the FE mechanism*   The F–N equation is currently used as a rough approximation to describe FE from CNTs. However, because of the nonmetal and

nanomaterial nature of CNTs, deviations from the F–N model for CNT field emitters have been observed. In addition, it is still unclear whether the sharpness of CNTs is their only advantage over other emitters, or whether the intrinsic properties of CNTs (e.g., occupied state level, Fermi level, tube chirality, diameter, and the eventual presence of defects) also influence their emission performance. For example, Bonard et al. [9] observed luminescence by the naked eye during FE from CNTs, which demonstrates that light emission is caused by electron transitions between different discrete energy levels participating in FE. FE energy distributions of individual MWCNTs also indicate that the position of peaks in the spectrum linearly depends on the extraction voltage [10], unlike the case of metallic emitters where the position stays in the vicinity of the Fermi level. The energy spread of electrons emitted from CNTs is very narrow ($\approx$0.2 eV), typically half that for metallic emitters [3]. These phenomena strongly suggest that the FE mechanism of CNTs may be distinctly different from the metallic continuum model in traditional metallic emitters (i.e., the F–N model). Therefore, an appropriate model that can precisely describe FE from CNTs is urgently required.

3. *Design and integration of FE devices* The facile assembly of nanometer-sized CNTs into a macroscopic form is an important step to their practical application. There are mainly two approaches for the integration of CNT field emitters. The first is the direct growth of CNT arrays or films on substrates by CVD [30,132–134]; the second is making a CNT paste by postsynthesis processing and then transplanting it onto substrates [135–138]. For the CVD process, a high growth temperature, small size, and complicated growth process limit their large-scale production and application. However, the transplant methods, such as screen printing [139,140], ink-jet printing [135,141], dip coating [142,143], spraying [144,145], electrophoretic deposition [146,147], and contact transfer [148,149] are attracting more interest owing to their low cost, simplicity, uniformity of emission, large size, mass production, and continuous manufacturing. Although the above integration techniques have prodigious advantages, some technical limitations still remain, including nonuniform dispersion of CNT powders, poor adhesion between substrates and the CNT layer, or the introduction of contamination and the possible destruction of the CNT pattern. As a consequence, simple and effective ways for assembling CNT field emitters, including positioning, density control, as well as a desirable combination between CNTs and substrates, still need to be developed.

4. *Cost efficiency of CNT-based field emitters* CNTs have attracted much interest for FE applications as described above; however, their costs have to be considered before the advent of satisfactory commercial products. Although mass production of agglomerated CNTs and aligned MWCNTs on the order of a few to several tens of kilograms per hour has already been realized [150,151], the low cost and scalable synthesis of high-quality, isolated SWCNTs with controllable electronic conductivity, chirality, good alignment, and predetermined diameters and lengths still presents great challenges. In addition, the development of low-cost and efficient processes for the integration of CNTs into FE devices is also important for the overall cost control. To realize these objectives, great efforts are needed to achieve breakthroughs on CNT manufacture and device assembly from both scientific and engineering respects.

REFERENCES

1. Schottky W, Concerning cold and hot electron discharge. *Zeitschrift Fur Physik*, 1923. 14: 63–106.
2. Fowler RH, Nordheim L, Electron emission in intense electric fields. *Proceedings of the Royal Society A: Mathematical, Physical and Engineering Sciences*, 1928. 119(781): 173–181.
3. Zhou G, Duan W, Field emission in doped nanotubes. *Journal of Nanoscience and Nanotechnology*, 2005. 5(9): 1421–1434.

4. Bockrath M, Cobden DH. et al., Single-electron transport in ropes of carbon nanotubes. *Science*, 1997. 275(5308): 1922–1925.

5. Tans SJ, Devoret MH. et al., Individual single-wall carbon nanotubes as quantum wires. *Nature*, 1997. 386(6624): 474–477.

6. Venema LC, Wildoer JWG. et al., Imaging electron wave functions of quantized energy levels in carbon nanotubes. *Science*, 1999. 283(5398): 52–55.

7. Collins PG, Zettl A, Unique characteristics of cold cathode carbon-nanotube–matrix field emitters. *Physical Review B*, 1997. 55(15): 9391–9399.

8. Bonard JM, Maier F. et al., Field emission properties of multiwalled carbon nanotubes. *Ultramicroscopy*, 1998. 73(1): 7–15.

9. Bonard JM, Stockli T et al., Field-emission-induced luminescence from carbon nanotubes. *Physical Review Letters*, 1998. 81(7): 1441–1444.

10. Fransen MJ, van Rooy TL, Kruit P, Field emission energy distributions from individual multiwalled carbon nanotubes. *Applied Surface Science*, 1999. 146(1–4): 312–327.

11. Zou R, Hu JQ. et al., Carbon nanotubes as field emitter. *Journal of Nanoscience and Nanotechnology*, 2010. 10(12): 7876–7896.

12. Gadzuk JW, Plummer EW, Field-emission energy-distribution (feed). *Reviews of Modern Physics*, 1973. 45(3): 487–548.

13. Lim SC, Lee K. et al., Field emission and application of carbon nanotubes. *Nano*, 2007. 2(2): 69.

14. De Heer WA, Chatelain A, Ugarte D, A carbon nanotube field-emission electron source. *Science*, 1995. 270(5239): 1179–1180.

15. Rinzler AG, Hafner JH. et al., Unraveling nanotubes: Field emission from an atomic wire. *Science*, 1995. 269(5230): 1550–1553.

16. Dai HJ, Hafner JH. et al., Nanotubes as nanoprobes in scanning probe microscopy. *Nature*, 1996. 384(6605): 147–150.

17. Saito Y, Carbon nanotube field emitter. *Journal of Nanoscience and Nanotechnology*, 2003. 3(1–2): 39–50.

18. Bonard JM, Kind H. et al., Field emission from carbon nanotubes: The first five years. *Solid-State Electronics*, 2001. 45(6): 893–914.

19. Chen GH, Shin DH. et al., Improved field emission stability of thin multiwalled carbon nanotube emitters. *Nanotechnology*, 2010. 21(1): 015704.

20. Lee LY, Lee SF. et al., Effects of potassium hydroxide post-treatments on the field-emission properties of thermal chemical vapor deposited carbon nanotubes. *Journal of Nanoscience and Nanotechnology,* 2011. 11(12): 11185–11189.

21. Nakahara H, Ichikawa S. et al., Carbon nanotube electron source for field emission scanning electron microscopy. *e-Journal of Surface Science and Nanotechnology*, 2011. 9: 400–403.

22. Rinzler AG, Hafner JH. et al., Unraveling nanotubes: Field emission from an atomic wire. *Science*, 1995. 269(5230): 1550–1553.

23. Saito Y, Hamaguchi K. et al., Field emission from multi-walled carbon nanotubes and its application to electron tubes. *Applied Physics A: Materials Science and Processing*, 1998. 67(1): 95–100.

24. Liu C, Tong Y. et al., Field emission properties of macroscopic single-walled carbon nanotube strands. *Applied Physics Letters*, 2005. 86(22): 223114.

25. Sato H, Takegawa H, Saito Y, Fabrication of carbon nanotubes array and its field emission property. *Journal of Vacuum Science and Technology*, 2004. 22(3): 1335–1337.

26. Stratakis E, Giorgi R. et al., Three-dimensional carbon nanowall field emission arrays. *Applied Physics Letters*, 2010. 96(4): 043110.

27. Bonard JM, Weiss N. et al., Tuning the field emission properties of patterned carbon nanotube films. *Advanced Materials*, 2001. 13(3): 184–188.

28. Murakami H, Hirakawa M. et al., Field emission from well-aligned, patterned, carbon nanotube emitters. *Applied Physics Letters*, 2000. 76(13): 1776–1778.

29. Xu XP, Brandes GR, A method for fabricating large-area, patterned, carbon nanotube field emitters. *Applied Physics Letters*, 1999. 74(17): 2549–2551.

30. Fan SS, Chapline MG. et al., Self-oriented regular arrays of carbon nanotubes and their field emission properties. *Science*, 1999. 283(5401): 512–514.

31. Bonard JM, Salvetat JP. et al., Field emission from single-wall carbon nanotube films. *Applied Physics Letters*, 1998. 73(7): 918–920.

32. Huarong L, Kato S, Saito Y, Effect of cathode–anode distance on field emission properties for carbon nanotube film emitters. *Japanese Journal of Applied Physics*, 2009. 48(1): 015007.

33. Huarong L, Saito Y, Influence of surface roughness on field emission of electrons from carbon nanotube films. *Journal of Nanoscience and Nanotechnology*, 2010. 10(6): 3983–3987.
34. LeMieux MC, Melburne C. et al., Self-sorted, aligned nanotube networks for thin-film transistors. *Science*, 2008. 321(5885): 101–104.
35. Liu H, Saito Y, Influence of surface roughness on field emission of electrons from carbon nanotube films. *Journal of Nanoscience and Nanotechnology*, 2010. 10(6): 3983–3987.
36. Nilsson L, Groening O. et al., Scanning field emission from patterned carbon nanotube films. *Applied Physics Letters*, 2000. 76: 2071.
37. Wang H, Li Z. et al., Synthesis of double-walled carbon nanotube films and their field emission properties. *Carbon*, 2010. 48(10): 2882–2889.
38. Bonard JM, Salvetat JP. et al., Field emission from carbon nanotubes: Perspectives for applications and clues to the emission mechanism. *Applied Physics A—Materials Science and Processing*, 1999. 69(3): 245–254.
39. Zhang YL, Zhang LL. et al., Synthesis and field emission property of carbon nanotubes with sharp tips. *New Carbon Materials*, 2011. 26(1): 52–55.
40. Jeong HJ, Jeong HD. et al., All-carbon nanotube-based flexible field-emission devices: From cathode to anode. *Advanced Functional Materials*, 2011. 21(8): 1526–1532.
41. Shearer CJ, Yu JX. et al., Highly resilient field emission from aligned single-walled carbon nanotube arrays chemically attached to *n*-type silicon. *Journal of Materials Chemistry*, 2008. 18(47): 5753–5760.
42. Min YS, Bae EJ. et al., Direct growth of single-walled carbon nanotubes on conducting ZnO films and its field emission properties. *Applied Physics Letters*, 2006. 89(11): 113116.
43. Bonard JM, Dean KA. et al., Field emission of individual carbon nanotubes in the scanning electron microscope. *Physical Review Letters*, 2002. 89(19): 197602.
44. Jeong HJ, Lim SC. et al., Edge effect on the field emission properties from vertically aligned carbon nanotube arrays. *Carbon*, 2004. 42(14): 3036–3039.
45. Wang XQ, Li L, Theoretical optimization for field emission current density from carbon nanotubes array. *Acta Physica Sinica*, 2008. 57(11): 7173–7177.
46. Wang XQ, Li L, Analytical optimization for field emission of carbon nanotube array. *Chinese Science Bulletin*, 2009. 54(10): 1801–1804.
47. Choi WB, Lee YH. et al., Carbon-nanotubes for full-color field-emission displays. *Japanese Journal of Applied Physics Part 1—Regular Papers Short Notes and Review Papers*, 2000. 39(5A): 2560–2564.
48. Lim SC, Choi YC. et al., Effect of gas exposure on field emission properties of carbon nanotube arrays. *Advanced Materials*, 2001. 13(20): 1563.
49. Dean KA, Chalamala BR, The environmental stability of field emission from single-walled carbon nanotubes. *Applied Physics Letters*, 1999. 75: 3017.
50. Hata K, Takakura A, Saito Y, Field emission microscopy of adsorption and desorption of residual gas molecules on a carbon nanotube tip. *Surface Science*, 2001. 490(3): 296–300.
51. Collazo R, Schlesser R, Sitar Z, Two field-emission states of single-walled carbon nanotubes. *Applied Physics Letters*, 2001. 78(14): 2058–2060.
52. Zhi CY, Bai XD, Wang EG, Enhanced field emission from carbon nanotubes by hydrogen plasma treatment. *Applied Physics Letters*, 2002. 81(9): 1690–1692.
53. Nilsson L, Groning O. et al., Carbon nano-/micro-structures in field emission: Environmental stability and field enhancement distribution. *Thin Solid Films*, 2001. 383(1–2): 78–80.
54. Dean KA, von Allmen P, Chalamala BR, Three behavioral states observed in field emission from single-walled carbon nanotubes. *Journal of Vacuum Science and Technology B*, 1999. 17(5): 1959–1969.
55. Dean KA, Chalamala BR, Current saturation mechanisms in carbon nanotube field emitters. *Applied Physics Letters*, 2000. 76(3): 375–377.
56. Kim C, Choi YS. et al., The effect of gas adsorption on the field emission mechanism of carbon nanotubes. *Journal of the American Chemical Society*, 2002. 124(33): 9906–9911.
57. Park N, Han S, Ihm J, Effects of oxygen adsorption on carbon nanotube field emitters. *Physical Review B*, 2001. 64(12): 125401.
58. Maiti A, Andzelm J. et al., Effect of adsorbates on field emission from carbon nanotubes. *Physical Review Letters*, 2001. 87(15): 155502.
59. Hwang YG, Lee YH, Adsorption of H_2O molecules at the open ends of single walled carbon nanotubes. *Journal of the Korean Physical Society*, 2003. 42: S267–S271.
60. Naieni AK, Yaghoobi P, Nojeh A, First-principles study of field-emission from carbon nanotubes in the presence of methane. *Journal of Vacuum Science and Technology B*, 2012. 30(2): 021803.
61. Li C, Fang G. et al., Effect of adsorbates on field emission from flame-synthesized carbon nanotubes. *Journal of Physics D—Applied Physics*, 2008. 41(19): 195401.

62. Wadhawan A, Stallcup RE, Perez JM, Effects of Cs deposition on the field-emission properties of single-walled carbon-nanotube bundles. *Applied Physics Letters*, 2001. 78(1): 108.

63. Dean KA, Chalamala BR , The environmental stability of field emission from single-walled carbon nanotubes. *Applied Physics Letters*, 1999. 75(19): 3017–3019.

64. Lim SC, Jeong HJ. et al., Field-emission properties of vertically aligned carbon-nanotube array dependent on gas exposures and growth conditions. *Journal of Vacuum Science and Technology a-Vacuum Surfaces and Films*, 2001. 19(4): 1786–1789.

65. Wadhawan A, Stallcup A. et al., Effects of O-2, Ar, and H-2 gases on the field-emission properties of single-walled and multiwalled carbon nanotubes. *Applied Physics Letters*, 2001. 79(12): 1867–1869.

66. Sheng LM, Liu P. et al., Effects of carbon-containing gases on the field-emission current of multiwalled carbon-nanotube arrays. *Journal of Vacuum Science and Technology A*, 2003. 21(4): 1202–1204.

67. Dai HJ, Wong EW, Lieber CM, Probing electrical transport in nanomaterials: Conductivity of individual carbon nanotubes. *Science*, 1996. 272(5261): 523–526.

68. Kim P, Shi L. et al., Thermal transport measurements of individual multiwalled nanotubes. *Physical Review Letters*, 2001. 87(21): 215502.

69. Lee H, Joak J. et al., High-current field emission of point-type carbon nanotube emitters on Ni-coated metal wires. *Carbon*, 2012. 50(6): 2126–2133.

70. Zhu W, Bower C. et al., Large current density from carbon nanotube field emitters. *Applied Physics Letters*, 1999. 75(6): 873–875.

71. Hu JQ, Bando Y. et al., Growth and field-emission properties of crystalline, thin-walled carbon microtubes. *Advanced Materials*, 2004. 16(2): 153.

72. Lim SC, Jeong HJ. et al., Extracting independently the work function and field enhancement factor from thermal-field emission of multi-walled carbon nanotube tips. *Carbon*, 2005. 43(13): 2801–2807.

73. Purcell ST, Vincent P. et al., Hot nanotubes: Stable heating of individual multiwall carbon nanotubes to 2000 K induced by the field-emission current. *Physical Review Letters*, 2002. 88(10): 105502.

74. Crespi VH, Chopra NG. et al., Anisotropic electron-beam damage and the collapse of carbon nanotubes. *Physical Review B*, 1996. 54(8): 5927–5931.

75. Doytcheva M, Kaiser M, de Jonge N, *In situ* transmission electron microscopy investigation of the structural changes in carbon nanotubes during electron emission at high currents. *Nanotechnology*, 2006. 17(13): 3226–3233.

76. de Jonge N, Bonard JM, Carbon nanotube electron sources and applications. *Philosophical Transactions of the Royal Society A: Mathematical, Physical and Engineering Sciences*, 2004. 362(1823): 2239–2266.

77. de Jonge N, Doytcheva M. et al., Cap closing of thin carbon nanotubes. *Advanced Materials*, 2005. 17(4): 451–455.

78. Xu Z, Bai XD, Wang EG, Geometrical enhancement of field emission of individual nanotubes studied by *in situ* transmission electron microscopy. *Applied Physics Letters*, 2006. 88(13): 133107.

79. Saito Y, Seko K, Kinoshita JI, Dynamic behavior of carbon nanotube field emitters observed by *in situ* transmission electron microscopy. *Diamond and Related Materials*, 2005. 14(11–12): 1843–1847.

80. Choi YS, Park KA. et al., Oxygen gas-induced lip–lip interactions on a double-walled carbon nanotube edge. *Journal of the American Chemical Society*, 2004. 126(30): 9433–9438.

81. Kuzumaki T, Sawada H. et al., Selective processing of individual carbon nanotubes using dual-nano-manipulator installed in transmission electron microscope. *Applied Physics Letters*, 2001. 79(27): 4580–4582.

82. Wang ZL, Gao RP. et al., *In situ* imaging of field emission from individual carbon nanotubes and their structural damage. *Applied Physics Letters*, 2002. 80(5): 856–858.

83. Tong Y, Lim SC. et al., Controlling field-emission patterns of isolated single-walled carbon nanotube rope. *Applied Physics Letters*, 2005. 87(4): 043114.

84. Charlier JC, Blase X, Roche S, Electronic and transport properties of nanotubes. *Reviews of Modern Physics*, 2007. 79(2): 677–732.

85. Wang QH, Setlur AA. et al., A nanotube-based field-emission flat panel display. *Applied Physics Letters*, 1998. 72: 2912.

86. Choi WB, Chung DS. et al., Fully sealed, high-brightness carbon-nanotube field-emission display. *Applied Physics Letters*, 1999. 75: 3129.

87. Kim JM, Choi WB. et al., Field emission from carbon nanotubes for displays. *Diamond and Related Materials*, 2000. 9(3–6): 1184–1189.

88. Zeng FG, Zhu C. et al., The fabrication and operation of fully printed carbon nanotube field emission displays. *Microelectronics Journal*, 2006. 37(6): 495–499.

89. Yi W, Su SG. et al., Secondary electron emission yields from MgO deposited on carbon nanotubes. *Journal of Applied Physics*, 2001. 89(7): 4091–4095.

90. Kim WS, Yi W. et al., Secondary electron emission from magnesium oxide on multiwalled carbon nanotubes. *Applied Physics Letters*, 2002. 81(6): 1098–1100.

91. Huang L, Lau SP. et al., Local measurement of secondary electron emission from ZnO-coated carbon nanotubes. *Nanotechnology*, 2006. 17(6): 1564–1567.

92. Lee J, Park J. et al., Double layer-coated carbon nanotubes: Field emission and secondary-electron emission properties under presence of intense electric field. *Journal of Vacuum Science and Technology B*, 2009. 27(2): 626–630.

93. Alam MK, Yaghoobi P, Nojeh A, Unusual secondary electron emission behavior in carbon nanotube forests. *Scanning*, 2009. 31(6): 221–228.

94. Nojeh A, Wong WK. et al., Scanning electron microscopy of field-emitting individual single-walled carbon nanotubes. *Applied Physics Letters*, 2004. 85(1): 112–114.

95. Alam MK, Yaghoobi P. et al., Secondary electron yield of multiwalled carbon nanotubes. *Applied Physics Letters*, 2010. 97(26): 261902.

96. Park JH, Son GH. et al., Screen printed carbon nanotube field emitter array for lighting source application. *Journal of Vacuum Science and Technology B*, 2005. 23(2): 749–753.

97. Saito Y, Uemura S, Hamaguchi K, Cathode ray tube lighting elements with carbon nanotube field emitters. *Japanese Journal of Applied Physics Part 2—Letters*, 1998. 37(3B): L346–L348.

98. Bonard JM, Stockli T. et al., Field emission from cylindrical carbon nanotube cathodes: Possibilities for luminescent tubes. *Applied Physics Letters*, 2001. 78(18): 2775–2777.

99. Lee S, Lm WB. et al., Low temperature burnable carbon nanotube paste component for carbon nanotube field emitter backlight unit. *Journal of Vacuum Science and Technology B*, 2005. 23(2): 745–748.

100. Huang JX, Chen J. et al., Field-emission fluorescent lamp using carbon nanotubes on a wire-type cold cathode and a reflecting anode. *Journal of Vacuum Science and Technology B*, 2008. 26(5): 1700–1704.

101. Fu W, Liu P. et al., Spherical field emission cathode based on carbon nanotube paste and its application in luminescent bulbs. *Journal of Vacuum Science and Technology B*, 2008. 26(4): 1404–1406.

102. Wei Y, Xiao L. et al., Cold linear cathodes with carbon nanotube emitters and their application in luminescent tubes. *Nanotechnology*, 2007. 18(32): 325702.

103. Zhang J, Yang J. et al., Stationary scanning x-ray source based on carbon nanotube field emitters. *Applied Physics Letters*, 2005. 86(18): 184104.

104. Cheng Y, Zheng J. et al., Dynamic radiography using a carbon-nanotube-based field-emission x-ray source. *Review of Scientific Instruments*, 2004. 75(10): 3264–3267.

105. Zhang J, Cheng Y. et al., A nanotube-based field emission x-ray source for microcomputed tomography. *Review of Scientific Instruments*, 2005. 76(9): 094301.

106. Sugie H, Tanemura M. et al., Carbon nanotubes as electron source in an x-ray tube. *Applied Physics Letters*, 2001. 78(17): 2578–2580.

107. Kim HS, Duy DQ. et al., Field-emission electron source using carbon nanotubes for x-ray tubes. *Journal of the Korean Physical Society*, 2008. 52(4): 1057–1060.

108. Calderon-Colon X, Geng H. et al., A carbon nanotube field emission cathode with high current density and long-term stability. *Nanotechnology*, 2009. 20(32): 325707.

109. Yue GZ, Qiu Q. et al., Generation of continuous and pulsed diagnostic imaging x-ray radiation using a carbon-nanotube-based field-emission cathode. *Applied Physics Letters*, 2002. 81(2): 355–357.

110. Liu Z, Yang G. et al., Carbon nanotube based microfocus field emission x-ray source for microcomputed tomography. *Applied Physics Letters*, 2006. 89(10): 103111.

111. de Jonge N, Allioux M. et al., Optical performance of carbon-nanotube electron sources. *Physical Review Letters*, 2005. 94(18): 186807.

112. de Jonge N, Brightness of carbon nanotube electron sources. *Journal of Applied Physics*, 2004. 95(2): 673–681.

113. de Jonge N, Lamy Y. et al., High brightness electron beam from a multi-walled carbon nanotube. *Nature*, 2002. 420(6914): 393–395.

114. Lim SC, Lee DS. et al., Field emission of carbon-nanotube point electron source. *Diamond and Related Materials*, 2009. 18(12): 1435–1439.

115. Riley DJ, Mann M. et al., Helium detection via field ionization from carbon nanotubes. *Nano Letters*, 2003. 3(10): 1455–1458.

116. Sheng LM, Liu P. et al., A saddle-field gauge with carbon nanotube field emitters. *Diamond and Related Materials*, 2005. 14(10): 1695–1699.

117. Huang JX, Chen J. et al., Bayard–Alpert ionization gauge using carbon-nanotube cold cathode. *Journal of Vacuum Science and Technology B*, 2007. 25(2): 651–654.

118. Yang YC, Qian L. et al., A low-vacuum ionization gauge with HfC-modified carbon nanotube field emitters. *Applied Physics Letters*, 2008. 92(15): 153105.

119. Xiao L, Qian L. et al., Conventional triode ionization gauge with carbon nanotube cold electron emitter. *Journal of Vacuum Science and Technology A*, 2008. 26(1): 1–4.

120. Wilfert S, Edelmann C, Field emitter-based vacuum sensors. *Vacuum*, 2012. 86(5): 556–571.

121. Suto H, Fujii C. et al., Fabrication of cold cathode ionization gauge using screen-printed carbon nanotube field electron emitter. *Japanese Journal of Applied Physics*, 2008. 47(4): 2032–2035.

122. Liu H, Nakahara H. et al., Ionization vacuum gauge with a carbon nanotube field electron emitter combined with a shield electrode. *Vacuum*, 2009. 84(5): 713–717.

123. Cai M, Detian L. et al., Latest development of ionization gauges with field emitters. *Journal of Vacuum Science and Technology*, 2011. 31(6): 732–738.

124. Dong CK, Myneni GR, Carbon nanotube electron source based ionization vacuum gauge. *Applied Physics Letters*, 2004. 84(26): 5443–5445.

125. Qian C, Qi H. et al., Fabrication of small diameter few-walled carbon nanotubes with enhanced field emission property. *Journal of Nanoscience and Nanotechnology*, 2006. 6(5): 1346–1349.

126. Bower C, Zhou O. et al., Fabrication and field emission properties of carbon nanotube cathodes. In *Amorphous and Nanostructured Carbon*, Sullivan JP. et al., eds., Materials Research Society: Warrendale, 2000. pp. 215–220.

127. Feng YT, Deng SZ. et al., Effect of carbon nanotube structural parameters on field emission properties. *Ultramicroscopy*, 2003. 95(1–4): 93–97.

128. Wang MS, Wang JY, Peng LM, Engineering the cap structure of individual carbon nanotubes and corresponding electron field emission characteristics. *Applied Physics Letters*, 2006. 88(24): 243108.

129. Wang YH, Lin J, Huan CHA, Macroscopic field emission properties of aligned carbon nanotubes array and randomly oriented carbon nanotubes layer. *Thin Solid Films*, 2002. 405(1–2): 243–247.

130. Charlier JC, Terrones M. et al., Enhanced electron field emission in B-doped carbon nanotubes. *Nano Letters*, 2002. 2(11): 1191–1195.

131. Golberg D, Dorozhkin PS. et al., Structure, transport and field-emission properties of compound nanotubes: CN x vs. BNC x (x <0.1). *Applied Physics A: Materials Science and Processing*, 2003. 76(4): 499–507.

132. Menda J, Ulmen B. et al., Structural control of vertically aligned multiwalled carbon nanotubes by radio-frequency plasmas. *Applied Physics Letters*, 2005. 87(17): 173106.

133. Tsai TY, Tai NH. et al., Growth of vertically aligned carbon nanotubes on glass substrate at 450 degrees C through the thermal chemical vapor deposition method. *Diamond and Related Materials*, 2009. 18(2–3): 307–311.

134. Siegal MP, Overmyer DL, Provencio PP, Precise control of multiwall carbon nanotube diameters using thermal chemical vapor deposition. *Applied Physics Letters*, 2002. 80(12): 2171–2173.

135. Maklin J, Mustonen T. et al., Inkjet printed resistive and chemical–FET carbon nanotube gas sensors. *Physica Status Solidi B*, 2008. 245(10): 2335–2338.

136. Fan ZJ, Wei T. et al., Fabrication and characterization of multi-walled carbon nanotubes-based ink. *Journal of Materials Science*, 2005. 40(18): 5075–5077.

137. Shin HY, Chung US. et al., Effects of bonding materials in screen-printing paste on the field-emission properties of carbon nanotube cathodes. *Journal of Vacuum Science and Technology B*, 2005. 23(6): 2369–2372.

138. Hunt CE, Glembocki OJ. et al., Carbon nanotube growth for field-emission cathodes from graphite paste using Ar–ion bombardment. *Applied Physics Letters*, 2005. 86(16): 163112.

139. Zou R, Zou G. et al., Improved emission uniformity and stability of printed carbon nanotubes in electrolyte. *Applied Surface Science*, 2009. 255(20): 8672–8675.

140. Kyung SJ, Park JB. et al., The effect of Ar neutral beam treatment of screen-printed carbon nanotubes for enhanced field emission. *Journal of Applied Physics*, 2007. 101(8): 083305.

141. Kordas K, Mustonen T. et al., Inkjet printing of electrically conductive patterns of carbon nanotubes. *Small*, 2006. 2(8–9): 1021–1025.

142. Lyth SM, Silva SRP. et al., Field emission from multiwall carbon nanotubes on paper substrates. *Applied Physics Letters*, 2007. 90(17): 173124.

143. Spotnitz ME, Ryan D, Stone HA, Dip coating for the alignment of carbon nanotubes on curved surfaces. *Journal of Materials Chemistry*, 2004. 14(8): 1299–1302.

144. Jeong HJ, Choi HK. et al., Fabrication of efficient field emitters with thin multiwalled carbon nanotubes using spray method. *Carbon*, 2006. 44(13): 2689–2693.
145. Lee YD, Lee KS. et al., Field emission properties of carbon nanotube film using a spray method. *Applied Surface Science*, 2007. 254(2): 513–516.
146. Cho J, Konopka K. et al., Characterisation of carbon nanotube films deposited by electrophoretic deposition. *Carbon*, 2009. 47(1): 58–67.
147. Jung SM, Hahn J. et al., Clean carbon nanotube field emitters aligned horizontally. *Nano Letters*, 2006. 6(7): 1569–1573.
148. Kumar A, Pushparaj VL. et al., Contact transfer of aligned carbon nanotube arrays onto conducting substrates. *Applied Physics Letters*, 2006. 89(16): 163120.
149. Zhu Y, Lim X. et al., Versatile transfer of aligned carbon nanotubes with polydimethylsiloxane as the intermediate. *Nanotechnology*, 2008. 19(32): 325304.
150. Zhang Q, Huang JQ. et al., Carbon nanotube mass production: Principles and processes. *Chemsuschem*, 2011. 4(7): 864–889.
151. Huang JQ, Zhang Q. et al., A review of the large-scale production of carbon nanotubes: The practice of nanoscale process engineering. *Chinese Science Bulletin*, 2012. 57(2–3): 157–166.

9 Nanocrystalline Diamond

Alexander Vul', Marina Baidakova, and Artur Dideikin

CONTENTS

9.1 INTRODUCTION

In present-day nanotechnologies, there has been a gradual shift from traditional, the so-called "top-down," processes to "bottom-up" methods, that is, processes based on assembly of nanosized building blocks to develop materials with required properties. Among such nanosized blocks, carbon nanomaterials are broadly assuming recognized significance. The current chapter addresses the technology of fabrication, methods of study, the structure and properties of the group of nanocrystalline (NC) diamond materials, which encompass NC diamond films, nanodiamond (ND) powders, produced by mechanical milling of microcrystalline synthetic diamonds, and powders/suspensions of NDs obtained by dynamic synthesis.

Several factors have generated interest in this group of materials. First, these materials retain certain unique properties inherent in bulk diamond: chemical stability, mechanical strength, high thermal conductivity, and optical transparency within a broad spectral range derived from the large bandgap. Second, the technology of production of these materials has either already been developed on an industrial scale or has been tested reliably under laboratory conditions, allows scaling, and results in the commercial availability of the materials. Third, the nontoxic nature, biocompatibility, and relatively simple methods of surface functionalization permit application of these materials in medicine. And, finally, various optical properties, in particular, the possibility of development of photoluminescent centers make NDs particularly attractive for preparation of biomarkers.

9.2   TYPES OF NC DIAMONDS AND METHODS OF THEIR FABRICATION

The term nanodiamond in present-day literature applies to several objects, among them are the following:

- Diamond nanocrystals obtained by milling of large crystals synthesized from graphite at high static pressures and temperatures in the presence of catalyst metals, the so-called high-pressure–high-temperature synthesis (HPHT).
- Nanometer-sized diamond powders and corresponding suspensions prepared by dynamic (detonation and laser) synthesis.
- Crystalline grains of NC diamond films prepared by chemical vapor deposition (CVD).
- ND crystals found to exist in meteorites.[*]

It appears only natural to start the discussion of the above types of ND and of methods of their preparation by a consideration of the phase diagram of carbon (Figure 9.1).

We see readily that the regions within which ND of the above three types can be produced differ from one another; indeed, the conditions in which dynamic synthesis is possible can be found in the region of thermodynamic stability of diamond, the HPHT synthesis is supported in the region of thermodynamic stability of diamond, but at the boundary of the thermodynamic stability regions of diamond and graphite, and, finally, the region favorable for CVD synthesis, in that of thermodynamic stability of graphite. Obviously enough, the conditions and mechanisms of synthesis favorable for these three types of NDs differ substantially, and these will be discussed in more detail later.

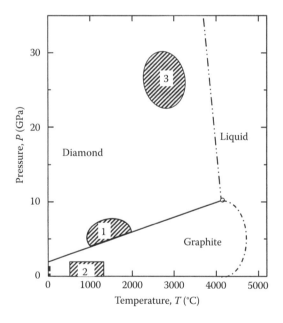

FIGURE 9.1   Phase diagram of carbon and regions of temperatures and pressures used in synthesis of various types of NDs. (1) Region of HPHT synthesis, (2) region of CVD synthesis, (3) region of dynamic synthesis, D—region of thermodynamic stability of diamond, G—region of thermodynamic stability of graphite, and L—region of thermodynamic stability of the liquid carbon phase.

---

[*] In this chapter, we are excluding from consideration nanodiamonds found in meteorites because their origin is still discussed primarily in the speculative context and their study would now hardly produce anything of scientific value. The reader interested specifically in this chapter can be referred to reviews.[1,2]

9.2.1 Synthesis of Diamond from Graphite in the Presence of Metal Catalysts

Publication of the first successful synthesis of diamond from graphite dates back to 1955, when the equipment designed at General Electric permitted reaching the pressures and temperatures necessary to drive the corresponding phase transition.[3] Depending on the metal catalyst employed, the required pressure and temperature may vary somewhat while remaining, as a general rule, in the ranges $P = 5.0–7.0$ GPa, $T = 1300–1800°C$.[4] Significantly, an increase in temperature and pressure brings about growth of the coefficient of graphite–diamond conversion and that of synthesis time, to grow the diamond crystals. In the current state-of-the-art technology, the HPHT synthesis of diamond from graphite provides reliable production of gem-quality crystals of sub-millimeter size.[5,6] Progress in HPHT diamond synthesis technology opened up a way to using macrosized crystals as starting material to produce NDs below 100 nm in size by mechanical milling. Despite the difficulties involved in the process of milling and subsequent fractionation to isolate nanosized particles, the problem of production of nanopowders from HPHT ND has been solved successfully in the recent years; the properties and application of such NDs will be considered in some details below.

9.2.2 Diamond Nanocrystals Obtained by Milling of HPHT Diamonds

In general, industrial HPHT synthesis of diamond produces crystals with size ranging from tens to hundreds of micrometers. Up to now, ordinary mechanical milling of as-grown HPHT diamond (20–50 μm) yields a very small and, thus, commercially expensive fraction of NDs.[7] However, a powder of commercial-type Ib-synthetic monocrystalline diamond fabricated for industrial polishing purposes can be used to produce monocrystalline material by crushing, purification, and precision grading to achieve a particle size distribution below 50 nm.[8]

What is more a modern milling process has recently opened an industrial scalable method that allows the efficient conversion of as-grown HPHT diamond microcrystals into spatially isolated nanoparticles with a mean size less than or equal to 10 nm.[9] The method consists of two stages of high-energy ball milling of commercially available diamond microcrystals. The first stage includes planetary milling with metallic beads with a nitrogen flow rate of 60 m^3/h and a high grinding pressure (8 bar) to produce particles having size below 2 μm. The second stage includes a planetary ball mill with hard alloy WC +6 wt% Co bowls. The method has been reported[9] to produce a pure concentrated aqueous colloidal dispersion of highly crystalline nanoparticles with a mean size less than or equal to 10 nm.

Taking into account demands on numerous applications of NC diamonds, in particular as bio-markers (see, e.g., Refs. 10,11) and success in preparation of ND powder from micrometer-sized HPHT- synthesized crystals, we can conclude on a current competition between different technologies for production of NC diamonds.

9.2.3 NC Diamonds Produced by Dynamic Synthesis

The idea underlying the dynamic synthesis of nanometer-sized diamonds essentially consists of providing the pressure and temperature needed to drive the graphite–diamond-phase transition by means of a shock (explosive) wave. Since both the pressure and temperature in a shock wave are characteristic of the region of thermodynamic stability of diamond, the time of synthesis is necessarily very short (as a rule, a few fractions of a microsecond), and, hence, the diamond crystallites produced usually remain nanometer sized.

The first announcement of the results achieved in dynamic synthesis was made in 1961 by DeCarli et al.,[12] who reported that diamond was detected in some carbon samples subjected to shock compression.

For the starting material in dynamic synthesis, by analogy with the HPHT approach, a mixture of graphite with a metal catalyst was chosen. The dynamic pressure (shock loading) can be produced by different methods, for instance:

- Subjecting the ampoule-containing graphite and the metal catalyst to an explosion wave set off close to this ampoule.[*][13]
- Impact on a carbon target containing the above-mentioned mixture of graphite with metal of a body flying with a velocity of several kilometer per second.[14]
- Interaction of a powerful laser pulse with a carbon target, the so-called laser synthesis.[15]

The next step in the development of the dynamic method is detonation synthesis, in which the starting material for the synthesis is not graphite but rather the carbon contained in the explosive itself. This method of ND production, called detonation synthesis, was discovered more than 40 years ago, but the appearance of generally available publications goes back to 1988, when the journals *Doklady Akademii Nauk*[16] and *Nature*[17] published articles currently referred to by most authors as the pioneering studies. The details of the breath-taking storytelling about the repeated discovery of the detonation ND synthesis by various research groups can be found in the historical review prepared by V. Danilenko, one of the pioneers in the method of detonation synthesis.[18]

The most frequently employed starting explosives in industrial-scale production of diamond by detonation synthesis are, as a rule, trinitrotoluene (TNT) and hexogen (also called RDX = Research Department Explosive), with detonation performed in a water medium to reach a higher productivity. The most complicated stage in the industrial process is chemical isolation of NC diamond from the detonation carbon produced in an explosion, which is actually a mixture of micro- and nanoparticles of graphite, various forms of sp^2-hybridized carbon and impurities originally contained in the explosive itself, and construction materials of the synthesis vessel. For details of the technology used in detonation diamond synthesis, the reader can be referred to Ref. 19 and Section 9.2.3.2 for the structure of the detonation diamond particles, their properties, and applications.

9.2.3.1   Laser Synthesis of NC Diamonds

Considered from a purely physical viewpoint, pulsed laser ablation (PLA) of a solid target is similar in many respects to dynamic synthesis of ND from graphite and, thus, has something in common with the detonation method. The PLA method, which is directly associated with the appearance of powerful rubidium lasers in the early 1960s, is currently widely employed in synthesis and modification of a broad range of nanomaterials.

The PLA-based production of diamond particles was first realized in 1992[20] when a graphite target was immersed in benzene and irradiated by nanosecond-range laser pulses. For this, the laser beam is focused on the graphite surface. Formation of diamond particles in laser ablation of graphite was demonstrated in a hydrogen atmosphere;[21] however, ablation of graphite immersed in various liquids was the most efficient method of all, and is currently the most widespread.[15,22–24] Single diamond particles of 3–6 nm in size were obtained using a low-power laser with 1.2 µs pulses.[15,22] Electron microscopy studies of such ND particles revealed a very high degree of twinning[15] and based on thermodynamic calculations, the authors concluded that twinning is a process energetically favorable for an increase of stability of ND particles within a certain particle size range. Another interesting feature is the observation of strong photoluminescence in the case where ND particles are passivated by polyethylene glycol. A diamond phase can be probably obtained by PLA without a graphite target as well.[23]

---

[*] Note that pilot production of diamond micrometer-sized powders by detonation synthesis was started by Dupont in the late 1970s, and in the Soviet Union, at the Institute of Problems in Chemical Physics, in 1982–1983, with graphite used as starting material in both cases.

An analysis of the above publications indicates a wide range of experimental conditions (laser parameters, target, and type of liquid) tested to determine the combination favorable for formation of diamond structures by PLA. It is believed that this process evolves formation of the so-called laser plume or a cloud of reaction products consisting of the evaporated substrate material and, partially, the surrounding liquid.[24] These evaporated substances form bubbles inside the liquid. As the amount of the evaporated material increases, the bubbles expand and, as the pressure and temperature reach a certain critical combination, they collapse. At the collapse of the bubbles, the temperature and pressure may reach in the range of thermodynamical stability of diamond.

9.2.3.2 Detonation NDs

Like all dynamic methods, the detonation-based synthesis of ND is basically a version of classical synthesis adapted to the high-pressure and temperature conditions providing the required thermodynamic stability of diamond. It is the short time during which these conditions have to be maintained in the shock wave in dynamic synthesis (not over a few microseconds) that is the main factor determining the size of the diamond nanocrystals fabricated by such methods.

One major feature of the detonation synthesis of NDs is that it does not require graphite or any other additional source of carbon to support it. It is the explosives themselves initiating the shock wave that acts as the source of the carbon to form diamond nanocrystals. For this purpose, their composition is chosen such as to ensure formation of free carbon in the chemical reactions running in the shock wave. This condition defines the so-called negative oxygen balance of the explosive composition.[25] The explosive composition customarily used in detonation synthesis of diamond is the TNT:RDX mixture taken in the 40:60–70:30 ratio.[26]

Obviously, the most essential factor accounting for the structural difference setting off detonation ND from all the other types of ND fabricated by dynamic techniques is the chemical composition of the medium in which diamond undergoes crystallization, as well as the dynamics of its cooling, and the latter depends on the conditions in which the explosion is run.

Selection of the optimum medium needed to sustain detonation synthesis of ND is governed by the need to prevent oxidation of the detonation products that can occur in contact with atmospheric oxygen, as well as to provide the maximum possible rate of cooling of the ND formed to preclude transition of the diamond to graphite. In industrial-scale production of detonation ND, the explosive charges are placed in carbon dioxide, water, or solid carbon dioxide ("dry ice").[19]

The charges used in industrial fabrication of detonation nanodiamond (DND) are exploded at the center of a strong, hermetically sealed vessel provided by a special device to discharge excess pressure, as well as by means to unload the products of the explosion containing ND. The dimensions of the vessel are determined by the maximum allowable mass of the explosive charge. Some vessels designed for large-scale applications can operate with explosive charge masses of up to 100 kg.

Depending on the used medium, the side products (detonation carbon) produced in detonation synthesis contain 20–70 wt% diamond. The remainder is a mixture of various structural forms of sp^2-carbon, the state of hybridized electronic orbitals characteristic of graphite. The yield of the diamond phase is the highest with carbon dioxide in the solid state, and the lowest, when it is in gaseous form. The medium most frequently used in industrial synthesis is water.

To isolate DND, the detonation carbon is treated with strong oxidizers. This method is based on the specific feature of carbon that becomes oxidized primarily in the sp^2-hybridized state. As for the carbon forming the diamond structure, it is in the sp^3-state and oxidizes at a substantially slower rate. The conditions of the treatment are chosen so as to reach the maximum possible difference between the oxidation rates of graphite and diamond. In industrial-scale production of ND, one predominantly employs liquid-phase oxidation, with mixtures of sulfuric acid and chromium anhydride or nitric acid, perchloric acid, as well as water solutions of nitric acid at elevated pressures and temperatures used as reagents.

Besides oxidation in the liquid phase, gas-phase treatment is also employed, in which the oxidizer is the oxygen of the air in the presence of catalysts, such as vanadium pentoxide, iron or manganese compounds, or boron anhydride. Another version of gas-phase oxidation is the treatment with ozone, which produces DND with the lowest content of sp^2-carbon. This method is, however, the most complicated and, besides, requires the highest energy consumption. While the techniques based on the use of hydrogen peroxide in the presence of nitric acid exhibit a high efficiency, they find application only in laboratory conditions because of the instabilities characteristic of the reagents involved.

Commercial DND isolated from detonation carbon is supplied in the form of suspensions and dry powders. This material usually contains 0.1–0.7 wt% graphite-like carbon, as well as, depending on the actual conditions of the synthesis procedure chosen, up to 1 wt% iron salts, silicon, and titanium oxides originating from the components of the equipment in which the synthesis and subsequent isolation of DND from the detonation carbon are performed.

Although for many practical purposes, the purity reached in industrial production of DND is quite satisfactory; new areas of application of this material, particularly those emerging in biology, medicine, or catalysis, require a substantially lower content of impurities. To remove iron salts from industrial DND, one can use hydrochloric acid as a medium combined with ultrasonic treatment. Particles of the inert oxides of titanium and silicon are removed by using hydrofluoric acid. Quite satisfactory results were obtained when centrifuging was involved. The residual content of ferromagnetic impurities in DND can be efficiently determined by electron paramagnetic resonance (EPR), a method permitting to analyze the residual iron down to levels of 0.1 ppm.[27]

A model for the structure of the detonation ND particles was first proposed in Ref. 28. The particle is formed of sp^3-hybridized carbon atoms making up a crystalline diamond core about 4.5 nm in size, which is partially coated by a shell of sp^2-hybridized carbon; the structure of the shell is a stack of several graphite layers curved in two directions (Figure 9.2). The shell curvature results from an alternation of the pentagonal and hexagonal elements making up the structure of the graphite layers in a pattern similar to that observed in larger fullerenes that are limited by the dimensions of the continuous fragments of these layers. The curved graphite layers are coated by amorphous carbon with inclusions of particles of crystalline graphite about 1 nm in size.

The shell is formed in the process of reverse diamond-to-graphite transition in the concluding stages of detonation synthesis, after the shock wave has passed and the pressure has dropped below the limit of thermodynamic stability of diamond. The thickness of the shell is largely determined by the conditions of the DND synthesis and in the course of DND isolation from detonation carbon, the thickness of the shell decreases. In the strongest regimes of sp^2-oxidation, the shell can be removed completely, except for separate single-layer sp^2-carbon islands, which result, as shown by calculations,[29] from natural reconstruction of the free surface of diamond nanoparticles.

As revealed by high-resolution transmission electron microscopy (HRTEM) studies, the diamond core has a distinct faceting of cuboctahedron with smoothed corners (Figure 9.3). The average

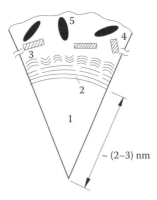

FIGURE 9.2 Model of structure of detonation carbon particle. (1) Diamond crystalline core, (2) onion-like shell, (3) set of graphene sheets, (4) nanographite grains, and (5) impurity particles.

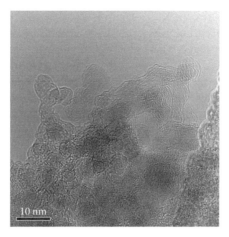

FIGURE 9.3 HRTEM image of highly purified DND powder.

size of a particle is 4.5 ± 0.5 nm that corresponds to that of a particle combining $N \approx 8000$ carbon atoms. This number may vary within the range of 5500–10,000 depending on the actual state of the faceting. The fraction of the carbon atoms located in this case on the surface of a DND particle amounts to 15–25%.

The information concerning the structure of the diamond core of a DND particle was refined in studies of diffraction of synchrotron radiation and by modeling calculations. It was established that the core contains at its center a single-crystal region 2.0–2.2 nm in size that is separated from the surface by a diamond layer with a distorted crystal structure and in which the defect concentration increases from the center to the surface of the particle (Figure 9.4).[30]

The presence of structural defects in a DND particle initiates stresses in its crystal lattice, a certain increase of the average length of the C–C valence bonds, and a slight decrease in the packing density of carbon atoms compared to that in bulk diamond.

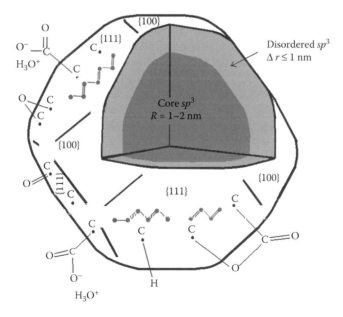

FIGURE 9.4 Model of structure of DND particle.

Dangling bonds related to structural defects in a DND particle have a nonzero spin and are revealed with the use of the EPR and NMR (nuclear magnetic resonance) techniques by measuring the spin-lattice relaxation times. Their number is estimated as 1–40 per one DND particle. Experiments performed with copper ions deposited on the surface of DND particles suggest that these defects are located inside the particle at a distance of 0.8–1.5 nm from its surface.[31]

As follows from elemental analysis, the main impurities in the structure of a DND particle are nitrogen (about 2 wt%) and hydrogen (about 1 wt%), which are assigned to the high concentration of these elements in the starting reagents employed in the detonation synthesis.

DND particles have a specific structural feature, namely, the EPR technique cannot detect a nitrogen impurity if it occupies in it a substitutional site. Only the indirect ^{15}N-NMR method is capable of corroborating its presence in the amount fitting the data obtained by elemental analysis.[32] One assumes conventionally that because of nitrogen atoms in the structure of a DND particle being closely spaced, their mutual influence results in the EPR line becoming so broad as to make its unambiguous identification impossible.[31]

A nitrogen impurity is also practically undetectable in the spectra of luminescence, which, when applied to bulk diamond, reveals the presence of charged NV (nitrogen vacancy) centers. Nevertheless, sintering DND particles at a high pressure and temperature transforms the nitrogen atoms they contain to charged NV–luminescence centers in concentrations of up to 10^{21}, that is, to levels inaccessible this far for any of the currently known methods of producing luminescence centers in diamond.[33]

The Raman spectrum of DND exhibits two maxima, a narrow one at 1326 cm^{-1} and a broad one at 1630–1640 cm^{-1}, which are detected against the background of a broad continuous luminescence band within a 500–900 nm interval, with a weakly pronounced peak close to 600 nm.[34] The narrow maximum derives from phonon vibrations of the diamond lattice. Its shift relative to the lattice vibrations of bulk diamond (1332.5 cm^{-1}) should be assigned to phonons being localized within a volume of the DND particle corresponding to a size of 4.5 nm. The line at 1630–1640 cm^{-1} is believed to be related both to sections formed of sp^2-hybridized carbon atoms on the surface of a DND particle and OH groups. The broad luminescence band of DND particles originates from structural defects of their surface. It is the surface characteristics that also account for the intense absorption of electromagnetic radiation by DND powders and suspensions over the visible spectral range. The dark color of the powders and water suspensions of 4.5 nm DND particles can be traced to the presence of defects in the form of one-dimensional chains of carbon atoms on their surface.[35]

The essential feature of NC materials, including DND, is that a sizable part of their atoms (up to 25%) is located at the surface that accounts for a large percentage of dangling valence bonds. It is saturation of these bonds that is the prerequisite for the chemical stability of the material.

The dangling valence bonds of carbon atoms on the surface of DND particles originate from the detonation synthesis and are saturated in interaction with the sp^2-phase of carbon. In the course of isolation of DND from detonation carbon and oxidation of the sp^2-phase, the structures combined of carbon atoms are replaced with functional groups formed in interaction with the oxidizing medium surrounding the particles. Among these functional groups are carboxyl (COOH$^-$) and hydroxyl (OH$^-$) groups along with protons (H+). The relative content of these groups depends on the composition of the oxidizing reagents employed and the conditions chosen for the oxidation. The presence of these groups on the surface of DND particles is revealed reliably by NMR[36] and can be derived from infrared (IR)[37] and Raman[38] spectra. The total amount of functional groups bound chemically to a DND particle is determined by the area of its surface free of sp^2-hybridized carbon.

Production of DND with a defined surface functionalization is a major technological challenge because it is the composition of surface functional groups that governs the properties required for the desired application. All the technologies developed so far for DND surface functionalization represent the various methods of substitution of the desired functional group for the primary set of DND functional groups obtained in the course of chemical purification.

DND functionalization proceeds, as a rule, in several steps. In the first stage, DND is treated in the specific conditions that would leave only one type of the primary functional group on its surface. For instance, to obtain a material with the surface containing the carboxyl group, the powder of commercial DND is heated in air at 400°C or treated with ozone at room temperature,[37] whereas if a material with hydrogen-saturated surface is required, the DND is treated with hydrogen plasma in the conditions approaching those accepted in synthesis of CVD diamond films.[39] To saturate the surface with hydroxyl groups, DND is treated with a mixture of sulfuric acid (H_2SO_4) and hydrogen peroxide (H_2O_2) in the presence of ferric ions (Fe^{+2}).[40] Annealing in vacuum or in argon atmosphere at temperatures above 800°C reduces the overall content of all three primary functional groups due to the broadening of the area of the surface of DND particles covered by sp^2-phase. After saturation with carboxyl groups, it should be replaced with the intermediate functional group that in its turn could be easily replaced for the final group that is required to obtain the desired properties of DND. An important intermediate functional group is represented by halogens, most often chlorine that forms C–Cl bonds. The chlorination is performed from gaseous phase at room temperature under ultraviolet (UV) irradiation. Also, chlorination is possible to execute from liquid phase using thionyl chloride ($SOCl_2$) at temperature near 70°C. Chlorine can be replaced easily in a subsequent step by aliphatic of aromatic amines bonded by an amine group (NH) that can be used further as an interlink to bioorganic molecules or polymers. Hydroxyl group on the surface of DND particles can be replaced by halogens or by amines. Using liquid-phase reactions, they could be easily replaced by the carboxyl group as well. This subsequently allows to link aliphatic chains to DND particles.

Dried, industrially produced DND represents the powder with particles that are linked into aggregates with a fractal structure (Figure 9.5), that is, self-similarity of the shape at variations of size at the range between 100 μm and 100 nm. The packing density of DND powder is usually between 1.0 and 1.5 g/cm^3, whereas pycnometry gives a value of approximately 3.2 g/cm^3 which is less than that for bulk diamond (3.54 g/cm^3). The specific surface of DND powder, measured by absorption of hydrogen, depends on the technique of DND purification and usually comprises about 300 m^2/g. The typical elemental composition of industrial DND includes 90 wt% of carbon, 5–10 wt% of oxygen, 1–3 wt% nitrogen, and about 1 wt% of hydrogen. Depending on the details of the DND synthesis, DND powders may contain up to 5 wt% of iron and up to 5 wt% of inert impurities (i.e., SiO_2 or TiO_2).

Commercial DND powders absorb water very easily. Aggregates of DND particles are able to absorb up to 60 wt% of water without losing their shape. However, only dry DND powder qualifies as an insulator. The conductance of DND powder can be observed after thermal treatment that produces a noticeable amount of sp^2-carbon. It is assumed that the aggregates in DND powder of size in the range 1–100 μm are linked together by electrostatic interaction that is commonly found for the majority of ultra-dispersed materials. These aggregates are disbanded when DND is immersed in water where the maximal size of DND aggregates does not exceed 0.8–1.0 μm.

FIGURE 9.5   SEM image of dried detonation ND powder.

Bonds inside DND aggregates with size in the range of 100–1000 nm are mostly because of the presence of metal impurities, primarily iron. Ferrous ions form in aqueous media bridge-like bonds between adjacent DND particles by interaction with dissociated negatively charged carboxyl groups that result in forming the aggregates of the mentioned size. Additional purification that completely removes the iron impurity allows reducing the size of DND aggregates in aqueous suspension up to 30–120 nm. DND aggregates of that size possess the solid structure that withstands destroying by ultrasonic treatment that is being effective for any other nanomaterial. It is supposed that these aggregates represent the polycrystalline structures of DND particles that are bonded by covalent C–C bonds formed during the late stages of detonation synthesis.

One of the techniques for disbanding strong DND aggregates and obtaining a suspension of free 4.5 nm particles is steering milling by hard spherical bodies of zirconia of size of about 30 μm in aqueous media.[41] The drawback of the method is the contamination of DND by the residuals of the milling material and partial graphitization of the particles due to local heating by the collisions with milling particles. Until recently, aggregation (and the presence of impurities) has significantly restricted the usage of DND in the areas that are sensitive to the grade of purity of the applied materials—the fine chemical technology, biology, and medicine. Recently developed methods for preparation of aqueous suspensions of free 4.5 nm DND particles require no mechanical milling and eliminate the contamination and unwanted modification of the DND surface. The basis of these techniques is the thermal annealing of dry DND powder after passing the preliminary purification in air[42] or in hydrogen atmosphere.[43] After annealing in air at a temperature of 420°C or in atmosphere of hydrogen at 800°C, the material is dispersed in deionized water under intensive ultrasonic treatment. The obtained aqueous suspension is applied to centrifugation to separate the fraction that contains free 4.5 nm DND particles from the pellet of the residual aggregates. Up to 20 wt% of the initial dry DND powder transforms into free condition. The yield can be significantly enhanced by repeatedly using the dried pellet.

Free DND particles in aqueous suspensions obtained by the mentioned techniques possess high absolute values of zeta potential (35–40 mV) that prevent their coagulation due to the mutual electrostatic repulsion in wide range of pH values of aqueous media. Particles of DND after annealing in hydrogen atmosphere possess positive zeta potential, whereas material that was annealed in air possesses a negative value of the zeta potential of free particles in aqueous suspension.

The challenge in obtaining high and stable value of zeta potential of DND particles in other liquid media, particularly in organic solvents, is still waiting its solution. That is why all suspensions of DND particles that are offered today by a number of vendors contain aggregates of size of about 10–60 nm instead of smaller, freely suspended diamond crystals.

The details underlying the mechanism of DND deaggregation today still remain largely unresolved. Functionalization of the surface of DND particles by carboxyl groups or hydrogen that takes place during annealing in corresponding atmosphere determines the value and sign of the zeta potential acquired by DND particles in aqueous media and turns up the condition of stable free status of 4.5 nm particles. Probably, it is the particular reason why one cannot obtain such small DND particles only by annealing in vacuum or in an argon atmosphere. However, the proper functionalization itself is insufficient for obtaining freely dispersed DND particles.[44] Apparently, the thermal annealing weakens the bonding inside hard aggregates of DND particles. Annealing does not eliminate the necessity of the intensive ultrasonic exposure for destroying those aggregates. Further studies of the details of structure of DND aggregates will help to establish a comprehensive model that could explain the process of emplacement of DND particles into free condition.

Free 4.5 nm DND particles demonstrate very interesting properties. While drying the aqueous suspension on smooth surfaces, the particles reveal the tendency to self-organize and form fanciful thread-like and fishnet-like structures. According to the results of self-consistent calculations based on density functional theory (DFT), the reason for these phenomena is in presence of the electric fields caused by the nonuniformity of distribution of the electronic density across the opposite facets of free particles of DND due to their small size.[45]

One more interesting property of DND particles is their ability to transform into onion-like carbon that are particles formed by multilayer closed shells of carbon atoms in sp^2-hybridization status also known to be described as multishell fullerenes.[46] DND particles transform into onions during annealing in an inert atmosphere or in vacuum at temperatures higher than approximately 1200°C. The mechanism underlying the transforming represents the sequential reconstruction of the groups of three atomic layers of sp^3-bonded carbon on the {111} facets of DND particles into couples of layers of sp^2-bonded carbon. Transformation of DND particles into carbon onions also occurs under electron beam irradiation with energy about 1 MeV. In this case, the reverse transformation of onions into diamond is also possible.[47] Onion-like structures are of particular interest to polymer composites, energy-storage devices, and are the suitable objects to study the magnetism in nanocarbons or field-effect emission.

9.2.4 CVD Synthesis of NC Diamond Films

Diamond can also be synthesized in its metastable regime. CVD growth of diamond films, at temperatures and pressures lower than that at HPHT synthesis of diamond from graphite, was suggested in 1960s.[48,49] Being in the region where diamond is metastable compared to graphite, CVD synthesis is driven by kinetics and not thermodynamics. The CVD diamond synthesis is usually performed using a small fraction of carbon (typically <5 wt%) in an excess of hydrogen. Decomposition of hydrocarbons (generally methane) is used as carbon precursors, besides aliphatic and aromatic hydrocarbons, alcohols, ketones, and solid polymers. The CVD diamond film growth can be broken down into three general stages: formation of carbon precursor and atomic hydrogen, nucleation process, and, finally, diamond growth. Diamond grown by CVD can fall into two categories: single and polycrystalline crystal. Single-crystal diamond is formed by homoepitaxial growth (see, e.g., Ref. 50), polycrystalline diamond films are typically formed in the case growth occurs on nondiamond substrate.

Because of the high surface energy of diamond, its nucleation on virgin substrates, in particular if the substance of the substrate does not have chemical affinity with carbon, is characterized by low nucleation densities (ca. 10^5 cm^{-2}) and long incubation times.[51]

As it was mentioned in Ref. 52, it is difficult to define particular forms of NC diamond films because of variation between laboratories and types of CVD reactors. However, the suppression or enhancement of renucleation processes can be chosen as that fundamental criterion. In conventional diamond growth, one uses a highly dilute concentration of hydrocarbon precursors, for example, methane in hydrogen. The so-called hydrogen-rich gas phase chemistry is known to reduce renucleation by the significantly higher etch rate of sp^2 phase (graphite) over sp^2 phase (diamond) in such a plasma. Thus, the suppression of renucleation results in an evolution of grain size from the small crystals at the seeded substrate to the film surface. The opposite type of ND films are those grown with a significant renucleation rate. It can be achieved by reducing the hydrogen concentration in the plasma to allow some sp^2 bonding to create new unepitaxial nucleation sites on the facets of growing crystals.

To enhance the renucleation process and formation of NC diamond films,[53] Gruen et al. recently suggested a new plasma-deposition process, which utilizes a high content of noble gas in plasma. Obviously, the renucleation process affects grain sizes of diamond films. Currently, the films grown with a low renucleation rate having the grain size 10–100 nm are termed as NC diamond films and those grown with a high renucleation rate are termed as ultrananocrystalline diamond (UNCD) films.*

Several methods have been used at the CVD growth for decomposition of hydrocarbons, activation plasma, and seeding of diamond precursors on a substrate (see, e.g., Refs. 55–58). Among the methods of seeding diamond precursors on a substrate at heteroepitaxial growth of CVD diamond

---

* The term "ultrananocrystalline" was suggested by Dieter M. Gruen for diamond films and particles with characteristic size of grains that are less than 10 nm.[54]

films, the seeding of detonation ND particles is the most effective.[59–60] A high density of nucleation sites (up to 10^{12} cm²) has been obtained on the substrate at the seeding by detonation NDs.[61] It can be concluded that there is no fundamental limitation for growing NC diamond film on nondiamond substrate and some applications of the films will be discussed below in Section 9.4.

9.3 METHODS FOR CHARACTERIZATION OF NC DIAMONDS

9.3.1 GENERAL REMARKS

It is appropriate to start consideration of the methods in use currently for characterization of ND by noting that the term diamond assumes the cubic crystalline form of sp^3-hybridized carbon with space group O_h^7. In this chapter, we are not going to consider forms of sp^3-hybridized carbon described by other symmetries, for example, hexagonal or rhombohedral, which are also referred to as "diamond." Falling in this category are, for instance, lonsdaleite called hexagonal diamond and superhard Z-carbon, representing a polymorphic phase of sp^3-hybridized carbon and exhibiting the elastic modulus and hardness comparable to those of the cubic diamond. The only exclusion we are making in this respect is n-diamond.[62]

Selecting a proper method of characterization of ND samples is governed by the need to obtain information on the average size of ND particles and their shape, as well as on the degree of aggregation and size distribution. Information on the sp^2-to-sp^3-ratio of hybridized atoms, the type and concentration of impurities in the bulk, and on the surface are also essential. There exist several methods employed in the characterization of optical properties of ND, an aspect of particular importance for CVD diamond films. Basic information related to the methods of ND characterization is listed in Table 9.1. It is divided into the following four main groups: diffraction-based methods, vibrational spectroscopy, resonance-based techniques, and imaging methods.

All the limitations imposed by the small ND size have great bearing on the applicability of a certain method or characterization technique. For instance, standard methods of data treatment employed in powder diffractometry are not suitable for investigation of ND materials, whereas the phonon confinement model (PCM), even complemented by taking particle size distribution and lattice defects into account, does not provide a comprehensive description of the results amassed in a Raman experiment. This means that data obtained by any one of the above methods on one sample and not corroborated by other techniques or models may not have a physical meaning at all. However, bearing in mind the above constraints on the applicability of these methods, they can be employed to characterize a series of samples to reveal trends in the variation of a parameter of interest initiated by an external action (e.g., temperature, pressure).

9.3.2 DIFFRACTION-BASED METHODS

The diffraction-based approach combines direct methods capable of evidencing diamond ordering of atoms in a crystal lattice and offering a possibility of estimating the degree of this ordering. Diffraction essentially involves a study of the angular distribution of the scattering intensity by the material of interest of radiation, for example, x-ray (including synchrotron) radiation, electron or neutron flux, and, generally speaking, Mössbauer γ-radiation. In all cases, the primary, most commonly, monochromatic beam is directed on the object under study and the resulting scattering pattern is analyzed. With radiation wavelength being usually not larger than 0.2 nm, which is comparable with atomic separations in matter (0.1–0.4 nm), the scattering of an incident wave can be identified with diffraction from atoms. The theory underlying the relation connecting the pattern of elastic scattering with spatial distribution of the scattering centers is the same for all radiations. However, because the physical nature of the interaction of matter with different kinds of radiations is different, both the pattern itself and the specific features of diffraction patterns are governed by the different characteristics of the atoms involved.

TABLE 9.1

Summary of Methods for Characterization of NC Diamond

Methods for Characterizations of NC Diamonds	Type of NC Diamonds	References
1 *Diffraction-based methods*		
X-ray and electron diffraction methods		
Powdered XRD	Nanometer-sized diamond powder	30, 64–66, 116, 117
GID	NC diamond films	71–74, 118
Electron diffractions	Nanometer-sized diamond powder	119
Small-angle scattering and reflectivity		
Small-angle x-ray and neutron scattering	Detonation NDs	69, 42, 76–77
X-ray and neutron reflectivity	NC diamond films	71, 78–80
2 *Vibrational spectroscopy*		
Raman spectroscopy	NC diamond films, detonation NDs	38, 80, 87, 88, 118, 120, 121
IR spectroscopy	Nanometer-sized diamond powder	37, 98, 101–106, 107, 109–110, 112, 122
HR-EELS	NC diamond films, nanometer-sized diamond powder	81–86, 123
3 *Electron spectroscopy*		
X-ray photoelectron spectroscopy (XPS)	Nanometer-sized diamond powder	121, 122, 124, 125, 126
Auger spectroscopy	Nanometer-sized diamond powder	124–125
Electron energy loss spectroscopy (EELS)	NC diamond films, detonation NDs	124, 127, 128, 129
4 *Spin resonance techniques*		
NMR	Nanometer-sized diamond powder	36, 31, 121, 130, 131, 132, 133
Electronic paramagnetic resonance (EPR)	Nanometer-sized diamond powder	31, 33, 134, 135, 136
5 *Imaging methods*		
Transmission electron microscopy (TEM)	Nanometer-sized diamond powder	47, 117, 119, 128, 129, 137
Scanning electron microscopy (SEM)	NC diamond films, nanometer-sized diamond powder	127, 138
Atomic force microscopy (AFM)	Nanometer-sized diamond powder	117, 127, 139
6 *Optical methods*		
Optical absorption	NC diamond films, nanometer-sized diamond powder	140, 141, 142
Photoluminescence	Nanometer-sized diamond powder	120, 143, 144
Dynamic light scattering (DLS)	Nanometer-sized diamond powder	42, 43, 140, 142, 141
7 *Chemical methods*		
Zeta-potential technique	Nanometer-sized diamond powder	42, 126
Titration	Nanometer-sized diamond powder	146, 147

9.3.2.1 X-Ray and Electron Diffraction Methods

Figure 9.6 visualizes typical x-ray diffractograms obtained from ND powders and CVD ND films measured in $(\theta, 2\theta)$ geometry. The graph (a) is plotted versus wave vectors q (along the abscissa axis), which are related to interplanar separations d_n through

$$q = (4\pi/\lambda) \sin(\theta) = 2\pi/d_n \tag{9.1}$$

where λ is the wavelength of the incident radiation, θ is the Bragg angle, and n is the reflection order coupled to the Miller indices through the relation $n = \sqrt{(h^2 + k^2 + l^2)}$ in the case of a cubic crystal. It should be noted that all ND samples produce diffraction maxima at the same wave vectors corresponding to interplanar separations in a diamond-type face-centered cubic lattice (FCC) with the lattice parameter $a_{vol} = 0.35667$ nm.[63]

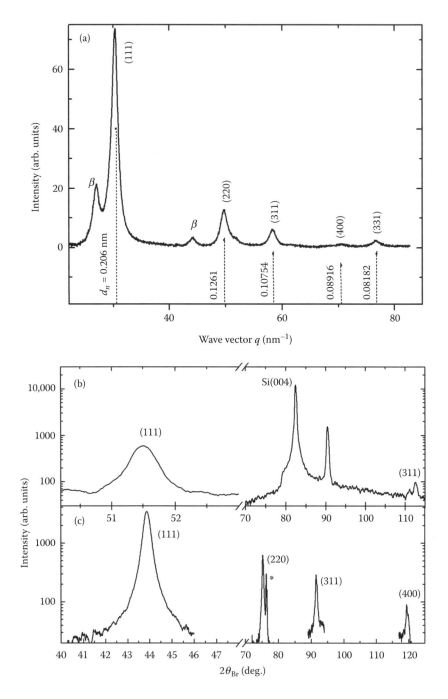

FIGURE 9.6 X-ray diffractograms of DND (a) and CVD diamond film on silicon substrate (b) and diamond particles produced from HPHT diamond by milling method (c). Positions for interplanar distances d_n are marked. The corresponding intensities are given by the lengths of the dashed lines. Positions of diffraction maxima are marked by corresponding Miller indexes. The asterisk identifies an impurity diffraction maximum. The (b, c) curves are received with CoK_α and CuK_α radiation.

A diffractogram of polycrystalline ND (i.e., excluding textured CVD films) should exhibit all the maxima specified in Ref. 63 within the range of attainable diffraction angles, and the more reflections can be found, the more complete is the information on ND structure that can be extracted from an analysis of the diffraction data. Therefore, one uses for measurements preferably short-wavelength synchrotron radiation[30,64] or hot neutrons.[65] An electron diffraction pattern also contains, as a rule, a large number of diffraction rings associated with different interplanar distances in the sample under study because it is obtained by diffraction of high-energy electrons (100 keV corresponds to a wavelength of 0.037 Å).

The appearance of diffraction peaks in addition to those typical of diamond is usually direct evidence for the carbon being crystallized to a nondiamond form[62] or for the presence of impurities in the sample.[66]

A particularly prominent example may serve the electron diffractograms of FCC carbon samples or of the so-called n-diamond.[62] Experimental studies reveal that the lattice parameter of n-diamond is close to that of cubic diamond, and that an electron diffraction pattern contains all the main reflections, even some of those forbidden for cubic diamond by extinction laws (e.g., (200), (222), and (420)). A recent analysis demonstrated that the model of hydrogen-doped diamond C_xH_{1-x} ($x = 0.04-0.19$), in which carbon and hydrogen atoms are distributed in a random manner over the lattice of cubic diamond, agrees well with the experimental diffractograms.[67] An analysis of diffraction data are capable of offering information on the ND structure, such as the lattice parameter, the average crystallite size (or, more precisely, the average coherence length that is also known as the domain size), type of their ordering (polycrystalline or textured), the presence of microstresses,[68,69] as well as detecting impurities with a sensitivity of usually higher than 1 wt% (depending on the scattering strength of a certain atom).[66]

Most of the methods employed in diffraction data treatment were developed for conventional crystalline materials, assuming infinite long-range order in the sample under study. For objects that are only a few nanometers in size, however, we can no longer assume long-range ordering because of the limited size of the nanocrystal. It was shown[70] using standard methods of data treatment assuming nanocrystallites to have uniform structure may produce wrong results and conclusions. This relates both to applying Bragg's equation to calculation of the lattice parameter and to the equation of Selyakov and Scherrer for estimation of crystallite size. The authors point out that a correctly developed analysis should be based on data obtained with the use of either "hot neutrons" or short-wave x-ray radiation available with synchrotron or neutron sources. Even then, reliable results will only be obtained if the interpretation is based on an accurate structural model that is also confirmed by other characterization methods. This statement can be exemplified by studies[65] in which the "apparent lattice parameter" approach was used to analyze neutron diffraction data obtained over a broad range of wave vectors. By invoking the "core-shell" model of DND particle structure, which demonstrated its validity in more than one respect and is based on assuming a perfect diamond core to be coated by a nondiamond shell, the existence of an additional layer of compressed diamond lattice between the core and the outer shell was demonstrated.

The structure of CVD diamond films obviously depends on both the specific features of the technology employed and the substrate type chosen and the composition and microstructure of the transition layers forming at the substrate/diamond film boundary. Ultrathin and fine-grained (about 5 nm) CVD films produced on a silicon substrate usually feature polycrystalline structure.[71] As the thickness of a film increases and the crystallites forming in them grow in size, they may acquire preferential orientation (texture).[72] Texture is characterized by a change in the intensity ratio among the diffraction maxima and can be directly visualized by two-dimensional (2D)-diffraction methods. To cite an example, by comparing the intensity ratio of the (220) and (111) diffraction maxima, it was shown[73] that the texture of the microwave plasma CVD films deposited on an (100) Si substrate behaves similarly with methane concentration and deposition temperature. At low values of these parameters, the texture is oriented along the <hhh> direction;

as these parameters grow, the crystallite orientation becomes random (0.7 vol% methane at 700°C and 0.5 vol% methane at 790°C), to convert to the <hh0> texture at still higher methane concentrations and deposition temperatures. An exactly opposite trend, that is, gradual enhancement of the <hhh> growth direction with increasing temperature, was observed by other researchers in CVD films deposited on titanium.[72] The explanation of this effect was found in studies[72,74] of how the deposition temperature influences the composition and microstructure of the transition layers forming at the interface between a titanium substrate and a thin diamond-like film. Such a study requires an analysis of the shape of both standard scans obtained in θ–2θ geometry and scans measured under grazing incidence.

The most comprehensive description of the x-ray grazing incidence diffraction (GID) method can be found in Ref. 75. In studies of CVD films, one employs a GID approach based on a slightly modified standard Bragg–Bretano (or θ–2θ) geometry. In the GID method, x-rays pass through a suitable slit system and subsequently fall on the sample at a fixed glancing angle (α), whereas the detector on the 2θ axis scans the x-ray diffraction (XRD) pattern. Application of GID geometry offers a possibility of noticeably increasing the intensity of the radiation diffracted from NC films. Earlier studies[72,74] reported application of the GID technique to studies of structural characteristics of the transition layers and of the carbide, hybrid, and other phases generated at the titanium surface during the GID process. The evolution of a peak intensity with glancing angle $\acute{\alpha}$ revealed by the various phases reflects the variation of the phase composition with depth.

9.3.2.2 Small-Angle Scattering and Reflectivity

Small-angle scattering and reflectivity based on both x-rays and neutrons are particularly important diffraction analytical techniques. An analysis of the shape of small-angle scattering curves can provide valuable insights concerning electron density distribution in a sample and, indirectly, the structure of both a single particle and of agglomerates. We are now going to briefly dwell on the results obtained by these methods in studies of DND and CVD diamond films.

The fractal structure of DND clusters was investigated by traditional methods, both on powders with x-rays[28,69,76] and on dilute suspensions of disaggregated 4 nm DND particles in neutron-scattering measurements.[77] The authors analyzed the dependence of the small-angle scattering amplitude I on the wave vector q. The peak position in the $I(q)$ curve allows determination of the typical size of a scatterer and the curve slope provides its fractal dimensionality. A comprehensive analysis revealed the fractal nature and dimensionality of isolated, 4 nm small DND particles and of their aggregates. Thus, the information obtained was employed in constructing a model of a detonation carbon particle[26] (see Figure 9.2) and permitted reconstruction of the shape of the primary aggregates[69] and determines the mass density of 4 nm isolated DND particles.[77] The small-angle scattering method applied to CVD films deposited on a substrate and having, as a rule, smooth interfaces is traditionally referred to as reflectivity approach.

In the reflectivity technique, the intensity of specularly reflected x-rays is measured as a function of the incidence angle. Since the reflection coefficient for x-rays is less than unity, it is totally reflected from the sample surface, up to the critical angle of incidence. This critical angle is proportional to the square root of electron density and, hence, of the mass density. The density measured in this way can be used to derive the sp^2/sp^3 ratio or, said otherwise, the extent to which a film is graphite like ($\rho = 2.25$ g/cm^3) or diamond like ($\rho = 3.51$ g/cm^3).[78] The results of a study of the density variation with the time of deposition of diamond-like films (converted to film thickness) in the initial stages of their growth can be found in Ref. 71.

Significantly, although x-ray radiation is insensitive to hydrogen content in a film, it is the hydrogen content that strongly affects such mechanical properties of films as their hardness and wear resistance. Scattering of neutrons is similar in many respects to that of x-rays, but, being scattered by nuclei and unpaired electrons, neutrons are much more sensitive to hydrogen content in a sample. In cases where the carbon density in a CVD film is known, for instance, from x-ray reflectivity studies, by measuring the critical scattering angle for neutrons, one can derive the concentration of hydrogen in it. Provided that the density of carbon atoms is known from x-ray study, the critical angle for neutron reflectivity is a measure of the concentration of hydrogen of ND films.[78,79] The reflectivity method can be readily applied to determination of the thickness of films and characterization of their surface roughness (see, e.g., Ref. 80).

9.3.3 Methods of Vibrational Spectroscopy

Vibrational spectroscopy is designed to study the structure of matter by analyzing the absorption, emission (e.g., Raman spectrocopy), and reflection spectra that are produced by quantum transitions between the vibration energy levels of molecules. The main methods employed in vibrational spectroscopy are IR spectroscopy and Raman spectroscopy. IR spectra are spectra of absorption and in their interpretation, one makes use of the concept of absorption bands. The presence of fluorescence and thermal emission makes obstacles for Raman measurements. Raman and IR spectroscopy, being complementary techniques, provide valuable insights into phase composition and surface termination of ND (see, e.g., Ref. 38).

Another vibrational method is the high-resolution electron energy loss spectroscopy (HR-EELS), which is operated in the 1–10 eV energy range and exhibits an energy resolution of better than 15 meV. HR-EELS is a method based on the use of inelastic scattering of low-energy electrons that was developed specially for measurement of vibrational spectra of surface groups. It is application of low-energy electrons that accounts for the surface sensitivity of this approach. In the first approximation, HR-EELS may be considered as an electron analog of Raman spectroscopy.

Recently, some works have been done on the HR-EELS spectrum of single-crystal diamond surfaces and polycrystalline diamond surface consisting of diamond grains in the micron to nano-size range deposited by CVD (see, e.g., Refs. 81–85) and of DND powder.[86]

9.3.3.1 Raman Spectroscopy

Raman spectroscopy is widely used as a diagnostic tool for the evaluation of diamond crystals and CVD diamond films. The technique is popular because each carbon allotrope displays a clearly identifiable Raman signature, it is nondestructive (when the correct laser irradiation parameters are chosen), requires little or no specimen preparation, and can be made confocal so that micrometer volumes can be sampled. Raman scattering from single-crystal, CVD diamond films, and ND has recently been reviewed in Refs. 87,88.

The Raman scattering of single-crystal graphite was found to contain a single narrow band (G-mode) at 1575 cm^{-1}, that is assigned to the planar E_{2g} mode of the Brillouin zone center.[89–91] However, sp^3-hybridized carbon, that is, single-crystal diamond, demonstrates a single first-order peak in the Raman spectrum, a narrow symmetric line at the frequency of 1332.5 cm^{-1} (with a peak width of about 2.0 cm^{-1}),[92] which derives from the transverse TO phonon of F_{2g} symmetry. At the same time, changes in the structure and size of the sp^2-hybridized phase become reflected in those of the spectrum near 1600 cm^{-1}.[93] Indeed, microcrystalline graphite demonstrates broadening and displacement of the so-called G-band at 1570 cm^{-1} toward higher frequencies. In addition, an additional feature at 1350 cm^{-1} appears, the D-band. The broadening and frequency shift of the "diamond" mode (1332.5 cm^{-1}) toward lower frequencies are observed to occur in Raman spectra at the transition from the bulk to NC diamond.[93,94]

Figure 9.7 shows a typical Raman spectrum of DND powder and those of microcrystalline diamond powder (the grain size of 200 μm) and microcrystalline graphite. One can clearly see the features mentioned above: the broadening and frequency shift of the diamond mode, as well as a new band due to microcrystalline graphite.[89]

The assignment for Raman peaks commonly observed in CVD diamond films is summarized in Ref. 87. In addition to the first-order Raman line at 1332 cm^{-1}, peaks appear, which for the most part reflect sp^2-bonded components of the films, either in the bulk or in the grain boundaries: 1150, 1350,

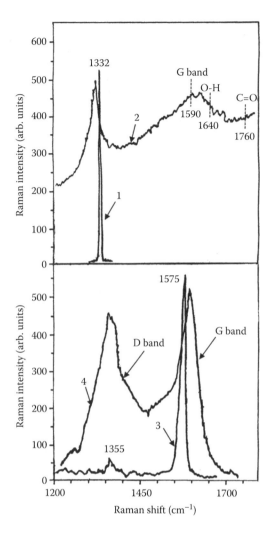

FIGURE 9.7 Raman spectra of different types of diamond (top) and different types of sp^2-hybridized carbon (bottom). (1) Microcrystalline diamond powder, (2) DND, (3) pyrolite, and (4) activated charcoal.

1480, and 1550 cm^{-1}. The peaks at 1350 and 1550 cm^{-1} are the D- and G-peaks of amorphous carbon. The peak at 1150 cm^{-1} has been attributed to NC diamond (see Ref. 87). Despite very convincing arguments,[88] the identification of the peaks at 1480 and 1150 cm^{-1} as due to trans-polyacetylene is still controversial. Comparing with silicon, the latter mode could correspond to the 650 cm^{-1} branch that is often observed in Raman from a-Si (that has been assigned to an Si–H bending mode), and it is also possible that this mode has more than one component. In some NC films, the substitution of deuterium for hydrogen in the growth mixture causes the peak to disappear, which supports the idea of it being connected with hydrogen.[95]

The Raman spectrum of DND (see Figure 9.7) consists of several characteristic features:[38] the first-order Raman mode of the cubic diamond lattice that is broadened and redshifted in DND (a peak at 1325 cm^{-1}) compared to bulk diamond (1332 cm^{-1}); a double-resonant D-band around 1400 cm^{-1} (at 325 nm excitation), resulting from disordered and amorphous sp^2-carbon, and a broad asymmetric peak between 1500 and 1800 cm^{-1}. Most often, the latter peak is labeled as the G-band and assigned to the in-plane vibrations of graphitic carbon. However, this peak significantly differs from the G-band of graphitic materials both in shape and position. Recently, the broad asymmetric

1640 cm^{-1} Raman peak of ND has assigned to a superposition of sp^2-carbon band at 1590 cm^{-1} with a peak of O–H bending vibrations at 1640 cm^{-1} C=O stretching vibrations, which were shown to be positioned at 1740 cm^{-1} in the Raman spectrum of oxidized ND, produce a shoulder on the 1640 cm^{-1} peak.[38] It has also been reported[34] that the shoulder observed at 1250 cm^{-1} is related to defects or crystalline particles of size less than 20 nm.

It is well known that the typical Raman spectrum contains basic information concerning the phase purity and crystalline perfection of the diamond material. But recent developments have shown that Raman spectroscopy can provide much more than this basic information. Surface-enhanced Raman spectroscopy, for example, has revealed the presence of new and unexpected structures on diamond surfaces.[87]

Size measurements using Raman spectroscopy are based on the analysis of the Raman frequency and Raman peak shape. For Raman scattering from finite-sized crystals, the wave vector of the vibration excitation is uncertain by a factor of $\Delta k = 2\pi/L$, where L is the crystal dimension. The authors of Refs. 96,97 independently investigated how the uncertainty of the wave vector will affect the Raman lines and suggest the phonon confinement model (PCM) that correlates the observed changes with the crystal size. The confinement of optical phonons in nanocrystals <10 nm results in asymmetrically broadened Raman lines, which are shifted toward lower wave numbers.

To date, the phonon confinement effects have not been explicitly detected for CVD diamond films[87] and results remained unsatisfactory in the case of DND.[98] To improve the agreement between the predictions of the model and experimental Raman spectra of DND, effects such as crystal size distribution, lattice defects, and the energy dispersion of the phonon modes were taken into consideration and incorporated into the PCM.[98] This work has shown that phonon wave vectors from small vibration domains lead to a broad shoulder peak at 1250 cm^{-1}, that is often observed in the Raman spectrum of DND. Although the agreement between experimentally obtained and calculated Raman spectra has been significantly improved, some limitations remain, as was pointed out in Ref. 98. The limitations imposed by the small ND size on the applicability of the PCM arise from the assumption that nanocrystals of 3–20 nm in size, showing extensive surface reconstruction and lattice defects, are assumed to have the phonon density of states of bulk diamond.

Because of the presence of several bands in the diamond and graphite spectra of the phonon density of states, some of which have frequency overlaps, the analysis of Raman spectra for determining the percentage of the sp^2-phase is very difficult. This task is also complicated by the influence of the $\pi-\pi^*$ resonance on the spectral amplitude, determined by the sp^2-phase.[99] The resonant enhancement of the sp^2- and sp^3-bonded portions of ND is very strongly dependent on the excitation wavelength, with the ratio $I_{(diamond)}/I_{(nondiamond)}$ increasing dramatically as the excitation wavelength decreases into the UV region.

Obviously, the application of multiple-wavelength excitations might prove very effective for characterization of ND by Raman spectral analysis. In disordered carbons, the G-peak position (G_{pos}) increases as the excitation wavelength decreases, from IR to UV, as indicated by Ferrari and Robertson.[88] The authors define G-peak dispersion (G_{disp}) as a function of the excitation wavelength

$$G_{disp} \text{ (cm}^{-1}/\text{nm)} = G_{pos}(244 \text{ nm}) - G_{pos}(514.5 \text{ nm})/(514.5 - 244) \text{ nm} \qquad (9.2)$$

to show that the dispersion is proportional to the degree of disorder. The G-peak does not disperse in graphite itself, in NC graphite, or in glassy carbon. In fact, G-peak dispersion is correlated with the sp^3-bonded fraction, density, and hardness.[80,88] Since the G-peak disperses only in more disordered carbons, this dispersion is related to the degree of disorder present in CVD films.

9.3.3.2 IR Spectroscopy

It is well known that defect-free, bulk diamond is transparent in the IR spectral region due to the high symmetry of its crystal lattice. Lowering of the C–C bond symmetry in a diamond crystal near its surface may give rise to the appearance of a band at 1332 cm^{-1}. This effect was observed in IR spectra of synthetic diamond with a specific surface area of 22 m^2/g after surface treatment.[100] The powder was pretreated by a mixture of sulfuric and nitric acids, with subsequent annealing in hydrogen environment. Ref. 101 discusses the size-dependent C=O stretching frequency (between 1680 and 1820 cm^{-1}) for ND particle diameter sizes from the 5 to 500 nm range.

IR spectroscopy was found to be very useful for studying DND because the surface of DND particles is usually saturated by various functional groups, which are covalently bound to them to form a dense molecular shell around a nanoparticle. Typical features of DND IR spectra can be found in Refs. 37,102–108. Most IR spectra of DND in the literature reveal the presence of water absorption bands, which consist of a broad composite band close to 3600 cm^{-1} and a somewhat weaker band, near 1620 cm^{-1}. The first of them derives from the symmetric and asymmetric stretch vibrations of water molecules, and the other, from the OH bend vibrations.[109] The disagreement with 1645 cm^{-1}, the value usually quoted in the literature, is usually assigned to adsorption effects.

By heating DND embedded in KBr and AgCl pellets in vacuum (140°C for 4 h), one succeeded in obtaining an IR spectrum of practically dehydrated DND.[109] The stretch vibration is practically suppressed in the spectrum, thus evidencing complete removal of water from the sample. Interestingly, the reverse process, that is, adsorption of water from air, takes only a few minutes to come to the end.

The OH bending vibration band (1620 cm^{-1}) does not vanish completely under dehydration, and one may therefore assume that this band masks the intrinsic IR spectrum of DND in this region.

The inert impurities usually occurring in DND (SiO$_2$, TiO$_2$) do not produce strong absorption lines in the (400–4000) cm^{-1} region. Some spectra may reveal a weak absorption band at 820 cm^{-1}. Disappearance of this band after fluorination permits assignment of this absorption to a TiO$_2$ impurity.[110]

The bending vibration band system in the (1600–1500) cm^{-1} region characteristic of aromatic carbon compounds[111] is not strong and can be masked by oxygen-containing functional groups. The CH stretch vibrations observed in aromatic systems near 3000 cm^{-1} are not usually present in DND spectra. This suggests that the amount of aromatic carbon left over in DND after the standard industrial acidic purification is fairly small.

Thus, the bands that can be assigned to intrinsic absorption of conventional (nonmodified) DND are: the group of maxima in the (2850–2950) cm^{-1} interval, the band at (1720–1750) cm^{-1}, the moderately strong band at 1640 cm^{-1} masked by water absorption, and a system of bands with frequencies of 1450, 1340, 1250, 1160, 1110, and 1050 cm^{-1} (see Figure 9.8[112]). The exact positions and the relative intensities of these bands may vary somewhat from one publication to another. [37,102–106,108,113]

The group of bands at (2850–2950) cm^{-1} is usually associated with CH stretch vibrations in aliphatic carbon chains (methyl and methylene groups). This assignment is borne out by the strong

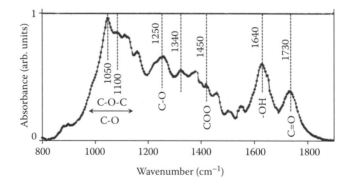

FIGURE 9.8   IR-absorption spectra of DND in the fingerprint spectral range 800–1900 cm^{-1}. Positions of the oxygen-containing groups are highlighted.

enhancement of intensity and broadening of these bands in DND treated in a hydrogen environment. A similar behavior is observed with the bands at 1450, 1250, and 1160 cm^{-1}, which are usually assigned to bending vibrations of CH$_2$ groups in a variety of environments.[103]

The band at (1720–1750) cm^{-1} is usually considered to belong to the ketone group with a secondary carbon atom. Reduction should transform it to the hydroxyl or methylene group. Some researchers believe that this group can transform to a carboxyl or anhydride group, to be accompanied by its shift by (20–40) cm^{-1} when treated by oxygen at temperatures above 400°C.[102,104]

This band (1720–1750) cm^{-1} could also derive from CO stretch vibrations in carboxyl groups. The presence of carboxyl groups is also corroborated by the presence of the complex bands at 1420 and 1340 cm^{-1}, which, however, could be masked by bending vibration lines of methyl groups.[106,113]

The vibration bands at 1110 and 1340 cm^{-1} may suggest the presence of ternary alcohol groups,[103] which become smeared in the course of reduction and disappear under fluorination.

A more accurate assignment of the absorption bands in DND IR spectra would require additional studies, including preparation of dehydrated samples[114] performing purposeful surface modification and application of different techniques of sample characterization.[115]

Table 9.1 summarizes the methods used for characterization of NC diamonds. Besides the methods mentioned above, we involved in the table additional methods, among them electron spectroscopy, spin resonance methods, optical methods, and methods involving visualization and chemical methods. We would like to emphasize that the complicated structure of NC diamonds' demands on using various methods for single object.[38,71,80,89]

9.4 APPLICATIONS OF NC DIAMONDS

There are many potential applications of NC diamond, and the applications (see Figure 9.9) are closely related to the various extreme physical and chemical properties they exhibit. Until the beginning of the twenty-first century, the main issue preventing the wide-scale use of CVD diamond films has been of economic nature because the main applications, coatings, were too expensive compared with existing alternatives. However, as higher power deposition reactors became standard, the cost for 1 carat (0.2 g) of CVD diamond fell below 1 U.S. $ in the year 2000. Along with that, growth rates of CVD diamond films increased from 0.05 μm/h to several tens of micrometers per hour at the time between 1995 and 2010. This technological development now allows using CVD diamond films for a wide variety of applications.[57] A similar situation can be identified for NC diamond powder, in particular DND, which has now become commercially available.[54]

The extreme hardness of diamond, combined with wear resistance, makes it an attractive material for use in cutting tools for machining nonferrous metals and abrasives. Indeed, HPHT industrial diamond is in use in this area since the 1960s. It is obvious, NC diamond primarily tried to use for the same purposes. NDs produced by milling of HPHT microcrystalline diamonds have been used as an ordinary abrasive material. A new application of ND with photoluminescent NV centers is the use as a biomarker.[148] In contrast to advanced ND production, NDs synthesized by laser methods are only starting to make their first steps from laboratories to industry.[116] The modern applications of the ND have been discussed in several recently published reviews on CVD films (see, e.g., Ref. 149 and detonation NDs[150]). Considering comprehensive reviews mentioned before, we would like to concentrate here only on main directions of several important points of NC diamond.

First, the main difference between NC and UNCD films has to be considered. The decrease of the crystal grain size from NC to UNCD films increases the number of grain boundary interfaces. This also results in a decrease of thermal conductivity and an increase of the optical absorption of UNCD films compared to NC films. At the same time, the friction coefficient of UNCD films is lower than that of NC films. Both types of CVD films can be used for MEMS (microelectromechanical systems), NEMS (nanoelectromechanical systems), and also for electrochemical applications.[149]

DND is applied as an abrasive for ultrafine mechanical polishing of hard surfaces of materials. Present-day polishing compositions based on detonation ND offer the possibility of obtaining

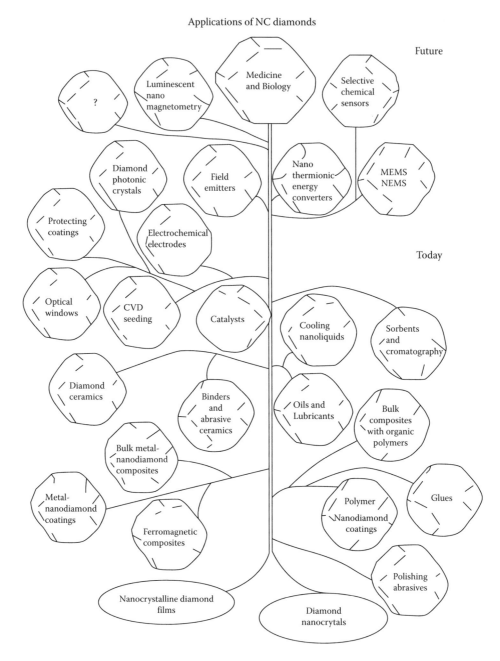

Applications of NC diamonds

FIGURE 9.9   Tree of applications of different types of NC diamond.

surfaces with an average roughness not in excess of 0.3 nm, which is a limitation to the techniques employed in mechanical polishing.[151] It is also possible to make abrasive tools based on a diamond ceramic prepared by sintering DND with glass-forming oxides at high pressure and temperature.[19] Another advantage to the use of DND in abrasive compositions is the improved removal of heat from the cutting region because of the high thermal conductivity of ND.

Adding DND into industrial lubricating oils has also emerged as a promising application due to the improving state of working metallic surfaces under friction. Adding DND to oils permitted, in some cases, reducing fuel consumption noticeably by up to 12% and increasing the service life of internal combustion engines. Intense research activities are currently dedicated to finding optimal conditions for the use of lubricants containing isolated 4 nm DND particles.[152]

One of the most promising areas of DND application in present-day mechanical engineering is the development of Cu-, Zn-, Sn-, Au-, Ag-, Cr-, or Ni-based composite metal coatings prepared by electroplating. The wear resistance of DND-containing platings increases up to 9× that can be used, for example, in aircraft engines, immersion pumps for use in oil wells, valves, stoppers, and pumps for chemical industry. For preparation of such platings, dry DND powder is directly poured into the water-based electrolyte of an electroplating bath. The incorporation of DND particles into the metallic base layer involves electrophoresis of the dielectric particles of ND, paralleled by galvanic deposition of the metal.[153]

Along with metal–ND coatings, the incorporation of DND particles enhances the parameters of bulk metallic compounds and ceramics. For instance, the incorporation of ND powders into an aluminum matrix opens the way of obtaining materials with a high thermal conductivity[154] and specific strength for use in aircraft engineering. In the atomic and electronic industry, DNDs are also used to manufacture nanoporous ceramic filters for biochemical engineering.[26]

Of particular interest for present-day industrial applications are DND composites based on organic polymers. The addition of DNDs (not exceeding 0.1–0.5 wt%) to the resin of car tires improves its elasticity, strength, and wear resistance, and also the force of friction on a smooth surface, even if the latter is coated by a water film.[155] It was demonstrated that introducing DND in a thermoplastic elastomer based on polypropylene improves both thermal stability elasticity and strength.[156] Progress has been reached in the application of polymer–ND composites based on polyimide and polyurethane with a DND content of up to 3 wt%. Such composites can be used in microelectronics as a material for insulating films possessing a low dielectric permittivity and a high thermal conductivity.[157] The high strength and thermal conductivity of such insulation are attractive to reduce the overall dimensions and increase the specific power of heavy-power small electric motors for use in transport facilities in the future. Adding DND to epoxy-based bonding agents significantly improves the strength of both the bond itself and, through adhesion, of the bonded materials as well. Another class of polymer–ND composites that has already gained broad popularity encompasses various cosmetics and skin-care products and several patents have been issued in this area.[158]

At the moment, one of the most important DND application areas is seeding for CVD diamond coatings. Layers of polycrystalline diamond produced by chemical deposition from a gas phase (CVD) are widely used as protective and antifriction coatings in engineering, high-precision instrumentation, optics, and in medicine to confer biological inertness to elements of the equipment used, surgical tools, and metallic and ceramic implants. The main condition for a crystalline CVD film to start growing on an arbitrary solid surface, it must have the maximal number of uniformly distributed centers of crystallization (precursors). By using DND as precursors in preparation of diamond CVD films, one can produce crystallization center densities on the substrate surface of 10^{12} cm^2 that is higher than those reached with any other diamond material.[32] One of the most impressive results of applying DND particles for CVD is the preparation of inverted bulk photonic crystals of polycrystalline CVD diamond.[141]

Hot-pressed granulated DND is used as a stationary phase in ion chromatography, which is indispensable in separation of ions of alkaline–earth and transition metals.[159] Recent years have highlighted a growth of interest in the use of DND as a carrier of metallic catalysts, in particular, of platinum group metals. Platinum film-coated DND grains manufactured for use as high-sensitivity sensors of carbon monoxide in the atmosphere operate by oxidizing CO to carbon dioxide.[119] Application of DND in biology and medicine is an area exhibiting the fastest growth that is strongly fueled by the nontoxic nature of DND. DND is also applied as a highly efficient absorbent, including binding of toxins in living organisms and high-efficiency stationary phase for chromatography of biologically active substances, including proteins.

The development of new methods for obtaining free 4.5 nm DND particles opens up new possibilities for targeted drug-delivery systems that overcome cell membranes, various biological markers, systems intended for determination of the concentration of bioactive agents in living organisms, and other applications.[160] Promising future of DND applications in electronic engineering concern their use in design of 2D electronic emitters and nanothermal energy converters. The possibility of forming charged luminescent NV centers inside the structure of DND particles opens the way for applying them as single photonic sources for future quantum-computing devices, quantum cryptography systems, and communication systems based on the entangled photonic states.[161] The first just developed devices employing the NV centers inside diamond nanoparticles allow measuring the magnetic fields with atomic resolution and sensitivity for the single spins at room temperature.[162] New MEMS sensors based on the mechanical resonance of a submicrometer-sized silicon rod with deposited DND particles with properly functionalized surface are able to detect the presence of single molecules in the environment.[163]

The broad field of the exciting and attractive future applications of DNDs documents that this unique material possesses all the possibilities to become an important part for future nanotechnology advancements.

9.5 CONCLUSIONS

Diamond has been well known from time immemorial, but development of HPHT and CVD synthesis in the beginning of the 1960s has opened the door for diamond crystals from jeweler's workshops to industry.

It is impossible to realize present-day industry without diamond crystals, microdiamond powder, and diamond films. List of applications of diamonds involves cutting of nonferrous metal and semiconductors, material of constructions, dry and wet grinding, and polishing.

Technology and properties of NC diamond powders and films, described in this chapter, open new horizons for this unique material. Applications of NC diamond have been recently discussed in several reviews.[110,150,151,164] In conclusion, we would like to emphasize the unsolved problems concerning NC diamond. First, we believe the important problem is still waiting for its solution, to understand the mechanism underlying impurities and defects formation in ND particle. It can be suggested that the solution of the problem can enable the creation of a scalable technology for formation of *pn*-junction using diamond nanoparticles as precursors.

The problem of similar importance is the technology of magnetic carbon materials. Surface functionalization of DND by metals[133] and paramagnetism of the edge states of π-electrons in nanographite, obtained from DND[165], are a promising way to the solution of this problem. Finally, namely the development of the technology for inserting photoluminescence impurity centers[33] Si-V[166] into diamond nanoparticles and also the functionalization of the surface of these particles determine the possibility to apply the NC diamonds for biomarkers and targeted drug-delivery systems.[160]

Among other tasks for future, we can highlight a technology for preparation of NC diamond particles with adjustable size distribution of grains in the range of 2–6 nm.

The list of these tasks requires a close collaboration of physicists, chemists, and experts in material science.

ACKNOWLEDGMENTS

The authors highly appreciate their colleagues from Ioffe Institute "nanodiamond group": A. E. Alexenskii, E. D. Eidelman, V. G. Golubev, M. A. Yagovkina, and V. Yu. Osipov for fruitful discussions.

Studies of Ioffe Institute "nanodiamond group" have been supported by the Russian Foundation for Basic Research, by the Ministry of Education and Science of the Russian Federation, and by programs of the Russian Academy of Sciences.

The experimental results by Ioffe Institute nanodiamond group cited in this chapter have been doing with using of equipment that was generously provided by Joint Research Center "Material science and characterization in advanced technology" of Ioffe Institute.

REFERENCES

1. Daulton, T.L., Nanodiamonds in the cosmos microstructural and trapped elements isotopic data, in *Synthesis, Properties and Applications of Ultrananocrystalline Diamond,* Gruen, D., Shenderova, O., Vul' A.Ya., eds., NATO Science Series II: Mathematics, Physics and Chemistry. Springer, Dordrecht, 192, 49, 2005.
2. Daulton, T.L., Extraterrestrial nanodiamonds in the cosmos, in *Ultrananocrystalline Diamond: Synthesis, Properties and Applications,* Shenderova, O. and Gruen, D., eds., William-Andrew Publisher, NY, Chapter 2, 23, 2006.
3. Bundy, F.P. et al., *Nature,* 176, 51, 1955.
4. Abbaschian, R., Zhu, H., and Clarke, C., *Diam. Relat. Mater.,* 14, 1916, 2005.
5. Liu, X. et al., *Cryst. Growth Des.,* 11, 3844, 2011.
6. Hu, M.H. et al., *J. Cryst. Growth,* 312, 2989, 2010.
7. Morita, Y. et al., *Small,* 4, 2154, 2008.
8. Treussart, F. et al., *Physica B,* 376–377, 926, 2006.
9. Boudou, J.-P. et al., *Nanotechnology,* 20, 235602, 2009.
10. Tisler, J. et al., *ACS Nano,* 3(7), 1959, 2009.
11. Chang, Y.-R. et al., *Nature Nanotechnol.,* 3(5), 284, 2008.
12. DeCarli, P.S. and Jamieson, J.C., *Science,* 133(3467), 1821, 1961.
13. Yamada, K. et al., *Carbon,* 37, 275, 1999.
14. Donnet, J.B. et al., *Diam. Relat. Mater.,* 9, 887, 2000.
15. Hu, S. et al., *Diam. Relat. Mater.,* 17, 142, 2008.
16. Lyamkin, A.I. et al., *Dokl. Akad. Nauk USSR,* 302, 611, 1988.
17. Greiner, N.R. et al., *Nature,* 333, 440, 1988.
18. Danilenko, V.V., *Phys. Solid State,* 46(4), 595, 2004.
19. Dolmatov, V.Y., *Russ. Chem. Rev.,* 76, 339, 2007.
20. Ogale, S.B. et al., *Solid State Comm.,* 84, 371, 1992.
21. Polo, M.C. et al., *Appl. Phys. Lett.,* 67, 485, 1995.
22. Sun, J. et al., *Appl. Phys. Lett.,* 89, 183115, 2006.
23. Hu, A. et al., *Diam. Relat. Mater.,* 18, 999, 2009
24. Pearce, S.R.J. et al., *Diam. Relat. Mater.,* 13, 661, 2004.
25. Danilenko, V.V., Nanocarbon phase diagram and conditions for detonation nanodiamond formation, in *Synthesis, Properties and Applications of Ultrananocrystalline Diamond,* Gruen, D., Shenderova, O., Vul' A. Ya., eds., NATO Science Series II: Mathematics, Physics and Chemistry. Springer, Dordrecht, 192, 181, 2005.
26. Baidakova, M. and Vul', A., *J. Phys. D: Appl. Phys.,* 40, 6300, 2007.
27. Alexenskii, A.E. et al., *J. Phys. Chem. Solids,* 63, 1993, 2002.
28. Baidakova, M.V. et al., *Phys. Solid. State,* 40, 715, 1998.
29. Barnard, A.S., *Analyst,* 134, 1751, 2009.
30. Hawelek, L. et al., *Diam. Relat. Mater.,* 17, 1186, 2008.
31. Shames, A.I. et al., *Diam. Relat. Mater.,* 29, 318, 2011.
32. Fang, X. et al., *J. Am. Chem. Soc.,* 131, 1426, 2009.
33. Baranov, P.G. et al., *Small,* 11, 1533, 2011.
34. Vlasov, I.I. et al., *Small,* 6(5), 687, 2010.
35. Pandey, K.C., *Phys. Rev. B,* 25, 4338, 1982.
36. Panich, A.M. et al., *Eur. Phys. J. B,* 52, 397, 2006.
37. Liang, Y. et al., *J. Colloid Interface Sci.,* 354, 23, 2011.
38. Mochalin, V. et al., *Chem. Mater.,* 21, 273, 2009.
39. Arnault, J.-C. et al., *Phys. Chem. Chem. Phys.,* 13, 11481, 2011.
40. Martin, R. et al., *Chem. Mater.,* 21, 4505, 2009.
41. Inaguma, M. et al., *Adv. Mater.,* 19, 1201, 2007.
42. Aleksenskii, A. et al., *Nanosci. Nanotechnol. Lett.,* 3, 68, 2011.

43. Williams, O. et al., *ACS Nano*, 4, 4824, 2010.
44. Xu, X. et al., *J. Solid State Chem.*, 178, 688, 2005.
45. Barnard, A.S., *J. Mater. Chem.*, 18, 4038, 2008.
46. Kuznetsov, V.L. et al., *Chem. Phys. Lett.*, 222, 343, 1994.
47. Banhart, F. and Ajayan, P.M., *Nature*, 382, 433, 1996.
48. Angus, J.C. et al., *J. Appl. Phys.*, 39(6), 2915, 1968.
49. Deryagin, B.V. et al., *Russ. Chem. Rev.*, 39(9), 783, 1970.
50. Lang, A.R. et al., *J. Cryst. Growth*, 200, 446, 1999.
51. Edelstein, R.S., *Diam. Relat. Mater.*, 8, 139, 1999.
52. Gouzman, I. et al., *Diam. Relat. Mater.*, 17, 1080, 2008.
53. Gruen, D.M., *Annu. Rev. Mater. Sci.*, 29, 211, 1999.
54. Gruen, D., Shenderova, O., Vul', A.Ya., eds., *Synthesis, Properties and Applications of Ultrananocrystalline Diamond,* NATO Science Series II: Mathematics, Physics and Chemistry. Springer, Dordrecht, 192, 2005.
55. Pierson, H.O., *Handbook of Carbon, Graphite, Diamond and Fullerenes*, Noyes Publications, Park Ridge, New Jersey, USA, 1993.
56. Plano, L.S., Growth of CVD diamond for electronic applications, in *Diamond: Electronic Properties and Applications,* Pan, L.S., Kania, D.R., eds., Kluwer Academic Publishers, Boston, Dordrecht, London, 62, 1995.
57. May, P.W., *Phys. Trans. R. Soc. Lond. A*, 358, 473, 2000.
58. Lopez, J.M. et al., *Appl. Surf. Sci.*, 185, 321, 2002.
59. Gordillo-Vazquez, F.J. and Albella, J.M., *Tech. Phys. Lett.*, 28, 787, 2002.
60. Chernov, V.V. et al., *Fullerenes, Nanotubes Carbon Nanostructures*, 20(4–7), 600, 2012.
61. Williams, O.A. et al., *Chem. Phys. Lett.*, 445, 255, 2007.
62. Konyashin, I., *Int. J. Refract. Met. Hard Mater.*, 24, 17, 2006.
63. Pattern 00-006-0675 ICCD database (http://www.iccd.com).
64. Hawelek, L. et al., *Diam. Relat. Mater.*, 17, 1186, 2008.
65. Palosz, B. et al., *Diam. Relat. Mater.*, 15, 1813, 2006.
66. Alexenskii, A.E. et al., *Phys. Solid State*, 39, 1007, 1997.
67. Wen, B. et al., *Chem. Phys. Lett.*, 516, 230, 2011.
68. Chen, P. et al., *Mater. Res. Bull.*, 39, 1589, 2004.
69. Ozerin, A.N. et al., *Crystallogr. Rep.*, 53, 60, 2008.
70. Palosz, B. et al., *Z. Kristallogr.*, 225, 588, 2010.
71. Hoffman, A., Mechanism and properties of nanodiamond films deposited by the DC–GD–CVD process, in *Synthesis, Properties and Applications of Ultrananocrystalline Diamond,* Gruen, D., Shenderova, O., Vul' A.Ya., eds., NATO Science Series II: Mathematics, Physics and Chemistry. Springer, Dordrecht, 192, 125, 2005.
72. Cappuccio, G. et al., *Appl. Phys. Lett.*, 69, 4176, 1996.
73. Windischmann, H. et al., *J. Appl. Phys.*, 69, 2231, 1991.
74. Braga, N.A. et al., *Diam. Relat. Mater.*, 18, 1065, 2009.
75. Tarey, R.D. et al., *Rigaku J.*, 4, 11, 1987.
76. Baidakova, M.V. et al., *Chaos, Solitons Fractals*, 10, 2153, 1999.
77. Avdeev, M.V. et al., *J. Phys. Chem. C*, 113(22), 9473, 2009.
78. Findeisen, E. et al., *J. Appl. Phys.*, 76, 4636, 1994.
79. Ozeki, K. et al., *Diam. Relat. Mater.*, 19, 489, 2010.
80. Singh, S.B. et al., *Surf. Coat. Technol.*, 203, 986, 2009.
81. Pehrsson, P.E. and Mercer, T.W., *Surf. Sci.*, 460, 74, 2000.
82. Michaelson, S.H. et al., *J. Appl. Phys.*, 104(8), 083527, 2008.
83. Hoffman, A. et al., *Surf. Sci.*, 602, 3026, 2008.
84. Akhvlediani, R. et al., *Surf. Sci.*, 604, 2129, 2010.
85. Azria, R. et al., *Prog. Surf. Sci.*, 86, 94, 2011.
86. Michaelson, S.H. et al., *Surf. Sci.*, 604, 1326, 2010.
87. Prawer, S. and Nemanich, R.J., *Phil. Trans. R. Soc. Lond. A*, 362, 2537, 2004.
88. Ferrari, A. C. and Robertson, J., *Philos. Trans. R. Soc. Lond. A*, 362, 2477, 2004.
89. Vul', A.Ya., Characterization and physical properties of UNCD particles, in *Ultrananocrystalline Diamond. Synthesis, Properties and Applications*, Gruen, D.M., Shenderova, O.A., eds., William Andrew Publishing, Norwich, New York, USA, 379, 2006.
90. Tuinstra, F. and Konig, J.L., *J.. Chem. Phys.*, 53, 1126, 1970.

91. Nemanich, R.J. and Solin, S.A., *Phys. Rev.,* 20, 392, 1979.
92. Solin, S.A. and Ramdas, A.K., *Phys. Rev. B*, 1, 1687, 1970.
93. Yoshikawa, M. et al., *Phys. Rev. B*, 46, 7169, 1992.
94. Obraztsov, A.N. et al., *Diam. Relat. Mater.*, 4, 968, 1995.
95. Pfeiffer, R. et al., *Appl. Phys. Lett.*, 82, 4149, 2003.
96. Richter, H. et al., *Solid State Commun.*, 39, 625, 1981.
97. Nemanich, R. J. et al., *Phys. Rev. B*, 23, 6348, 1981.
98. Osswald, S. et al., *J. Am. Chem. Soc.*, 128, 11635, 2006.
99. Leeds, S.M. et al., *Diam. Relat. Mater.*, 7, 233, 1998.
100. Ando, T. et al., *Chem. Soc. Faradey Trans.,* 89, 1783, 1993.
101. Tu, J.-S. et al., *J. Chem. Phys.*, 125, 174713, 2006.
102. Kulakova, I. I., *Phys. Solid State*, 46, 636, 2004.
103. Krüger, A. et al., *Carbon*, 43, 1722, 2005.
104. Shenderova, O. et al., *Diam. Relat. Mater.*, 15, 1799, 2006.
105. Petrov, I. et al., *Carbon*, 29, 665, 1991.
106. Jiang, T. and Xu, K., *Carbon*, 33, 1663, 1995.
107. Mironov, E. et al., *Diam. Relat. Mater.,* 11, 872, 2002.
108. Ji, S., Jiang, T., Xu, K., and Li, S., *Appl. Surf. Sci.*, 133, 231, 1998.
109. Chung, P.-H. et al., *Diam. Relat. Mater.*, 15, 622, 2006,
110. Aleksenskiy, A., Baidakova, M., Osipov, V., and Vul', A., The fundamental properties and characteristics of nanodiamonds, in: *Nanodiamonds: Applications in Biology and Nanoscale Medicine*, Ho, D., ed., Springer, New York, 55, 2009.
111. Nakanishi, K., *In Infrared Absorption Spectroscopy: Practical.* Holden-Day, USA, 233 , 1962.
112. Osipov, V. Yu. et al., *Diam. Relat. Mater.,* 20, 1234, 2011.
113. Zhu, Y.W. et al., *Phys. Solid State*, 46, 681, 2004.
114. Larionova, I. et al., *Diam. Relat. Mater.*, 15, 1804, 2006.
115. Korolkov, V. V. et al., *Diam. Relat. Mater.*, 16, 2129, 2007.
116. Panich, A.M. et al., *Diam. Relat. Mater.*, 23, 150, 2012.
117. Jee, A.Y. and Lee, M., *Curr. Appl. Phys.*, 9, E144, 2009.
118. Buijnsters, J.G. et al., *J. Appl. Phys.,* 108, 103514, 2010.
119. Vershinin, N.N. et al., *Fullerenes, Nanotubes Carbon Nanostructures*, 19, 63, 2011.
120. Shiryaev, A.A. et al., *J. Phys. Conden. Matter,* 18, L493, 2006.
121. Panich, M. et al., *J. Phys. Chem. C*, 114(2), 774, 2010.
122. Shenderova, O. et al., *J. Phys. Chem. C*, 115(39), 19005, 2011.
123. Hirai, H. et al., *Diam. Relat. Mater.*, 8,1703, 1999.
124. Dementjev, A. et al., *Diam. Relat. Mater.*, 16, 2083, 2007.
125. Zemek, J. et al., *Diam. Relat. Mater.*, 26, 66, 2012.
126. Zeppilli, S. et al., *Diam. Relat. Mater.*, 19, 846, 2010.
127. Pichot, V. et al., *J. Phys. Chem. C*, 114, 10082, 2010.
128. Lu, Y.-G. et al., *Diam. Relat. Mater.*, 23, 93, 2012.
129. Turner, S. et al., *Adv. Funct. Mater.,* 19, 2116, 2009.
130. Panich, A.M. et al., *Fullerenes, Nanotubes Carbon Nanostructures*, 20(4–7), 579, 2012.
131. Levin, E.M., *Phys. Rev. B*, 77(5), 054418, 2008.
132. Dubois, M. et al., *Solid State Nucl. Mag.*, 40, 144, 2011.
133. Panich, A.M. et al., *J. Phys. D: Appl. Phys.,* 44(12), 125303, 2011.
134. Orlinskii, S.B. et al., *Nanosci. Nanotechnol. Lett.*, 3(1), 63, 2011.
135. Ilyin, I.V. et al., *Fullerenes, Nanotubes Carbon Nanostructures*, 19(1–2), 44, 2011.
136. Osipov, V.Yu. et al., *J. Phys. Condens. Matter.,* 24, 225302, 2012.
137. Chang, L.Y. et al., *Nanoscale*, 3, 958, 2011.
138. Souza, F.A. et al., *Chem. Vapor Deposit.*, 18, 159, 2012.
139. Gaebel, T. et al., *Diam. Relat. Mater.*, 21, 28, 2012.
140. Aleksenskii, A.E. et al., *Phys. Solid State*, 54(3), 578, 2012.
141. Kurdyukov, D.A. et al., *Nanotechnology*, 23, 015601, 2012.
142. Vul', A. Ya. et al., *Diam. Relat. Mater.*, 20, 279, 2011.
143. Bradley, R.S. et al., *Diam. Relat. Mater.*, 19, 314, 2010.
144. Grichko, V. et al., *Nanotechnology,* 19, 225201, 2008.
145. Gibson, N. et al., *Diam. Relat. Mater.*, 18, 620, 2009.
146. Schmidlin, L. et al., *Diam. Relat. Mater.*, 22, 113, 2012.

147. Zhukov, A. N. et al., *Colloid J.*, 74(4), 463, 2012.
148. Faklaris, O. et al., *J. Eur. Opt. Soc. – Rap. Publ.,* 4, 09035, 2009.
149. Williams, O.A. et al., *Diam. Relat. Mater.,* 17, 1080, 2008.
150. Mochalin, V.N. et al., *Nat. Nanotechnol.*, 7, 11, 2012.
151. Zhu, Y. et al., *Chin. Particuol.*, 2(4), 153, 2004.
152. Chou, C.C. et al., *J. Mater. Process. Technol.*, 201, 542, 2008.
153. Mandich, N.V. and Dennis, J.K., *Met. Finish.*, 99, 117, 2001.
154. Mizuuchi, K. et al., *Composites: Part B*, 42, 825, 2011.
155. Wanga, Y.-X. et al., *Appl. Surf. Sci.*, 257, 2058, 2011.
156. Jahromi, H. and Katbab, A.A., *Appl. Polym. Sci.*, 125(3), 1942, 2012.
157. Shenderova, O. et al., *Diam. Relat. Mater.,* 16, 1213, 2007.
158. US patent 7,294,340 issues nanodiamonds for cosmetics, November 13, 2007.
159. Nesterenko, P.N. et al., *J. Chromatogr. A*, 1155, 2, 2007.
160. Schrand, A.M. et al., *Crit. Rev. Solid State Mater. Sci.*, 34, 18, 2009.
161. Aharonovich, I. et al., *Nano Lett.*, 9(9), 3191, 2009.
162. Maze, R. et al., *Nature*, 455, 644, 2008.
163. Kulha, P. et al., *Vacuum*, 84, 53, 2010.
164. Vul', A.Ya., Aleksenskiy, A.E., and Dideykin, A.T., Detonation nanodiamonds: Technology, properties and applications, in *Nanosciences and Nanotechnologies*, Kharkin, V.N., Bai, C., and Kim, S.-C., eds., *Encyclopedia of Life Support Systems (EOLSS), Developed under the Auspices of the UNESCO*, EOLSS Publishers, Oxford, UK, 2009.
165. Enoki, T. et al., *Chem. Asian J.,* 4(6), 796, 2009.
166. Grudinkin, S.A. et al., *J. Phys. D: Appl. Phys.,* 45, 062001, 2012.

10 Carbon Onions

Yuriy Butenko, Lidija Šiller, and Michael R. C. Hunt

CONTENTS

10.1 INTRODUCTION

Carbon onions are a member of the family of nanometer-scale graphite-like all-carbon allotropes, the emergence of which was catalyzed by the Nobel Prize-winning discovery of the first member, the fullerene, by Kroto et al. in 1985.[38] Initially, carbon onions were observed by Iijima in 1980,[33] and were brought to popular attention by the experiments of Ugarte in 1992.[83] Structurally, they consist of concentric spherically closed carbon shells and receive their name from the close resemblance between their nanoscale structure and the more familiar concentric layered structure of an onion. Closely related to carbon onions is a class of material known as onion-like carbons (OLCs), which include polyhedral nanostructures such as ideal nested fullerenes. This material, rather than ideal spherical carbon onions, can be currently produced in macroscopic quantities, and, hence, be used for future applications.

In this chapter, we discuss the structure of carbon onions and provide an overview of the various methods by which carbon onion-like materials may be produced. Also, some of the most important physical properties of carbon onion-like structures are reviewed with an emphasis on those properties, which may be relevant for future applications of this material.

10.2 CARBON ONIONS: STRUCTURE AND BONDING

It is the ability of carbon atoms to form covalent bonds with different orientations and bond strength that gives rise to the diverse range of all-carbon structures observed in nature. This is described in quantum chemistry in terms of the different hybridization states that carbons display. A free atom of carbon has an electronic structure $1s^2 2s^2 2p^2$ and, when forming a single molecule or a solid material, we can linearly combine the outermost $2s$ and $2p$ atomic orbitals (valence orbitals) to form the hybrid orbitals, which are the basic building blocks of carbon materials. Such combination of orbitals involves the formal promotion of an electron from the $2s$ to the unoccupied $2p$ orbital; however, the energy cost associated with this (≈ 4 eV) is recovered by the energy liberated by the formation of chemical bonds. The $2s$ orbital may be combined (hybridized) with one, two, or all three $2p$ orbitals, which leads to the three different geometric relationships observed for the covalent bonds among carbon atoms. Depending on the state of hybridization, carbon–carbon bonds can span one, two, or three dimensions:

1. *sp Hybridization* occurs when a single $2p$ orbital hybridizes with the $2s$ orbital. This enables the formation of two strong bonds with neighboring carbon atoms in a linear, one-dimensional arrangement with an angle of 180° between them.
2. *sp^2 Hybridization* involves the combination of two $2p$ orbitals with the $2s$ orbital. Three hybrid orbitals make an angle of 120° with respect to each other in a two-dimensional plane, whereas the remaining electron remains in a $2p$ orbital (typically defined as the $2p_z$ orbital) oriented perpendicular to the plane.
3. *sp^3 Hybridization* results when all three orbitals hybridize to form four hybrid orbitals. These hybrid orbitals are directed along the sides of a tetrahedron with an angle of $\approx 109°$ between them.

The capability of carbon atoms to form different formal hybridization states from sp to sp^3 is the reason for the surprising flexibility of graphite sheets. On the basis of a formal sp^2 hybridization state, an sp^2 bond graphite sheet should only display a planar geometry. However, a variety of graphite-based nanostructures, such as carbon onions and nanotubes, display a curvature. This deformation arises from the possibility of forming an admixture between sp^2 and sp^3 hybridized states (which we can denote by $sp^{2+\delta}$), which has a slightly higher excitation energy than the pure sp^2 hybridized state and has a nonplanar arrangement of hybrid orbitals. The energy cost associated with a curvature is recovered by closing dangling bonds that result from the resulting closed structure. Simple bending of a graphite sheet, although sufficient for the formation of a carbon nanotube,[62] cannot lead to the formation of a fully closed cage structure. To create an all-carbon cage, an additional curvature must be realized through the insertion of nonhexagonal rings of carbon atoms into the graphite net.

Positive curvature may be introduced within a graphite sheet through the incorporation of pentagonal rings of carbon atoms—the structure formed by a single pentagon surrounded by hexagonal rings of carbon atoms subtends a solid angle of $^\pi/_6$. We can neglect smaller rings of carbon atoms in this discussion because of the large bond strain resulting from their incorporation into an sp^2-bonded graphitic network. The number of pentagonal rings that need to be incorporated into a graphite sheet to form a closed shell may be found from Euler's theorem for polyhedra[25]:

$$f + v = e + 2$$

where f is the number of faces of the polyhedron, v is the number of vertices, and e is the number of edges. For a polyhedron containing f_h hexagonal faces and f_p pentagonal faces, we obtain the following relations:

$$f = f_h + f_p$$
$$2e = 6f_h + 5f_p$$
$$3v = 6f_h + 5f_p$$

and hence

$$6(f + v - e) = p = 12$$

which demonstrates that any all-carbon closed cage containing only hexagonal and pentagonal rings must possess 12 pentagonal rings and may contain an arbitrary number of hexagonal faces.

In addition to single-layer closed-shell all-carbon polyhedra, known as fullerenes, remarkable nested structures have been observed. It is possible to produce structures in which fullerenes can be concentrically enclosed within one another to produce what are known as multishell fullerenes or carbon onions. The first observation of carbon onions was in the product of arc discharge experiments.[33] Later, in 1992, Ugarte demonstrated the transformation of particles of fullerene soot into spherical carbon onions when exposed to an intense electron beam.[83] Ugarte's discovery sparked extensive interest in these unique carbon structures and stimulated many researchers to explore new approaches for synthesis and characterization of this unique material. A high-resolution transmission electron microscopy (HRTEM) image of a typical carbon onion structure, in this case, synthesized by annealing nanometer-sized diamond particles,[81] is shown in Figure 10.1 (see also Section 10.3.2).

The carbon onions produced in Ugarte's experiments have almost perfect spherical shape, and the size of the innermost shell is approximately equal to the size of the most abundant fullerene, C_{60}. Fullerene molecules, which contain $60n^2$ carbon atoms (where n is a nonzero positive integer), display an icosahedral shape. The ideal structure of a fullerene molecule containing 240 carbon atoms is shown in Figure 10.2. The icosahedral shape of these fullerenes originates from the local curvature introduced by the 12 pentagons present in their structure. This icosahedral shape becomes clearly visible as the size of fullerene molecules increases.

A carbon onion composed of four fullerenes (C_{60n^2}, where $n = 1, 2, 3,$ and 4) is shown in Figure 10.3. The icosahedral structure is very clear, with planar facets lined by pentagonal rings. This facetted structure is clearly at odds with the observation of nearly spherical carbon onions, such as that shown in Figure 10.1. The distance between the shells in this onion C_{60n^2} is approximately 3.5 Å, which is slightly larger than the interplanar separation observed in bulk crystals of graphite (3.354 Å).

An explanation for the contradiction between the observed spherical morphology of carbon onions and the evident icosahedral shape of fullerenes consisting of hexagonal arrays of carbon atoms incorporating 12 pentagons to ensure closure was advanced by Terrones et al.[76–78] They suggested that Stone–Wales-type transformations,[72] which convert four adjacent hexagons into two pentagons and two heptagons, can be applied throughout the icosahedral onions to smooth their

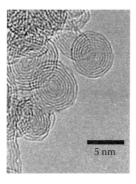

5 nm

FIGURE 10.1   HRTEM images of carbon onions produced by annealing ultradispersed NDs at 1700°C. The diameter of the innermost shell of the onion is larger than the diameter of the fullerene C_{60}. (Adapted from Tomita, S. et al., 2002a, *Carbon* **40**, 1469–1474.)

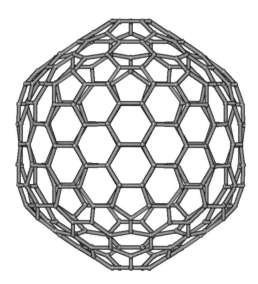

FIGURE 10.2 A fullerene molecule containing 240 carbon atoms has planar facets linked by pentagonal rings of carbon atoms. The faceting increases as the size of the fullerene becomes larger.

structure, producing quasi-spherical molecules. In such pentagon–heptagon pairs, the positive curvature of the pentagonal ring is offset by a similar negative curvature associated with the heptagonal ring. As a result, only local changes to the curvature are caused by such Stone–Wales pairs. It can also be seen, by consideration of Euler's theorem, that a closed carbon shell (or polyhedron) will remain closed after the introduction of a pentagon–heptagon pair.

Rings containing more than six and seven carbon atoms can be included to a closed carbon shell to create its curvature. Figure 10.4a shows a C_{240} fullerene (I_h) molecule consisting of pentagonal and hexagonal rings, which is the second innermost shell of an ideal icosahedral carbon onion. Figure 10.4b presents an O_h isomer of the C_{240} molecule, which consists of not only

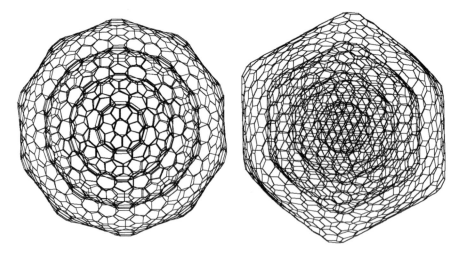

FIGURE 10.3 Two different projections of a carbon onion consisting of four fullerene molecules, each of which contains $60n^2$ carbon atoms, where n is the number of the fullerene shell in the carbon onion. The carbon onion has evident icosahedral shape due to the presence of 12 pentagons in each of the constituent fullerene shells.

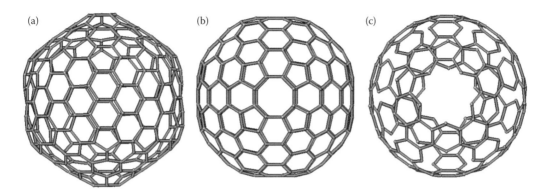

FIGURE 10.4 Fullerene molecules obtained via energy minimization: (a) fullerene C_{240} (I_h) with 12 vertices composed of 12 pentagonal rings; (b) O_h isomer of C_{240} containing 24 pentagons, 92 hexagons, and 6 octagons; the molecule has a more spherical shape; (c) spherical molecule C_{180} obtained by removing pentagons from the icosahedral C_{240} isomer.

pentagonal and hexagonal rings of carbon but also includes octagonal rings. The O_h isomer is clearly much more spherical than its icosahedral counterpart. The spherical shape exhibited by carbon onions can also originate from other defects, such as holes in the fullerene-like shells. Figure 10.4c presents an example of such a molecule, which was obtained by removing pentagons from the C_{240} (I_h) molecule. Therefore, one can conclude that the shape of carbon onions depends on the presence of different ring structure types and defects in the fullerene-like shells from which they are constructed.

In contrast to the spherical carbon onions observed in the first experiments by Ugarte,[83] OLC particles were subsequently produced with polyhedral facets, more closely matching the polyhedral structures predicted from the consideration of nested fullerene structures described above. These polyhedral onion-like particles were synthesized by vacuum heat treatment of carbon soot[23] and diamond nanoparticles.[41] Figure 10.5 presents HRTEM images of the polyhedral OLC particles produced in the experiments of Kuznetsov et al. The range of synthesis methods available has led to the production of different types of OLC. In addition to their shape, such carbon onions can be characterized by other parameters, such as the number of concentric shells, the spacing between adjacent shells, the size of the innermost shell, and the presence of different types of defects.

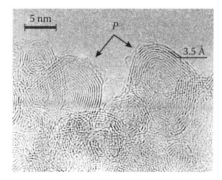

FIGURE 10.5 HRTEM images of polyhedral onion-like particles produced by annealing NDs at 1500°C for approximately 1 h (marked in the figure by letter *P*). Dark contrast lines in the figure correspond to graphite-like shells. The distance between lines is approximately 3.5 Å. (Adapted from Kuznetsov, V.L. et al., 1994a, *Chem. Phys. Lett.* **222**, 343–348.)

10.3 SYNTHESIS

10.3.1 EARLY METHODS FOR CARBON ONION SYNTHESIS AND TRANSFORMATION UNDER ELECTRON BEAM IRRADIATION

The first carbon onions were synthesized in arc discharge experiments by Iijima.[33] It was demonstrated that the condensation of carbon vapor results in the formation of not only single-shell fullerene molecules but also of multishell fullerenes (i.e., carbon onions). It was proposed that quasi-spherical spiral-like particles could be intermediates for the formation of carbon onions via condensation of carbon vapor.[39,91] The close value of the diameter of the innermost shells of these carbon onions to the diameter of the C_{60} fullerene molecule suggested the idea that another possible candidate for growth centers for carbon onions can be fullerenes themselves.[86]

The transformation of carbon nanoparticles to carbon onions can be stimulated by exposure to an intense beam of electrons. Indeed, Ugarte's experiment[83,84] demonstrated for the first time that closed, curved carbon nanostructures can be produced not only by the condensation of carbon vapor but also by the transformation of condensed carbon nanoparticles. The other very important aspect of this work was the realization that electron irradiation not only has a destructive effect on carbon materials but can actually be used to synthesize new carbon nanostructures. This discovery opened a completely new field of carbon materials research related to stability and transformation of carbon nanostructures. It was found that not only carbon soot can be converted into carbon onions but also diamond crystals. Qin and Iijima demonstrated that carbon onions can form on the surfaces of 1–3 µm diamond crystals under intense electron beam irradiation (150 A/cm²).[97] Growth of carbon onions (which consisted of between 4 and 10 closed graphitic shells) proceeded from the inner shell because of the high mobility of carbon atoms induced by electron irradiation. However, further irradiation resulted in the destruction of the carbon onions. Transformation of nanometer-sized diamond crystals (3–10 nm in diameter) to carbon onions under electron irradiation (20–40 A/cm²) was later observed by Roddatis et al.[60] In this case, complete transformation of nanodiamond crystals occurred, starting from their surfaces and proceeding inward.

The transformation of diamond to graphitic structures is perhaps not very surprising because graphite is more thermodynamically stable than diamond under normal conditions. Therefore, it was a great surprise when the opposite transformation was observed under electron irradiation. In 1996, Banhart and Ajayan reported on the formation of a diamond nanocrystal inside a carbon onion particle under intense energetic electron irradiation (1250 keV, 200 A/cm²), at a temperature of ≈700°C.[3] An HRTEM image of such a carbon onion containing an encapsulated diamond nanoparticle is shown in Figure 10.6. It was found that high-energy electrons are capable of mobilizing carbon atoms through thermally activated jumps, ballistic knock-on displacements. They can also induce mass loss by sputtering (i.e., ejection of atoms from the carbon onion). All these processes cause dynamic migration of interstitials, which together with mass loss are responsible for the contraction of the whole particle.[4,5,7,59,89,90] This contraction can be clearly observed in the HRTEM micrograph of Figure 10.7, by measuring the spacing between graphitic shells.[7] It was concluded that initial diamond nucleation occurs in the core of carbon onions under extremely high pressures caused by this contraction and the high temperature resulting from electron irradiation. Subsequent growth of the diamond crystal proceeds at low pressure because of the generation of defects under irradiation in the presence of which a phase transition from graphite to diamond could lead to a decrease of the free energy of the system.[7,98] Transformation of graphitic structures to diamond under electron irradiation generated a new understanding of the nature of the direct graphite–diamond transition. As a result, in 1997, Zaiser and Banhart developed a new formalism to predict the phase stability under irradiation for the graphite–diamond system.[89] This work also demonstrated that carbon onions can actually serve as nanocapsules for different materials or conduct confined reactions. It was found that metals, such as cobalt, nickel, iron, molybdenum, and gold, can be encapsulated inside carbon onions under electron irradiation.[6,8,46,75,85] Carbon onions

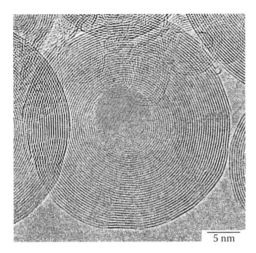

FIGURE 10.6 Spherical carbon onion with a monocrystalline diamond core 10 nm in diameter. The core shows the lattice fringes of diamond with a separation of 2.06 Å. (Adapted from Banhart, F., Ajayan, P.M., 1996, *Nature,* **382**, 433–435.)

containing cobalt metal crystals are shown in the HRTEM image of Figure 10.8 at different electron beam radiation doses. The radiation promotes contraction of the onions, which forces metal atoms to migrate outward through the shells.[6]

The pioneering experiments discussed in this section generated new ideas and concepts that significantly improved our knowledge of the physics and chemistry of carbon nanomaterials. The development of techniques for carbon onion synthesis by electron irradiation enabled investigators to establish an understanding of properties of these remarkable materials. Carbon onions produced by electron irradiation of carbon nanomaterials have a nearly perfect spherical form and a very

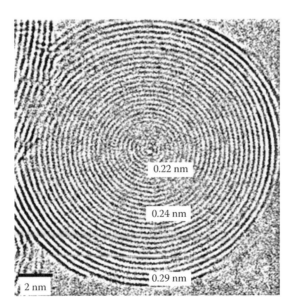

FIGURE 10.7 Contraction of carbon onions under electron irradiation can be deduced from the decreasing distances between layers upon moving from the surface of the carbon onion toward the center. The carbon onion was formed under electron irradiation at 700°C. (Adapted from Banhart, F., 1999, *Rep. Prog. Phys.* **62**, 1181–1221.)

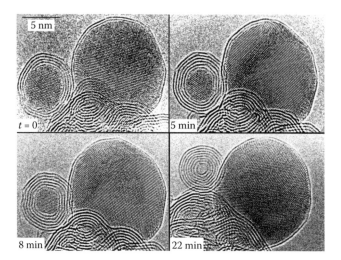

FIGURE 10.8 HRTEM images showing the behavior of two cobalt crystals covered by different number of graphitic shells under electron irradiation at 1000 K. It can be seen that while the smaller crystal on the left side is displaced rather rapidly, the larger crystal with initially one and then two graphitic shells shows almost no shrinking. (Adapted from Banhart, F., Redlich, Ph., Ajayan, P.M., 1998, *Chem. Phys. Lett.* **292**, 554–560.)

well-ordered structure. However, although highly perfect, the small synthesis yield made it necessary to explore new synthetic pathways for large-scale production of carbon onions.

10.3.2 Synthesis of Onion-Like Carbon as Alternative Material

The electron irradiation experiments, performed by Ugarte in 1992,[83] encouraged investigators to explore alternatives to electron irradiation for the formation of carbon onion structures and the simplest of which was the heat treatment of preformed carbon nanostructures. During the first 2 years after Ugarte's original experiment, it was shown that the transformation of nanostructures in carbon soot into carbon onions could occur through heating under vacuum at 2250°C for 1 h,[23] and that diamond nanoparticles (so-called nanodiamonds [NDs]) annealed at 1500°C for approximately 1 h[13,41,42] in vacuum also resulted in carbon onions. However, these onions have significantly less perfect structures and, therefore, are usually referred to as OLC rather than simply carbon onions. Figure 10.9 shows HRTEM micrographs of the products resulting from annealing NDs at different temperatures in vacuum.[43] It can be seen that graphitization starts from the diamond surfaces, proceeds toward the center of the nanodiamond particles, and eventually leads to the formation of OLC particles at higher temperatures.

The temperature-induced graphitization observed in these experiments resulted in the formation of closed multishell graphitic cages (carbon onions) from structurally different carbon precursors. Annealing at elevated temperatures helps overcome the kinetic barrier associated with graphitization. It was unexpected that heat treatment did not lead to the formation of planar graphite sheets because graphite is considered to be thermodynamically the most stable form of carbon under normal conditions.[11] The formation of symmetric, low-entropy, stable closed carbon multishell cages instead of planar graphite during carbon vapor condensation or heat-induced transformation of carbon nanoparticles does raise a question about stability of carbon structures on the nanoscale.[9,58] Decreasing the size of a particle to the nanoscale substantially increases the contribution of the surface energy of the particle to its total energy. The formation of flat graphite nanocrystals would result in the presence of carbon atoms with dangling bonds at their edges. Such carbon atoms have high energy in comparison with those with fully saturated bonds. The ability of graphitic sheets to curl and form closed carbon cages provides a possibility to eliminate these dangling bonds at

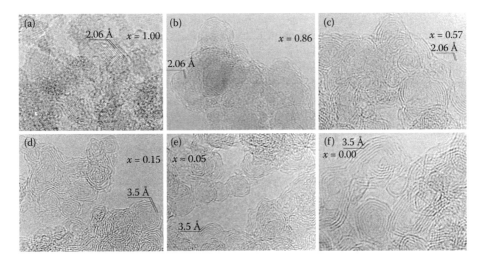

FIGURE 10.9 HRTEM micrographs of a UDD sample annealed under vacuum at (a) 1170 (1 h), (b) 1420 (1.4 h), (c) 1600 (1 h), (d) 1800 (1 h), (e) 1900 (1 h), and (f) 2140 K (0.5 h). The dark straight contrast lines in micrographs (a), (b), and (c) correspond to the (111) crystallographic diamond layers. The distance between these lines is 2.06 Å. The dark curved lines in (b) through (f) correspond to the (0002) crystallographic graphite layers. The distance between these lines is approximately 3.5 Å. The diamond weight fractions (x) of the samples are also presented within each image. (Adapted from Kuznetsov V.L. et al., 2001, *Chem. Phys. Lett.* **336**, 397–404.)

the cost of energy associated with bond strain and curvature and reduces the surface energy of the carbon nanoparticles significantly.

It is interesting to note that carbon onions produced by heat transformation of carbon soot[23,86,87] and NDs[13,41,42] have innermost shells with different diameters. Thermal transformation of nanostructured soot results in the formation of carbon onions with large internal shells in comparison to using NDs as the precursor.[86] This is the result of different mechanisms involved in the formation of carbon onions. The formation of carbon onions from carbon soot involves the transformation of amorphous carbon to initially curved graphite layers and subsequently the formation of closed carbon cages and consequent growth of multiwall cages around them. Such a mechanism implies growth outward from the center of the carbon onions. In contrast, when annealing NDs at elevated temperatures, graphitization starts from the diamond surfaces and progresses toward centers of nanodiamond crystals.[13,41,42,45] As a result, the formation of carbon onions occurs from a higher-density material and results in the growth of carbon onions with small internal shells. Because carbon onions have higher free energy than graphite, further annealing at sufficiently high temperatures for a long enough time should eventually lead to the formation of flat graphite crystals.

Synthesis of carbon onions through heat treatment of soot by de Heer and Ugarte[23] represented the first approach in producing carbon onions in macroscopic quantities. This is an important requirement for many further characterization steps and for any potential applications of this material. These carbon onions consist of hollow carbon onions, with between 2 and 8 graphitic shells, and their production in macroscopic quantities by this experimental approach enabled their examination by ultraviolet–visible (UV–Vis) and Raman spectroscopy. It was proposed that carbon onions could be a possible carrier of the 2175 Å interstellar absorption bump.[23,30,88] Raman spectra of this material revealed pronounced differences to other graphitic materials.[2]

Vacuum-annealing NDs with diameters in the range of 2–20 nm produced by detonation allowed the production of several grams of OLC at a time.[13,41,42,45] The synthesis of OLC in macroscopic quantities has allowed the possibility of exploring the physical and chemical characteristics of this material in a much more thorough manner. As a result of larger-scale production, it has been shown, for example, that because of an efficient optical limiting action, OLCs are good candidates for photonic applications.[37] OLC has also been shown to demonstrate high selectivity and catalytic activity in the oxidative dehydrogenation of ethylbenzene to styrene.[35,74,92] For this process, it has been demonstrated that OLC performs better as a catalyst than industrially used catalysts. The properties of OLC produced by annealing ND have been investigated by x-ray emission spectroscopy,[54] Raman spectroscopy,[52] electron energy-loss spectroscopy,[79] electron-spin resonance,[80] x-ray diffraction,[81] and UV–Vis absorption spectroscopy.[82]

The synthesis of carbon onions or OLC by heat treatment of nanostructures can be easily scaled to an industrial level. However, these techniques also have disadvantages: in case of NDs, the strong aggregation of the starting material leads to a final OLC product that mainly consists of aggregates of carbon onions bonded by curved graphitic layers and having different shapes and containing structural defects (see Figure 10.9d through f). Similar problems occur when producing OLC by annealing soot; the resulting carbon onions form aggregates and have irregular shape and structure.

10.3.3 Recent Development in Methods of Carbon Onion Synthesis

Three major challenges have been identified for the synthesis of carbon onions: (i) the quantity of carbon onion material produced, (ii) control of the structure of the carbon onions and of the level and nature of the defects, and (iii) purity, that is, production of carbon onions that do not contain other forms of carbon. Of these, isolation of carbon onions from other carbon materials is probably the most difficult task because carbon onions do not have such high solubility in organic solvents as fullerene molecules form strong bonds or aggregates with other carbon structures. Difficulties in obtaining deagglomerated and pure carbon onions, especially having perfect spherical shape and highly controlled structure, have prompted researchers to develop new synthetic techniques for this material.

Cabioc'h et al.[16–18] developed a method based on carbon ion implantation into a metal matrix (Ag, Cu), resulting in onions with typical diameters in the 3–15 nm range. Sufficient quantities could be produced for investigation of their optical, electronic, and tribological properties.[19–21,56] Fourier transform infrared (FTIR) spectroscopy[19] measurements on these carbon onions demonstrated that the most stable state for the onions consists of concentric spheres of fullerenes. The electronic properties of the onions were characterized by spatially resolved electron energy loss spectroscopy (EELS) in transmission,[36,56,71] and reflection mode.[31]

Sano et al.[1,64,65] reported the production of several milligrams of carbon onions with diameters in the 4–36 nm range, using an arc discharge between two graphite electrodes submerged in water or liquid nitrogen. It has been shown that depending on the synthesis conditions, the carbon nanoparticles produced by this method contain carbon onions with a spherical shape along with defective carbon onions and a minimum amount of amorphous carbon or carbon nanotubes. These carbon onions were characterized by Raman spectroscopy[61] and by UV–Vis absorption spectroscopy.[22] New Raman peaks were observed in spectra of the carbon onions compared with Raman spectra from highly oriented pyrolytic graphite (HOPG). The appearance of these new features could be explained by the curvature of the graphitic walls of the carbon onions.[61]

A new approach to the synthesis of carbon onions was proposed by Du et al.[26] They used radio-frequency and microwave plasma to produce OLC particles from coal and carbon black. Optimization of synthesis conditions in their experiments enabled Du et al. to produce carbon onions with a minimum of structural defects. The growth of carbon onions with a diameter of about 30 nm was achieved by Shimizu et al. by the application of inductively coupled microplasma.[69] However, both these plasma-based techniques resulted in the formation of carbon onions mixed with other carbons, which is a significant issue as the problem of carbon onions purification remains unsolved.

Carbon onions can also serve as nanocapsules for a range of different materials. Special interest is associated with carbon onions with metal cores because these particles can, for example, be used for the production of magnetic liquids or magnetic devices for information storage. Recently, synthetic techniques based on chemical vapor deposition (CVD) have been used to produce metal particles encapsulated by several graphitic layers.[29,63,93] CVD used for synthesis of metal particles encapsulated in carbon is based on the decomposition of carbon-containing substances on the surface of small metal catalyst particles that may themselves be produced by the decomposition of metal-containing molecules in the gas phase. For example, Sano et al. reported the synthesis of large quantities of iron particles covered by several graphitic layers through the decomposition of ferrocene in pure hydrogen.[63] Wang et al. synthesized iron particles encapsulated in graphitic layers using acetylene and cyclohexane as carbon sources.[93] The CVD method is indeed widely used for the synthesis of carbon nanotubes. Carbon onions can be considered as a zero-dimensional relative of carbon nanotubes; therefore, the mechanism of the formation of the carbon onions containing metal particles should be similar to that for carbon nanotubes produced by CVD. One of the possible routes for the synthesis of carbon onions with encapsulated metal particles involves the initial formation of metal nanoparticles in the hot zone of a CVD reactor by the decomposition of a metal-containing precursor compound; these metal particles catalyze the thermal decomposition of carbon-containing gases at their surfaces; carbon produced as a result of this decomposition dissolves in the metal particles up to a point of saturation; eventually, when such a metal particle enters the cold zone of the reactor, carbon dissolved in the metal precipitates and forms graphitic layers on and around the metal particles. As a result of this series of transformations, metal particles encapsulated in carbon onions are formed. A schematic illustration of the main stages of this process is presented in Figure 10.10.

Another interesting example of the synthesis of metal particles covered by graphitic layers is that of Xu et al.[94,95] Xu et al. synthesized carbon nanocapsules containing different metals by striking an arc discharge between graphite electrodes in water solutions containing salts of the chosen metals. Another potentially attractive approach could enable the formation of a wider range of materials by filling the preopened carbon onions. This method does no longer rely on the filling material being compatible with the process by which the carbon onions are produced. It has been shown by Butenko et al. that carbon onions can be opened by oxidative treatment in carbon dioxide,[15] and it is reasonable to assume that these opened onions can be subsequently used for the preparation of nanocapsules for different substances through controlled filling.

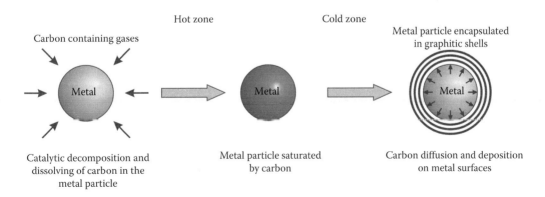

FIGURE 10.10 Schematic showing the stages of synthesis of a metal particle encapsulated in graphitic shells during the CVD process.

10.4 PROPERTIES AND APPLICATIONS

10.4.1 Electronic and Electrical Properties of Carbon Onions

The electrical conductivity of carbon allotropes spans an enormous range from insulating to (semi) metallic, and this can be generally understood with reference to the hybridization state of the carbon atoms, as discussed in Section 10.2.1, and the resultant localization or delocalization of electrons within the carbon solid. For a homogeneous carbon allotrope, bonds will be of σ- or π-character: electrons in the σ-bonds are strongly localized between carbon atoms and, hence, cannot contribute to the electrical conductivity. However, π-bonds on neighboring pairs of carbon atoms can overlap and become spatially delocalized. If partially occupied by electrons, such delocalized orbitals facilitate electrical conduction. Conversely, in the absence of a part-filled π-orbital, an undoped all-carbon material is typically an insulator (such as diamond that has sp^3-bonding; hence, no π-electrons and a band gap larger than 5 eV).

Graphite as a semimetal is a conductor with a low density of charge carriers at the Fermi level arising from band overlap and shows a highly anisotropic conductivity: the conductivity parallel to the graphite planes can be an order of magnitude or more larger than that perpendicular to them, depending on the quality of the graphite. The wrapping of graphitic sheets into nanoparticles, such as carbon onions, can have a significant impact on electronic structure. At the simplest level, the imposition of finite size and periodic boundary conditions can lead to discrete and well-defined molecular electronic states, such as observed in small fullerenes. In OLC, the molecule-like electronic behavior is suppressed 2003,[50] possibly because of the presence of curved graphitic planes connecting the onion-like structures. In addition to the constraints directly imposed by finite size, the introduction of pentagons, heptagons, and octagonal rings into hexagonal graphitic network necessary to introduce curvature (as discussed in Section 10.2.2) and shell closure results in partial rehybridization of the carbon atoms leading to bonds that have a character between the sp^2 bonds in a graphite sheet and the sp^3 bonds of diamond.[43] The precise details of bonding will depend on the precise curvature of the graphene sheet, which in turn depends on the exact structure of the carbon onion, with a deficit of π-electron density (resulting from the admixture of sp^3 hybridization) increasing as the curvature of the sheet increases.[43]

10.4.1.1 Electronic Properties of Carbon Onions

The electronic structure of OLC samples produced by annealing NDs has been investigated by means of x-ray photoemission spectroscopy (XPS),[14,50] soft x-ray emission spectroscopy, and x-ray absorption spectroscopy (XAS).[53,54] XPS, although a technique that is primarily associated with determining the elemental composition and chemical state of solid materials through the measurement of the binding energies of core electrons, can provide insight into the hybridization state and bonding of a carbon material. Moreover, the presence of loss structures associated with the excitation of plasmons and interband transitions provides valuable information regarding molecular orbital/band structure and electron density. Valence band photoemission spectroscopy is a complementary technique, in which low-energy photons are used to directly probe the full valence band, and hence, the electronic structure of solids. The occupied electronic states are also accessible through soft x-ray emission spectroscopy in which a core hole is created and an x-ray photon emitted when the core hole is subsequently filled by a valence electron. The resulting spectrum is proportional to the projection of the density of occupied valence-electron states for the element at which the core hole is created. Thus, rather than the complete density of states observed in the valence band photoemission experiment, x-ray spectroscopy yields a local partial density of occupied valence states (LPDOS). XAS is, by contrast, a technique by which the *unoccupied* valence states can be accessed. In this technique, the absorption of an x-ray photon causes the promotion of a core electron into the unoccupied part of the valence band manifold, the relative magnitude of absorption reflecting the matrix-element-weighted unoccupied density of states.

Annealing of ultradisperse diamond (UDD) nanoparticles leads to a variety of structures, depending on the precise annealing temperature, as shown in Figure 10.9 (Kuznetsov et al. 1994b, also see Section 10.3.2). Analysis of the C 1s x-ray photoemission line from intermediates of ND transformation prepared by thermal treatment at 1420 and 1600 K and then exposed to the atmosphere (Figure 10.11) shows sp^2- and sp^3-bonded carbon in addition to the presence of oxygen-containing groups.[14] The sp^2 component to the C 1s line is significantly increased in relative contribution to the sample annealed at higher temperature, which is consistent with increased nanoparticle graphitization. Indeed, one of the great advantages of photoemission spectroscopy is that it is possible to determine the sp^2:sp^3 ratio, at least semiquantitatively. Increased sp^2 hybridization and graphitic structural organization are also reflected in the presence of a significant number of delocalized π-electrons for samples subject to higher temperature annealing, as demonstrated by the observation of a plasmon loss feature at 290.8 eV binding energy associated with the collective motion of the π-electrons, which is absent for the less well-graphitized sample.

Figure 10.12 shows valence band spectra of UDD samples annealed at five different temperatures.[14] The spectra contain two prominent peaks at 2.8 ± 0.1 and 7.5 ± 0.1 eV and a shoulder at 4.2 ± 0.1 eV. The first peak at 2.8 ± 0.1 eV is related to π-bonding in graphitic materials,[50,53,70]

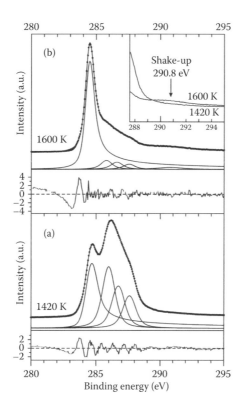

FIGURE 10.11   X-ray photoemission spectra showing the C1s line of ND annealed at (a) 1420 K and (b) 1600 K. The spectra were obtained in normal emission geometry at a photon energy of 1486.6 eV. The dots are experimental data and the solid lines are the fit components into which the spectra were decomposed. The resulting fit is superimposed on the data. The inset shows the appearance of the π plasmon peak at 290.8 eV in the spectrum of the sample annealed at 1600 K. The residuals from the curve fitting (in units of standard deviation) are displayed in the bottom panel beneath the spectra. (Adapted from Butenko, Yu. V. et al., 2005, *Phys. Rev. B* **71**, 075420.)

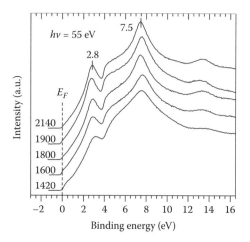

FIGURE 10.12 Evolution of valance band spectra as a function of annealing temperature for ND annealed at 1420, 1600, 1800, 1900, and 2140 K (temperatures in the Figure are presented in kelvin). Spectra were obtained in normal emission geometry using a photon energy of 55 eV. (Adapted from Butenko, Yu. V. et al., 2005, *Phys. Rev. B* **71**, 075420.)

and is seen to increase in intensity with annealing temperature, once again reflecting increased graphitization as the annealing temperature is increased.[14] A powerful aspect of valence band photoemission spectroscopy is that the intensity in the valence band close to the Fermi level is related to the density of conduction electron states that govern the transport properties of the material. Examination of this region of the valence band spectrum in the annealed ND material[14] shows an increase in density of states at the Fermi level for samples prepared at 1600, 1800, and 1900 K compared with that for a well-graphitized sample prepared at 2140 K (Figure 10.13). This behavior has been associated with an accumulation of different types of defects in the curved

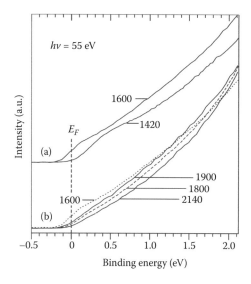

FIGURE 10.13 Evolution of the valance band spectra near the Fermi level (E_F) as a function of annealing temperature for samples annealed at (a) 1420 and 1600 K; (b) 1600, 1800, 1900, and 2140 K (temperatures in the Figure are presented in kelvin). Spectra were obtained in normal emission geometry using a photon energy of 55 eV. (Adapted from Butenko, Yu. V. et al., 2005, *Phys. Rev. B* **71**, 075420.)

graphite layers during graphitization of diamond.[14] The faceting of carbon onions at the highest annealing temperature (Figure 10.9) is associated with the reduction of the defects required for the spherical carbon onion shape (discussed in Section 10.2.2) and is reflected in the valence band photoelectron spectra.

10.4.1.2 Resistivity

The electrical resistivity of films of OLC produced by annealing UDD has been measured by a four-probe direct current (DC) method as a function of temperature between 4 and 300 K.[43] The use of four-probe measurements is particularly important in samples such as these, in which the magnitude of the contact resistance is unknown and may be significant. It has been found that UDD samples heated at temperatures lower than 1170 K, which do not exhibit any surface graphitic species, have very high values of resistivity, above 10^9 Ωcm.[43] As the graphitization of the ND material occurs with increasing annealing temperature, the resistivity of the samples drops significantly. For UDD annealed at temperatures above 1600 K, the resistivity at 300 K lies in the range 0.2–0.5 Ωcm and is comparable with that of carbon black. For all samples annealed at temperatures higher than 1170 K, the electrical resistivity shows a temperature dependence typical of systems with a variable-range hopping conductivity.[43,51] For typical graphitic materials (carbon black, graphitized soot, and graphite powder), three-dimensional hopping is usually observed; however, the hopping conductivity in the UDD annealing products displays a significantly lower dimensionality (Figure 10.14), which can be directly related to the conduction pathways available, which range from one-dimensional carbon chains in samples annealed at 1420 K to defective two-dimensional layers for high-temperature annealing treatments.[43]

10.4.1.3 Field Emission

Field emission, also known as Fowler–Nordheim tunneling, is the process by which electrons tunnel through the work function barrier at a surface in the presence of a high electric field and are emitted into the vacuum. Fowler–Nordheim theory is generally used to quantitatively describe the

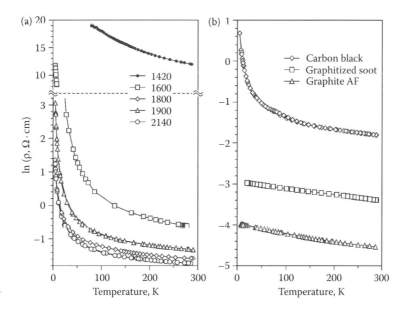

FIGURE 10.14 Temperature dependence of the resistivity of UDD annealing products and carbon comparison standards. (Adapted from Kuznetsov, V.L. et al., 2001, *Chem. Phys. Lett.* **336**, 397–404.)

field emission current density as a function of the electric field. This quantum-mechanical tunneling process is an important mechanism for thin barriers. The tunneling probability is

$$\Theta = \exp\left(-\frac{4}{3} \frac{\sqrt{2qm^*}}{\hbar} \frac{\phi_B^{3/2}}{E} \right)$$

where E is the applied electric field (which can be related to the applied voltage V by $E = V/d$, where d is the distance over which the voltage is applied), q is the carrier charge, m^* is effective mass of the charge carrier, ϕ_B is the barrier height, and $\hbar$ is the Planck h constant divided by 2π. The tunneling current is obtained from the product of the carrier charge, q, velocity, v, and density, n with the tunneling probability. The velocity, v, is taken as the average velocity with which the carriers approach the barrier. The net tunnel current density J is then given by

$$J = qvn\Theta$$

Ni-filled onion-like graphitic-particle films have shown electron field emission with a low threshold voltage.[48] These films were prepared by cosputtering of Ni and graphite and subsequent thermal annealing in vacuum. Figure 10.15a shows the electron field emission characteristics for an onion-like graphitic-particle film and an amorphous carbon film prepared by sputtering. A field emission current is detected from the OLC film at an applied voltage above 400 V and increases exponentially with the applied voltage, although there is no evidence of substantial field emission from the corresponding amorphous carbon film. In Figure 10.15b, the corresponding Fowler–Nordheim plot is shown; it can be seen that a straight line, representing Fowler–Nordheim behavior is obtained over a wide current range. It has been suggested[48] that the main reason for field emission from the onion-like graphitic-particle films is because of the presence of nanometer-scale protrusions that lead to a strong local electric field enhancement reinforced by the nonuniform resistivity of the OLC films that further enhances local electric fields. As can be seen from the derivation of the Fowler–Nordheim equation presented above, an increased local electric field will produce a greater tunneling probability and hence field emission current for a given applied voltage, leading to a low-threshold voltage value.

10.4.1.4 Carbon Onions as an Energy Material for Supercapacitor Electrodes

Electrochemical double-layer capacitors, often called supercapacitors, can store energy via reversible interfacial electrosorption of ions, and are used to power portable electronic equipment, hybrid electronic vehicle, and other devices. Supercapacitors have fast-charging and discharging rates and the ability to sustain millions of cycles. They bridge the gap between the batteries that offer high energy densities but are slow and between conventional electrolytic capacitors that are fast but have low energy densities.

OLC is an attractive material for electrical energy storage in regard to high-rate, high-power applications. The first report on electrochemical behavior of carbon onions in aqueous electrolyte was reported by Bushueva.[12] A more extended study of the electrochemical performance of carbon onions has been studied in electrical double-layer capacitors with organic electrolyte.[57] It was reported that carbon onion cells are able to deliver the stored energy under a high current density with a capacitance twice that of the one obtained with multiwalled carbon nanotubes.[57] It has been shown[55] that microsupercapacitors based on OLC, produced by several-micrometer-thick layer of nanostructured carbon onions, have a power that is comparable to electrolytic capacitors but the capacitance is four orders of magnitude higher and energies per volume are one order of magnitude higher. The discharge rates were also increased,[49,55] compared to conventional supercapacitors that are based on porous carbons, such as activated carbon. Improved properties of carbon onions as a

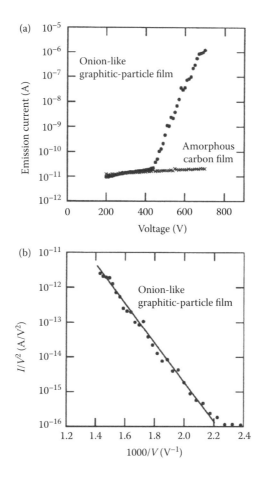

FIGURE 10.15 (a) Electron field emission characteristics for an onion-like graphitic-particle film and an amorphous carbon film prepared by sputtering. (b) Corresponding Fowler–Nordheim plot. (Adapted from Mamezaki, O. et al., 2000, *Jpn. J. Appl. Phys.* **39**, 6680–6683.)

material for supercapacitors, either as the main component or as a conductive additive, have been assigned to the good ordering of carbon onions,[27] surface with a pronounced positive curvature of carbon onion nanoparticles that is suitable to accommodate larger charges compared to negative curvature,[34] lack of use of any organic binders, and polymer separators that improve the mobility of the ions.[55] The important aspect is that the pure outer surface facilitates ion transport in contrast to mass transport limitations commonly encountered in porous carbon materials.

One way of manipulating the properties of carbon onions as supercapacitor electrodes is by heating of ND detonation soot by changing the sp^3/sp^2 ratio.[49] In addition, it might be worth investigating the electrochemical properties of OLC by manipulating carbon onions by the direct method of covalent functionalization.[10]

10.4.2 Optical Properties of Carbon Onions

10.4.2.1 Transmittance and Absorption

Since the discovery of carbon onions, it has been believed that they are a component of the interstellar dust and that they contribute to the strong absorption band centered at a wavelength of 217.5 nm.[23,30,40,87,88] Optical transmission spectroscopy measurements in the wavelength range of

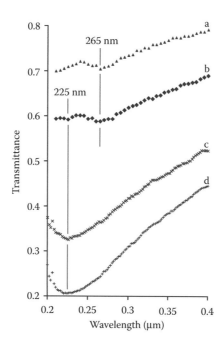

FIGURE 10.16 Optical transmittance of carbon onion layers obtained by the implantation of 120 keV carbon ions into Ag held at 500°C at various fluences: (a) 5×10^{16}, (b) 10^{17}, (c) 2×10^{17}, and (d) 3×10^{17} ions/cm². (Adapted from Cabioc'h, T. et al., 2000, *Eur. Phys. J. B* **18**, 535–540.)

0.2–1.2 μm were performed by Cabioc'h et al.[20] on carbon onions produced by carbon ion implantation into silver thin films and deposited on a silica surface. This method of production was chosen because after ion implantation into a 400-nm-thick silver film, the silver can be evaporated in vacuum (at 850°C, 10^{-5} Pa) leaving carbon onions weakly bond to the silica. Silica is almost transparent in the wavelength range of 0.2–1.2 μm, which enables measuring the optical transmittance of carbon onions supported by silica.[20] It is also possible by varying the fluence of the carbon ion implantation to change the size of the carbon onions, for example, increasing the fluence of 120 keV carbon ions from 5×10^{16} ions/cm² to 3×10^{17} ions/cm² produces an increase in carbon onion diameter from 4 to 8 nm.[20] Optical transmittance spectra of carbon onion layers deposited on silica substrates in this manner reveal two absorption peaks as shown in Figure 10.16. Minima in the optical transmittance are found to occur at 220–230 and 265 nm and are attributed to the presence of carbon onions and residual disordered graphitic carbon, respectively.[20] The transmittance decrease, which arises with increasing ion fluence during carbon onion production, has been explained by the increase of the average thickness of the carbon layer.

10.4.2.2 Electromagnetic Shielding in Onion-Like Carbons and Their Composites

Recently, there have been several reports that OLC obtained by annealing detonation NDs provides electromagnetic (EM) shielding over the microwave spectral band from 2 to 37 GHz[47,66] and, in addition, efficient attenuation of the EM spectrum over the frequency range of 12–230 THz as compared to the detonation ND starting material.[68] Another group of authors[28] reported high dielectric loss in the 2–18 GHz frequency range for carbon onions synthesized by DC arc discharge in the water method and carbon onions encapsulating Fe nanoparticles synthesized by CVD.

There have been a variety of mechanisms suggested as contributors to the broadband EM attenuation displayed by OLC, including electrical conductivity that can be tuned by the hierarchical assembly of the primary carbon onion particles and inner shell defects.[44] It has been observed that thermally treated NDs display a wide range of diameters over a broad length scale, such as sp^2 patches between defects (holes) within a single graphitic nanoshell, onion aggregates, and conglomerates of the OLC aggregates, which could be the additional reason for the wide band attenuation of the EM spectrum.[44,68] However, until now, the detailed microscopic mechanism of attenuation still remains unknown.

It has also been reported that dispersed nanodiamonds (DNDs) are efficient UV absorbers,[67] and on the basis of this observation, Shenderova et al.[68] suggest that the combination of the OLC and DND in a nanocomposite material could provide effective protection from EM radiation over a rather wide spectral range. There is still a need for the optimization of the thickness of the film and composition, and hence, further studies would be desirable.

10.4.3 TRIBOLOGICAL PROPERTIES: FRICTION AND LUBRICATION

Carbon onions have significant potential for the reduction of friction and wear and consequently can increase the lifetime of mechanical bearings. There are several ways by which frictional forces may be measured, the most common of which is the ball-on-disk or pin-on-disk tribometer. In this device, a flat, pin, or sphere is mounted on a stiff elastic arm, which is brought into contact with a sample and loaded with a precisely known weight. This ensures a nearly fixed contact point and thus, a stable position in the friction track. The relative friction coefficient is commonly determined by rotating the disk with respect to the probe (although reciprocating action is also possible with certain designs) and measuring the deflection of the elastic arm or by directly measuring the change in torque. Wear coefficients may be found from the amount of material lost in the test, which, for example, can be determined by scanning electron microscopy (SEM) or profilometry. This simple approach allows studies to be conducted of friction and wear for various material combinations, with or without lubrication.

Researchers at National Aeronautics and Space Administration (NASA) have investigated the tribological properties of carbon nanoonions as additives for aerospace applications.[24,73] So far, carbon onions have demonstrated superior lubrication properties when compared with some other conventional lubricants.[32,96] The tribological properties of carbon onions prepared by heat treatment of diamond clusters or particles at 1730°C using an infrared radiation furnace in argon at atmospheric pressure have been examined by ball-on-disc-type friction testing of the type described above using a silicon wafer and a steel ball.[32] The carbon onions, which are spread on the silicon wafer without adhesive, exhibit stable friction coefficients lower than 0.1 both in air and in vacuum at room temperature. It is found that the larger carbon onions prepared from diamond particles show low friction on the rough surface of silicon discs. The good solid lubrication displayed by carbon onions is assumed[32] to be because of the quasi-spherical structure of the carbon onions (diameter ~10 nm) material that is believed to behave as a film of nanometer scale. An additional contribution to the lubrication displayed by the carbon onions may arise from the absence of dangling bonds on the surface of well-graphitized carbon onions; when carbon atoms arrange perfectly so that the outermost shell contains no defects, weak intermolecular (van der Waals) bonding with the counterpart material is expected to occur.[32] Finally, the closed-shell structure of the carbon onion can be expected to be highly stable, leading to high mechanical strength, wear resistance, and thermal tolerance.[32]

10.5   SUMMARY

The highly symmetrical structure of carbon onions suggests that this material has unique properties with various significant potential applications. However, the synthesis of well-separated carbon onions with controlled structure in large quantities and free from other graphitic "impurities" is an extremely challenging task. Several research groups have taken up this challenge and explored a variety of approaches for the synthesis of this remarkable material, leading to the production of carbon onions characterized by a range of structures and degrees of perfection. In particular, depending on the exact method and conditions of synthesis, carbon onions can vary in shape, number of graphitic shells, lattice spacing between shells, diameter of the internal shell, and in presence and density of different types of defects. Despite the difficulties in the synthesis of carbon onions, they have already demonstrated great potential for several applications. Among the most promising applications of carbon onions is their use as additives to lubricants due to their excellent tribological properties, which arise in part from their highly symmetric shape. The electronic properties of carbon onions suggest that they may be successfully applied as a material for EM shielding equipment, field emission electrodes, and for photonic devices.

Perhaps, the most important results of the intensive research in the field of carbon onions are the establishment of a fundamental understanding and the development of new concepts. A better understanding of the diamond-to-graphite transformations has significantly broadened our ability to manipulate carbon nanostructures and to understand processes occurring under extremes of heat, pressure, and irradiation. The exciting idea of applying carbon onions as a nanocapsule for reactions and/or protection of species is an area still waiting to be fully explored with significant potential for impact in biology and medicine.

REFERENCES

1. Alexandrou, I., Wang, H., Sano, N., Amaratunga, G.A.J., 2004, Structure of carbon onions and nanotubes formed by arc in liquids, *J. Chem. Phys.* **120**, 1055–1058.
2. Bacsa, W.S., de Heer, W.A., Ugarte, D., Châtelain, A., 1993, Raman spectroscopy of closed-shell carbon particles, *Chem. Phys. Lett.* **211**, 346–352.
3. Banhart, F., Ajayan, P.M., 1996, Carbon onions as nanoscopic pressure cells for diamond formation, *Nature*, **382**, 433–435.
4. Banhart, F., 1997a, The transformation of graphitic onions to diamond under electron irradiation, *J. Appl. Phys.* **81**, 3440–3445.
5. Banhart, F., Füller, T., Redlich, Ph., Ajayan, P.M., 1997, The formation, annealing and self-compression of carbon onions under electron irradiation, *Chem. Phys. Lett.* **269**, 349–355.
6. Banhart, F., Redlich, Ph., Ajayan, P.M., 1998, The migration of metal atoms through carbon onions, *Chem. Phys. Lett.* **292**, 554–560.
7. Banhart, F., 1999, Irradiation effects in carbon nanostructures, *Rep. Prog. Phys.* **62**, 1181–1221.
8. Banhart, F., Charlier, J.-C., Ajayan, P.M., 2000, Dynamic behaviour of nickel atoms in graphitic networks, *Phys. Rev. Lett.* **84**, 686–689.
9. Barnard, A.S., Russo, S.P., Snook, I.K., 2003, Coexistence of bucky diamond with nanodiamond and fullerene carbon phases, *Phys. Rev. B* **68**, 073406.
10. Brieva, A.C., Jager, C., Huisken, F., Siller, L., Butenko, Yu. V., 2009, A sensible route to covalent functionalisation of carbon nanoparticles with aromatic compounds, *Carbon* **47**, 2812–2820.
11. Bundy, F.P., Bassett, W.A., Weathers, M.S., Hemley, R.J., Mao, H.K., Goncharov, A. F., 1996, The pressure–temperature phase and transformation diagram for carbon; updated through 1994, *Carbon* **34**, 141–963.
12. Bushueva, E.G., Okotrub, A.V., Galkin, P.S., Kuznetsov, V.L., Moseenkov, S.I., 2006, *Electrochemical Supercapacitors Based on Carbon Materials, Nanocarbon and Nanodiamond Conference*, St. Petersburg, Russia, pp. 11–15.
13. Butenko, Yu. V., Kuznetsov, V.L., Chuvilin, A.L. et al., 2000, Kinetics of the graphitization of dispersed diamonds at "low" temperatures, *J. App. Phys.* **88**, 4380–4388.
14. Butenko, Yu. V., Krishnamurthy, S., Chakraborty, A.K. et al., 2005, Photoemission study of onion-like carbons produced by annealing nanodiamonds, *Phys. Rev. B* **71**, 075420.
15. Butenko, Yu. V., Chakraborty, A.K., Peltekis, N. et al., 2008, Potassium intercalation of opened carbon onions, *Carbon* **46**, 1133–1140.
16. Cabioc'h, T., Riviere, J.P., Delafond, J., 1995, A new technique for fullerene onion formation, *J. Mater. Sci.* **30**, 4787–4792.
17. Cabioc'h, T., Girard, J.C., Jaouen, M., Denanot, M.F., Hug, G., 1997, Carbon onions thin film formation and characterization, *Europhys. Lett.* **38**, 471–475.

18. Cabioc'h, T., Jaouen, M., Denanot, M.F., Bechet, P., 1998a, Influence of the implantation parameters on the microstructure of carbon onions produced by carbon ion implantation, *Appl. Phys. Lett.* **73**, 3096–3098.

19. Cabioc'h, T., Kharbach, A., Le Roy, A., Riviere, J.P., 1998b, Fourier transform infra-red characterization of carbon onions produced by carbon-ion implantation, *Chem. Phys. Lett.* **285**, 216–220.

20. Cabioc'h, T., Camelio, S., Henrard, L., Lambin, P.H., 2000, Optical transmittance spectroscopy of concentric-shell fullerenes layers produced by carbon ion implantation, *Eur. Phys. J. B* **18**, 535–540.

21. Cabioc'h, T., Thune, E., Riviere, J.P. et al., 2002, Structure and properties of carbon onion layers deposited onto various substrates, *J. Appl. Phys.* **91**, 1560–1567.

22. Chhowalla, M., Wang, H., Sano, N., Teo, K.B.K., Lee, S.B., Amaratunga, G.A.J., 2003, Carbon onions: Carriers of the 217.5 nm interstellar absorption feature, *Phys. Rev. Lett.* **90**, 155504.

23. de Heer, W.A., Ugarte, D., 1993, Carbon onions produced by heat treatment of carbon soot and their relation to the 217.5 nm interstellar absorption feature, *Chem. Phys. Lett.* **207**, 480–486.

24. Delgado, J.L., Herranz, M.A., Martin, N., 2008, The nano-forms of carbon, *J. Mat. Chem.* **18**, 1417–1426.

25. Dresselhaus, M.S., Dresselhaus, G., Eklund, P.C., 1996, *Science of Fullerenes and Carbon Nanotubes*, Academic Press, San Diego.

26. Du, A.B., Liu, X.G., Fu, D.J., Han, P.D., Xu, B.S., 2007, Onion-like fullerenes synthesis from coal, *Fuel* **86**, 294–298.

27. Fulvio, P.F., Mayes, R.T., Wang, X., Mahurin, S. M., Bauer, J.C., Presser, V., Mc Donough, J., Gogotsi, Y., Dai, S., 2011, "Brick-and-mortar" self assembly approach to graphitic mesoporous carbon nanocomposites, *Adv. Funct. Mater.* **21**, 2208–2215.

28. Ge, A.Y., Xu, B.S., Wang, X.M., Li, T.B., Han, P.D., Liu, X.G., 2006, Study on electromagnetic property of nano onion-like fullerenes, *Acta Phys. Chim. Sinica* **22**, 203–208.

29. He, C.N., Shi, C.S., Du, X.W., Li, J.J., Zhao, N.Q., 2008, TEM investigation on the initial stage growth of carbon onions synthesized by CVD, *J. Alloys Compd.* **452**, 258–262.

30. Henrard, L., Lambin, P.H., Lucas, A.A., 1997, Carbon onions as possible carriers of the 2175 Angstrom interstellar absorption bump, *Astrophysics. J.* **487**, 719–727.

31. Henrard, L., Malengreau, F., Rudolf, P. et al., 1999, Electron-energy-loss spectroscopy of plasmon excitations in concentric-shell fullerenes, *Phys. Rev. B* **59**, 5832–5836.

32. Hirata, A., Igarashi, M., Kaito, T., 2004, Study on solid lubricant properties of carbon onions produced by heat treatment of diamond clusters or particles, *Tribol. Int.* **37**, 899–905.

33. Iijima, S., 1980, Direct observation of the tetrahedral bonding in graphitized, *J. Cryst. Growth* **5**, 675–683.

34. Huang, J., Sumpter, B.G., Meunier, V., Yushin, G., Portet, C., Gogotsi, Y., 2010, Curvature effects in carbon nanomaterials: Exohedral versus endohedral supercapacitors, *J. Mater. Res.* 25, 1525–1531.

35. Keller, N., Maksimova, N.I., Roddatis, V.V. et al., 2002, The catalytic use of onion-like carbon materials for styrene synthesis by oxidative dehydrogenation of ethylbenzene, *Angew. Chem. Int. Ed.* **41**, 1885–1888.

36. Kociak, M., Henrard, L., Stéphan, O., Suenaga, K., Colliex, C., 2000, Plasmons in layered nanospheres and nanotubes investigated by spatially resolved electron energy-loss spectroscopy, *Phys. Rev. B* **61**, 13936–13944.

37. Koudoumas, E., Kokkinaki, O., Konstantaki, M. et al., 2002, Onion-like carbon and diamond nanoparticles for optical limiting, *Chem. Phys. Lett.* **357**, 336–340.

38. Kroto, H.W., Heath, J.R., O'Brien, S.C. et al., 1985, C_{60}— Buckminsterfullerene, *Nature* **318**, 162–163.

39. Kroto, H.W., McKay, K., 1988, The formation of quasi-icosahedral spiral shell carbon particles, *Nature* **331**, 328–331.

40. Kroto, H.W., 1992, Carbon onions introduce new flavor to fullerene studies, *Nature* **359**, 670–671.

41. Kuznetsov, V.L., Chuvilin, A.L., Butenko, Yu. V., Titov, V.M., 1994a, Onion-like carbon from ultradisperse diamond, *Chem. Phys. Lett.* **222**, 343–348.

42. Kuznetsov, V.L., Chuvilin, A.L., Moroz, E.M. et al., 1994b, Effect of explosions on the structure of detonation soots: Ultradisperse diamond and onion carbon, *Carbon* **32**, 873 882.

43. Kuznetsov, V.L., Butenko, Yu. V., Chuviln, A.L., Romanenko, A.I., Okotrrub, A.V., 2001, Electrical resistivity of graphitized ultra-disperse diamond and onion-like carbon, *Chem. Phys. Lett.* **336**, 397–404.

44. Kuznetsov, V.L., Butenko, Yu. V., 2005, in: D. Guen, O. Shenderova, A. Vul (eds.), *Synthesis, Properties and Applications of Ultrananocrystalline Diamond (NATO Science Series)*, Springer, Amsterdam, p. 199.

45. Kuznetsov, V.L. Butenko, Yu. V., 2006, Diamond phase transition at nanoscale, in: Olga A. Shendorova and Dieter M. Gruen (eds.), *Ultra Nanocrystalline Diamond, Synthesis, Properties and Applications*, William Andrew Publishing, Norwich, New York, NY, Ch. 13, pp. 405–475.

46. Li, J., Banhart, F., 2005, The deformation of single, nanometer-sized metal crystals in graphitic shells, *Adv. Mater.* **17**, 1539–1542.

47. Maksimenko, S.A., Rodinova, V.N., Slepyan, Ya. G. et al., 2007, Attenuation of electromagnetic waves in onion-like carbon composites, *Diam. Relat. Mater.* **16**, 1231–1235.

48. Mamezaki, O., Adachi, H., Tomita, S., Fujii, M., Hayashi, S., 2000, Thin films of carbon nanocapsules and onion-like graphitic particles prepared by the cosputtering method, *Jpn. J. Appl. Phys.* **39**, 6680–6683.

49. McDonough, J.K., Frolov, A.I., Presser, V., Niu, J., Miller, C.H., Ubieto, T., Fedorov, M.V., Gogotsi, Y., 2012, Influence of the structure of carbon onions on their electrochemical performance in supercapacitor electrodes, *Carbon* **50**, 3298–3309.

50. Montalti, M., Krishnamurthy, S., Chao, Y. et al., 2003, Photoemission spectroscopy of clean and potassium-intercalated carbon onions, *Phys. Rev. B* **67**, 113401–113404.

51. Mott, N.F., Davis, E.A., 1979, *Electron Processes in Noncrystalline Materials*, Claredon Press, Oxford.

52. Obraztsova, E.D., Fujii, M., Hayashi, S., Kuznetsov, V.L., Butenko, Yu. V., Chuvilin, A.L., 1998, Raman identification of onion-like carbon, *Carbon* **36**, 821–826.

53. Okotrub, A.V., Bulusheva, L.G., Kuznetsov, V.L., Butenko, Yu.V., Chuvilin, A.L., Heggie, M.I., 2001, X-ray emission studies of the valence band of nanodiamonds annealed at different temperatures, *J. Phys. Chem. A* **105**, 9781–9787.

54. Okotrub, A.V., Bulsheva, L.G., Kuznetsov, V.L., Vyalikh, D.V., Poyguin, D.V., 2005, Electronic structure of diamond/graphite composite nanoparticles, *Eur. Phys. J. D* **34**, 157–160.

55. Pech, D., Brunet, M., Durou, H., Huang, P., Mochalin, V., Gogotsi, Y., Taberna, P.-L., Simon, P., 2010, Ultrahigh-power micrometre-sized supercapacitors based on onion-like carbon, *Nat. Nanotechnol.* **5**, 651–654.

56. Pichler, T., Knupfer, M., Golden, M.S., Fink, J., Cabioc'h, T., 2001, Electronic structure and optical properties of concentric-shell fullerenes from electron-energy-loss spectroscopy in transmission, *Phys. Rev. B* **63**, 155415.

57. Portet, C., Yushin, G., Gogotsi, Y., 2007, Electrochemical performance of carbon onions, nanodiamonds, carbon black and multiwalled nanotubes in electrical double layer capacitors, *Carbon* **45**, 2511–2518.

58. Raty, J.-Y., Galli, G., Bostedt, C., van Buuren, T.W., Terminello, L.J., 2003, Quantum confinement and fullerene-like surface reconstructions in nanodiamonds, *Phys. Rev. Lett.* **90**, 037401.

59. Redlich, P.H., Banhart, F., Lyutovich, Y.U., Ajayan, P.M., 1998, EELS study of the irradiation-induced compression of carbon onions and their transformation to diamond, *Carbon* **36**, 561–563.

60. Roddatis, V.V., Kuznetsov, V.L., Butenko, Yu.V., Su, D.S., Schlögl, R., 2002, Transformation of diamond nanoparticles into carbon onions under electron irradiation, *Phys. Chem. Chem. Phys.* **4**, 1964–1967.

61. Roy, D., Chhowalla, M., Wang, H. et al., 2003, Characterisation of carbon nano-onions using Raman spectroscopy, *Chem. Phys. Lett.* **373**, 52–56.

62. Saito, R., Dresselhaus, G., Dresselhaus, M.S., 1998, *Physical Properties of Carbon Nanotubes,* Imperial College Press, London.

63. Sano, N., Akazawa, H., Kikuchi, T., Kanki, T., 2003, Separated synthesis of iron-included carbon nanocapsules and nanotubes by pyrolysis of ferrocene in pure hydrogen, *Carbon* **41**, 2159–2162.

64. Sano, N., Wang, H., Chhowalla, M., Alexandrou, I., Amaratunga, G.A.J., 2001, Nanotechnology—Synthesis of carbon "onions" in water, *Nature* **414**, 506.

65. Sano, N., Wang, H., Alexandrou, I. et al., 2002, Properties of carbon onions produced by an arc discharge in water, *J. Appl. Phys.* **92**, 2783–2788.

66. Shenderova, O., Tyler, T., Cunningham, G. et al., 2007a, Nanodiamond and onion-like carbon polymer nanocomposites, *Diam. Relat. Mater.* **16**, 1213–1217.

67. Shenderova, O., Grichko, V., Hens, S., Walch, J., 2007b, Detonation nanodiamonds as UV radiation filter, *Diam. Relat. Mater.* **16**, 2003–2008.

68. Shenderova, O., Grishko, V., Cunningham, G., Moseenkov, S., McGuire, G., Kuznetsov, V., 2008, Onion-like carbon for terahertz electromagnetic shielding, *Diam. Relat. Mater.* **17**, 462–466.

69. Shimizu, Y., Sasaki, T., Ito, T., Terashima, K., Koshizaki, N., 2003, Fabrication of spherical carbon via UHF inductively coupled microplasma CVD, *J. Phys. D: Appl. Phys.* **36**, 2940–2944.

70. Skytt, P., Glans, P., Mancini, D.C. et al., 1994, Angle-resolved soft-x-ray fluorescence and absorption study of graphite, *Phys. Rev. B* **50**, 10457–10461.

71. Stöckli, T., Bonard, J.-M., Châtelain, A., Wang, Z.L., Stadelmann, P., 2000, Plasmon excitations in graphitic carbon spheres measured by EELS, *Phys. Rev. B* **61**, 5751–5759.

72. Stone, A.J., Wales, D.J., 1986, Theoretical studies of icosahedral C_{60} and some related species, *Chem. Phys. Lett.* **128**, 501–503.

73. Street, K.W., Marchetti, M., Vandar Wal, R.V., Tomasek, A.J., 2004, Evaluation of the tribological behaviour of nano-onions in Krytox 143AB. *Tribol. Lett.* **16**, 143–149.

74. Su, D.S., Maksimova, N., Delgado, J.J. et al., 2005, Nanocarbons in selective oxidative dehydrogenation reaction, *Catal. Today* **102**, 110–114.

75. Sun, L., Banhart, F., 2006, Graphitic onions as reaction cells on the nanoscale, *Appl. Phys. Lett.* **88**, 193121(3).

76. Terrones, H., Terrones, M., Hsu, W.K., 1995, Beyond C_{60}: Graphite structures for the future, *Chem. Soc. Rev.* **24**, 341–350.

77. Terrones, M., Hsu, W.K., Hare, J.P., Kroto, H.W., Terrones, H., Walton, D.R.M., 1996, Graphitic structures: From planar to spheres, toroids and helices, *Phil. Trans. R. Soc. Lond. A* **354**, 2025–2054.

78. Terrones, H., Terrones, M., 1997, The transformation of polyhedral particles into graphitic onions, *J. Phys. Chem. Solids* **58**, 1789–1796.

79. Tomita, S., Fujii, M., Hayashi, S., Yamamoto, K., 1999, Electron energy-loss spectroscopy of carbon onions, *Chem. Phys. Lett.* **305**, 225–229.

80. Tomita, S., Sakurai, T., Ohta, H., Fujii, M., Hayashi, S., 2001, Structure and electronic properties of carbon onions, *J. Chem. Phys.* **114**, 7477–7482.

81. Tomita, S., Burian, A., Dore, J.C., LeBolloch, D., Fujii, M., Hayashi, S., 2002a, Diamond nanoparticles to carbon onions transformation: X-ray diffraction studies, *Carbon* **40**, 1469–1474.

82. Tomita, S., Hayashi, S., Tsukuda, Y., Fujii, M., 2002b, Ultraviolet–visible absorption spectroscopy of carbon onions, *Phys. Solid State* **44**, 450–453.

83. Ugarte, D., 1992, Curling and closure of graphitic networks under electron-beam irradiation, *Nature* **359**, 707–709.

84. Ugarte, D., 1993a, Formation mechanism of quasi-spherical carbon particles induced by electron bombardment, *Chem. Phys. Lett.* **207**, 473–479.

85. Ugarte, D., 1993b, How to fill or empty a graphitic onion, *Chem. Phys. Lett.* **209**, 99–103.

86. Ugarte, D., 1994, High-temperature behaviour of "fullerene black", *Carbon* **32**, 1245–1248.

87. Ugarte, D., 1995a, Onion-like graphitic particles, *Carbon* **33**, 989–993.

88. Ugarte, D., 1995b, Interstellar graphitic particles generated by annealing of nanodiamonds and their relation to the 2175-Angstrom peak carrier, *Astroph. J.* **443**, L85–L88.

89. Zaiser, M., Banhart, F., 1997, Radiation-induced transformation of graphite to diamond. *Phys. Rev. Lett.* **79**, 3680–3683.

90. Zaiser, M., Lyutovich, Y.U., Banhart, F., 2000, Irradiation-induced transformation of graphite to diamond: A quantitative study, *Phys. Rev. B.* **62**, 3058–3064.

91. Znang, Q.L., O'Brien, S.C., Heath, J.R. et al., 1986, Reactivity of large carbon clusters: Spheroidal carbon shells and their possible relevance to the formation and morphology of soot, *J. Phys. Chem.* 90, 525–528.

92. Zhang, J., Su, D.S., Zhang, A.H., Wang, D., Schlogl, R., Hebert, C., 2007, Nanocarbon as robust catalyst: Mechanistic insight into carbon-mediated catalysis, *Angew. Chem. Inter. Edit.* **46**, 7319–7323.

93. Wang, X.M., Xu, B.S., Uu, X.G., Guo, J.J., Ichinose, H., 2006, Synthesis of Fe-included onion-like fullerenes by chemical vapor deposition, *Diam. Relat. Mater.* **15**, 147–150.

94. Xu, B.S., Guo, J.J., Wang, X.M., Liu, X.G., Ichinose, H., 2006, Synthesis of carbon nanocapsules containing Fe, Ni or Co by arc discharge in aqueous solution, *Carbon*, **44**, 2631–2634.

95. Xu, B.S., 2008, Prospects and research progress in nano onion-like fullerenes, *New Carbon Mater.* **23**, 289–301.

96. Yao, Y., Wang, X., Guo, J., Yang, X., Xu, B., 2008, Tribological property of onion-like fullerenes as lubricant additive, *Mater. Lett.* **62**, 2524–2524.

97. Qin, L.C., Iijima, S., 1996, Onion-like graphitic particles produced from diamond, *Chem. Phys. Lett.* **262**, 252–258.

98. Bar-Yam, Y., Moustakas, T.D., 1989, Defect-induced stabilization of diamond films, *Nature* **342**, 786–787.

11 Carbide-Derived Carbons

Yair Korenblit and Gleb Yushin

CONTENTS

11.1 INTRODUCTION

The synthesis of carbon can be arranged through many different techniques, including chemical-vapor deposition (CVD; plasma enhanced,[1,2] low pressure,[3] laser ablation,[4,5] and arc discharge[6,7]), hydrothermal synthesis,[8] electrolysis,[9] and various other techniques.[10] The temperatures needed to synthesize carbonaceous materials can vary from subzero[9] to thousands of degree centigrade,[11] although the pressure can vary from low pressures (as when making amorphous carbon coatings[3]) to extremely high pressures (e.g., diamond synthesis[11]). With thermal energy being the main source of energy for bond breaking and atomic mobility (such as in CVD), different synthesis conditions are required for the various allotropes of carbon.

The development of novel methods to synthesize carbon structures and to allow for strict control of the resulting structure is of key importance for the advancement of technology and for providing scientists with new means to study various phenomena. Through strict control of material structure, the necessary properties for various applications can be obtained. For example, in the case of electrochemical capacitors, it is vital to control the pore size distribution, surface area, and conductivity to obtain high-energy and power-density devices. For the storage of gases, such as hydrogen or methane, the pore size, pore volume, and surface area are also of key importance, among other parameters. A synthesis method that allows for such tight control relies on the extraction of metals from metal carbides, resulting in what is often termed carbide-derived carbon (CDCs). The extraction of the metal can be accomplished through several routes, including hydrothermal etching,[12–14] high-temperature treatment in halogens,[12,15–38] and vacuum decomposition.[29,39] The transformation from carbide to carbon is manifested as a carbon growth front growing from the outer layer of the carbide inward. During the synthesis of CDC, a variety of structures may form, including amorphous,[40] various crystal sizes of graphite,[18,41] nanotubes,[18] fullerene-like structures,[18] graphitic ribbons,[18] carbon onions,[18] and nano- to microcrystalline diamond.[18,42] In essence, the majority of known carbon allotropes may be produced via the halogene treatment of metal carbides.

The morphology of the carbon structure is dictated by the carbide from which it originates and, because of the transformation mechanism, this method can provide control on the sub-nm level. The morphology retention or conformality provides a great advantage over CVD and similar techniques. It allows for simplified and strict design of the shape and structure of the carbon, providing engineers and scientists an extremely powerful tool for the advancement of nanotechnology. Applications that have already benefited from the development of CDC will be discussed further in this chapter including electrochemical energy storage, gas storage, tribological applications, catalyst supports, and biomedical applications. Many promising results have been obtained in these areas through the synthesis of various carbon structures. The most recent results will be discussed below.

11.2 SYNTHESIS, STRUCTURE, AND PROPERTIES OF CDC

11.2.1 HISTORICAL PERSPECTIVE

Chlorine treatment of carbides has become the most commonly used method of producing CDC.[43] This method has come a long way since its humble beginnings. In the early twentieth century, the main purpose of the chlorination of metal carbides was to produce silicon tetrachloride and the remnants were not given much thought. In 1918, Hutchins patented the process entailing the flow of dry chlorine gas over hot (>1000°C) silicon carbide[44]:

$$SiC_{(s)} + 2Cl_{2(g)} \rightarrow SiCl_{4(g)} + C_{(s)} \tag{11.1}$$

The resulting silicon tetrachloride vapor was retrieved in a condenser, and the carbon was not further considered. In 1953, a U.S. patent addressing the remaining carbon was filed and the patent was awarded in 1956.[45] This patent regarded the remaining carbon as an unwanted waste product and devised a method of removing it without having to open the system and retrieve the carbon by passing air through the hot chamber to form, presumably, CO and CO_2. This facilitated large-scale production efficiency and maintained that the carbon was of no significant importance.

In 1959, Mohun filed a patent (and publicly presented the findings) for a similar process to that of Hutchins,[44] but with a paradigm shift. The end means of chlorinating the metal carbides was to retrieve carbon, not the metal chloride. This "novel" carbon was described as "highly adsorbent and catalytically active." It was distinguished clearly from previously produced activated carbons by calling it a "mineral active carbon" that was different from amorphous carbons, such as hard carbons, soft carbons, and carbon black, often produced from organic precursors. In 1962, Mohun's idea was patented and along with it a description of many different carbides and various properties of the resulting carbon, including sorption ability and crystallinity. It was found that the heat of adsorption for organic vapors was very high (comparable to commercial activated carbons of the time) and the conductivity of these carbons was improved over activated carbons. Unlike previously prepared carbons from organic precursors, the produced carbon was presumed to be a much purer carbon as it was not contaminated by a large variety of functional groups as those attached to organic-precursor-derived carbons.

Mohun's experiments included chlorine treatment of silicon carbide, aluminum carbide, boron carbide, titanium carbide, zirconium carbide, and zirconium carbonitride, generally at temperatures

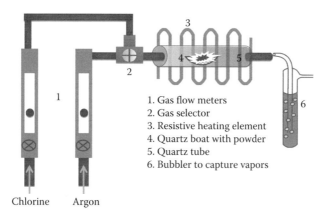

3

4   5

2

1

1. Gas flow meters
2. Gas selector
3. Resistive heating element
4. Quartz boat with powder
5. Quartz tube
6. Bubbler to capture vapors

6

Chlorine   Argon

FIGURE 11.1 Schematic of the apparatus used in the synthesis of CDC through the halogen treatment of metal carbides.

above 500°C in a reactor made of mullite. Various methods of reaction product removal were examined, including vacuum at elevated temperature, flowing nitrogen, or air at elevated temperature or by wet chemistry methods (e.g., by using water, sodium hydroxide, ammonium hydroxide, sodium hydrogen sulfate, hydrochloric acid, and hydrogen peroxide). Many samples revealed formation of a lamellar structure at low magnification (e.g., 350×) that was seen in other carbons, but unique to carbon with a low crystalline order. Another characterization technique, magnetic resonance, indicated no measurable resonance, again, unlike other carbons at that time. Diffraction studies indicated that unlike prior amorphous carbons exposed to temperatures above 400°C, the majority of carbons synthesized by Mohun were mostly x-ray amorphous and showed promise with respect to their ability to adsorb various gases and chemicals from solution (e.g., acetone, benzene, *m*-xylene, iodine, and phenol). These various characteristics of the "mineral active carbon" distinguished it from commercial carbons and created the foundation for future developments in this area.

As time progressed, more studies on this unique carbon came about. These studies included investigations of the kinetics of the metal-removal process,[48,49] the use of various metal carbides (including the ones not previously mentioned, such as niobium carbide, calcium carbide, and molybdenum carbide),[31,37,48,50–61] effect of using other halogens as metal-etching agents,[62–64] the impact of crystal structure[65] and pore structure,[30,52,56,61,66–75] and various other aspects of the conversion process.[76,77] Over time, the term "carbide-derived carbon" has become fairly popular and will be used throughout this chapter. A diagram of a commonly used variant of the setup used to synthesize CDCs is illustrated in Figure 11.1. The system usually entails the flow of an inert gas used to assist in removing moisture and oxidizing gases from the system before removal of the metal by flowing chlorine or other halogens. The exhaust gas passes through a bubbler (to avoid any back stream of oxidizing gases) and a scrubber to neutralize chlorine and other toxic or dangerous chemicals.

As more studies were completed into the development of CDCs, their use in gas sorption, filtration, energy storage, and many other applications have become sufficiently interesting that large quantities can now be purchased through companies such as Y-Carbon Inc. (USA) and Skeleton Technologies Inc. (Estonia). Continued investigations have provided engineers and scientists an improved understanding of the mechanisms, thermodynamics, and kinetics of the process of CDC formation, resulting in better control over the design and processing. Such studies are discussed in the following sections.

11.2.2 Thermodynamics

To better understand the underlying mechanisms for the formation of CDC materials, several studies under thermodynamic equilibrium have been undertaken for CDC synthesis through halogen treatment of SiC, Ti_3AlC_2, Ti_3SiC_2, B_4C, Ti_2AlC, TiC, and ZrC.[68,78–83] The majority of thermodynamic data associated with these studies were performed using a Gibbs energy minimization program. These calculations provided interesting insight into the nature of carbide-to-carbon transition and enabled a first idea on what experimental conditions to choose for an actual experiment, such as the temperature. However, it is important to note that such software assumes a closed system, ideal gases, and an equilibrium state. In reality, synthesis systems (Figure 11.1) entail the continuous flow of nonideal gases and equilibrium is not attained. These assumptions make it so that, in addition to various other complicating factors (such as the effect of structure on stability and kinetics), experiments may not comport with thermodynamic predictions.

Despite these limitations, there are lessons to be learned from such calculations. After examination of various studies,[78] a tri-region segmentation can be made with respect to chlorine treatment, temperature, and likely products. For example, Dash[78] examined the products and reactants in systems involving ZrC (Figure 11.2), TiC, B_4C, and SiC in a temperature range from 0°C to 1200°C with a 10:1 mol ratio of chlorine to carbide. In the region below 450°C (region I), the formation of CCl_4 is stable and CDC formation is not. Thus, for CDC synthesis, this may be a region to avoid. The following is an example of such a reaction:

$$ZrC_{(s)} + 4Cl_{2(g)} \rightarrow ZrCl_{4(s,g)} + CCl_{4(s,g)} \tag{11.2}$$

In the temperature region between 450°C and 750°C (region II), CCl_4 is still stable, but an increasing amount of carbon becomes stable as the temperature is increased, or alternatively, if the concentration of chlorine is reduced. A reaction illustrating this region is

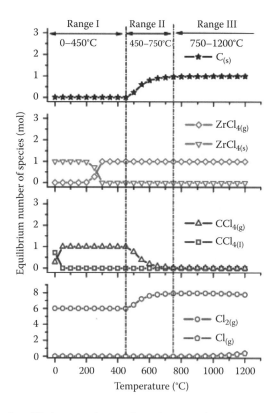

FIGURE 11.2 Calculated equilibrium reaction products for a reaction of 10 mol of Cl_2 with 1 mol of ZrC. (Adapted from Dash, R.K. In *Department of Materials Science and Engineering, Dr. of Philosophy,* Drexel University, Philadelphia, PA; 2006.)

$$ZrC_{(s)} + [2(2 - x)]Cl_{2(g)} \rightarrow ZrCl_{4(s,g)} + xC_{(s)} + (1 - x)CCl_{4(s,g)} \tag{11.3}$$

Finally, at the range above 750°C (region III), the formation of CCl_4 is no longer stable and the efficient synthesis of CDC can be attained through

$$ZrC_{(s)} + 2Cl_{2(g)} \rightarrow ZrCl_{4(g)} + C_{(s)} \tag{11.4}$$

It may seem somewhat intuitive that increasing the chlorine content may provide an impetus for the formation of CCl_4, but thermodynamic modeling provides researchers with an estimate for the effect of chlorine content on the reaction mechanisms and appropriate temperature windows. For example, range II can be shifted to higher temperatures by simply increasing the chlorine content per 1 mol of carbide.[80]

The predictions made by the modeling in some studies, where synthesis of CDC appears unlikely,[79] may turn out to be inaccurate, as was the case when examining the effect of adding hydrogen to chlorine in the conversion process.[84] As the thermodynamic calculations indicate, the progression from stability to instability is a function of temperature, pressure, and ratios of gases to solids. Although some limited studies may not observe the formation of CDC under certain conditions, as in Ref. 84, a variation of the process parameters will indicate that the implied stability through thermodynamic calculations merely required a change in condition. For example, in Ref. 29, hydrochloric acid was also found to form CDC but it was not observed until a vast increase in etching time and temperature was employed. One reason behind the need to increase operating temperatures and etching time may simply be because of a decreased driving force. A contrast between the Gibbs energy can be seen in the following two possible reactions[85]:

$$SiC + 2Cl_2 \rightarrow SiCl_4 + C \quad \Delta G° = -434.1 \text{ kJ mol}^{-1} \tag{11.5}$$

$$SiC + 4HCl \rightarrow SiCl_4 + C + 2H_2 \quad \Delta G° = -25 \text{ kJ mol}^{-1} \tag{11.6}$$

Although Ischenko[84] provides slightly different numbers, the change in driving force is clearly visible and may account for such a drastic change in processing conditions to be taken to observe CDC formation. Simply changing the chlorine to hydrogen ratio from 2:0.6 to 2:1 resulted in significant changes and the resulting carbons had different crystal structure and very different mechanical properties.[42] Low halogen-to-carbon ratios often leave behind unreacted carbide and possible contaminants such as CCl_4. However, they may also allow for the synthesis of more disordered carbon that could conceivably be beneficial for certain applications.[86]

Computational thermodynamics also enables an estimate of different metal chloride species that are formed during the chlorine treatment. For example, there are a variety of possible reaction products for the chlorine treatment of iron carbide[87]:

$$2Fe_3C + 9Cl_2 \rightarrow 6FeCl_3 + 2C \quad \text{or} \tag{11.7}$$

$$Fe_3C + 3Cl_2 \rightarrow 3FeCl_2 + C \tag{11.8}$$

In addition to these reactants, Fe_2Cl_6 and Fe_2Cl_4 are possible in the gas phase. If solid iron salts (such as $FeCl_2(s)$) are formed, it may be necessary to remove them at very high temperatures (>1000°C); however, as the calculations show, increasing the chlorine content shifts the reaction toward the formation of Fe_2Cl_6 (g) avoiding the issue of forming salts in the pores.

11.2.3 KINETICS

The kinetics of CDC formation are not easily generalized as there is a dependency on thickness, porosity, and experimental setup. The majority of studies have reported a linear growth rate for CDC films,[29,32,42,88,89] though there are studies indicating a large deviation from linear growth at longer times for the chlorine treatment or increased thicknesses.[90,91] Films have been synthesized on templates ranging from sintered SiC, monolithic SiC films, from fibers to whiskers, and other geometries to CDC thicknesses in the 100s of micrometers on millimeter-thick samples.[29,92] For short times and thin layers, the carbon layer grows in a linear fashion and can be characterized through a simple equation[88]

$$d_{thickness} = k_l t \tag{11.9}$$

where d is the thickness of the carbon layer, t is the time, and k_l is the linear rate constant (commonly 1–10 μm/h), which is a function of various parameters, including experimental setup, gas mixture, precursor structure and morphology, temperature, reaction products, and more. The time and thickness limitations for such a fitting equation are to be determined on a case-by-case basis to verify its applicability. Under these constraints, the conversion seems to be reaction controlled, rather than gas diffusion controlled that would result in linear-parabolic or parabolic growth kinetics.[93] Only for a sufficiently thick sample or treatment times, the kinetics deviate significantly from a linear correlation and diffusion control becomes more prominent.

During conversion to CDC, in many cases, micropores appear to facilitate the transport of halogens to the carbide–CDC interface and the extraction of the resulting reaction products. If this situation develops and rapid transport is facilitated, a surface reaction-controlled regime is made possible and a time-independent linear rate constant is observed, as can be seen in Figure 11.3. The conversion has been shown to allow complete transformation of carbide fibers,[89,94] films,[95] plates,[96] and powders[80] to CDC. Under certain conditions, such as thick CDC layers or perhaps increased flow rates, the linear reaction kinetics are no longer prominent and parabolic growth kinetics may be observed.[97] As would be expected, increasing the chlorine flow rate has been shown to increase the carbide conversion rate, but not significantly enough to warrant the use of pure chlorine.[91] However, caution must be employed when considering a diluting agent for economic reasons because at least in one study, a 22% decrease in growth rate was found when adding 1% hydrogen to the gas mix.[98]

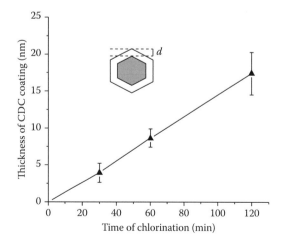

FIGURE 11.3 Kinetics data plot for chlorine treatment of SiC showing the thickness of the CDC coating formed at 700°C as a function of time. (Adapted from Cambaz, Z.G. et al., *Journal of the American Ceramic Society* **89**, 509–514, 2006.)

From the various findings, it seems that a linear-parabolic model would best fit the conversion of carbides changing based on flow rate, temperature, experimental setup, geometry of the carbide, thickness of CDC, and chlorine treatment time. For example, in at least one study, it appeared that a lower synthesis temperature decreased the linear growth region.[43]

11.2.4 CARBON: ORDER AND MORPHOLOGY

Mohun's early examination of CDC revealed that the majority of his samples indicated no perceptible or very weak (002) x-ray diffraction peaks up until 1250°C, a feature that distinguished them from the carbons studied before that point.[47] He estimated that the structures forming this amorphous carbon should correspond only to 2–3 aromatic layers (graphitic sheets). Mohun was able to discover some very exciting properties of CDCs and in the following decades, through continued studies and new and improved techniques, much more was discovered. Through the use of transmission electron microscopy (TEM), a myriad of carbon structures was discovered in CDCs, including amorphous carbon,[40] various crystal sizes of graphite,[18,41] nanotubes,[18] fullerene-like structures,[18] graphitic ribbons,[18] onions,[18] nano- to microcrystalline diamond,[18,42] and even nano-needles[18] in a proverbial amorphous carbon haystack.

By varying the process parameters and precursors, it is possible to select and control the resulting nanostructures. For example, by introducing the proper concentration of hydrogen into the chlorine-treatment process of SiC, it is possible to induce the formation of nanodiamond (5–10 nm) at fairly low temperatures (<1000°C) and ambient pressures.[42] The choice of carbide may dictate the propensity of the CDC to form a specific carbon structure. Nanobarrels were observed to form in significant fraction in Al_4C_3CDC. These multiwalled structures were observed by various researchers.[99–101]

Welz et al.[18] found a large variety of structures by looking at different polytypes of SiC varying the chlorine treatment temperatures and chlorine treatment time. At very low temperatures (500–600°C), carbon showed a very amorphous nature, even after long chlorine treatment, but fullerene-like structures were observed. Once temperatures were increased to 1000°C, graphitization became pronounced and, as time progressed and temperatures increased, more ordering was observed. It was suggested that initially small clusters of six-membered rings began to coalesce through the formation of networks, and as thermal energy was supplied, graphene sheets began to form. Along with turbostratic carbon, with time, more layers were added and the correlation between them strengthened. The lattice spacing, somewhat larger than graphite's 0.33 nm (≈0.35 nm), began to reduce with annealing, and 0.334 nm was attained upon complete conversion to carbon, and with sufficient time and temperature. The studies resulted in the observation of bent graphitic sheets or nanoribbons, as well as nano-needles, carbon onions, curved graphitic shapes, tubular structures, nanotubes, and diamond (cubic, hexagonal, and n-diamond, up to 800 nm in size[102]). The researchers described the formation of onions as an intermediate transformation product of the nanodiamond and, with sufficient temperature and time, the onions converted to graphite.

11.2.5 CONSERVATION OF SHAPE

One key feature of CDCs is shape preservation, referenced as "conformality." The conformality stems from the ability to use a metal carbide of desired shape and converts it to carbon, while maintaining the general shape and volume of the original preform.[43] Conformality has been reported to range from the hundreds of nanometers scale[29,92] to the macroscale.[96,103] This characteristic seems to have first been reported for a TiC powder that maintained its grain size and shape after chlorine treatment.[43] The trait can be shown quite elegantly in the TEM study in Figure 11.4 that demonstrates the gradual conversion of a β-SiC whisker to CDC.[29] As can be seen, on chlorine treatment, a thin film of CDC forms on the surface of the whisker and grows inward. With sufficient chlorine treatment time and temperature (Figure 11.4c), the entire whisker can be converted into CDC.

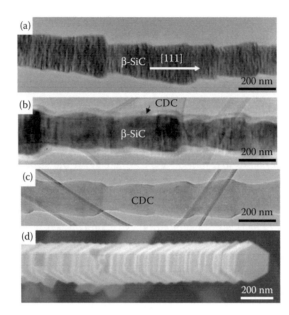

FIGURE 11.4 TEM (a–c) and SEM (d) images demonstrating the gradual conversion of a β-SiC whiskers to carbon at various stages of the transformation: (a) as received, (b) treated in chlorine gas at 700°C and 1200°C, (c) shape, and (d) size are largely maintained. (Adapted from Cambaz, Z.G. et al., *Journal of the American Ceramic Society* **89**, 509–514, 2006.)

A recent study compared CDC particles converted through chlorine treatment and via a mixture of chlorine gas and an oxidizing agent (CO$_2$). Despite the etching of the CDC, the particle size of the original carbide, the CDC, and the etched CDC maintained a relatively constant particle size as shown in Figure 11.5.[104]

This conformality is extremely useful in designing devices based on carbon, or as a multistep process (carbide → carbon → final material system) where it would be beneficial to begin with a

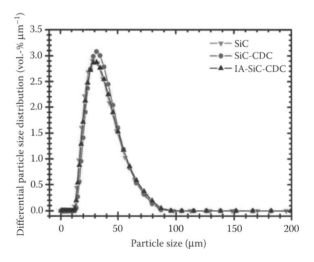

FIGURE 11.5 Particle size distributions of SiC and SiC–CDC (nonactivated, activated) from SiC—37 μm particles chlorinated at 900°C. Activation was carried out via CO$_2$ at 900°C. (Adapted from Schmirler, M., Glenk, F., and Etzold, B.J.M. *Carbon* **49**, 3679–3686, 2011.)

metal carbide template or preform. However, caution should be taken, as the conformality is an approximate descriptor and is not impervious to failure under certain conditions and scales. For example, the high porosity created during conversion (e.g., 52–83%[78]) can result in a partial collapse of the carbon structure, along with cracks, possible surface area and pore volume reduction, and other possible deleterious effects. Such collapses were seen for micrometer-sized powder from Ti_3AlC_2, Ti_2AlC, and $Ti_2AlC_{0.5}N_{0.5}$[105] and possibly with B_4C at higher temperatures (e.g., 1800°C).[106] In general, during halogen treatment, the volumetric changes can be minimized through proper care under a large range of sizes and shapes.[32,79,94,107]

11.2.6 Porosity

CDC has the ability to create a very uniform and controlled pore structure with the appropriate processing conditions. An example of such a produced pore structure is shown in Figure 11.6. In this study,[108] TiC was treated in chlorine gas at a variety of temperatures and the resulting pore size evolution was studied via sorption studies. The pore size distribution at both 1000°C and 500°C indicates strong control over pore size distribution as afforded by the CDC when compared to other carbon synthesis methods (e.g., activation of pyrolyzed organic precursors). In addition, a gradual increase in the average pore size is possible by increasing the chlorine treatment temperature.

The porosity created within CDC is a function of many variables, including processing parameters (gas mixtures, temperature), initial carbide chemistry and porosity, and post-halogen treatment. The gas mixtures have been shown to affect the type of carbon forming during chlorine treatment and affect the resulting pore structure. For example, Gogotsi et al.[42] observed that by adding hydrogen to chlorine gas, formation of diamond became more favorable. The initial carbide chemistry and crystal structure strongly affect the distribution of atoms throughout the metal carbide and, as a result, control the resulting pore structure and density of the resulting CDC. The temperature affects the mobility of carbon atoms liberated from bonding to metal atoms, as well as the mobility

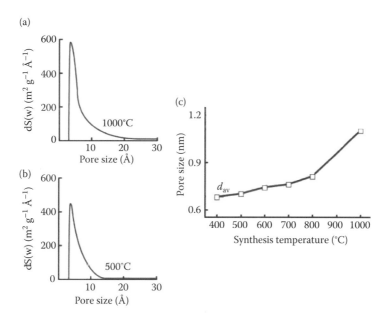

FIGURE 11.6 Pore size distributions calculated assuming slit-shaped pores for TiC–CDC synthesized at (a) 1000°C and (b) 500°C (fitted DFT distribution curves). A plot of the average pore size, d_{av}, versus the synthesis temperature (c) shows the pore size increase from about 0.7 nm to about 1.0 nm. (Adapted from Chmiola, J. et al., *Angewandte Chemie-International Edition* **47**, 3392–3395, 2008.)

of metal chloride species, both of which may, in turn, affect the developing porosity. These parameters must be controlled to obtain the desired pore volume, surface area, and pore size distribution. For example, in the boron carbide system alone, it was shown that by varying the chlorine treatment temperature, one could obtain surface areas ranging from 2200 m^2/g (600°C) down to 30 m^2/g (1800°C) and pore volumes from 1.08 to 0.20 cm^3/g.[106] However, we note that, in general, porosities more than 50 vol.% along with surface areas higher than 1000 m^2/g are commonly observed when synthesizing CDC.[78,86,109–111]

Early studies on CDC already indicated that the method used to remove trapped chlorine could significantly impact the porosity. Mohun used a variety of methods to remove the chlorine and reaction products, including exposing CDC to vacuum, gas flow, and employing wet purification methods.[47] Following studies observed similar effects of purification on increasing the CDC porosity. Among them, one study found that even when cooling the samples in argon after chlorine treatment, chlorine atoms can account for 20–45 wt.%.[81] Such large amounts of remaining chlorine have been attributed to a degradation in the specific surface area (SSA) and were more noticeable when chlorine treatment was performed at lower temperatures.[68,81]

In addition to the porosity developed during chlorine treatment and the steps to remove any remnants (such as during cooling, hydrogen flowing, or other methods mentioned by Mohun), there are additional treatments that may be applied to further develop the pore structure. The majority of reported steps often involve physical or chemical activation where an agent (e.g., O_2, CO_2, or KOH) is used to remove carbon atoms often resulting in an increase of the average pore size along with pore volume.[33,104,112–114]

A general trend suggested by Presser et al.[43] allows for an estimate of the resulting pore structure based on the initial carbide's structure. Some studies showed that carbides with the NaCl structure (e.g., TiC, VC, and ZrC) or wurtzite/zinc blend structure (SiC) result in a fairly uniform pore structure, attributed to the first neighbor separation uniformity in the original carbide. However, the two-tiered nearest-neighbor separation existing in rhombohedral (B_4C) and orthorhombic (Mo_2C) structures results in a larger, less uniform, pore structure.[60,82,106]

The temperature at which chlorine treatment is carried out has a clear effect on a variety of CDC characteristics, as has been intimated in prior sections of this chapter. Because temperature is one of the most convenient parameters to change in the experimental setup for CDC, many such studies on various carbide precursors have been done and, generally, the average pore size and the pore size distribution increase as the temperature of the chlorine treatment is increased. This has been seen for TiC,[115] ZrC,[115] VC,[110] WC,[116] chromium carbides,[117] Al_4C_3,[101] Mo_2C,[60] SiCN,[72] and others. Higher chlorine treatment temperatures increase the mobility of carbon atoms that facilitates the formation of larger-sized graphene fragments and multiwalled structures, which, in turn, result in the formation of larger pores.[22] Larger and straighter pores facilitate the removal of trapped metal chlorides. In addition, the chlorides rapidly produced at high temperatures may cause expansion of CDC particles, particularly those produced from layered carbides.[105,118,119] These phenomena contribute to increasing the total pore volume in CDC. The pore structure control offered by the synthesis methods of CDC has earned the materials the reputation of materials with tunable porosity. Some CDC materials display pore uniformity on the level of, and higher than certain carbon nanotube variations, and on the level of zeolites[22] and zeolite-templated carbons.[120–125]

The SSA of CDC is controlled by various parameters. This includes the type of carbide precursors, their particle size,[126] initial pore size distribution,[86,127] chlorine treatment temperature, halogen flow rate, cooldown gas, post synthesis treatment, and cleaning/purification procedures post synthesis, to name a few. Many carbides show a pseudo-bell-shaped variation[60,61,128,129] in surface area as a function of the chlorine treatment temperature, where at lower temperatures, an increase in surface area is seen and at a certain maximum, a decrease in surface area is seen with further increasing temperatures. The decrease in the surface area at elevated temperatures can be linked to the formation of graphene ribbon structures and the structural closing of internal pores. There are cases where this trend is not seen within the range of temperatures studied,[28,116] but it is likely that if higher temperatures were examined, the decrease in surface area would have been observed, too. The temperature window where this is seen may vary for different carbide precursor and different experimental setup and conditions but is in general correlated with the onset of significant graphitization.

The observed reduction in surface area with increased treatment temperatures can be attributed to a variety of causes. These factors include the trapping of reaction products (e.g., metal chlorides[97]) and increased crystallization at elevated temperatures. The trapping may be more prominent for lower temperature synthesis conditions along with the possible etching of carbon (CCl_4, see Section 11.2.2) leading to structure collapse, whereas graphitization and other ordering phenomena may occur at higher temperatures. As the SSA is measured using a probing gas (such as N_2, Ar, CO_2, etc.), there is a limitation for the accuracy of the pore size distribution technique used by most researchers because these gases cannot necessarily probe the smallest pores and their different interactions with the exposed surfaces. Moreover, different gases will give different results based on the pore shape and the used gas adsorption model, the gas molecule size, pore morphology, and functional groups present in CDC, among various possible variables. A word of caution is then provided that not only should the SSA and (apparent) pore size distributions be studied, but also *in situ* studies of the materials in their designated applications to obtain an understanding of the application-related, "effective" porosity. Furthermore, the use of diffraction methods, such as small angle neutron[130,131] or small angle x-ray scattering[115] along with immersion calorimetry,[132] offers unique insights.

11.2.7 CATALYST-MODIFIED CDC STRUCTURES

There are a variety of situations in which formation of ordered structures may be desirable, as will be seen in Section 11.3. The synthesis of such ordered structures in CDC often requires elevated temperatures beyond those required for initiation of the reaction between the halogen and metal. It turns out that by adding a catalyst to the reacting metal carbides, ordered structures can be produced at lower temperatures. A variety of studies on the impact of catalysts on CDC formation have been undertaken.[34,75,99,133–136] Catalysts that have been examined include CeO_2/Pt, Ru(III), Fe(II), Fe(III), Ni, and Co. Kockrick et al.[136] employed a polymeric SiC precursor mixed with CeO_2/Pt catalysts to create CDC materials containing catalytically active nanoscale CeO_2/Pt. The study found that in the CDC containing catalytic particles, the thermal stability was lowered in differential thermal analysis studies due to activity in combustion of the CeO_2/Pt nanoparticles, thus, lowering the minimum temperature from 588°C to 378°C. A study on SiC and the effect of catalysts[133] involved the formation of CDC films on top of SiC for investigations of catalyst impact on structure and tribological applicability. Chlorine treatment was carried out between 1000°C and 2000°C, and it was found that adding iron catalyst did not change the film thickness but the ordering of the CDC was increased. TEM studies showed that the CDC films included multiwalled carbon nanotube (MWCNT) and onion-like carbon (OLC) in samples synthesized at 1000°C, unlike the mainly amorphous carbon and graphitic crystals seen in the CDC without the presence of a catalyst under similar conditions. At 1200°C, MWCNT and OLC were not visible, and so, there is likely a temperature window in which the catalysts are most effective. As a result, the wear rate was improved by the presence of the CNT and OLC reinforcing the CDC matrix. The effect of iron and ruthenium catalysts on CDC derived from biomorphic TiC and SiC was investigated by Kormann et al.[34] These studies found a linear increase with catalyst (Ru(III), Fe(II), and Fe(III)) concentration on TiC and SiC etching rate. Although the TiC etching rate was independent of the temperature even in the presence of a catalyst, SiC showed a significant temperature dependence of the etching rate. However, for SiC, with a catalyst present, there was only a very small dependence on temperature. The SSA of both carbides showed the highest values at intermediate temperatures, but they reported a very low SSA (<250 m²/g). Contrary to earlier studies, they found ordered structures forming even at low temperatures as a result of catalyst addition.

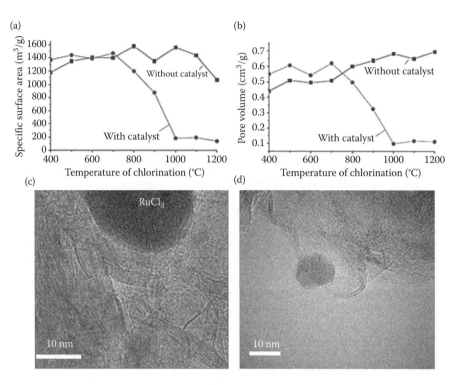

FIGURE 11.7 BET–SSA (a) and pore volume (b) of TiC–CDC as a function of chlorine treatment temperature and catalyst inclusion in the TiC precursor. The catalyst consisted of a mixture of CoCl₂, NiCl₂, and FeCl₃ solutions in ethanol resulting in ≈73 mg of each chloride per gram of carbide. (Adapted from Leis, J. et al., *Carbon* **40**, 1559–1564, 2002; Yushin, G., Nikitin, A. and Gogotsi, Y. in *Nanomaterials Handbook*. 239–282, CRC Press, Boca Raton, FL, 2006.) TEM micrographs of CDC obtained from TiC/Ru(III) after chlorine treatment at (c) 1200°C and (d) from TiC/Ru(III)/Fe(III) after chlorine treatment at 1200°C. (Adapted from Kormann, M. et al., *Carbon* **47**, 2344–2351, 2009.)

In general, addition of catalysts to the carbide precursors has been shown to significantly impact the resulting CDC structure (Figure 11.7) even at concentrations as low as 0.1 wt.%. Addition of catalysts often results in increased ordering, formation of ordered crystallites such as nanobarrels, MWCNT and OLCs, reduced SSA, and increased mesoporosity.

11.2.8 HALOGEN SELECTION

Because chlorine is a fairly low cost and relatively easy to work with halogen, only a few studies on the synthesis of carbon from nonchlorine halogens have been undertaken.[64,138–140] Babkin et al.[138] studied the differences between the resulting CDCs when varying the halogen from chlorine to bromine and iodine. It seemed that the total pore volume, as studied via benzene adsorption, was approximately the same for a series of ZrC–CDC samples. It turned out, however, that the sorption capacities for different halogens varied significantly between samples: samples synthesized via bromine treatment and iodine treatment, respectively, showed an increased sorption ability for those gases. The authors considered this as evidence that the porous structure of CDC was affected by the size of the halogen and evolving zirconium tetrahalogenide molecule, as well as by surface termination of the CDC. It is important to keep in mind that the volatile reaction products are not necessarily of a single type as often discussed (say, SiCl₄), and the distribution of reaction products may also affect the porosity. For example, when studying fluorine treatment of various carbides, a large number of different reaction products were identified by infrared spectroscopy and gas chromatography, including CF_4, C_2F_6, C_3F_8, and more.[64] Another important reason for the use of different halogens entails the reactivity of the halogen with the carbide. Batisse et al.[139] showed that when using F_2 on SiC thin films to create CDC films, the resulting film is not as homogeneous as when using a less-aggressive XeF_2 that appears to react only with the silicon atoms, rather than both the silicon and carbon. The temperature at which CDC formation can take place is also highly dependent on the halogen. For example, XeF_2 can react with SiC to form CDC at temperatures below 130°C, whereas hundreds of degrees higher are usually required for the synthesis of SiC–CDC using chlorine treatment.[139]

11.3 APPLICATIONS

11.3.1 SUPERCAPACITORS

Supercapacitors are electrochemical energy storage devices that can provide higher power densities and cyclability as compared to conventional battery chemistries, while having significantly higher energy densities as compared to solid-state capacitors. Commercial devices are sold by several companies and often consist of activated carbon electrodes in a solution of tetraethylammonium tetrafluoroborate in acetonitrile or propylene carbonate. To store energy, the devices rely on the electrostatic formation of the electrochemical double layer (EDL) by ions in solution across the surface of the active material. Relying on the EDL formation to provide the mechanism for energy storage enables the very high power density along with cyclability. However, the faradaic reactions in batteries are attributed to a reduced charge–discharge rate, efficiency, and cyclability while yielding a significantly higher energy density.

Supercapacitors show potential in a variety of areas. These include power management, electric vehicles, hybrid–electric vehicles, sensors, communication devices, renewable energy storage, emergency door actuation, cordless devices, on-chip power storage, and many more.[95,120,141–143] The breadth of available electrolytes and salts available for use in supercapacitors make them extremely attractive and versatile. They have larger temperature-operating windows than many battery systems and a variety of new materials are already being commercialized in an effort to improve their performance and use.

The design of electrode materials for electrochemical capacitors as currently understood dictates that certain material characteristics must be present. For increased specific capacitance (tied to the specific energy), it is understood that high surface area along with micropores and possibly disordered carbon should be present.[28] It is, however, also important to note that because of the limited charge-screening ability of very thin carbon pore walls, there is a limit to the maximum surface area useful for supercapacitor applications.[144] However, it is necessary to have a pore structure that can allow for rapid ion transport, either by having large mesopores or straight small mesopores, to have high power density. CDCs have shown significant promise with respect to energy density and templated CDCs have provided extremely promising results with respect to both specific energy and power.[86,129] Templated CDC involves the use of a sacrificial template to dictate the morphology of the carbide precursor so as to confer upon it desirable attributes. Korenblit et al. were able to attain one of the highest reported specific capacitances in organic electrolytes by first templating SiC on a sacrificial silica template and then using chlorine to create a templated CDC in which a combination of straight ordered mesopores and random micropores generated a large pore volume. The use of these materials resulted in a specific capacitance of around 170 F/g[86] and very high capacitance retention at elevated current densities (<3% loss of specific capacitance at currents as high as 17,500 mA/g) as demonstrated in Figure 11.8.

Among the latest developments in CDC synthesis and supercapacitor construction is the use of an electrospun TiC nano-felt that allows for the synthesis of flexible, binder-free CDC electrodes for supercapacitors resulting in high power electrodes (Figure 11.9).[145] However, while yielding a very high power handling ability, the volumetric energy and power density of such fiber-based devices

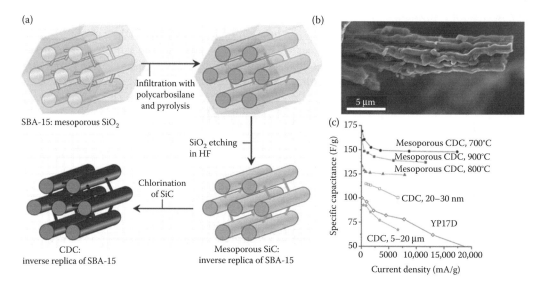

FIGURE 11.8 Aligned mesoporous-templated silicon carbide-derived carbon synthesis and application. (a) Schematic illustration of the fabrication of CDC with aligned mesopores, (b) SEM micrograph showing the morphology of the produced particles, (c) capacitance retention versus current density of CDC with aligned mesopores produced at 700°C, 800°C, and 900°C in comparison to that of activated carbon commonly used in commercial electrochemical double-layer capacitors (YP17D) and CDC produced from micrometer-sized and nanosized SiC powder at 900°C.

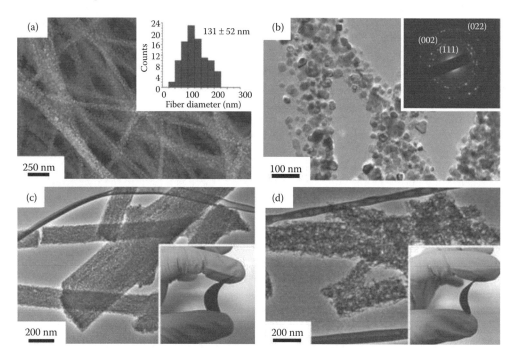

FIGURE 11.9 SEM (a) and TEM (b–d) images of electrospun TiC nanofibers (CDC precursor). Each fiber consisted of TiC nanocrystallites. The inset in (b) shows the electron diffraction pattern along the (111)-, (200)-, and (220)-planes of cubic TiC crystal with d values of 2.50, 2.19, and 1.55 Å, respectively. TEM images of the TiC–CDC nano-felts after chlorine treatment at 400°C (c) and 600°C (d). The insets in (c) and (d) demonstrate the flexibility of the developed TiC–CDC binder-free electrodes. (Adapted from Presser, V. et al. *Advanced Energy Materials* **1**, 423–430, 2011.)

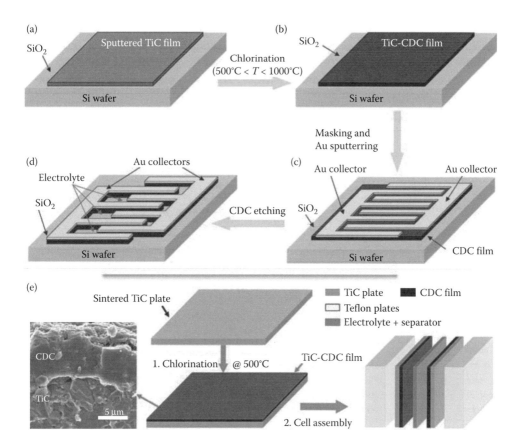

FIGURE 11.10 (a–d) Schematic of the fabrication of a micro-supercapacitor integrated onto a silicon chip based on the bulk CDC film process. Standard photolithography techniques can be used for fabricating CDC capacitor electrodes (oxidative etching in oxygen plasma) and deposition of gold current collectors. (e) Schematic of CDC synthesis and electrochemical test cell preparation. Ti is extracted from TiC as $TiCl_4$, forming a porous carbon film. Two TiC plates with the same CDC coating thickness ranging from 1 to 200 mm are placed face to face and separated by a polymer fabric soaked with electrolyte. An SEM micrograph shows a representative image of the CDC/TiC interface. (Adapted from Chmiola, J. et al., *Science* **328**, 480–483, 2010.)

remain moderate. An additional development is the concept of making on-chip type supercapacitors by deposition of CDC films on substrates that can then be etched using standard photolithography techniques as seen schematically in Figure 11.10.[146] Various materials have been tested as potential supercapacitor electrodes and some examples are given in Table 11.1.

The high surface area and micropores resulting from the halogene treatment of various carbide precursors have resulted in fairly high capacitances (>100 F/g) in organic electrolytes. Many carbides have been tested in supercapacitors and have included TiC,[30] chromium carbide,[117] SiC,[86] VC,[128] and CaC_2,[56] in various electrolytes. The energy and power densities achieved by using CDCs are some of the highest reported in both organic and aqueous electrolytes. They show significant promise for the advancement of commercial EDLC for high-performance supercapacitors.

11.3.2 Gas Storage

The storage of gases in a solid sorbent is of significant importance for society. Whether it is for their capturing, so as to prevent environmental harm (e.g., CO_2, SO_2), for storing the gas for later use as

TABLE 11.1

Impact of CDC Precursor, Electrolyte (AC, Acetonitrile; PC, Propylene Carbonate), and Halogen Treatment Temperature (among Various Other Parameters) on Pore Structure and Specific Capacitance

CDC Precursor	BET SSA (m^2/g)	Temperature (°C)	Electrolyte	Maximum Capacitance (F/g)	References
TiC powder	1436	900	1 M $(C_2H_5)_4N(BF_4)$ in PC	121	147
TiC powder	1595	800	PYR_{14}-TFSI	130 (at 100°C)	145
TiC fiber mats	1188	1000	1 M H_2SO_4 and 1.5 M $(C_2H_5)_4N(BF_4)$ in AN	120 (organic) 135 (aqueous)	148
WC powder	1580	1000	1 M $(C_2H_5)_3CH_3NBF_4$ in AN	132	142
TiC powder	1627	950/800 (two step)	2 M H_2SO_4 in H_2O	196	36
B_4C powder	1461	1000	6 M KOH in H_2O	177	149
SiO_2- templated SiC powder	2914	800	1 M $(C_2H_5)_4N(BF_4)$ in AN	100	150
Al_4C_3 powder	1470	≈500 (varied for cation vs. anion affecting the maximum capacitance)	1 M $(C_2H_5)_3CH_3NBF_4$ in PC	110	60
Mo_2C powder	1811	700	1 M $(C_2H_5)_3CH_3NBF_4$ in AN	143	128
VC powder	1305	900	1 M $(C_2H_5)_3CH_3NBF_4$ in AN	133	151
SiC powder	1234	1000	1 M $(C_2H_5)_4NBF_4$ in AN	129	151
TiC powder	1627	950/800	1 M $(C_2H_5)_4NBF_4$ in AN	152	151
SiC powder	1234	1000	2 M H_2SO_4 in H_2O	153	151

an energy source (e.g., H_2) or for medical applications (e.g., NO).[152] The Brunauer–Emmett–Teller (BET) surface areas provided by CDC can be larger than 2000 m^2/g and contain a large volume of micropores. Table 11.2 shows the great breadth of available physical characteristics from different CDC precursors and their ability to adsorb various gases. The ability of CDC to offer variable porosity makes these materials promising for the storage of hydrogen and other gases. Dash[78] showed that small micropores are much more efficient for hydrogen storage at atmospheric pressure and that microporous CDC can outperform metal-organic frameworks (MOFs), MWCNT, and SWCNT in their hydrogen storage capacity. Similarly, Presser et al.[153] concluded that the surface area and total pore volume were not sufficient measures to indicate CO_2 uptake capacity, and that, depending on the applied pressure, the micropore volume was an important indication for CO_2 adsorption capacity of TiC–CDC. Just like for the storage of H_2,[154] the pore volume of micropores smaller than 1 nm is of particular importance.

CDCs provide precise control over the pore size distribution, surface area, and surface functionality that could control the affinity of the material for certain gases, thus providing the material with significant promise for hydrogen storage and other gases. Seredych et al.[156] showed that by adding nitrogen groups to the surface of CDC and maintaining sufficient microporosity, while assuring that pores are not so small as to cause steric hindrance, TiC–CDC can most efficiently remove H_2S for environmental protection applications. Through the use of different activation methods, the pore structure and functional groups can be further modified, for example, by using CO_2 or KOH to etch the carbon. Yeon et al.[157] were able to obtain surface areas of over 3300 m^2/g and an uptake of methane as high as 18.5 wt.% and Sevilla et al.[114] showed that by using KOH activation on a ZrC–CDC, an increase in hydrogen uptake of 63% was possible along with varying pore volume and surface area and attaining surface areas of 2800 m^2/g and 1.47 cm^3/g pore volumes. Templated SiC–CDCs were also shown to provide higher methane and hydrogen adsorption as compared to ordered mesoporous carbons (CMK-3) and microporous SiC–CDC.[158]

TABLE 11.2

Impact of CDC Precursor and Halogen Treatment Temperature (among Various Other Parameters) on the Resulting Pore Structure and Gas Sorption Ability

CDC Precursor	BET SSA (m^2/g)	Chlorine Treatment Temperature (°C)	Pore Volume (cm^3/g)	Average Pore Size (nm)	Gas Adsorbed ≈1 atm (Unless Otherwise Noted)	References
Nano-TiC	434	200	0.36	0.58	CO_2, 0°C 2.74 mol/kg	153
Nano-TiC	1920	1000	1.61	0.96	CO_2, 0°C 3.70 mol/kg	153
Micro-TiC	1669	1000	0.75	0.78	CO_2, 0°C 5.45 mol/kg	153
Micro-TiC	≈1500	800	≈0.8	≈0.9	H_2, −196°C 330 cm^3/g, 3 wt.% CH_4, 25°C 46 cm^3/g, 3.1 wt.%	30
Micro-SiOC	2593	1200	1.40	2	60 bar H_2, −196°C 5.5 wt.% 60 bar CH_4, 25°C 21.5 wt.%	109
Micro-B_4C	2012	800	0.99	0.97	H_2, −196°C 213 cm^3/g, 1.91 wt.%	155
Micro-TiC	1943	800	0.94	1.00	H_2, −196°C 336 cm^3/g, 3.02 wt.%	155
Micro-ZrC	1388	600	0.65	0.97	H_2, −196°C 287 cm^3/g, 2.58 wt.%	155
Micro-SiC	1279	1200	0.49	0.97	H_2, −196°C 287 cm^3/g, 2.10 wt.%	155
Micro-ZrC	1880	800	0.93	NA	H_2, −196°C 1.9 wt.%	114
Micro-ZrC (with CDC activation in KOH)	2770	800 (900 activation temperature)	1.47	NA	H_2, −196°C 2.7 wt.%	114

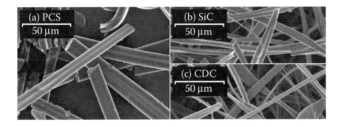

FIGURE 11.11 SEM micrographs of (a) electrospun PCS-3500 fibers, (b) SiC fibers after pyrolysis, and (c) CDC fibers after chlorine treatment. (Adapted from Rose, M. et al., *Carbon* **48**, 403–407, 2010.)

Another method to create novel CDC-based sorption materials has recently been reported and entails electrospinning of polycarbosilane (PCS) with subsequent pyrolysis and chlorine treatment (fibers shown in Figure 11.11).[94] The resulting fibers demonstrate great potential for the adsorption of hydrogen. The surface area was over 3000 m^2/g and hydrogen sorption measurements at 77 K indicated adsorption of 2.75 wt.% at a pressure of only 1 bar. Combined with this high performance is the ability to integrate such fibers into textiles and air filters, providing significant benefit over other particulate sorption media.

11.3.3 LITHIUM-ION BATTERIES

The commercial proliferation of lithium-ion batteries has initiated an avalanche of intensive research into candidate materials for the advancement of the technology. Some of the first studies on the utility of CDC as lithium-ion anode materials focused on the diffusion characteristics of lithium ions in CDC and estimated that the mechanism for the diffusion was Li diffusion along pore walls and accumulation in pores post-cluster formation. The examined materials included SiC, TiC, and Mo_2C–CDC and evinced a diffusion coefficient between 10^{-9} and 10^{-7} cm^2/s that was pore size dependent.[159] Sakamoto et al. examined TiC–CDC at 600°C, 1000°C, and 1200°C as anode materials versus lithium metal and $LiCoO_2$. The samples showed large irreversible capacities during the initial cycling because of high SSAs of these materials. The most disordered carbon chlorinated at 600°C demonstrated the highest specific capacity of ≈290 mAh/g.[160] There have not been many studies related to the use of CDC in Li-ion battery anodes or cathodes,[159–163] but initial studies have indicated some promise for fundamental investigations where pore size and particle size control are desired.

11.3.4 CDC as a Catalyst Support

High surface area carbon materials are a promising substrate for researchers developing future generations of catalyst supports, which may be of great utility in areas such as fuel cells, metal–air batteries, and catalytic reforming. Zeolites and other materials are used for catalyst supports because of their pore structure and design flexibility[164] among various other characteristics, but may not be suitable for certain applications, for example, because of the lack of a sufficiently high electronic conductivity.

The design flexibility of CDC, such as pore structure, surface area, conductivity, and other parameters make it a material of interest for catalyst support applications. Kockrick et al.[136] used an inverse microemulsion method to incorporate ceria and platinum nanoparticles into PCS that was later pyrolyzed and then chlorine treated to produce CeO_2/Pt on a high surface area CDC and showed complete oxidation capability of methane to carbon monoxide and hydrogen. This method for the introduction of catalysts into the CDC structure provides for a high level of control. Other methods for impregnation with catalysts have been demonstrated as well.[165] Ersoy et al.[165] used a method that involved the dispersion of Pt throughout the CDC by introduction of bulk Pt into the reaction zone of the chlorine treatment and attributed the dispersion to gas-phase transport of Pt_3Cl_3 to the site of decomposition and deposition of the Pt. The deposition of Au catalyst on CDC with pore-limited size control showed very promising performance.

11.3.5 Tribological Applications

CDC has shown promising results in the area of tribology. CDC can provide friction coefficients as low as 0.03 and wear rates as low as 10^{-9} mm³/Nm.[167] As a result, CDC can have potential uses in a variety of tribological applications. Erdemir et al.[167] showed that CDC films are self-lubricating and capable of providing a very low friction coefficient to sliding surfaces. The findings also indicated that by removing trapped chlorine products and performing the friction studies in dry air, further reduction in friction coefficients can be obtained. The temperature and chlorine-gas mixtures have also been indicated as important parameters, as these will affect the resulting carbon structures and, in turn, the friction and wear resistance. Bae et al.[168] observed an increase in wear rate, but a decrease in the friction coefficient, with increasing chlorine treatment temperatures of SiC to a certain transition temperature at which point the wear rate began to decrease. Gao et al.[169] also confirmed the need for removing trapped chlorine and found evidence that commercial graphite had a reduced tribological performance, as compared to SiC–CDC against steel. The importance of metal carbides for sliding contact applications because of their high wear resistance, strength, and chemical stability is well known. However, the work done on CDC film formation will assist in reducing their friction coefficients and the advancement of these materials as indicated by the research on their tribological properties[19,20,38,90,133,167–175] and an R&D100 award associated with a CDC sliding seal.

11.3.6 Biomedical Applications

The ability to control the pore structure and modify the surface chemistry of CDC has also demonstrated benefit in the area of biomedical engineering. Sepsis syndrome can result because of the inflammatory response to an inciting event such as a microbiologically confirmed infection. The body's response is to release an excessive amount of cytokines, which can result in sepsis syndrome, associated with mortality rates above 40%.[176] CDC materials have been targeted as possible sorbents to remove the cytokines from plasma. Yushin et al.[177] and Yachamaneni et al.[176] studied the applicability of Ti_2AlC and Ti_3AlC_2–CDC to the removal of cytokines (e.g., TNF-α, IL-6, and IL-1b) and were able to demonstrate a very high removal efficiency. The studies revealed that the large size of the cytokines (see Figure 11.12) necessitated the presence of mesopores for sufficient exposure of the surface area to the cytokines for adsorption purposes. In these studies, mesoporous CDCs were synthesized from layered ternary carbides (the so-called MAX phases). The results indicated that up to 99.7% of TNF-α could be removed in a 1 h time frame when the samples were post-annealed in Ar at 800°C for 5 h and could further reach 100% when the annealing environment was changed to NH_3, indicating the importance of surface chemistry. The smaller IL-6 cytokine adsorption was much faster and all samples in Yachamaneni's study removed 100% in 60 min except for samples annealed in chlorine. The smallest cytokine, IL-1b, was the first to be removed and reached 100% in 5 min for all samples. CDC derived from a polymer-derived Si–C–N ceramic with hierarchical porosity was recently introduced as a facile method to obtain potentially monolithic filters for efficient cytokine removal.[178]

In addition to the ability of CDC to remove cytokines from blood plasma, another study on the ability of CDC to effectively kill various bacteria showed further promise for biomedical applications.[71] Often, it is desired to remove chlorine from produced CDC but, in this case, it was found that

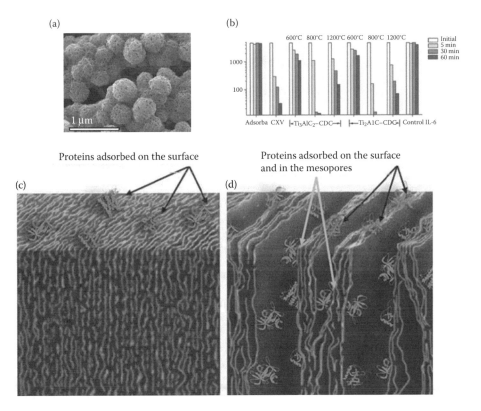

FIGURE 11.12 Physical characterization and diagram of mesoporous CDC and its impact on cytokine removal. (a) Scanning electron micrograph of the synthesized mesoporous CDC. (b) Superior performance of mesoporous CDC (800°C) as compared to other materials with respect to the ability to remove cytokines from human blood plasma: concentration of IL-6 in the plasma solution initially and after 5, 30, and 60 min of adsorption. (c,d) Schematics of protein adsorption by porous carbons: (c) surface adsorption in microporous carbon. Small pores do not allow proteins to be adsorbed in the bulk of carbon particles; (d) adsorption in the bulk of mesoporous CDC. Large mesopores are capable to accommodate a larger fraction of the proteins. (Adapted from Yushin, G. et al. *Biomaterials* **27**, 5755–5762, 2006.)

the ability of CDC to hold up to 60 wt.% of chlorine was very beneficial. Chlorine-loaded CDC was able to release chlorine into water over a prolonged period of time, rather than simply add a large amount of chlorine in one batch as would be done, for example, with chlorine tablets. The CDC efficiently killed the *Bacillus anthracis* and *Escherichia coli* bacteria and indicated a capacity to perform the task over time.[71]

11.3.7 MEMBRANE APPLICATIONS

Thin film membranes have many potential and current applications, among them purification/separation of gases and fluids (e.g., hydrocarbon fractionation, water desalination, and biological fluids). Although polymer membranes dominate in the industry, it is highly desirable to design and synthesize new materials of higher chemical and thermal stability.[179] In addition to increasing the thermal and chemical stability as compared to using polymer membranes, the use of CDCs allows for the tailoring of various other important parameters such as pore size distribution, surface chemistry, which in turn controls affinity for various materials of interest (e.g., via hydrophilicity). The research entailed the deposition of a TiC precursor on a preexisting macroporous ceramic membrane, followed by chlorine treatment at 350°C. The results are shown in Figure 11.13. This method of producing a dual-ceramic macroporous–CDC membrane was the first of its kind and through future developments may result in the creation of many interesting and useful membranes.

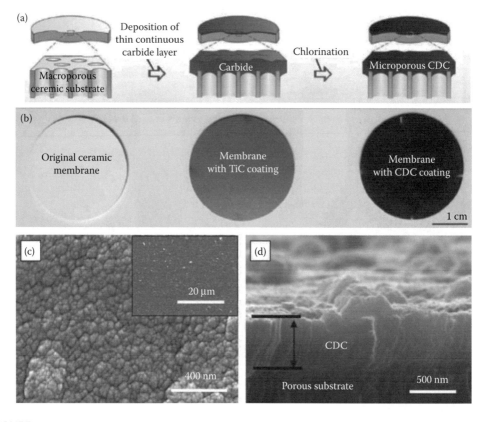

FIGURE 11.13 CDC thin film formation: (a) schematic of the process flow, (b) optical images of the thin film and corresponding processing stages. SEM micrographs of TiC coating on an anodic substrate after chlorine treatment: (c) top view, (d) cross section of the membrane depicting the substrate pores covered by an approximately 500-nm-thick CDC layer. (Adapted from Hoffman, E.N. et al., *Materials Chemistry and Physics* **112**, 587–591, 2008.)

11.4 CONCLUSIONS

This chapter has discussed the diversity and potential of CDCs. The available precursor carbides, chlorine treatment temperatures, postsynthesis treatment methods, presynthesis template methods, synthesis configurations, and many other parameters give rise to the diversity of chemistries and morphologies found in CDCs. These parameters are studied so that scientists and engineers can have a better understanding of the mechanisms involved in the synthesis of CDC and to provide them with model materials to study other phenomena such as electroadsorption.

Although we have been using porous carbon for the benefit of society for thousands of years, the development and study of pore structure and properties in CDC have only been actively pursued in the past 20 years. In this short time frame, we have developed an ability to produce structures ranging from graphene and nanotubes to onions and nanobarrels, all the way to graphite. We have managed to create materials that can be used for tribological applications, the sorption of gases and chemicals for water treatment and fuel distribution, advanced electrochemical energy storage technologies, and materials that can be used to fight disease. CDCs have shown great promise in a short time frame. Continued research and development will enable the application of these materials more effectively and to applications that will only become known to us in the years to come.

ACKNOWLEDGMENTS

This work was partially supported by US Army Research Office (grant W911NF-12-1-0259), Petroleum Research Fund managed by the American Chemical Society (grant PRF # 49045-DNI10) and Exide, Inc.

REFERENCES

1. Wang, J.J. et al. Synthesis of carbon nanosheets by inductively coupled radio-frequency plasma enhanced chemical vapor deposition. *Carbon* **42**, 2867–2872, 2004.
2. Malesevic, A. et al. Synthesis of few-layer graphene via microwave plasma-enhanced chemical vapour deposition. *Nanotechnology* **19**(30), 305604, 2008.
3. Kajdos, A., Kvit, A., Jones, F., Jagiello, J., and Yushin, G. Tailoring the pore alignment for rapid ion transport in microporous carbons. *Journal of the American Chemical Society* **132**, 3252, 2010.
4. Scott, C.D., Arepalli, S., Nikolaev, P., and Smalley, R.E. Growth mechanisms for single-wall carbon nanotubes in a laser-ablation process. *Applied Physics A-Materials Science and Processing* **72**, 573–580, 2001.
5. Rode, A.V., Luther-Davies, B., and Gamaly, E.G. Ultrafast ablation with high-pulse-rate lasers. Part II: Experiments on laser deposition of amorphous carbon films. *Journal of Applied Physics* **85**, 4222–4230, 1999.
6. Gamaly, E.G. and Ebbesen, T.W. Mechanism of carbon nanotube formation in the arc-discharge. *Physical Review B* **52**, 2083–2089, 1995.
7. Hutchison, J.L. et al. Double-walled carbon nanotubes fabricated by a hydrogen arc discharge method. *Carbon* **39**, 761–770, 2001.
8. Wei, L., Sevilla, M., Fuertes, A.B., Mokaya, R., and Yushin, G. Hydrothermal carbonization of abundant renewable natural organic chemicals for high-performance supercapacitor electrodes. *Advanced Energy Materials* **1**, 356–361, 2011.
9. Matveev, A.T., Goldberg, D., Novikov, N.P., Klimkovich, L.L., and Bando, Y. Synthesis of carbon nanotubes below room temperature. *Carbon* **39**, 137–158, 2001.
10. Szabó, A. et al. Synthesis methods of carbon nanotubes and related materials. *Materials (1996–1944)* **3**, 3092–3140, 2010.
11. Akaishi, M., Kanda, H., and Yamaoka, S. Synthesis of diamond from graphite carbonate systems under very high-temperature and pressure. *Journal of Crystal Growth* **104**, 578–581, 1990.
12. Gogotsi, Y.G. et al. In *22nd Annual Conference on Composites, Advanced Ceramics, Materials, and Structures*, vol. 19, 87–94, Cocoa Beach, Boca Raton, FL; 1998.
13. Gogotsi, Y.G. et al. Structure of carbon produced by hydrothermal treatment of beta-SiC powder. *Journal of Materials Chemistry* **6**, 595–604, 1996.
14. Gogotsi, Y. Nanostructured carbon coatings. *Nanostructured Films and Coatings NATO Science Series* **78**, 25–40, 2000.

15. Dai, C.L., Wang, X.Y., Huang, Q.H., and Li, J. Porous carbide derived carbon. *Progress in Chemistry* **20**, 42–47, 2008.
16. Gogotsi, Y.G. Formation of carbon coatings on carbide fibers and particles by disproportionation reactions. *Advanced Multilayered and Fibre-Reinforced Composites* **43**, 217–230, 1998.
17. Jeon, I.D., McNallan, M.J., and Gogotsi, Y.G. Formation of carbon coatings on silicon carbide by reactions in halogen containing media. *Fundamental Aspects of High Temperature Corrosion* **96**, 256–268, 1997.
18. Welz, S., McNallan, M.J., and Gogotsi, Y. Carbon structures in silicon carbide derived carbon. *Journal of Materials Processing Technology* **179**, 11–22, 2006.
19. Carroll, B., Gogotsi, Y., Kovalchenko, A., Erdemir, A., and McNallan, M.J. Effect of humidity on the tribological properties of carbide-derived carbon (CDC) films on silicon carbide. *Tribology Letters* **15**, 51–55, 2003.
20. Carroll, B., Gogotsi, Y., Kovalchenko, A., Erdemir, A., and McNallan, M.J. Tribological characterization of carbide-derived carbon (CDC) films in dry and humid environments, in *Nanostructured Materials and Coatings for Biomedical and Sensor Applications*, vol. 102 (eds. Y.G. Gogotsi and I.V. Uvarova), 119–130, Dordrecht, The Netherlands; 2003.
21. Chen, L.L., Ye, H.H., Gogotsi, Y., and McNallan, M.J. Carbothermal synthesis of boron nitride coatings on silicon carbide. *Journal of the American Ceramic Society* **86**, 1830–1837, 2003.
22. Gogotsi, Y. et al. Nanoporous carbide-derived carbon with tunable pore size. *Nature Materials* **2**, 591–594, 2003.
23. Erdemir, A. et al. Effects of high-temperature hydrogenation treatment on sliding friction and wear behavior of carbide-derived carbon films. *Surface and Coatings Technology* **188**, 588–593, 2004.
24. Nikitin, A. and Gogotsi, Y. Nanostructured carbide-derived carbon, in *Encyclopedia of Nanoscience and Nanotechnology*, vol. 7 (ed. H.S. Nalwa), 553–574, American Scientific Publishers, Valencia, CA; 2004.
25. Zinovev, A.V. et al. Coating of SiC surface by thin carbon films using the carbide-derived carbon process. *Thin Solid Films* **469–70**, 135–141, 2004.
26. Gogotsi, Y. et al. Tailoring of nanoscale porosity in carbide-derived carbons for hydrogen storage. *J. Am. Chem. Soc.* **127**, 16006–16007, 2005.
27. Yushin, G., Dash, R.K., Gogotsi, Y., Jagiello, J., and Fischer, J.E. Carbide-derived carbons: Effect of pore size on hydrogen uptake and heat of adsorption. *Advanced Functional Materials* **16**, 2288–2293, 2006.
28. Chmiola, J. et al. Anomalous increase in carbon capacitance at pore sizes less than 1 nanometer. *Science* **313**, 1760–1763, 2006.
29. Cambaz, Z.G., Yushin, G.N., Gogotsi, Y., Vyshnyakova, K.L., and Pereselentseva, L.N. Formation of carbide-derived carbon on beta-silicon carbide whiskers. *Journal of the American Ceramic Society* **89**, 509–514, 2006.
30. Dash, R. et al. Titanium carbide derived nanoporous carbon for energy-related applications. *Carbon* **44**, 2489–2497, 2006.
31. Kim, H.S., Singer, J.P., Gogotsi, Y., and Fischer, J.E. Molybdenum carbide-derived carbon for hydrogen storage. *Microporous and Mesoporous Materials* **120**, 267–271, 2009.
32. Kormann, M., Gerhard, H., and Popovska, N. Comparative study of carbide-derived carbons obtained from biomorphic TiC and SiC structures. *Carbon* **47**, 242–250, 2009.
33. Osswald, S. et al. Porosity control in nanoporous carbide-derived carbon by oxidation in air and carbon dioxide. *Journal of Solid State Chemistry* **182**, 1733–1741, 2009.
34. Kormann, M., Gerhard, H., Zollfrank, C., Scheel, H., and Popovska, N. Effect of transition metal catalysts on the microstructure of carbide-derived carbon. *Carbon* **47**, 2344–2351, 2009.
35. Cheng, G., Long, D.H., Liu, X.J., and Ling, L.C. Fabrication of hierarchical porous carbide-derived carbons by chlorination of mesoporous titanium carbides. *New Carbon Materials* **24**, 243–249, 2009.
36. Wang, H.L. and Gao, Q.M. Synthesis, characterization and energy-related applications of carbide-derived carbons obtained by the chlorination of boron carbide. *Carbon* **47**, 820–828, 2009.
37. Janes, A., Thomberg, T., Kurig, H., and Lust, E. Nanoscale fine-tuning of porosity of carbide-derived carbon prepared from molybdenum carbide. *Carbon* **47**, 23–29, 2009.
38. Choi, H.J., Bae, H.T., and Lim, D.S. Tribology of carbon layers fabricated from SiC exposed to different H(2)/Cl(2) gas mixtures. *Journal of Ceramic Processing Research* **10**, 330–334, 2009.
39. Forbeaux, I., Themlin, J.M., Charrier, A., Thibaudau, F., and Debever, J.M. Solid-state graphitization mechanisms of silicon carbide 6H-SiC polar faces. *Applied Surface Science* **162**, 406–412, 2000.
40. Adu, K.W. et al. Morphological, structural, and chemical effects in response of novel carbide derived carbon sensor to NH(3), N(2)O, and air. *Langmuir* **25**, 582–588, 2009.
41. Dimovski, S., Nikitin, A., Ye, H.H., and Gogotsi, Y. Synthesis of graphite by chlorination of iron carbide at moderate temperatures. *Journal of Materials Chemistry* **14**, 238–243, 2004.

42. Gogotsi, Y., Welz, S., Ersoy, D.A., and McNallan, M.J. Conversion of silicon carbide to crystalline diamond-structured carbon at ambient pressure. *Nature* **411**, 283–287, 2001.
43. Presser, V., Heon, M., and Gogotsi, Y. Carbide-derived carbons—From porous networks to nanotubes and graphene. *Advanced Functional Materials* **21**, 810–833, 2011.
44. Hutchins, O. Method for the production of silicon tetrachloride, US Patent 1271713, July 9, 1918.
45. Andersen, J.N. Silicon tetrachloride manufacture, US Patent 2739041, March 20, 1956.
46. Mohun, W.A. In *Proceedings of the 4th Biennial Conference on Carbon 443–453*, Pergamon Press, New York, USA.
47. Mohun, W.A. Mineral active carbon and process, USPTO 3,066,099, 1962.
48. Kirillova, G.F., Meerson, G.A., and Zelikman, A.N. Kinetics of chlorination of titanium and niobium carbides. *Izvestiya vuzov, Tsvetnaya Metallurgiya* **3**, 90–96, 1960.
49. Orekhov, V.P. et al. Kinetics of chlorination of zirconium carbide briquets. *Journal of Applied Chemistry of the USSR* **42**, 230–237, 1969.
50. Fedorov, N.F., Ivakhnyuk, G.K., and Samonin, V.V. Mesoporous carbon adsorbents from calcium carbide. *Journal of Applied Chemistry of the USSR* **54**, 2253–2255, 1981.
51. Fedorov, N.F., Ivakhnyuk, G.K., and Gavrilov, D.N. Adsorbent carbons from carbides of group iv–vi transition-metals. *Journal of Applied Chemistry of the USSR* **55**, 243–246, 1982.
52. Samonin, V.V., Ivakhnyuk, G.K., and Fedorov, N.F. Mechanism of the formation of characteristic pore types in carbon from calcium carbide. *Journal of Applied Chemistry of the USSR* **59**, 1378–1381, 1986.
53. Samonin, V.V. et al. Carbon adsorbent from carbonized calcium carbide. *Journal of Applied Chemistry of the USSR* **59**, 1022–1023, 1986.
54. Ivakhnyuk, G.K. et al. Carbon enriched calcium carbide and possibility of its application. *Zhurnal Prikladnoi Khimii* **60**, 852–856 (in Russian), 1987.
55. Ivakhnyuk, G.K., Samonin, V.V., and Fedorov, N.F. Study of properties of carbon derived from calcium carbide in the presence of nitrogen. *Zhurnal Prikladnoi Khimii* **60**, 1413–1415 (in Russian), 1987.
56. Dai, C.L., Wang, X.Y., Wang, Y., Li, N., and Wei, J.L. Synthesis of nanostructured carbon by chlorination of calcium carbide at moderate temperatures and its performance evaluation. *Materials Chemistry and Physics* **112**, 461–465, 2008.
57. Zheng, L.P., Wang, Y., Wang, X.Y., An, H.F., and Yi, L.H. The effects of surface modification on the supercapacitive behaviors of carbon derived from calcium carbide. *Journal of Materials Science* **45**, 6030–6037, 2010.
58. Zheng, L.P. et al. Surface N, O functionalization of calcium carbide-derived carbon and its electrochemical performance. *Acta Chimica Sinica* **68**, 2516–2522, 2010.
59. Zheng, L.P. et al. The preparation and performance of calcium carbide-derived carbon/polyaniline composite electrode material for supercapacitors. *Journal of Power Sources* **195**, 1747–1752, 2010.
60. Thomberg, T., Janes, A., and Lust, E. Energy and power performance of electrochemical double-layer capacitors based on molybdenum carbide derived carbon. *Electrochimica Acta* **55**, 3138–3143, 2010.
61. Leis, J., Arulepp, M., Kaarik, M., and Perkson, A. The effect of Mo(2)C derived carbon pore size on the electrical double-layer characteristics in propylene carbonate-based electrolyte. *Carbon* **48**, 4001–4008, 2010.
62. Kuriakose, A.K. and Margrave, J.L. Kinetics of reactions of elemental fluorine with zirconium carbide + zirconium diboride at high temperatures. *Journal of Physical Chemistry* **68**, 290, 1964.
63. Kuriakose, A.K. and Margrave, J.L. Kinetics of reaction of elemental fluorine. 2. Fluorination of hafnium carbide + hafnium boride. *Journal of Physical Chemistry* **68**, 2343, 1964.
64. Schumb, W.C. and Aronson, J.R. The fluorination of carbides. *Journal of the American Chemical Society* **81**, 806–807, 1959.
65. Boehm, H.P. and Warnecke, H.H. Structural parameters and molecular-sieve properties of carbons prepared from metal carbides. *Carbon* **13**, 548–548, 1975.
66. Fedorov, N.F., Ivakhnyuk, G.K., and Gavrilov, D.N. Porous structure of carbon adsorbents made from titanium carbide. *Journal of Applied Chemistry of the USSR* **55**, 40–43, 1982.
67. Babkin, O.E., Ivakhnyuk, G.K., and Fedorov, N.F. Porous structure of carbon adsorbents made from zirconium carbide. *Journal of Applied Chemistry of the USSR* **57**, 463–467, 1984.
68. Hoffman, E.N., Yushin, G., Barsoum, M.W., and Gogotsi, Y. Synthesis of carbide-derived carbon by chlorination of Ti$_2$AlC. *Chemistry of Materials* **17**, 2317–2322, 2005.
69. Yushin, G., Dash, R., Jagiello, J., Fischer, J.E., and Gogotsi, Y. Carbide-derived carbons: Effect of pore size on hydrogen uptake and heat of adsorption. *Advanced Functional Materials* **16**, 2288–2293, 2006.
70. Janes, A., Kurig, H., Thomberg, T., and Lust, E. Advanced nanostructured carbon materials for electrical double layer capacitors, in *Functional Materials and Nanotechnologies: Fm&Nt-2007*, vol. 93. (eds. A. Sternberg and I. Muzikante), Riga, Latvia; 2007.

71. Gogotsi, Y. et al. Bactericidal activity of chlorine-loaded carbide-derived carbon against *Escherichia coli* and *Bacillus anthracis*. *Journal of Biomedical Materials Research Part A* **84A**, 607–613, 2008.

72. Yeon, S.H. et al. Carbide-derived-carbons with hierarchical porosity from a preceramic polymer. *Carbon* **48**, 201–210, 2010.

73. Kormann, M. and Popovska, N. Processing of carbide-derived carbons with enhanced porosity by activation with carbon dioxide. *Microporous and Mesoporous Materials* **130**, 167–173, 2010.

74. Bae, J.S., Nguyen, T.X., and Bhatia, S.K. Influence of synthesis conditions and heat treatment on the structure of Ti(3)SiC(2)-derived carbons. *Journal of Physical Chemistry C* **114**, 1046–1056, 2010.

75. Popovska, N. and Kormann, M. Processing of porous carbon with tunable pore structure by the carbide-derived carbon method. *Jom* **62**, 44–49, 2010.

76. Vasilenko, B.D. In *Mintsvetmetzoloto*, Moscow, USSR; 1956.

77. Babkin, O.E., Ivakhnyuk, G.K., Lukin, Y.N., and Fedorov, N.F. Study of structure of carbide derived carbon by XPS. *Zhurnal Prikladnoi Khimii* **57**, 1719–1721 (in Russian), 1984.

78. Dash, R.K. In *Department of Materials Science and Engineering, Dr. of Philosophy,* Drexel University, Philadelphia, PA; 2006.

79. Gogotsi, Y.G., Jeon, I.D., and McNallan, M.J. Carbon coatings on silicon carbide by reaction with chlorine-containing gases. *Journal of Materials Chemistry* **7**, 1841–1848, 1997.

80. Dash, R.K., Yushin, G., and Gogotsi, Y. Synthesis, structure and porosity analysis of microporous mesoporous carbon derived from zirconium carbide. *Microporous and Mesoporous Materials* **86**, 50–57, 2005.

81. Yushin, G.N. et al. Synthesis of nanoporous carbide-derived carbon by chlorination of titanium silicon carbide. *Carbon* **43**, 2075–2082, 2005.

82. Dash, R.K., Nikitin, A., and Gogotsi, Y. Microporous carbon derived from boron carbide. *Microporous and Mesoporous Materials* **72**, 203–208, 2004.

83. Hoffman, E.N. In *Materials Science and Engineering,* Drexel University, Philadelphia, PA; 2006.

84. Ischenko, V. et al. The effect of SiC substrate microstructure and impurities on the phase formation in carbide-derived carbon. *Carbon* **49**, 1189–1198, 2011.

85. Jeon, I.D., McNallan, M.J., and Gogotsi, Y.G. Formation of carbon coatings on silicon carbide by reactions in halogen containing media, in *Proceedings of the Symposium on Fundamental Aspects of High Temperature Corrosion*, vol. 96–126 (eds. D.A. Shores, R.A. Rapp, and P.Y. Hou), 256–268, The Electrochemical Society, Pennington, NJ, USA; 1996.

86. Korenblit, Y. et al. High-rate electrochemical capacitors based on ordered mesoporous silicon carbide-derived carbon. *Acs Nano* **4**, 1337–1344, 2010.

87. Dimovski, S., Nikitin, A., Ye, H.H., and Gogotsi, Y. New carbon based materials for electrochemical energy storage systems: Batteries, supercapacitors and fuel cells, *NATO Science Series II: Mathematics, Physics and Chemistry* **229**, 399–410, 2006.

88. Ersoy, D.A., McNallan, M.J., and Gogotsi, Y. Carbon coatings produced by high temperature chlorination of silicon carbide ceramics. *Materials Research Innovations* **5**, 55–62, 2001.

89. Chen, L., Behlau, G., Gogotsi, Y., and McNallan, M.J. Carbide derived carbon (CDC) coatings for tyranno ZMI SiC fibers, in *27th International Cocoa Beach Conference on Advanced Ceramics and Composites: A*, Cocoa Beach, Florida, vol. 24 (eds. W.M. Kriven and H.T. Lin), 57–62, 2003.

90. Lee, A. In *Civil and Materials Engineering*, University of Illinois at Chicago, Chicago, IL; 2005.

91. Becker, P., Glenk, F., Kormann, M., Popovska, N., and Etzold, B.J.M. Chlorination of titanium carbide for the processing of nanoporous carbon: A kinetic study. *Chemical Engineering Journal* **159**, 236–241, 2010.

92. Chen, X.Q. et al. Carbide-derived nanoporous carbon and novel core-shell nanowires. *Chemistry of Materials* **18**, 753–758, 2006.

93. Presser, V. and Nickel, K.G. Silica on silicon carbide. *Critical Reviews in Solid State and Materials Sciences* **33**, 1–99, 2008.

94. Rose, M., Kockrick, E., Senkovska, I., and Kaskel, S. High surface area carbide-derived carbon fibers produced by electrospinning of polycarbosilane precursors. *Carbon* **48**, 403–407, 2010.

95. Heon, M. et al. Continuous carbide-derived carbon films with high volumetric capacitance. *Energy and Environmental Science* **4**, 135–138, 2011.

96. Yeon, S.H., Knoke, I., Gogotsi, Y., and Fischer, J.E. Enhanced volumetric hydrogen and methane storage capacity of monolithic carbide-derived carbon. *Microporous and Mesoporous Materials* **131**, 423–428, 2010.

97. Lee, A., Zhu, R.Y., and McNallan, M. Kinetics of conversion of silicon carbide to carbide derived carbon. *Journal of Physics-Condensed Matter* **18**, S1763–S1770, 2006.

98. Ersoy, D.A., McNallan, M.J., and Gogotsi, Y. High temperature chlorination of SiC for preparation of tribological carbon films, in *High Temperature Corrosion and Materials Chemistry* (eds P.Y. Hou, M.J. McNallan, R. Oltra, E.J. Opila, and D. Shores), The Electrochemical Society, Inc., Pennington, NJ, 324–333; 1998.

99. Perkson, A. et al. Barrel-like carbon nanoparticles from carbide by catalyst assisted chlorination. *Carbon* **41**, 1729–1735, 2003.
100. Jacob, M. et al. Synthesis of structurally controlled nanocarbons—In particular the nanobarrel carbon. *Solid State Sciences* **5**, 133–137, 2003.
101. Leis, J., Perkson, A., Arulepp, M., Kaarik, M., and Svensson, G. Carbon nanostructures produced by chlorinating aluminium carbide. *Carbon* **39**, 2043–2048, 2001.
102. Welz, S. *Identification of Carbon Allotropes in Carbide Derived Carbon Using Electron Microscopy*, University of Illinois at Chicago, Chicago; 2003.
103. Schmirler, M. et al. Fast production of monolithic carbide-derived carbons with secondary porosity produced by chlorination of carbides containing a free metal phase. *Carbon* **49**, 4359–4367, 2011.
104. Schmirler, M., Glenk, F., and Etzold, B.J.M. In-situ thermal activation of carbide-derived carbon. *Carbon* **49**, 3679–3686, 2011.
105. Hoffman, E.N., Yushin, G., El-Raghy, T., Gogotsi, Y., and Barsoum, M.W. Micro and mesoporosity of carbon derived from ternary and binary metal carbides. *Microporous and Mesoporous Materials* **112**, 526–532, 2008.
106. Kravchik, A.E., Kukushkina, J.A., Sokolov, V.V., and Tereshchenko, G.F. Structure of nanoporous carbon produced from boron carbide. *Carbon* **44**, 3263–3268, 2006.
107. Portet, C., Yushin, G., and Gogotsi, Y. Effect of carbon particle size on electrochemical performance of EDLC. *Journal of Electrochemical Society* **155**(7), A531–A536, 2008.
108. Chmiola, J., Largeot, C., Taberna, P.L., Simon, P., and Gogotsi, Y. Desolvation of ions in subnanometer pores and its effect on capacitance and double-layer theory. *Angewandte Chemie-International Edition* **47**, 3392–3395, 2008.
109. Vakifahmetoglu, C., Presser, V., Yeon, S.H., Colombo, P., and Gogotsi, Y. Enhanced hydrogen and methane gas storage of silicon oxycarbide derived carbon. *Microporous and Mesoporous Materials* **144**, 105–112, 2011.
110. Janes, A., Thomberg, T., and Lust, E. Synthesis and characterisation of nanoporous carbide-derived carbon by chlorination of vanadium carbide. *Carbon* **45**, 2717–2722, 2007.
111. Chmiola, J. In *Materials Science and Engineering,* Drexel University, Philadelphia, PA; 2009.
112. Sevilla, M. and Mokaya, R. Activation of carbide-derived carbons: A route to materials with enhanced gas and energy storage properties. *Journal of Materials Chemistry* **21**, 4727–4732, 2011.
113. Portet, C., Lillo-Rodenas, M.A., Linares-Solano, A., and Gogotsi, Y. Capacitance of KOH activated carbide-derived carbons. *Physical Chemistry Chemical Physics* **11**, 4943–4945, 2009.
114. Sevilla, M., Foulston, R., and Mokaya, R. Superactivated carbide-derived carbons with high hydrogen storage capacity. *Energy and Environmental Science* **3**, 223–227, 2010.
115. Laudisio, G. et al. Carbide-derived carbons: A comparative study of porosity based on small-angle scattering and adsorption isotherms. *Langmuir* **22**, 8945–8950, 2006.
116. Tallo, I., Thomberg, T., Kontturi, K., Janes, A., and Lust, E. Nanostructured carbide-derived carbon synthesized by chlorination of tungsten carbide. *Carbon* **49**, 4427–4433, 2011.
117. Thomberg, T., Kurig, H., Janes, A., and Lust, E. Mesoporous carbide-derived carbons prepared from different chromium carbides. *Microporous and Mesoporous Materials,* **141**, 88–93, 2011.
118. Yushin, G. et al. Synthesis of nanoporous carbide-derived carbon by chlorination of titanium silicon carbide. *Carbon* **44**, 2075–2082, 2005.
119. Hoffman, E., Yushin, G.N., Barsoum, B.M., and Gogotsi, G. Synthesis of nanoporous carbide-derived carbon by chlorination of titanium aluminum carbide. *Chemical Materials* **17**, 2317–2322, 2005.
120. Korenblit, Y. et al. *In-situ* studies of ion transport in microporous supercapacitor electrodes at ultra-low temperatures. *Advanced Functional Materials* **22**(8), 1655–1662, April 2012.
121. Kajdos, A., Kvit, A., Jones, F., Jagiello, J., and Yushin, G. Tailoring the pore alignment for rapid ion transport in microporous carbons. *Journal of American Chemical Society* **132**, 3252, 2010.
122. Portet, C. et al. Electrical double-layer capacitance of zeolite-templated carbon in organic electrolyte. *Journal of the Electrochemical Society* **156**, A1–A6, 2009.
123. Itai, H., Nishihara, H., Kogure, T., and Kyotani, T. Three dimensionally arrayed and mutually connected 1.2-nm nanopores for high-performance electric double layer capacitor. *Journal of the American Chemical Society* **133**, 1165–1167, 2011.
124. Nishihara, H. et al. A possible buckybowl-like structure of zeolite templated carbon. *Carbon* **47**, 1220–1230, 2009.
125. Kyotani, T., Nagai, T., Inoue, S., and Tomita, A. Formation of new type of porous carbon by carbonization in zeolite nanochannels. *Chemistry of Materials,* **9**, 609–615, 1997.
126. Portet, C., Yushin, G., and Gogotsi, Y. Effect of carbon particle size on electrochemical performance of EDLC. *Journal of Electrochemical Society* **155** (7), A531–A536, 2008.

127. Rose, M. et al. Hierarchical micro- and mesoporous carbide-derived carbon as a high-performance electrode material in supercapacitors. *Small* **7**, 1108–1117, 2011.

128. Thomberg, T., Janes, A., and Lust, E. Energy and power performance of vanadium carbide derived carbon electrode materials for supercapacitors. *Journal of Electroanalytical Chemistry* **630**, 55–62, 2009.

129. Rose, M. et al. Hierarchical micro- and mesoporous carbide-derived carbon as a high-performance electrode material in supercapacitors. *Small* **7**, 1108–1117, 2011.

130. He, L.L. et al. Small-angle neutron scattering characterization of the structure of nanoporous carbons for energy-related applications. *Microporous and Mesoporous Materials* **149**, 46–54, 2012.

131. Nguyen, T.X. and Bhatia, S.K. Characterization of accessible and inaccessible pores in microporous carbons by a combination of adsorption and small angle neutron scattering. *Carbon* **50**, 3045–3054, 2012.

132. Centeno, T.A. and Stoeckli, F. The assessment of surface areas in porous carbons by two model-independent techniques, the DR equation and DFT. *Carbon* **48**, 2478–2486, 2010.

133. Jeong, J.H., Bae, H.T., and Lim, D.S. The effect of iron catalysts on the microstructure and tribological properties of carbide-derived carbon. *Carbon* **48**, 3628–3634, 2010.

134. Leis, J., Perkson, A., Arulepp, M., Nigu, P., and Svensson, G. Catalytic effects of metals of the iron subgroup on the chlorination of titanium carbide to form nanostructural carbon. *Carbon* **40**, 1559–1564, 2002.

135. Kaarik, M., Arulepp, M., Karelson, M., and Leis, J. The effect of graphitization catalyst on the structure and porosity of SiC derived carbons. *Carbon* **46**, 1579–1587, 2008.

136. Kockrick, E. et al. CeO(2)/Pt catalyst nanoparticle containing carbide-derived carbon composites by a new *in situ* functionalization strategy. *Chemistry of Materials* **23**, 57–66, 2011.

137. Yushin, G., Nikitin, A., and Gogotsi, Y. Carbide derived carbon, in *Nanomaterials Handbook*. (ed. Y. Gogotsi), 239–282, CRC Press, Boca Raton, FL, 2006.

138. Babkin, O.E., Ivakhnyuk, G.K., and Fedorov, N.F. Porous structure of carbon adsorbents from zirconium carbide. *Zhurnal Prikladnoi Khimii* **57**, 504–508 (in Russian), 1984.

139. Batisse, N. et al. Fluorination of silicon carbide thin films using pure F(2) gas or XeF(2). *Thin Solid Films* **518**, 6746–6751, 2010.

140. Watanabe, M. et al. Thermal reaction of polycrystalline SiC with XeF2. *Journal of Vacuum Science and Technology A* **23**, 1638–1646, 2005.

141. Liu, F., Gutes, A., Laboriante, I., Carraro, C., and Maboudian, R. Graphitization of *n*-type polycrystalline silicon carbide for on-chip supercapacitor application. *Applied Physics Letters* **99**(11), 3, 2011.

142. Largeot, C., Taberna, P.L., Gogotsi, Y., and Simon, P. Microporous carbon-based electrical double layer capacitor operating at high temperature in ionic liquid electrolyte. *Electrochemical and Solid State Letters* **14**, A174–A176, 2011.

143. Simon, P. and Gogotsi, Y. Materials for electrochemical capacitors. *Nature Materials* **7**, 845–854, 2008.

144. Barbieri, O., Hahn, M., Herzog, A., and Kotz, R. Capacitance limits of high surface area activated carbons for double layer capacitors. *Carbon* **43**, 1303–1310, 2005.

145. Presser, V. et al. Flexible nano-felts of carbide-derived carbon with ultra-high power handling capability. *Advanced Energy Materials* **1**, 423–430, 2011.

146. Chmiola, J., Largeot, C., Taberna, P.L., Simon, P., and Gogotsi, Y. Monolithic carbide-derived carbon films for micro-supercapacitors. *Science* **328**, 480–483, 2010.

147. Sun, G. et al. Significantly enhanced rate capability in supercapacitors using carbide-derived carbons electrode with superior microstructure. *Journal of Solid State Electrochemistry* **16**, 1263–1270, 2012.

148. Tallo, I., Thomberg, T., Janes, A., and Lust, E. Electrochemical behavior of alpha-tungsten carbide-derived carbon based electric double-layer capacitors. *Journal of the Electrochemical Society* **159**, A208–A213, 2012.

149. Oschatz, M. et al. A cubic ordered, mesoporous carbide-derived carbon for gas and energy storage applications. *Carbon* **48**, 3987–3992, 2010.

150. Latt, M. et al. A structural influence on the electrical double-layer characteristics of Al(4)C(3)-derived carbon. *Journal of Solid State Electrochemistry* **14**, 543–548, 2010.

151. Fernandez, J.A., Arulepp, M., Leis, J., Stoeckli, F., and Centeno, T.A. EDLC performance of carbide-derived carbons in aprotic and acidic electrolytes. *Electrochimica Acta* **53**, 7111–7116, 2008.

152. Morris, R.E. and Wheatley, P.S. Gas storage in nanoporous materials. *Angewandte Chemie-International Edition* **47**, 4966–4981, 2008.

153. Presser, V., McDonough, J., Yeon, S.H., and Gogotsi, Y. Effect of pore size on carbon dioxide sorption by carbide derived carbon. *Energy and Environmental Science* **4**, 3059–3066, 2011.

154. Gogotsi, Y. et al. Importance of pore size in high-pressure hydrogen storage by porous carbons. *International Journal of Hydrogen Energy* **34**, 6314–6319, 2009.

155. Gogotsi, Y. et al. Tailoring of nanoscale porosity in carbide-derived carbons for hydrogen storage. *Journal of the American Chemical Society* **127**, 16006–16007, 2005.

156. Seredych, M., Portet, C., Gogotsi, Y., and Bandosz, T.J. Nitrogen modified carbide-derived carbons as adsorbents of hydrogen sulfide. *Journal of Colloid and Interface Science* **330**, 60–66, 2009.

157. Yeon, S.H. et al. Enhanced methane storage of chemically and physically activated carbide-derived carbon. *Journal of Power Sources* **191**, 560–567, 2009.

158. Kockrick, E. et al. Ordered mesoporous carbide derived carbons for high pressure gas storage. *Carbon* **48**, 1707–1717, 2010.

159. Kotina, I.M. et al. Study of the lithium diffusion in nanoporous carbon materials produced from carbides. *Journal of Non-Crystalline Solids* **299–302**, 815–819, 2002.

160. Sakamoto, K.M., *Master's thesis*, Naval Postgraduate School, Monterey, CA; 2011.

161. Kotina, I.M. et al. Lithium in nanoporous carbon materials produced from SiC. *Hydrogen Materials Science and Chemistry of Carbon Nanomaterials* **172**, 391–398, 2004.

162. Kotina, I.M. et al. The phase composition of the lithiated samples of nanoporous carbon materials produced from carbides. *Journal of Non-Crystalline Solids* **299**, 820–823, 2002.

163. Kotina, I.M. et al. Study of the lithium diffusion in nanoporous carbon materials produced from carbides. *Journal of Non-Crystalline Solids* **299**, 815–819, 2002.

164. Davis, M.E. Ordered porous materials for emerging applications. *Nature* **417**, 813–821, 2002.

165. Ersoy, D.A., McNallan, M.J., and Gogotsi, Y. Platinum reactions with carbon coatings produced by high temperature chlorination of silicon carbide. *Journal of the Electrochemical Society* **148**, C774–C779, 2001.

166. Jie Niu, J., Presser, V., Karwacki, C.J., and Gogotsi, Y. Ultrasmall gold nanoparticles with the size controlled by the pores of carbide-derived carbon. *Materials Express* **1**, 259–266, 2011.

167. Erdemir, A. et al. Synthesis and tribology of carbide-derived carbon films. *International Journal of Applied Ceramic Technology* **3**, 236–244, 2006.

168. Bae, H.T., Choi, H.J., Jeong, J.H., and Lim, D.S. The effect of reaction temperature on the tribological behavior of the surface modified silicon carbide by the carbide derived carbon process. *Materials and Manufacturing Processes* **25**, 345–349, 2010.

169. Gao, F., Lu, J.J., and Liu, W.M. Tribological behavior of carbide-derived carbon coating on SiC polycrystal against SAE52100 steel in moderately humid air. *Tribology Letters* **27**, 339–345, 2007.

170. Ersoy, D.A., McNallan, M.J., Gogotsi, Y., and Erdemir, A. Tribological properties of carbon coatings produced by high temperature chlorination of silicon carbide. *Tribology Transactions* **43**, 809–815, 2000.

171. Carroll, B., Gogotsi, Y., Kovalchenko, A., Erdemir, A., and McNallan, M.J. Tribological characterization of carbide-derived carbon layers on silicon carbide for dry friction applications, in *Euro Ceramics Viii, Pts 1–3*, vol. 264–268. (eds. H. Mandal and L. Ovecoglu), Trans Tech Publications Inc., Durnten-Zurich, Switzerland, 465–468, 2004.

172. Choi, H.J. et al. Sliding wear of silicon carbide modified by etching with chlorine at various temperatures. *Wear* **266**, 214–219, 2009.

173. Sui, J.A., Zhang, Y.J., Ren, S.F., Rinke, M., and Lu, J.J. *Super-Hydrophobic and Self-Lubricating Carbon Coating on Ti(3)SiC(2)*, Springer, Berlin, Heidelberg, Germany; 2009.

174. Bae, H.T., Jeong, J.H., Choi, H.J., and Lim, D.S. Fabrication and tribological characterization of gradient carbon layer derived from SiC–TiC composites. *Journal of the Ceramic Society of Japan* **118**, 1150–1153, 2010.

175. Sui, J. and Lu, J. Formulated self-lubricating carbon coatings on carbide ceramics. *Wear* **271**, 1974–1979, 2011.

176. Yachamaneni, S. et al. Mesoporous carbide-derived carbon for cytokine removal from blood plasma. *Biomaterials* **31**, 4789–4794, 2010.

177. Yushin, G. et al. Mesoporous carbide-derived carbon with porosity tuned for efficient adsorption of cytokines. *Biomaterials* **27**, 5755–5762, 2006.

178. Presser, V. et al. Hierarchical porous carbide-derived carbons for the removal of cytokines from blood plasma. *Advanced Healthcare Materials* **1**, 796–800, 2012.

179. Hoffman, E.N., Yushin, G., Wendler, B.G., Barsoum, M.W., and Gogotsi, Y. Carbide-derived carbon membrane. *Materials Chemistry and Physics* **112**, 587–591, 2008.

12 Templated and Ordered Mesoporous Carbons

Pasquale F. Fulvio, Joanna Gorka, Richard T. Mayes, and Sheng Dai

CONTENTS

12.1 INTRODUCTION

Porous carbon materials have attracted considerable attention in recent years, in particular, for applications, such as energy storage,[1,2] catalysis,[3] and separation/filtration.[4,5] As a common definition according to the International Union of Pure and Applied Chemistry (IUPAC), pores are classified on the basis of their diameter as micropores (<2 nm), mesopores (between 2 and 50 nm), and macropores (>50 nm).[6,7] Various synthesis methods have been developed for the preparation of carbons with one or more types of pores to yield complex and sometimes ordered hierarchical pore structures.[8–14] Simultaneously, the search for the most suitable carbon precursors that allow for the formation of carbons with well-defined pore structures, combined with good thermal stability, well-defined surface functional groups, or high electrical conductivities for targeted applications, has been one of the most active topics of research in the field of carbon materials. Although some carbon precursors of natural origin or synthetic organic molecules can develop micro-, meso-, or macropores during thermal treatment, other carbon sources require chemical activation or templates to generate such pores.[15–17]

Templates can be defined as hard or soft templates.[9–14] The former refers to solid matter having interconnected pore systems, such as some zeolites,[18,19] ordered mesoporous silica (OMS),[20–25] or colloidal silica particles, whereas, soft templates relate to soft matter, such as surfactant micelles, polymers, and other organic molecules, exhibiting liquid crystalline properties.

Despite sacrificing the template and introducing additional synthesis steps, templating offers the advantage of properly controlling both pore geometry and size, as well as enabling use of a wider spectrum of starting carbon precursors, ranging from small natural or synthetic organic molecules to polymers.[9–13]

12.2 HARD-TEMPLATED MICROPOROUS, MESOPOROUS, AND HIERARCHICAL CARBONS

12.2.1 Nanostructured Hard-Template Preparation: OMSs and Zeolites

Many hard templates, for example, of OMS, are prepared from the hydrolysis and condensation of silica precursors (e.g., silicon alkoxides, sodium silicates, etc.) in the presence of liquid crystalline phases of alkyl-ammonium surfactants as shown in Figure 12.1a (i.e., MCM-41, MCM-48)[20] or polyethylene oxide–polypropylene oxide–polyethylene oxide (PEO–PPO–PEO; PEO: polyethylene oxide; PPO: poly(p-phenylene oxide)) triblock copolymers (i.e., SBA-15, SBA-16, KIT-6, FDU-1).[21–25] This synthesis proceeds through an induced self-assembly of the inorganic frameworks in the presence of the liquid crystal template (LCT). After condensation and removal of the organic phase, a porous material is obtained. In general, only long-range periodicity in the mesopore structures is observed in OMSs, and these have amorphous pore walls.[20–22,24,25] A similar synthesis mechanism for the direct preparation of porous carbons through self-assembly will be discussed in this section.

Crystalline aluminosilicate zeolites are prepared via a hydrothermal synthesis using a silica source, in the presence of a small quaternary ammonium molecule.[18,19,26] The inorganic framework organizes around the molecular template instead of the LCT, as illustrated in Figure 12.1b. After removing the organic template, the resulting materials exhibit well-defined crystalline phases and periodic micropore structures.[18,19]

12.2.2 Zeolite-Templated Microporous Carbons

Microporous carbons have been prepared as inverse replicas of zeolites.[27–32] Zeolite-templated carbons (ZTCs) are prepared by filling the micropores of zeolites, such as zeolite *Y* and *X13*, with a suitable carbon source, for example, furfuryl alcohol, followed by carbonization of composites and dissolution of the siliceous framework. As a result of the well-defined three-dimensional (3D) crystal structure of most starting zeolite materials, ZTCs exhibit amorphous carbon walls having a periodic arrangement of micropores.[27–32] The ZTCs exhibit the same periodicity as the starting microporous templates in the range of 1.0–2.0 nm, as shown in Figure 12.2.[27–32]

Whenever zeolites with disconnected two-dimensional (2D) pore system are used as templates, for example, zeolite *L*, the carbon structure collapses after dissolution of the template, resulting in randomly oriented nonporous carbon rods. This scenario is illustrated in Figure 12.3.[32]

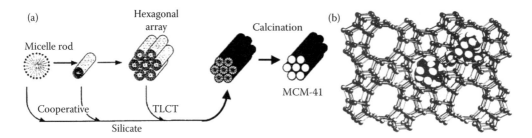

FIGURE 12.1 General liquid crystal templating mechanism for the preparation of OMS (a). The transformation from micelle aggregates into an ordered liquid crystalline phase is induced by changes in surfactant concentration, as in mechanism 1. The required surfactant concentration to obtain a specific liquid crystal phase can be induced by the silicate–surfactant ratios in the synthesis gels according to mechanism 2. (From Beck, J. S. et al., *Journal of the American Chemical Society*, 114, 10834, 1992. With permission.) (b) Structure of the TPA+-ZSM-5 composite, showing how the molecular tetrapropylammonium cation template is located in the crystalline zeolite channels. The size and geometry of the cavities of the negatively charged zeolite framework are determined by the cation template used. (From Schüth, F., *Angewandte Chemie International Edition*, 42, 3604, 2003. With permission.)

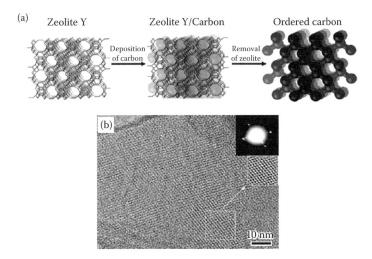

FIGURE 12.2 Synthesis scheme for the preparation of microporous ZTC from zeolite *Y* as a template (a) and high-resolution transmission electron microscopy (HRTEM) image of the periodic microporous carbon inverse replica (b) obtained following this procedure. (From Lee, J. et al., *Advanced Materials,* 18, 2073, 2006. With permission.)

12.2.3 COLLOIDAL SILICA TEMPLATING

Dispersed colloidal silica particles of various sizes[33–36] and colloidal silica crystals (opals)[37–39] have been used as templates of porous carbons, with spherical pores having narrow pore size distributions (PSDs). By coating monodisperse colloidal silica particles or crystals with a suitable carbon precursor, followed by carbonization and etching of the silica, porous carbon particles can be obtained. The diameters of the mesopores are determined by the size of the silica particles. Because

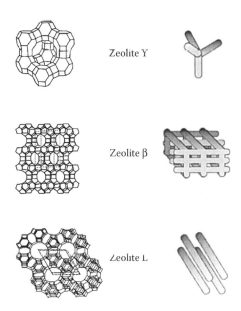

FIGURE 12.3 Illustration of the pore connectivity in various types of zeolite materials used as templates of microporous carbons. Pore interconnections are a requirement for the preparation of microporous ZTCs with the same pore periodicity as the zeolite template. (From Johnson, S. A. et al., *Chemistry of Materials,* 9, 2448, 1997. With permission.)

of the large variety of colloidal particles with various sizes that are commercially available, carbons with pores having diameters in the mesopore range from approximately 5 to 50 nm, as well as in the macropore range (i.e., above 50 nm) have been reported. Templating via colloidal silica is also suitable for the preparation of ordered macroporous carbons[37–39] and polymers.[40]

The first reported colloid-templated carbons exhibited high specific surface area and a large mesopore volume,[41] but with broad PSD owing to uncontrolled aggregation of the silica colloids. To prevent the latter, surfactants were used as stabilizing agents.[42] However, to obtain spherical mesopores accurately resembling the size of the silica colloids, suitable carbon precursors that can uniformly fill the voids between all silica particles are required.[43] An example of a suitable carbon source is mesophase pitch,[33–35] which is characterized by a softening point. Optimum heating and stabilization temperatures before carbonization were found to assure a complete coverage of the surface of silica colloids with the carbon precursor. In this manner, carbons with spherical pores having similar diameters as those of the silica colloidal particles are obtained,[33–35,43] as shown in Figure 12.4a.

Initial theoretical studies were later confirmed experimentally indicating that complete filling of pores of a colloidal crystal template with carbon precursor may yield carbons with pore volumes as large as 1.36 cm^3/g.[36] Whenever a thin carbon film from a phenolic resin was uniformly cast on the surface of a silica colloidal crystal, as illustrated in Figure 12.4b, the total pore volume reached 9 cm^3/g after etching the silica template.[36,44]

For many applications, nanoporous carbons with a monolithic shape are required. Monoliths can be prepared from powder processing using polymeric binders and additional carbonization stages. Precise control over particle size and morphology is needed to successfully prepare a monolith. Because of the difficulties associated with controlling all these properties, the development of techniques to prepare monolithic carbon structures is desired. The use of colloidal silica was also shown to facilitate the preparation of carbon monoliths, and monolithic colloidal silica templates can be fabricated by evaporation of the solvent off the colloidal silica suspension and by compressing the dry powder into desired shape and size.[45] The prepared carbon inverse replicas of the templates retained the monolithic shape even after chemical removal of silica. In addition to colloidal silica particles, metal nanoparticles of interest for catalysis, such as silver, were also dispersed within the starting silica monolith template.[45] In this manner, silver-supported mesoporous carbon monoliths were obtained (Figure 12.5).

Besides carbon powders and monoliths with interconnected pores, dispersed porous carbon hollow spheres can be prepared by coating well-dispersed silica particles with thin organic coatings, followed by carbonization and silica etching.[46–49] For example, amine-functionalized silica colloids with various diameters were coated with glucose using hydrothermal conditions.[47] The final carbon spheres were microporous and 10 nm thick and the diameters comparable with that of the starting silica. Another advancement was achieved by replacing conventional silica spheres by gold–silica core-shell particles to prepare carbon spheres with a single Au nanoparticle trapped in each sphere (Figure 12.6a).[47] Mesoporous carbon hollow spheres were prepared by first forming a mesoporous silica layer on Au–silica particles. The final carbon hollow sphere was roughly 24 nm thick and had disordered mesopores resembling those of the thin porous silica layer and containing Au nanoparticles. When 400 nm diameter Au–silica shell particles were coated with dopamine, 4 nm thick microporous carbon hollow spheres with 15 nm gold particles were formed (Figure 12.6b).[49]

Compared with the various templating methods, some advantages associated with the colloidal silica templating method are the commercial availability of colloidal particles of various sizes, and of carbon precursors, such as mesophase pitches, organic monomers, polymers, and phenolic resins. Hence, the colloidal silica and opal templating approach are attractive for the possible large-scale production of nanoporous carbons.

12.2.4 OMS Templating

Ordered mesoporous carbons (OMCs) templated by OMSs have been extensively studied, especially the CMK-1[50] and CMK-3[51] carbons templated by MCM-48[20] and SBA-15[21,22] silicas, respectively.

Because of the nature of this templating synthesis, which involves introduction of a carbon precursor into pores of the template followed by carbonization and template dissolution, inverse carbon replicas of the OMS templates are obtained. Although carbon replicas of MCM-48 often undergo a symmetry change[50,52,53] (Figure 12.7) and the MCM-41 replication results in disordered carbon rods, the SBA-15 replication permits exact inverse replicas (Figure 12.8).[51] The observed differences in replication of the aforementioned OMSs are caused by structural differences of these templates. For instance, the MCM-41 and MCM-48 (*P6 mm* and *Ia3d* symmetry, respectively) OMSs are

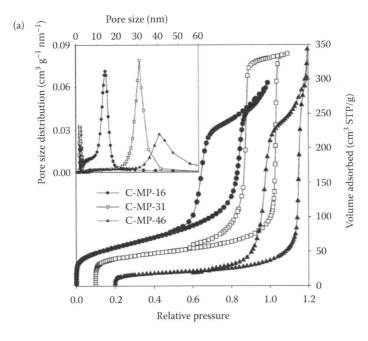

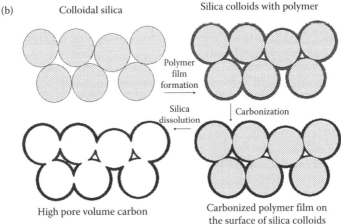

FIGURE 12.4 Nitrogen adsorption isotherms at 77 K for pitch-based carbons, offset horizontally in increments of $P/P_0 = 0.1$, prepared with colloidal silicas of various diameters as templates (a). Inset shows the calculated PSDs. (From Gierszal, K. P. et al., *Journal of Materials Chemistry,* 16, 2819, 2006.) By increasing the diameter of the silica particles, the mesopores of the final carbons were systematically enlarged. The synthesis scheme for highly interconnected carbon thin films obtained after carbonization of phenolic resin-coated silica colloids and silica dissolution (b). Carbons with pore volumes approaching 9 cm³/g can be obtained by this method. (From Gierszal, K. P. et al., *Journal of the American Chemical Society,* 128, 10026, 2006. With permission.)

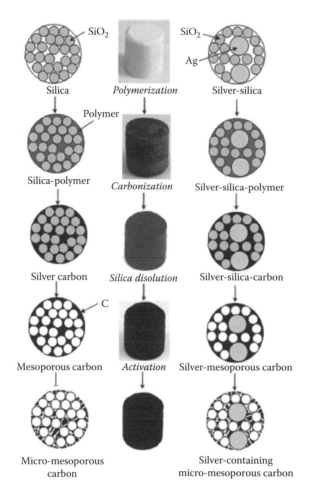

FIGURE 12.5 Carbon monolith synthesis using resorcinol–crotonaldehyde polymer as carbon precursor. The polymerization took place in the pores of the large silica monoliths. Silver nanoparticles could also be dispersed in the silica template and transferred to the final templated carbon materials. The carbons retained the monolithic shape of the silica template and resisted chemical activation with potassium hydroxide (KOH). After activation, carbons exhibited microporous–mesoporous structures. (From Jaroniec, M. et al., *Chemistry of Materials,* 20, 1069, 2008. With permission.)

synthesized by using cationic surfactants as soft templates, which afford purely mesoporous materials. In the case of SBA-15 (*P6 mm*), because of its fine interconnecting pores that create a 3D porous system, it affords a stable and exact inverse replica, CMK-3, with hexagonal structure composed of ordered carbon rods interconnected by irregular carbon threads.[8] An analogous structure of CMK-3 is CMK-5, which consists of ordered carbon pipes instead of carbon rods formed because of an incomplete filling of the ordered mesopores of SBA-15 with a carbon precursor.[54,55] The mechanism of formation for CMK-3 and CMK-5 is shown in Figure 12.9.

Contrary to colloidal silica templating where the pore size is determined by the diameter of the silica colloids used, the pore widths of the hard-templated OMCs depend on the pore wall thickness of the selected OMS template.[56] The mesopore width of SBA-15 will then influence the wall thickness of the resulting CMK-3 carbon. Hence, control over the mesopore size in OMS hard-templated carbons is possible by tailoring the adsorption and structural properties of the starting template (Figure 12.10). Small variations in the expected pore widths and pore wall thicknesses of the final OMCs result from shrinkage of the carbonaceous materials inside the silica template

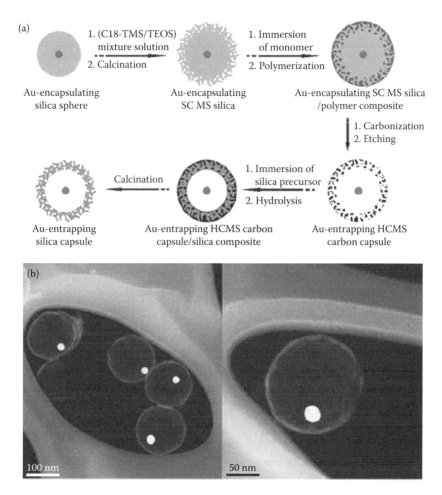

FIGURE 12.6 (a) Illustration of the preparation of mesoporous Au–mesoporous carbon and silica capsules using glucose as carbon source. (From Kim, J. Y. et al., *Chemical Communications*, 790, 2003.) (b) TEM images of Au nanoparticles in microporous carbon hollow spheres prepared from dopamine as the carbon precursor. The carbon spheres were as large as the Au–silica template spheres, approximately 400 nm in diameter, and the carbon walls were ≈4 nm thick. (From Liu, R. et al., *Angewandte Chemie-International Edition*, 50, 6799, 2011.)

during thermal treatments. Hence, the OMC inverse replicas can display slightly larger mesopore widths and thinner carbon walls depending on the carbon source and carbonization temperature.

Other OMSs obtained in the presence of triblock copolymers, which were successfully used as hard templates for the synthesis of carbons, are SBA-16 with *Im3m* symmetry[57] and KIT-6 with *Ia3d* symmetry.[25] Although the latter is a large pore analog of MCM-48, SBA-16 is composed of a 3D array of ordered spherical cages and each cage is connected to eight neighboring cages through small apertures. Similarly, to SBA 15, both mesostructures possess additional irregular micropores in the siliceous mesopore walls. In comparison to a 3D cubic bicontinuous arrangement of cylindrical mesopores in KIT-6, the replication process of SBA-16 is challenging because of its cage-like structure. In the case of SBA-16, its carbon inverse replica (*Im3m* symmetry) consists of interconnected spherical carbon particles. Although several carbon precursors have been successfully used for the replication of SBA-15 and KIT-6, some difficulties were reported for SBA-16 nanocasting.[58–60] For instance, the carbon replicas of SBA-16 prepared from sucrose could be only obtained for pores exceeding 6 nm that excludes this material for a number of applications.[58]

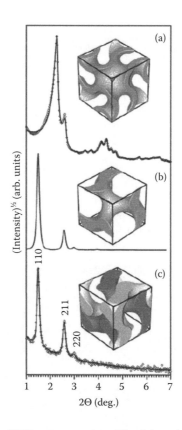

FIGURE 12.7 Small-angle powder XRD patterns and models of the unit cells for MCM-48 template (a) for a single-carbon subframework enantiomer (b) and the CMK-1 carbon. Diamonds show experimental data and solid lines show the calculated diffraction patterns. The MCM-48 silica has a double-gyroid pore structure (*Ia3d* symmetry), with two disconnected pore systems. The lack of connectivity between both systems causes the symmetry to change in the resulting carbon inverse replica, CMK-1. The symmetry for some of the segments of the CMK-1 structure is compatible to the $I4_132$ space group. (From Solovyov L. A. et al., *Journal of Physical Chemistry B,* 106, 12198, 2002. With permission.)

In an attempt to improve electronic conductivity, carbon precursors, such as furfuryl alcohol,[59–63] pitch,[35,64–66] polyacrylonitrile (PAN),[67] and polyaniline,[68,69] have been employed to obtain graphitic carbons, which could retain, at least partially, originally ordered mesostructures during the graphitization process ($\approx$2000°C). Doping the carbon with heteroatoms, such as nitrogen,[67–71] was also investigated. For instance, nitrogen-doped carbons with partial graphitic ordering were obtained from acetonitrile[70,71] after thermal treatments at relatively low temperatures (below 1000°C). Sulfur-doped carbons obtained by polymerization and carbonization of thiophene in SBA-15 have also been studied,[72] leading to some conductivity enhancement.

Graphitic carbons were also obtained by *in situ* oxidative polymerization of pyrrole in the SBA-15 pores in the presence of iron chloride ($FeCl_3$).[73–75] The resulting iron species after polymerization acted as a catalyst for the formation of graphitic domains;[73,74] the amount of nanocasted carbon and the extent of graphitization were found to be dependent on the catalyst loading within the silica host.[74] Further studies of this system showed that some doping with nitrogen remained in the graphitic framework after thermal treatments up to 1000°C.[73] Additional studies suggested that a small fraction of the iron catalyst was deeply embedded in the carbon matrix and consequently protected from leaching.[75] The same work reported that carbonization and removal of silica template with a sodium hydroxide solution afford an amorphous carbon-containing super paramagnetic iron oxide (Fe_3O_4) nanoparticles with traces of nonoxidized α-Fe (ferrite). In addition to the study of carbons

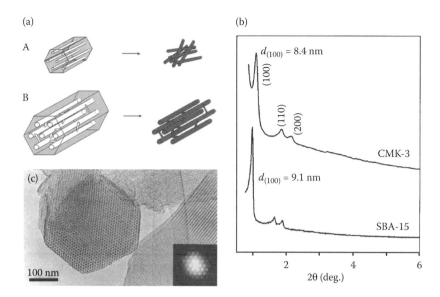

FIGURE 12.8 Synthesis mechanism for disconnected carbon rods using MCM-41 and for CMK-3 porous carbon inverse replica of SBA-15. (a) Powder XRD patterns of SBA-15 and carbon inverse replica CMK-3 (b) and TEM image with selected area diffraction pattern, inset. (c) As a result of the small interconnecting pores in the SBA-15 structure, a stable carbon inverse replica, having the same *P6 mm* symmetry as the starting template, is obtained. (From Ryoo, R. et al., *Advanced Materials,* 13, 677, 2001. With permission; Jun, S. et al., *Journal of the American Chemical Society,* 122, 10712, 2000. With permission.)

with graphitic domains,[59,64,71] particle morphology,[63,76] and the preparation of carbon thin films for fuel cell applications,[77] CMK-3-supported nanoparticles have been studied widely. Oxides of tin, iron, nickel, and manganese,[78–84] and metal nanoparticles of iron, tin, platinum, platinum–ruthenium,[75,85,86] for example, have been introduced to the external surfaces of CMK-3 by postsynthesis deposition methods (insipient wetness impregnation of metal chlorides and thermal treatments) or within the carbon walls by mixing the particle sources with the carbon precursors.

So far, CMK-5 carbons have been much less explored than CMK-3. One possible reason is the difficulty of finding suitable precursors that permit a strict control of the pore wall thickness to obtain stable carbon replicas. There are reports indicating the possibility of preparing CMK-5 with controlled pore wall thickness by using different methods of furfuryl alcohol polymerization[55,87,88] and ferrocene[89] as carbon precursors. Also, there are reports on the incorporation of highly dispersed nanoparticles of platinum[54,89] and cobalt[88] into CMK-5 to enhance its electrochemical and

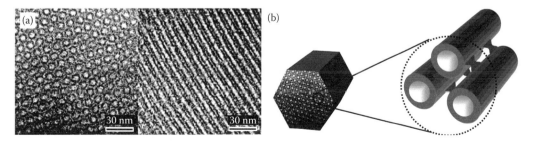

FIGURE 12.9 TEM image and selected area electron diffraction of CMK-5 material (a) and the proposed structure having a bimodal distribution of mesopores (b). A pore system is formed by the tubular structure of the carbon walls and the second system of pores results from the interconnected carbon pipes. (From Kruk, M. et al., *Chemistry of Materials,* 15, 2815, 2003. With permission.)

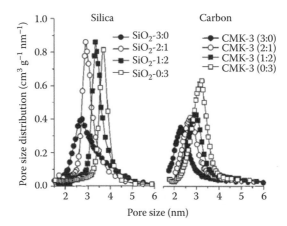

FIGURE 12.10 Pore size control in OMS-templated OMC. The 2D hexagonal OMS templates with various pore diameters were prepared by varying ratios of hexadecyltrimethylammonium bromide (HTAB) to polyoxyethylene hexadecyl ether-type surfactants. (From Lee, J. S. et al., *Journal of the American Chemical Society,* 124, 1156, 2002. With permission.)

catalytic properties. In the case of cobalt-containing CMK-5, magnetic properties too are added. It has been shown that the nanocasting synthesis of CMK-5 by using furfuryl alcohol together with cobalt, iron, and nickel salts[88] resulted in carbon nanopipe materials only in the case of cobalt. In the case of iron and nickel, only CMK-3 (rod-type structure) carbons have been reported.

A major step in the preparation of tubular carbons has been made by using chemical vapor deposition (CVD) of acetylene in the MCM-48 channels.[90] Following this route, relatively stable carbon inverse replicas with ultrathin pore walls have been obtained. In the case of SBA-15, the stable ultrathin carbon films have been reported only for carbon–silica nanocomposites.[91] An aromatic carbon precursor, 2,3-dihydroxynaphthalene (DHN), which reacts with silanols present on the silica surface, was used to form homogeneous ultrathin carbon films but they were unstable after silica dissolution. These carbon films obtained by carbonization of DHN attached to the siliceous pore walls were hydrophobic and electrically conductive such as graphitic carbons.[91] When a nickel catalyst was added to the SBA-15/DHN nanocomposites, partially stable CMK-5 carbons with a peapod distribution of Ni nanoparticles throughout the tubular films were obtained after carbonization and silica dissolution.[92] Despite the lower molecular weight of DHN compared to mesophase pitch and phenolic resins, the ability of DHN to form chemical bonds with the silica surface[91,93,94] as well as its easy graphitization makes this precursor attractive for further investigations.

Such results illustrate the enormous possibilities to tailor the structure of hard-templated OMCs using OMS templates. OMCs having pore symmetries previously known almost exclusively for OMSs were reported for the first time. Also, carbons with mesopores below 10 nm and narrow PSDs with precise control over the pore diameters with geometries other than spherical were made possible through this method.

12.3 SOFT-TEMPLATED MESOPOROUS CARBONS

In addition to the hard-templating synthesis of OMCs, a soft-templated method for the preparation of OMCs, without the silica scaffolds, was reported by Dai et al. in 2004. This direct soft-template synthesis of OMC thin films used resorcinol–formaldehyde as the carbon source and a polystyrene-*block*-poly(4-vinylpyridine) (PS-P4VP) diblock copolymer as the soft template.[95] Resorcinol monomers can interact with the P4VP segment of the copolymer via hydrogen bonds. The spin-coated films were preorganized into well-ordered mesostructured films with the assistance of amphiphilic PS-P4VP self-assembly through solvent annealing and then polymerized *in situ* with formaldehyde vapor. On carbonization under N_2 atmosphere, OMC films with a hexagonal structure and large pore size of *ca.* 34 nm were obtained (Figure 12.11).

OMC films were later reported by Tanaka et al. (COU-1) through a similar route, but using triblock copolymer Pluronic F127 as a template.[96] For this synthesis, triethyl orthoacetate (EOA) and resorcinol–formaldehyde resins are used as carbon precursors. However, the ordered mesostructure may undergo partial collapse during the high-temperature carbonizations.[97]

A major advance in the soft-templated synthesis of OMCs was made by using all commercially available, low-cost raw materials, namely, phenolic resins and a PEO–PPO–PEO triblock copolymer (e.g., Pluronic F127, P123, and F108) to prepare OMC films, powders, disordered flexible fibers, and monoliths[98–108] with the hexagonal *P6 mm* structure and also free-standing carbon membranes[109] with disordered mesopores (Figure 12.12). In one synthesis, low-molecular-weight and water-soluble phenolic resins (resols), which are polymerized from phenol and formaldehyde under alkaline conditions, first mix with the Pluronic PEO–PPO–PEO in ethanol.[101,102] The homogeneous solution is poured into dishes to evaporate the solvent. It induces the organic–organic self-assembly of resols with amphiphilic block copolymers by hydrogen bonds to form ordered mesostructures. In the synthesis reported by Dai et al., phloroglucinol–formaldehyde,[99] resorcinol–formaldehyde,[100] or phloroglucinol–glyoxal[108] are polymerized in the presence of Pluronic triblock copolymers. The *in situ* polymerization with formaldehyde is followed by microphase separation of a polymer nanocomposite owing to the strong hydrogen bond and electrostatic interactions with the hydrophilic blocks of the F127 template. The obtained gels are then redispersed in tetrahydrofuran and ethanol, then cast as films on different substrates. The drying steps involve similar structural organization

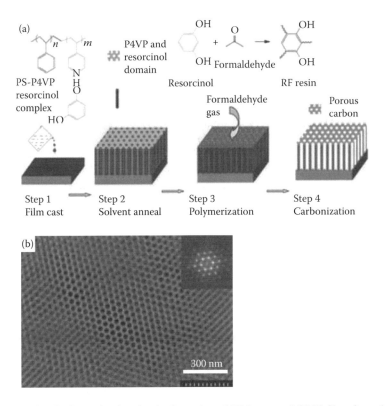

FIGURE 12.11 (a) Synthesis mechanism for the formation of 2D hexagonal OMC films from the casting and microphase separation of PS-P4VP/resorcinol composite, followed by formaldehyde cross-linking and carbonization. (b) Z-contrast image and electron diffraction of the highly ordered 2D hexagonal structure of the final films, 300 nm scale bar. The pore diameter is 33.7 ± 2.5 nm and the pore wall thickness is 9.0 ± 1.1 nm. (From Liang, C. D. et al., *Angewandte Chemie-International Edition*, 43, 5785, 2004. With permission.)

FIGURE 12.12 Photographic images of (a) flexible mesoporous carbon fibers (From Liang, C. D. et al., *Journal of the American Chemical Society*, 128, 5316, 2006. With permission.) and (b) free-standing mesoporous carbon membranes (From Wang, X. Q. et al., *Carbon,* 48, 557, 2010. With permission.).

during solvent annealing as in the other method. Because of the different chemical and thermal stability between the resins and triblock copolymers, the templates can be removed by thermal treatments at 350°C in an inert atmosphere, resulting in ordered mesoporous polymer frameworks. When heating at a high temperature above 600°C, polymers are converted into homologous carbon frameworks. By simply varying the surfactant-to-resol ratio or choosing different block copolymers synthesis, porous carbons with various mesostructures have been obtained. Figure 12.13 illustrates this solvent evaporation-induced self-assembly (EISA) synthesis procedure.

Following this concept, other PEO-containing block copolymers can also be used as templates to propagate the pore structures, such as custom-made diblock copolymer poly(ethylene oxide)-*b*-polystyrene (PEO-*b*-PS),[104] poly(ethylene oxide)-*b*-poly(methyl methacrylate) (PEO-*b*-PMMA),[110] ABC triblock copolymer poly(ethylene oxide)-*b*-poly(methylmethacrylate)-*b*-polystyrene (PEO-*b*-PMMA-*b*-PS),[111] and reverse triblock copolymer $PO_{53}EO_{136}PO_{53}$.[105,112] The mesostructures can be extended to 3D face-centered cubic *Fm3m* and *Fd3m* symmetries.

As previously demonstrated for hard-templated OMCs, the soft-templated synthesis method also allows for the preparation of materials with highly disperse catalytically active ceramic,[113] oxide,[114–116] and metallic nanoparticles[117–120] present in the frameworks. Heteroatom-doped carbons have been prepared by the self-assembly of organosilanes and phenolic resins[121] or by replacing the latter with melamine-based monomers.[122] Surfaces can be further modified by using phosphoric,[123–125] boric,[125] and sulfuric acid[126] by postsynthesis grafting or "one-pot" synthesis methods. The functionalized carbon surfaces exhibited high surface acidities for dehydration reactions.[123,124] Because of the thick carbon walls of the soft-templated OMCs, these have also been found to withstand high-temperature graphitization,[119,127,128] fluorination,[129] chemical activation, and surface modifications including diazonium and dipolar cycloaddition reactions.[109,130–135]

Furthermore, although carbons prepared by the EISA method exhibit pore sizes in the range of 2–5 nm, those from microphase separation are typically larger, above 6 nm.[99,100,108,109] The ability to prepare mesoporous carbons with different pore sizes and with narrow PSDs makes both synthetic pathways complementary, thus offering a broader range of materials to be chosen based on the desired applications.

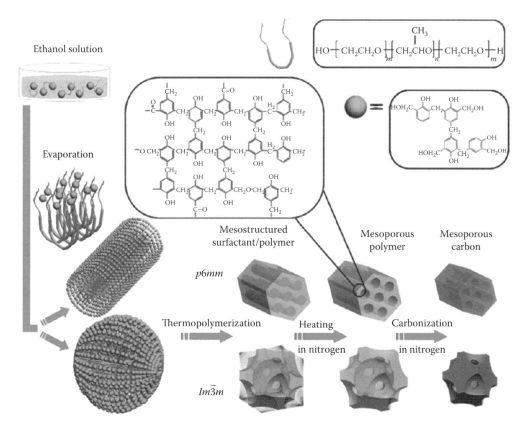

FIGURE 12.13 Proposed mechanism for the induced self-assembly synthesis of mesoporous polymers and OMCs from a phenolic resin and triblock copolymers. (From Meng, Y. et al., *Angewandte Chemie-International Edition*, 44, 7053, 2005. With permission.)

12.3.1 PORE SIZE CONTROL

Inspired from the pore size control of silica, initial work on pore size regulation of the mesoporous carbons synthesized from a soft-template approach mainly focuses on selecting surfactants with varying lengths of the hydrophobic block segments. Generally, larger hydrophobic segments occupy more space, translating to larger mesopores. For example, the pore size of mesoporous carbon FDU-16 increases from 2.8 to 4.3 nm when replacing the triblock copolymer F108 template $(EO_{132}PO_{50}EO_{132})$ with F127 $(EO_{106}PO_{70}EO_{106})$, which contains a longer hydrophobic PPO block.[101] Diblock copolymers, such as PEO-*b*-PS and PEO-*b*-PMMA, are inclined to form a compact packing of globular micellar aggregation and consequently favor large pore sizes in mesoporous carbons. By using the custom-made diblock copolymers PEO_{125}-*b*-$PMMA_{144}$ and PEO_{125}-*b*-PS_{230}, mesoporous carbons with pore sizes of 10.5 and 23.0 nm, respectively, are obtained.[104,136] Large pore mesoporous carbons can also be prepared under acidic conditions by using resorcinol–formaldehyde and template blends of F127 and polyoxyethylene (20) cetyl ether $(C_{156}H_{453}EO_{20}$, Brij 58), or polyoxyethylene (20) stearyl ether $(C_{18}H_{37}EO_{20}$-OH, Brij 78).[137] By changing the ratios of Brij 58 and Brij 78 to F127, the mesopores double in diameter, increasing from ≈4 to 8 nm. Since Brij polymers interact with the hydrophilic segments of F127, the average size and micropore volume also increase at higher Brij to F127 ratios.[137]

The temperature and pH for the *in situ* polymerization and self-assembly of phloroglucinol–formaldehyde with Pluronic F127, in combination with the mass ratios used largely change the mesopore size and other adsorption parameters in mesoporous carbons.[138] For instance, mesopores are nearly twice as large for the same mass ratios of carbon precursors to F127 when lowering the temperature from 55°C or 40°C to 15°C. At the lowest temperature investigated, the mesopores of final carbons were as large as 16 nm.[138]

The addition of organic swelling agents also plays a significant role in expanding the pore sizes of mesoporous silica. Only a disordered mesostructure is observed when hydrocarbon molecules, for example, 1,3,5-trimethylbenzene (TMB), heptane, and hexane, are added in the phenol–formaldehyde/copolymer P123[98] or phloroglucinol–formaldehyde/copolymer P123 systems.[139] Ordered mesostructures can be obtained in the former system by employing decane and hexadecane as swelling agents; however, the expansion effects are quite limited.[98] The only successful preparation of ordered ultra-large-mesopore carbons via a pore-swelling approach was reported for the homopolymer h-PS$_{49}$–diblock copolymer PEO$_{125}$-b-PS$_{230}$ templated system.[136] Since the chain length of homopolymer h-PS$_{49}$ ($\approx$5100 g/mol) is much shorter than that of a PS segment ($\approx$15,000 g/mol) of the PEO$_{125}$-b-PS$_{230}$, the affinity between the hydrophobic PS segments with the same chain composition is much stronger. Therefore, the homopolymer can solubilize at the core microdomains of the diblock copolymer, and, thus, swelling of the hydrophobic core occurs. The pores of the resulting mesoporous carbon can continuously increase from 22.9 to 37.4 nm, while preserving the mesostructure. In a large excess of h-PS$_{49}$ (h-PS$_{49}$/PEO$_{125}$-b-PS$_{230}$ > 20 wt.%), even larger pore diameters are obtained (40–90 nm), but at the expense of losing the uniformity of the mesopores and forming disordered foam-type pores.

Until now, pore size control of OMCs based on templates and organic additives remains limited, mostly because of the shrinkage of the phenolic resin frameworks during pyrolysis. The latter largely decreases the final pore size in carbons with increasing pyrolysis temperatures. For instance, the pore width of FDU-15 reduces from 7.1 nm in the carbonaceous samples prepared at 350°C, to 4.3 nm after heating at 800°C.[101] Consequently, attempts to lessen the framework shrinkage have been made. Liu et al. demonstrated a triconstituent coassembly route to prepare well-ordered mesoporous carbon–silica nanocomposites. In their system, the phenol–formaldehyde resol along with tetraethyl orthosilicate (TEOS) was used to coassemble with triblock copolymer F127 via an EISA strategy.[139] After removal and carbonization of triblock copolymer, ordered mesoporous carbon–silica composites were obtained. The presence of rigid silicates that act as a "reinforcing-steel-bar" dramatically inhibits framework shrinkage during pyrolysis. The pore size of carbon–silica composites is $\approx$6.7 nm, which is much larger than that ($\approx$2.9 nm) of ordered mesoporous carbon from similar synthetic process without silica addition. Mesoporous carbon was obtained by hydrofluoric (HF) etching of silica in the carbon–silica composites. This process builds pores in the carbon pore walls, and the dimension can be simply adjusted by tuning hydrolysis and condensation degrees of silicates before coassembly. The highly ordered mesoporous carbon shows a very large specific surface area ($\approx$2400 m^2/g^{-1}) and relatively large mesopores ($\approx$6.7 nm). Another recent example of reinforced nanocomposite materials was that for mesoporous carbon–silica with silica from two different sources.[116] By mixing colloidal silica particles and concomitantly adding TEOS to the carbon synthesis gels, composites having mesopore diameters ranging from 7.1 to 15.4 nm were prepared. After silica dissolution, the large mesopores of the F127 template were preserved, with additional mesopores left by the silica colloidal particles and micropores left from the small particles from TEOS hydrolysis. Also, soft-templated mesoporous carbons prepared in the presence of alpha-alumina (α-Al$_2$O$_3$) nanosheets (ANS)[140] were found to exhibit ordered cylindrical mesopores perpendicular to the alumina (001) facets (Figure 12.14) with pores as large as 14 nm.[141] The thermally stable ANS substrate may have prevented shrinkage of the coating carbon film during pyrolysis.

Carbon materials with large mesopore widths were also reported for carbon-graphitic nanoparticle composites made by a novel "brick and mortar"[142] soft-templating method as shown in Figure 12.15.[143] These nanocomposites were prepared by systematically replacing the resorcinol–formaldehyde "mortar" in the synthesis gels of soft-templated carbons with graphitic nanostructures "bricks," namely onion-like carbons (OLCs) and carbon black (CB). After carbonizations at 850°C, carbons with larger mesopore volumes and widths were obtained with increasing brick/mortar

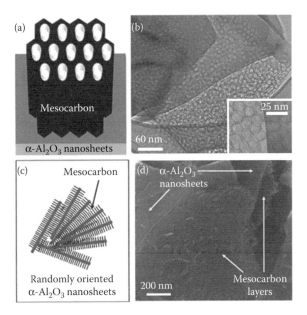

FIGURE 12.14 Simple model of ordered mesoporous carbon–α-alumina nanosheet composites supported by HRTEM and SEM images. Carbon layers with mesopores oriented perpendicularly to the (001) facets of α-alumina nanosheets (panels a and b). Higher magnification of the mesopores is shown in the lower inset of (b). Several nanosheets with layers of mesoporous carbon between them (panels c and d). The exposed layers of carbon in (d) reveal the presence of mesopores oriented perpendicularly to the underlying nanosheets. (Reproduced with permission from Gorka, J. et al., *Nanoscale*, 2, 2868, 2010.)

ratios as shown by the calculated PSDs. Besides improved thermal stability, the resorcinol-based mesoporous carbon provided sufficient electronic conductivity for electronic percolation between the graphitic OLC and CB particles, resulting in highly improved electronic conductivity of the mesoporous carbons without high graphitization temperatures, making these very attractive for future electrochemical applications, such as supercapacitors or electrocatalyst supports.

12.3.2 HIERARCHICAL PORE STRUCTURE

Preparing carbons with hierarchical pore structure is done by impregnation of preformed macroporous structures, such as silica colloids, with the carbon precursor gels, followed by precursor carbonization and macropore template dissolution. Although in hard-templated carbons, two particle sizes of the hard template are simultaneously required for the colloidal imprinting method;[33] the soft-templating method makes this procedure much simpler and broadens the selection of templates for the larger mesopores and macropores.

For example, hierarchically macro-/mesoporous carbons were fabricated by combining a surfactant-templating organic resol self-assembly with a colloidal-crystal templating approach (Figure 12.16).[144] Monodispersed silica colloids with particle sizes of 240, 320, and 450 nm are first used for the generation of ordered *fcc* colloidal crystal arrays. The organic–organic self-assembly of the resins and triblock copolymer F127 occurs in the interstices between silica particles. The rigid silica spheres prevent the shrinkage of the mesostructure during thermosetting and carbonization procedure, resulting in large mesopore sizes (≈11 nm). Finally, silica spheres were etched by HF solution to yield hierarchically ordered macro-/mesoporous carbons with tunable macropore sizes of 230–430 nm and 3D interconnecting mesopores of 30–65 nm. These interconnecting pores between macropores result from the inability of the carbon precursor to penetrate the areas where the silica spheres touch in the colloidal crystal lattice.

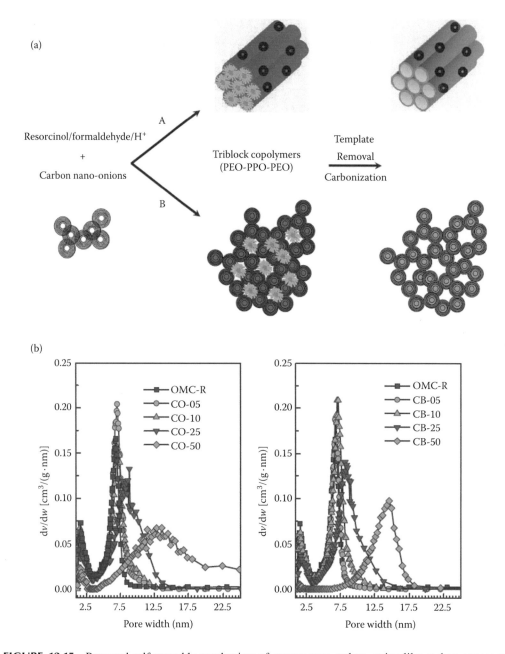

FIGURE 12.15 Proposed self-assembly mechanism of mesoporous carbon–onion-like carbon nanocomposites ("brick and mortar"). The onion-like carbon as well as other carbon nanostructures such as CB is dispersed in the final OMC framework, whereas a mesoporous OLC framework bound by a thin carbon film from the resorcinol–formaldehyde carbon is obtained in excess of the nanoparticles (a). Calculated PSD curves (b) for OLC and CB/OMC nanocomposites. The mesopore diameters increased with increasing OLC and CB contents. (From Fulvio, P. F. et al., *Advanced Functional Materials,* 21, 2208, 2011. With permission.)

This dual hard-silica colloid/soft-block copolymer-templating approach was also adopted to prepare OMCs with large interconnected channels.[145] However, the SiO$_2$ colloids used were small, ≈8 nm in diameter, and the mass ratio of SiO$_2$/phenol was low, >7 wt.%. The hydrophilic SiO$_2$ nanoparticles and phenolic resin coassemble with Pluronic F127 to form the organic–inorganic composite walls. Because the diameters of the SiO$_2$ nanoparticles are larger than the thickness

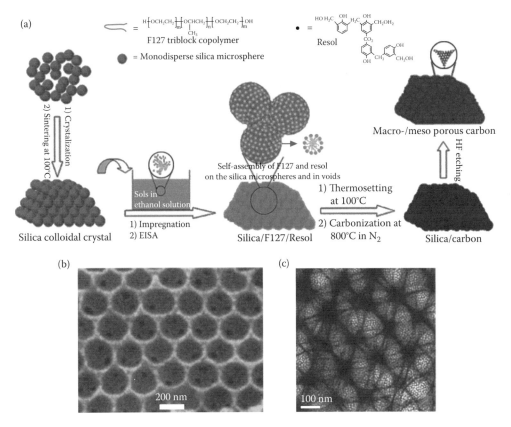

FIGURE 12.16 Colloidal crystal templating of soft-templated mesoporous carbon. (a) The pores between silica colloidal particles are infiltrated with the phenolic resin–triblock copolymer gel. After carbonization and silica dissolution, a mesoporous–macroporous hierarchical carbon is obtained. SEM image of the final carbon, showing the three-dimensional ordered macroporous structure, 3DOM (inverse replication of the large-diameter colloidal silica template) (b) and TEM image (c) showing the carbon mesopores. (From Deng, Y. H. et al., *Chemistry of Materials,* 19, 3271, 2007. With permission.)

of the pore walls, neighboring channels of the 2D hexagonal mesostructure are interconnected after removal of the SiO_2. The reported mesoporous carbons exhibited surface areas of $\approx$750 m^2. g^{-1} and bimodal pores of $\approx$3.4 and 7.6 nm. When larger SiO_2 colloids, 20 or 50 nm diameters, were used,[115,116,146] bimodal distributions of mesopores were obtained after SiO_2 etching.[115,116,146] The addition of TEOS to the previous synthesis gels afforded micropores after treatment with HF.[116,146] The final carbons could be further activated with steam and CO_2 resulting in carbons with pore volumes of $\approx$6 cm^3/g and specific surface areas exceeding 2000 m^2/g.[146]

However, the use of hard templates adds an extra step to eliminate the templates, complicating the preparation procedures. To facilitate the preparation of hierarchical soft-templated carbons, Dai et al. reported an all-organic polymerization-induced spinodal decomposition route to obtain mesoporous–macroporous carbon monoliths.[147] According to this route, prepolymerized mixtures of phloroglucinol–formaldehyde resin and copolymer F127 were dissolved in glycolic solvents and immediately transferred into a glass tube. Mesopores were formed by the self-assembly of the resin with the triblock copolymer (Figure 12.17). As the polymerization of the resin progressed by heating in air bath, it became less miscible with glycolic solvent, resulting in a polymerization-induced phase separation, with the glycolic solvent trapped in the segregated gel phase. As a result, macropores formed from solvent evaporation in the solvent-rich phase during thermal treatments. The sizes of the resulting macropores were in the range of 1–3 μm and varied with reaction temperatures,

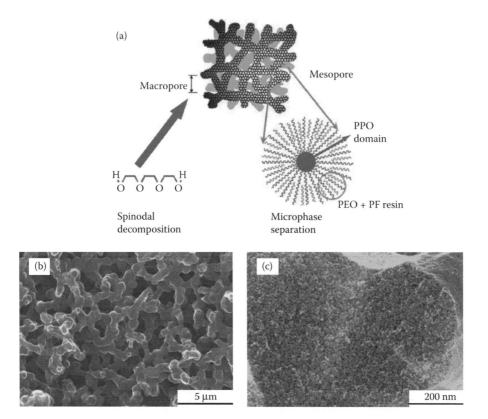

FIGURE 12.17 Illustration of the preparation of a bimodal mesoporous–macroporous carbon by dual-phase separation. The macropores are formed from the spinodal decomposition of glycolic solvents (a). Bicontinuous structure, framework structure, and the large macropores left by the solvent after annealing and carbonization (b). The carbon walls display large amounts of mesopores (c) templated by the triblock copolymer. (From Liang, C. D. et al., *Chemistry of Materials,* 21, 2115, 2009. With permission.)

prepolymerization times, and solvent compositions. The diameter of the mesopores ($\approx$8.0 nm) was independent of the spinodal decomposition process.

A similar polymerization-induced phase separation was reported for blends of Pluronic F127 and P123[148] triblock copolymer templates. By carrying out the synthesis under hydrothermal conditions at 100°C, a fast polymerization of the phenolic resin precursor induced a macrodomain separation of continuous and interconnected phenolic resin/Pluronic-rich and water-rich phases. After drying in air, the aqueous-phase domains left interconnected macropores. The resulting carbons had an ordered 2D hexagonal mesostructure of the mesopore (width $\approx$3 nm) and a 3D system of $\approx$3 μm irregular macropores.

Despite the benefits introduced by the aforementioned methods, these still require the use of additional solvents, polymer blends, or hydrothermal conditions. All these factors increase costs and the energy input significantly. Hence, another major advancement to the synthesis of hierarchical carbons with meso/macroporous structures was reported by Mayes et al.[108] in which formaldehyde was substituted by glyoxal in the self-assembly synthesis with phloroglucinol and F127. OMCs with additional macropores with average sizes of $\approx$200 nm were obtained without further steps added to the original OMC synthesis with formaldehyde.[108] This synthesis did not require hydrothermal synthesis, the need for a secondary volatile solvent, or additional triblock copolymers, making it more attractive from both environmental and economic perspectives.

Recently, Xue et al.[149,150] improved this EISA method by replacing the planar substrate with bulk polyurethane (PU) foams. The PU foam scaffolds have a hydrophilic surface. Most importantly, its unique 3D interconnecting macroporous architecture provides large voids and interfaces for self-assembly of the ordered mesostructures. As presented in Figure 12.18,[149] when the PU foam is impregnated with the ethanolic solution, the precursors, for example, resols with/without silicate oligomers and triblock copolymer F127, together with the ethanol solvent, are infused into the inter-connecting 3D networks and large macropore voids by capillary and wetting driving forces. During the solvent evaporation, the precursors can coat the struts of the PU foams and form a uniform layer of ordered mesostructured composites on the PU foam skeletons. It is noted that triblock copolymer templates and more than 97% of PU foam can be removed during carbonization and the primitive macroporous architecture with the 3D interconnecting struts is retained to form hierarchical macro/mesoporous carbon–silica or carbon materials.

PPO–PEO–PPO reverse triblock copolymers are an important family of commercial surfactants that have the hydrophilic PEO block located between the hydrophobic PPO blocks. It has been suggested that the reverse copolymers facilitate the formation of interconnected micelles[151,152] in which the two outer PPO blocks of a chain participate in two different micelles or aggregation.[153,154] Their unique micelle structures make them promising candidates to prepare porous carbons with hierarchical structures. As previously mentioned, using a larger-chain reverse triblock copolymer $PO_{53}EO_{136}PO_{53}$ template resulted in an OMC phase with cage-like pores (cubic $Fd3m$ symmetry), thus having a bimodal PSD of small interconnecting pores (3.2–4.0 nm) and larger mesopores (5.4–6.9 nm).[112] However, for the short-chain reverse block copolymer PO_{15}–EO_{22}–PO_{15}, monolithic

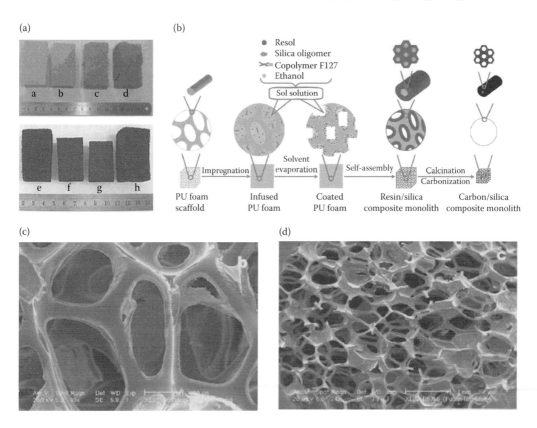

FIGURE 12.18 Photographic images of PU foam before (top) and carbon–silica monoliths (bottom) (a). Illustration of the procedure used for the impregnation of PU foam with the resorcinol–silica–F127 gel to prepare ordered mesoporous carbon–silica composites (b). The silica added to the polymer frameworks prevents the shrinkage of the PU–nanocomposites during carbonizations. SEM images of the resin–PU foam nanocomposite (c) and the carbon–silica material after carbonization. (d) The large macropores from the PU foam are present in the final mesoporous carbon–silica monolith. (From Xue, C. F. et al., *Advanced Functional Materials,* 18, 3914, 2008. With permission.)

carbon aerogels with macro- and micropores were prepared.[155] Hydrolysis and condensation of the resorcinol and formaldehyde resol were conducted in an acid water/ethanol solution of PO_{15}–EO_{22}–PO_{15} block copolymer. Low water concentrations induced the formation of a macroscopic polymer-rich phase containing water nanodroplets. Resin colloids formed at those hydrophilic nanodomains. As polymerization of the resin continued, some PO_{15}–EO_{22}–PO_{15} was entrapped within the grown resin colloids. Further resin growth provided the contact between colloidal particles, resulting in an interconnected carbon framework after pyrolysis at 800°C. Although the triblock copolymer-rich phase left open macropores, larger macropores were formed between the aggregated particles, whereas decomposition of the copolymer entrapped within the resin colloids left micropores in the final carbon walls. The carbon monoliths had specific surface and micropore areas of 725 and 470 m^2/g, respectively, with a trimodal PSD: two types of macropores having 100 nm, 2.4 μm, and micropores.

12.4 CONCLUSIONS

Carbon materials science has progressed tremendously from simply carbonizing cellulosic biomass or transforming coal precursors into porous materials. The ability to generate pore sizes that span the micropore to macropore range, while maintaining well-defined mesopores continuously motivates research on these materials. Recent developments in nanoporous carbon syntheses further allowed for the preparation of pure carbons and carbon-based nanocomposites having the most varied morphologies, such as thin films, free-standing membranes, monoliths, and powders. The easiness of preparation of carbons with tailored surface groups further broadens the fields of application for these novel materials. Emerging fields, such as catalysis, for the production of fuels from renewable biomass sources, gas- or liquid-phase separations, and energy storage and conversion (e.g., supercapacitors, Li-ion batteries) will greatly benefit from the consistency of properties found in such carbon materials. The latter properties result from their tailorable and uniform PSDs and pore wall structure, as well as from the reproducibility of these parameters.

ACKNOWLEDGMENTS

P. F. F. and S. D. acknowledge support by the Fluid Interface Reactions, Structures and Transport (FIRST) Center, an Energy Frontier Research Center funded by the U.S. Department of Energy, Office of Science, and Office of Basic Energy Sciences.

REFERENCES

1. Zhang, L. L. et al., *Chemical Society Review,* 38, 2520, 2009.
2. Wang, D. W. et al., *Angewandte Chemie-International Edition,* 47, 373, 2008.
3. Upare, D. P. et al., *Korean Journal of Chemical Engineering,* 28, 731, 2011.
4. Saufi, S. M. et al., *Carbon,* 42, 241, 2004.
5. Ismail, A. F. et al., *Journal of Membrane Science,* 193, 1, 2001.
6. Sing, K. S. W. et al., *Pure Applied Chemistry,* 57, 603, 1985.
7. Kruk, M. et al., *Chemistry of Materials,* 13, 3169, 2001.
8. Ryoo, R. et al., *Advanced Materials,* 13, 677, 2001.
9. Lee, J. et al., *Advanced Materials,* 18, 2073, 2006.
10. Lu, A. H. et al., *Advanced Materials,* 18, 1793, 2006.
11. Wan, Y. et al., *Accounts of Chemical Research,* 39, 423, 2006.
12. Liang, C. D. et al., *Angewandte Chemie-International Edition,* 47, 3696, 2008.
13. Wan, Y. et al., *Chemistry of Materials,* 20, 932, 2008.
14. Xia, Y. D. et al., *Nanoscale,* 2, 639, 2010.
15. Mohamed, A. R. et al., *Renewable and Sustainable Energy Reviews,* 14, 1591, 2010.
16. Titirici, M. M. et al., *Chemical Society Review,* 39, 103, 2010.
17. White, R. J. et al., *Chemical Society Review,* 38, 3401, 2009.

18. Davis, M. E. et al., *Chemistry of Materials,* 4, 756, 1992.
19. Cundy, C. S. et al., *Chemical Reviews,* 103, 663, 2003.
20. Beck, J. S. et al., *Journal of the American Chemical Society,* 114, 10834, 1992.
21. Zhao, D. Y. et al., *Science,* 279, 548, 1998.
22. Zhao, D. Y. et al., *Journal of the American Chemical Society,* 120, 6024, 1998.
23. Matos, J. R. et al., *Journal of the American Chemical Society,* 125, 821, 2003.
24. Yu, C. Z. et al., *Chemical Communications,* 575, 2000.
25. Kleitz, F. et al., *Chemical Communications,* 2136, 2003.
26. Schüth, F., *Angewandte Chemie International Edition,* 42, 3604, 2003.
27. Ma, Z. X. et al., *Carbon,* 40, 2367, 2002.
28. Ma, Z. X. et al., *Chemistry of Materials,* 13, 4413, 2001.
29. Ma, Z. X. et al., *Chemical Communications,* 2365, 2000.
30. Rodriguez-Mirasol, J. et al., *Chemistry of Materials,* 10, 550, 1998.
31. Kyotani, T. et al., *Chemistry of Materials,* 9, 609, 1997.
32. Johnson, S. A. et al., *Chemistry of Materials,* 9, 2448, 1997.
33. Li, Z. J. et al., *Journal of the American Chemical Society,* 123, 9208, 2001.
34. Li, Z. J. et al., *Chemistry of Materials,* 15, 1327, 2003.
35. Li, Z. J. et al., *Chemical Communications,* 1346, 2002.
36. Gierszal, K. P. et al., *Journal of the American Chemical Society,* 128, 10026, 2006.
37. Zakhidov, A. A. et al., *Science,* 282, 897, 1998.
38. Stein, A. et al., *Chemistry of Materials,* 20, 649, 2008.
39. Yoon, S. B. et al., *Journal of the American Chemical Society,* 127, 4188, 2005.
40. Johnson, S. A. et al., *Science,* 283, 963, 1999.
41. Han, S. J. et al., *Carbon,* 37, 1645, 1999.
42. Han, S. J. et al., *Chemical Communications,* 1955, 1999.
43. Gierszal, K. P. et al., *Journal of Materials Chemistry,* 16, 2819, 2006.
44. Gierszal, K. P. et al., *Journal of Physical Chemistry C,* 111, 9742, 2007.
45. Jaroniec, M. et al., *Chemistry of Materials,* 20, 1069, 2008.
46. Kim, M. et al., *Nano Letters,* 2, 1383, 2002.
47. Kim, J. Y. et al., *Chemical Communications,* 790, 2003.
48. Zhang, Y. X. et al., *Journal of Materials Chemistry,* 21, 3664, 2011.
49. Liu, R. et al., *Angewandte Chemie-International Edition,* 50, 6799, 2011.
50. Ryoo, R. et al., *Journal of Physical Chemistry B,* 103, 7743, 1999
51. Jun, S. et al., *Journal of the American Chemical Society,* 122, 10712, 2000.
52. Kaneda, M. et al., *Journal of Physical Chemistry B,* 106, 1256, 2002.
53. Solovyov, L. A. et al., *Journal of Physical Chemistry B,* 106, 12198, 2002.
54. Joo, S. H. et al., *Nature,* 412, 169, 2001.
55. Kruk, M. et al., *Chemistry of Materials,* 15, 2815, 2003.
56. Lee, J. S. et al., *Journal of the American Chemical Society,* 124, 1156, 2002.
57. Sakamoto, Y. et al., *Nature,* 408, 449, 2000.
58. Guo, W. et al., *Carbon,* 43, 2423, 2005.
59. Kim, T.-W. et al., *Journal of Materials Chemistry,* 15, 1560, 2005.
60. Fuertes, A. B. et al., *Electrochimica Acta,* 50, 2799, 2005.
61. Fuertes, A. B. et al., *Journal of Power Sources,* 133, 329, 2004.
62. Fuertes, A. B., *Microporous and Mesoporous Materials,* 67, 273, 2004.
63. Fuertes, A. B., *Journal of Materials Chemistry,* 13, 3085, 2003.
64. Kim, T. W. et al., *Angewandte Chemie-International Edition,* 42, 4375, 2003.
65. Vix-Guterl, C. et al., *Journal of Materials Chemistry,* 13, 2535, 2003.
66. Gierszal, K. P. et al., *Journal of Physical Chemistry B,* 109, 23263, 2005.
67. Kruk, M. et al., *Microporous and Mesoporous Materials,* 102, 178, 2007.
68. Vinu, A. et al., *Microporous and Mesoporous Materials,* 109, 398, 2008.
69. Vinu, A. et al., *Chemical Letters,* 36, 770, 2007.
70. Xia, Y. et al., *Chemistry of Materials,* 17, 1553, 2005.
71. Xia, Y. et al., *Advanced Materials,* 16, 1553, 2004.
72. Shin, Y. et al., *Advanced Functional Materials,* 17, 2897, 2007.
73. Fuertes, A. B. et al., *Journal of Materials Chemistry,* 15, 1079, 2005.
74. Yang, C.-M. et al., *Chemistry of Materials,* 17, 355, 2004.
75. Lee, J. et al., *Carbon,* 43, 2536, 2005.

76. Xia, Y. et al., *Chemistry of Materials,* 18, 140, 2005.
77. Lin, M.-L. et al., *The Journal of Physical Chemistry C,* 112, 867, 2008.
78. Fan, J. et al., *Advanced Materials,* 16, 1432, 2004.
79. Zhu, S. et al., *Advanced Functional Materials,* 15, 381, 2005.
80. Minchev, C. et al., *Microporous and Mesoporous Materials,* 81, 333, 2005.
81. Huwe, H. et al., *Carbon,* 45, 304, 2007.
82. Cao, Y. et al., *Journal of Solid State Chemistry,* 180, 792, 2007.
83. Dong, X. P. et al., *Journal of Physical Chemistry B,* 110, 6015, 2006.
84. Li, H. et al., *Materials Letters,* 60, 943, 2006.
85. Grigoriants, I. et al., *Chemical Communications,* 921, 2005.
86. Ding, J. et al., *Electrochimica Acta,* 50, 3131, 2005.
87. Lu, A.-H. et al., *Carbon,* 42, 2939, 2004.
88. Park, I.-S. et al., *Journal of Materials Chemistry,* 16, 3409, 2006.
89. Lei, Z. et al., *The Journal of Physical Chemistry C,* 112, 722, 2008.
90. Lo, A. Y. et al., *Thin Solid Films,* 498, 193, 2006.
91. Nishihara, H. et al., *Carbon,* 46, 48, 2008.
92. Fulvio, P. F. et al., *European Journal of Inorganic Chemistry,* 2009, 605, 2009.
93. Kamegawa, K. et al., *Carbon,* 35, 631, 1997.
94. Kamegawa, K. et al., *Journal of Colloid and Interface Science,* 159, 324, 1993.
95. Liang, C. D. et al., *Angewandte Chemie-International Edition,* 43, 5785, 2004.
96. Tanaka, S. et al., *Chemical Communications,* 2125, 2005.
97. Li, H. Q. et al., *Carbon,* 45, 2628, 2007.
98. Zhang, F. Q. et al., *Chemistry of Materials,* 18, 5279, 2006.
99. Liang, C. D. et al., *Journal of the American Chemical Society,* 128, 5316, 2006.
100. Wang, X. Q. et al., *Langmuir,* 24, 7500, 2008.
101. Meng, Y. et al., *Chemistry of Materials,* 18, 4447, 2006.
102. Meng, Y. et al., *Angewandte Chemie-International Edition,* 44, 7053, 2005.
103. Zhang, F. Q. et al., *Journal of the American Chemical Society,* 127, 13508, 2005.
104. Deng, Y. H. et al., *Journal of the American Chemical Society,* 129, 1690, 2007.
105. Huang, Y. et al., *Chemistry-an Asian Journal,* 2, 1282, 2007.
106. Yan, Y. et al., *Chemical Communications,* 2867, 2007.
107. Zhang, F. Q. et al., *Journal of the American Chemical Society,* 129, 7746, 2007.
108. Mayes, R. T. et al., *Journal of Materials Chemistry,* 20, 8674, 2010.
109. Wang, X. Q. et al., *Carbon,* 48, 557, 2010.
110. Deng, Y. et al., *Journal of Materials Chemistry,* 18, 91, 2008.
111. Zhang, J. Y. et al., *Chemistry of Materials,* 21, 3996, 2009.
112. Huang, Y. et al., *Angewandte Chemie-International Edition,* 46, 1089, 2007.
113. Zhu, Q. et al., *Journal of Power Sources,* 193, 495, 2009.
114. Patel, M. N. et al., *Journal of Materials Chemistry,* 20, 390, 2010.
115. Gorka, J. et al., *Journal of Physical Chemistry C,* 112, 11657, 2008.
116. Jaroniec, M. et al., *Carbon,* 47, 3034, 2009.
117. Li, J. S. et al., *Microporous and Mesoporous Materials,* 128, 144, 2010.
118. Gupta, G. et al., *Chemical Materials,* 21, 4515, 2009.
119. Shao, Y. Y. et al., *Journal of Power Sources,* 195, 1805, 2010.
120. Wang, X. Q. et al., *Adsorption-Journal of the International Adsorption Society,* 15, 138, 2009.
121. Gorka, J. et al., *Journal of Physics Chemistry C,* 114, 6298, 2010.
122. Kailasam, K. et al., *Chemistry of Materials,* 22, 428, 2010.
123. Fulvio, P. F. et al., *Catalysis Today,* 186, 12, 2012.
124. Mayes, R. T. et al., *Physical Chemistry Chemical Physics,* 13, 2492, 2011.
125. Zhao, X. C. et al., *Chemistry of Materials,* 22, 5463, 2010.
126. Hou, K. K. et al., *Journal of Colloid and Interface Science,* 377, 18, 2012.
127. Liang, C. et al., *Journal of American Chemical Society,* 128, 5316, 2006.
128. Liang, C. D. et al., *Journal of American Chemical Society,* 131, 7735, 2009.
129. Fulvio, P. F. et al., *Chemistry of Materials,* 23, 4420, 2011.
130. Gorka, J. et al., *Carbon,* 46, 1159, 2008.
131. Wang, X. Q. et al., *Chemistry of Materials,* 22, 2178, 2010.
132. Jin, J. et al., *Carbon,* 48, 1985, 2010.
133. Liang, C. D. et al., *Chemical Materials,* 21, 4724, 2009.

134. Wang, X. Q. et al., *Chemical Materials,* 20, 4800, 2008.
135. Liang, C. D. et al., *European Journal of Organic Chemistry,* 586, 2006.
136. Deng, Y. H. et al., *Chemistry of Materials,* 20, 7281, 2008.
137. Choma, J. et al., *Adsorption-Journal of the International Adsorption Society,* 16, 377, 2010.
138. Gorka, J. et al., *Colloids and Surfaces A-Physicochemical and Engineering Aspects,* 352, 113, 2009.
139. Liu, R. L. et al., *Journal of the American Chemical Society,* 128, 11652, 2006.
140. Suchanek, W. L. et al., *Crystengcomm,* 12, 2996, 2010.
141. Gorka, J. et al., *Nanoscale,* 2, 2868, 2010.
142. Szeifert, J. M. et al., *Chemistry of Materials,* 21, 1260, 2009.
143. Fulvio, P. F. et al., *Advanced Functional Materials,* 21, 2208, 2011.
144. Deng, Y. H. et al., *Chemistry of Materials,* 19, 3271, 2007.
145. Liang, Y. R. et al., *Langmuir,* 25, 7783, 2009.
146. Gorka, J. et al., *Carbon,* 49, 154, 2011.
147. Liang, C. D. et al., *Chemistry of Materials,* 21, 2115, 2009.
148. Huang, Y. et al., *Chemical Communications,* 2641, 2008.
149. Xue, C. F. et al., *Advanced Functional Materials,* 18, 3914, 2008.
150. Xue, C. F. et al., *Nano Research,* 2, 242, 2009.
151. Mortensen, K. et al., *Macromolecules,* 27, 5654, 1994.
152. Wang, Q. Q. et al., *Langmuir,* 21, 9068, 2005.
153. Alexandridis, P. et al., *Journal of Physical Chemistry,* 100, 280, 1996.
154. Alexandridis, P. et al., *Langmuir,* 14, 2627, 1998.
155. Gutierrez, M. C. et al., *Journal of Materials Chemistry,* 19, 1236, 2009.

13 Oxidation and Purification of Carbon Nanostructures

Sebastian Osswald and Bastian J. M. Etzold

CONTENTS

13.1 INTRODUCTION

The oxidation of carbon materials has been explored for decades and has become a powerful route to alter the properties of carbonaceous materials, particularly coal and graphite. Until recently, the oxidation process had two main functions: (1) increasing surface area and porosity by removing carbon atoms from the surface (activation), and (2) formation of organic functional groups on the carbon surface (functionalization). With the rise of nanotechnology, however, researchers have recognized a third application of the oxidation process: (3) intelligent design and modification of carbon nanostructures, by utilizing the selectivity of oxidation toward specific structural elements at the nanoscale.

The large number of oxidation methods existing today is commonly divided into dry chemistry (gas phase; also used for "physical" activation) and wet chemistry (liquid phase) approaches (Table 13.1). Liquid-phase oxidation is based on treatments in concentrated or diluted acids and/or bases. One advantage of liquid-phase oxidation is the ability to control the purification process and reaction rates by adjusting the concentrations of reactants. Simultaneous dissolution of metal impurities parallel to oxidation of undesired carbon phases is another advantage of this approach. Unfortunately, liquid-phase oxidation techniques suffer from several disadvantages. Although being effective oxidation reactants, most acids and bases are environmentally harmful and require corrosion-resistant equipment for processing, transport, and storage. In other cases, oxidation with acids, salts, or oxides may introduce foreign elements and may lead to additional impurities, such as nitrogen- and sulfur-containing compounds, chlorine, or chromium. These impurities alter the

TABLE 13.1

Comparison of Dry Chemistry- and Wet Chemistry-Based Oxidation Methods

	Wet Chemistry Methods	**Dry Chemistry Methods**
Reactant	Acid: HNO_3, H_2SO_4, H_3PO_4, HCl, H_2O_2 Basic: KOH, NaOH, KMnO, NH_4OH	Gas: air, O_2, O_3, CO_2, H_2O vapor
Benefits	• Very homogeneous • High selectivity with respect to sp^2 and sp^3 carbon • Lower temperature	• High process control using *in situ* characterization • Fast and inexpensive • Easy to scale up • No filtration/separation required
Limitations	• Require extensive washing and corrosion-resistant equipment • Environmentally harmful and expensive waste disposal • Little process control during treatment: high sample loss/reduced selectivity	• Require high temperatures • Not able to remove metal impurities • Use of aggressive substances or supplementary catalysts • May require additional supplements (e.g., catalyst, oxidation inhibitors)

intrinsic characteristics of the carbon material and therefore need to be removed through additional purification steps. Especially, heavy metal compounds cannot be ignored and typically result in a costly waste disposal process.

To overcome these issues, a variety of alternative, gas-based purification techniques have been developed. These dry chemistry approaches use high-temperature reactions in gaseous oxidizers, such as air, oxygen (O_2), ozone (O_3), or carbon dioxide (CO_2) [1–7]. A comparison of the advantages and disadvantages of both methods is shown in Table 13.1. Although oxidation has been used extensively to modify and activate macroscopic forms of carbon, a comparison of experimental data is difficult as many of the studies reported in literature provide only a qualitative description of the oxidation kinetics and often lack a thorough investigation of the physical and chemical processes that take place at the nanoscale. For example, various types of carbon are often simply referred to as "coke" or "carbon soot," although these materials were synthesized under very different conditions and are characterized by distinct differences in crystal structure, purity, and surface terminations.

The structural diversity of carbon at the nanoscale exceeds that of all other materials. Some of these nanostructures, particularly carbon nanotubes (CNT), have been studied extensively for decades and are generally quite well understood; however, other carbon nanomaterials, including nanodiamond (ND) and onion-like carbon (OLC), have received less attention but offer similar potential for a variety of applications. To study their properties and open avenues for widespread utilization, one must provide materials of high purity and well-defined composition.

Although oxidation has been used widely to purify carbon materials, carbon–oxygen reactions have also been shown to drastically alter the physiochemical properties of nanostructures, particularly their wettability and adsorption/desorption characteristics. Moreover, oxidation potentially induces damage to carbon nanomaterials or even destroys the structures under improper conditions. To fully utilize the selectivity of the oxidation process at the nanoscale, a comprehensive understanding of the chemical and physical nature of a material and the structure dependence of the oxidation kinetics is required. For the latter, one must systematically study the interactions of the different carbon nanostructures with gaseous and liquid oxidizer, monitor changes in structure and composition, and analyze the reaction kinetics in greater detail.

13.2 MECHANISM AND REACTION KINETICS OF CARBON OXIDATION

The interaction between oxygen and carbon is of great technological importance and one of the most studied reactions [22,24]. It consists of two primary steps: (1) activation of molecular oxygen at the carbon surface (physisorption), and (2) stabilization of the activated oxygen by formation of covalent bonds with the carbon atoms (chemisorption).

With the emergence of nanotechnology, graphite has received renewed interest as it is considered an important "model material" for a variety of carbon nanomaterials. Two-dimensional (2D) graphite layers (graphene) are the basic building block for many carbon-based nanostructures and have been used to explain their exceptional physical properties. For example, the interactions between the graphite basal planes were used to explain bundling effects in CNTs and to analyze wall–wall interactions in multiwall nanostructures and carbon onions. Therefore, many of the properties of carbon nanomaterials, including thermal stability and chemical reactivity, are closely related to that of the graphitic crystal lattice.

13.2.1 Mechanism of Carbon Oxidation

The oxidation of carbon materials consists of several individual steps, of which each, if ranked the slowest, can potentially become the rate-controlling process of the reaction. A typical, noncatalyzed oxidation is subject to the following reaction steps:

1. Diffusion of oxygen to the outer surface of carbon particle (film diffusion).
2. Diffusion of oxygen through pores/cavities of carbon particle.
3. Adsorption of oxygen onto the carbon surface (physisorption).
4. Formation of oxygen–carbon bonds (chemisorption).
5. Breaking of carbon–carbon bonds.
6. Desorption of reaction products from the carbon surface.
7. Diffusion of reaction products through pores/cavities of carbon particle.
8. Diffusion of reaction products away from outer surface of carbon particle (film diffusion).

The contribution of each step to the overall reaction depends on the conditions, and can be controlled by temperature, partial pressure of the oxidant, active (exposed) surface area, amount of catalyst impurities and surface functionalities, diffusion constants, and, if applicable, the concentration of defects.

Depending on the oxidation temperature, one can distinguish between three distinct reaction regimes. At low temperatures (corresponding with low oxidation rates), the concentration of the oxidant is essentially the same everywhere in and on the sample. This is referred to as the chemical regime. The reaction rates are largely determined by the intrinsic reactivity of a carbon material, although different parts of the structure may react at different rates. For most carbon materials, the chemical regime ranges from approximately 300°C to 600°C, and rate-controlling processes are steps 3–6. The second regime is characterized by internal mass transport, which is the diffusion of reactant and reaction products through internal pores and cavities, where reaction rates are determined by steps 2 and/or 7. The diffusion of oxidants and reactants is strongly affected by the pore structure of a material, especially at the nanoscale. The third regime is dominated by the external mass transfer. The reaction rate is so high that nearly all of the oxidant is consumed on the outer surface of the material. This region is controlled by steps 1 and/or 8. In most cases, oxidation reactions occur mainly in the chemical regime and are, therefore, in a first approximation, independent from the diffusion of oxidants and reactants. The rate-controlling steps are physisorption (3) and chemisorption of oxygen (4), and the breaking of the carbon–carbon bonds (5). Using low-temperature thermal desorption spectroscopy, Ulbricht et al. [8] showed that the binding energy of oxygen on defect-free graphitic surfaces ranges between 0.12 and 0.19 eV. In their experiments, the reaction of molecular oxygen with graphitic surfaces did not produce any reactants like carbon monoxide or carbon dioxide, suggesting the dominant interaction to be of van der Waals type. This assumption was confirmed by density functional theory calculations which revealed the absence of any kind of charge transfer between the oxygen molecules and the carbon surface as a result of the energy mismatch between the unoccupied states of molecular oxygen and the valence band of graphite [9]. Therefore, any reaction between oxygen and graphitic surfaces requires a transformation of unoccupied states. Physisorption onto defective sites, such as edges or vacancies, lowers the energy of the unoccupied states in oxygen molecules, and allows oxidation (chemisorption) of graphitic materials at temperatures below 600°C (chemical regime) [8]. Only temperatures above 700°C facilitate the removal of carbon atoms from defect-free basal planes [10].

Different reaction mechanisms have been proposed for the oxidation of graphitic surfaces, the two most common being the Eley–Rideal mechanism (ERM) [11] and the Langmuir–Hinshelwood mechanism (LHM) [8]. The latter consists of two distinct steps: (1) physisorption of oxygen on the graphitic surface, and (2) diffusion of the physisorbed species to the defective sites, where carbon–oxygen bonds are formed. The ERM process is characterized by a direct collision of oxygen molecules with the surface [11]. Both the physics of reaction mechanism and the adsorption energies are dependent on the structure of the graphitic surface [11]. For example, initial studies suggested that the oxidation of graphite follows the LHM [11], whereas oxidation of CNTs occurs mainly because of the direct collision of oxygen molecules with defective sites and edge atoms [9]. More recent studies using scanning tunneling microscopy (STM) showed that reaction kinetics on the atomic level appear to be more complex because both size and shape of the defects strongly affect the oxidation mechanism, leading to energetically different adsorption sites [12]. Diffusion-controlled oxidation mechanisms have also been studied by several groups [13]. In the case of carbon nanomaterials, size and surface effects are dominant and, therefore, are of particular importance for diffusion processes.

13.2.2   KINETICS AND THERMODYNAMICS OF THE CARBON–OXYGEN REACTION

The reaction of molecular oxygen with carbon can be described by the three basic equations:

$$C(s) + O_2(g) = CO_2(g) \tag{13.1}$$

$$2C(s) + O_2(g) = 2CO(g) \tag{13.2}$$

$$2CO(g) + O_2(g) = 2CO_2(g) \tag{13.3}$$

One also needs to consider the Boudouard equation as an intermediate state:

$$C(s) + CO_2(g) = 2CO(g) \tag{13.4}$$

The temperature-dependent standard Gibbs energy change ($\Delta G°$) for each reaction is shown in Figure 13.1 and given by

$$\Delta G° = \Delta H - T\Delta S \tag{13.5}$$

where T is the temperature and ΔH and ΔS are the enthalpy and entropy change of the reaction, respectively [14]. The first three reactions, Equations 13.1 through 13.3, are exothermic, whereas the Boudouard reaction, Equation 13.4, is endothermic. Spontaneous reactions are characterized by a negative $\Delta G°$. The more negative the $\Delta G°$, the higher is the yield of reaction products. A positive $\Delta G°$ does not predicate that a reaction is improbable, but that the yield of reaction products is rather low. According to Figure 13.1, the Gibbs free energy changes in Equations 13.1 through 13.3 intersect at approximately 1000 K.

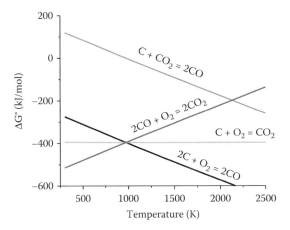

FIGURE 13.1 Standard Gibbs free energy change of Equations 13.1 through 13.4 as a function of the temperature. Below the interception point at ≈1000 K formation of CO_2 is thermodynamically favored.

To the right of the interception point (i.e., at higher temperatures), $\Delta G°$ of Equation 13.2 passes below the others, suggesting the formation of CO to be the dominant process. To the left of the interception point, at temperatures below 1000 K, formation of CO_2 is thermodynamically favored [14,15]. Because the Gibbs free energy change of Equation 13.1 has a large negative value, the reaction occurs with a large driving force even at low oxygen partial pressures, and is the predominant process in the oxidation regime of interest (chemical regime). Given the stoichiometry of reaction (1), the rate of depletion for carbon and molecular oxygen is equal to the rate of formation of CO_2. In many cases, the reaction rate approach can be described in the studied regime by a power law:

$$-\frac{d[C]}{dt} = -\frac{d[O_2]}{dt} = \frac{d[CO_2]}{dt} = k[O_2]^n \tag{13.6}$$

where [C] and $[O_2]$ are the concentration of the reactants carbon and oxygen, respectively; $[CO_2]$ is the concentration of carbon dioxide formed during the reaction, k is the reaction rate constant, and the index n characterizes the order of the reaction. The temperature dependence of the reaction rate constant k is given by the Arrhenius equation:

$$\ln k = \ln k_0 - \frac{E_a}{R} \cdot \frac{1}{T} \tag{13.7}$$

where E_a is the activation energy of the oxidation (in kJ/mol), T is the temperature (in K), R is the universal gas constant, and k_0 is a preexponential constant often referred to as frequency factor. The activation energy can be understood as a measure of the energy barrier to activate the chemical reaction between carbon and oxygen. Nevertheless, in a simplified reaction rate approach, the activation energy often summarizes all temperature dependencies, including that of diffusion coefficients or sorption equilibrium constants, if these are not accounted for separately. The frequency factor reflects the probability of reactants involved to effectively interact with each other with sufficiently high energy. Therefore, it is also a measure of the total number of molecules that possess the activation energy. The rate of reactant depletion and reaction product formation can, therefore, be determined by directly measuring the concentration of the reaction products or by determining the decrease in the concentration of the reactants [16]. The change in concentration of the reactant is typically measured by analyzing the weight change of the carbon samples during oxidation.

Following the simplified power law in Equation 13.6 and assuming a first-order reaction ($n = 1$), the weight loss during oxidation is given by

$$\frac{dm}{dt} = -k \cdot m^c \tag{13.8}$$

where m is the sample weight, t is the oxidation time, and k is the reaction rate constant for a given temperature (isothermal reactions) [17]. The index c is used to account for particle shape and/or surface area available for oxidation. For example, oxidation reactions that are characterized by a contracting volume or a contracting area apply $c = 2/3$ and $c = 1/2$, respectively. However, in the case of formal phase boundary reactions, we can assume $c = 1$ [18]. The weight change during oxidation is often expressed in terms of the weight fraction, α, which for a highly porous material ($c = 1$, no shape affects) is defined as

$$\alpha = -\frac{m_0 - m}{m_0} = k \cdot t \tag{13.9}$$

Using Equations 13.7 and 13.9, one can then estimate the activation energy for the oxidation reaction by determining the slope of the Arrhenius curve according to

$$\ln \alpha = \ln k_0 + \ln t - \frac{E_a}{R} \cdot \frac{1}{T} \tag{13.10}$$

Because $[\ln k_0 + \ln t]$ is constant for a fixed temperature, plotting $\ln \alpha$ versus $1/T$ leads to a linear relationship, known as Arrhenius plot. The activation energy E_A can be calculated from the slope of the curve, whereas the interception with the y-axis determines the frequency factor k_0. Under nonisothermal conditions, activation energy and frequency factor of the oxidation reactions can be determined, for example, by the Achar–Brindley–Sharp–Wendeworth (ABSW) equation:

$$\ln \left[\frac{1}{F(\alpha)} \cdot \frac{d\alpha}{dT} \right] = \ln \left(\frac{k_0}{\beta} \right) - \frac{E_a}{R} \cdot \frac{1}{T} \tag{13.11}$$

where β is the heating rate, $d\alpha/dT$ is the oxidation reaction rate, and $F(\alpha)$ is a differential function describing the reaction mechanism. At lower oxidation temperatures (<600°C), the chemical reaction is the rate-controlling step and the oxidation process is well described by a first-order reaction using the random nucleation and growth mechanism (Mampel unimolecular law) [19], where $F(\alpha) = 1 - \alpha$ [13,20]. The kinetic parameters are then determined by plotting $\ln [d\alpha/dT/(1 - \alpha)]$ versus $1/T$.

For both isothermal and nonisothermal oxidation, changes in the slope of the Arrhenius-type curves indicate a change in the reaction mechanism. At low temperatures, reactions are controlled by the chemical process itself, whereas at elevated temperatures (>600°C) oxidation reactions are increasingly dominated by the diffusion of the active species. The critical temperature that separates both steps depends on several factors, including surface area and pore structure of the material [21], and has been neither well defined nor exactly determined [15]. Other critical factors are the partial pressure of the oxidizing agents, the flow and/or agitation rates used during the experiments, and mechanical stress applied to the material [15].

The solid–gas reaction between molecular oxygen and graphite has been described using both ERM [11] and LHM [8]; however, in most studies one can assume the latter to be the dominant process. The oxidation of a graphite layer (graphene) proceeds therefore through two distinct steps:

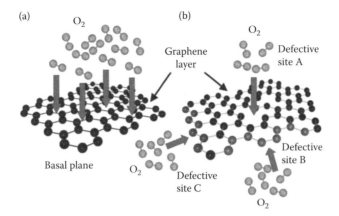

FIGURE 13.2 Reaction of molecular oxygen with graphite layer. (a) Physisorption of oxygen on the basal plane. (b) Chemisorption of oxygen at defective sites: vacancies (site A), armchair (site B), and zigzag edges (site C).

(1) physisorption of molecular oxygen onto the basal plane, and (2) diffusion of molecular oxygen to defective sites [25].

The physisorption of oxygen onto the basal plane is of van der Waals type and occurs without any charge transfer (Figure 13.2a) [8]. The diffusion of physisorbed oxygen to defective sites then lowers the kinetic barrier of the oxygen–graphite reaction, enabling a charge transfer (chemisorption). There exist various types of defects which are energetically different. In general, one distinguishes between vacancies and edge sites (armchair, zigzag) as illustrated in Figure 13.2b.

The presence of defects leads to a reduction of the adsorbed oxygen at the surface and formation of various activated oxygen species, as O^-, O^{2-}, O_2^{2-}, and O_2^-. The O^{2-} ion is the most stable surface species and preferentially formed according to

$$O_2(ads) + C_{defect} + 2e^- \rightarrow O_2^{2-}(ads) \tag{13.12}$$

$$O_2^{2-}(ads) + 2e^- \rightarrow 2O^{2-}(ads) \tag{13.13}$$

The oxide ions then lead to the formation of oxygen–carbon surface groups:

$$O^{2-}(ads) + C_{defect} \rightarrow CO(ads) + 2e^- \tag{13.14}$$

The most common surface functionalities formed during the oxidation of graphite are carboxyl and carbonyl groups. The exact nature of the functional groups depends on the temperature, the concentration of the reactants and reaction products, and the reactivity of the defective site which varies for different types of defects. The decomposition of carboxylic acids and lactones occurs at low temperatures (<300°C) and leads mainly to desorption of carbon dioxide, whereas phenols and quinones favor formation of carbon monoxide (CO), but decompose only at elevated temperatures (>600°C). The exact decomposition/desorption temperatures strongly depend on the carbon material and values reported in literature vary substantially, as discussed in greater detail in the following sections.

13.2.3 Oxidation of Carbon Nanostructures and Effect of Metal Impurities

Although substantial progress has been made in both synthesis and purification of carbon nanomaterials, bulk samples still contain large amounts of impurities, such as amorphous carbon and metal catalyst, alongside the desired nanostructures. Moreover, the as-produced carbon nanomaterials are typically composed of mixtures of different nanostructures all of which are structurally different and possess different reactivity. Even within an individual structure category (e.g., CNTs), significant variations in size, shape, and number of defects often exist. Because the oxidation kinetics are closely related to structural features, reaction rates and activation energies are expected to differ for these distinct forms of carbon, which is an important factor for oxidation-based purification or surface functionalization. Analogous to graphite or carbon soot, the oxidation of most carbon nanostructures is often described by a simple first-order reaction rate approach, with respect to the carbon component in Equation 13.6. Therefore, depending on the temperature, the oxidation of carbon nanomaterials can be understood as an overlap of several first-order oxidation reactions resulting from different nanostructures present in the sample.

CNT, ND, fullerenes, OLC (also called carbon onions or carbon nano-onions), nanocrystalline graphite, amorphous carbon, and carbide-derived carbon (CDC) are all considered carbon nanomaterials, but their shape, structure, and chemical stability vary largely. In addition, various types of lattice defects, surface functionalities, and impurities resulting from the synthesis process further differentiate carbon nanomaterials, even within the same structure category. Although small differences in size and shape may be neglected for bulk materials, they become significant at the nanoscale and drastically change the properties and chemical reactivity of a material. The structural and compositional complexity of carbon nanomaterials determines their oxidation behavior [26]. The resistance against oxidation generally increases with the level of graphitization and is lowest for highly disordered and amorphous structures. However, a proper analysis of the oxidation kinetics requires detailed information on structure, size, and surface chemistry, as well as type and number of defects. In general, there exists a distinct difference in oxidation onset temperature between amorphous carbon ($\approx 350°C$) and other graphitic or diamond-like carbon phases ($>400°C$), allowing for a selective oxidation (purification) of amorphous species from the samples.

Figure 13.3 shows the oxidation behavior of small- and large-diameter CNTs by monitoring the intensity ratio between the D and G band Raman features (I_D/I_G) over a wide temperature range. The intensity ratio between disorder-induced D band and the graphitic G band is commonly used to evaluate the purity and structural ordering in graphitic (sp^2) carbons. Small-diameter CNTs (single-walled CNTs: SWCNTs, double-walled CNTs: DWCNTs) typically exhibit a lower I_D/I_G ratio as compared to large-diameter CNTs (multiwalled CNTs: MWCNTs), suggesting a higher structural

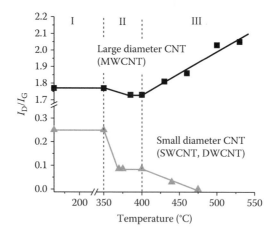

FIGURE 13.3 Changes in the I_D/I_G Raman intensity ratio of small- and large-diameter CNTs during nonisothermal oxidation in air. (From S. Osswald, E. Flahaut, Y. Gogotsi, *Chem. Mater.* 2006, *18*, 1525; S. Osswald, M. Havel, Y. Gogotsi, *Journal of Raman Spectroscopy* 2007, *38*, 728.)

perfection and/or a lower content of amorphous carbon (region I). The latter, amorphous carbon, is often present as a by-product of the synthesis of carbon nanomaterials and its content varies depending on the synthesis type and conditions. The oxidation process can be divided into three temperature regions. At temperatures below $\approx 350°C$ (region I), oxidation of carbon does not occur. Region II is characterized by the removal of amorphous and disordered carbon, but temperatures remain insufficient for the oxidation of the actual carbon nanomaterial (e.g., CNTs). Above $\approx 400°C$ (region III), both amorphous carbon and nanostructures are oxidized simultaneously. The transition between region I and II typically occurs around 350–370°C. The transition temperature between region II and III may vary from sample to sample, and depends on the relative amount of amorphous carbon and the structural properties of the nanomaterial.

The removal of amorphous carbon from the small-diameter CNT samples often results in a large decrease in I_D/I_G as compared to MWCNTs. The majority of the D band intensity in the Raman spectrum of small-diameter CNTs can be ascribed to amorphous carbon, whereas the D band of larger CNTs (e.g., MWCNTs) originates primarily from structural defects. At higher temperatures (region III), the I_D/I_G of large-diameter CNTs increases due to formation of defects in the wall structures of the tube. In contrast, the I_D/I_G ratio of small-diameter CNTs decreases which can be explained by the preferred or more rapid oxidation of defective tubes. The carbon–carbon bonds in small-diameter CNTs are subject to higher strain as compared to larger CNTs because of an increase in curvature perpendicular to the tube axis. However, large-diameter tubes exhibit a higher number of defects, which also reduce the oxidation resistance. The separation of graphitic carbon from the diamond phase is a more complex process, but can be achieved around 425–430°C without significant sample loss, mainly because of notable differences in the rate constants of graphitic and diamond-like carbon. Although oxidation in air was found suitable for selective removal of amorphous and/or graphitic carbon, the high reactivity of O_2 molecules may lead to enhanced sample loss and limited process control, particularly during oxidation-based activation of carbon nanomaterials. In these cases, milder oxidizers, such as CO_2, can be applied in a similar fashion using slightly higher oxidation temperatures.

Arguably, the most significant factor for the oxidation behavior of carbon nanostructures is the presence of metal particles. In many experiments, the energy required for the oxidation of carbon materials is significantly lower than the activation energy predicted by theoretical calculations [29]. Catalytic reactions are one possibility to lower the activation energy [30]. Catalysts enable alternative reaction pathways by interacting with the reactants (O_2) and forming an intermediate state (typically oxides), which subsequently yields the reaction product (CO_2). Most of the existing industrial production techniques require addition of metal catalysts to promote formation and growth of carbon nanomaterials. Unfortunately, current purification methods do not allow a complete removal of the metal impurities without modifying the nanostructures or changing the composition of the sample. These catalytic impurities enable reactions that would be thermodynamically impossible in their absence, or serve to accelerate reactions rates by several orders of magnitude. The catalyst itself is not consumed during the reaction and may participate in multiple catalytic cycles. Therefore, even small amounts of catalysts in the sample can significantly affect the oxidation reaction, and sample-to-sample variations in the amount and nature of these catalyst impurities often inhibit an exact determination of the reaction kinetics. Figure 13.4 illustrates the effect of Fe on the oxidation behavior of carbon nanomaterials. Thermogravimetric analysis (TGA) of metal-free carbon black (Pure Black®) with and without the 1:1 addition of UD50 (i.e., detonation soot with a 25% content of sp^3 NDs) shows that while Pure Black resists oxidation at temperatures below $\approx 600°C$, the mixed powder (Fe content ≈ 0.65 wt.%) is fully oxidized before reaching 600°C (Figure 13.4a).

To prove that the observed decrease in oxidation temperature results from a catalytic process rather than from internal heating effects related to the exothermic oxidation of amorphous carbon, we performed TGA under similar conditions using a mixture (1:1) of Pure Black and Fe-free TiC-CDC. Figure 13.4b reveals that the oxidation of the carbon mixture occurs in two distinct temperature ranges. Below 600°C only TiC-CDC (≈ 50 wt.%) is oxidized, suggesting that the reaction heat

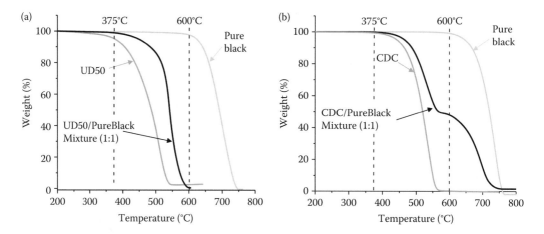

FIGURE 13.4 Thermogravimetric analysis of metal-free pure black with and without an addition (mixture 1:1) of UD50 (a) and 600°C-TiC-CDC (b). As-received UD50 and TiC-CDC are shown for comparison.

created by oxidation of amorphous carbon between 375°C and 600°C does not significantly affect the activation energy of the carbon black. This clearly demonstrates that, in the case of ND powders containing Fe impurities, catalytic processes dominate the oxidation reactions.

A large number of different inorganic materials, ranging from noble metals to salts and oxides of alkali and alkaline earth metals, were found to catalyze the reaction of carbon with gaseous reactants such as O_2 or CO_2 at elevated temperatures. The common property of these materials is that they are able to form several oxidation states in the temperature range of interest (300–800°C). Although catalytic reactions are not yet fully understood and are the subject of ongoing research, it is generally agreed that catalytic processes result from oxidation/reduction cycles on the carbon surface. The effects of catalysts on the oxidation temperature can be significant. For example, lead (Pb), copper (Cu), silver (Ag), iron (Fe), platinum (Pt), and nickel (Ni) were found to lower the ignition temperature of graphite powder from 740°C to 382°C, 570°C, 585°C, 593°C, 602°C, and 613°C, respectively. In all cases, the concentration of the metal in the sample was <0.2 wt.% [31]. Finally, it should be noted that nonmetals may also act as catalyst during nanomaterial oxidation [32].

Structural features, lattice defects, surface chemistry, and metal impurities simultaneously affect the oxidation behavior of carbon nanomaterials, making a differentiation between the individual contributions challenging, if not impossible. It is, therefore, important to note that the measured activation energies, reaction rate constants, and frequency factors of carbon nanomaterials only reflect the oxidation behavior of the sample as a whole, but not that of a particular nanostructure. A direct comparison between different carbon nanostructures, or similar nanostructures produced by different synthesis techniques, is thus extremely difficult.

Finally, one must consider the effects of the characterization tools and the measurement technique used to investigate the oxidation process. There exists a variety of thermal analysis and structural characterization tools that are routinely used to study physical and chemical changes in carbon nanomaterial during oxidation. Structural characterization techniques, such as Raman spectroscopy or electron microscopy, can be used to optimize the oxidation process. These methods provide detailed structural information about carbon samples before, during, or after the oxidation, and can, thus, help prevent inadvertent damage to the nanostructures. Although generally considered nondestructive, these characterization techniques irradiate the samples using electron beams, lasers, or other intense radiation sources, increasing the energy of the probed system. While the higher energy state may be insufficient to modify structure or composition of the sample under standard conditions, it may affect the reaction kinetics and the oxidation behavior of nanomaterials in the temperature range of interest.

Other common techniques used to study oxidation reactions are thermogravimetric methods, which probe the weight loss of the sample and investigate the reaction kinetics by monitoring changes in the concentration of reactants and reaction products. Such measurement techniques are very sensitive to heating rates and mass transport, and may be affected by experimental parameters such as sample size, size of the heating chamber, and gas flow used during analysis, to mention a few.

In summary, the oxidation behavior of carbon nanomaterials depends on a series of factors. In addition to the structural characteristics of the nanostructures themselves, one has to account for contributions from catalytic impurities, but must also consider potential effects resulting from the experimental setup and the measurement process itself. Therefore, literature data on oxidation behavior of carbon nanomaterials vary widely. Specific examples will be discussed in greater detail in the following sections.

13.3 APPLICATIONS OF CARBON NANOMATERIAL OXIDATION

13.3.1 PURIFICATION

The majority of the existing carbon nanomaterial synthesis methods yield bulk samples that contain significant amounts of amorphous carbon and comprise a variety of carbon nanostructures co-produced under selected growth conditions. Furthermore, the majority of these techniques require metal catalysts (e.g., Fe, Co, Ni, Mg) to promote nanostructure growth and, as a consequence, the as-produced materials typically contain significant amounts of metal impurities [33–36]. Figures 13.5a and 13.5b show a schematic and a transmission electron micrograph (TEM) of a double-walled carbon nanotube (DWCNT) sample, respectively. The as-produced DWCNTs contain noticeable amounts of amorphous carbon and catalyst particles, which may be covered by carbon shells. In addition, the mixture contains several single- and triple-wall CNTs alongside the bundled DWCNTs, all of which are subject to a certain diameter distribution [33–38].

It is important to note that there exists no single purification process which would remove all impurities and separate a particular class of nanostructure from the complex mixtures. Other well-known carbon nanomaterials that can be produced on an industrial scale suffer from similar constraints. Detonation ND, also referred to as nanocrystalline diamond, ultra-dispersed diamond, or simply, ND, is another promising carbon nanomaterial that has attracted the interest of the scientific community. A simple production process using expired munitions for the detonation synthesis allows for moderate material costs. The diamond-bearing soot obtained from the detonation contains a mixture of ND particles, nondiamond carbon, as well as metals, metal oxides, and other impurities coming from the detonation chamber or the explosives in use (Figure 13.5c–e). Until now, because of this complex composition, the purification treatment remains the most complicated and expensive stage of the ND production accounting for up to 40% of the material cost [39].

The process of purification of carbon nanomaterials may not only simply refer to the removal of amorphous carbon and metal catalyst, but may also aim to achieve structural selectivity and size control. Removal of amorphous carbon and most catalyst particles has been achieved by using a combination of HCl treatments (for catalyst removal) and exposure to acidic and basic reactants (removal of undesired carbon species) [40–42]. The most common purification agents are HCl [43]; HNO_3 [28,43–52], H_2SO_4 [28,47,48,53], and mixtures thereof [28,50,52,54,55]; H_2SO_4/H_2O_2 [43,44]; NH_4OH/H_2O_2 [43,44]; $H_2SO_4/KMnO_4$ [43]; chlorine and ammonia water [56]; $(NH_4)_2S_2O_8$ [57]; and dilute aqueous O_3 solution [58,59]. In the following paragraphs, we will discuss the purification of two common, but structurally different carbon nanomaterials in greater detail, CNTs and ND, which consist primarily of sp^2- and sp^3-carbon, respectively. Elaborated techniques are used to purify other carbon nanomaterials in similar or slightly adjusted processes. Studies on the effect of liquid-phase oxidation on the structural integrity of MWCNTs revealed that, although effective in the removal of amorphous carbon and metal impurities, oxidation using acids such as HNO_3 or H_2SO_4 can lead to nanotube degradation in the form of tube shortening and side-wall damage [44,51,55].

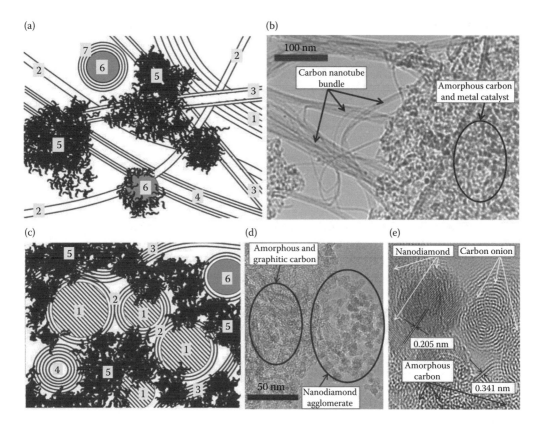

FIGURE 13.5 (a) Schematic of as-produced SWCNT sample showing common impurities, such as catalyst particles and/or amorphous carbon, and simultaneous presence of SWCNTs, DWCNTs, and MWCNTs. (Notation: (1) nanotube bundle, (2) SWCNT, (3) DWCNT, (4) MWCNT, (5) amorphous carbon, (6) metal catalyst, (7) fullerene shell). (b) HRTEM image of DWCNT sample showing CNT bundles and a large number of catalyst particles surrounded by amorphous carbon. (c) Schematic of ND detonation soot showing coexistence of different nanostructures and corresponding HRTEM images of as-produced ND showing. (d) Large amounts of sp^2 carbon and agglomerates of ND crystals and (e) presents of carbon onion, ND and amorphous carbon. (Notation: (1) ND crystal, (2) graphitic shell, (3) graphite ribbon, (4) carbon onion, (5) amorphous carbon, (6) metal catalyst).

In contrast, treatments in basic mixtures (e.g., NH$_4$OH/H$_2$O$_2$) were found to maintain the structural integrity of the MWCNTs, while being efficient in the removal of amorphous carbon and metal oxide impurities [44]. Using Raman spectroscopy, Osswald et al. [28] monitored the oxidation behavior of MWCNTs in concentrated HNO$_3$ and H$_2$SO$_4$ to determine the optimal purification conditions. In their study, they used the intensity ratio between the D and G band Raman features to evaluate the content of disordered carbon and wall defects in the samples. They concluded that treatment in a mixture of HNO$_3$/H$_2$SO$_4$ (1:1) is favorable for improved MWCNTs solubility, but that gas-phase oxidation is more suitable for selective removal of amorphous carbon impurities. MWCNTs have also been purified using treatments in chlorine (Cl$_2$) and ammonia (NH$_3$) water to remove catalyst particles and nontubular carbon like amorphous carbon [56]. Although the treatment did not alter the integrity of the tube structure, some defects were observed on the outermost walls of the MWCNTs.

Menna et al. studied the oxidation of SWCNTs in a solution of concentrated H$_2$SO$_4$ and HNO$_3$ [48]. Using resonant Raman spectroscopy, they showed that the diameter of the SWCNTs is the most important factor determining the resistance of the SWCNTs toward oxidation. Similar results were reported by Li et al., who achieved purity levels up to 98% [52]. SWCNTs were also purified using an aqueous solution of H_2O_2 [60–62]. The treatment not only resulted in a removal of amorphous carbon, but also led to the preferred oxidation of small-diameter CNTs, providing a possible pathway for size control in SWCNT samples.

Although effective in the removal of amorphous carbon and catalyst impurities, liquid-phase oxidation techniques suffer from a variety of limitations. Most importantly, majority of the oxidation agents are environmentally harmful requiring proper handling and storage. This is of particular importance for aggressive oxidizers, such as acids and bases, which require special corrosion-resistant equipment. Oxidation with acids and bases is also known to introduce foreign elements, particularly nitrogen- and sulfur-containing compounds, chlorine, and chromium, to mention a few, which alter the properties and reactivity of the nanomaterial sample and therefore need to be removed in additional purification steps.

Alternatively, gas-phase purification, which employs high-temperature reactions with gases, such as air, O_2, O_3, or CO_2, has also been studied extensively [1–7,63–65]. Li et al. used air oxidation in the temperature range of 480–750°C to remove amorphous carbon from MWCNTs [66]. Under these conditions, the authors also observed the opening of the tube caps and localized damage to the wall structure. Air oxidation of MWCNTs between 5 min (with air flow) and 40 min (static air) at 760°C in a rotating tube furnace was reported to be a suitable high yield (>40%) purification method, allowing for a selective removal of amorphous carbon from MWCNT samples [67]. Lower oxidation temperatures required relatively long treatment times, whereas higher treatment temperatures resulted in high sample losses. Wang et al. employed Au catalyst-assisted oxidation in O_2 to selectively remove the caps from MWCNTs and to control their length and defect concentration [50]. Smith and coworkers [68] purified SWCNTs in air, O_2, and CO_2. They reported that the increase in selectivity toward the different species is inversely proportional to the oxidative severity of the reactant, making CO_2 the most selective oxidant for the separation of SWCNTs from amorphous carbon impurities and nanotube fragments. Purification of a bulk SWCNT sample at 600°C for 1 h in a tube furnace using CO_2 confirmed the suitability of CO_2 oxidation for selective removal of the most reactive components (e.g., amorphous carbon, tube fragments, heavily damaged tubes). At the same time, the authors noted a slight weight increase in the sample due to the oxidation of the remaining catalyst, which points out one of the major limitation of gas-phase purification: unlike oxidation in HCl or HNO_3, most gaseous oxidizers do not allow for a removal of metal impurities.

Gas-phase oxidation was also used to selectively remove metallic SWCNTs using a combination of *in situ* and postsynthesis processing [69]. *In situ* oxidation was conducted by introducing small amounts of oxygen during the synthesis step, whereas the postsynthesis oxidation consisted of a separate treatment exposing the sample to air at 400°C. The combination of *in situ* and postsynthesis oxidation improves the selectivity of the oxidation process toward small-diameter (<1.5 nm) SWCNTs, which are subject to higher tube curvature, and large-diameter SWCNT (>2.0 nm) that typically contain higher defect concentrations. Because the SWCNTs in the diameter range of 1.5–2.0 nm are primarily semiconducting, the oxidation method can be used to yield samples with high contents of semiconducting SWCNTs.

Although purification using gas-phase oxidation was found to selectively remove amorphous carbon, highly damaged tubes, or small-diameter CNTs, the resulting sample loss can be substantial without suitable process control [63,70–73]. Therefore, sufficient understanding of the oxidation behavior of the individual nanomaterials is needed to determine the optimal purification conditions that allow for selective removal of impurities without loss of, or damage to the desired nanostructures. This is difficult to realize for acid purification or high-temperature vacuum annealing, but feasible for oxidation treatments using gases, such as in air, O_2, and CO_2. Therefore, gas-phase oxidation, particularly air oxidation, is considered the most suitable purification route [74].

Other nanomaterials that can also be produced on an industrial scale suffer from similar constraints. Producers of detonation ND employ liquid-phase oxidizers, such as HNO_3 [75,76], $HClO_4$, H_2SO_4, H_2O_2 [77], and mixtures thereof; or aqueous and acidic solutions of $NaClO_4$, CrO_3, or $K_2Cr_2O_7$ to selectively remove amorphous and graphitic carbon from ND powders [41,42]. The treatments utilize differences in reactivity of diamond and nondiamond species toward oxidation; however, the required oxidation times and oxidation temperatures depend on the used reactant(s) and the desired purity levels [41,42]. In addition, several dry chemistry techniques, including air oxidation, catalyst-assisted oxidation [78], oxidation using boric anhydride as an inhibitor of diamond oxidation [79], and ozone-enriched air oxidation [80] have successfully been used to remove nondiamond phases from ND powders. Although being effective purification methods, both catalyst-assisted oxidation and oxidation using boric anhydrite require additional substances to be added to the ND, thus risking further sample contamination. Ozone-enriched air oxidation is based on blowing an ozone–air mixture through the detonation soot at temperatures between 150°C and 400°C [80]. This treatment was also reported to eliminate effectively nondiamond carbon species from ND. However, like acid-based purification, the above processes also require the use of either toxic and aggressive substances, or supplementary catalysts, which result in an additional contamination or a significant loss of the diamond phase. Moreover, while some of these techniques discussed above are already in use, none of these individual processes can provide the purity levels required for most ND applications, leading to the dark-gray color of commercially available powders [81].

Recently, air purification of ND powders was shown to be a simple, inexpensive, environmentally friendly, and scalable method to remove selectively nondiamond carbon from the detonation soot. Although first thought of as being unfeasible [79], Osswald et al. demonstrated purity levels as high as 96 wt% without significant loss of the diamond phase. Cataldo et al. studied the oxidation of ND in an air flow and found that the ND sample is stable toward oxidation at temperatures below 450°C, but burns quite suddenly above 450°C [82]. However, no changes in content of diamond or nondiamond carbon were reported.

13.3.2 Activation

Nanoporous carbon consists of interconnected cells and networks of highly amorphous carbon and is characterized by nanometer-sized pores which generate a high specific surface area (SSA) and are commonly referred to as nanostructures. They are of great technological importance with applications ranging from gas separation and storage, to water/air filtration, to catalyst supports, to lithium-ion batteries, and supercapacitors [83–86]. In particular, activated carbons, aerogels, templated carbons, and, more recently, CDC [87] have been investigated and can now be produced and activated on industrial scale [88]. Although activated carbons are widely used and considered the most powerful conventional adsorbents, despite extensive studies and improvements in the activation process, little control over the pore structure has been achieved [89–91]. In the following section, we will focus our analysis on the activation of CDCs, which, unlike other nanoporous carbons, can be manufactured with a large variety of porosities and, thus, offer great potential for future application.

Activation using oxidation treatments has become a routine technique to adjust the properties of porous carbons by modifying their microstructure under controlled environmental conditions [92,93]. During the years, a variety of activation methods have been developed, including chemical activation in bases (NaOH and KOH) [94,95] or acids (HNO_3, H_2SO_4, H_3PO_4) [96], and physical activation in air, CO_2, or steam [1–7]. The properties of the activated carbons strongly depend on the activation process and choice of activation conditions. Even minor changes in activation temperature, activation time, or reactant concentration for the same oxidizing agent can lead to significant differences in the resulting pore texture, surface chemistry, and microstructure [97,98].

The activation process is a crucial step toward a successful utilization of CDCs. Although large pore volumes can be achieved by selecting precursors with high metal-to-carbon ratios, pore growth and structural collapse may occur when precursors with low carbon contents are used [99]. From an economical point of view, it is also more suitable to use low-cost precursors, such as TiC or SiC, and to employ postsynthesis activation to adjust the microstructure to obtain the desired structural characteristics.

Of the potential activation techniques, physical activation is particularly beneficial due to the use of inexpensive gaseous reactants and process simplicity. Particularly, activation in air, O_2, steam, or CO_2 have been studied widely [2,98,100]. CO_2 is known to be a relatively mild oxidizer, allowing for higher control over changes in microporosity during activation [1,4,5,7]. However, air activation is a low-energy/low-cost process because of the higher reactivity, requiring lower activation energy as compared to CO_2 or steam [3,100,101]. However, the process is more difficult to control and great care must be taken to avoid excessive sample loss. Furthermore, it should be noted that activation processes that maximize pore volume and surface area of carbon materials also lead to a broadening of the pore size distribution (PSD), which may limit the control over the porosity. Ultimately, the optimal activation conditions for each sample depend on the nature of the carbon material of interest, its porosity, degree of graphitization, and the surface chemistry [3].

Studies on activation of TiC-CDC synthesized via chlorine treatment at 600°C (600°C-TiC-CDC) in CO_2 and 1000°C (1000°C-TiC-CDC) in air in air revealed that both oxidants can be used to control the microporosity in CDCs (Figure 13.6). Although air oxidation is suitable for partially graphitized samples, such as 1000°C-TiC-CDC, activation in CO_2 allows sufficient control over the porosity of highly amorphous and less oxidation resistant 600°C-TiC-CDC. The higher reactivity of O_2 would lead to a rapid oxidation of amorphous species and would limit the control over porosity development. CO_2 is a milder oxidizer, but requires higher oxidation temperatures. The observed shifts in pore size (Figure 13.6a) result mainly from a decrease in the pore volume of smaller pores.

Activation usually yields a general trend toward larger pores, thus decreasing the pore volume related to micropores (<1.5 nm) and increasing mesopore volume (>1.5 nm). Oxidation of 1000°C-TiC-CDC in air was found to provide only moderate control over the SSA, reaching a

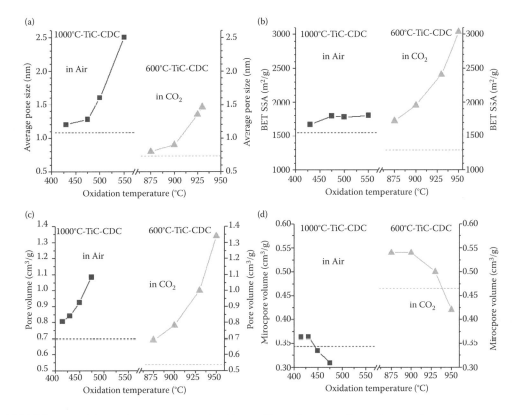

FIGURE 13.6 Changes in average pore size (a), specific surface area (b), total pore volume (c), and micropore volume (d) upon oxidation of 1000°C-TiC-CDC and 600°C-TiC-CDC in air (6 h) and CO_2 (2 h), respectively. The dashed lines represent the values of nonactivated samples.

maximum value of ≈ 1800 m^2/g, whereas CO_2 activation of 600°C-TiC-CDC significantly increases the SSA, reaching values up to 3000 m^2/g or higher (Figure 13.6b). Both activation agents (air and CO_2) lead to an increase in the total pore volume compared to the nonactivated samples (Figure 13.6c). Although the relative increase of total pore volume and micropore volume is larger for 600°C-TiC-CDC, both air and CO_2 activation exhibit similar trends. The total pore volume significantly increases with activation temperature (Figure 13.6c). The micropore volume shows only a small increase at low temperature (6% for 1000°C-TiC-CDC and 17% for 600°C-TiC-CDC), but decreases at higher activation temperatures following a two-step process, with initial formation of new micropores followed by the reconstitution and enlargement of those micropores into mesopores (Figure 13.6d). The optimum conditions for oxidation of 600°C-TiC-CDC in CO_2 with respect to a high SSA, development of micropores, and low weight loss are, therefore, low oxidation temperatures (≈ 875°C) and long oxidation times (>8 h) [102]. High temperatures and longer oxidation times lead to a larger total pore volume, but are unfavorable for the formation of micropores. The average pore size shifts toward higher values with increasing oxidation time and oxidation temperature. At constant activation temperature, oxidation in CO_2 enlarges small pores with time, thus decreasing micropore and increasing mesopore volume. Similar trends were found for oxidation at different temperatures with constant treatment time.

The ability to control the pore structure of CDCs is of great importance for a large number of applications, especially for adsorption and storage of gases such as hydrogen and methane [102–104]. Studies on activated 1000°C-TiC-CDC (3 h at 475°C) and 600°C-TiC-CDC (8 h at 875°C) revealed a 14% and 40% increase in H_2-uptake, respectively, compared to nonactivated samples [102]. The increase is larger for CO_2-activated 600°C-TiC-CDC because of a better porosity development, that is, 40–50% increase in SSA and a significantly larger micropore volume. Similar improvements were achieved for methane uptake in activated TiC-CDC. While micropores (<1.5 nm) are of great importance for hydrogen/methane storage [105,106] and supercapacitor [107,108] applications, larger average pore sizes (1.5–5.0 nm) are very attractive for sorption of biological molecules and toxins, which are usually larger in size compared to the electrolyte ions or hydrogen molecules used in most energy-related applications.

When discussing pore size influences, these fundamental structure–property relationships are usually obtained by adjusting homogeneously the pore size within the carbon particle, resulting in a constant pore size distribution throughout the particle. Nevertheless, Hegedus et al. postulated in 1980 that optimal pore structures for heterogeneous catalysis vary within particles [109]. It was discovered that a local variation of micropore and mesopore content is superior to a homogeneous distribution if mass transport is involved in parallel to a reaction or absorption [110–118]. The combination of the CDC synthesis with the pore widening during the oxidation presents a promising pathway for the establishment of local pore size variations. The extraction of the metal atoms from carbide precursors follows a shrinking core-like conversion pathway. Hence, as the reaction progresses, the reaction front migrates from the outside of the particle to the particle center [119,120]. When CO_2 is used as an oxidizing agent to widen the pores, the oxidation reaction can take place in parallel to the extraction since the reactants or products do not interfere with each other. Therefore, CO_2 can function as a parallel reactant to Cl_2, carrying out an *in situ* activation of the CDC during synthesis [121]. Following a shrinking core model, CDC formation starts at the outer edge of the carbide particles and continues toward the center. Thus, carbon located on the outer parts is oxidized over a longer period of time as compared to the inner regions. This leads to different levels of pore widening. The largest pore sizes will be encountered at the outer edge of the particles and smallest close to the center. The approach has been demonstrated using SiC-CDC, but can be applied to other CDCs as well [121].

To improve visualization of the changes in local pore size, SiC was converted into CDC under parallel activation. The reaction was stopped after 2.5 h, resulting in a final carbide conversion of 25%. The resulting nanoporous carbon core shell structure covering the remaining carbide was then activated for an additional 2.5 h without reactive extraction. Figure 13.7 compares the pore size

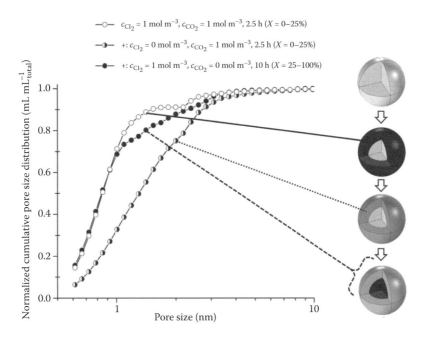

FIGURE 13.7 Normalized pore size distributions (N_2-sorption, deconvolution via QSDFT: quenched solid density functional theory) for SiC powder with parallel and subsequent CO_2 activation and Cl_2 extraction.

distribution of both core shell structures and shows that the mean pore size of the shell increased from 1.0 to 1.6 nm. In a subsequent reaction, the carbide core of the core shell structure was converted to CDC with smaller pore size without parallel activation ($c_{Cl_2} = 0, c_{CO_2} = 1 \text{ mol/m}^3$). The resulting pore size distribution as given in Figure 13.7 shows a clear bimodal distribution of the core and the shell and the expected ratio of approximately 25% pore volume in the shell and 75% in the core. Thus, the *in situ* activation offers the possibility to synthesize nanoporous carbons with a local pore size variation. It was further demonstrated that pore size distribution can be controlled by changing the ratio of the extraction rate to the activation rate [121]. Available pore structures range from the border case of homogeneous pore widening within the particle, to core shell structures, such as those obtained by activation in series. The reaction rate ratio is influenced by the CO_2 and Cl_2 concentrations and the reaction temperature, all of which can be controlled in the course of the reaction.

In situ activation is also beneficial from an economical point of view. Combining the two high-temperature steps lowers the total process time and reduces the overall material cost. For example, for TiC-CDC at the same reaction time, the specific surface area is increased by 52% and the pore volume by 66% when the CDC is activated in parallel.

13.3.3 FUNCTIONALIZATION

Arguably, the most important characteristic of nanomaterials is their high surface-to-volume ratio. Because nanomaterial properties are strongly influenced by surface properties, alterations in the chemical surface functionality can have significant effects. Depending on the nanomaterial and synthesis method used, a carbon surface can be largely free of functional groups, as is the case for high-quality graphene or nanotubes; it can be covered by a single functionality, as is the case for hydrogen or halogen terminations; or it can exhibit a mixed functionalization, commonly comprising functional groups of the main elements of hydrocarbon chemistry, that is, containing oxygen, hydrogen, nitrogen, and sulfur.

Properties directly affected by the surface chemistry include the wetting behavior, adsorption/attraction, surface charges, basicity/acidity, and chemical reactivity. Examples include the increased biocompatibility of CNTs with increasing water solubility, a property which can be achieved by PEGylating (PEG: polyethylenglycol) CNTs [122]; the covalent incorporation of nanomaterials into a polymer matrix, which results in good dispersions and direct uptake of loads by the reinforcement material [123]; and the transformation of nanoporous carbons to solid acids upon sulfonation (e.g., with 4-benzenediazoniumsulfonate) and its use in catalysis [124].

The advanced surface chemistries mentioned in these examples are not a result of the actual synthesis, but rather must be introduced after the synthesis using additional treatments. Oxidation, currently the most common method used for carbon functionalization, introduces a variety of oxygen-containing functional groups to carbon surfaces. These oxygen-modified surfaces can serve as a starting point for further, more advanced functionalization steps. Common follow-on treatments include the following [125,126]:

- Activation of carboxylic groups with thionyl chloride and the subsequent reaction with amine-terminated reactants (e.g., $PEG-NH_2$) or esterification with alcohols. A direct reaction with terminal amines is also possible in this case.
- Carbodiimide-activated direct esterification with alcohols.
- Introduction of azide functionalization by reaction with sodium azide followed by subsequent reduction with lithium aluminum hydride to obtain amino-functionalization. Reaction with sodium azide can also be followed by the subsequent reaction of azides with triplebonds in click chemistry reactions.
- Direct reaction of epoxy functionalities with amine-terminated reactants in a nucleophilic ring-opening reaction.
- Formation of amides and carbamate esters by reaction with isocyanates.
- Surface termination with hydrophilic phenolic functionalities by reduction with lithium aluminum hydride.

The selection of a suitable chemical route depends primarily on two factors: (1) the availability of a counter functionality for the desired reactant, and (2) the influence of the posttreatment on the desired properties of the carbon material (e.g., electrical conductivity, stability, wettability).

Although a subsequent modification of the oxidized surface is often desired, the oxygen-containing surface groups themselves may be beneficial or even crucial for selected applications. A good example of this is the catalytic activity of C–O groups for the oxidative dehydrogenation of ethylbenzene to styrene [127]. High-temperature oxygen groups are alkaline and, if positioned close to one other, can dehydrogenate reactants through the formation of phenolic OH groups. In combination with the dissociation of oxygen on the basal graphite plane, the phenolic groups are reoxidized to quinones through the desorption of water [128]. Thus, the high-temperature oxygen groups enable a Mars–van Krevelen reaction mechanism. Interestingly, the active oxygen groups can be generated *in situ* under reaction conditions of 515°C in an oxygen-containing atmosphere, leading to an initial delay in activity [127]. This delay is not observed if the active functional groups are introduced before reaction via oxidation at 570°C.

The various methods of carbon oxidation discussed in Section 13.1 result in a variety of functional surface groups. These include carboxylic acids, anhydrides, lactones, carbonyls, aldehydes, esters, and phenols. The quantity and type of functionalities formed vary depending on the oxidation procedure and the prevailing reaction conditions. For example, wet chemical oxidation with nitric acid tends to yield more carboxylic acids, whereas gas-phase oxidation yields more anhydrides, lactones, phenols, and carbonyls [129].

Identifying the nature and quantity of functional groups is a critical step toward a successful application of carbon nanomaterials. Commonly used methods include Fourier-transformed infrared spectroscopy (FTIR) of compacted carbon/potassium bromide mixtures, temperature-programmed

TABLE 13.2

Desorption Temperature and Off-Gas Species and FTIR Bands for the Most Common Oxygen Surface Functionalities

	TPD		
	Desorbing Species	Temperature Range (°C)	FTIR
Carboxylic acid	CO_2	100–400 [129–132]	1120–1200
			1665–1760
			2500–3300 [129,131–133]
Carboxylic anhydride	CO, CO_2	350–600 [129–132]	980–1300
			1780–1840 [129,133]
Cyclic anhydride			1740–1760 [133]
Lactone	CO_2	400–650 [129,130,132]	1160–1370
			1675–1790 [129,131,132]
Phenols	CO	600–700 [129–132]	1000–1220
			2500–3620 [129,132]
Carbonyl, quinones	CO	700–1000 [129,130,132]	1100–1500
			1550–1680 [129,132]
Ethers	CO	600–700 [129,131]	1000–1300 [129,132,133]

desorption (TPD) in inert atmosphere or vacuum conditions eventually combined with off-gas analytic (H_2O, CO, CO_2), x-ray photoelectron spectroscopy (XPS), Boehm titration, solid-state nuclear magnetic resonance (NMR), and point-of-zero-charge measurements. Typical FTIR band ranges for different oxygen surface functionalities are given in Table 13.2. By identifying new peaks or slight changes in peak positions after oxidation, the impact of the oxidation procedure on surface functionalization can be qualitatively determined. A quantitative analysis, however, is much more challenging. The total oxygen content must first be determined by elemental analysis methods. The fraction of each functionality can then be estimated from the temperature-dependent mass loss and off-gassing during TPD. However, in TPD, only carboxylic acid groups can be clearly separated from the other groups, which overlap in broad peaks at temperatures above 350°C during desorption (see Table 13.2). Boehm titration, in contrast, allows different acidic and alkaline groups (carboxylic acids, lactones and phenols, and reaming weaker acids) to be quantified. Unfortunately, deriving quantitative information with Boehm titration can be a tedious procedure, because small amounts of atmospheric CO_2 dissolved in the titration solutions often falsify the results.

13.4 EXEMPLARY CARBON NANOMATERIALS

13.4.1 CARBON NANOTUBES

As mentioned earlier, CNTs are commonly divided into SWCNTs and MWCNTs. SWCNTs are basically a single graphene layer rolled into a hollow cylinder, whereas MWCNTs comprise two, three, or more concentrically arranged cylinders. CNTs can be understood as individual molecules or quasi one dimensional (1D) crystals with translational periodicity along the tube axis. Their diameters range from 0.4 nm to tens of nanometers, with lengths reaching several hundred micrometers or more. The tube ends are open or capped by hemispherical fullerenes and cones. One can further distinguish between metallic and semiconducting CNTs. Although MWCNTs are metallic, approximately two-thirds of all possible SWCNTs are semi-conducting and only one-third are metallic [134]. Currently, there exist three primary synthesis techniques: laser ablation, arc discharge, and chemical vapor deposition (CVD). Owing to their confined dimensions, CNTs exhibit exceptional electronic, mechanical, and optical properties. For details on synthesis, properties, and application of these unique nanostructures, the reader is hereby referred to two comprehensive reviews by Dai et al. [135] and Terrones et al. [136].

Soon after the knowledge of their existence became widespread in 1992, researchers began to explore the potential of the oxidation process for the purification, modification, and functionalization of CNTs. As early as 1993, researchers reported on the thinning and opening of MWCNTs via heat-treatment in air [137,138] and CO_2 [139]. Oxidation was found to lead to a partial or complete removal of the nanotube caps and a stripping of the outer layers of the MWCNTs. It was concluded that the curved caps exhibit higher reactivity because of higher strain and presence of pentagonal rings. Once the caps were removed, the edges of the MWCNT sidewalls were exposed and further eroded as the oxidation proceeded. However, follow-up studies revealed that oxidation of CNTs does not always start at the tips of the tubes. The reaction mechanism is rather complex and controlled by several factors, including CNT curvature, helicity, and the possible presence of structural defects, such as pentagons and heptagons, which affect both initiation and propagation of the oxidation reaction [140].

Since then, numerous studies have focused their efforts on a more detailed investigation of the oxidation mechanism in CNTs [58,67,68,141–146]. For example, Park et al. performed density functional theory (DFT) calculations to demonstrate that the desorption energy barrier of a C–O pair at the nanotube edge is higher than that of amorphous carbon; thus, selective oxidation of amorphous carbon and other nontubular carbon species become feasible [67]. Similar studies on the reaction mechanism of SWCNTs were also conducted for ozone [146] and water [147]. Experimental studies on the oxidation behavior of SWCNT samples in oxygen, air, and carbon dioxide revealed that the oxidation process contained several individual reaction steps stemming from the presence of different carbon species [68]. The oxidation temperature of the individual peaks increased in going from oxygen to air to carbon dioxide. Also, the increase in selectivity toward the different carbon species was found to be inversely proportional to the oxidative severity of the reactant, making carbon dioxide a very selective oxidant for removal of nontubular carbon impurities. Zhang et al. investigated the oxidation behavior of SWCNT and its dependence on both tube diameter and electronic structure, revealing that the oxidation resistance of small-diameter CNTs not only increases with tube diameter, but also modifies its electronic properties [148]. Li et al. oxidized SWCNTs in dilute aqueous ozone solution and concluded that the reaction pathway follows a two-step process. At the onset of the oxidation, C=C bonds are first converted to hydroxyl groups, which are then transformed to carbonyl and carboxyl groups [58]. Singh et al. [144] investigated the oxidation kinetics of MWCNTs in air. They found that the oxidation follows different mechanisms, depending on the oxidation temperature. At lower temperatures (<350°C), the reaction is chemically controlled and based on the oxidation of the metal catalyst and the reaction of access oxygen with the MWCNT, thus shortening the tubes as the oxidation proceeds. Between 400°C and 450°C, the oxidation becomes diffusion controlled, beginning with the opening of the nanotubes and continuing along the tube axis following the cylinder model. Oxidation above 500°C results in a breakdown of the MWCNTs into smaller parts, which leads to a powder-like oxidation behavior. This also results in a lower activation energy and an increased oxidation. However, it should be noted that the samples contain catalyst particles which strongly affect the activation energies of individual oxidation steps. Savage et al. [141] investigated the process of UV light-assisted oxidation on the oxidation behavior of SWCNT and MWCNTs. They proposed two possible mechanisms by which UV radiation may enhance the oxidation of CNTs. The first contribution may arise from a reduced energy barrier for the adsorption of photo-generated singlet oxygen. The second contribution could stem from the presence of defects in the wall structure of the CNTs that may facilitate the formation of local electron-rich and electron-deficient sites, which in turn further the adsorption of oxygen molecules on the CNTs. The effects of tube diameter, chirality, and the number of walls on the O_2 adsorption process were found to be small in comparison to the importance of lattice defects [142]. Osswald et al. compared the oxidation behavior of SWCNTs and DWCNTs under isothermal and nonisothermal conditions using Raman spectroscopy [27]. It was concluded that structural characterization techniques, such as Raman spectroscopy, can be used to maximize the CNT yield by optimizing the selectivity of the oxidation process for the various CNT samples. Using DFT calculations, the oxidation of small-diameter CNTs was also investigated as a means to the formation of graphene nanoribbons [143].

Mechanisms of liquid-phase oxidation of CNTs have been studied in much less detail. Kanai et al. [149] reported on first-principles calculation of NO_3-SWCNT interactions and provided an atomistic model for the oxidation process.

With a better understanding of the oxidation mechanism, many researchers have used oxidation for the modification of CNTs, in particular for development of high-yield purification techniques. A recent review by Hou et al. [40] provides a detailed overview on existing purification methods. A large number of studies found in literature report on the use of gas-phase oxidizers, such as air [28,63,66–69,71–73,144,150–153], oxygen [50,64,68,69,150], carbon dioxide [68], ozone [59,151], mixtures of Ar, O_2, and H_2O [154,155], and steam [156].

Purification of SWCNTs was achieved using isothermal treatments in air and oxygen at temperatures ranging from 250°C to 500°C [27,63–65,155,157]. Dementev et al. developed a nonisothermal oxidation process, referred to as dynamic oxidation, which exposed SWCNTs to a wider temperature range, but shorter oxidation times [158]. Oxidation has also been used to control both the diameter and the length distribution of SWCNTs. Several studies showed that during oxidation, diameter is the most important factor determining the reactivity of the SWCNTs, allowing for selective removal of small-diameter [48,60] and semiconducting tubes [62]. Light-assisted oxidation in an aqueous solution of H_2O_2 led to selective removal of semiconducting SWCNT in a specific diameter range, suggesting an enhanced selectivity of the oxidation process on controlled light exposure [61].

Purification of DWCNTs, the smallest member of the MWCNT family, was realized using treatments in air at temperatures between 350°C and 550°C [159]. Air oxidation of MWCNTs between 450°C and 750°C was shown not only to successfully remove amorphous carbon impurities and enable the opening of CNT caps, but may also lead to the introduction of defects to the wall structure [66,153]. Several studies have identified temperatures around 400°C to be optimal for removing amorphous carbon from MWCNT samples without damaging the tube structure [28,160]. Osswald et al. proposed that rapid oxidation at higher temperatures (≈550°C), referred to as flash oxidation, may be a suitable alternative to conventional purification methods, as it yields similar purification levels, but simultaneously introduces a certain number of wall defects that are necessary for subsequent functionalization. Several groups have scaled their oxidation experiments to the bulk level and demonstrated that gas-phase oxidation is a suitable technique for industrial CNT purification [67,68].

A large number of studies on the oxidation of CNTs employed liquid-phase oxidizers, such as acids and bases, to modify structure and composition of CNT samples. The most common reactants investigated include HCl [43], HNO_3 [28,43–52], H_2SO_4 [28,47,48,53], and mixtures thereof [28,50,52,54,55]; H_2O_2 [62]; H_2SO_4/H_2O_2 [43,44]; NH_4OH/H_2SO_2 [43,44]; $H_2SO_4/KMnO_4$ [43]; $(NH_4)_2S_2O_8$ [57]; chlorine and ammonia water [56]; and dilute aqueous O_3 solution [28,58]. Similar to gas-phase oxidation, utilization of acids and bases can lead to the formation of wall defects and extensive tube damage under improper oxidation conditions [55]. Rosca et al. investigated the effect of HNO_3 oxidation on the structure and properties of MWCNTs. In their study, the weight loss increased in a nearly linear relation to oxidation time. During the first 6–9 h of oxidation in concentrated HNO_3, MWCNTs suffer only minor damage. After ≈24 h of oxidation, small diameter MWCNTs are removed almost completely. After 48 h of oxidation, the remaining MWCNTs were highly fragmented and covered by amorphous carbon. Therefore, although acid oxidation removes amorphous carbon impurities in the early stages of the oxidation treatment, it also leads to the formation of amorphous carbon through severe shortening and thinning of CNTs for prolonged treatment times [45]. Datsyuk et al. compared the effect of liquid-phase oxidation in acids (HNO_3 and H_2SO_4/H_2O_2) and bases (NH_4OH/H_2SO_2) on MWCNTs and noted that in contrast to acid treatment, base oxidation yielded a complete removal of amorphous carbon and metal oxide impurities, while maintaining the structural integrity of the MWCNTs [44]. Li et al. purified SWCNT using HNO_3 and H_2SO_4 and showed that optimized acid mixtures enable purity levels as high as 98 wt.% SWCNTs [52].

Although the oxidation process allows for a controlled removal of amorphous carbon and other impurities, the oxidation also affects both structure and properties of CNTs. Zhang et al. showed that acid-purified MWCNT exhibit higher oxidation resistance as compared to untreated CNT [161]. Spitalsky et al. [43] reported on the oxidation of MWCNT using hydrochloric acid (HCl), nitric acid (HNO_3), acidic permanganate ($H_2SO_4/KMnO_4$), piranha (H_2SO_4/H_2O_2), and an NH_4OH/H_2O_2 mixture. They discovered that the mechanical properties of the treated MWCNT films increase with the power of the oxidizing agent and that treatment in HNO_3, H_2SO_4/H_2O_2, or HCl yield films with narrow pore size distributions and high surface conductivities. Oxidation of MWCNTs in chlorine and ammonia water has also been demonstrated. The surface functionalization resulting from the treatment resulted in improved dispersion in polar solvents, particularly water, ethanol, acetone, chloroform, and DMF [56].

Collins et al. [150] and Zaporotskova et al. [162] analyzed the electronic properties of SWCNTs before and after exposure to oxygen. Their study revealed that the electronic properties of SWCNTs can be reversibly modified by small concentrations of adsorbed oxygen and that, on exposure, semiconducting SWCNT can be converted to metallic nanotubes. Tantang et al. [47] used treatments in HNO_3 (60%) and H_2SO_4 (98%) to increase the electrical conductivity of SWCNT electrodes. The formation of wall defects and subsequent saturation with COOH (HNO_3 and H_2SO_4 treatment) and SO_3H (H_2SO_4 treatment) functional groups increase or stabilize the p-doping of the SWCNTs because of their electron-accepting character. DFT calculations were found to support these assumptions [163]. Oxidation was also found to enhance the field emission performance of vertically aligned MWCNT films [151]. Oxidation in air and ozone yielded an eightfold increase in the emission current. Air oxidation in the temperature range of 480–750°C was found to improve the performance of MWCNT electrodes in electric double-layer capacitors (EDLC) because of a resulting increase in surface area, a decrease in CNT aggregation, and a reduction of the micropore volume [66].

Zhao et al. demonstrated that an oxidation-based surface modification of SWCNTs is a promising way for preparing chemically selective CNTs. Formation of carboxylate groups on the tube surface led to a reversible response of the optical absorption properties of semiconducting SWCNTs to pH changes in aqueous solutions [164]. Similar effects on the gas-sensing properties of MWCNTs were observed by Xu et al. [165]. Kyotani et al. reported on the selective functionalization of the inner walls of alumina template-grown MWCNT using treatments in HNO_3, thus rendering the inner surface hydrophilic, while the outer surface maintained its hydrophobic character [46]. Studies conducted by Pavese et al. [54] and Stobinski et al. [51] investigated changes in the wettability of SWCNT and MWCNT after oxidation, and demonstrated that acid oxidation lowers the contact angle of water with the CNTs and change their behavior from super-hydrophobic to hydrophilic.

Marques et al. [49] used hydrothermal oxidation in HNO_3 at elevated pressure (0.5 MPa) and temperature (120–200°C) to functionalize SWCNTs and to control both the type and amount of surface functional groups. Their results showed that the degree of surface functionalization is correlated with the HNO_3 concentration by a first-order exponential function. Lebron-Colon et al. [53] used dye-assisted photo-oxidation to functionalize SWCNTs. Their study revealed that photo-oxidized SWCNT contain up to 40% more oxygen (11.3 at.%) bound to the tube surface (chemisorption), as compared to H_2SO_4-treated samples (6.7 at.%), primarily in the form of carboxylic, carboxylate, and ester groups.

Figure 13.8 shows changes in I_D/I_G of MWCNTs after various oxidation treatments in comparison to as-received and vacuum annealed (1800°C) MWCNT. Air oxidation at temperatures around 400°C removes amorphous carbon impurities without damaging the CNTs, while air oxidation at temperatures above 500°C and acid treatments increase the number of defects, leading to lower and higher I_D/I_G values, respectively. Vacuum annealing heals defects in the wall structure, as indicated by the substantial reduction in I_D/I_G.

In summary, oxidation provides great potential for the modification of CNT samples. A large variety of gaseous and liquid reactants have been explored and were used to selectively remove

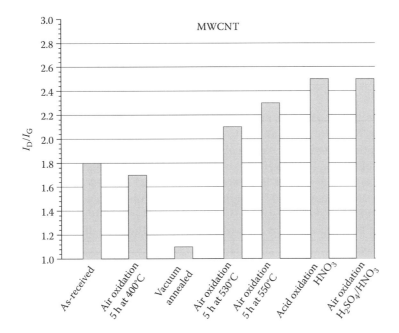

FIGURE 13.8   Changes in the I_D/I_G band intensity ratio of MWCNT after air oxidation and acid treatment under various conditions. Values of as-received and vacuum-annealed MWCNTs are shown for comparison.

amorphous carbon and other nontubular carbon impurities, and to control the defect density and surface functionalization of the nanotubes.

13.4.2   GRAPHENE

The pioneering work of Novoselov and Geim on graphene has had a great impact on the scientific community and has also generated much interest [166]. The most remarkable properties of graphene include the 2D geometry of the crystal itself, its extraordinary electrical and mechanical properties, and its high surface area [166]. The oxidation of graphene is of great interest, as it allows for an introduction of surface functionalization, a controlled alteration of electronic properties, an increase in specific surface area, and the production of graphene from graphite via graphite oxide. Although the oxidation of graphite has been studied extensively, graphene oxidation remains today a relatively new, unexplored field.

Oxidation also plays an important role in the chemical exfoliation synthesis of graphene. Graphene oxide (GO), the precursor for low-cost, large-scale production of graphene, is fabricated by liquid-phase oxidation of graphite [167] that induces delamination and separation of the layered graphite structure. In this process, it is important that functional oxygen surface groups be created without removing carbon atoms from the structure. This process has been known by the scientific community since 1859, when D. C. Brodie demonstrated that graphite oxide can be produced with potassium chlorate and fuming nitric acid [168]. Staudenmeier [169] as well as Hummers and Offeman [170] developed less hazardous process methods. In the Hummers method, graphite is oxidized with H_2SO_4 and HNO_3 to a blue stage, which then forms graphite hydrogen sulfate intercalated with H_2SO_4 and HSO_4^- ions. This compound is subsequently oxidized with $KClO_3$ and $KMnO_4$ to form transparent flakes. Internal mass transfer limitations were observed during the oxidation of 400 µm graphite flakes, which showed slower oxidation rates than 45 µm flakes [171]. GO contains exfoliated sheets and multilayer platelets, with an increased layer distance of 0.7 nm. The intercalated platelets can be further exfoliated through ultrasonic treatment [172–174] or through heat-induced (thermal) expansion [171,175]. In his original research paper, Brodie mentioned the extreme thinness of graphite oxide. However, it was not until 100 years later that Boehm et al. proved for the first time the existence of single-layer carbon sheets [167,168,176,177]. For further purification, dialysis can be used to remove ions from the synthesis suspension. Sedimentation and centrifugation can also be employed to further separate the oxidized sheets from the nonoxidized particles.

The exact structure and surface functionalization of GO is still debated and is at the center of ongoing research. Since Hofmann and Holst published their initial model in 1939, advanced characterization techniques, including elemental analysis, ion exchange experiments, base titrations, and methylation of carboxyl groups [178], have been employed to further study the complex surface structure of GO and the model has been significantly revised [179–184]. Today, mainly the Lerf-Model [185,186] and Dékány-Model [187] are used [174,188] (see Figure 13.9). Hydroxyl and epoxy groups are the most prevalent oxygen functionalities in GO [189,190] and carboxylic groups can also occasionally be found along the sheet edges [189,191]. The main differences between the two models are listed below:

Lerf Model:

- C-O bonds are believed only to be present in the carboxylic groups located on the sheet edges.
- Nonfunctionalized aromatic and functionalized aliphatic ring areas exist in parallel.
- The basal plan is primarily flat with some wrinkles because of distorted tetrahedral configurations of carbons with OH groups.

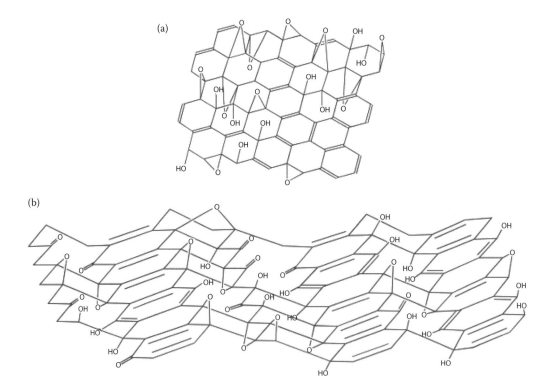

FIGURE 13.9 Schematic of the surface functionalization of graphene oxide (a) Lerf model and (b) Dékány model. (From T. Szabo et al., *Chemistry of Materials* 2006, *18*, 2740.)

Dékány Model:

- C-O bonds can also be present in the form of ketones/quinones in the basal plane.
- Flat carbon ribbons connected by C-C bonds alternate with ribbons of functionalized cyclohexane chairs.
- A slight tilting angle between the boundary layer of both ribbon domains can result in a wrinkling of the basal plane.

It has been experimentally and theoretically shown that the oxidation of graphite and graphene cannot lead to 100% carbon functionalization while maintaining a sheet-like structure [192,193]. This is different from hydrogenation or fluorination reactions, where a 100% functionalization can also be achieved for single layers. The rich surface functionalization of graphene oxide can be an ideal starting point for further functionalization [125], for example, for covalent grafting of graphene in polymer composites [194].

While the oxidation and subsequent delamination of graphite is already an important process for the production of graphene, the oxidation of graphene itself is also of great technological importance, as it is one of the primary methods to alter the surface functionalization and material properties. When discussing graphene oxidation, the influence of the number of graphene layers on the transition to graphite is of high importance. In the last century, gas-phase oxidation has been by far the most studied chemical transformation involving graphite. As interest in the topic rose in recent years, studies in single- and few-layer graphene oxidation also began to appear in the literature.

The most abundant reactant in gas-phase oxidation is atmospheric molecular oxygen. Molecular oxygen has a low sticking coefficient on the basal plane of graphite layers, and thus reacts mainly with surface defects and edge sites [195]. According to TEM and STM studies, 850°C molecular oxygen first begins to react with natural defects and then oxidation subsequently continues along the (002) graphite plane [23,196–200]. Hence, monolayer pits are formed along the graphene plane. Although the in-plane etching rate for gas-phase oxidation is determined by classical kinetic parameters, such as the reactant concentration and temperature, the overall reaction rate depends on the number of defects in the layer. Also, the initial number of defects determines the resulting pit density. For graphite, "synthetic" defects were successfully introduced by ion bombardment [11,201]. The literature reports a faster etching of multilayer pits compared to monolayer ones, which can be explained by the subsequent reaction of the underlying layer with the product carbon monoxide [23,201]. If the reaction temperature rises above 875°C, oxidation occurs not only along defects, but also along planes [200]. Angle-resolved x-ray absorption spectroscopy of air-oxidized HOPG demonstrated that the oxidation process and pronounced pit growth do not result in a heavily functionalized surface [195]. Most carbon atoms that build up a carbon bond are removed during the oxidation process to the gas phase. Thus, the remaining graphitic carbon is nearly nonaltered in its semimetallic electrical properties.

Studies on single- and multilayer graphene showed that three- or more layered materials behave comparably to graphite [202]. For bi- or monolayer graphene, however, an increased oxidation rate was observed. The oxidation-related changes in graphite are defect-induced, whereas those of single-layer graphene already become random at temperatures much lower than 450°C.

The influence of defects becomes more pronounced when graphene oxide and reduced graphene oxide (rGO) are oxidized using gas-phase reactions. GO itself is highly defective. These defects can be helpful initiators for a subsequent gas-phase oxidation or chemical functionalization. It has been shown that steam oxidation of GO and rGO can take place at temperatures as low as 200°C [203]. In accordance with the number of defect sides, the etching rate of GO was observed to be significantly higher than that of rGO. Atomic force microscopy (AFM) images showed the development of a porous network during GO steaming. The decrease in the graphene surface fraction (corresponding to an increase in porosity) with increasing reaction time is shown in Figure 13.10. Thus, in principal, materials of interest can undergo a series of oxidation and reduction cycles; first,

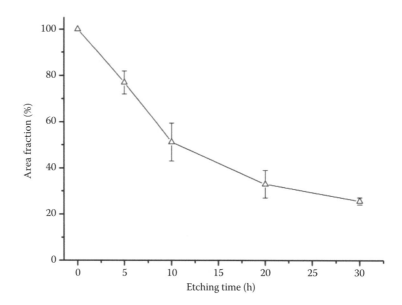

FIGURE 13.10   Decreasing area fraction of GO (= increasing porosity) with increasing steaming time, determined by AFM. (Reprinted with permission from T. H. Han et al., *Journal of the American Chemical Society*, 133, 15264, Copyright 2011, American Chemical Society.)

a liquid-phase oxidation of graphite can be used to induce delamination. The highly functionalized surface of the produced GO must then be chemically reduced to obtain reduced graphene oxide. Subsequent gas-phase oxidation allows monolayer pits to grow from the defects, thus increasing the surface area. However, the basal plane is only slightly functionalized during this oxidation step. This repetitive cycle of oxidation and reduction could potentially be an economically interesting production method for high surface area 2D carbon materials.

In addition to the introduction of porosity via gas-phase oxidation, a wet chemical approach, commonly referred to as chemical activation, has also been studied for GO [204]. The impregnation of GO with KOH and its subsequent heating to 800°C at 400 Torr (Ar atmosphere) resulted in a material with a specific surface area of 3100 m^2/g. Interestingly, the treatment also resulted in a restructuring of the 3D structure of the material. The material exhibited pores from 0.6 to 5 nm, low oxygen and hydrogen content, and 98% sp^2 hybridization, yielding an excellent performance as electrode material for supercapacitor in combination with several different electrolytes.

Depending on future applications of the material, a trade-off between increasing surface functionalization and maintaining of graphene-like properties will become important. This is already the case for the application of oxidized graphene as a TEM support for biological purposes [205]. Although graphene shows promising properties as a TEM support as it is transparent to electrons, it lacks the hydrophilicity necessary for biological molecules. Thus, oxygen functionalization is necessary to introduce a hydrophilic character. Nevertheless, it is important that the graphene substrate retains its electrical conductivity; hence, the degree of oxygen functionalization must be kept sufficiently low. This was realized by a slow reaction in molecular oxygen atmosphere between 200°C and 300°C for 2 h [202,205]. The resulting material showed minimal structural damage but sufficient hydrophilic character to study a tobacco mosaic virus adsorbed on the graphene support.

Oxygen surface functionalization is also necessary for the production of stiff and strong freestanding graphene membranes, also known as graphene paper by flow-directed assembly [206–208]. If nonfunctionalized graphene is used, the membranes become porous and highly brittle. Finally, the oxygen-containing functional groups can also be used as starting point for further, more advanced surface functionalizations.

Similar to graphite, there exist several methods that can be used to lower the oxidation temperature of graphene, including the use of catalysts [209] or employment of more reactive gas-phase species. If, for example, atomic (dissociated) oxygen is used for graphite oxidation, monolayer pit growth can already occur at room temperature [210,211]. At this low temperature, hexagonal monolayer pits are formed. In contrast, circular pits are observed if molecular oxygen is used at a higher temperature [210]. Ozonization of single- and multilayer graphene at room temperature led to a transition from a highly crystalline to a nanocrystalline graphene layer within minutes, and finally to an amorphous material [212]. Raman analysis reveals that the transition from a nanocrystalline to an amorphous material occurs after approximately 7 min of ozone treatment at these temperatures. Figure 13.11 shows the I_D/I_G ratio for different ozonization times. The ratio is a commonly used measure for the degree of crystallinity. A perfect graphene sheet shows no D-band; hence, it has a ratio of zero. The ratio increases later on with decreasing crystallinity. After 6 min, the size of the crystallites L_a derived from the I_D/I_G ratio is ≈25 nm. An explanation for the higher reactivity of single-layer graphene compared to multilayer graphene could be the interaction with oxide substrates, resulting in wrinkling. Thus, curvature is introduced, increasing the resulting sp^3 hybridization character. Trenches accumulating at the bottom of the graphene folds after ozone treatment support this assumption. Similar ozonization times were reported in a study where the change in ozone-treated graphene conductance was used for NO$_2$ gas sensing [213]. The sensor principle is based on the charge transfer from the adsorbed molecules to graphene, and on local electrostatic gating effects [213,214]. The electrical response can be amplified by the addition of surface functional groups, which increase the adsorbate binding energy [215–217]. The introduction of ethers, epoxides, and carbonyls to CVD grown graphene via ozone treatment at room temperature resulted in an eight times higher time response of NO$_2$ than untreated graphene [213]. The sensitivity was decreased from 10 ppm for pristine graphene, to below 200 ppb (theoretically 1.3 ppb) after ozonization. An optimum ozonization time of 70 s was observed, compromising an increase in the density of surface functional groups and a simultaneous decrease in the sheet conductivity [213,218,219].

A similar increase in electrochemical sensitivity to deoxyribonucleic acid (DNA) detection was observed for electrochemically oxidized (anodized) graphene [220]. Oxygen functional groups were

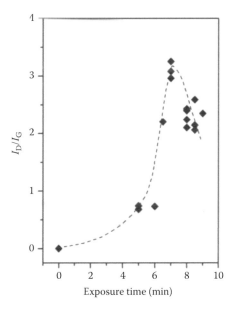

FIGURE 13.11 Time-dependent Raman analysis of graphene ozone treatment. (Data obtained from H. Tao et al., *Journal of Physical Chemistry C* 2011, *115*, 18257.)

introduced using a three-electrode setup in a phosphate buffer solution and applying a potential of 2 V versus Ag/AgCl. The anodized graphene was shown to be superior at detecting mixtures or single solutions of nucleic acids, uric acids, dopamine, and ascorbic acids, compared to boron-doped diamond, CNTs or glassy carbon [220].

Oxidation can also be of interest for unzipping graphene to obtain nanoribbons. These ribbons are half-metals and, the band gap can be tuned by adjusting the ribbon width [221,222]. For exfoliated GO, fault lines and cracks are already observed and may serve as a starting point for unzipping the plane to ribbons. It is has been proposed that the alignment of epoxy groups induces strain and initiates cracks [190]. If, after the initial crack has been formed, a new epoxy group is attached during further oxidation, the unzipping process can continue and lead to the formation of nanoribbons.

In summary, the oxidation of graphene and GO holds great promise for a variety of applications. The ability to control the nature and density of surface functionalities by the oxidation process is a crucial step for controlling material properties between highly conductive graphene and insulating graphene oxide, and is therefore decisive for the success of subsequent applications. Reaction time, temperature, pressure, oxidation agent, and concentration are the main tools used to control the ensuing degree of functionalization.

13.4.3 NANODIAMOND

Nanocrystalline diamond or ultradispersed diamond, often simply referred to as "nanodiamond," is another carbon nanomaterial with promising properties for a widespread range of applications. Although it has received much less attention as compared to other nanostructures such as CNTs or graphene, the interest in this unique material is increasing rapidly [223]. ND can be produced either as thin film using CVD techniques or as powder via detonation of carbon-containing explosives, such as trinitrotoluene (TNT) and hexogen (RDX) in a steel chamber [41]. Although ND films are undoubtedly attractive as biocompatible, smooth, and wear-resistant coatings, it is ND powder that has the potential to achieve truly widespread use on a scale comparable to that of CNTs. In this work, the term "nanodiamond" refers to powders produced by detonation synthesis (i.e., detonation ND). These powders are composed of aggregates of primary particles with an average size of ≈ 5 nm, each consisting of a diamond core partially or completely covered by layers of graphitic and/or amorphous carbon. The current interest in ND stems from the fact that these nanocrystals combine a highly active surface, featuring a variety of chemically reactive moieties, with the favorable properties of macroscopic diamonds. Such properties include an extreme hardness and high Young's modulus, chemical stability, biocompatibility, high thermal conductivity, and electrical resistivity, to name a few. Moreover, ND is one of the few nanomaterials that can be produced in large commercial quantities at a reasonable cost. A simple production process using expired amunition as the energy and carbon source for the detonation results in moderate synthesis costs, leading to numerous applications. Currently, ND is used in composite materials [41], as additive in cooling fluids [224], lubricants [225], and electroplating baths [39]. However, a large number of other promising applications, including drug delivery, stable catalyst support, and transparent coatings for optics, still remain underexplored.

Like most bulk synthesis methods, detonation yields a material that contains not only a variety of carbon nanostructures, but also amorphous carbon and other impurities such as metals and metal oxides [39]. In most cases, less than 30% of the as-produced soot consists of diamond-like (sp^3) carbon, making the purification process a crucial step and often expensive process on the route for a successful ND application [39]. As for other carbon nanostructures, treatments in concentrated acids and other liquid oxidizers have been used to remove nondiamond species from detonation product. These purification techniques are based on different reactivity of ND and nondiamond carbon species toward various oxidizers. Common purification agents include HNO_3 [75,76], H_2SO_4, and mixtures thereof at elevated temperatures and pressures; KOH/KNO_3; Na_2O_2; HNO_3/H_2O_2 under pressure [77]; $HClO_4$; and CCl_4, with HNO_3 being the most common oxidizer in use. In addition, oxidizing agents, such chromic anhydride (CrO_3) or potassium dichromate ($K_2Cr_2O_7$) may be added to further enhance the oxidation power of these liquids [223]. Sushchev et al. purified ND in aqueous solution of HNO_3 at temperatures between 230°C and 240°C under increased pressure (6–10 MPa) for 20–30 min [76]. Their experiments revealed the process of removal of up to 99.5% of nondiamond species. After washing and removal of salts, the material showed a residual ash content of only 0.5%. Purification at milder conditions was found to be unsuitable. Under low-temperature conditions, thermal energy and solubility of nitrogen oxides are not sufficient for breaking the C=C bonds. The oxidation rate increases to a noticeable level when reaching temperatures around 160–170°C. Under these conditions, HNO_3 decomposes and forms free radicals, which then initiate the oxidation reaction. The effect of the increased pressure is twofold. First, depending on the concentration of the acid, the pressure is needed to ensure a liquid-phase process ($\approx$7–8 MPa). Second, the pressure affects the oxidation rate, which increases with increasing pressure. The concentration of the acid should be at least 50–57%. During the reaction, HNO_3 is consumed and the concentration decreases, which reduces the oxidation rate. The initial concentration must therefore be sufficiently high to ensure high oxidation rates throughout the purification process. According to Sushchev et al. [76], the liquid-phase oxidation of ND in HNO_3 contains the following steps:

1. Solvation of ND particles by the H_2O/HNO_3 mixture, which causes swelling and slow homogenization of the mixture.
2. Diffusion of solvent (H_2O) and oxidizer (HNO_3) into ND agglomerates and removal of adsorbed heteroatoms from the ND surface.
3. Thermal decomposition of nitric acid and formation of free radicals capable of oxidizing carbon fragments.
4. Etching of carbon starting at defective sites and interparticle bonds. The total surface area available for oxidation increases and becomes saturated with reaction products.
5. Etch removal of the loose surface, gasification, and removal of surface oxidation products in the liquid-phase volume.
6. Formation of surface oxygen-containing functional groups and surface reconstruction that offsets the excessive free energy of the ND particles.

During the oxidation process, the outer surface of the ND particles changes continuously. To minimize mass transport limitations during the reaction, the mixture should be agitated. Although the high-pressure/high-temperature purification of ND using HNO_3 is economically and environmentally the most promising liquid-phase process, it raises a number technical difficulties and safety concerns. Most importantly, it requires corrosion resistance equipment that can withstand high pressures and temperatures and is subject to a costly waste disposal process. As a consequence, the purification process currently makes up for 40% of the total ND material cost and the purification of 1 kg of detonation soot may require about 35 L of concentrated acid [226]. Finally, oxidation with acids, salts, or oxides often introduces impurities, such as nitrogen- and sulfur-containing compounds, chlorine, or chromium, which need to be removed using additional and subsequent purification steps.

To overcome these challenges, several gas-phase purification processes have been developed. In particular, air oxidation has been reported to be a simple and inexpensive technique for ND purification, as it does not require any toxic substances or additional catalysts [81,227,228]. Depending on the temperature, oxidation in air may be used to remove water and other functional species from the ND surface, to etch amorphous and graphitic carbon impurities [81], and ultimately to reduce the ND crystal size through oxidation of the sp^3 carbon phase [229,230].

Gubarevich et al. purified ND powders using catalyst-assisted oxidation by bubbling air through a water suspension of ND in the presence of catalyst to remove nondiamond carbons [231]. The need for an additional catalyst, which can be expensive and may contaminate the ND sample, is the main disadvantage of this method. Chiganov et al. reported on air oxidation to selectively remove nondiamond carbon from ND at temperatures between 300°C and 550°C [232]. In this process, boric anhydride is used as an inhibitor of diamond oxidation. However, similar to catalyst-assisted oxidation, the process further contaminates the sample and boron is extremely difficult to remove from ND. Osswald et al. used *in situ* Raman spectroscopy to optimize the air purification process and selectively remove graphitic carbons from ND powders, eliminating the need for additional contaminants. The optimal temperature range for oxidation of the ND samples investigated was found as 400–430°C. Depending on the ND sample, 5 h oxidation at 425°C increased the content of sp^3-bonded to 94–96% [81]. Other studies reported higher purification temperatures for the selective removal of nondiamond carbon. Tyurnina et al. recommended oxidation temperatures as high as 550°C, without significant loss of the diamond phase, suggesting potential differences in structure and composition of commercially available ND powders [233]. Alternatively, ozone-enriched air has been proposed for ND purification. Pavlov et al. purified the ND-bearing detonation soot by blowing an ozone–air mixture through the sample at temperatures 150–400°C [234]. A similar process was used by Petrov et al. using slightly lower treatment temperatures (150–200°C), but longer oxidation times [235]. Although these processes require the use of ozone, which is toxic, aggressive, and difficult to handle, they lead to a unique surface chemistry and provide other benefits such as reduced agglomeration [236].

Oxidation in air has also been employed to modify the average crystal size in ND. It was found that small NDs (<10 nm) exhibit higher oxidation rates during thermal treatment as compared to larger crystals present in the samples (the diameter distribution ranges from 3 up to 50 nm), thus shifting the average ND crystal size to higher values [229,237]. Oxidation in air provides a pathway to remove selectively small ND crystals and to effectively narrow the crystal size distribution in detonation ND powders. Mohan et al. demonstrated that the size of selected ND crystals can be reduced by oxidation. Treatments for 2 h at 500°C in air [238] allowed for a reduction of the size of larger ND particles in the size range 20–50 nm. However, it should be mentioned that ND investigated by the authors were milled synthetic type Ib diamond powders with a median size of 35 nm. Gaebel et al. investigated the air oxidation behavior of individual ND crystal using atomic force microscopy (AFM) [230]. Similar to Mohan et al., their studies used microdiamond powder with reduced grain sizes (0–0.1 µm). Etch rates of 1, 5, and 10 nm/h were measured at 500°C, 550°C, and 600°C, respectively. No noticeable size dependence of the oxidation kinetics was found in the size range investigated (>20 nm).

13.4.4 ONION-LIKE CARBON

OLCs, also known as carbon nano-onions (CNO), multishelled fullerenes, or simply carbon onions, were discovered by Ugarte in 1992 when submitting carbon soot to intense electron beam radiation [239]. Newer methods, especially vacuum annealing of ND powder, also allow for OLC production on an industrial scale. Indeed, Kuznetsov et al. reported that the vacuum annealing of NDs above 1200°C also leads to OLC [240]. Thermal annealing in an inert atmosphere (e.g., helium or argon) rather than in a vacuum is also possible and has been demonstrated [241]. More recent synthesis efforts include arc discharge [242] and CVD [243–245]. The onions are built up of several, concentric, sp^2-hybridized carbon shells. Similar to fullerenes, the bonding character shows some sp^3-hybridization due to the bending of the sp^2-plane. Nevertheless, this is less pronounced in OLC as their diameters are larger than that of most fullerenes. The size of OLC depends on the synthesis procedure. For example, onions derived by arc discharge (AD-OLC) are approximately 20 nm in diameter (20–30 shells), whereas the annealing of nanodiamonds (ND-OLC) leads to onions around 5 nm (6–8 shells) [241]. The combination of a high degree of graphitization, a shell-like structure which allows for intercalations, and a small size make OLC an interesting carbon nanomaterial. In contrast to fullerenes, OLC is insoluble in both polar and unpolar solvents and highly electrically conductive. Hence, surface functionalization will most likely be a key step for future applications. Nevertheless, compared to other carbon nanomaterials, such as CNTs or graphene, only a small number of studies focusing on the functionalization of OLCs are found in the literature.

Oxidation and the resulting oxygen surface groups are crucial for subsequent functionalization steps. Palkar et al. presented one of the most complete studies on OLC oxidation and studied the difference in reactivity between AD-OLC and ND-OLC [241]. The largest difference between the two materials is the resulting onion size, which is larger in AD-OLC, as discussed above. Thus, the relative available outer surface area of AD-OLC is smaller than that of ND-OLCs. It is argued that arc discharge-produced onions display a higher degree of graphitization. Raman analysis showed a more pronounced G-band and lower I_D/I_G ration for AD-OLC; however, the size difference between the two onion types makes a direct comparison difficult. Radicals were probed by electron paramagnetic resonance and were only observed in ND-OLCs. Thus, ND-OLCs seem to be more defective than AD-OLCs. The reactivity toward molecular oxygen was studied by TGA in air. The results are redrawn, to provide more clarity, in Figure 13.12 (upper graph). Interestingly, the larger and less defective AD-OLC started to react at approximately 500°C, whereas the ND-OLC began to oxidize at 700°C. The authors also reported that annealing AD-OLC at 2300°C for 1 h increases the upper stability limit up to 700°C. Thus, for AD-OLC, which is quenched rapidly in the water bath after synthesis, the thermodynamical optimum was probably not reached because of kinetic restrictions. This is different for ND-OLC, which, in contrast, was cooled down slowly from the annealing temperature. Further details, such as a characterization of the annealed AD-OLC, have not been published in the literature as yet.

Palkar et al. also studied the wet chemical oxidation of both kinds of OLC with nitric acid to induce carboxylic groups [241]. Onions prepared by both routes were oxidized by heating for 48 h in reflux in (i) nitric acid (3 M), and (ii) a 1:1 mixture of concentrated nitric and sulfuric acid. Raman and TGA analysis in air, showing the desorption of carboxylic groups below 400°C and before air oxidation of the carbon, were used to confirm the functionalization after the acid treatment. The TGA curves are redrawn in Figure 13.12 (lower graph). Surprisingly, the trends displayed by oxidation resistance were the exact opposite of those demonstrated by the air oxidation. AD-OLC did not

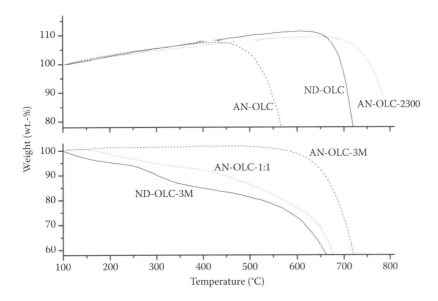

FIGURE 13.12 TGA of as is AN-OLC and ND-OLC, arc discharge onions annealed at 2300°C (AN-OLC-2300), onions heated in reflux for 48 h in 3 M nitric acid (AN-OLC-3M, ND-OLC-3M) or 1:1 mixture of concentrated sulfuric and nitric acid (AN-OLC-1:1). (Data obtained from A. Palkar et al., *Chemistry —An Asian Journal* 2007, 2, 625.)

show functionalization (no mass loss up to 400°C) after the weaker treatment in nitric acid (3 M), whereas ND-OLC showed a mass loss of approximately 15 wt.% up to 400°C, typical for carboxylic groups. Under harsher conditions (a 1:1 mixture of concentrated H_2SO_4 and HNO_3), AD-OLC was oxidized completely and no residue remained. Additionally, surface functionalization was observed for AD-OLC (with an approximately 8 wt.% mass loss at 400°C). The difference in the air and wet chemical oxidations could be explained by the different reaction rates, and thus a transition from the kinetic to the thermodynamic regime. Owing to the high-temperature difference, the wet chemical oxidation should be much slower than that of the air oxidation. If the air oxidation reaction approaches the conditions required for spontaneity, the thermodynamic stability and not the kinetic reactivity is the dominant factor. The slower wet chemical oxidation is assumed to take place in a kinetic region, before reaching the thermodynamic equilibrium, and thus probes the kinetic reactivity. Therefore, it could be possible that the smaller ND-OLC is more reactive (i) because of higher surface area and (ii) because of more defects. The ND-OLC could be thermodynamically more stable because of the long cooling period after annealing.

Another detailed study on ND-OLC oxidation is presented by Plonska-Brzezinska et al. using ozone as the oxidation agent [246]. The onions were ozonized after suspension/being suspended in tetrachloromethane for 10 h at room temperature with a 9 wt.% ozone in air mixture. High solubility of the onions in polar solvents was observed after ozone oxidation. Raman analysis gives an I_D/I_G ratio of 1.14 after ozonization. Compared to reference data for ND-OLC [247], the degree of graphitization is rather high. Thus, it seems that the graphitic structure is retained during ozone treatment. Boehm titration revealed that no carboxylic acids, but rather a large number of lactones (3.7 g/kg) and hydroxylic groups (0.7 g/kg), were introduced by the treatment. In comparison, an ND-OLC oxidized using 3 M nitric acid (oxidation parameters not further specified) shows carboxyl groups (0.7 g/kg), fewer lactones (1.1 g/kg) but no hydroxyl groups. The ozonides on the surface allow for further tuning of the surface oxygen functionalization as they can be reduced or oxidized. Reductive cleavage with dimethyl sulfide should lead to aldehydes and cleavage with sodium bromide leads to alcohol. An oxidative cleavage with hydrogen peroxide and sodium hydroxide should result in carboxylic acids [246].

In most of the small number of studies currently available on OLC functionalization, oxidizing acids are used in the initial reaction stage to introduce carboxyl groups to the surface. The ensuing surface functionalization and the change in material properties after oxidizing acid treatment have yet not received much attention. Heating the substrate in a reflux of 3 M nitric acid for 48 h is the main method used to introduce carboxylic groups [248,249], but introduction via a short treatment of 10 min duration in a 3:1 mixture of concentrated sulfuric and nitric acid (temperature not given) has also been reported [250,251]. It seems that wet chemical oxidation conditions have not yet been optimized. Detailed studies on acid oxidation strength, acid concentrations, and mixture and reaction temperatures with a quantitative and qualitative analysis (in situ or subsequent) of the surface functionalization and change in material properties are necessary. Based on this, the first step could be optimized regarding the time and energy necessary, the amount of surface groups introduced, and the type of surface groups required. One of the main reactions used for further functionalization is the reaction of surface carboxyl groups with terminating amine groups. Solubility of onions in nonpolar solvents could be achieved, for instance, through the direct amination of oxidized AD-OLC with octadecylamin, whereas PEGylation with diamine-terminated PEG1500-NH2 could be carried out to increase the water solubility [249]. Palkar et al. reported that the amidation of oxidized AD-OLC with 4-aminopyridine as functionalization agent resulted in a water solubility of 2 g/L [250]. Raman analysis showed a slight increase in the I_D/I_G ratio for the final functionalized material, which could be because of the oxidizing acid treatment. The amount of aminopyridin grafted to the surface was estimated by TGA analysis to be one functional group per 120 carbon atoms. Considering the size of AD-OLC and the number of carbon atoms at the surface compared to the bulk, this value seems to be overestimated. It was shown by Luszczyn et al. that oxidized ND-OLC can also be used to construct self-assembled onion monolayers by the amination reaction [251]. Therefore, amine group-terminated alkanethiol was generated on a gold surface. After the oxidized onions are brought in contact with the gold surface, the carboxylic groups react with the amine, and the onion is thereby covalently grafted by an amide. Because the OLC shows further unreacted carboxylic groups, they can be employed for further functionalization of the grafted onion. This was shown for Biotin (vitamin H), which also reacts to form an amide bond with the onion and can subsequently act as a receptor for the protein avidin, while allowing the protein to retain its biological activity. Biosensors are an important application of these self-assembled monolayers [251]. Wet chemical oxidation is also the starting point for coating ND-OLC by the polymerization of anilin with polyanilin (PANI), which is a conducting polymer. Dense layers of approximately 1 to 4 nm of PANI were wrapped around the onions and resulted in solubility in protic solvents. The resulting coating could potentially be of interest to biomedical and electrochemical applications [248].

Butenko et al. showed that defects introduced by oxidation can result in better accessibility of the interplanar graphitic space OLCs [252]. ND-OLC was oxidized in carbon dioxide at 750°C for 1 h, cooled in nitrogen gas, and then analyzed for potassium intercalation. After the nitrogen cooling step, no oxygen was measured by XPS analysis. The XPS analysis results also showed no sp^3 carbon. It is assumed that surface oxygen desorbs and the carbon rearranges to sp^2 during slow cooling in the presence of nitrogen. Nevertheless, if only the Boudouard reaction is taking place, it is also possible that no oxygen sticks to the surface during CO_2 oxidation. The potassium intercalation was followed by XPS. In the oxidized form of OLC, potassium could be intercalated, something which was not observed for the untreated sample. Hence, it seems that the oxidation opened up one or more outer shells which made pathways available for intercalation [252].

13.5   CONCLUSION

Owing to its versatility, ease of use, and scalability, oxidation has become one of the most frequently used modification methods applied to carbon nanostructures. Known for the purpose of activation and surface functionalization of carbonaceous bulk materials, such as graphite, liquid- and gas-phase oxidation offers even greater possibilities at the nanoscale. Oxidation plays a crucial role in carbon nanomaterial purification as it utilizes the variations in the reactivity of different nanostructures and carbonaceous impurities toward oxidation. Carbon nanomaterials exhibit exceptionally high surface-to-volume ratio, and surface modifications, such as oxidation, were found to have high impact on bulk properties. For example, on proper surface functionalization, graphene changes from a semiconductor with zero band gap to an isolator. Moreover, for covalent grafting of carbon nanomaterial to matrixes (e.g., for polymer reinforcement) or complex surface functionalization (e.g., binding of protein receptors), oxidation is commonly the first functionalization step, and influences all subsequent treatments. Mechanisms known from bulk carbon oxidation can be transferred partially to the nanoscale, but need to be adapted to account for higher structural complexity and higher reactivity. To be able to fully exploit the potential of nanostructure oxidation, more advanced quantitative analytic tools and *in situ* characterization methods are required. To date, differences in reactivity reported on oxidation of carbon nanostructures arise mainly from variations in the starting material. Therefore, structure and composition of the respective nanomaterials must be known to allow for proper data interpretation and enable a comprehensive study of the oxidation behavior. This is critical to both understanding the fundamentals of the oxidation process in the laboratory and to utilizing oxidation on industrial scale, where the oxidation process may need to be adapted if the raw material properties change. Because of its universal use and the growing number of applications, it is believed that oxidation will remain the main process for the purification and modification of carbon nanomaterials. Further breakthroughs in understanding the fundamental mechanisms that take place during oxidation will allow us to provide carbon nanomaterials of high purity and well-controlled composition, and thus pave the way for a widespread application of these unique nanostructures in our daily lives.

REFERENCES

1. P. T. Williams, A. R. Reed, *Biomass & Bioenergy* **2006**, *30*, 144.
2. F. Rodríguez-Reinoso, M. Molina-Sabio, M. T. González, *Carbon* **1995**, *33*, 15.
3. F. Rodriguez-Reinoso, M. Molina-Sabio, *Carbon* **1992**, *30*, 1111.
4. J. Pastor-Villegas, C. J. Duran-Valle, *Carbon* **2002**, *40*, 397.
5. M. V. Navarro, N. A. Seaton, A. M. Mastral, R. Murillo, *Carbon* **2006**, *44*, 2281.
6. M. J. Munoz-Guillena, M. J. Illan-Gomez, J. M. Martin-Martinez, A. Linares-Solano, C. Salinas-Martinez de Lecea, *Energy Fuels* **1992**, *6*, 9.
7. Z. H. Huang, F. Y. Kang, J. B. Yang, K. M. Liang, R. W. Fu, A. P. Huang, *Journal of Materials Science Letters* **2002**, *22*, 293.
8. H. Ulbricht, G. Moos, T. Hertel, *Surface Science* **2003**, *532*, 852.
9. P. Giannozzi, R. Car, G. Scoles, *The Journal of Chemical Physics* **2003**, *118*, 1003.
10. Z. Klusek, Z. Waqar, P. K. Datta, W. Kozlowski, *Corrosion Science* **2004**, *46*, 1831.
11. S. M. Lee, Y. H. Lee, Y. G. Hwang, J. R. Hahn, H. Kang, *Physical Review Letters* **1999**, *82*, 217.
12. H. P. Chang, A. J. Bard, *Journal of the American Chemical Society* **1991**, *113*, 5588.
13. P. Z. Gao, H. J. Wang, Z. H. Jin, *Thermochimica Acta* **2004**, *414*, 59.
14. A. I. Demidov, I. A. Markelov, *Russian Journal of Applied Chemistry* **2005**, *78*, 707.
15. K. S. Goto, K. H. Han, G. R. Saintpierre, *Materials Science and Engineering* **1987**, *88*, 347.
16. R. Brukh, S. Mitra, *Journal of Materials Chemistry* **2007**, *17*, 619.
17. Y. Yin, J. G. P. Binner, T. E. Cross, S. J. Marshall, *Journal of Materials Science* **1994**, *29*, 2250.
18. E. Illekova, K. Csomorova, *Journal of Thermal Analysis and Calorimetry* **2005**, *80*, 103.
19. A. B. Phadnis, V. V. Deshpande, *Thermochimica Acta* **1983**, *62*, 361.
20. Y. Yang, F. He, M. Wang, *Tansu* **1998**, *1*, 2.
21. H. W. Chang, S. K. Rhee, *Carbon* **1978**, *16*, 17.
22. G. R. Hennig, *Science* **1965**, *147*, 733.
23. E. L. Evans, R. J. M. Griffiths, J. M. Thomas, *Science* **1971**, *171*, 174.
24. R. T. Yang, C. Wong, *Science* **1981**, *214*, 437.
25. R. Schlogl, G. Loose, M. Wesemann, *Solid State Ionics* **1990**, *43*, 183.
26. I. A. Valuev, G. E. Norman, B. R. Shub, *Russian Journal of Physical Chemistry B* **2011**, *5*, 156.
27. S. Osswald, E. Flahaut, Y. Gogotsi, *Chem. Mater.* **2006**, *18*, 1525.
28. S. Osswald, M. Havel, Y. Gogotsi, *Journal of Raman Spectroscopy* **2007**, *38*, 728.
29. D. W. McKee, in *Chemistry and Physics of Carbon*, Vol. 16, edited by P. L. Walker, Jr. and P. A. Thrower (Marcel Dekker, New York, **1981**).
30. M. A. Vannice, *Kinetics of Catalytic Reactions,* Springer, New York, **2005**.
31. D. W. McKee, *Carbon* **1970**, *8*, 623.
32. Y. Kobayashi, M. Sano, *Chemcatchem* **2010**, *2*, 397.
33. L. J. Ci, Z. L. Rao, Z. P. Zhou, D. S. Tang, Y. Q. Yan, Y. X. Liang, D. F. Liu et al., *Chemical Physics Letters* **2002**, *359*, 63.
34. E. Couteau, K. Hernadi, J. W. Seo, L. Thien-Nga, C. Miko, R. Gaal, L. Forro, *Chem. Phys. Lett.* **2003**, *378*, 9.
35. M. Daenen, R. D. de Fouw, B. Hamers, P. G. A. Janssen, K. Schouteden, M. A. J. Veld, *The Wondrous World of Carbon Nanotubes. A Review of Current Carbon Nanotube Technologies* Eindhoven University of Technology, **2003**.
36. E. Flahaut, R. Bacsa, A. Peigney, C. Laurent, *Chemical Communications* **2003**, *12*, 1442.
37. Y. Ando, X. L. Zhao, *New Diamond and Frontier Carbon Technology* **2006**, *16*, 123.
38. M. Kusaba, Y. Tsunawaki, *Thin Solid Films* **2006**, *506*, 255.
39. V. Y. Dolmatov, *Russian Chemical Reviews* **2001**, *70*, 607.
40. P.-X. Hou, C. Liu, H.-M. Cheng, *Carbon* **2008**, *46*, 2003.
41. V. Y. Dolmatov, *Ultradisperse Diamonds of Detonation Synthesis: Production, Properties and Applications,* State Polytechnical University, St. Petersburg, **2003**.
42. D. M. Gruen, O. A. Shenderova, A. Y. Vul, in *NATO Science series. Series II: Mathematics, Physics and Chemistry, Vol. 192*, Springer, Dordrecht, **2005**, p. 401.
43. Z. Spitalsky, C. Aggelopoulos, G. Tsoukleri, C. Tsakiroglou, J. Parthenios, S. Georga, C. Krontiras, D. Tasis, K. Papagelis, C. Galiotis, *Materials Science and Engineering B-Advanced Functional Solid-State Materials* **2009**, *165*, 135.
44. V. Datsyuk, M. Kalyva, K. Papagelis, J. Parthenios, D. Tasis, A. Siokou, I. Kallitsis, C. Galiotis, *Carbon* **2008**, *46*, 833.

45. I. D. Rosca, F. Watari, M. Uo, T. Akaska, *Carbon* **2005**, *43*, 3124.
46. T. Kyotani, S. Nakazaki, W. H. Xu, A. Tomita, *Carbon* **2001**, *39*, 782.
47. H. Tantang, J. Y. Ong, C. L. Loh, X. C. Dong, P. Chen, Y. Chen, X. Hu, L. P. Tan, L. J. Li, *Carbon* **2009**, *47*, 1867.
48. E. Menna, F. Della Negra, M. Dalla Fontana, M. Meneghetti, *Physical Review B* **2003**, *68*, 193412.
49. R. R. N. Marques, B. F. Machado, J. L. Faria, A. M. T. Silva, *Carbon* **2011**, *48*, 1515.
50. L. Wang, L. Ge, T. E. Rufford, J. L. Chen, W. Zhou, Z. H. Zhu, V. Rudolph, *Carbon* **2011**, *49*, 2022.
51. L. Stobinski, B. Lesiak, L. Kover, J. Toth, S. Biniak, G. Trykowski, J. Judek, *Journal of Alloys and Compounds* **2010**, *501*, 77.
52. Y. Li, X. B. Zhang, J. H. Luo, W. Z. Huang, J. P. Cheng, Z. Q. Luo, T. Li, F. Liu, G. L. Xu, X. X. Ke, L. Li, H. J. Geise, *Nanotechnology* **2004**, *15*, 1645.
53. M. Lebron-Colon, M. A. Meador, D. Lukco, F. Sola, J. Santos-Perez, L. S. McCorkle, *Nanotechnology* **2011**, *22*, 455707.
54. M. Pavese, S. Musso, S. Bianco, M. Giorcelli, N. Pugno, *Journal of Physics-Condensed Matter* **2008**, *20(47)*, 474206.
55. B. Scheibe, E. Borowiak-Palen, R. J. Kalenczuk, *Materials Characterization* **2010**, *61*, 185.
56. J. M. Yuan, X. H. Chen, Z. F. Fan, X. G. Yang, Z. H. Chen, *Carbon* **2008**, *46*, 1266.
57. J. Xie, M. N. Ahmad, H. Bai, H. Li, W. Yang, *Science China-Chemistry* **2010**, *53*, 2026.
58. M. H. Li, M. Boggs, T. P. Beebe, C. P. Huang, *Carbon* **2008**, *46*, 466.
59. H. Naeimi, A. Mohajeri, L. Moradi, A. M. Rashidi, *Applied Surface Science* **2009**, *256*, 631.
60. F. Simon, A. Kukovecz, H. Kuzmany, in *Molecular Nanostructures* **2003**, *685*, 185.
61. M. Yudasaka, M. Zhang, S. Iijima, *Chemical Physics Letters* **2003**, *374*, 132.
62. Y. Miyata, Y. Maniwa, H. Kataura, *Journal of Physical Chemistry B* **2006**, *110*, 25.
63. S. Gajewski, H. E. Maneck, U. Knoll, D. Neubert, I. Dorfel, R. Mach, B. Strauss, J. F. Friedrich, *Diamond and Related Materials* **2003**, *12*, 816.
64. R. Sen, S. M. Rickard, M. E. Itkis, R. C. Haddon, *Chemistry of Materials* **2003**, *15*, 4273.
65. A. Anson-Casaos, M. Gonzalez, J. M. Gonzalez-Dominguez, M. Teresa Martinez, *Langmuir* **2011**, *27*, 7192.
66. C. S. Li, D. Z. Wang, T. X. Liang, X. F. Wang, J. J. Wu, X. Q. Hu, J. Liang, *Powder Technology* **2004**, *142*, 175.
67. Y. S. Park, Y. C. Choi, K. S. Kim, D. C. Chung, D. J. Bae, K. H. An, S. C. Lim, X. Y. Zhu, Y. H. Lee, *Carbon* **2001**, *39*, 655.
68. M. R. Smith, S. W. Hedges, R. LaCount, D. Kern, N. Shah, G. P. Huffman, B. Bockrath, *Carbon* **2003**, *41*, 1221.
69. B. Yu, P. X. Hou, F. Li, B. L. Liu, C. Liu, H. M. Cheng, *Carbon*, *48*, 2941.
70. W. Zhou, Y. H. Ooi, R. Russo, P. Papanek, D. E. Luzzi, J. E. Fischer, M. J. Bronikowski, P. A. Willis, R. E. Smalley, *Chemical Physics Letters* **2001**, *350*, 6.
71. S. Bandow, M. Takizawa, K. Hirahara, M. Yudasaka, S. Iijima, *Chemical Physics Letters* **2001**, *337*, 48.
72. E. Borowiak-Palen, T. Pichler, X. Liu, M. Knupfer, A. Graff, O. Jost, W. Pompe, R. J. Kalenczuk, J. Fink, *Chemical Physics Letters* **2002**, *363*, 567.
73. W. Z. Li, J. G. Wen, M. Sennett, Z. F. Ren, *Chemical Physics Letters* **2003**, *368*, 299.
74. A. Suri, K. S. Coleman, *Carbon* **2011**, *49*, 3031.
75. G. Post, V. Y. Dolmatov, V. A. Marchukov, V. G. Sushchev, M. V. Veretennikova, A. E. Sal'ko, *Russian Journal of Applied Chemistry* **2002**, *75*, 755.
76. V. G. Sushchev, V. Y. Dolmatov, V. A. Marchukov, M. V. Veretennikova, *Journal of Superhard Materials* **2008**, *30*, 297.
77. T. M. Gubarevich, V. F. Pyaterikov, I. S. Larionova, V. Y. Dolmatov, R. R. Sataev, A. V. Tyshetskaya, L. I. Poleva, *Journal of Applied Chemistry of the USSR* **1992**, *65*, 2075.
78. T. M. Gubarevich, R. R. Sataev, V. Y. Dolmatov, in *5th All-Union Meeting on Detonation, Vol. 1*, Krasnoyarsk USSR, **1991**, pp. 135.
79. A. S. Chiganov, *Physics of the Solid State* **2004**, *46*, 595.
80. E. V. Pavlov, Y. A. Skryabin, *Method for Removal of Impurity of Non-Diamond Carbon and Device for Its Realization*, **1994**: Russia.RU2019502
81. S. Osswald, G. Yushin, V. Mochalin, S. Kucheyev, Y. Gogotsi, *Journal of the American Chemical Society* **2006**, 128, 11635.
82. F. Cataldo, A. P. Koscheev, *Fullerenes Nanotubes and Carbon Nanostructures* **2003**, 11, 201.
83. H. Marsh, F. Rodriguez-Reinoso, *Activated Carbon*, Elsevier, Oxford, **2006**.
84. R. Setton, P. Bernier, S. Lefrant, *Carbon Molecules and Materials*, Taylor & Francis, London, **2002**.

85. C. K. Lim, *Advances in Chromatography* **1992**, 32, 1.
86. J. Lee, J. Kim, T. Hyeon, *Advanced Materials* **2006**, 18, 2073.
87. G. Yushin, A. Nikitin, Y. Gogotsi, in *Nanomaterials Handbook* (Ed.: Y. Gogotsi), CRC Taylor and Francis Group, Boca Raton, London, **2006**, pp. 239.
88. A. Linares-Solano, D. Lozano-Castelló, M. A. Lillo-Ródenas, D. Cazorla-Amorós, in *Chemistry and Physics of Carbon* Vol. 30, edited by R. L. Radovic (CRC Press, **2008**, p. 1).
89. S. Villar-Rodil, F. Suarez-Garcia, J. I. Paredes, A. Martinez-Alonso, J. M. D. Tascon, *Chemistry of Materials* **2005**, *17*, 5893.
90. D. Mowla, D. D. Do, K. Kaneko, in *Chemistry and Physics of Carbon, Vol. 28*, **2003**, 229.
91. O. Ioannidou, A. Zabaniotou, *Renewable & Sustainable Energy Reviews* **2007**, *11*, 1966.
92. A. Linares-Solano, D. Lozano-Castelló, M. A. Lillo-Ródenas, D. Cazorla-Amorós, in *Chemistry and Physics of Carbon*, Vol. 30, edited by R. L. Radovic (CRC Press, **2008**, p. 1).
93. H. Marsh, F. R. Reinoso, *Activated Carbon, Vol. 1*, Elsevier Science, Oxford, **2006**.
94. M. J. Illan-Gomez, A. Garcia-Garcia, C. Salinas-Martinez de Lecea, A. Linares-Solano, *Energy Fuels* **1996**, *10*, 1108.
95. M. A. Lillo-Rodenas, J. P. Marco-Lozar, D. Cazorla-Amoros, A. Linares-Solano, *Journal of Analytical and Applied Pyrolysis* **2007**, *80*, 166.
96. Y.-R. Nian, H. Teng, *Journal of The Electrochemical Society* **2002**, *149*, A1008.
97. A. Ahmadpour, D. D. Do, *Carbon* **1996**, *34*, 471.
98. J. A. Maciá-Agulló, B. C. Moore, D. Cazorla-Amorós, A. Linares-Solano, *Carbon* **2004**, *42*, 1367.
99. E. N. Hoffman, G. Yushin, T. El-Raghy, Y. Gogotsi, M. W. Barsoum, *Microporous and Mesoporous Materials* **2008**, *112*, 526.
100. G. Q. Lu, D. D. Do, *Carbon* **1992**, *30*, 21.
101. E. A. Dawson, G. M. B. Parkes, P. A. Barnes, M. J. Chinn, *Carbon* **2003**, *41*, 571.
102. S. Osswald, C. Portet, Y. Gogotsi, G. Laudisio, J. P. Singer, J. E. Fischer, V. V. Sokolov, J. A. Kukushkina, A. E. Kravchik, *Journal of Solid State Chemistry* **2009**, *182*, 1733.
103. Y. Gogotsi, C. Portet, S. Osswald, J. M. Simmons, T. Yidirim, G. Laudisio, J. E. Fischer, *International Journal of Hydrogen Energy* **2009**, *34*, 6314.
104. S. H. Yeon, S. Osswald, Y. Gogotsi, J. P. Singer, J. M. Simmons, J. E. Fischer, M. A. Lillo-Rodenas, A. Linares-Solanod, *Journal of Power Sources* **2009**, *191*, 560.
105. V. Presser, J. McDonough, S. H. Yeon, Y. Gogotsi, *Energy and Environmental Science* **2011**, *4*, 3059.
106. G. Yushin, R. Dash, J. Jagiello, J. E. Fischer, Y. Gogotsi, *Advanced Functional Materials* **2006**, *16*, 2288.
107. C. Largeot, C. Portet, J. Chmiola, P.-L. Taberna, Y. Gogotsi, P. Simon, *Journal of the American Chemical Society* **2008**, *130*, 2730.
108. J. Chmiola, G. Yushin, Y. Gogotsi, C. Portet, P. Simon, P. L. Taberna, *Science* **2006**, *313*, 1760.
109. L. L. Hegedus, *Industrial & Engineering Chemistry Product Research and Development* **1980**, *19*, 533.
110. M. O. Coppens, S. Gheorghiu, P. Pfeifer, *Studies in Surface Science and Catalysis* **2005**, *156*, 371.
111. S. Gheorghiu, M. O. Coppens, *AIChE Journal* **2004**, *50*, 812.
112. E. Johannessen, G. Wang, M. O. Coppens, *Industrial and Engineering Chemistry Research* **2007**, *46*, 4245.
113. G. Wang, M. O. Coppens, *Industrial and Engineering Chemistry Research* **2008**, *47*, 3847.
114. G. Wang, E. Johannessen, C. R. Kleijn, S. W. de Leeuw, M. O. Coppens, *Chemical Engineering Science* **2007**, *62*, 5110.
115. N. Hansen, T. Kerber, J. Sauer, A. T. Bell, F. J. Keil, *Journal of the American Chemical Society* **2010**, *132*, 11525.
116. J. A. Swisher, N. Hansen, T. Maesen, F. J. Keil, B. Smit, A. T. Bell, *Journal of Physical Chemistry C* **2010**, *114*, 10229.
117. F. J. Keil, *Chemie Ingenieur Technik* **2010**, *82*, 881.
118. F. J. Keil, C. Rieckmann, *Hungarian Journal of Industrial Chemistry* **1993**, *21*, 277.
119. T. Knorr, M. Kaiser, F. Glenk, B. J. M. Etzold, *Chemical Engineering Science* **2012**, *69*, 492.
120. V. Presser, M. Heon, Y. Gogotsi, *Advanced Functional Materials* **2011**.
121. M. Schmirler, F. Glenk, B. J. M. Etzold, *Carbon* **2011**, *49*, 3679.
122. Z. Liu, S. Tabakman, K. Welsher, H. Dai, *Nano Research* **2009**, *2*, 85.
123. V. N. Mochalin, I. Neitzel, B. J. M. Etzold, A. Peterson, G. Palmese, Y. Gogotsi, *ACS Nano* **2011**, *5*, 7494.
124. P. Krawiec, E. Kockrick, L. Borchardt, D. Geiger, A. Corma, S. Kaskel, *The Journal of Physical Chemistry C* **2009**, *113*, 7755.
125. K. P. Loh, Q. Bao, P. K. Ang, J. Yang, *Journal of Materials Chemistry* **2010**, *20*, 2277.
126. V. N. Mochalin, O. Shenderova, D. Ho, Y. Gogotsi, *Nature Nanotechnology* **2012**, *7*, 11.

127. D. Su, N. I. Maksimova, G. Mestl, V. L. Kuznetsov, V. Keller, R. Schlögl, N. Keller, *Carbon* **2007**, *45*, 2145.
128. G. Mestl, N. I. Maksimova, N. Keller, V. V. Roddatis, R. Schlögl, *Angewandte Chemie International Edition* **2001**, *40*, 2066.
129. J. L. Figueiredo, M. F. R. Pereira, M. M. A. Freitas, J. J. M. Orfao, *Carbon* **1999**, *37*, 1379.
130. V. Z. Radkevich, T. L. Senko, K. Wilson, L. M. Grishenko, A. N. Zaderko, V. Y. Diyuk, *Applied Catalysis A: General* **2008**, *335*, 241.
131. T. G. Ros, A. J. Van Dillen, J. W. Geus, D. C. Koningsberger, *Chemistry—A European Journal* **2002**, *8*, 1151.
132. U. Zielke, K. J. Hüttinger, W. P. Hoffman, *Carbon* **1996**, *34*, 983.
133. H. P. Boehm, *Carbon* **2002**, *40*, 145.
134. A. Jorio, R. Saito, G. Dresselhaus, M. S. Dresselhaus, *Philosophical Transactions of the Royal Society of London. Series A. Mathematical, Physical and Engineering Sciences* **2004**, *362*, 2311.
135. H. J. Dai, *Accounts of Chemical Research* **2002**, *35*, 1035.
136. M. Terrones, *Annual Review of Materials Research* **2003**, *33*, 419.
137. P. M. Ajayan, T. W. Ebbesen, T. Ichihashi, S. Iijima, K. Tanigaki, H. Hiura, *Nature* **1993**, *362*, 522.
138. T. W. Ebbesen, P. M. Ajayan, *Nature* **1992**, *358*, 220.
139. S. C. Tsang, P. J. F. Harris, M. L. H. Green, *Nature* **1993**, *362*, 520.
140. N. Yao, V. Lordi, S. X. C. Ma, E. Dujardin, A. Krishnan, M. M. J. Treacy, T. W. Ebbesen, *Journal of Materials Research* **1998**, *13*, 2432.
141. T. Savage, S. Bhattacharya, B. Sadanadan, J. Gaillard, T. M. Tritt, Y. P. Sun, Y. Wu, S. Nayak, R. Car, N. Marzari, P. M. Ajayan, A. M. Rao, *Journal of Physics-Condensed Matter* **2003**, *15*, 5915.
142. M. Grujicic, G. Cao, A. M. Rao, T. M. Tritt, S. Nayak, *Applied Surface Science* **2003**, *214*, 289.
143. Y. F. Guo, L. Jiang, W. L. Guo, *Physical Review B*, **2010**, *82, 115440.
144. A. K. Singh, X. M. Hou, K. C. Chou, *Corrosion Science* **2010**, *52*, 1771.
145. J. G. Coroneus, B. R. Goldsmith, J. A. Lamboy, A. A. Kane, P. G. Collins, G. A. Weiss, *Chemphyschem* **2008**, *9*, 1053.
146. W.-L. Yim, J. K. Johnson, *Journal of Physical Chemistry C* **2009**, *113*, 17636.
147. A. M. da Silva, Jr., G. M. A. Junqueira, C. P. A. Anconi, H. F. Dos Santos, *Journal of Physical Chemistry C* **2009**, *113*, 10079.
148. Y. F. Zhang, Z. F. Liu, *Journal of Physical Chemistry B* **2004**, *108*, 11435.
149. Y. Kanai, V. R. Khalap, P. G. Collins, J. C. Grossman, *Physical Review Letters* **2010**, *104*, 66401.
150. P. G. Collins, K. Bradley, M. Ishigami, A. Zettl, *Science* **2000**, *287*, 1801.
151. S. C. Kung, K. C. Hwang, I. N. Lin, *Applied Physics Letters* **2002**, *80*, 4819.
152. E. Flahaut, A. Peigney, C. Laurent, A. Rousset, *Journal of Materials Chemistry* **2000**, *10*, 249.
153. J. M. Moon, K. H. An, Y. H. Lee, Y. S. Park, D. J. Bae, G. S. Park, *Journal of Physical Chemistry B* **2001**, *105*, 5677.
154. I. W. Chiang, B. E. Brinson, A. Y. Huang, P. A. Willis, M. J. Bronikowski, J. L. Margrave, R. E. Smalley, R. H. Hauge, *Journal of Physical Chemistry B* **2001**, *105*, 8297.
155. I. W. Chiang, B. E. Brinson, R. E. Smalley, J. L. Margrave, R. H. Hauge, *Journal of Physical Chemistry B* **2001**, *105*, 1157.
156. G. Tobias, L. D. Shao, C. G. Salzmann, Y. Huh, M. L. H. Green, *Journal of Physical Chemistry B* **2006**, *110*, 22318.
157. C. M. Yang, K. Kaneko, M. Yudasaka, S. Iijima, *Nano Letters* **2002**, *2*, 385.
158. N. Dementev, S. Osswald, Y. Gogotsi, E. Borguet, *Journal of Materials Chemistry* **2009**, *19*, 7904.
159. S. Osswald, E. Flahaut, H. Ye, Y. Gogotsi, *Chemical Physics Letters* **2005**, *402*, 422.
160. K. Behler, S. Osswald, H. Ye, S. Dimovski, Y. Gogotsi, *Journal of Nanoparticle Research* **2006**, *8*, 615.
161. X. X. Zhang, C. F. Deng, R. Xu, D. Z. Wang, *Journal of Materials Science* **2007**, *42*, 8377.
162. I. V. Zaporotskova, N. G. Lebedev, L. A. Chernozatonskii, *International Journal of Quantum Chemistry* **2004**, *96*, 149.
163. C. C. Wang, G. Zhou, J. Wu, B. L. Gu, W. H. Duan, *Applied Physics Letters* **2006**, *89, 173130.
164. W. Zhao, C. H. Song, P. E. Pehrsson, *Journal of the American Chemical Society* **2002**, *124*, 12418.
165. M. Xu, Z. Sun, Q. Chen, B. K. Tay, *International Journal of Nanotechnology* **2009**, *6*, 735.
166. A. K. Geim, K. S. Novoselov, *Nature Materials* **2007**, *6*, 183.
167. H.-P. Boehm, *Angewandte Chemie International Edition* **2010**, *49*, 9332.
168. B. C. Brodie, *Philosophical Transactions of the Royal Society of London* **1859**, *149*, 249.
169. L. Staudenmaier, *Berichte der Deutschen Chemischen Gesellschaft* **1898**, *31, 1481–1487.
170. W. S. Hummers Jr, R. E. Offeman, *Journal of the American Chemical Society* **1958**, *80*, 1339.

171. M. J. McAllister, J. L. Li, D. H. Adamson, H. C. Schniepp, A. A. Abdala, J. Liu, M. Herrera-Alonso et al., *Chemistry of Materials* **2007**, *19*, 4396.

172. S. Stankovich, R. D. Piner, X. Chen, N. Wu, S. T. Nguyen, R. S. Ruoff, *Journal of Materials Chemistry* **2006**, *16*, 155.

173. X. Sun, Z. Liu, K. Welsher, J. Robinson, A. Goodwin, S. Zaric, H. Dai, *Nano Research* **2008**, *1*, 203.

174. S. Park, R. S. Ruoff, *Nature Nanotechnology* **2009**, *4*, 217.

175. H. C. Schniepp, J.-L. Li, M. J. McAllister, H. Sai, M. Herrera-Alonso, D. H. Adamson, R. K. Prud'homme, R. Car, D. A. Saville, I. A. Aksay, *The Journal of Physical Chemistry B* **2006**, *110*, 8535.

176. H.-P. Boehm, W. Scholz, *Justus Liebigs Annalen der Chemie* **1966**, *691*, 1.

177. H. P. Boehm, W. Scholz, *Zeitschrift für Anorganische und Allgemeine Chemie* **1965**, *335*, 74.

178. U. Hofmann, R. Holst, *Berichte der Deutschen Chemischen Gesellschaft (A and B Series)* **1939**, *72*, 754.

179. G. Ruess, *Monatshefte für Chemie/Chemical Monthly* **1947**, *76*, 381.

180. A. Clause, R. Plass, H. P. Boehm, U. Hofmann, *Zeitschrift für Anorganische und Allgemeine Chemie* **1957**, *291*, 205.

181. M. Mermoux, Y. Chabre, A. Rousseau, *Carbon* **1991**, *29*, 469.

182. T. Nakajima, A. Mabuchi, R. Hagiwara, *Carbon* **1988**, *26*, 357.

183. T. Nakajima, Y. Matsuo, *Carbon* **1994**, *32*, 469.

184. F. Cataldo, *Fullerenes, Nanotubes and Carbon Nanostructures* **2003**, *11*, 1.

185. H. He, T. Riedl, A. Lerf, J. Klinowski, *The Journal of Physical Chemistry* **1996**, *100*, 19954.

186. A. Lerf, H. He, M. Forster, J. Klinowski, *Journal of Physical Chemistry B* **1998**, *102*, 4477.

187. T. Szabo, O. Berkesi, P. Forgo, K. Josepovits, Y. Sanakis, D. Petridis, I. Dekany, *Chemistry of Materials* **2006**, *18*, 2740.

188. W. Cai, R. D. Piner, F. J. Stadermann, S. Park, M. A. Shaibat, Y. Ishii, D. Yang, A. Velamakanni, S. J. An, M. Stoller, J. An, D. Chen, R. S. Ruoff, *Science* **2008**, *321*, 1815.

189. G. Eda, M. Chhowalla, *Advanced Materials* **2010**, *22*, 2392.

190. J. L. Li, K. N. Kudin, M. J. McAllister, R. K. Prud'homme, I. A. Aksay, R. Car, *Physical Review Letters* **2006**, *96*, 176101-1–176101-4.

191. L. J. Cote, F. Kim, J. Huang, *Journal of the American Chemical Society* **2009**, *131*, 1043.

192. D. W. Boukhvalov, M. I. Katsnelson, *Journal of Physics Condensed Matter* **2009**, *21*, 344205-1–344205-12.

193. C. Hontoria-Lucas, A. J. Lopez-Peinado, J. D. D. Lopez-Gonzalez, M. L. Rojas-Cervantes, R. M. Martin-Aranda, *Carbon* **1995**, *33*, 1585.

194. S. Stankovich, D. A. Dikin, G. H. B. Dommett, K. M. Kohlhaas, E. J. Zimney, E. A. Stach, R. D. Piner, S. T. Nguyen, R. S. Ruoff, *Nature* **2006**, *442*, 282.

195. F. Atamny, J. Blöcker, B. Henschke, R. Schlögl, T. Schedel-Niedrig, M. Keil, A. M. Bradshaw, *Journal of Physical Chemistry* **1992**, *96*, 4522.

196. B. H. Chen, J. P. Huang, L. Y. Wang, J. Shiea, T. L. Chen, L. Y. Chiang, *Journal of the Chemical Society—Perkin Transactions* **1998**, *1*, 1171

197. A. Tracz, G. Wegner, J. P. Rabe, *Langmuir* **1993**, *9*, 3033.

198. H. Chang, A. J. Bard, *Journal of the American Chemical Society* **1991**, *113*, 5588.

199. X. Chu, L. D. Schmidt, *Surface Science* **1992**, *268*, 325.

200. J. R. Hahn, *Carbon* **2005**, *43*, 1506.

201. J. R. Hahn, H. Kang, S. M. Lee, Y. H. Lee, *Journal of Physical Chemistry B* **1999**, *103*, 9944.

202. L. Liu, S. Ryu, M. R. Tomasik, E. Stolyarova, N. Jung, M. S. Hybertsen, M. L. Steigerwald, L. E. Brus, G. W. Flynn, *Nano Letters* **2008**, *8*, 1965.

203. T. H. Han, Y. K. Huang, A. T. L. Tan, V. P. Dravid, J. Huang, *Journal of the American Chemical Society* **2011**, *133*, 15264.

204. Y. Zhu, S. Murali, M. D. Stoller, K. J. Ganesh, W. Cai, P. J. Ferreira, A. Pirkle et al., *Science* **2011**, *332*, 1537.

205. R. S. Pantelic, J. W. Suk, Y. Hao, R. S. Ruoff, H. Stahlberg, *Nano Letters* **2011**, *11*, 4319.

206. D. A. Dikin, S. Stankovich, E. J. Zimney, R. D. Piner, G. H. B. Dommett, G. Evmenenko, S. T. Nguyen, R. S. Ruoff, *Nature* **2007**, *448*, 457.

207. T. J. Booth, P. Blake, R. R. Nair, D. Jiang, E. W. Hill, U. Bangert, A. Bleloch, M. Gass, K. S. Novoselov, M. I. Katsnelson, A. K. Geim, *Nano Letters* **2008**, *8*, 2442.

208. A. K. Geim, *Science* **2009**, *324*, 1530.

209. D. W. McKee, *Carbon* **1970**, *8*, 623.

210. C. Wong, R. T. Yang, B. L. Halpern, *The Journal of Chemical Physics* **1983**, *78*, 3325.

211. P. Solís-Fernández, J. I. Paredes, A. Cosío, A. Martínez-Alonso, J. M. D. Tascón, *Journal of Colloid and Interface Science* **2010**, *344*, 451.

212. H. Tao, J. Moser, F. Alzina, Q. Wang, C. M. Sotomayor-Torres, *Journal of Physical Chemistry C* **2011**, *115*, 18257.
213. M. G. Chung, D. H. Kim, H. M. Lee, T. Kim, J. H. Choi, D. K. Seo, J. B. Yoo, S. H. Hong, T. J. Kang, Y. H. Kim, *Sensors and Actuators, B: Chemical* **2012**, 166–167.
214. S. Capone, A. Forleo, L. Francioso, R. Rella, P. Siciliano, J. Spadavecchia, D. S. Presicce, A. M. Taurino, *Journal of Optoelectronics and Advanced Materials* **2003**, *5*, 1335.
215. Y. Xuan, Y. Q. Wu, T. Shen, M. Qi, M. A. Capano, J. A. Cooper, P. D. Ye, *Applied Physics Letters* **2008**, *92*, 013101-1–013101-3.
216. J. A. Robinson, E. S. Snow, û. C. Badescu, T. L. Reinecke, F. K. Perkins, *Nano Letters* **2006**, *6*, 1747.
217. G. Lee, B. Lee, J. Kim, K. Cho, *Journal of Physical Chemistry C* **2009**, *113*, 14225.
218. N. Leconte, J. Moser, P. Ordejón, H. Tao, A. Lherbier, A. Bachtold, F. Alsina, C. M. Sotomayor Torres, J. C. Charlier, S. Roche, *ACS Nano* **2010**, *4*, 4033.
219. J. Moser, H. Tao, S. Roche, F. Alzina, C. M. Sotomayor Torres, A. Bachtold, *Physical Review B—Condensed Matter and Materials Physics* **2010**, *81*, 205445-1–205445-6.
220. C. X. Lim, H. Y. Hoh, P. K. Ang, K. P. Loh, *Analytical Chemistry* **2010**, *82*, 7387.
221. Y. W. Son, M. L. Cohen, S. G. Louie, *Nature* **2006**, *444*, 347.
222. L. Yang, C. H. Park, Y. W. Son, M. L. Cohen, S. G. Louie, *Physical Review Letters* **2007**, *99*, 186801-1–186801-4.
223. O. A. Shenderova and D. M. Gruen. *Ultrananocrystalline Diamond Synthesis Properties and Applications*, William Andrew Publishing, New York, **2006**, p. 293.
224. J. L. Davidson, D. T. Bradshaw, C09 K 5/00 ed., Vanderbilt University, USA, **2005**, p. 18.
225. V. E. Red'kin, *Chemistry and Technology of Fuels and Oils* **2004**, *40*, 164.
226. J. R. Maze, P. L. Stanwix, J. S. Hodges, S. Hong, J. M. Taylor, P. Cappellaro, L. Jiang, M. V. G. Dutt, E. Togan, A. S. Zibrov, A. Yacoby, R. L. Walsworth, M. D. Lukin, *Nature* **2008**, *455*, 644.
227. A. S. Chiganov, *Physics of the Solid State* **2004**, *46*, 620.
228. Kulakova, II, *Physics of the Solid State* **2004**, *46*, 636.
229. S. Osswald, M. Havel, V. Mochalin, G. Yushin, Y. Gogotsi, *Diamond and Related Materials* **2008**, *17*, 1122.
230. T. Gaebel, C. Bradac, J. Chen, J. M. Say, L. Brown, P. Hemmer, J. R. Rabeau, *Diamond and Related Materials* **2012**, *21*, 28.
231. T. M. Gubarevich, R. R. Sataev, V. Y. Dolmatov, in *Proceedings of the 5th All-Union Meeting on Detonation, Vol. 1*, Krasnoyarsk USSR, **5–15 August 1991**, pp. 135.
232. A. S. Chiganov, G. A. Chiganiv, Y. V. Tushko, A. M. Staver, *Vol. No. RU2004491*, Russia, **1993**.
233. A. V. Tyurnina, I. A. Apolonskaya, Kulakova, II, P. G. Kopylov, A. N. Obraztsov, *Journal of Surface Investigation-X-Ray Synchrotron and Neutron Techniques* **2010**, *4*, 458.
234. E. V. Pavlov, J. A. Skrjabin, *Vol. RU2019502*, Russia, **1994**.
235. I. Petrov, O. Shenderova, V. Grishko, V. Grichko, T. Tyler, G. Cunningham, G. McGuire, *Diamond and Related Materials* **2007**, *16*, 2098.
236. O. Shenderova, A. Koscheev, N. Zaripov, I. Petrov, Y. Skryabin, P. Detkov, S. Turner, G. Van Tendeloo, *Journal of Physical Chemistry C* **2011**, *115*, 9827.
237. S. Gordeev, S. Korchagina, *Journal of Superhard Materials* **2007**, *29*, 124.
238. N. Mohan, Y. K. Tzeng, L. Yang, Y. Y. Chen, Y. Y. Hui, C. Y. Fang, H. C. Chang, *Advanced Materials* **2010**, *22*, 843.
239. D. Ugarte, *Nature* **1992**, *359*, 707.
240. V. L. Kuznetsov, A. L. Chuvilin, Y. V. Butenko, I. Y. Mal'kov, V. M. Titov, *Chemical Physics Letters* **1994**, *222*, 343.
241. A. Palkar, F. Melin, C. M. Cardona, B. Elliott, A. K. Naskar, D. D. Edie, A. Kumbhar, L. Echegoyen, *Chemistry—An Asian Journal* **2007**, *2*, 625.
242. N. Sano, H. Wang, M. Chhowalla, I. Alexandrou, G. A. J. Amaratunga, *Nature* **2001**, *414*, 506.
243. L. S. Chen, C. J. Wang, *Advanced Materials Research* **2012**, *490–495*, 3211.
244. C. N. He, N. Q. Zhao, X. W. Du, C. S. Shi, H. Ding, J. J. Li, Y. D. Li, *Scripta Materialia* **2006**, *54*, 689.
245. Y. Shimizu, T. Sasaki, T. Ito, K. Terashima, N. Koshizaki, *Journal of Physics D-Applied Physics* **2003**, *36*, 2940.
246. M. E. Plonska-Brzezinska, A. Lapinski, A. Z. Wilczewska, A. T. Dubis, A. Villalta-Cerdas, K. Winkler, L. Echegoyen, *Carbon* **2011**, *49*, 5079.
247. J. K. McDonough, A. I. Frolov, V. Presser, J. Niu, C. H. Miller, T. Ubieto, M. V. Fedorov, Y. Gogotsi, *Carbon* **2012**, *50*, 3298.

248. M. E. Plonska-Brzezinska, J. Mazurczyk, B. Palys, J. Breczko, A. Lapinski, A. T. Dubis, L. Echegoyen, *Chemistry—A European Journal* **2012**, *18*, 2600.
249. A. S. Rettenbacher, B. Elliott, J. S. Hudson, A. Amirkhanian, L. Echegoyen, *Chemistry—A European Journal* **2006**, *12*, 376.
250. A. Palkar, A. Kumbhar, A. J. Athans, L. Echegoyen, *Chemistry of Materials* **2008**, *20*, 1685.
251. J. Luszczyn, M. E. Plonska-Brzezinska, A. Palkar, A. T. Dubis, A. Simionescu, D. T. Simionescu, B. Kalska-Szostko, K. Winkler, L. Echegoyen, *Chemistry—A European Journal* **2010**, *16*, 4870.
252. Y. V. Butenko, A. K. Chakraborty, N. Peltekis, S. Krishnamurthy, V. R. Dhanak, M. R. C. Hunt, L. Šiller, *Carbon* **2008**, *46*, 1133.

14 Hydrothermal Synthesis of Nano-Carbons

Masahiro Yoshimura and Jaganathan Senthilnathan

CONTENTS

14.1 INTRODUCTION

Water is an inevitable component of hydrothermal systems and both temperature and pressure have a great influence on the resulting properties of water. When water attains supercritical state, its surface tension approaches nearly zero. At this point, the distinction between gas and liquid phase breaks down and water effectively occupies the pores of the material and facilitates the hydrothermal process [1]. Hydrothermal processing can be defined as "any homogeneous (nanoparticles) or heterogeneous (bulk materials) reaction in the presence of aqueous solvents or mineralizers under high pressure (above 1 atmospheric) and temperature (above a room temperature) conditions to dissolve and recrystallize (recover) materials that are relatively insoluble under ordinary conditions" [1, p.1]. Hydrothermal synthesis of materials indeed has a longstanding history. Schafthaul was the first person who used hydrothermal treatment to prepare fine particles of quartz in a Papin's digester during 1845 [2]. In the early 1900s, more than 150 mineral species, including diamond, were synthesized by hydrothermal methods [3]. Since 1940s, the early stage of hydrothermal research was conducted by several groups in the United States, Europe, and Japan. They mostly focused on crystal growth of artificial Zeolite and Quartz as indicated in a review by Somiya [4]. During the years between 1950s and 1970s, hydrothermal processes have become a facile method mostly used in the area of geology and mineralogy [5–7]. However, ever since the late 1970s, material scientists adopted hydrothermal processes to prepare various compounds with controlled size, shape, and composition.

During the late 1980s, solvothermal reaction has been defined as a high-temperature reaction in a closed system in the presence of a solvent (generally nonaqueous) [8a,b]. A solvothermal process can also be termed as glycothermal, ammonothermal, carbonathermal, lyothermal, alchothermal, or hydrothermal depending on the nature of the used solvent. Recently, nonaqueous solution processes have also been developed which generates a small amount of H_2O during the process. This small amount of water then becomes a major reactant in the actual solution process. Hydrothermal and solvothermal process might also relate to supercritical fluid process, molten salt, and/or ionic liquid process [1,9].

Hydrothermal research nucleated in Japan, the United States, and other parts of the world, and this opened a new avenue for hydrothermal research in the area of materials research [4–7,10–17]. The evolutionary trends of hydrothermal processing of various materials are given in Table 14.1 [9]. Yoshimura et al. have proposed a novel concept of soft solution process (SSP), for fabricating ceramic and other materials as a more effective and environmental friendly approach and the end products have no disparity with hydrothermal or other synthetic routes [18–21]. Although this term has a broader meaning, it covers a part of the hydrothermal research and refers mainly to any solution processing at or near the ambient conditions. Hydrothermal process has been widely investigated and used for the synthesis of a wide range of materials, especially various metal oxides and nonoxides, such as diamond, carbon, selenides, tellurides, sulfides, fluorides, nitrides, and aresenides. Hydrothermal technologies have evolved as a very powerful tool in materials processing because it is versatile and environmentally benign. Hydrothermal treatment is, thus, highly suitable for advanced materials synthesis starting from bulk single crystals to fine and ultrafine crystals, including nanocrystals or nanoparticles with a controlled size and morphology [9,22]. Recently, hydrothermal process has also emerged as a promising technique for the preparation of carbon material from various sources and importantly carbides, organic compounds, and biomass as described subsequently in more detail.

14.2   DIFFERENT FORMS OF CARBONS

The important allotropes of carbons are amorphous carbon, diamond, and graphite. However, other allotropes, such as fullerenes, nanotubes, and graphene, have also been recognized recently. Carbon atoms form bonds in three different configurations (sp, sp^2, and sp^3). Many disordered forms of carbon have structures based on the graphite lattice of sp^2-bonded carbon [23]. Apart from its conventional structure, various other forms of carbons, such as onion-like carbon, diamond-like carbon, disordered graphite, filamentous carbon, nanotubes, nanocells, nanowire, nonofiber, nanochains, hallow carbon, carbon sheet, and carbon spheres, have also been reported. Properties of different forms of carbon are dissimilar from each other and classifications of these carbons have been established by various research groups [24–40].

14.2.1   CARBONS FROM CARBIDES

From early 1970 to middle of 1990, only a selected few research groups have worked on the formation of carbon from carbides, but no significant breakthrough was achieved during this period [8,41–46]. Reichle and Kickl observed the formation of carbon and metal oxide by air oxidation (800–1200°C) of TiC single crystals using O_2/Ar and CO_2/CO mixtures [47]. Formation of carbon from TaC, TiC, NbC, SiC, ZrC, HfC, WC, Cr_3C_2, and Ti_3SiC has been extensively studied by Shimada et al. [30,42,44,48–53]. During hydrothermal oxidation of TiC, ZrC, and HfC, a carbon-rich layer has been observed by varying the range of temperature and pressure. Similarly, formation of amorphous carbon was reported by most of the researchers and in some cases hexagonal diamonds were also observed [8,42,44,50–52,54–57]. Initial attempts had been made to form a carbon from powdered, sintered, and fibrous SiC (Tyranno©) by hydrothermal condition and the final products showed solid SiO_2 along with the following gas species: CO_2, CO, and CH_4 [41].

$$SiC + 2H_2O \rightarrow SiO_2 + CH_4 \qquad (14.1)$$

TABLE 14.1:
Development of Hydrothermal Processing of Materials

Area	Periods	Example, Materials Applied
Hydrometallurgy	1900	Sulphate ore and oxide ore
Crystal synthesis, and growth	1940	Quartz, oxides, sulfides, fluorides, layered compounds
Fire crystals with controlled composition, size, and shape	1970	PZT, ZrO_2, PSZ, $BaTiO_3$, hydroxyapatite
Whiskers	1980	Hydroxyapatite, Mg-sulfate, SiC, Si_3N_4, Al_2O_3, ZrO_2, and K-titanate
Crystalline films (thin, thick)	1980	$BaTiO_3$, $LiNbO_3$, ferrite, carbon, and $LiNiO_2$
Hydrothermal etching	1980	Oxides and non-oxides
Hydrothermal machining	1980	Oxides and non-oxides
Combination with electro-, photo-, mechano-, and electrochemical, etc.	1970–80	Synthesis, alternation, coating, and modification
Organic- or biomaterials	1980	Organic hybrid nanoparticles: Co_3O_4: C_9COOH; Fe_3O_4: $C_{17}COOH$. Bioceramics: Hydroxyapatite$[Ca_{10}(PO_4)_6(OH)_2]$ Hydrolysis, wet-combustion, extraction, polymerization, decomposition, and remediation
Solvothermal process	1980	CeO_2, $PbTiO_3$, Synthesis, extraction, and reaction
Continuous process	1990	Synthesis extraction and decomposition TiO_2, ZrO_2, γ-AlOOH, γ-Al_2O_3, $BaTiO_3$, $Ca_{0.8}Sr_{0.2}Ti_{1-x}Fe_xO_3$, ZnO, $CoFe_2O_4$, $(Y_{2.7}Tb_{0.3})Al_5O_{12}$, $LiMn_2O_4$, In_2O_3, SnO_2, $Y_2Al_5O_{12}$, $La_{n+1}Ni_nO_{3n+1}$, α-Fe_2O_3, $LiFePO_4$, $BaZrO_3$, $Ce_xZr_{1-x}O_2$, $Ni(OH)_2Co_xNi_{1-x}(OH)_2$, $NiCo_2O_4$, CeO_2, $Ca_{10-x}Mg_y(PO_4)_6(OH)_2$, and $Ca_{3-y}Mg_y(HPO_4)_2(PO_4)_{2-2x/3}$
Nanomaterial synthesis	1992	Oxides: Cu_2O, BeO, ZnO, Al_2O_3, TiO_2, ZrO_2, HfO_2, TeO_2, α-Fe_2O_3, β-Fe_2O_3, γ-Fe_2O_3, γ-MnO_2, La_2O_3, Mn_3O_4, PbO, VO_2, GeO_2, Ge_2O_3, SnO_2, V_2O_3, V_2O_5, In_2O_3, Bi_2O_3, CdO, CoO, Co_3O_4, HgO, $BaZrO_3$, $BaFe_{12}O_{19}$, $LiMn_2O_4$, and $LiCoO_2$ Mixed oxides: $In_3Sb_5O_{12}$, $BaTiO_3$, $PbTiO_3$, $CoFe_2O_4$, $ZnFe_2O_4$, $ZnAl_2O_4$, Fe_2CoO_4, La_3SbO_{12}, $Pr_3Sb_5O_{12}$ and $Yb_3Sb_5O_{12}$
Hydrothermal synthesis of various forms of carbon from carbides	1994	Graphene, CNTs (bucky tube), fullerenes, (bucky ball), carbon onions (bucky onions), filamentous carbon, nanocell, nanobeads and nanodiamond etc.
Patterning	2000	$PbTiO_3$, Synthesis and fixing
Microwave-assisted hydrothermal synthesis	2009	$BiFeO_3$, $CsAl_2PO_6$, $ZnFe_2O_4$, $NiFeO_4$, $MnFe_2O_4$, Co Fe_2O_4, $BaTiO_3$, $SrTiO_3$, $BaZrO_3$, $PbZrO_3$, $SrZrO_3$, $Ba_{0.5}Sr_{0.5}TiO_3$ and carbonaceous materials from organic waste

Sources: Reproduced from M. Yoshimura and K. Byrappa, *J. Mater. Sci.*, 43, 2085–2103, 2008; K. Byrappa and M. Yoshimura, *Handbook of Hydrothermal Technology: Hydrothermal Technology for Nanotechnology—A Technology for Processing of Advanced Materials,* 2nd Edition, Elsevier Amsterdam, p 615, 2013.

$$SiC + 4H_2O \rightarrow SiO_2 + CO_2 + 4H_2 \qquad (14.2)$$

$$SiC + 3H_2O \rightarrow SiO_2 + CO + 3H_2 \qquad (14.3)$$

14.2.2 Formation of Amorphous Carbon from Carbides

Gogotsi and Yoshimura first reported the formation of carbon from SiC fiber by hydrothermal process [58]. In this study, amorphous SiC fibers were used for the hydrothermal process at 300–800°C and a fixed pressure of 100 MPa. The uniqueness of this study was that the carbon film was not deposited on the surface of the SiC, but that the SiC was transformed into carbon conformally. The scanning electron microscopy (SEM) images of hydrothermal corrosion of SiC fiber at different temperatures are given in Figure 14.1. Hydrothermal corrosion of SiC fibers at 300°C showed no change in strength or any surface damage. When the temperature was increased beyond 300°C and up to 400°C, diffusion of water into an oxycarbidic phase occurred. Figure 14.1b shows only slight surface damage at this temperature. Hydrothermal corrosion at 500°C led to a very drastic decrease in the fiber diameter (Figure 14.1e) and further increase in temperature (600°C for 25 h) caused

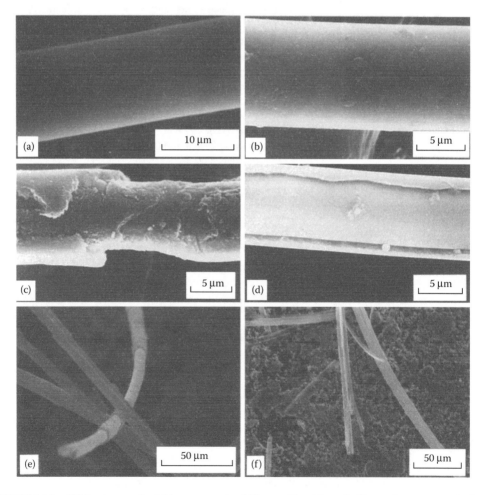

FIGURE 14.1 SEM micrograph of (a) an as-received fiber and LoxM grade fibers after water corrosion at (b–d) 450°C (e) 500°C, and (f) 600°C. (Reproduced from Y. Gogotsi and M. Yoshimura, *J. Mater. Sci. Lett.* 3, 395–399, 1994.)

complete degradation of the fibers (Figure 14.1f) [58,59]. Fourier transform infrared (FTIR) and Raman spectra confirmed the formation of amorphous or microcrystalline carbon and the following reaction mechanisms was proposed [58–61]:

$$SiC + 2H_2O \rightarrow SiO_2 + C + 2H_2 \qquad (14.4)$$

The silica which is formed during the hydrothermal process of SiC is readily dissolved in water. This is, however, a very different situation for other metal oxides as most of them are insoluble in water; thus, for other carbides, both carbon and metal oxide formations occur during hydrothermal reaction [62]. The type of carbide and the carbide-to-water ratio are the two most important controlling factors for the formation of carbon in hydrothermal reaction [63].

In the case of SiC, at lower water and SiC molar ratios, both carbon and silica are deposited on the surface. At moderate molar ratios, formation of carbon was observed on the surface and silica dissolutes into the solution. At higher molar ratios, carbon reacts with water and forms CO/CO_2. The trend was similar to WC, TaC, NbC, and TiC, except for B_4C for which carbon formation has never been reported [63].

14.2.3 Formation of Diamond by Hydrothermal Process

Extensive research has been carried out on artificial synthesis of diamond. The direct conversion of graphite to diamond was carried out under very harsh conditions (120 kbar and 3300 K) by electric flash heating technique [64,65]. Wentorf et al. exposed various nondiamond carbons including organic compounds to a range of pressures (95–150 kbars) and temperatures (1300–3000°C). Under these conditions, most of these materials turn partially to diamond within 0.2–55 min. In those studies, formation of diamonds depends on the nature of carbonaceous material used and compounds which contain benzene rings (naphthalene, anthracene, or chrysene) form graphite. Similarly, compounds which contain comparatively large amounts of nitrogen also favor the formation of graphite [66]. Transformation of glassy carbon to diamond through well-crystallized graphite was reported at pressure above 9 GPa and 3000°C [67]. The evidence for the presence of graphite, poorly crystallized carbons, and diamond-like structures in hydrothermally corroded SiC raises the question about the hydrothermal behavior and optimum condition for diamond formation. For diamond synthesis, it is necessary to identify the conditions under which diamond will be nucleated and the formation of graphite and/or other nondiamond phases will be inhibited. In contrast to such high-pressure and high-temperature conditions, hydrothermal conditions (several hundred degrees Celcius and several MPa) have been applied to synthesize diamond. Roy et al. studied the hydrolysis of β-SiC powder at 140 MPa and 800°C for 48 h and found that this leads to the formation of a sharp Raman band at 1330 cm⁻¹. However, additional experimental results did not show sufficient evidence for presence of diamond, and this might be because of small quantity (few nanometers) of diamond crystal present in the sample [49]. The formation of diamond has been carried out with α and β-SiC powder. Twenty percentage of HF was used for the removal of SiO_2 and hot $HClO_4$ acid was used for the removal of nondiamond carbon. The addition of hot $HClO_4$ acid was to overcome the interference of nondiamond carbon during the analysis. IR absorption clearly showed a band at 1310 cm⁻¹ corresponding to sp³ carbon and shoulder peak at 1280 cm⁻¹ indicating the presence of diamond (Figure 14.2) [49,58,68,69]. TEM and selected area electron diffraction (SAD) pattern of carbon films after hydrothermal treatment at 700°C under 100 MPa and HF etching are given in Figure 14.3 [23,70]. Most recently, Yamasaki et al. have reported that fine particles of diamond can be synthesized using chlorinated hydrocarbon, such as dichloromethane and 1,1,1-trichloroethane, at 573 K and 1 GPa for 24 h under alkaline (10 M NaOH solution) hydrothermal conditions [71]. Similarly, the above studies were carried out in the presence of diamond seeds at 300°C and pressure 1 GPa [72–74]. Synthetic diamonds are produced in an artificial process which involves very

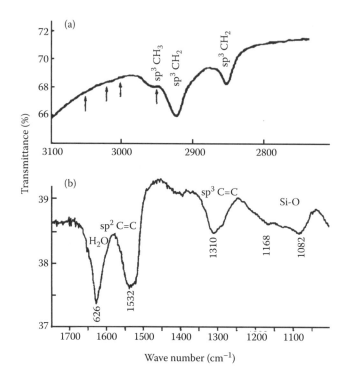

FIGURE 14.2 FTIR spectra of CH_n, (a) and carbon (b) regions recorded after leaching β-SiC powders in HF (a) hydrothermal treatment for 5 h at 750°C under 100 MPa pressure (the predicted positions of sp^2-CH, groups are marked with arrows), (b) hydrothermal treatment for 5 h at 700°C under 100 MPa pressure (Ni-NiO buffer). (Reproduced from Gogotsi et al., *J. Mater. Chem.*, 6(4), 595–604, 1996.)

high temperature and pressure and it is widely known as high-pressure and high-temperature diamonds (HPHT) or chemical vapor deposition (CVD) diamonds.

14.2.3.1 Identification Diamond from Carbides

Gogotsi and Yoshimura have carried out extensive study on formation of diamond on the surface of SiC by hydrothermal process. On the basis of their studies, the nature of the carbon (sp^3, sp^2, and sp) formation purely depends on a structure of the SiC precursor and single instrument analysis provides insufficient information about the types of bonding and structure of carbon [23]. Identification of diamond was complicated by the presence of metal oxide and carbide phases. In specific, XRD

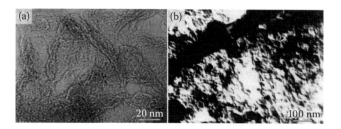

FIGURE 14.3 TEM images of carbon films after hydrothermal treatment at 700°C under 100 MPa and etching in HF. (a) Region producing only graphite rings on SAD patterns; (b) region producing graphite rings and spots with d = 2.70, 2.53, 2.37, 2.06–2.08, 1.92, and 1.65 Å. Some of the crystals that were bright on the dark-field image for a reflection with d = 2.06 Å are marked with arrows. (Reproduced from Gogotsi et al., *J. Mater. Chem.* 6(4), 595–604, 1996.)

analysis of hydrothermally corroded SiC showed the strongest lines of graphite, (002) and (004), are superimposed on the peaks of α-quartz. Similarly, the (220) reflection of diamond coincides with the (222) reflection of β-SiC, leaving only one intense peak at a d-value of 2.06 Å for detecting diamond. Thus, one cannot draw any conclusion about the carbon structure on the basis of XRD studies alone. FTIR spectroscopy is a useful tool for analysis of C–H bonding, but it is difficult to detect pure carbon or diamond by this method as pure diamond does not have any IR active modes. However, we note that Raman spectroscopy is the simplest and most powerful technique for identifying carbon allotropes. The phonon-confinement effect plays a substantial role leading to the shift and to the strong broadening of a fundamental diamond mode, usually observed at 1332 cm^{-1} [23,41,49,75]. It is possible that keeping its integral intensity constant, this line would have too low an amplitude to be extracted from the graphite background even in the case of comparable contents of diamond and graphite phases in the sample. In particular, it has been reported that the Raman cross-section of graphite is much higher than that of crystalline diamond and hence sp^2-bonded regions dominate the Raman spectra [23,41,58,76] and the Raman response of hexagonal diamond is many times smaller than that of cubic diamond [77]. By now, the formation of diamond from SiC by hydrothermal process has been established by different research groups [8,43,49,58,78,79].

14.2.3.2 Formation of Disordered Carbon by Hydrothermal Treatment

Szymanski et al. have reported diamond crystal growth under hydrothermal conditions [78]. At very high temperature and pressure (7.7 GPa and 2200°C), the growth of natural diamond was increased in the presence of diamond seed crystal-embedded graphite [79]. On the contrary, the deposition of nondiamond carbon was observed by interaction of diamond single crystals with water at 5.5 GPa and 1000°C and the complete dissolution of diamond was also observed at 5 GPa and 1500°C [80]. Several other groups have also reported the formation of diamond by hydrothermal process [49,67,81,82], but it was always questioned whether growth was seeded or not, because the nucleation of diamond is always more difficult than the continuing growth. Hydrothermal corrosion of β-SiC was carried out in the presence of organic compounds (malonic acid, glycolic acid, citric acid, sucrose, formic acid, acetic acid, ascorbic acid, succinic acid, and malic acid) at 100–200 MPa and 600–850°C. In a closed system, the organic compound increases the atomic hydrogen, C and H radical concentrations. This creates a highly reducing environment in the system which facilitates the SiC dissociation and forms different polymorphs of carbon at very low temperature and pressure [83,84]. Two types of carbon phases are formed in this process, one is glassy or disordered graphitic carbon and the other is nano- or micrometer-sized spherical-shaped carbon particles (Figure 14.4).

The behavior of diamond under hydrothermal conditions has been studied in detail and no significant change was observed at 750°C and 500 MPa. With further increase in temperature (850°C) and reaction time (94 h), formation of graphite was observed and this was further confirmed by combination of TEM and SAD analysis which is given in Figure 14.5 [69,85].

14.2.4 Hydrothermal Synthesis of Filamentous Carbons

Hollow filamentous carbons can be divided into two different classes: (a) The layers of graphene encompassing the walls of the tubes are finite in length, that is, they terminate on both the inner and outer surfaces of the tubes with a conical geometry of arbitrary conical angle [86–90]. (b) Filamentous carbons comprising continuous cylinders of graphene whose lengths are limited only by the filament length or possible defects (i.e., nanotubes). They have very small inside diameters often related to the dimensions of fullerene molecules and can be single or multiwalled where each cylindrical layer can have its own unique structure and orientation [90–92].

Deviations from the above classification are also reported in the literature [93,94]. Synthesis of filamentous carbon from Cr_3C_2 was studied under hydrothermal condition at 100–200 MPa and 350–800°C. The formation of spherical and filamentous carbon was observed in the presence of organic compounds (ascorbic acid, malonic acid, glycolic acid, oxalic acid, citric acid, and stearic acid) at

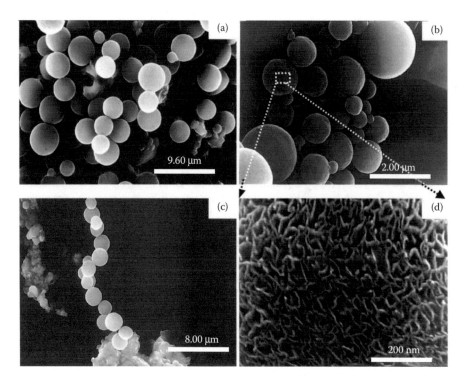

FIGURE 14.4 Scanning electron images of the spherical-shaped carbon particles (a) and (c) individual spherules as well as linked chains, (b) carbon spherules having characteristic surface pattern, (d) enlarged image of a particle surface. (Reproduced from Basavalingu et al. *Carbon*, 39, 1763–1767, 2001.)

temperatures above 600°C (Figure 14.6). At this temperature, supercritical fluids generated by the dissociation of organic compounds have great influence on the decomposition of Cr_3C_2. Carbon particles formed by this method were solid curved filaments with a mean diameter of 50–100 nm [30].

Synthesis of filamentous carbon from paraformaldehyde by high temperature (600°C) and high pressure has been reported [60]. Formation of filamentous carbon in C-H-O-Ni, C-H-O-Fe, and

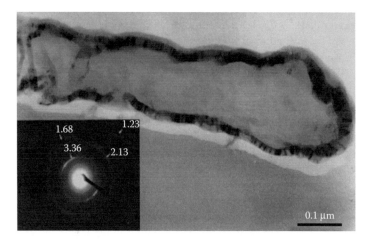

FIGURE 14.5 Bright-field TEM micrograph and SAD pattern of a typical graphite particle in microcrystalline diamond powder after hydrothermal treatment for 94 h at 850°C under 100 MPa. (Reproduced from Gogotsi et al. *Diamond Rel. Mater.* 7, 1459–1465, 1998.)

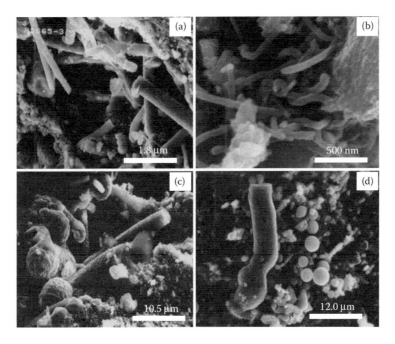

FIGURE 14.6 SEM images showing (a–c) filamentous particles (d) filamentous and spherical particles. (Reproduced from Basavalingu et al., *J. Mater. Sci.*, 43, 2153–2157, 2008.)

C-H-O-Co systems at ambient pressure (1 atm) showed low graphitization and conical layer morphology [95]. At high pressure, (100 MPa) the formation of filamentous carbon are different from that of lower pressures, [60,96]. Formation of bamboo-like carbon filaments and spherical-shaped carbon was also reported [84,92,97]. The nanocells showed diameters lesser than 100 nm and outer diameters ranging from 15 to 100 nm. The internal cavities of nanocell diameters were in the range of 10–80 nm. Perfect spherical morphology of nongraphitic carbon was prepared by heating of mesitylene at 700°C in a closed cell [97].

14.2.5 Hydrothermal Synthesis of Graphitic Carbons

Graphite with different crystallinity in natural conditions has been reported by many researchers [98–101]. For example, Buseck and Huang have reported that the transformation of carbonaceous material into graphite is a continuous process which begins with conversion of the carbonaceous material into coaly material, followed by transitional stage, and finally it becomes graphite [102].

Graphitization is greatly influenced by various factors, such as temperature, duration of the transformation, structural difference of organic precursors, pressure, and mineral assemblages [98–100,103]. It is still very difficult to establish the transformation of coal to graphite. Coal contains a large amount of hydrogen and oxygen and possible formation of graphite from coal is given as follows [103]:

$$\text{Coal } (C, H, O)_n \rightarrow \text{Graphite } (C) + \text{Gas } (CO_2 + H_2 + CH_4 + H_2O)_n \qquad (14.5)$$

French and Rosenberg studied the stability of siderite ($FeCO_3$) in CO_2 and disordered graphite at about 450°C at 1 kbar which was formed through the breakdown reaction of siderite [104]. Hirano et al. synthesized graphite from coke which was prepared from polyvinylchloride in the presence of $CaCO_3$ at 900°C and 9.8 kbar pressure. Graphitization was observed in the absence of $CaCO_3$ even at temperature elevated up to 1400°C [105]. Formation of graphite from bituminous coal occurred in the presence of Li_2CO_3/Ni metal catalyst at water vapor pressure around 0.5–5 kbar. Graphite crystallization showed a positive sloping below 1 kbar and a negative orientation beyond 1 kbar [103]. Similarly, graphite, hydrocarbons (CH_4 and C_2H_6), and solid black residue were obtained in the presence of calcite and hydrogen at 500°C [106]. Under alkaline hydrothermal conditions, hydrogen and chlorine were decomposed from the halogenated hydrocarbon and graphite-like materials were reported [107]. Unlike diamond, graphite is a good electron conductor because of delocalization of the π (sp^2) bond electrons.

In addition to graphite and diamond, new carbon structures such as nanocell [108], nanochain [31], and graphite polyhydral crystals [32] have also been reported. The SEM images of different graphite polyhedral structure from glassy carbon are given in Figure 14.7. Olivary (olive-shaped) carbon particles (OCPs) with a diameter of ~1.5–2 μm at the middle and a length of ~3–4 μm were synthesized by pyrolysis of acetone with Zn as the catalyst at 600°C. In this study, Mg, Ni, Fe, Cu, Zn, and Cd powder were used as catalysts and Zn powder showed excellent olivary carbon compared with other metals [109]. Large pore-ordered mesoporous carbons were prepared in the presence of silica (large pore silica KIT-6 material), Pluronic P-123, and *n*-butanol solution [110].

Cross-linked polyvinyl alcohol (PVA) nano-cables were formed in the presence of Cu ions by a one-step hydrothermal approach at 200°C [111]. Synthesis of metal-free submicro meter graphitic carbon plates from glucose with controllable thickness has been reported and H_2SO_4 used for the enhancement of carbonization [34]. Furthermore, a low amount of glucose and a high amount of

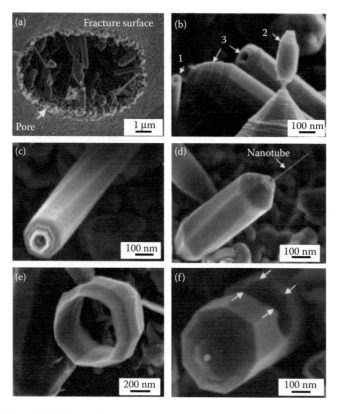

FIGURE 14.7 SEMs of GPCs found in pores of glassy carbon (a) Fracture surface, showing carbon nanotubes and GPCs growing in the pore. (b) Carbon nanotube (1), double cone (2), and microrods (3), which are typical structures. (c) Twisted rod with a heptagonal cross section. (d) Twisted GPC with a protruding nanotube. (e) Faceted ring that might be formed by pullout of the core structure. (f) Twisted rod that has a notch from crossing another GPC growing from the other side of the pore, which was removed when the pore fractured. Arrows mark edges of the interrupt. (Reproduced from Y. Gogotis et al., *Science* 290, 317–320, 2000.)

TABLE 14.2
Hydrothermal Carbonization in the Absence of Metal Catalyst/Composite

Precursors	Temp. (°C)	Pres. (MPa)	Time (h)	Nature of Carbon	Ref.
Formaldehyde	650–750	100	312	Filamentous carbon	[60]
Mesitylene	700	28	3	Spherules carbon	[97]
Carbon soot	800	100	—	MWCNTs	[108]
Amorphous carbon	600	100	—	Carbon nanocell	[31]
Glucose	190		6	Carbon sheet	[34]
Glucose/sodium dodecyl sulfate	190		6	Carbon sheet	[112]
Polyethyleneglycol	160	—	20	MWCNTs	[33]
SWCNT	200–800	100	0.5–48	MWCNTs/polyhedral	[35]
Wood powder	400–800	—	2	Liquid wood	[114]
Cellulose	220–250	—	2–4	Carbon sphere	[115]
Acrylic acid/glucose	190	—	16	Amorphous carbon	[116]
Palm kernel shells/glucose	180–190	—	3–6	Activated carbon	[117]
Glassy carbon	750	100	24	MWCNTs	[118]
Sugar	190–1000	—	5	Spherules carbon	[119]
Apple residue	400–800	—	2	Amorphous carbon	[120]
Anthracite coal	450–900	20	2	Ball like carbon	[121]
Fructose	120–140	0.2–0.7	—	Carbon sphere	[122]
Oak wood/char	525–775	29	60	Activated carbon	[123]
Cyclodextrins	160	—	2–16	Carbon sphere	[124]
Glucose	500	—	12	Microsphere	[125]
Saw dust	110–800	—	1–6	Microporous carbon	[126]
Starch	600	—	12	Carbon microsphere	[127]
Swine manure	275–350	7–28	2	Solid char	[128]
Ethylenetriamine/polyethyleneglycol	150–180	—	24	MWCNTs	[129]
Starch	180	—	12	Carbon microspheres	[40]

sodium dodecyl sulfate were used to produce a thin graphitic carbon-nano-sheet [112]. Nitrogen-doped hollow carbon microspheres with graphitic carbon shells have been reported by direct pyrolysis of solid melamine–formaldehyde resin [113]. Synthesis of various forms of carbon in the presence and absence of metal catalyst and composites is given in Tables 14.2 and 14.3.

14.2.6 CARBON NANOTUBE

14.2.6.1 Formation of Nanotube in the Absence of Catalysts

Mutliwalled, hollow CNTs composed of well-ordered concentric graphitic layers were synthesized in the absence of metal catalyst from amorphous carbon in presence of water at 800°C and 100 MPa [108]. Amorphous carbon was transformed into pristine multiwalled CNTs yielding diameters in the range of ~10 nm with a length of ~100 nm. TEM revealed that the polygonal particles had polyhedral shapes with a hollow core and multiwalled concentric graphene layers (Figure 14.8). Similarly, formation of multiwalled carbon nanocell (<100 nm) from amorphous carbon can be achieved at 600°C by hydrothermal treatment [108].

Formation of multiwalled carbon nanotubes (MWCNTs) by hydrothermal process has been carried out using mixed aqueous solution of diethylenetriamine and polyethyleneglycol (PEG) in basic condition. The MWCNTs were formed in the temperature range of 150–180°C in 24 h reaction time [151]. The hydrothermal formation of MWCNTs has been carried out in the presence of Cl_2 gas using aqueous NaOH with dichloromethane and Li as the starting materials at 150–160°C for 24 h. Nanotubes produced in this manner were about 60 nm in diameter and 2–5 μm long [152].

TABLE 14.3

Hydrothermal Carbonization in the Presence of Metal Catalyst/Composite

Precursors	Catalyst/ Composite	Temp. (°C)	Pres. (MPa)	Time, (h)	Nature of Carbon	Ref.
Bituminous coal	Ni/Li	450	100		Graphite	[103]
Polyvinyl chloride (coke)	CaCO$_3$	900	980	4	Graphite	[105]
Acetone	Zn	600	—	12	Olivary carbon	[109]
Poly(vinyl-alcohol)	Cu	200	—	72	Nanocable	[111]
Melamine–formaldehyde resin	N	150–700	—	5	Carbon microspheres	[113]
Ethylene glycol	Ni	730–800	52–90	0.5–24	Graphite tube	[143]
Ethylene glycol	Ni	730–800	60–100	24	Graphite tube	[144]
Starch/PVA	Ag	180		72	Carbon microcable	[145]
Glucose	Si/SiO$_2$	750		4	Core/shell structure	[146]
Glucose	Te	160–200	—	—	Nanowire	[37]
Glucose	Pd/Pt/Au	550	—	1	Carbon nanofibers	[38]
Glucose/starch	Ru/Pt	100–900	—	14	Carbon sphere	[147]
Starch	Ag	160–500	—	12	Carbon nanofiber	[148]
Starch/rice-grain	Fe	200	—	12	Carbon sphere	[149]
Saccharide	Pt/Ru	900	—	3	Carbon microsphere	[150]
Ethanol	Mg/NiCl$_2$	650	—	8	Hallow carbon	[39]
Sugar	Pt	190–1000	—	12	Carbon spherules	[130]
Sucrose	Pt/Pd	600		10	Carbon sphere	[131]
Furfural	Pb	190	—	14	Carbon sphere	[132]
Glucose/vinylimidazole	Si	190	—	16	Carbon spheres	[133]
Glucose/polyvinylpyrrolidone	Pt	750	—	3	Mesoporous carbon	[134]
Glucose	SiO$_2$/SnO$_2$	180–500	—	4	Carbon hollow sphere	[135]
Glucose	Fe$_3$O$_4$	190	—	12–15	Carbon nanospindles	[136]
Phenol formaldehyde	Ag	160	—	4	Carbon nanospheres	[137]
Glucose	TiO$_2$	180	—	4	Carbon spheres	[138]
Acetylene, benzene, and ethylene	Fe, Co, Ni	1950–2600	—	—	Bamboo-like carbon	[139]
Glucose	Ag	160–180	—	4–20	Carbon microsphere	[140]
Fullerenes	Ni	200–800	100	48	MWCNTs	[141]
Divinylbenzene and styrene tris(allyl)borane	B	650	125		Amorphous carbon	[142]

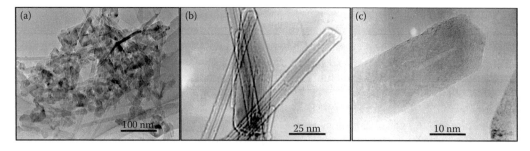

FIGURE 14.8 Electron micrographs after treatment showing multiwall nanotubes and graphitic nanoparticles (a); detail at higher magnification showing the hollow cores (b); and HRTEM lattice fringe image of several hydrothermal multiwall nanotubes (c). (Reproduced from J.M. Calderon-Moreno and M. Yoshimura, *J. Am. Chem. Soc.*, 123(4), 741–742, 2001.)

MWCNTs were also synthesized at 160°C without adding catalysts using polyethylene glycol as the carbon source. The diameters of the MWCNTs are much smaller than that of those prepared by high-temperature hydrothermal method [33].

14.2.6.2 Formation of Nanotubes in the Presence of Catalysts

Open end and closed MWCNTs with a large inner diameter of 20–800 nm were produced using polyethylene/water mixtures in the presence of Ni at 700–800°C under 60–100 MPa [143]. A very important feature of closed nanotubes is their ability to retain water and gases (CO, CO_2, H_2, H_2O, and CH_4) encapsulated at supercritical conditions (Figure 14.9). The size of the Ni catalyst particles influences the carbon structure and, in consequence, flake-like graphite or carbon tubules were formed. Carbon tubes were also synthesized in the absence of water, but these tubes have multiple internal caps. The carbon tubes which were produced in the presence of water have only very few internal obstructions and a large inside diameter (70 nm to 1.3 μm) [90,144,153]. The excellent wettability of the graphitic inner tube walls by the aqueous liquid and the mobility of this liquid in the nanotube channels were also observed (Figure 14.10). Also, the stability of single-wall carbon

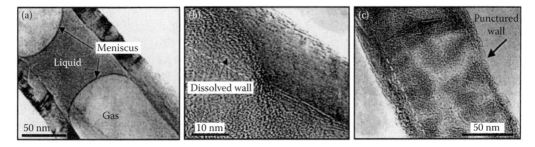

FIGURE 14.9 TEM micrographs showing water trapped in closed tubes. (a) The meniscus shows a good wettability of carbon with water (contact angle is <5°). Fast heating with the electron beam to high temperatures results in chemical reactions between the tube wall and water-based supercritical fluid leading to (b) dissolution and (c) puncture of the wall. (Reproduced from Gogotsi et al. *J. Mater. Res.*, 15(12), 2591–2594, 2000.)

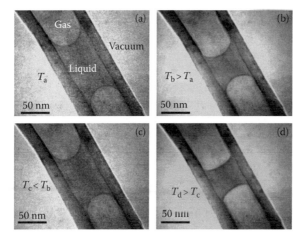

FIGURE 14.10 TEM micrograph sequence of a typical carbon nanotube portion showing the reversible volume contraction/expansion of a liquid entrapment upon heating/cooling achieved by manipulating the illuminating electron beam. (a) Initial shape of liquid at temperature T_a, (b) inclusion gets thinner on heating at $T_b > T_a$, (c) liquid returns to its initial size on cooling at $T_c < T_b$, (d) heating is repeated ($T_d > T_c$), resulting in a renewed contraction of the liquid volume. (Reproduced from Gogotsi et al. *Appl. Phys. Lett.* 79, 1021–1023, 2001.)

nanotubes (SWCNTs) was investigated under hydrothermal conditions at 100 MPa pressure, from 30 min to 48 h in the temperature range from 200°C to 800°C.

Similarly, C_{60} was converted into MWCNTs above 700°C via amorphous carbon under hydrothermal conditions [119]. SWCNTs under hydrothermal conditions can only survive mild- and short-term treatment in high-temperature, high-pressure water. With time, SWCNTs gradually transform into MWCNTs and polyhedral graphitic nanoparticles [35]. Both carbon filaments and MWCNTs are produced using ferrocene, Fe, or Fe/Pt nanocrystals at supercritical toluene at 600°C and 12.4 MPa. In this study, toluene serves as both the carbon source for nanotube formation and as a solvent [154]. Recently, direct synthesis of MWCNTs/CdS, MWCNT/ZnO, Ti/CNT, FeO/MWCNTs, and CNT-F-doped SnO_2 nanocomposites under hydrothermal process has been used for various applications [155–159].

14.2.7 Hydrothermal Synthesis of Graphene from Graphene Oxides

Graphene is a two-dimensional atomic monolayer of sp^2-hybridized carbon atoms. It has attracted significant interest in the recent year because of its extraordinary electrical and mechanical properties [160]. Hydrothermal synthesis of metal–graphene composite was achieved by reducing the graphene oxide to graphene [161–163]. ZnS nanodots have been deposited on the surface of graphene by reducing graphene oxide in the presence of reducing agent Na_2S [161]. Similarly, CdS-graphene oxide microparticle was prepared through a one-pot hydrothermal method and by using sodium thiosulfate-reducing agent [162]. Similarly, TiO_2-reduced graphene oxide has been formed in the presence of glucose, a reducing agent by hydrothermal process [163]. Synthesis of graphene-CdS and graphene-ZnS from graphene oxides has been reported by one-step solvothermal method [164]. Commercial grade TiO_2 has been used for the synthesis of TiO_2/graphene nanotube composites from graphene oxide under alkaline conditions [165]. Micro- and nanometer-sized Co-Al-layered double hydroxides were grown on the surface of graphene oxide, and these oxides surface prevent the restacking of the graphene nanosheets [166].

14.3 HYDROTHERMAL SYNTHESIS OF CARBONACEOUS MATERIALS

Hydrothermal carbonization is an effective method and has been extensively studied for the synthesis of carbonaceous material from different organic sources, especially from biomass. Pyrolysis is one of the most promising thermochemical processes for the conversion of biomass mainly into liquid (bio-oil), gaseous (gaseous C1–C4 hydrocarbons), and solid (char) products [167,168].

The high temperature which is required in the pyrolysis process has a critical influence on the properties of the char [168]. Conversely, hydrothermal degradation of biomass is initiated by hydrolysis which exhibits lower activation energy than the pyrolytic decomposition reactions and in hydrothermal processes, biomass components decompose at lower temperature compared to pyrolytic process [168]. For example, hemicelluloses decompose between 200–400°C and 180–220°C in pyrolytic and hydrothermal treatment, respectively [168–170]. The resulting carbon mass (carbon efficiency, CE) at the end of the process is much higher when compared to other methods such as anaerobic digestion, fermentation, and combustion techniques (Figure 14.11) [171]. Hydrothermal carbonization has low toxicological effect and no pre-drying steps, which are necessary for other carbon synthesis methods and required for hydrothermal carbonization. Hydrothermal carbonization is also free from undesired products, such as CO_2, CO, and odor or noise pollutions.

14.3.1 Hydrothermal Carbonization of Biomass and Carbohydrates

Biomass is mainly derived from animal and vegetable sources. Lignocellulosic biomass is a plant biomass which is composed of cellulose, hemicellulose, and lignin. The cellulose content in the lignocellulosic varies depending on the nature of the biomass [36]. The rate of degradation or decomposition

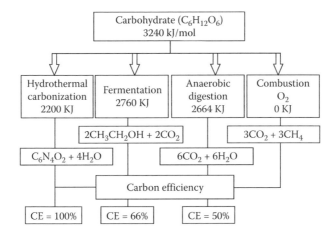

FIGURE 14.11 Carbon efficiency comparison of different carbonaceous process. *Note*: Carbon efficiency (CE) is defined as the relative amount of carbon from the starting product bound in the final product. This is analogous to the group efficiency of green chemistry. (From M. M. Titirici et al., *New J. Chem.*, 31(6), 787–789, 2007.)

of cellulose is much higher when compared to lignin and hemicelluloses [36,172,173]. Bergius et al. reported the transformation of cellulose into a coal-like material by hydrothermal process [174]. Berl and Schmidt have carried out methodical investigations of various biomasses by hydrothermal process at 150–350°C [175]. Bobleter et al. studied the hydrothermal degradation of cellulosic matter and hemicellulose under 200 to 275°C [176]. During hydrothermal treatment at low temperatures, cellulose has been converted into sugars and hydroxyl-methyl-furfural. At high temperatures, cellulose has been converted into glucose and cellobiose; similarly, hemicellulose has been converted into xylose and arabinose [177]. At low temperature (200°C), easily hydrolyzable polysaccharides are converted into soluble products. The conversion rate was further increased when the temperature rose to 260°C [177]. Cellulose (wood powder) impregnated with phenol in acidic condition at 150°C forms liquefied wood. Wood ceramic has been prepared from liquefied wood at various carbonization temperatures (400°C, 500°C, 650°C, and 800°C). When the temperature increases beyond 250°C, dehydration and de-polymerization of cellulose take place and with further increase in temperature beyond 400°C, condensation aromatic polynuclear structures start (Figure 14.12) [114].

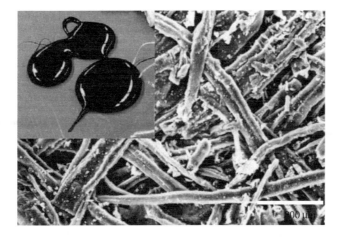

FIGURE 14.12 SEM images of liquefied wood and wood ceramic (400°C). (Reproduced from T. Hirose et al., *J. Mater. Sci.*, 36, 4145–4149, 2001.)

Formation of functionalized carbonaceous materials, such as saccharides, starch, sucrose, and glucose, has been observed by hydrothermal carbonization of cellulose at 220–250°C [115]. The formation of this material follows essentially the path of a hydrolysis followed by dehydration and fragmentation into soluble (monomer) products [115]. Shu-Hong Yu and Titirici extensively studied the hydrothermal carbonization of biomass and the mechanisms of char formation. The chemical reactions of biomass (e.g., saccharides, cellulose, lignins) involved in the hydrothermal process are complex in nature and most of the reports focus on initial steps of the reaction, such as hydrolysis, dehydration, and decarboxylation. Recently, formation of polymerization and aromatization in cellulose hydrothermal carbonization has been described, too [36,178,179].

Hydrothermal treatment of glucose, sucrose, or starch in aqueous solution at 150–350°C gives rise to water-soluble organic substances and a carbon-rich solid product [115]. Highly reactive carbon (surface area ~2700 m^2/g and micro-pore range 0.7–2.0 nm) was formed by hydrothermal carbonization using furfural, glucose, starch, cellulose, and eucalyptus sawdust. In this study, KOH was used as an activating agent [180]. Both ordered microporous and mesoporous functional carbonaceous materials were formed from D-fructose in the presence of copolymer surfactant Pluronic F127 by hydrothermal carbonization at 130°C [181]. Preparation of nitrogen-doped carbonaceous material has been demonstrated using D(+)-glucosamine hydrochloride by hydrothermal treatment at 180°C and the nitrogen content in the material was unaffected up to 750°C [182].

The presence of metal ions can effectively accelerate hydrothermal carbonization of various organic compounds, and an understanding of the formation of carbon materials in the presence of metal catalyst is not completely understood. Recent studies showed that carbonaceous materials can combine with metal nanoparticles and form distinctive structure and properties [37,38,110,111,145, 146,183–188]. Pt/Ru/carbon materials (carbon spheres) were prepared by hydrothermal carbonization process using glucose and starch as a precursor materials.

The precursor materials were thermally treated under argon atmosphere initially at 550°C for 4 h and thereafter at 900°C for 3 h [147]. Carbon nanocable/nanofiber was prepared from starch solution in the presence of Ag catalyst at 160–500°C for 12 h reaction time. In this study, various carbonaceous nanostructures, such as nanocables, hollow tubes, and hollow spheres, were demonstrated [148]. Hydrothermal carbonization of starch and rice grains has been carried out in the presence of Fe^{2+} and iron oxide (Fe_2O_3) nanoparticles at 200°C. Fe^{2+} ions and Fe_2O_3 nanoparticles showed significant changes in the morphology of the formed carbon nanomaterials [115]. Carbon-rich composite nanocables were prepared using glucose as starting materials in the presence of Te-nanowire. Removal of Te-nanowire from the nanocables produced uniform and functionalized carbonaceous nanofibers [37]. A well-defined silver/carbon microcables were produced by a one-step hydrothermal reaction at 180°C for 96 h using a mixture of $AgNO_3$, PVA, and a glucose-based saccharide [145]. Synthesis of Si/SiO_x/C nanocomposites was reported by hydrothermal carbonization of glucose in the presence of Si nanoparticles at 200–750°C [146].

14.4 APPLICATIONS OF HYDROTHERMAL CARBON

The production of nanostructured carbon materials by hydrothermal processes from natural precursors is one of the most attractive subjects in material science today. Carbon materials prepared from hydrothermal process are currently being used in various fields of research including environmental, electrical, chemical, and biomedical fields. In environmental application, carbon is mainly used as a sorbent material for the removal of heavy metal ions (CrO_4^{2-}, Pb^{2+}, and Cd^{2+}) from water and wastewater [116,117]. Carbon nanocoils prepared from saccharides (sucrose, glucose, and starch) with a support of Pt/Ru nanoparticles exhibit a high catalytic activity for the electro-oxidation of methanol in an acid medium [150]. Similarly, electro-oxidation of ethanol and methanol was carried out with electro catalysts, such as Pd/CHC (coin-like hollow carbon), Pt/HCS (hard carbon spherules), and Pt/Pd/CMS (carbon microspheres) in acidic and alkaline media

[39,130,131]. The conversion of CO to CO_2 at low temperatures was reported using Pd-, Pt-, and Au-loaded carbon nanofiber catalyst [38]. Selective hydrogenation of phenol to cyclohexanol was achieved using Pb-supported carbon spheres [132]. Silicon template mesoporous carbon sphere containing imidazole group is used as a catalyst for esterification, Knoevenagel, and Aldol reactions [133]. Pt-embedded microporous carbon (Pt/C/MC) nanoparticles showed methanol-tolerant behavior and considerable stability for oxygen electroreduction [134]. Recent study showed that the anodes made of carbon incorporated with Si/SiO_2, SnO_2, and Fe_3O_4 improved the overall electrochemical performance, such as cycle life and storage capacity of lithium-ion batteries [146,135,136]. Ag-embedded phenol formaldehyde resin core/shell nanospheres displayed strong green luminescent properties at the wavelength of 340 nm. Their application to *in vivo* bioimaging on human lung cancer cells has been demonstrated [137]. Anatase TiO_2 carbon hollow sphere (CHS) prepared by hydrothermal process displayed enhanced photocatalytic activity compared to commercially available P-25 TiO_2. Kinetic studies of TiO_2/CHS have been demonstrated against the degradation methylene blue in aqueous solution under UV light. The leaf-shaped ferric oxide which was impregnated in MWCNTs is used as catalyst for the electro-oxidation of ascorbic acid [155]. MWCNTs/CdS nanocomposites which were synthesized hydrothermally via direct growth of CdS nanoparticles on the functionalized MWCNT surface showed higher photocatalytic activity under visible light [156]. Similarly, MWCNT/ZnO nanocomposite was used as a photocatalyst for the removal of cyanide from water [157]. Hydrothermally synthesized rice-shaped Ti/CNT hybridized material which was used as the working electrode of dye-sensitized solar cells showed highest photoconversion efficiency compared to anatase TiO_2 [158]. Dye-sensitized solar cells performance was enhanced by coating of MWCNTs on conductive glass which was annealed in an Ar atmosphere [159].

14.5 FUTURE PERSPECTIVE

Hydrothermal technology has emerged as a most powerful tool because of its environmental-friendly approach. Hydrothermally synthesized carbons have been applied in various fields of research including electronics, nanotechnology, mechanical, biotechnology, and biomedical. Conversion of biomass and waste materials into alternative energy sources open new potentials for future research. A considerable amount of research is still required to fully understand the carbonization process of biomass and polysaccharides.

ACKNOWLEDGMENTS

The authors thank Professor Yury Gogotsi (Drexel University) for his encouragement and support. In addition, helpful discussion with Dr. Chin-Chiang Weng is gratefully acknowledged.

REFERENCES

1. K. Byrappa and M. Yoshimura, *Handbook of Hydrothermal technology: A Technology for Crystal Growth and Materials Processing*, Noyes Publications, New York, 2001.
2. K. F. E. Schafthaul, Die neuesten geologischen Hypothesen und ihr Verhältnis zur Naturwissenschaft überhaupt. *Gelehrte Anzeigen*, Koniglich Bayerische Akademie der Wissenschaften, 20, 557, 1845.
3. G. W. Morey and P. Niggli, The hydrothermal formation of silicates, a review. *J. Am. Chem. Soc.*, 35, 1086–1130, 1913.
4. S. Somiya, Historical developments of hydrothermal works in Japan, especially in ceramic science. *J. Mater. Sci.*, 41, 1307–1318, 2006.
5. G. W. Morey, Hydrothermal synthesis. *J. Am. Ceram. Soc.*, 36, 279–285, 1953.
6. R. Roy and O. F. Tuttle, *Physics and Chemistry of the Earth*. Pergamon Press, London, 1, 138, 1956.
7. K. Byrappa and M. Yoshimura, *Handbook of Hydrothermal Technology: Hydrothermal Technology–Principles and Application*, 2nd Edition, Elsevier, Amsterdam, p. 1, 2013.

8. (a) G. Demazeau, Solvothermal reaction: An original route for the synthesis of novel materials *J. Mater. Sci., 43*, 2104–2114, 2008. (b) G. Demazeau, V. Gonnet, V. Solozhenko, B. Tanguy, and H. Montigaud, C.R. *Acad. Sci.*, 320 (II b), 419, 1995.

9. M. Yoshimura and K. Byrappa, Hydrothermal processing of materials: Past, present and future. *J. Mater. Sci.*, 43, 2085–2103, 2008.

10. A. Rabenau, The role of hydrothermal synthesis in preparative chemistry. *Angewandte Chemie International Edition*, 24(12), 1026–1040, 1985.

11. E. Givagirov, K. S. Bagdasaroy, V. A. Kuznetsov, I. N. Demianets, and A. N. Lobachev, Growth from solutions In: *Modern Crystallography III*, M. Cardona, P. Fulde, and H.-J. Queisser, Editors, Springer Verlag, Berlin, Heidelberg, 353–414, 1984.

12. M. Yoshimura and S. Somiya, Fabrication of dense non-stabilized ZrO_2 ceramics by hydrothermal reaction sintering. *Bull. Am. Ceram. Soc.*, 59(2), 246, 1980.

13. E. Tani, M. Yoshimura, and S. Somiya, Effect of mineralizers on the crystallization of solid solutions in the system ZrO_2-CeO_2 under hydrothermal conditions. *Yogyo-Kyokai-Shi*, 90, 195–201 (Jpn.) 1982.

14. E. P. Stamburgh, Hydrothermal processing for advanced ceramic oxides. *Proceedings ICCTE, 86*, Technology shaping our future, 71–75, 1986.

15. S. Komarneni, E. Fregeau, E. Breval, and R. Roy, Hydrothermal preparation of ultrafine ferrites and their sintering. *J. Am. Ceram. Soc.*, 71(1), 26C–C28, 1988.

16. S. Somiya, Hydrothermal preparation of fine powders, *Advanced Ceramics III* (Elsevier), 207, 243, 1990.

17. S. Somiya, Powders: Hydrothermal preparation. In: *Concise Encyclopedia of Advanced Ceramic Materials*, R. J. Brook Editors. Pergamon Press, Oxford, 375–377, 1991.

18. M. Yoshimura, Importance of soft solution processing for advanced inorganic materials. *J. Mater. Res.*, 13(4), 796–802, 1998.

19. M. Yoshimura, W. L. Suchanek, and K. Byrappa, Soft solution processing—A strategy for one-step processing of advanced inorganic materials. *MRS Bulletin Special Issue*, 25(9), 17–25, 2000.

20. M. Yoshimura, Powder less processing for nano-structured bulk ceramics: Realization of direct fabrication from solutions and/or melts. *J. Ceram. Soc. Japan*, 114(11), 888–895, 2006.

21. M. Yoshimura, Soft solution processing, concept and realization of direct fabrication of shaped ceramics (Nano-crystals, whiskers, films, and/or patterns) in solutions without post-firing. *J. Mater. Sci.*, 41(5), 1299–1306, 2006.

22. M. Yoshimura, Feature and future of hydrothermal reactions for synthesis/preparation of nano-materials with desired shape, size and structures. *The 2nd International Solvothermal and Hydrothermal Association Conference, ISHA*, China National Conventional Center, Beijing, China, PL-4, 7–8, 2010.

23. Y. Gogotsi, P. Kofstad, M. Yoshimura, and K. G. Nickel, Formation of sp^3-bonded carbon upon hydrothermal treatment of SiC. *Diamond Rel. Mater.*, 5, 151–162, 1996.

24. S. Iijima, Helical microtubules of graphitic carbon. *Nature*, 354, 56–58, 1991.

25. M. Inagaki, Discussion of the formation of nanometric texture in spherical carbon bodies. *Carbon*, 35, 711–713, 1997.

26. P. Serp, R. Feurer, P. Kalek, Y. Kihn, J. L. Faria, and J. L. Figueiredo, A chemical vapor deposition process for the production of carbon nanospheres. *Carbon*, 39, 621–626, 2001.

27. M. Washiyama, M. Sakai, and M. Inagaki, Formation of carbon spherules by pressure carbonization relation to molecular structure of precursor. *Carbon*, 26, 303–307, 1988.

28. M. Inagaki, M. Washiyama, and M. Sakai, Production of carbon spherules and their graphitization. *Carbon*, 26, 169–172, 1988.

29. S. Tomita, T. Sakurai, H. Ohta, M. Fujii, and S. Hayashi, Structure and electronic properties of carbon onions. *J. Chem. Phys.*, 114, 7477–7483, 2001.

30. B. Basavalingu, P. Madhusudan, A. S. Dayananda, K. Lal, K. Byrappa, and M. Yoshimura, Formation of filamentous carbon through dissociation of chromium carbide under hydrothermal conditions. *J. Mater. Sci.*, 43, 2153–2157, 2008.

31. J. M. Calderon-Moreno, T. Fujino, and M. Yoshimura, Carbon nanocells grown in hydrothermal fluids. *Carbon*, 39, 618–621, 2001.

32. Y. Gogotsi, J. A. Libera, N. Kalashnikov, and M. Yoshimura, Graphite polyhedral crystals. *Science*, 290, 317–320, 2000.

33. W. Wang, J. Y. Huang, D. Z. Wang, and Z. F. Ren, Low-temperature hydrothermal synthesis of multiwall carbon nanotubes. *Carbon*, 43, 1317–1339, 2005.

34. M. Liu, C. Wang, and X. Wang, Interface-facilitated hydrothermal synthesis of sub-micrometre graphitic carbon plates. *J. Mater. Chem.*, 21, 15197–15200, 2011.

35. S. S. Swamy, J. M. Calderon-Moreno, and M. Yoshimura, Stability of single wall carbon nano tube under hydrothermal condition. *J. Mater. Res.*, 17(4), 734–737, 2002.

36. M. M. Titirici, R. J. White, C. Falco, and M. Sevilla, Black perspectives for a green future: Hydrothermal carbons for environment protection and energy storage. *Energy Environ. Sci.*, 5, 6796–6822, 2012.

37. H. S. Qian, S. H. Yu, L. B. Luo, J. Y. Gong, L. F. Fei, and X. M. Liu, Synthesis of uniform Te@Carbon rich composite nanocables with photoluminescent property and carbonaceous nanofibers by hydrothermal carbonization of glucose. *Chem. Mater.*, 18, 2102–2108, 2006.

38. H. S. Qian, M. Antonietti, and S. H. Yu, Hybrid "Golden Fleece": Unique approaches for synthesis of uniform carbon nanofibres and silica nanotubes embedded/confined with high population of noble metal nanoparticles and their catalytic performance. *Adv. Funct. Mater.*, 17, 637–643, 2007.

39. D. S. Yuan, C. W. Xu, Y. L. Liu, S. Z. Tan, X. Wang, Z. D. Wei, and P. K. Shen, Synthesis of coin-like hollow carbon and performance as Pd catalyst support for methanol electrooxidation. *Electrochem. Commun.*, 9, 2473–2478, 2007.

40. S. Ratchahat, N. Viriya-empikul, K. Faungnawakij, T. Charinpanitkul, and A. Soottitantawat, Synthesis of carbon microspheres from starch by hydrothermal process. *Sci. J. UBU*, 1(2), 40–45, 2010.

41. M. Yoshimura, J. Kase, and S. Somiya, Oxidation of Si_3N_4 and SiC by high temperature High pressure water vapor; *2nd International Conference on Ceramic Components for Engine*, Apr. 14–17, 529–536, *Proceedings of the Second International Symposium*, Edited by W. Bunk and H. Hausner, Verlag Deutsche Keramische Gesellscheft, 1986.

42. S. Shimada and T. Ishii, Oxidation kinetics of zirconium carbide at relatively low temperatures, *J. Am. Ceram. Soc.*, 73, 2804–2808, 1990.

43. R. C. DeVries, Synthesis of diamond under metastable conditions. *Annu. Rev. Mater. Sci.*, 17, 161–187, 1987.

44. S. Shimada, M. Inagaki, and K. Matsui, Oxidation kinetics of hafnium carbide in the temperature range of 480–600°C. *J. Am. Ceram. Soc.*, 75, 2671–2678, 1992.

45. R. C. De-Vries, R. Roy, S. Somiya, and J. Yamada, A review of highest phase systems pertinent to diamond systems. *Trans. Mater. Res. Sot. Jpn.*, B19, 641–667, 1994.

46. Y. Gogotsi and M. Yoshimura, Degradation of SiC-based fibers in high-temperature, high-pressure water. *J. Mater. Sci. Lett.*, 13, 395–399, 1994.

47. M. Reichle and J. J. Kickl, Investigation about the high temperature oxidation of titanium carbide (in Ger.). *J. Less-Common Met.*, 27, 213–236, 1972.

48. Y. Gogotsi, Y. Tanabe, E. Yasuda, and M. Yoshimura, Effect of oxidation and hydrothermal corrosion on strength of SiC fibres. In: *Advanced Materials '93, I/A: Ceramics, Powders, Corrosion and Advanced Processing*, N. Mizutani, Editor, Elsevier, Amsterdam, Netherlands, 1994.

49. R. Roy, D. Ravichandran, A. Badzian, and E. Breval, Attempted hydrothermal synthesis of diamond by hydrolysis of β-SiC powder. *Diamond Rel. Mater.*, 5, 973–976, 1996.

50. S. Shimada, F. Yunazar, and S. Otani, Oxidation of hafnium carbide and titanium carbide single crystals with the formation of carbon at high temperatures and low oxygen pressures. *J. Am. Ceram. Soc.*, 83, 721–728, 2000.

51. S. Shimada, Formation and mechanism of carbon-containing oxide scales by oxidation of carbides (ZrC, HfC, TiC). *Mater. Sci. Forum.*, 369–372, 377–384, 2001.

52. S. Shimada, A thermo analytical study on the oxidation of ZrC and HfC powders with formation of carbon. *Solid State Ionics*, 149, 319–326, 2002.

53. H. Zhang, V. Presser, C. Berthold, K. G. Nickel, X. Wang, C. Raisch, T. Chasse, L. He, and Y. Zhou, Mechanisms and kinetics of the hydrothermal oxidation of bulk titanium silicon carbide. *J. Am. Ceramic Soc.*, 93, 1148–1155, 2010.

54. S. Shimada, M. Nishisako, M. Inagaki, and K. Yamamoto, Formation and microstructure of carbon containing oxide scales by oxidation of single crystals of zirconium carbide. *J. Am. Ceram. Soc.*, 78, 41–48, 1995.

55. S. Shimada, K. Nakajima, and M. Inagaki, Oxidation of single crystals of hafnium carbide in a temperature range of 600–900°C. *J. Am. Ceram. Soc.*, 80, 1749–1756, 1997.

56. S. Shimada, M. Yoshimatsu, M. Inagaki, and S. Otani, Formation and characterization of carbon at the ZrC/ZrO_2 interface by oxidation of ZrC single crystals. *Carbon*, 36, 1125–1131, 1998.

57. S. Shimada, Microstructural observations of ZrO_2 scales formed by oxidation of ZrC single crystals with formation of carbon. *Solid State Ionics*, 101–103, 749–753, 1997.

58. Y. Gogotsi and M. Yoshimura, Formation of carbon films on carbides under hydrothermal conditions. *Nature*, 367, 628–630, 1994.

59. Y. Gogotsi and M. Yoshimura, Low temperature oxidation, hydrothermal corrosion and their effects on properties of SiC (Tyranno) fibers. *J. Am. Ceram. Soc.*, 78, 1439–1450, 1995.

60. Y. Gogotsi and K. G. Nickel, Formation of filamentous carbon from paraformaldehyde under high temperatures and pressures. *Carbon*, 36(7), 937–942 1998.

61. Y. Gogotsi, Interactions of non-oxide structural ceramics with high-temperature gaseous media. Dr. Sci. Thesis, Kiev, Institute for Problems of Materials Science, 1995.

62. G. Yushin, A. Nikitin, and Y. Gogotsi, Carbide derived carbon. In: *Nanomaterials Handbook,* Y. Gogotsi, Editor, CRC Press, Boca Raton, FL, 2006.

63. N. S. Jacobson, Y. Gogotsi, and Y. Yoshimura, Thermodynamic and experimental study of carbon formation on carbides under hydrothermal conditions. *J. Mater. Chem.*, 5 (4) 595–601, 1995.

64. F. P. Bunty, H. T. Hall, H. M. Strong, and R. H. Wentorf, Man-made diamonds. *Nature*, 176, 51–55, 1955.

65. F. P. Bundy, Direct conversion of graphite to diamond in static pressure apparatus. *J. Chem. Phys.* 38, 631–643, 1963.

66. R. H. Wentorf, The behavior of some carbonaceous materials at very high pressures and high temperatures. *Phys. Chem.*, 69, 3063–3069, 1965.

67. S. I. Hirano, K. Shimono, and S. Naka, Diamond formation from glassy carbon under high pressure and temperature conditions. *J. Mater. Sci.*, 17, 1856–1862, 1982.

68. K. E. Spear, A. W. Phelps, and W. B. White, Diamond polytype and their vibrational spectra. *J. Mater Res.*, 5, 2277–2285, 1990.

69. Y. Gogotsi, A. Kailer, and K. G. Nickel, transformation of diamond to graphite. *Nature*, 401, 663–664, 1999.

70. Y. Gogotsi, K. G. Nickel, D. Bahloul-Hourlier, T. Merle-Mejean, G. E. Khomenko, and K. P. Skjerlie, Structure of carbon produced by hydrothermal treatment of -SiC powder. *J. Mater. Chem.*, 6(4), 595–604, 1996.

71. K. Yokosawa, S. Korablov, K. Tohji, and N. Yamasaki, The possibility of diamond sintering by hydrothermal hot-pressing. *Water Dynamics: 3rd International Workshop on Water Dynamics, Nov.*, Sendai, Japan, 2005.

72. S. Korablov, K. Yokosawa, D. Korablov, K. Tohji, and N. Yamasaki, Hydrothermal formation of diamond from chlorinated organic compounds. *Mater. Lett.*, 60, 3041–3044, 2006.

73. N. Yamasaki, K. Yokosawa, S. Korablov, and K. Tohjt, Synthesis of diamond particles under alkaline hydrothermal conditions. *Solid State Phenomena*, 114, 271–276, 2006.

74. S. Korablov, K. Yokosawa, T. Sasaki, D. Korablov, A. Kawasaki, K. Ioku, H. Ishida, and N. Yamasaki, Synthesis of diamond from a chlorinated organic substance under hydrothermal conditions. *J. Mater. Sci.*, 42, 7939–7949, 2007.

75. S. Osswald, V. N. Mochalin, M. Havel, G. Yushin, and Y. Gogotsi, Phonon confinement effects in the Raman spectrum of nano-diamond. *Phys. Rev.-B*, 80, 075419, 2009.

76. M. Yoshikawa, N. Nagai, M. Matsuki, H. Fukuda, G. Katagiri, H. Ishida, A. Ishitani, and I. Nagai, Raman scattering from sp^2 carbon clusters. *Phys. Rev.-B*, 46, 7169–7174, 1992.

77. M. Frenklach, R. Kematick, D. Huang, W. Howard, K. E. Spear, A. W. Phelps, and R. Koba, Homogeneous nucleation of diamond powder in the gas phase. *J. Appl. Phys.*, 66, 395–400, 1989.

78. A. Szymanski, E. Abgarowicz, A. Bakon, A. Niedbalska, R. Slalcinski, and J. Sentek, Diamond formed at low pressures and temperatures through liquid phase hydrothermal synthesis. *Diamond Rel. Mater.*, 4, 234–235, 1995.

79. S. Yamaoka and M. Akaishi, Growth of diamond under HP/HT water condition. In: *Proceedings of the 6th NIRIM International Symposium Advanced Nat. (ISAM 99)*, pp. 1–2, Tsukuba, Jpn., 1999.

80. N. A. Bendeliani, T. D. Varfolomeeva, A. N. Glushko, N. A. Nikolaev, V. N. Slesarev, and S. V. Volkov, Interaction of diamonds with high-pressure water fluid. *Kristallografia*, 40(2), 380–381, 1995.

81. X. Z. Zhao, R. Roy, K. A. Cherian, and A. Badzian, Hydrothermal growth of diamond in metal-C-H$_2$O systems. *Nature*, 385, 513–515, 1997.

82. D. Ravichandran and R. Roy, Growth of diamond on diamond substrates in presence of an alkali and metal under hydrothermal conditions. *Mat. Res. Bull.*, 31, 1075–1082, 1996.

83. I. Schmidt and C. Benndorf, Mechanisms of low temperature growth of diamond using halogenated precursor gases. *Diamond Rel. Mater.*, 7, 266–271, 1998.

84. B. Basavalingu, J. M. Calderon-Moreno K. Byrappa, Y. Gogotsi, and M. Yoshimura, Decomposition of silicon carbide in the presence of organic compounds under hydrothermal conditions. *Carbon*, 39, 1763–1767, 2001.

85. Y. Gogotsi, T. Kraft, K. G. Nickel, and M. E. Zvanut, Hydrothermal behavior of diamond. *Diamond Rel. Mater.*, 7, 1459–1465, 1998.

86. E. Boellaard, P. K. De-Bokx, A. J. H. M. Kock, and J. W. Geus, The formation of filamentous carbon on iron and nickel catalysts: III. Morphology. *J. Catal.*, 96, 481–90, 1985.

87. M. Audier and M. Coulon, Kinetic and microscopic aspects of catalytic carbon growth. *Carbon*, 23(3), 317–323, 1985.
88. N. A. Kiselev, J. Sloan, D. N. Zakharov, E. F. Kukovitskii, J. L. Hutchison, J. Hammer, and A. S. Kotosonov, Carbon nanotubes from polyethylene precursors: Structure and structural changes caused by thermal and chemical treatment revealed by them. *Carbon*, 36(7), 1149–1157, 1998.
89. V. D. Blank, E. V. Polyakov, B. A. Kulnitskiy, A. A. Nuzhdin, Y. L. Alshevskiy, U. Bangert, A. J. Harvey, and H. J. Davock, Nanocarbons formed in a hot isostatic pressure apparatus. *Thin Solid Films*, 346, 86–90, 1999.
90. J. A. Libera and Y. Gogotsi, Hydrothermal synthesis of novel carbon filaments. *J. Am. Ceram. Soc.*, 82(11), 2942–2943, 1999.
91. M. Endo, S. Iijima, and M. S. Dresselhaus, Editors, *Carbon Nanotubes*, New York: Elsevier Science Ltd, 35–110, 1996.
92. K. Tanaka, T. Yamabe, and K. Fukui, Editors, *The Science and Technology of Carbon Nanotubes*, Amsterdam: Elsevier, 143–152, 1999.
93. R. T. K. Baker, Catalytic growth of carbon filaments. *Carbon*, 27(3), 315–323, 1989.
94. M. Endo, K. Takeuchi, K. Kobori, K. Takahashi, H. W. Kroto, and A. Sarkar, Pyrolytic carbon nanotubes from vapor-grown carbon fibers. In: *Carbon Nanotubes*, M. Endo S. Iijima, and M.S. Dresselhaus, Editors, Pergamon, Oxford, pp. 27–35, 1996.
95. G. A. Jablonski, F. W. Geurts, A. J. Sacco, and R. R. Biederman, Carbon deposition over Fe, Ni, and Co foils from CO-H-CH-H_2O, CO-CO_2, CH_4-H_2, and CO-H_2-H_2O gas mixtures: I. Morphology. *Carbon*, 30(1), 87–98, 1992.
96. Y. Gogotsi, J. A. Libera, and M. Yoshimura, Formation of nanostructured carbons under hydrothermal conditions. In: *Perspective of Fullerene Nanotechnology*, E. Osawa, Editor, Kluwer Academic Publisher, Dordrecht, p. 252, 2002.
97. V. G. Pol, M. Motiei, A. Gedanken, J. M. Calderon-Moreno, and M. Yoshimura, Carbon spherules: Synthesis, properties and mechanistic elucidation. *Carbon*, 42, 111–116, 2004.
98. M. Tagiri and S. Tsuboi, Mixed carbonaceous material in mesozoic shales and sandstones from the Yamizo mountain system, Japan. *J. Jpn. Assoc. Mineral. Petrol. Econ. Geol.*, 74, 47–56, 1979.
99. G. F. Wang, Carbonaceous material in the Ryoke metamorphic rocks, Kinki district, Japan. *Lithos*, 22, 305–316, 1989.
100. T. Itaya, Carbonaceous material in pelitic schists of the Sanbagawa metamorphic belt in central Shikoku, Japan. *Lithos*, 14, 215–224, 1981.
101. S. I. Hirano, K. Nakamura, and S. Somiya, Graphitization of carbon in presence of calcium compounds under hydrothermal condition by use of high gas pressure apparatus. In: *Hydrothermal Reactions for Materials Science and Engineering*, S. Somiya, Editor, Elsevier, London, 331–336, 1989.
102. P. R. Buseck and B. J. Huang, Conversion of carbonaceous material to graphite during metamorphism. *Geochim. Cosmochim. Acta*, 49, 2003–2016, 1985.
103. M. Tagiri and T. Oba, Hydrothermal syntheses of graphite from bituminous coal at 0.5–5 kbar water vapor pressure and 300–600°C. *J. Jpn. Assoc. Mineral. Petrol. Econ. Geol.*, 81, 260–271, 1986.
104. B. M. French and P. E. Rosenberg, Siderite ($FeCO_3$): Thermal decomposition in equilibrium with graphite. *Sci.*, 147, 1283–1284, 1965.
105. S. I. Hirano, K. Nakamura, and S. Somiya, Graphitization of carbon in the presence of calcium compounds under hydrothermal condition by use of high gas pressure apparatus. *International Conference on High Pressure, 4th, Proceedings*, Kyoto, Japan 25th Nov., 418–423, 1974.
106. A. A. Giardini, C. A. Salotti, and J. F. Lakner, Synthesis of graphite and hydrocarbons by reaction between calcite and hydrogen. *Science*, 159, 317–319, 1968.
107. N. Yamasaki, K. Hosoi, K. Yanagisawa. *J. Chem. Soc. Japan*, 11, 1909–1911, 1998.
108. J. M. Calderon-Moreno and M. Yoshimura, Hydrothermal processing of high quality multiwall nanotubes from amorphous carbon. *J. Am. Chem. Soc.*, 123(4), 741–742, 2001.
109. T. Luo, L. S. Gao, J. W. Liu, L. Y. Chen, J. M. Shen, L. C. Wang, and Y. T. Qian, Olivary particles: Unique carbon microstructure synthesized by catalytic pyrolysis of acetone. *J. Phys. Chem. B*, 109, 15272–15277, 2005.
110. T. W. Kim and L. A. Solovyov, Synthesis and characterization of large-pore ordered mesoporous carbons using gyroidal silica template. *J. Mater. Chem.*, 16, 1445–1455, 2006.
111. J. Y. Gong, L. B. Luo, S. H. Yu, H. S. Qian, and L. F. Fei, Synthesis of copper/cross-linked poly (vinyl alcohol) (PVA) nanocables via a simple hydrothermal route. *J. Mater. Chem.*, 16, 101–105, 2006.
112. M. Liu, Y. Yan, L. Zhang, X. Wang, and C. Wang, Hydrothermal preparation of carbon nanosheets and their supercapacitive behavior. *J. Mater. Chem.*, 22, 11458–11461, 2012.

113. F. Ma, H. Zhao, L. Sun, Q. Li, L. Huo, T. Xia, S. Gao, G. Pang, Z. Shi, and S. Feng, A facile route for nitrogen-doped hollow graphitic carbon spheres with superior performance in supercapacitors. *J. Mater. Chem.*, 22, 13464–13468, 2012.

114. T. Hirose, T. X. Fan, T. Okabe, and M. Yoshimura, Effect of carbonizing temperature on the basic properties of wood ceramics impregnated with liquefied wood. *J. Mater. Sci.*, 36, 4145–4149, 2001.

115. M. Sevilla, and A. B. Fuertes, The production of carbon materials by hydrothermal carbonization of cellulose. *Carbon*, 47, 2281–2289, 2009.

116. R. Demir-Cakan, N. Baccile, M. Antonietti, and M. M. Titirici, Carboxylate-rich carbonaceous materials via one-step hydrothermal carbonization of glucose in the presence of acrylic acid. *Chem. Mater.*, 21, 484–490, 2009.

117. Y. J. Xu, G. Weinberg, X. Liu, O. Timpe, R. Schlogl, and D. S. Su, Nano-architecturing of activated carbon: New facile strategy for chemical functionalization of the surface of activated carbon. *Adv. Funct. Mater.*, 18, 3613–3619, 2008.

118. Y. Gogotsi, S. Dimovski, and J. A. Libera, Conical crystals of graphite, *Carbon*, 40, 2263–2284, 2002.

119. W. L. Suchanek, J. A. Libera, Y. Gogotsi, and M. Yoshimura, Behavior of C60 under hydrothermal conditions: Transformation to amorphous carbon and formation of carbon nanotubes. *J. Solid State Chem.*, 160, 184–188, 2001.

120. B. Y. Zhao, T. Hirose, T. Okabe, M. Yoshimura, and K. A. Hu, Apple residue-new starting material for high surface area carbon. *J. Mater. Science Lett.*, 21, 333–336, 2002.

121. J. Qiu, Y. Li, Y. Wang, C. Liang, T. Wang, and D. Wang, A novel form of carbon micro-balls from coal. *Carbon*, 41, 767–772, 2003.

122. C. Yao, Y. Shin, L. Q. Wang, C. F. Windisch, W. D. Samuels, B. W. Arey, C. Wang, W. M. Risen, and G. J. Exarhos, Hydrothermal dehydration of aqueous fructose solutions in a closed system. *J. Phys. Chem. C*, 111, 15141–15145, 2007.

123. F. Salvador, M. J. Sanchez-Montero, and C. Izquierdo, C/H$_2$O reaction under supercritical conditions and their repercussions in the preparation of activated carbon. *J. Phys. Chem. C*, 111, 14011–14020, 2007.

124. Y. Shin, L. Q. Wang, I. T. Bae, B. W. Arey, and G. J. Exarhos, Hydrothermal syntheses of colloidal carbon spheres from cyclodextrins, *J. Phys. Chem. C*, 112, 14236–14240, 2008.

125. Y. Mi, W. Hu, Y. Dan, and Y. Liu, Synthesis of carbon micro-spheres by a glucose hydrothermal method. *Mat. Lett.*, 62, 1194–1196, 2008.

126. F. Cheng, J. Liang, J. Zhao, Z. Tao, and J. Chen, Biomass waste-derived microporous carbons with controlled texture and enhanced hydrogen uptake. *Chem. Mater.*, 20, 1889–1895, 2008.

127. M. T. Zheng, Y. L. Liu, Y. Xiao, Y. Zhu, Q. Guan, D. S. Yuan, and J. X. Zhang, An easy catalyst-free hydrothermal method to prepare monodisperse carbon microspheres on a large scale, *J. Phys. Chem. C*, 113, 8455–8459, 2009.

128. B. J. He, Y. Zhang, Y. Yin, T. L. Funk, and G. L. Riskowski, Operating temperature and retention time effects on the thermochemical conversion process of swine manure, *Trans. ASAE*, 43(6), 1821–1825, 2000.

129. S. Manafi, M. B. Rahaei, Y. Elli, and S. Joughehdoust, High-yield synthesis of multi-walled carbon nanotube by hydrothermal method. *Can. J. Chem. Eng.*, 88(3), 283–286, 2010.

130. R. Z. Yang, X. P. Qiu, H. R. Zhang, J. Q. Li, W. T. Zhu, Z. X. Wang, X. J. Huang, and L. Q. Chen, Monodispersed hard carbon spherules as a catalyst support for the electrooxidation of methanol. *Carbon*, 43, 11–16, 2005.

131. C. W. Xu, L. Q. Cheng, P. K. Shen, and Y. L. Liu, Methanol and ethanol electrooxidation on Pt and Pd supported on carbon microspheres in alkaline media. *Electrochem. Commun.*, 9, 997–1001, 2007.

132. P. Makowski, R. D. Cakan, M. Antonietti, F. Goettmann, and M. M. Titirici, Selective partial hydrogenation of hydroxy aromatic derivatives with palladium nanoparticles supported on hydrophilic carbon. *Chem. Commun.*, (8), 999–1001, 2008.

133. R. Demir-Cakan, P. Makowski, M. Antonietti, F. Goettmann, and M. M. Titirici, Hydrothermal synthesis of imidazole functionalized carbon spheres and their application in catalysis. *Catal. Today*, 150, 115–118, 2010.

134. Z. H. Wen, J. Liu, and J. H. Li, Core/Shell Pt/C nanoparticles embedded in mesoporous carbon as a methanol-tolerant cathode catalyst in direct methanol fuel cells. *Adv. Mater.*, 20, 743–747, 2008.

135. X. W. Lou, C. M. Li, and L. A. Archer, Designed synthesis of coaxial SnO$_2$@carbon hollow nanospheres for highly reversible lithium storage. *Adv. Mater.*, 21, 2536–2539, 2009.

136. W. M. Zhang, X. L. Wu, J. S. Hu, Y. G. Guo, and L. J. Wan, Carbon coated Fe$_3$O$_4$ Nanospindles as a superior anode material for lithium-ion batteries. *Adv. Funct. Mater.*, 18, 3941–3946, 2008.

137. S. R. Guo, J. Y. Gong, P. J. M. Wu, Y. Lu, and S. H. Yu, Biocompatible, luminescent silver Large-scale synthesis and application for *in vivo* bioimaging. *Adv. Funct. Mater.*, 18, 872–879, 2008.

138. Y. Ao, J. Xu, D. Fu, and C. Yuan, A simple method for the preparation of titania hollow sphere. *Catal. Commun.*, 9, 2574–2577, 2008.

139. V. V. Kovalevski and A. N. Safronov, Pyrolysis of hollow carbons on melted catalyst. *Carbon*, 36(7–8), 963–968, 1998.

140. X. M. Sun and Y. D. Li, Colloidal carbon spheres and their core/shell structures with noble-metal nanoparticles. *Angew. Chem. Int. Ed.*, 43, 597–601, 2004.

141. Q. Wang, H. Li, L. Q. Chen, and X. J. Huang, Monodispersed hard carbon spherules with uniform nanopores. *Carbon*, 39, 2211–2214, 2001.

142. T. Yogo, H. Tanaka, S. Naka, and S. I. Hirano, Synthesis of boron-dispersed carbon by pressure pyrolysis of organoborane copolymer. *J. Mater. Sci.*, 25, 1719–1723, 1990.

143. Y. Gogotsi, J. A. Libera, and M. Yoshimura, Hydrothermal synthesis of multiwall carbon nanotubes. *J. Mater. Res.*, 15(12), 2591–2594, 2000.

144. Y. Gogotsi, N. Naguib, and J. A. Libera, *In situ* chemical experiments in carbon nanotubes. *Chem. Phys. Lett.*, 365, 354–360, 2002.

145. L. B. Luo, S. H. Yu, H. S. Qian, and J. Y. Gong, Large scale synthesis of uniform silver@carbon rich composite (carbon and cross-linked PVA) sub-microcables by a facile green chemistry carbonization approach. *Chem. Commun.*, (7), 793–795, 2006.

146. Y. S. Hu, D. C. Rezan, M. M. Titirici, J. O. Muller, R. Schlogl, M. Antonietti, and J. Maier, Superior lithium storage performance of Si@SiOx/C nanocomposite as novel anode material for lithium-ion batteries. *Angew. Chem., Int. Ed.*, 47, 1645–1649, 2008.

147. M. M. Tusi, M. Brandalise, O. V. Correa, A. O. Neto, M. Linardi, and E. V. Spinacé, Preparation of PtRu/ carbon hybrids by hydrothermal carbonization. *Proc. Mat. Res.*, 10(2), 171–175, 2007.

148. S. H. Yu, X. Cui, L. Li, K. Li, B. Yu, M. Antonietti, and H. Coelfen, From starch to carbon/metal hybrid nanostructures: Hydrothermal metal catalyzed carbonization. *Adv. Mater.*, 16, 1636–1640, 2004.

149. X. J. Cui, M. Antonietti, and S. H. Yu, Structural effects of iron oxide nanoparticles and iron ions on the hydrothermal carbonization of starch and rice carbohydrates. *Small*, 2(6), 756–759, 2006.

150. M. Sevilla, G. Lota, and A. B. Fuertes, Saccharide-based graphitic carbon nanocoils as supports for Pt/ Ru nanoparticles for methanol electrooxidation. *J. Power Sources*, 171, 546–511, 2007.

151. S. Manafi, M. B. Rahaei, Y. Elli, and S. Joughehdoust, High yield synthesis of multi-walled carbon nanotube by hydrothermal method. *Can. J. Chem. Eng.*, 88(3), 283–286, 2010.

152. S. Manafi, H. Nadali, and H. R. Irani, Low temperature synthesis of multi-walled carbon nanotubes via a sonochemical/hydrothermal method. *Mat. Lett.*, 62, 4175–4176, 2008.

153. Y. Gogotsi, J. A. Libera, A. Güvenç-Yazicioglu, and C. M. Megaridis, In-situ multi-phase fluid experiments in hydrothermal carbon nanotubes. *Appl. Phys. Lett.*, 79(7), 1021–1023, 2001.

154. D. C. Lee, F. V. Mikulec, and B. A. Korgel, Carbon nanotube synthesis in supercritical toluene. *J. Am. Chem. Soc.*, 126, 4951–4957, 2004.

155. K. Ding, Hydrothermal synthesis of leaf shaped ferric oxide particles onto multi-walled carbon nanotubes (MWCNTs) and it's catalysis for the electrooxidation of ascorbic acid. *Int. J. Electrochem. Sci.*, 4, 943–953, 2009.

156. T. Peng, P. Zeng, D. Ke, X. Liu, and X. Zhang, Hydrothermal preparation of multiwalled carbon nanotubes (MWCNTs)/CdS nanocomposite and its efficient photocatalytic hydrogen production under visible light irradiation. *Energy Fuels*, 25, 2203–2210, 2011.

157. T. A. Saleh, M. A. Gondal, and Q. A. Drmosh, Preparation of a MWCNT/ZnO nanocomposite and its photocatalytic activity for the removal of cyanide from water using a laser. *Nanotech.*, 21(49), 495705, 2010.

158. T. Charinpanitkul, P. Lorturn, W. Ratismith, N. Viriya-empikul, G. Tumcharern, and J. Wilcox, Hydrothermal synthesis of titanate nanoparticle/carbon nanotube hybridized material for dye sensitized solar cell application. *Mat. Res, Bull.*, 46, 1604–1609, 2011.

159. S. Siriroj, S. Pimanpang, M. Towannang, W. Maiaugree, S. Phumying, W. Jarernboon, and V. Amornkitbamrung, High performance dye-sensitized solar cell based on hydrothermally deposited multiwall carbon nanotube counter electrode. *Appl. Phys. Lett.*, 100, 243303, 2012.

160. A. K. Geim and K. S. Novoselov, The rise of graphene. *Nat. Mater.*, 6, 183–191, 2007.

161. L. Xue, C. Shen, M. Zheng, H. Lu, N. Li, G. Ji, L. Pan, and J. Cao, Hydrothermal synthesis of graphene-ZnS quantum dot nanocomposites. *Mat. Lett.*, 65, 198–200, 2011.

162. J. Chu, X. Li, and J. Qi, Hydrothermal synthesis of CdS microparticles-graphene hybrid and its optical properties. *Cryst. Eng. Comm.*, 14, 1881–1884, 2012.

163. J. Shen, B. Yan, M. Shi, H. Ma, N. Li, and M. Ye, One step hydrothermal synthesis of TiO_2-reduced graphene oxide sheets. *J. Mater. Chem.*, 21, 3415–3421, 2011.

164. P. Wang, T. Jiang, C. Zhu, Y. Zhai, D. Wang, and S. Dong, One-Step, solvothermal synthesis of graphene-CdS and graphene-ZnS quantum dot nanocomposites and their interesting photovoltaic properties. *Nano Res.*, 3(11), 794–799, 2010.

165. S. D. Perera, R. G. Mariano, K. Vu, N. Nour, O. Seitz, Y. Chabal, and K. J. Balkus, Hydrothermal synthesis of graphene-TiO$_2$ nanotube composites with enhanced photocatalytic activity. *ACS Catal.*, 2, 949–956, 2012.

166. S. Huang, G. Zhu, C. Zhang, W. W. Tjiu, Y. Y. Xia, and T. Liu, Immobilization of Co-Al layered double hydroxides on graphene oxide nanosheets: Growth mechanism and super capacitor studies. *ACS Appl. Mater. Interfaces*, 4, 2242–2249, 2012.

167. S. Stephanidis, C. Nitsos, K. Kalogiannis, E. F. Iliopoulou, A. A. Lappas, and K. S. Triantafyllidis, Catalytic upgrading of lingo-cellulosic biomass pyrolysis vapour: Effect of hydrothermal pre-treatment of biomass. *Catal. Today*, 167(1), 37–45 2011.

168. J. A. Libra, K. S. Ro, C. Kammann, A. Funke, N. D. Berge, Y. Neubauer, M. M. Titirici, C. Fühner, O. Bens, J. Kern, and K. H. Emmerich, Hydrothermal carbonization of biomass residuals: A comparative review of the chemistry, processes and applications of wet and dry pyrolysis. *Biofuels*, 2(1), 89–124, 2011.

169. O. Bobleter, Hydrothermal degradation of polymers derived from plants. *Prog. Polym. Sci.*, 19, 797–841 1994.

170. M. G. Groenli, G. Varhegyi, and C. Di-Blasi, Thermo gravimetric analysis and devolatilization kinetics of wood. *Ind. Eng. Chem. Res.*, 41, 4201–4208, 2002.

171. M. M. Titirici, A. Thomas, and M. Antonietti, Back in the black: Hydrothermal carbonization of plant material as an efficient chemical process to treat the CO$_2$ problem? *New J. Chem.*, 31(6), 787–789, 2007.

172. Y. S. Sun, X. B. Lu, S. T. Zhang, R. Zhang, and X. Y. Wang, Kinetic study for Fe(NO$_3$)$_3$ catalyzed hemi-cellulose hydrolysis of different corn stover silages. *Bioresour. Technol.*, 102, 2936–2294, 2011.

173. M. P. Pandey and C. S. Kim, Lignin depolymerization and conversion: A review of thermochemical methods. *Chem. Eng. Technol.*, 34(1), 29–41, 2011.

174. F. Bergius, *Die Anwendung hoher drucke bei chemischen vorgängen und eine nachbildung des entstehungsprozesses der steinkohle*; Verlag Wilhelm Knapp: Halle an der Saale. *Germany*, 58, 1913.

175. E. Berl and A. Schmidt, Über die entstehung der kohlen. II. Die Inkohlung von cellulose und lignin in neutralem medium. *Justus Liebigs Annalen Chemie.*, 493, 97–123, 1932.

176. O. Bobleter, R. Niesner, and M. Rohr, The hydrothermal degradation of cellulosic matter to sugars and their fermentative conversion to protein. *J. Appl. Poly. Sci.*, 20, 2083–2093 1976.

177. G. Bonn, R. Concin, and O. Bobleter, Hydrothermolysis—A new process for the utilization of biomass. *Wood Sci. Technol.*, 17, 195–202, 1983.

178. C. Falco, F. P. Caballero, F. Babonneau, C. Gervais, G. Laurent, M. M. Titirici, and N. Baccile, Hydrothermal carbon from biomass: Structural differences between hydrothermal and pyrolyzed carbons via ^{13}C solid state NMR. *Langmuir*, 27, 14460–14471, 2011.

179. M. M. Titirici, M. Antonietti, and N. Baccile, Hydrothermal carbon from biomass: A comparison of the local structure from poly to monosaccharides and pentoses/hexoses. *Green Chem.*, 10, 1204–1212, 2008.

180. M. Sevilla, A. B. Fuertes, and R. Mokaya, High density hydrogen storage in superactivated carbons from hydrothermally carbonized renewable organic materials. *Energy Environ. Sci.*, 4, 1400–1410, 2011. (123).

181. S. Kubo, R. J. White, N. Yoshizawa, M. Antonietti, and M. M. Titirici, Ordered carbohydrate-derived porous carbons. *Chem. Mater.*, 23, 4882–4885, 2011.

182. L. Zhao, N. Baccile, S. Gross, Y. Zhang, W. Wei, Y. Sun, M. Antonietti, and M. M. Titirici, Sustainable nitrogen-doped carbonaceous materials from biomass derivatives. *Carbon*, 48, 3778–3787, 2010.

183. B. B. Hu, K. Wang, L. Wu, S. H. Yu, M. Antonietti, and M. M. Titirici, Engineering carbon materials from the hydrothermal carbonization process of biomass. *Adv. Mater.*, 22, 813–828, 2010.

184. Y. C. Liang, M. Hanzlik, and R. Anwander, Periodic mesoporous organosilicas: Mesophase control via binary surfactant mixtures. *J. Mater. Chem.*, 16, 1238–1253, 2006.

185. M. M. Titirici, A. Thomas, and M. Antonietti, Aminated hydrophilic ordered mesoporous carbons. *J. Mater. Chem.*, 17, 3412–3418, 2007.

186. J. Wang, J. C. Groen, W. Yue, W. Zhou, and M. O. Coppens, Facile synthesis of ZSM-5 composites with hierarchical porosity. *J. Mater. Chem.*, 18, 468–474, 2008.

187. B. Deng, A. W. Xu, G. Y. Chen, R. Q. Song, and L. P. Chen, Synthesis of copper-core/carbon-sheath nanocables by a surfactant-assisted hydrothermal reduction/carbonization process. *J. Phys. Chem. B*, 110, 11711–11716, 2006.

188. X. M. Sun, and Y. D. Li, Ag Controlled synthesis, characterization, and assembly. *Langmuir*, 21, 6019–6024, 2005.

189. K. Byrappa and M. Yoshimura, *Handbook of Hydrothermal Technology: Hydrothermal Technology for Nanotechnology—A Technology for Processing of Advanced Materials*, 2nd Edition, Elsevier Amsterdam, p 615, 2013.

15 Carbon Nanomaterials for Water Desalination by Capacitive Deionization

P. Maarten Biesheuvel, Slawomir Porada, Albert van der Wal, and Volker Presser

CONTENTS

15.1 INTRODUCTION

The availability of affordable clean water is one of the key technological, social, and economical challenges of the twenty-first century. Clean water, acknowledged as a basic human right by the United Nations (General Assembly GA/10967), is still unavailable to one of seven people worldwide. Because of increasing groundwater extraction, salt water ingress in wells and aquifers continues. As a consequence, interest in desalination technologies is steadily growing. During the years, several desalination technologies have been developed among which distillation, reverse osmosis, and electrodialysis are the most commonly known and the most widespread.[1] There is an increased interest in rendering these technologies more energy and cost efficient both for the deionization of seawater as well as for brackish water. On the basis of the fact that there is more brackish water than fresh water in the world, it is evident that it is particularly attractive to utilize the large brackish water resources for human consumption and residential use, agriculture, and industry.

Capacitive deionization (CDI) has emerged as a robust, energy-efficient, and cost-efficient technology.[1] CDI is particularly attractive for desalination of brackish water with a low or moderate salt content. Brackish water is defined as that between potable water (freshwater) and seawater and, thus, contains 0.5–30 mass‰ of dissolved salt, whereas the salt content of seawater ranges between 30 and 50 mass‰. CDI for brackish water is energy efficient due to the fact that the salt ions, which are the minority compound, are removed from the mixture, not the water. Furthermore, energy release during electrode regeneration (ion release) can be utilized to charge a neighboring cell in the ion-electrosorption step. As will be explained later in detail, a CDI cycle consists of an ion (electro)adsorption, or charging, step to purify the water, where ions are adsorbed in anode/cathode porous carbon electrode pairs, followed by a step in which the ions are released, that is, are desorbed from the electrodes, and thus the electrodes are regenerated. When we neglect further energy-consuming processes (like pumps, regulating power electronics, etc.), we may write as a first approximation for the energy consumption of water desalination (with W denoting energy) as: $W_{CDI} = W_{deionization} - W_{regeneration} = W_{charge} - W_{discharge}$.[2] Energy consumption in CDI will be discussed in detail in Section 15.6 of this chapter.

With most of the research efforts dealing with porous carbon electrodes dedicated to capacitive energy storage devices, CDI has been somewhat overlooked for decades. In fact, the application of porous carbon electrodes for water desalination has been documented since the 1960s, when it was called "electrochemical demineralization" or "electrosorb process for desalting water."[3–7] However, it never attracted significant research initiatives nor has it led to commercially viable technologies in the past. During the last couple of years, the situation has changed completely: academic interest in this technology has increased exponentially (Figure 15.1), and companies are successfully marketing their commercial CDI technologies (Figure 15.2).

The mechanism of CDI is schematically shown in Figure 15.2. CDI uses pairs of oppositely placed porous carbon electrodes which store ions upon applying an electrical voltage difference. Such electrodes can be utilized in stacks of n pairs. The ions are harvested from the water flowing through a "spacer channel" between the two electrodes and are stored in the pores inside the carbon material. This process is based on the formation of an electrical double layer (EDL)—the cornerstone of capacitive energy storage and the mechanism by which salt ions are immobilized and selectively extracted from brackish water. After some time, all of the accessible surface area is saturated by electrosorbed ions and the storage capacity of the device is reached. To regenerate the carbon electrodes, the ions are released again by reducing or even reversing the cell voltage. In this manner, a small stream enriched in ions is produced and the electrodes regain their initial ion uptake capacity. Without chemical reactions, this process is purely physical in nature and withholds the potential of a long service life and low maintenance.

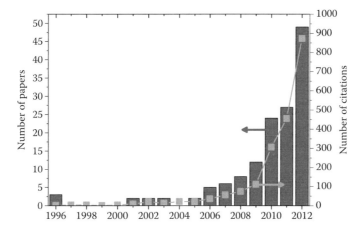

FIGURE 15.1 Number of papers on CDI published annually since 1996 and the corresponding number of annual citations of these publications (status June 2013; data extracted from Web of Knowledge Citation Report).

Compared to the classic work of the twentieth century, various new technologies are considered for CDI today, such as the inclusion of ion-exchange thin membrane barriers in front of the electrodes,[8–12] optimized operational modes such as stop-flow operation during ion release,[13] salt release at reversed voltage,[12] constant-current operation,[14] energy recovery from the desalination/release cycle,[14] and flow-through electrodes where the water is directed head-on through the electrodes.[15,16]

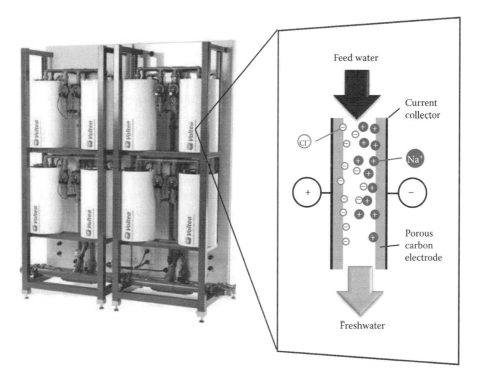

FIGURE 15.2 CDI is based on a flow-through cell design where influenced by the electrical field, cations migrate from the spacer channel into the negatively charged porous carbon electrode (cathode) while anions migrate into the anode. Shown also is a commercial CDI system with 8 modules, each consisting of 12 stacks, which each consist of 24 cell pairs (CapDI CT 8, Voltea B.V., The Netherlands).

On the materials side, new materials and design strategies for novel and improved electrodes continue to emerge. Fundamentally, even the very basic question of the "best" material remains unanswered. Naturally, the choice of electrode material largely depends on required performance (desalting capacity, initial and final salt concentration), system requirements (flow rate, stack configuration), and cost consideration (efficiency, material costs, lifetime). Although the choice of material is in theory not limited to carbon, only carbon is used in practice. Carbon development has emerged from "just" using activated carbons known from energy storage or sorption applications, to the synthesis of CDI optimized structures, including ordered mesoporous carbons[17] and metal-oxide/carbon mixtures.[18,19]

In this chapter, we summarize the basics and the theory behind CDI. Alongside with reporting on carbon materials used for CDI, we provide guidelines and strategies for a rational design of porous carbon electrodes for desalination applications.

In Sections 15.2 through 15.6, the terminologies macropores and micropores are based on that used in porous electrode theory,[20–23] with the term "macropores" denoting the electrolyte-filled continuous interparticle space in between carbon particles, serving as transport pathways for ion transport across the electrode, whereas the term "micropores" is used for all the pore space within the carbon particles (intraparticle porosity; see Figure 15.5b). In Section 15.7, the formal IUPAC terminology for porous material characterization is used where macro-, meso-, and micropores are distinguished on the basis of the pore sizes in a porous material.[24]

15.2 EXPERIMENTAL APPROACHES TO OPERATION AND TESTING OF CDI

15.2.1 ELECTROCHEMICAL EXPERIMENTATION VERSUS DESALINATION USING A TWO-ELECTRODE CELL PAIR

For the characterization and testing of CDI electrodes, two different approaches have been described in literature. The first approach is to perform electrochemical (EC) analysis, in general, based on a setup consisting of a working, counter, and reference electrode. In this approach, only current and voltage signals are measured, and not the actual change of salt concentration, and the water containing the salt ions is not necessarily flowing along or through the porous carbon electrode (which is the working electrode). All three electrodes are typically different, with only the working electrode made of the CDI material to be tested.

In a different approach, water containing salt flows along or through a cell pair consisting of two porous carbon electrodes usually made from the same material and equal in mass and dimensions, between which a cell voltage is applied and no third (i.e., reference) electrode is used. The degree of desalination (see Section 15.2.3) as well as the current responding to a certain cell voltage (see Sections 15.2.3 and 15.2.4) can be measured in this setup. In the following sections, we discuss this second approach, where water is desalinated using cell pairs consisting of two porous electrodes. Section 15.5 discusses one mode of EC testing using a symmetric porous electrode cell pair.

The use of the first approach (EC testing using a three-electrode setup) finds its roots in similar testing of single electrodes for supercapacitor applications, and in the study of (planar) electrodes where faradaic reactions take place. Instead, in CDI ideally no faradaic reactions take place, and capacitive processes are most important. In such a three-electrode experiment, the voltage of the electrode under study is measured relative to a reference electrode, as function of the current running between the working and counter electrode. Several well-known techniques are available to modulate the voltage and current signals and two of the most often used for CDI are: (1) cyclic voltammetry (CV) where the (working) electrode potential is swept in time following a triangular pattern between a lower and higher set point in voltage and the resulting current is plotted in a CV diagram; and (2) electrochemical impedance spectroscopy (EIS), where a small voltage signal, sinusoidally changing over time, is varied in a large window of frequencies, and the responding current (also sinusoidally varying, but with a time lag) is analyzed to construct Nyquist plots and other representations that allow one to derive the magnitude of resistances and capacities in the electrode.

Although these techniques are well established,[25] their application to porous electrodes is more complicated and analysis requires understanding of distributed charge/resistance networks.[26-28] Classic analysis of data obtained from these methods assumes that the local resistances in the electrode are constant, for example, not changing as a function of time and position in the electrode. However, this condition is not met in water of a salt concentration in the brackish water regime, as we will discuss in Section 15.4.3, and therefore we argue in Section 15.5 that such experiments (also when used for CDI in brackish water) should be based on high-ionic strength conditions of the electrolyte (e.g., >0.5 M). Mathematical methods to analyze CV and EIS curves obtained in low salinity solutions need first to be developed for porous electrodes, before these methods can be used to characterize CDI electrodes directly in brackish water.

Therefore, in this chapter, we focus on the second experimental approach discussed earlier, namely testing of a symmetric two-porous electrode cell, using the cell voltage applied between the two electrodes as the primary electrical signal. We note that this second approach is also more reminiscent of actual operation and desalination using CDI, which is another reason to use this approach. In the next three sections, we first discuss different CDI cell geometries (Section 15.2.2), followed by different approaches to testing desalination performance (Section 15.2.3), and third, different modes of applying an electrical driving force (Section 15.2.4).

15.2.2 Geometries for CDI Testing Based on a Two-Electrode Layout

Most experimental work on CDI uses two porous carbon film electrodes with a typical thickness between 100 and 500 μm, placed parallel to one another in such a way that a small planar gap is left in between the electrodes through which water can flow along the electrodes. A typical electrode for laboratory scale experiments is in the range of 5×5 cm^2 to 10×10 cm^2. Such electrodes can be constructed either as freestanding thin films, or be coated directly onto a flexible current collector such as graphite foil.[9,29-31] It is possible to test a single cell pair, or to construct a stack of multiple cell pairs. In that case, each current collector layer is contacting two porous electrodes (one on each side) and, thus, the sequence of anode–cathode is reversed from one cell to the next in the stack. Typically, each electrode is the same as the other in a cell pair, though cell pairs with electrodes based on different synthesis conditions have been described.[32-37] Recently,[38] a different kind of asymmetric cell geometry was introduced, namely to increase the cathode electrode mass relative to the anode electrode mass by factors of two and three (and vice versa), and compare desalination performance with the symmetric system where the two electrodes have equal masses.

In the published work on CDI, the water usually flows through a slit in between the two parallel porous electrodes, from one side to the other. This slit can be an open channel, typically at least 1 mm in thickness, or can be constructed from a spacer material, being a porous thin layer, of thickness typically between 100 and 300 μm. The geometry is normally not such that a purely one-dimensional flow pattern arises, but instead water flows from one edge of a square channel to an exit point at the opposite corner,[39] or water flows from a hole in the center of a square cell and then radially outward to leave the cell on all four sides;[40] the direction of this flow pattern can also be reversed.

This approach, with water flowing along the electrodes is called "flow-through capacitor technology" in the patent literature, distinguishing this approach from supercapacitor-like technology where the electrolyte is stagnant and the box containing electrodes and electrolyte can be closed off. However, in some papers,[15,41] the wording is quite different from this convention. There, a distinction is made between the above approach (called "flow-by" mode) and the "flow-through" mode, where the latter refers to an operational mode where water is pumped straight through the electrodes.[15,16]

Another approach is called "parametric pumping."[6,13,42,43] In this approach, two separate streams are produced from different exit points: a freshwater stream from one end of the device and a brine stream from the other end, with the feed water being injected about halfway along the length of the relatively long spacer (water) channel. In parametric pumping, although water is being fed continuously through the inlet hole in the middle of the column, the cell voltage between the two electrodes is applied during a certain period of time and turned off during another period of time. During the period of applying the cell voltage, a valve is opened on one end of the channel (valve A) and another valve closed on the opposite end of that channel (valve B), allowing desalinated water to come out at end A. During the period of zero cell voltage, valve A is closed and valve B is opened so that brine can be extracted through B.

A new design employs movable carbon rod electrode wires,[44] thereby avoiding the sequential production of freshwater and brine from the same device for different periods of time. Instead, the freshwater and brine streams are separated at all times. Thus, in contrast to the standard mode for CDI, it does not have the disadvantage of the required precise switching of the effluent stream into freshwater and into brine, and the inadvertent mixing of just-produced water with the untreated water that may occur right after switching. In the wire-based approach, cell pairs are constructed from wires, or thin rods, with anode wires positioned close to cathode wires. An array of such wire pairs is lowered into the water and upon applying a voltage difference between the anode and cathode wires, salt ions are adsorbed into their counter electrodes. The wire approach requires no spacer layer as long as the wires are sufficiently rigid. The assembly of wires is lifted from the compartment (or stream) that is desalinated, and immersed into another water stream, upon which the cell voltage is reduced again to zero and salt is released. After salt release, the procedure can be repeated; thus, two continuous streams are obtained, one in which the feed water is steadily converted into freshwater and one where feed water continuously becomes more saline. The saline stream (into which adsorbed salt is released) can be of a different source as the water to be desalinated. When there is only one source of feed water, then by upfront splitting the feed water stream in a certain ratio of volumes, the water recovery of the process is automatically determined. Here, water recovery (WR) is the ratio of freshwater volume over inlet volume, an important determinant to analyze performance of a water desalination process.

For all such setups, it is important to study dynamic equilibrium. CDI is a cyclic process, and each cycle is broken down in a well-defined series of steps (flow rate adjustment, voltage adjustment, etc.). After the commencement of a new experiment with the fixed conditions that determine the cycle, it will take a few cycles before the desalination performance of each new cycle is the same as the one before, that is, the salt effluent concentration versus time profile has become the same for each cycle. This situation is called the limit cycle, or dynamic equilibrium, and this condition is typically reached after two or three cycles. Not taking degradation mechanisms into account and as long as the operational settings do not change, dynamic equilibrium should continue *ad infinitum*. It is important to report experimental results of dynamic equilibrium, as the first few cycles after start of a new experiment can give quite markedly different desalination performance, due to transient effects of salt accumulation in the cell and electrodes. For instance, experiments showing results of the very first cycle can be influenced by effects such as chemisorption of ions in the carbon on first contact, and will not be desorbed when the voltage is reversed.[9,45,46] Such transient, start-up effects are interesting to study in their own respect, but for CDI it is more important to present results of dynamic equilibrium behavior because CDI is meant to run continuously for many cycles, and the performance of start-up cycles is usually not representative for such continuous operation.

The water used in CDI experiments can have very different compositions—ranging from analytical grade water with specified amounts of ions, to the complex compositions of brackish water found in nature and industry. First of all, it is important to decide whether or not to do experiments in oxygen-saturated water. In an experiment with oxygen-containing water, it is advisable to make sure that the oxygen content is known and kept constant. The second situation of oxygen-free (anaerobic) water can be obtained by using a nitrogen blanket in the water storage (recycle) vessel, or by bubbling nitrogen gas through this tank. Typically, results for CDI in anaerobic water are experimentally more reproducible.

In general, we can differentiate between the following types of water:

1. Real water (diluted sea water, tap water, ground water, waste water from agriculture, or industrial sources). This water contains many different ions, monovalent as well as

divalent, and with some ions being amphoteric (i.e., their charge dependent on pH, such as HCO_3^- or $H_2PO_4^-$).

2. Water of a synthetic composition simulating a "real" water source, but which is free of organic pollutants, solid particles, and so on.
3. Water containing only a single salt solution, such as NaCl or KCl. This is most commonly the choice for laboratory-scale experiments in the literature.

When "real" water is chosen (option 1), one must decide on what pretreatment to use to remove particulate matter, biological species, and organic pollutants. Because ionic mixtures are used in options 1 and 2, the effluent freshwater produced must be analyzed using offline individual ion detection, such as inductively coupled plasma in combination with optical emission spectroscopy or mass spectroscopy. However, in option 3, when using only single salt solutions, the measurement of conductivity is sufficient. In single salt solutions, we have to consider that the diffusion coefficients of the anion and cation may be (almost) equal (KCl) or are different (NaCl).

These three types of water have its own advantages and disadvantages and the choice must be made based on the objectives of the actual study. For basic CDI experiments to determine, for example, the salt adsorption capacity of an electrode material, option 3 using a single salt solution is most straightforward to analyze, because the desalination performance can be followed in time and online using a conductivity meter. We note, however, that because of the complexity of natural or industrial water compositions, CDI performance under real conditions may vary greatly from what has been determined based on "clean" single salt solution experiments.

15.2.3 "SINGLE-PASS EXPERIMENTS" VERSUS "BATCH-MODE EXPERIMENTS"

For all CDI cell designs, to measure the actual water desalination by CDI, we need to measure the change of ion concentration over time. This can be done by taking water samples and analyzing the ion composition. This is also the required procedure for studies with ion mixtures, such as for most real water sources or complex artificial mixtures.[23] Only if a single salt solution (such as NaCl or KCl) is used, simple online measurement of the water conductivity suffices.

For the layout of a CDI experiment and in particular for the location of the conductivity probe, that is, where the conductivity is actually measured, two methods are possible. In the single pass (SP) method (Figure 15.3a), water is fed from a storage vessel and the salinity (conductivity) of the water leaving the cell is measured directly at the exit of the cell or stack.[12,38,47,48] In this case, the measured effluent salinity starts to drop after applying the cell voltage, because desalination is ongoing, and later on the effluent (measured) salinity rises again to the inlet value, because the electrodes have reached their adsorption capacity. The effluent water is either discarded or can be recycled to a reservoir container. This reservoir needs to be large such that concentrations here only change very slightly within the adsorption half of the cycle, say less than 1%, to make sure that the influent concentration remains virtually constant during the cycle. The total amount of removed salt molecules can be calculated from numerical integration of the effluent concentration versus time data (taking the difference with the feed concentration, and multiplying by the water flow rate Φ).[12,14,23,38,40,47]

In another common approach, called the batch-mode (BM) method (Figure 15.3b), the recycling reservoir is much smaller, and the water conductivity in this container is measured.[9,17,49] The volume of the recycling reservoir needs to be small because otherwise the change in salinity is low and cannot be measured accurately. In this experiment, the measured salinity drops steadily and does not have a minimum; instead it levels off at a final, low, value. The difference in salinity between the initial and final situation can be multiplied by the total water volume in the whole system to calculate the amount of ions removed from the water.

The analysis by the BM method (Figure 15.3b) is simpler than by the SP method (Figure 15.3a). However, one problem is that the equilibrium salt adsorption is measured for a different (final)

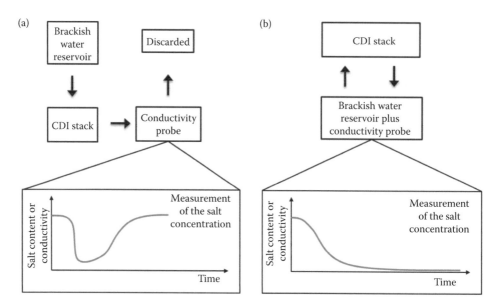

FIGURE 15.3 Schematic representation of two designs for CDI experiments. (a) Single-pass experiment (SP-method): The water conductivity is measured at the exit of the stack, or cell, and the outflow solution is discarded or recycled to a large container. (b) Batch-mode experiment (BM-method): The conductivity is measured in a (small) recycle beaker.

reservoir concentration in each experiment, and this value is unknown *a priori*. Thus, in the BM method it is difficult to compare data, for example, as function of cell voltage only. The reason is that with increasing cell voltage, desalination increases and thus the salt concentration in the system decreases. And thus, in this experimental design, it is difficult to make a parametric comparison between theory and experiment. This problem does not arise when using the SP method because here we can measure all properties (effluent salinity, electrical current) at well-defined values of the salinity of the feed stream. This is one advantage of the SP method, the other being that it is also more reminiscent of a real CDI application with water to be treated only passing through the device once, instead of being recycled multiple times, which will be less efficient.

15.2.4 CONSTANT-VOLTAGE VERSUS CONSTANT-CURRENT OPERATION OF CDI

Practically all published work using one of the above experimental procedures for CDI uses a constant cell voltage which is applied for the duration of the ion adsorption step and which is reduced abruptly during the ion desorption step where salt is released. The lower voltage level does not necessarily have to be equal to zero, for example, cycling between 1.5 V (adsorption) and 0.5 V (desorption) has been described in literature,[42] whereas Ref. 12 presents results of desalination cycles where ion desorption occurs at a reversed voltage.

The advantage of working with a reverse voltage during ion desorption is that the stored salt is released more quickly, allowing for a shorter desorption step and for higher WR. Another advantage is that the electrode regions can be more effectively depleted of the counter-ions, allowing for more ion adsorption in the next cycle (but only when membranes are utilized).

An operational mode quite different from applying constant voltage (CV) is constant-current operation (CC). As sketched in Figure 15.3a, in the CV mode, the effluent water first decreases in salinity level, and then the salinity increases again. This may, however, not be the most practical operational mode for actual devices when the production of freshwater with a constant composition over time is required. In CC operation, the effluent salt concentration level remains at a constant value, namely at a constant low value during adsorption, and at a constant high value during desorption. Desorption

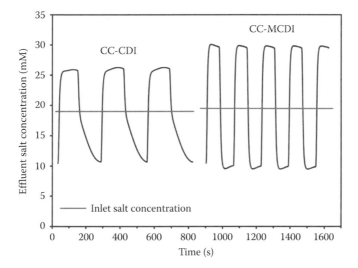

FIGURE 15.4 Effluent salt concentration in several cycles at dynamic equilibrium for constant-current operation of CDI and membrane CDI (MCDI). Inlet salinity of NaCl 20 mM, flow rate 7.5 mL/min per cell. Each cell has an electrode area of 33 cm^2. During adsorption, a current of 37 A/m^2 is applied, while during ion desorption the current is −37 A/m^2. The current is reversed from positive to negative when the cell voltage reaches the upper limit of 1.6 V (after which we switch to the ion desorption step until we reach a zero cell voltage). Ions are adsorbed when the effluent salt concentration is below the inlet value of 20 mM (which is represented by the horizontal solid lines), while ion desorption leads to effluent concentrations above 20 mM.

can be regulated by setting the cell voltage to zero (or a low voltage) during this step as in CV operation, or by using a reverse current. However, note that to generate a constant effluent concentration, CC operation works only when membranes are placed in the cell (MCDI; Figure 15.4) because in CDI (thus, without membranes) and when the voltage is still low at the start of a new adsorption step, even when the current is constant, the salt adsorption rate changes over time. This is because of the fact that in CDI in the electrode the electrical current is partially compensated by counter-ion adsorption and for the other part by co-ion desorption.[12,15,40] The second effect decreases at high voltages and then, the current is directly proportional to water desalination rate, but this is not yet the case at low cell voltages. This is why in CC-CDI the effluent salinity only drops slowly before reaching the desired constant level, see Figure 15.4 (left part). For CC operation in combination with membranes (CC-MCDI), constant levels of the effluent salt concentration are quickly reached because the co-ions are kept within the electrode structure; thus, MCDI is required to obtain a constant effluent concentration in CC operation.[14] Another advantage of CC operation is that it is easy to adjust the effluent salt concentration level by adjusting the current as control parameter.[14]

15.3 CONCEPTUAL APPROACHES TO UNDERSTAND THE PHENOMENON OF CDI

The study of CDI requires conceptual approaches to rationalize experimental data. Two related questions in this respect are:

1. How to understand the fundamental phenomenon that porous electrodes adsorb salt under the application of an external voltage?
2. How to quantify experimental data for CDI, that is, what is the right theoretical modeling framework?

In the next sections, we highlight various conceptual approaches that can be used to understand CDI experimental results.

15.3.1   Understanding of CDI Using the Concept of Operational Voltage Windows

To describe the behavior and the performance of porous carbon electrodes in a CDI cell, one approach is based on a general description of EC processes, and points out the importance that each electrode's potential must be positioned appropriately relative to a reference potential, or within a voltage window, required to have optimized ion adsorption and minimal faradaic, parasitic electrode reactions.[15,50,51] Instead, if the potentials are not chosen correctly, then ion adsorption is not optimized.

Because the values of this potential depend on the material's potential of zero charge (PZC), modifying the PZC either by oxidation or reduction of the carbon materials can improve the resulting CDI performance. This can be done by reducing the positive electrode in a way that its PZC is shifted negatively and likewise the negative electrode can be oxidized to positively shift its PZC. As a result, after applying a voltage difference to a previously shortened CDI system, both electrodes will work in the voltage window such that expulsion of co-ions is limited, and adsorption of counterions is dominant, due to introduced potential shifts of both electrodes in opposite directions. The electrical potential of both electrodes can also be optimized by the use of a third electrode (reference electrode) which can lead to a higher charge efficiency and salt adsorption capacity.[41]

15.3.2   Modeling Based on Charge Transfer between the Electrodes

In a different approach, based on concepts from the field of colloid science and electrokinetics, classical electrostatic double-layer (EDL) theory for capacitive, ideally polarizable, electrodes is used to describe the charge- and salt-voltage characteristics of the cell. In a first approximation, it is assumed that the charge is only related to electrons (in the carbon electrode) and ions (in the aqueous phase); that is, charge due to chemical ion adsorption or pseudo-capacitance effects is neglected, and, thus, when the material is not charged, there is a zero voltage drop across the EDL. Electrokinetic modeling based on the Nernst–Planck equation can be used to quantify the dynamics of the process. Although neglected in a first approximation, chemical charge of the electrodes can be readily incorporated in a more detailed model.[22]

In this approach, the focus is not on how voltage windows are chosen relative to a reference electrode, but on how much charge is transferred from one electrode to the other, and how this impacts ion concentrations inside the porous electrodes (EDL structure) and the resulting local voltage drops across the EDL. This chapter follows this approach and we discuss it in more detail in Section 15.4. Using similar concepts, Section 15.5 describes EC testing based on a pair of carbon electrodes where only purely capacitive processes occur. Detailed transport and EDL modeling taking a similar approach but using a two-dimensional geometry of the pores and of the spacer channel is provided by Jeon and No.[52]

15.3.3   Isotherm-Based Modeling

A third approach for CDI modeling is to quantify experimental data for salt adsorption in the electrodes as function of salt concentration in the external bath (recycle volume) using one of several adsorption isotherms, such as those based on the Langmuir or Freundlich equation.[45] From the fitted parameters such as equilibrium constant K, useful information can be extracted on the interaction energy between ion and substrate. The fitted isotherms can also be used to predict adsorption at other values of the reservoir ionic strength.

Problematic in this approach is that it does not describe the fact that in CDI anions and cations are separated into their respective counter electrode when they are removed from the water. Instead, isotherm-based modeling describes the adsorption of the whole salt molecule as if onto one and the same sample of carbon material. Consequently, the impact of several parameters cannot be included in this description, such as the role of the cell voltage. This approach does neither describe how much charge is required for a certain salt adsorption (i.e., the charge efficiency is not described, see Section 15.4) nor how this approach can be extended to describe asymmetric electrodes (unequal mass of anode to cathode).[38]

15.3.4 MEMBRANE CAPACITIVE DEIONIZATION AND THE ROLE OF ION-EXCHANGE MEMBRANE BARRIERS

One of the most promising developments in CDI is to include ion-exchange membranes (IEMs) in front of the electrodes (MCDI). IEMs can be placed in front of both electrodes, or just in front of one. Because of their high internal charge, they will allow easy access for one type of ion (the counter-ion) and block access for the ion of equal charge sign as the membrane, the co-ion. The membrane can be further modified to include selectivity between ions of the same sign class: for instance, between nitrate and chloride (both anions).[53] The membranes can be included as stand-alone films of thicknesses between 50 and 200 μm, or can be coated directly on the electrode with a typical coating thickness of 20 μm.[11]

To describe the functioning of the IEMs, theory from the field of charged membranes must be adapted for MCDI to describe the voltage–current relationship and the degree of transport of the co-ions. This implies that (in contrast to most membrane processes) the theory must be made dynamic (time dependent) because it has to include the fact that across the membrane the salt concentrations on either side of the membrane can be very different, and change in time.[12] This means that approximate, phenomenological approaches based on (constant values for) transport (or transference) numbers or permselectivities are inappropriate, and that instead a microscopic theory must be used. An appropriate theory includes as input parameters the membrane ion diffusion coefficient D_{mem} and a membrane charge density X.

The theory required for MCDI has been introduced in Ref. 12 assuming linearized profiles in ion concentration and electrical potential across the membrane. These simplifications imply that accumulation effects in the membrane are neglected. Irrespective of these assumptions, the model was shown to work excellently to describe a large data set for MCDI operation using constant-voltage operation[12] and also for constant-current operation.[14]

A more simplified model was presented in Ref. 10, where the membrane was assumed to be perfectly permselective toward the counter-ion, and the salt concentration in the macropores of the electrode assumed to be unvarying in time. A basic element in the modeling of the membranes in MCDI is that in the membrane the cation concentration is different from the anion concentration, with the difference compensated by the fixed membrane charge density, X. Except for this difference, the same ion transport model can be used as in free solution (Nernst–Planck equation), thus, with ions moving under the influence of a concentration gradient and because of an electrical field (electromigration). At the edges of the membranes, a Donnan potential difference develops between the outside solution and inside the membrane. For more information on MCDI, see Section 15.4.3, where a porous electrode is modeled which has an ideally permselective membrane layer in front.

15.4 THEORY

In this section, we will discuss several important theoretical concepts helpful to describe ion transport and ion storage properties of carbon electrodes for CDI. The literature on ion and water transport in porous media is vast, and we do not attempt to review this field entirely (cf. Refs. 54, 55 for an introduction with references). Instead, we describe our perspective on the most pertinent topics for carbon electrode performance and will do so by introducing certain mathematical routes that have shown to be effective tools in describing equilibrium and transport properties in CDI.

15.4.1 EDL Models

15.4.1.1 Introduction

In this section, we focus on describing salt adsorption and charge storage in the micropores within the carbon particles. Basically, the EDL is the concept describing that across an interface (in our case, the carbon/electrolyte interface in the pores inside the carbon particles) there can be charge separation, with some excess charge in one phase (i.e., the electronic charge in the carbon matrix) locally charge-compensated by charge in the other phase (due to the ions in the electrolyte-filled pores). These two components of charge sum up to zero, that is, as a whole, the EDL is uncharged. The concept of the EDL dates back to Helmholtz,[56,57] who assumed in the nineteenth century that all surface charge (either electronic charge in a conductor, or chemically bound surface charge) is directly charge-compensated by countercharge adsorbed to ("condensed onto") the surface; put in other words, the condensed layer of counter-ions directly compensates the surface charge. In this context, counter-ions are not to be mistaken with co-ions: counter-ions are the ions of opposite charge as that of the surface (which in our case is the electronic charge in the carbon, which can be of both positive and negative sign) and which are, therefore, attracted into the micropores in the carbon upon introducing charge in the electrode. In contrast, co-ions have the same charge sign as the surface and will be repelled away from the surface.

If the Helmholtz model would hold, this would be an ideal situation for CDI: for every electron transferred from one electrode to the other, one cation would be transferred into the cathode to compensate the negative electronic charge there (assuming that all ions are monovalent), while one anion would be transferred into the anode to compensate for the positive electronic charge there, and as a result effectively one full salt molecule would be removed from the spacer channel. Thus, the charge efficiency Λ (to be defined below) would be unity: one salt molecule is removed for each electron transferred from one to the other electrode. Note here that in CDI the definition of anode and cathode is based on the charging step (when we adsorb salt): the cathode is where the cations go to during charging. This is opposite to the definition in battery and supercapacitor applications, where anode and cathode are defined based on where the ions go during *dis*charging.

Unfortunately, the Helmholtz model insufficiently describes the EDL structure in porous CDI electrodes. For real systems, we must first consider the diffuse layer theory as developed by Gouy and Chapman (GC)[58–60] and second the concept of an inner (or compact) layer in between the electrode (the carbon matrix) and the diffuse layer; this inner layer is also called Stern layer or Helmholtz layer. Combination of the diffuse layer and (what is now called) Stern layer dates back to Stern in 1924.[61] The diffuse layer considers that ions *do not* condense in a plane right next to the surface, but remain diffusively distributed in the proximity of the surface. This diffuse layer does not have a precise width, but instead ion concentrations progressively decay with increasing distance from the surface. The Debye length λ_D is a characteristic distance for the counter-ion concentration and potential to decay by a factor e (~2.7) (in the low-voltage limit of the theory and for a planar single surface). For a NaCl solution of 10 mM ionic strength, the Debye length is approximately λ_D ~ 3.1 nm at 20°C. As a rule of thumb, we can consider the diffuse layer to have ended after two or three times the Debye length.

The above approach, as depicted graphically in Figure 15.5 assumes that the diffuse layer extending from one surface is not "overlapping" with that of a nearby opposite surface. If, however, this is the case, then the diffuse layer does have a finite extension of half of the distance between the two surfaces. This half-space contains all diffuse countercharge. Of course, we have the same situation for cylindrical pores, or pores of other geometries with a finite space for the diffuse layer to form. This situation of "EDL overlap" will typically be the situation for the micropores (∅ < 2 nm) in activated carbon particles as the average pore size is often *smaller* than the Debye length.

In the GC model (Figure 15.5a), the ions are not surface adsorbed in a condensed layer as considered by Helmholtz, but remain in solution because of their thermal motion. At equilibrium, the ion concentration profiles can be described in first approximation by the Boltzmann distribution,

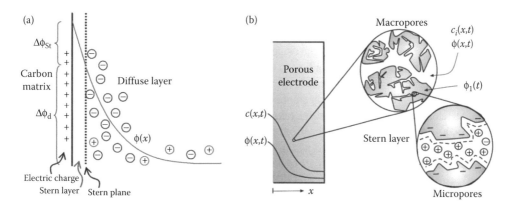

FIGURE 15.5 Models for charge and ion storage in porous CDI electrodes. (a) Structure of the electrical double layer (EDL) according to the GCS theory for a single planar EDL. (b) Two-porosity-model for the electrode.[20] Both macro- and micropores are electrolyte-filled, and in both we assume locally averaged ion concentrations (within the micropores described by the Donnan model). The large continuous interparticle pores (macropores) transport salt across the electrode thickness and are charge neutral, with salt concentration c, while in the (intraparticle) micropores the excess ionic charge is charge-compensated by electrical charge located in the carbon matrix.

which is on the basis of the Poisson–Boltzmann equation. One of the key predictions of the GC theory for the diffuse layer is that to compensate surface charge, two routes are available. The first is counter-ion adsorption in the diffuse layer (the diffuse part of the EDL). The second option is co-ion desorption, which implies that ions that were close to the surface in the absence of charge are now being expelled because they have the same charge sign as the surface charge that builds up. This effect is disadvantageous for CDI, because now for each electron transferred between the two electrodes only an extra, say, 0.6 or 0.8 ions are estimated to be stored in each electrode (relative to the situation without charge) and (with the same process also occurring in the other electrode) there are only 0.6 or 0.8 salt molecules removed from the water flowing through the spacer channel (we consider a monovalent salt only in the present analysis). Thus, the charge efficiency Λ will drop significantly below 100% to 60–80%.

This is by far not yet the most unfavorable situation, because the charge efficiency can drop even further, all the way to zero. Zero salt adsorption is possible because salt (electro)adsorption (and thus CDI) is a nonlinear process, requiring relatively high voltages. This is because of the fact that for low voltages (e.g., for diffuse layer voltages below two times the thermal voltage, V_T, which is $V_T = k_B T/e = RT/F \approx 25.7$ mV at room temperature) we approach the Debye–Hückel limit where for each electron transferred we have half a counter-ion adsorbed and half a co-ion desorbed. Consequently, in the full electrode pair (consisting of the two porous electrodes) we have no salt removal from the water channel. Data in the literature quantitatively underpin this effect, namely that with increasing cell voltage not only (1) the charge increases relatively linearly (as expected), but (2) also the charge efficiency increases, from zero to unity.[40] As salt adsorption is the product of these two parameters, it increases more than linearly with increasing cell voltage.

In CDI, we typically work at cell voltages above 1.0 V, thus, we expect that for symmetrical cell design the EDL-voltage will be above 0.5 V and thus about 20-times the "thermal voltage," which brings us very far from the Debye–Hückel limit. For such high voltages, GC theory would predict that for each extra electron transferred between the anode and cathode, we are very close to removing a complete salt molecule from the water, because in each electrode we have only counter-ion adsorption and no longer any co-ion desorption because all co-ions initially present are already expelled and there are none left. Thus, we expect the charge efficiency to be close to $\Lambda \sim 1$. Unfortunately, this is not the case. The origin of this difference is the importance of the Stern layer, which is the thin dielectric layer envisioned to be in between the charged surface and the start of the diffuse layer. This "start of the diffuse layer" (or Stern *plane*) can in first approximation be considered as the closest-approach-plane for the centers of the ions to the charged carbon surface, with the thickness of the Stern *layer* corresponding to the (hydrated) radius of the ion, for example, 0.3 nm (see Figure 15.5a). Across this dielectric layer, the voltage drop can be very high. Our estimates are that it is not improbable that 80% of the applied voltage drops across this layer for the microporous carbons currently used for CDI. This does not imply that 80% of the energy invested will be lost in this way, as the energy associated with the Stern layer is simply stored and can be completely recovered when the cell is discharged again. Thus, the Stern layer does not contain charge but it is a dielectric layer. Also, it is important to realize that in the simplest GCS model as depicted in Figure 15.5a, the intersection of Stern and diffuse layer, which is the Stern plane, neither contains any charge nor (electro-)adsorbed ions. Instead, the Stern plane only denotes the closest approach distance for ions to come to the surface.

The two concepts discussed above, namely the diffuse layer of diffusely distributed ions in the electrolyte-filled pores, and the dielectric Stern layer separating that layer from the carbon matrix structure, will be used in the further theory. Note that the Donnan model, discussed in Section 15.4.1.3, assumes that the diffuse layer has properties independent of the distance from the Stern plane, but otherwise this "Donnan part" of the model is qualitatively similar to the diffuse GC layer.

15.4.1.2 Gouy-Chapman–Stern Theory for Nonoverlapping EDLs

EDL theory is normally applied to derive the relation between stored charge (density) and the voltage difference across the EDL, that is, from the conducting electrode material ("matrix," or "carbon" in our case) to outside the EDL. "Outside the EDL" is defined as a position sufficiently far away from the surface that local charge neutrality is regained and where the concentration of cations equals that of the anions. This distance depends on many factors but, for example, for a 10 mM salt solution a distance of 15 nm will be more than sufficient. EDL theory is an equilibrium theory and equilibrium formally assumes that no further changes occur over time. One might think that such a theory cannot be applied for a dynamic process like CDI where—by definition—transport rates of ions into the EDLs are always significant. Still, in the EDL layer of a few nanometers in thickness, local equilibrium can be assumed: although there are significant ion transport fluxes across the electrode during CDI, the charge–voltage (and salt-voltage) relations that describe ion concentrations in the micropores are very well described by assuming local equilibrium in the EDL layers. Formal derivations for the validity of this assumption are given in Ref. 62, but the main reason is simply the very short (nm-scale) typical thickness of the EDL across which equilibration will be very fast, relative to the time scale of seconds and minutes in a full CDI cell with typical dimensions of the electrode layers of more than 100 μm and full cell dimensions of tens of centimeters.

In contrast to what most literature on EDL theory may indicate, for the application of CDI, we are less interested in the stored charge and more in the total amount of ions stored in the EDL. This "salt storage" has been studied in much less detail than the classical charge–voltage relationship of EDLs. Interestingly, the same EDL model that predicts the charge versus voltage relation also predicts the salt versus voltage relation. An important effect is that at low voltages across the EDL, the capacity (or, capacitance) for charge storage is nonzero (i.e., the slope of the charge versus voltage curve is nonzero), but the capacity for salt storage will be zero. This effect implies that CDI is inherently a nonlinear process: voltages are required across the diffuse part of the EDL that are at least a few times the thermal voltage (V_T, which is ~25.7 mV at room temperature) to have desalination.[21]

The equilibrium ratio of salt removal (in a 1:1 salt solution) versus charge using a two-electrode setup is called the charge efficiency, Λ. Experiments reported in Ref. 40 show that at low cell voltages (the applied voltage between the two electrodes), the charge efficiency Λ approaches zero, and only approaches unity, which is the theoretical maximum, at cell voltages above $V_{cell} = 1.0$ V. Note that "charge" in this definition is the charge Σ expressed in the same dimension as salt removal, either in moles, moles/area, or moles/gram electrode, and so forth. To convert to charge Σ_F into the more common dimension of C (Coulomb) per area or per gram, we must multiply by Faraday's number, $F = 96,485$ C/mol. In the sections below, both Σ and Σ_F are used. Also the pore charge density, to be discussed below, σ_{mi}, is expressed in mol/m^3 = mM, and can likewise be multiplied by F to obtain a volumetric charge density in C/m^3.

Several of the above concepts, such as the difference between ion adsorption and charge storage, are depicted in Figure 15.6, conceptually based on a classical figure that was first presented by Soffer and Oren in 1983.[42] In Figure 15.6, we show the adsorption into the carbon pores of cations and anions from a monovalent (NaCl) solution, as a function of the electrode charge. Charge and adsorption are defined per unit carbon micropore volume which is estimated here to be ~0.75 mL/g. The data are obtained from recalculating data for charge and salt adsorption in a two-electrode setup.[12,40] To construct Figure 15.6, both theoretically (here based on the modified Donnan model to be discussed in Section 15.4.1.3) and experimentally, the assumption of symmetry must be made, which implies that the anion adsorption (cation desorption) in the anode equals the cation adsorption (anion desorption) in the cathode; this also implies that the applied cell voltage is assumed to be equally divided between both electrodes. Our experimental work reported in Ref. 38 suggests that this assumption is very appropriate, at least for electrodes based on activated carbon powders that have not undergone chemical modification to introduce a permanent chemical charge. Figure 15.6 shows, how around zero charge, the counter-ion adsorption increases at the same rate as the co-ion desorption (when the electrode is charged), and thus the net salt adsorption will be zero in this range. Only at high (either negative or positive) surface charge, are we in the limit that for each increment in charge, an equal increment in counter-ion adsorption can be expected. This is the ideal situation that for each additional electron transferred between the electrodes, an additional salt molecule is removed from solution. Figure 15.6 also suggests that it may be favorable to operate CDI not by discharging from a charge of say 100 C/mL back to 0 C/mL before starting a new cycle, but that a cycle where we go from, for instance, +100 C/mL charge back to +25 C/mL charge (with reversed numbers in the opposite electrode) may be advantageous.

Many models are available to describe the structure of the EDL at planar surfaces and in the pores of electrodes. The literature is extensive and we will not attempt a review. In the context of CDI and supercapacitors, advanced EDL models describing ion removal and charge have been set up.[63–68] In the next two sections, we will summarize two more simple mean-field approaches that can be used in two important limits, namely (1) the limit that the typical pore size (radius) is much

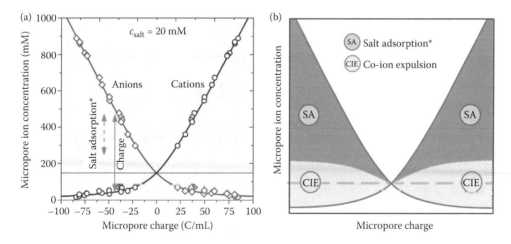

FIGURE 15.6 (a, b) Typical curves for anion and cation equilibrium adsorption in the micropores of activated carbon electrodes, based on data from Ref. 12, expressed as moles of ions adsorbed per volume of micropores. Theoretical lines based on the modified-Donnan model. Note that the micropore charge is always higher than the salt adsorption. (*Salt adsorption relative to the situation of zero charge.)

larger than the Debye length, where we use the Gouy–Chapman–Stern (GCS) theory, and (2) the opposite limit where the pore size is small relative to the Debye length. In the latter limit, the EDLs overlap strongly and we can use a "modified Donnan" (mD) approach. These two models have the important property that they are mathematically sufficiently tractable to be implemented readily in larger scale transport models, as we will show in Sections 15.4.2 and 15.4.3. More complicated EDL models may not allow one to do this.

Thus, in case we have a porous material with pores sufficiently large compared to the EDL thickness we can assume that the EDLs do not overlap and we can use the classical GCS theory developed for a single ("isolated"), planar (flat) electrode. This limit is approached both in high salt concentration (short Debye length), or for very large pores (macropores). In that case, the general Boltzmann relationship gives the ion concentration at position x away from the carbon surface, as function of φ, the dimensionless potential relative to that in the neutral macropores, where the ion (salt) concentration is $c_{salt,mA}$, as

$$c_j(x) = c_{salt,mA} \cdot \exp\left(-z_j \cdot \phi(x)\right) \tag{15.1}$$

with z_j the ionic charge number, and φ defined as the dimensional voltage V divided by V_T. Equation 15.1 is valid for ions as point charges in the mean-field approximation, and can be integrated across the diffuse layer to obtain the surface charge density, σ, in mol/m^2,

$$\sigma = 4\lambda_D c_{salt,mA} \cdot \sinh\left(\tfrac{1}{2} \cdot \Delta\phi_d\right) \tag{15.2}$$

where λ_D is the Debye length, given by $\lambda_D = 1/\kappa$, with the inverse Debye length, κ, given by

$$\kappa^2 = \frac{2F^2 c_{salt,mA}}{\varepsilon_r\varepsilon_0 RT} = 8\pi\lambda_B c_{salt,mA} N_{av} \tag{15.3}$$

with $\varepsilon_r\varepsilon_0$ the dielectric permittivity of water ($= 78 \cdot 8.854 \times 10^{-12}$ C/Vm), R the universal gas constant, T temperature, λ_B the Bjerrum length, given by $\lambda_B = F^2/(4\pi\varepsilon_r\varepsilon_0 RT \cdot N_{av})$, and N_{av} Avogadro's number. The Bjerrum length is 0.72 nm at room temperature. The surface charge density, σ, is defined in moles of charges per unit area of the electrolyte/carbon interface and can be multiplied by F to obtain a surface charge density in C/m^2.

From Equation 15.1, we can also derive the total concentration of ions stored in the EDL, given by

$$w = 8\lambda_D c_{salt,mA} \cdot \sinh^2\left(\tfrac{1}{4} \cdot \Delta\phi_d\right) \tag{15.4}$$

Next, we first define the charge efficiency, Λ, as an experimentally accessible parameter for any CDI cell geometry, as given by

$$\Lambda = \frac{\Gamma_{salt}}{\Sigma} \tag{15.5}$$

where Γ_{salt} is the removed amount of salt on applying a cell voltage (e.g., in moles per gram of all electrodes together) and Σ is the total charge transferred (obtained from Σ_F in C/gram divided by F). This equation can be used for any CDI experiment, as long as the current (which can be integrated in time to obtain Σ_F) and the salt removal are measured (see Section 15.2.3 for a description of various methods). Thus, Equation 15.5 is equally valid for symmetric as for asymmetric CDI cells, and likewise is valid for electrodes where chemical reactions ("pseudo-capacitance") will play a role.

The charge efficiency is an equilibrium property; thus, measurements require sufficient time for desalination to come to an end after a certain cell voltage has been applied. To adequately describe the salt/charge ratio while the CDI process is continuing, one option is to use the term "dynamic charge efficiency."[14] Typically, we define charge efficiency for an experiment where we step up the cell voltage from zero to a certain value. However, it is also possible to go from a certain nonzero cell voltage $V_{cell,1}$ to a second value, $V_{cell,2}$ and base Λ on the resulting change in Σ and Γ.[42] One can also develop alternative definitions, such as a differential charge efficiency, which can be defined in many different ways, for example, as the slope in a plot of Λ versus V_{cell}, or the slope in Λ versus σ, and so on. In a transport model, one can use a differential charge efficiency to relate the ions flux into an electrode, to the current, as in Ref. 69. In this work, we focus on the equilibrium charge efficiency Λ and use it to describe an experiment with a cell voltage of zero, $V_{cell} = 0$, as reference.

The charge efficiency cannot only be measured, but can also be predicted based on EDL theories. In this section, we do this on the basis of the "single-plane" GCS theory. Additional to the assumptions underlying the GCS theory, we will also assume that we describe an experiment with two electrodes and that the value of the diffuse layer voltage, $\Delta\phi_d$, is the same in each electrode; thus, we assume ideal symmetry of the CDI cell pair.[38] In that case, we can use (1) $\Sigma = \sigma \cdot a$, where a is the specific electrode surface area in m^2 per gram of electrode, and (2) $\Gamma_{salt} = w \cdot a$, based on the fact that for a symmetrical cell pair the total adsorption of ions in one electrode equals the total adsorption of *salt* molecules in the whole pair (at least for a symmetric monovalent salt). Combining these definitions with Equations 15.2, 15.4, and 15.5 results in (see Ref. 40)

$$\Lambda = \tanh \frac{\Delta\phi_d}{4} \tag{15.6}$$

Equation 15.6 predicts that according to the GCS model, the charge efficiency Λ only depends on the diffuse layer voltage, $\Delta\phi_d$. Finally, in the GCS-model we must consider the Stern layer voltage, $\Delta\phi_{St}$, which relates directly to the charge, σ, according to

$$\sigma \cdot F = C_{St} \cdot \Delta\phi_{St} \cdot V_T \tag{15.7}$$

with C_{St} the Stern layer capacity (in F/m^2). To complete the calculation of the equilibrium GCS model, we need to relate the applied cell voltage V_{cell} to $\Delta\phi_d$ and $\Delta\phi_{St}$. Assuming a symmetric cell, this relation is $V_{cell}/(2 \cdot V_T) = |\Delta\phi_d + \Delta\phi_{St}|$. Thus, we have assumed throughout this section that the EDL structure in the one electrode equals that in the other electrode except for the obvious difference in sign. For electrodes made of porous carbon, and for not very concentrated NaCl solutions, this assumption finds strong experimental support in the work described in Ref. 38. One reason for this assumption to be correct may be the very similar size of the anion and cation in NaCl solutions, and for other ionic systems, this assumption may no longer be correct.

This finalizes our treatment of the GCS model valid for a nonoverlapping, planar, EDL structure. In the next section, we consider the modified-Donnan model, valid in the opposite limit, where the EDL thickness is much larger than the typical pore size.

15.4.1.3 Modified Donnan Theory for Fully Overlapped EDLs

In the previous section, we discussed the GCS model applicable when EDLs are not overlapping. However, applying this theory to experimental data using standard electrodes made of porous activated carbons, we found that at high cell voltages this theory predicts co-ion expulsions from the EDL that are beyond the amount of co-ions initially present in an electrode. This anomaly is because of the fact that GCS theory cannot be applied to micropores where the Debye length is of the order of, or larger than, the pore size. This problem is resolved when the mD model is applied for CDI. The mD approach is valid in the limit of strongly overlapped EDLs, and was shown to work well to describe many data sets for salt adsorption and charge storage in CDI (see Refs. 14, 22, 38).

The mD model can be used to theoretically describe the equilibrium salt adsorption and charge in microporous carbons, and assumes that the EDLs inside the carbon particles are strongly overlapping, to the point that the potential in the micropores becomes constant (i.e., does not vary with position in the pore). This will be a valid assumption when the Debye length is much larger than typical micropore sizes, which are in the 1–2 nm range. In this limit, it is possible to make the "Donnan" assumption that the electrolyte inside the carbon particles has a constant electrical potential. Obviously, this is an approximation of the detailed structure of the EDL in microporous carbons,[68,70] but the Donnan approach has the advantage of being mathematically simple and it has shown to accurately describe data both for charge and salt adsorption. Because of its simplicity, it can easily be included in porous electrode mass transport theory,[20,22,23] while it will not predict larger co-ion expulsions than physically possible. The latter prediction, as we mentioned before, is a severe complication of the GCS theory when applied to micropores.

The "pure" Donnan approach, however, does not describe well various data sets for salt adsorption and charge in microporous carbons. Instead, we must make two quite natural modifications to the classical Donnan approach.[70,71] The first is to consider the Stern layer located in between the electron charge and the ions; the presence of this layer reflects that the ion charge cannot come infinitely close to the electron charge, for instance because of the (hydrated or dehydrated) ion size, or because the electron charge is not exactly located at the edge of the carbon material, or because of an atomic "roughness" of the carbon/electrolyte interface. The second modification to the pure Donnan approach is to include a chemical attraction energy for the ion, an energy gained when the ion transfers from outside to inside the carbon particles, described by a term μ_{att}. Thus, in this way we consider an additional, nonelectrostatic, attraction for the ion to go into a pore. This term also reflects that also uncharged carbons adsorb some salt.[72]

The mD model containing these two modifications (see Refs. 14, 23, 38, 47) is described by the following equations. First, the concentration of ion j in the micropores inside the carbon particle is given by

$$c_{j,mi} = c_{salt,mA} \cdot \exp\left(-z_j \cdot \Delta\phi_d + \mu_{att}\right) \tag{15.8}$$

where "mi" stands for the intraparticle (micro)pores (the pores inside a carbon particle), and "mA" for the macropores, being the transport pathways outside the particles (see Figure 15.5b). In the macropores, the anion and cation concentration are equal (local electroneutrality) because we only consider a monovalent salt solution, and thus $c_{j,mA}$ can be replaced by the macropore salt concentration, $c_{salt,mA}$, which will be a function of time and position (depth) within the electrode, just like $c_{j,mi}$. In Equation 15.8, $z_j = +1$ for the cation and $z_j = -1$ for the anion, while $\Delta\phi_d$ is the Donnan electrostatic potential difference between micro- and macropores, that is, between inside and outside the carbon particle. It may be noticed that Equation 15.8 is very similar to Equation 15.1 in Section 15.4.1.2 except for the extra term μ_{att}, with the Donnan potential now replacing the diffuse layer potential, but using the same symbol, $\Delta\phi_d$. Note that Equation 15.8 will be used for the whole micropore volume, whereas Equation 15.1 was only used to describe the ion concentration in a plane at distance x from the electrode surface, and was integrated over x to obtain the charge–voltage and salt–voltage relations given by Equations 15.2 and 15.4. In the mD model, this integration is not made, and instead, summing up Equation 15.8 for both ions directly gives the total ion density in the pores:

$$c_{ions,mi} = c_{cation,mi} + c_{anion,mi} = 2 \cdot c_{salt,mA} \cdot \exp(\mu_{att}) \cdot \cosh(\Delta\phi_d) \tag{15.9}$$

Although it is possible to consider the fact that μ_{att} will be different for anions and cations (and may depend on other factors), in this work we will assume that they are the same for each ion. In future work, it may be possible to derive from dedicated experiments with different salt solutions the individual values of μ_{att} for each ion separately.

The local ionic micropore charge density, σ_{mi}, follows from Equation 15.8 as

$$\sigma_{mi} = c_{cation,mi} - c_{anion,mi} = -2 \cdot c_{salt,mA} \cdot \exp(\mu_{att}) \cdot \sinh(\Delta\phi_d) \tag{15.10}$$

This volumetric ionic micropore charge density (dimension mM = mol/m³), σ_{mi}, relates to the Stern layer potential difference, $\Delta\phi_{St}$, according to

$$\sigma_{mi} \cdot F = -C_{St,vol} \cdot \Delta\phi_{St} \cdot V_T \tag{15.11}$$

where $C_{St,vol}$ is a volumetric Stern layer capacity (in F/m³). To compare with experiment, we must recalculate pore concentrations to the measurable parameters of equilibrium charge $\Sigma_F (= \Sigma \cdot F)$ and salt adsorption Γ_{salt} (relative to salt adsorption at zero applied voltage, i.e., at $V_{cell} = 0$ V) per gram of electrodes as given by Refs. 14, 23, 38, and 47:

$$\Gamma_{salt} = \tfrac{1}{2} \cdot p_{mi}/\rho_e \cdot (c_{ions,mi} - c^0_{ions,mi}), \quad \Sigma_F = -\tfrac{1}{2} \cdot F \cdot p_{mi}/\rho_e \cdot \sigma_{mi} \tag{15.12}$$

where superscript "0" refers to the total ion adsorption at a cell voltage of $V_{cell} = 0$, and where ρ_e is the electrode density (mass per unit total electrode volume), and p_{mi} the micropore volume relative to the total electrode volume. Note that the ratio p_{mi}/ρ_e is equal to the micropore volume per gram of carbon powder (e.g., measured by nitrogen adsorption) times the mass fraction of carbon powder relative to the total electrode mass (e.g., 85%). By combining Equations 15.9 through 15.12, we can derive that in the mD model the charge efficiency, Λ, which is the measurable ratio in the two-electrode cell pair of equilibrium salt adsorption Γ_{salt} over charge Σ, relates to the Donnan potential $\Delta\phi_d$ according to

$$\Lambda = \tanh\frac{\Delta\phi_d}{2} \tag{15.13}$$

for an experiment where the initial and final salt concentrations outside the EDL are the same (Figure 15.3a). Equation 15.13 is slightly, but notably, different from the prediction based on the GCS model as given by Equation 15.6.

It is important to note that Equations 15.12 and 15.13, again, assume symmetry: the double-layer structure in the cathode is equal to that in the anode, except for the difference in the sign of charge; thus, μ_{att} must be the same for the cation and the anion. More general theories including differences in μ_{att} as well as the natural charge of the carbon (dependent on local pH) and pseudo-capacitance effects (e.g., the quinone to hydroquinone transition of oxidized carbons) should be developed in the future. Extensions of Equation 15.12 for electrodes with unequal masses between the two electrodes are given in Ref. 38. Finally, as for the GCS model, for equilibrium, the applied cell voltage V_{cell}, defined as a positive number, relates to $\Delta\phi_d$ and $\Delta\phi_{St}$ according to $V_{cell}/(2 \cdot V_T) = |\Delta\phi_d + \Delta\phi_{St}|$.

15.4.2 SIMPLIFIED DYNAMIC CDI TRANSPORT MODEL FOR BATCH-MODE EXPERIMENT

In this section, we describe a simple dynamic CDI process model valid for the batch-mode experiment (see Section 15.3.2), where the water leaving the CDI cell is flowing to a small reservoir (recycle vessel) and from there fed back into the CDI cell (cf. Figure 15.3b). Assuming a low desalination "per pass," we can make the assumption that throughout the cell the salt concentration is the same everywhere (though it will decrease in time), and is the same as in the recycle vessel. The conductivity in the recycle vessel is measured, not the effluent salt concentration from the CDI cell. As we will show, this model can be written as a single ordinary differential equation (ODE) coupled to an algebraic equation (AE), two equations which can be solved jointly quite accurately even in simple spreadsheet-software, using the forward Euler method (see Box 15.1). This model assumes symmetry at several points, that is, that the EDL structure in the anode is the same as that in the cathode, except for the obvious difference in sign. The present model describes CDI and does not include the membranes, required for a description of membrane-CDI (MCDI).

We will develop this transport model on the basis of the modified Donnan (mD) model (Section 15.4.1.3) as this seems to be the most suitable EDL model for CDI. If necessary, it can also be derived for the GCS model.

The ODE that is solved is a cell balance for the ionic charge density in the intraparticle micropores, σ_{mi}, which relates to the current density J running from electrode to electrode, as given by

$$v_{mi} \frac{d\sigma_{mi}}{dt} = J \cdot A \qquad (15.14)$$

where J is defined in mol/m²/s, and where A is the electrode area (i.e., the area of the spacer channel covered by the porous electrode, either anode or cathode), and v_{mi} is the micropore volume in all electrodes of the same sign (i.e., all anodes, or all cathodes). The volume v_{mi} excludes the volume of the interparticle macropores in the electrode, which is included in v_{tot}. To obtain the current as measured externally in A (Ampère), we can multiply $J \cdot A$ by Faraday's constant, F. The current J depends on the driving force for transport, $\Delta\varphi_{tr}$, which is the applied cell voltage minus the voltages in the EDLs in both electrodes, and it will depend on the resistance, to which there are contributions both in the spacer channel and within the electrode. The outside resistance (in the spacer) can be considered as a "linear resistance" as all charge effectively has to transfer from one electrode across the channel to the other. Within the electrode, however, the situation is very different because charge will be stored in a distributed fashion, and the ion concentration profile increasingly penetrates into the electrode: initially, there is hardly any resistance in the electrode as ions will be stored in the outer layers of the electrode, right next to the interface with the spacer. It is only with the passing of time that the ions will need to travel a constantly increasing distance to reach unsaturated pores. To describe all these effects, porous electrode theory is required which will be treated in the following section.

In the simplified transport model, we lump the resistance inside the electrode with that in the spacer channel. To describe the current–voltage relation, we will not assume a constant resistance but include how the resistance increases when the salt concentration goes down. We assume that at each moment in time the salt concentration within the electrode macropores (the transport pathways in the electrode) is the same as in the spacer channel. Therefore, we can use a single relation between J, $\Delta\phi_{tr}$, and c (with c being the salt concentration, assumed equal in spacer channel, macropores, and recycle vessel), resulting for J in

$$J = k \cdot c \cdot \Delta\phi_{tr} \qquad (15.15)$$

where k is an effective total transport coefficient (in m/s) and $\Delta\phi_{tr}$ is the dimensionless voltage drop that drives transport, that is, the voltage between the start of the EDLs in the one electrode and that in the other electrode. This voltage drop, $\Delta\phi_{tr}$, relates to the cell voltage, V_{cell}, and the EDL voltages according to

$$V_{cell}/V_T = (\Delta\phi_{St} + \Delta\phi_d)_{anode} + \Delta\phi_{tr} - (\Delta\phi_{St} + \Delta\phi_d)_{cathode} = \Delta\phi_{tr} + 2 \cdot |\Delta\phi_{St} + \Delta\phi_d| \quad (15.16)$$

where the second equality is based on the assumption of symmetry of the EDL structure in anode and cathode. Vertical lines denote that the absolute value of the EDL voltages is used. By combining Equations 15.10, 15.11, and 15.15 with Equation 15.16, we obtain a relation for J as function of σ, c, and V_{cell} according to

$$J = kc \left\{ \frac{V_{cell}}{V_T} - \frac{2 \cdot \sigma_{mi} \cdot F}{C_{St,vol} \cdot V_T} - 2 \cdot \text{asinh} \frac{\sigma_{mi}}{2 \cdot c \cdot \exp(\mu_{att})} \right\} \tag{15.17}$$

which can be substituted directly in Equation 15.14.

The required AE in the model follows from two equations, first of all, a balance stating that the total number of moles of salt molecules in the system is conserved,

$$v_{tot} c_0 + v_{mi} c_{ions,mi,0} = v_{tot} c + v_{mi} c_{ions,mi} = v_{mi} \gamma \tag{15.18}$$

where v_{tot} is the water volume in the whole system, including the recycle vessel, tubing, spacer channel, and macropores, but excluding micropores. Volumes v_{mi} and v_{tot} as well as area A must either be based on a single cell, or on the full CDI stack. In Equation 15.18, subscript "0" refers to time zero, the moment that the voltage is applied. To understand Equation 15.18, one must realize that the total amount of moles of *ions* in the micropores of *one electrode*, $v_{mi} \cdot c_{ions,mi}$, equals the number of moles of *salt* molecules in the micropores of *both* electrodes taken together. As shown in Equation 15.18, we describe the total amount of salt molecules in the system, divided by the micropore volume, v_{mi}, by the parameter γ.

The second required equation in the model is based on combining Equations 15.9 and 15.10 and gives

$$c_{ions,mi} = \sqrt{\sigma_{mi}^2 + (2 \cdot c \cdot \exp(\mu_{att}))^2} \tag{15.19}$$

Next, Equations 15.18 and 15.19 can be combined to result in

$$c = \frac{\sqrt{b^2 - 4\beta(\sigma_{mi}^2 - \gamma^2)} - b}{2\beta}, \quad \beta = (2\exp(\mu_{att}))^2 - \left(\frac{v_{tot}}{v_{mi}}\right)^2, \quad b = 2\gamma \frac{v_{tot}}{v_{mi}} \tag{15.20}$$

Thus, we have a model based on a single ODE, Equation 15.14, expressing micropore charge density, σ_{mi}, as function of current J, together with auxiliary relations for the salt concentration in the recycle volume, c, and for current density, J. Just as in Refs. 38 and 47, we will not use a constant Stern capacity to fit the data, but use a function where $C_{St,vol}$ increases with increasing charge, a classical observation in the colloid literature,[47,73,74] which we can describe empirically by using $C_{St,vol} = C_{St,vol,0} + \alpha \cdot \sigma_{mi}^2$. Next, we will show how this set of equations can be solved in simple spreadsheet software (example sheet available upon request from the authors) and we show a comparison of theoretical results with example data.

To solve this model, total saltwater volume v_{tot}, micropore volume (of all electrodes of one sign), v_{mi}, and electrode geometrical area A (of all electrodes of one sign) are obtained from the experimental geometry and carbon analysis (e.g., pore volume in mL/g from nitrogen sorption, times the mass of porous carbon in the electrode, results in v_{mi}). From fitting of the mD model to equilibrium data we obtain parameter estimates for $C_{St,vol}$, α, and μ_{att}. Fitting to the time-dependent part of the data, the only remaining parameter is then the transfer coefficient, k, to fit all possible data sets.

BOX 15.1   SOLVING THE SIMPLIFIED DYNAMIC CDI MODEL USING SPREADSHEET SOFTWARE

A simple procedure allows us to solve Equations 15.14, 15.17, and 15.20 as function of time in standard spreadsheet software, using the Euler Forward method.

Step 1. Assign columns for the parameters t (s), c (mM), σ_{mi} (mM), and J (mol/m^2/s).

Step 2. Make a separate list of constants: k, V_{cell}, v_{tot}, $C_{St,vol}$, α, F, μ_{att}, v_{mi}, β, and A.

Step 3. Make a list of time values from 0 to a certain final time, with time steps Δt.

Step 4. For time zero we know that (1) c is equal to $c_{salt,0}$ and (2) that $\sigma_{mi} = 0$ and $c_{ions,mi} = c_{ions,mi,0}$ from Equation 15.19. Based on these parameters, we can calculate γ, b, and c from Equations 15.18 and 15.20. Also, J can be calculated at time zero using Equation 15.17.

Step 5. A value for σ_{mi} at the next moment (at the next "time-line") follows from Euler Forward's method based on Equation 15.14: $\sigma_{mi,i} = \sigma_{mi,i-1} + J_{i-1} \cdot \Delta t \cdot A/v_{mi}$ where subscript "i" refers to the actual time-line, and "$i-1$" refers to the previous time-line.

Step 6. For the actual time-line "i" we now calculate c using Equation 15.20, based on σ_{mi} from the same time-line "i," and next we calculate J using Equation 15.17, based on c and σ_{mi} from the same time-line "i."

Step 7. We copy the equations for c, σ_{mi}, and J from Steps 4 and 5 down to the final time.

Step 8. We can now plot the salt concentration c as function of time, t.

Step 9. It is important to compare results for various values of the timestep Δt, and only when Δt is small enough that it does not influence the outcome of the calculation, we can then use the calculation outcome (we find good results with $\Delta t = 1$ s).

In Figure 15.7, we present example of data obtained in the same stack of $N = 8$ cells as described in Refs. 14, 38, 40, and 47. The data in Figure 15.7 are obtained using activated carbon electrodes (Norit Supersorb 50) with electrodes each of mass 0.58 g and with each cell having $A = 33.8$ cm^2 projected (geometric, outer) surface area, using a spacer of $L_{sp} = 250$ μm thickness and an open spacer porosity of $p_{sp} = 0.5$, after compression of the stack. Of the electrode mass, 85 wt% is activated carbon (AC), and the micropore volume of pores <1 nm measured by nitrogen sorption is 0.217 mL/g AC. The water flow rate is 60 mL/min for the stack.

We present data at an initial salt concentration of $c_{salt,0} = 5$, 10, and 20 mM, and at cell voltages of $V_{cell} = 0.8$ and 1.2 V. We have the following geometrical input parameters for the whole system (stack level calculation): $v_{tot} = 200$ mL, $v_{mi} = 0.86$ mL, and $A = 270$ cm^2. We can fit the theory to the data quite well using the following set of fitting parameters: $C_{St,vol} = 210$ MF/m^3, $\alpha = 19.2$ F · m^3/mol^2, $\mu_{att} = 1.5$ kT, and $k = 1.5$ μm/s. Note that in Figure 15.7 we have shifted the experimental data, to the left by 20 s for all data sets, necessary because of the slight delay between transport of water from the stack to the recycle vessel where the conductivity is measured. From the fit of theory to data, we can rather confidentially estimate the system transport coefficient as $k \sim 1.5$ μm/s. Comparing this with an ideal spacer transfer coefficient of $2*D*p_{sp}/L_{sp} \sim 7$ μm/s ($D_{avg} \sim 1.7 \times 10^{-9}$ m^2/s for NaCl), we can estimate that the resistance in the electrode is ~2 μm/s (assuming simple addition of resistances, R, with $R \propto 1/k$), and thus the resistance in the electrode is more than 3× higher than in the spacer channel. This is a very tentative analysis, and for a full analysis the CDI system must be modeled in a more rigorous fashion, including porous electrode theory, as will be explained in the next section.

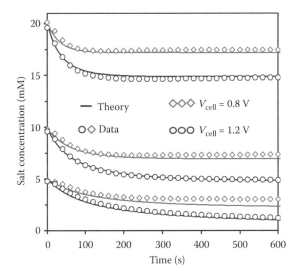

FIGURE 15.7 Salt concentration versus time in batch-mode CDI-experiment (see Figure 15.3b) for three values of initial salt concentration, and in each case for two values of the cell voltage. Lines: theory, points: experimental data.

15.4.3 Porous Electrode Theory for Membrane-CDI with an Ion-Selective Blocking Layer

In this section, we present a detailed model for transport and storage of ions in a porous electrode, based on a monovalent salt solution. This is an extension classic porous-electrode theory which assumes a constant and high salt concentration in the transport pores (macropores) and a linear EDL capacity, resulting in classical transmission line theory.[27,28] Here, we do not make these assumptions. This has the consequence that instead of a single partial differential equation (PDE), two PDEs must be solved simultaneously, together with additional algebraic equations. This complete model describes as a function of the depth in the electrode (x) and time (t) a total number of four coupled variables: the concentration in the interparticle macropores, c_{mA}, the potential there, φ_{mA}, the charge density in the intraparticle micropores, σ_{mi}, and the net salt adsorption in the micropores. The latter variable will be described using an effective salt concentration, c_{eff}, which is a summation over all pores of the total ion concentration (divided by 2 to get a salt concentration), defined per unit total electrode volume, thus:

$$c_{eff} = p_{mA}c_{mA} + \tfrac{1}{2}p_{mi}c_{ions,mi} \tag{15.21}$$

Note that all these parameters depend on depth x and time t, and that it is only after sufficient time that all parameters level off to their final, equilibrium, value when gradients through the electrode become zero. Equilibrium properties are described by the set of equations given for the mD model in Section 15.4.1.3.

Capacitive charging leading to desalination has been discussed in the literature,[20–22] where both the GCS and mD model are used and both a single monovalent salt solution is considered, as well as mixtures of salts including ions of multiple valencies.[23,40] Here, we will assume *a priori* that we only have a monovalent salt with both ions having the same diffusion coefficient, D. Note that this is an effective diffusion coefficient for transport in the macropores, which contains a contribution of pore tortuosity. We will not consider transport outside the electrode (thus, the Biot number is set to infinity),[20,21,23] but we will consider the presence of an ion-selective membrane layer between spacer and electrode to simulate MCDI. The present model can be combined with the CDI transport model of the previous section, to obtain a full (M)CDI transport model which describes ion adsorption and transport into an electrode in detail, thereby distinguishing between spacer and electrode resistances. This important extension, however, must be left for future work.

In the calculations in this section, in the spacer channel, we assume a constant salt concentration, c_{sp}. The membrane in front of the electrode is assumed to be 100% permselective, that is, only counter-ions can go through; we also neglect a transport resistance within the membrane. In that case, the total potential drop from spacer across the membrane (including the two EDL Donnan edges of the membrane) to a point in the macropores located at the edge with the membrane is

$$\Delta\phi_{mem} = \ln\frac{c_{sp}}{c_{mA}} \qquad (15.22)$$

which is based on a cation-exchange membrane (otherwise a minus sign must be added), with c_{mA} the macropore salt concentration in the electrode, right next to the membrane. Within the electrode the salt mass balance is given by

$$\frac{\partial c_{eff}}{\partial t} = p_{mA}D\frac{\partial^2 c_{mA}}{\partial x^2} \qquad (15.23)$$

with x the position across the electrode, $0 < x < L$, where L is the electrode thickness, and D is the ion diffusion coefficient in the interparticle macropores. Furthermore, the micropore charge balance is given by

$$p_{mi}\frac{\partial\sigma_{mi}}{\partial t} = 2p_{mA}D\frac{\partial}{\partial x}\left(c_{mA}\frac{\partial\phi_{mA}}{\partial x}\right) \qquad (15.24)$$

At each position in the electrode, the macropore potential ϕ_{mA} is related to that in the carbon matrix, ϕ_1, according to

$$\Delta\phi_d + \Delta\phi_{St} = \phi_1 - \phi_{mA} \qquad (15.25)$$

with expressions for $\Delta\phi_d$ and $\Delta\phi_{St}$ given by Equations 15.10 and 15.11. Boundary conditions are as follows: at the backside of the electrode, we have $dc_{mA}/dx = 0$ and $d\phi_{mA}/dx = 0$. Initial conditions are a certain value for the macropore concentration c_{mA}, and with $\sigma_{mi} = 0$ everywhere in the electrode, we can use Equations 15.10 and 15.22 to determine c_{eff} at time zero. To simulate a real experiment where we apply a certain cell voltage between two electrodes (with the spacer channel in between), we make a step change in voltage between that in the spacer channel ϕ_{sp} (just outside the membrane) relative to that in the carbon matrix, ϕ_1. This difference $\phi_{sp} - \phi_1$ is quickly stepped up at time zero from zero to $\Delta\phi_{step}$. In the carbon matrix, we assume a constant potential ϕ_1; thus, we neglect a possible electrical resistance in the carbon, in the current collectors, and in the connecting wires. At the front side of the electrode (where it contacts the membrane), the potential ϕ_{mA} relates to that in the spacer channel according to: $\phi_{mA} = \phi_{sp} + \Delta\phi_{mem}$, with $\Delta\phi_{mem}$ a function of c_{mA} at that position (see Equation 15.22).

The final boundary condition follows from the assumption of a perfectly selective cation-exchange membrane in front of the electrode, and thus the total anion number in the electrode is conserved, that is, the anion flux through the membrane is zero. This implies that at the membrane/electrode edge ($x = 0$) we have as a final boundary condition: $dc_{mA}/dx|_{x=0} = c_{mA,x=0}*d\phi_{mA}/dx|_{x=0}$. This is in principle a valid boundary condition. However, we find that this is numerically troublesome and it is better to use instead a different constraint, namely that of total anion conservation in the whole electrode: $\int_0^L c_{mA}(p_{mA} + p_{mi} \cdot 2c_{mA}e^{2\cdot\mu_{att}}/(\sigma + \sqrt{\sigma^2 + (2c_{mA}e^{\mu_{att}})^2}))dx = $ constant, where we have implemented Equation 15.10 in Equation 15.8 for the anion concentration in the micropores ($z_i = -1$).

Next we show an example calculation for the macropore salt concentration c_{mA} as a function of position x and time t during ion adsorption after a step change of $\Delta\phi_{step} = -30$, for $\mu_{att} = 0$, $p_{mi} = p_{mA} = 0.30$ and $C_{St,vol} = 0.15$ GF/m³. The value of $\Delta\phi_{step} = -30$ corresponds to applying a negative voltage of −770 mV to the electrode relative to the midplane (i.e., $V_{cell} = 1.44$ V). As Figure 15.8a shows, after applying this voltage, macropore concentrations increase in time. This is because anions cannot leave the electrode region and as they are expelled from the micropores they must end up in the macropores, which therefore also adsorb extra cations, compared to the situation without the membrane.[10,12] In Figure 15.8b, we show the effect of increasing the electrode voltage, not simply back to $\Delta\phi_{step} = 0$, but further, to an opposite polarity of $\Delta\phi_{step} = 30$, that is, ions are now desorbed at reversed voltage, a very effective operational mode in MCDI, to allow for fast ion release, and enhanced ion (electro-)adsorption in the next cycle.[12] As can be observed, the salt concentration in the macropores drops tremendously during this stage, and gets very close to zero (to reach a value of about 30 µM), making further ion transport in the electrode problematic (low conductivity). In a final calculation (not shown), the voltage is reduced to zero and the macropore salt concentration increases again to the starting value without noteworthy gradients across the electrode in this stage (i.e., the concentration increases equally rapid everywhere).

Other important parameters that follow from the calculation are salt adsorption and charge. The micropore charge increases to beyond 1000 mM during charging, slightly below the value for the cation concentration in the micropores (with the anions in the micropores at a concentration of 1.4 mM). These numbers for the micropore concentration imply that the macropore-to-micropore Donnan potential was never higher than ~3.3 thermal voltage units, or about $\Delta\phi_d$ ~ 85 mV during charging. During discharge at reversed voltage, the micropore charge goes from positive to negative, to reach a minimum value of σ_{mi} ~ −40 mM.

In our view, these effects of strong transients in the ion concentrations across the electrode are very important in the study of the dynamics of CDI and MCDI. Simple analysis based on RC networks or other theories that assume a constant resistance in the electrode will fail completely

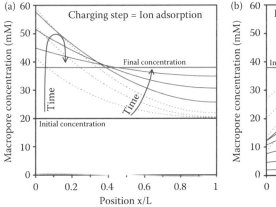
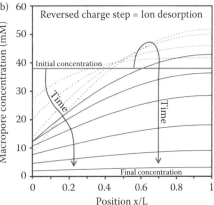

FIGURE 15.8 Calculation results of 1D dynamic porous electrode model for an electrode with a selective cation exchange membrane layer in front (located at $x = 0$) that is perfectly blocking for co-ions. (a) Macropore salt concentration profiles as function of time (direction of arrows) during application of −770 mV voltage to the electrode relative to the bulk (spacer channel) outside the membrane, and (b). After subsequent increase of the potential to +770 mV.

for (M)CDI, because as we show here ion concentrations (and thus resistances for ion transport) actually change dramatically across the electrode, and are very time dependent as well.

An important assumption in this modeling approach is that ion transport from macropore to micropore is sufficiently fast to be at equilibrium, that is, transport resistances are due to ion diffusion and electromigration across the thickness of the electrode. An extended model accounting for such a local transport resistance can be based on describing the individual ion adsorption fluxes into the carbon micropores j_i by $j_i = k_\rightarrow \cdot c_{mA} \cdot \exp(-z_i \cdot \alpha \cdot \Delta\phi_d) - k_\leftarrow \cdot c_{mi,i} \cdot \exp(z_i \cdot (1 - \alpha) \cdot \Delta\phi_d)$, where α is a transfer coefficient ($0 < \alpha < 1$) and the kinetic adsorption and desorption constants, $k_\rightarrow$ and $k_\leftarrow$, relate to μ_{att} according to $\mu_{att} = \ln(k_\rightarrow / k_\leftarrow)$. For high values of the kinetic constants, or low values of the flux j_i, the equilibrium Donnan model is recovered.

15.5 ELECTROCHEMICAL ANALYSIS OF CDI

EC analysis is an essential tool to evaluate the performance of porous electrode systems. Many options are available for what current or voltage signals to apply in EC analysis, and for what experimental system to test. The experimental system is often one of the following two options. The first option is to test a single porous electrode in combination with a counter- and a reference electrode ("three electrode setup"). The other option is to test two porous electrodes as one another's counter electrode. The use of a reference electrode is not obligatory in this second option, and solely the cell voltage difference between the two electrodes is externally controlled or measured ("two electrode setup"). This second option resembles actual CDI operation (likewise for supercapacitors), and we will, therefore, discuss this EC analysis method in this chapter.

The main question is then: can we use the standard toolbox of EC test methods (galvanostatic cycling, cyclic voltammetry, impedance spectroscopy, etc.) also for realistic CDI conditions? The problem here is that CDI is often applied for water of a low ionic strength which is initially below 200 mM (and more often even below 20 mM), and this value decreases even further during desalination. These are concentrations much lower than those used in standard EC analysis where the salt concentration is typically 1 M or higher. So, does it make sense to use the same methods and the same analysis approach when we do EC analysis for water of say 10 mM ionic strength?

Of course, one can still do the same type of experiment, and using an appropriate theory we can obtain important information on local resistances and capacities. However, applying the standard theoretical tools developed for testing at high ionic strength, based on constant RC theory for porous electrodes,[27,28] seems undesirable as these methods assume a constant capacity and resistance as well as equal ion diffusion coefficients, and for water desalination neither of these assumptions is very much applicable; instead, capacities are nonlinear but more importantly resistances for ion transport are highly place- and time-dependent, as illustrated in the previous Section 15.4.3. Applying a theory that does not account for these effects will easily lead to erroneous conclusions. As mentioned in Section 15.4.3, this implies that the transmission line theory completely fails and we need to revert to solving a porous electrode model based on several coupled PDEs.[20–23] Analytical solutions are not readily available for this problem and thus it seems that EC analysis in low ionic strength aqueous solutions will require novel dedicated software making use of exact porous electrode theory, for example, along the lines of that presented in Section 15.4.3.

Is EC analysis for CDI then not relevant, using the current analytical methods? On the contrary, EC analysis can be a very important tool to find out details of the origin of resistances and the magnitude of storage capacities, but to be able to use standard theoretical tools, we need to test at high ionic strength conditions, and ideally with a salt with equal ion diffusion coefficients, such as KCl.

First, let us show how to derive the required porous electrode RC theory ("transmission line-theory") from our results in Section 15.4.3. At high ionic strength conditions, the diffuse (or, Donnan) layer potential will be sufficiently low such that we are in the low-potential limit where we have no net salt (electro-)adsorption in the EDLs (see Equation 15.4). This is the fundamental reason that we can hope to have no gradients in salt concentration inside and outside the electrode. Making that assumption allows us to modify Equation 15.24 to

$$\frac{\partial \sigma_{mi}}{\partial t} = \frac{2p_{mA}cD}{p_{mi}} \frac{\partial^2 \phi_{mA}}{\partial x^2} \tag{15.26}$$

where c is the constant salt concentration in macropores and outside the electrode. Implementing $\sigma_{mi} \cdot F = -C_{St,vol} \cdot \Delta\phi_{St} \cdot V_T$ and $(\Delta\phi_d) + \Delta\phi_{st} = \phi_1 - \phi_{mA}$ assuming ϕ_1 (the electrode potential) not to vary, we can rewrite Equation 15.26 to

$$\frac{\partial^2 \phi_{mA}}{\partial x^2} - RC \frac{\partial \phi_{mA}}{\partial t} = 0 \tag{15.27}$$

which is the classical equation for the transmission-line model, where according to our porous electrode theory the dimensionless group RC is given as the product of $R = V_T/2Fp_{mA}cD$ for the resistance per unit length (in Ω m), and $C = C_{St,vol}p_{mi}$ for the volumetric capacitance (in F/m³). Equation 15.27 can be solved analytically for all kinds of applied voltage or current signal.[26]

In this work, we will use Equation 15.27 for an experiment where we apply a step-change in cell voltage and measure the resulting current signal (abbreviated here as SW-CAM, for "square-wave chronoamperometry") for porous electrodes made by a standard casting technique (see Ref. 47; Norit Supersorb-50 activated carbon, Norit B.V., The Netherlands). We will analyze a two porous-electrode setup, with a spacer layer in between the electrodes (electrode thickness $L = 340$ μm, electrode mass density $\rho_e = 0.50$ g/mL). Assuming symmetry, we only need to solve Equation 15.27 for one electrode, whereas across the half-spacer layer we then have a constant resistance, $R_{tot,ext}$.

Boundary conditions for Equation 15.27 are that at the inner edge of the electrode ($x = L$) we have $d\phi_{mA}/dx = 0$, whereas at the electrode/spacer interface we have

$$\frac{I_e \cdot R_{tot,ext}}{V_T} = \phi^* - \phi_{mA}\Big|_{x=0} = -\frac{L}{Bi} \frac{\partial \phi_{mA}}{\partial x}\Big|_{x=0} \tag{15.28}$$

where I_e is the current density (in A/m²) and the Biot number describes the ratio in electrode resistance over external resistance, given by $Bi = RL/R_{tot,ext}$, where the electrode resistance per unit length, R (in Ω m), is given above, and $R_{tot,ext}$ is the external resistance over half of the channel (in Ω m²). In Equation 15.28, ϕ^* is the dimensionless voltage step change applied to the half-cell. The analytical solution for the current density I_e as function of time after applying a step change in voltage is[26]

$$I_e(t) = I_{e,0} \cdot \sum_{n=0}^{\infty} \frac{2 \cdot Bi}{Bi^2 + Bi + \lambda_n^2} \exp\left(-\lambda_n^2 \cdot t/RCL^2\right) \tag{15.29}$$

where the values of λ_n are the roots of the transcendental equation $\lambda_n \cdot \tan\lambda_n = Bi$, and where the initial current density is

$$I_{e,0} = V_{cell}/2R_{tot,ext} \tag{15.30}$$

where V_{cell} is the voltage applied between the two electrodes.

The advantage of the SW-CAM technique is that based on a set of curves at various values of V_{cell} we obtain a data set which can be analyzed in various ways to make several "checks and balances" to analyze step by step whether the assumptions underlying the analysis are correct. We will discuss these various steps next. As we show, using SW-CAM, we can independently and robustly derive values for R, C, and Bi, and we can analyze these numbers to evaluate the system.

First, the starting data of current density I_e versus time t (see Figure 15.9a) must fulfill several properties, namely that I_e starts off high and must drop to zero, both in the first step where we apply a voltage difference between the two electrodes (charging step), as well as in the second step where the voltage is reduced to zero again (discharging). At each switching moment, the current must instantly step up or down, as indeed observed in Figure 15.9a. The higher the voltage, the higher the currents must be.

Next, we integrate the current signal with time, and after dividing by the electrode thickness we obtain the charge σ^* (per unit total electrode volume), as plotted in Figure 15.9b versus cell voltage.

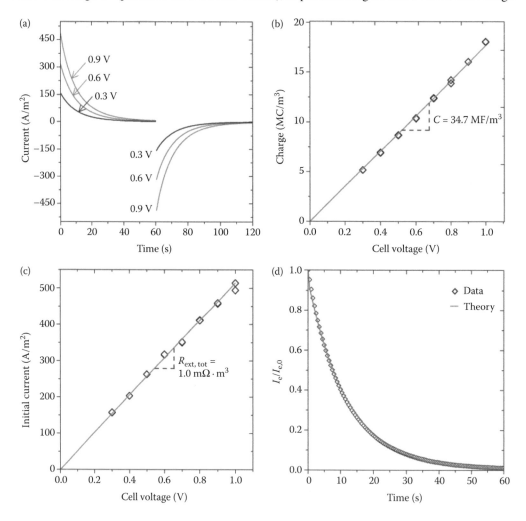

FIGURE 15.9 EC analysis using square wave chrono-amperometry of a two-porous electrode cell filled with 0.5 M KCl solution. (a) Transient current signals after applying a cell voltage step change, V_{cell}, and again after the voltage is reduced to zero. (b) Electrode charge in C/m^3. From the slope, we can derive the capacity C. (c) Initial current versus V_{cell}. (d). Mastercurve of dimensionless current versus time for the data sets of panel (a) and comparison with theory (Bi = 0.15). Mastercurve of dimensionless current versus time for the data set of panel (a) (diamonds) and comparison with theory (gray line, Bi = 0.15).

Two requirements can here be checked for validity: first of all whether the charge increases in proportion to voltage, implying that the capacity, C, does not depend on voltage, a basic requirement in the transmission line theory. Second, the total charge, σ^*, transported in the one direction during the charging step must be close to the total charge transported back in the discharging step, and thus the two values derived for charge σ^* at each voltage must be the same. Both these requirements can be observed in Figure 15.9b to hold true. Taking the slope of σ^* versus V_{cell} and multiplying by a factor 2, we obtain the value for the capacity C required in the transmission line theory, which in this case is $C = 34.7$ MF/m^3. We can also multiply this number by the electrode mass density ρ_e and obtain for the capacity ~70 C/g, which is a fairly standard number for the single-electrode capacity. This number can be divided by 4 to obtain the actual *system* capacity such as reported, for example, in Refs. 40 and 75. For 500 mM NaCl solution and a different electrode material, a value of 23 C/g system capacity was measured in Ref. 75, whereas for the same system $C = 19$ C/g was reported for a salinity of 20 mM salt.

Next, we can analyze the initial current $I_{e,0}$. As Equation 15.30 shows, we expect $I_{e,0}$ to be proportional to V_{cell}, and indeed Figure 15.9c shows this to be true. From the slope of $I_{e,0}$ versus V_{cell}, we can derive the external resistance to be $R_{tot,ext} = 1.0$ mΩ m^2. This linear resistance, derived from the initial current, can be due to an ionic resistance (spacer channel) or electrical resistance (wires, current collector, etc.), but in any case it must be outside the porous electrode, and can therefore also be due to a contact resistance, or a resistance in the wires or current collector. The analysis given below suggests that the ionic resistance in the spacer channel cannot account for the observed value of the linear resistance $R_{tot,ext}$ and thus it is likely because of an electronic effect.

Having established these two numbers, $R_{tot,ext}$ and C, the only parameter yet to be derived is the resistance R of the porous electrode. But, first, as shown in Figure 15.9d, we test a final prediction of the theory, namely that we can collapse all curves of I_e versus t onto a common plot by dividing the current I_e by the initial current $I_{e,0}$ (see Equation 15.29). This renormalization has been made in Figure 15.9d and shows that both for the charging and the discharging step a perfect collapse of all curves is obtained. Now, fitting the theory, Equation 15.29, to this collapsed set of data curves gives the following result. Namely, that for Bi above 0.15, we do not get a good fit, with the theoretical curve first dropping too quickly and around 10–30 s, dropping off too slowly (not shown). For Bi ≤ 0.15, the fit is actually quite good, but not perfect. There is no further influence of the value of Bi on the theoretical curve (and thus neither on the quality of the fit) when Bi ≤ 0.15, because for such low Bi-numbers, the resistance in the system is now completely in the linear part (described by $R_{ext,tot}$). This has the consequence that also a simple exponential curve, $I_e(t)/I_{e,0} = \exp(-t/\tau)$, describes the data just as well, with $\tau = R_{ext,tot}CL = 11.5$ s. The choice for Bi = 0.15, is interesting (a lower value would describe the data just as well) because this value follows exactly when we take ideal ionic conductivities in the macropores and neglect tortuosity effects; that is, we apply $R = V_T/2Fp_{mA}cD$ with $p_{mA} = 0.30$, $c = 500$ mM, and $D = 2 \times 10^{-9}$ m^2/s, which results in an electrode volumetric resistance of $R = 0.44$ Ω m. It is difficult to imagine that R can theoretically be even lower, because we already assumed perfect equilibration of the voltage through the carbon matrix, that is, zero electrical resistance in the porous electrode. Thus, the only remaining volumetric resistance is in the electrolyte-filled pores and $R = 0.44$ Ω m must then be the lowest possible resistance, unless unaccounted for effects of enhanced ion transport play a role.

If indeed, apparently, the ionic resistance in the electrode can be described by the theoretical formula based on the free diffusion coefficient, then for the spacer channel we may expect the same, and thus we can calculate the spacer layer contribution to $R_{tot,ext}$ as $R_{tot,ext,th} = V_T L_{half-spacer}/2Fp_{spacer}cD$, which assuming $p_{spacer} \sim 0.50$ results in $R_{tot,ext,th} \sim 0.066$ mΩ m^2 ($L_{half-spacer}$ is 250 µm) which is about 15 times below the experimental value of $R_{ext,tot}$, suggesting that in this particular experiment the linear external resistance is in the electrical circuit/current collector, not in the ionic solution in between the electrodes.

In conclusion, SW-CAM allows us to accurately test the properties of capacitive porous carbon electrodes and calculate the electrode capacity and the various contributions to the observed resistance. In this case, the linear (external) resistance determines the total resistance and analysis suggests that we can assign this resistance to the external electrical circuit, while we can also tentatively conclude that the distributed (volumetric) resistance within the electrode may be close to the ideal value based on an ion transport resistance only determined by the free solution ion diffusion coefficients. This finalizes our exposition of the derivation of the various constants in the transmission line theory based on the SW-CAM technique. In conclusion, the SW-CAM technique is a robust, precise, and very informative method to perform EC analysis on two-electrode capacitive cells in aqueous solutions.

15.6 ENERGY: THERMODYNAMIC MINIMUM AND ACTUAL ENERGY CONSUMPTION IN CDI

Desalination of water is a process in which the ionic content of the water is demixed, and thus the entropy of the system decreases. This implies that we always need an energy input. Desalination implies that we go from a single starting solution (brackish water, volume flow rate $\phi_{vol,in}$, salt concentration c_{in}) to at least two separate product streams, one being the fresh (dilute) water ($\phi_{vol,d}$, c_d), and the other being the concentrate, or brine ($\phi_{vol,c}$, c_c); the two product streams are separated in time to enable a charge/discharge = adsorption/desorption = purify/regeneration cyclic mode of operation. Desalination performance is often defined by the freshwater concentration and by the water recovery WR, which is the ratio of $\phi_{vol,d}$ over $\phi_{vol,in}$.

Water desalination leads to an increase of the system free energy, ΔG. This requires an energy input which is at least equal to this value ΔG. This is the minimum energy input for desalination, independent of the chosen process. In practice, the energy requirement will be significantly higher. The objective of material and engineering studies in water desalination is to design a process for which the energy input is as close as possible to ΔG. To describe ΔG as a function of the desalination performance (recovery and fresh water concentration), the general equation, valid for ideal thermodynamics for the ions in the water, is

$$\Delta G = G_{fresh} + G_{brine} - G_{brackish}, \quad G_i = RT\phi_{vol,i}c_i \ln c_i \tag{15.31}$$

where c is the total dissociated ion concentration in the water. For instance, for a monovalent fully dissociated salt like NaCl, c is twice the salt concentration, c_{salt}. For water, ideal thermodynamics closely describes the exact energy consumption, even up to the concentration of sea water ($c_{salt} \sim 500$ mM). Equation 15.31 predicts that for the desalination of sea water with a recovery WR = 0.6 we need exactly 1 kWh per m³ of produced fresh water (when $c_{salt,d} = 5$ mM).

Instead of Equation 15.31, there are several equivalent equations, perhaps the most elegant being

$$\Delta G = RT\phi_{vol,d}(c_{in} - c_d)\left[\frac{\ln\alpha}{1-\alpha} - \frac{\ln\beta}{1-\alpha}\right] \tag{15.32}$$

where $\alpha = c_{in}/c_d$ and $\beta = c_{in}/c_c$.[76,12] The recovery, WR, relates to concentrations according to WR = $(c_c - c_{in})/(c_c - c_d)$.

Another angle into the topic of energy consumption in CDI is to give data and theory for actual operation, as plotted in Figure 15.10 both for CDI and MCDI. Here, we present data for desalination of an NaCl solution of a starting concentration of $c_{salt} = 20$ mM at a cell voltage of $V_{cell} = 1.2$ V. The presented data are obtained in constant-voltage mode which implies that the effluent salt concentrations are varying in time and it is not straightforward to directly compare with the minimum energy consumption as can be calculated by Equations 15.31 or 15.32. This is more straightforward

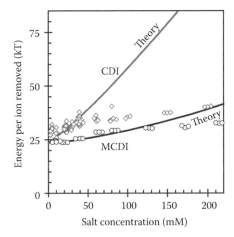

FIGURE 15.10 Energy consumption, data, and theory for constant-voltage CDI and membrane-CDI (V_{cell} = 1.2 V, half-cycle-time 500 s). (Adapted from Zhao, R. et al., *Energy & Environmental Science*, 5, 9520, 2012.)

with constant-current operation (see Ref. 14). Instead, here, we present data for the actual energy consumption per ion removed, which is calculated from the electrical energy input during desalination (ion adsorption step), as the current in Ampère multiplied by the cell voltage in V integrated over the duration of the adsorption step (500 s), and divided by the total ion removal in one cycle. Note that this is a gross energy input during charging—a large part of which can be recovered from the energy release during discharging the cell. Interestingly, we see in Figure 15.10 that for MCDI the energy consumption is somewhat lower than for CDI. For the lower values of ionic strength, a typical number for energy consumption in practice seems to be 25 kT per ion removed. This recalculates to 25 RT per mol of ions removed, and thus to ≈125 kJ per mol of monovalent salt molecules removed.

15.7 MATERIALS FOR CDI

15.7.1 Introduction

Just as in the field of capacitive energy storage devices, essential in any CDI system is the porous electrode. In both fields carbon is the material of choice for developing and making porous electrodes. In the following sections, we will review several carbon materials used for CDI, ranging from conventional activated carbons to highly tunable carbide-derived carbons and highly ordered structures such as carbon nanotubes (CNTs) or graphene. Figure 15.11 provides a selection of various carbons used for CDI applications.

The performance of electrodes for CDI not only relates to the total pore volume, pore size, and pore connectivity, but must also take into account electronic conductivity, EC stability, and cost considerations. The following list summarizes the most important requirements for CDI electrode materials and their influence on CDI performance (see Ref. 82):

1. Large ion-accessible specific surface area
   - The salt electrosorption capacity is related to the surface area.
   - However, not the entire surface area calculated from experimental methods may be accessible to ions.
2. High (electro) chemical stability over the used pH and voltage range (no oxidation, etc.)
   - Important to ensure longevity and system stability.

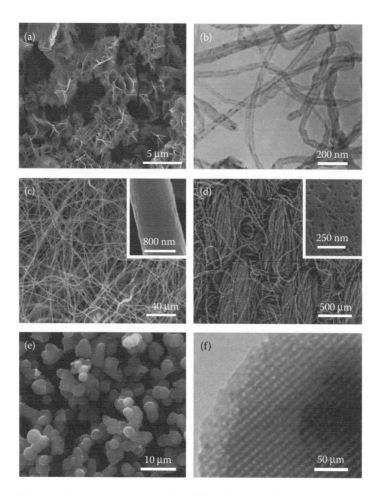

FIGURE 15.11   Selection of carbon materials used for CDI. (a) Graphene (Adapted from Li, H. B. et al., *Journal of Materials Chemistry*, 19, 6773, 2009.); (b) multi-walled carbon nanotubes (Adapted from Zhang, D. et al., *Materials Chemistry and Physics*, 97, 415, 2006.); (c) electrospun fibers (Adapted from Wang, M. et al., *New Journal of Chemistry*, 34, 1843, 2010.); (d) activated carbon cloth (Adapted from Oh, H. J. et al., *Thin Solid Films*, 515, 220, 2006.); (e) carbon aerogel (Adapted from Li, J. et al., *Journal of Power Sources*, 158, 784, 2006.); and (f) ordered mesoporous carbon (Adapted from Peng, Z. et al., *Journal of Physical Chemistry C*, 115, 17068, 2011.).

3. Fast ion mobility within the pore network
   - Bottlenecks or very small pores pose diffusional limitations and limit the kinetics.
   - This not only concerns the porosity within carbon particles, but also the pore structure of the entire CDI electrode considering, for example, interparticle distances and electrode thickness.
4. High electronic conductivity
   - Metallic or metal-like electronic conductivity ensures that the entire electrode surface of all particles is charged without large voltage gradients within the carbon.
   - Only a high electronic conductivity ensures a low energy dissipation and low heating.
5. Low contact resistance between the porous electrode and the current collector
   - A low interfacial resistance is required to avoid a large voltage drop from the electrode to the current collector.
6. Good wetting behavior
   - Hydrophilicity ensures that the entire pore volume is participating in the CDI process.

7. Low costs and scalability
   - Cost considerations are important for large-scale applications.
8. Good processability
   - Shapeable into film electrodes: either based on compacted powders, fibers, or monoliths.
9. Large (natural) abundance and low CO_2 footprinting
   - Availability and environmental impact considerations not only affect cost considerations, but also sustainability concerns.
10. High bio-inertness
    - Biofouling needs to be avoided for long-term applications with surface of brackish water.

Particularly important yet difficult to accomplish is combining a very high specific surface area (#1) and high ion mobility (#3). A smaller pore size and a larger total number of such small pores translate to a larger specific surface area (SSA; defined as surface area per mass). However, more small pores bring along transportation limitations, and steric hindrance as also the number of constrictions increases and the pore walls become more curved. The final and ultimate limit to the pore size is the bare ion size: 1.16 Å for Na^+ and 1.67 Å for Cl^-. These numbers increase when we consider solvated ion sizes: 3.58 Å for sodium and 3.31 Å for chloride.[83] Commonly, the majority of pores of most porous carbons are still significantly larger. Larger pores provide better transport pathways; however, they also decrease the total specific surface area.

With the great importance of the pore structure and its influence on the desalination performance, we first have to clarify the pore size terminology. The International Union of Pure and Applied Chemistry (IUPAC) defines pores strictly according to their size as follows (Refs. 84, 85; see also Figure 15.12):

- *Macropores*: larger than 50 nm
- *Mesopores*: between 2 and 50 nm
- *Micropores*: smaller than 2 nm

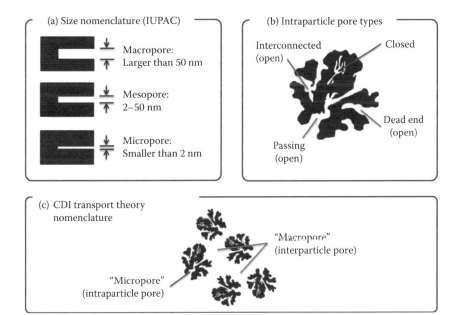

FIGURE 15.12 Pore nomenclature according to IUPAC (a), classification of intraparticle pores (b), and nomenclature in porous media transport theory (c).

Because the term "micro" pores may incorrectly be associated with "micrometer" sized pores, some authors have preferred the term "nanopores" for pore diameters smaller than 2 nm (e.g., Ref. 86). It is important to note that the IUPAC classification is independent of the choice of the porous material (carbon, metal, metal oxide . . .), the kind of pores (closed-pore, open pore . . .), or *where* the pore is actually located (inside a particle versus between particles). As schematically shown in Figure 15.12, open pores may be dead-end-pores (also called semi-open; that is a pore with only one open end which cannot contribute to percolation), interconnected (i.e., connected with another pore but not necessarily to the entire pore volume), or passing—and only closed pores cannot contribute to desalination.

It is important to note that the commonly used term "average pore size" is insufficient to describe a complex pore structure, and it is even misleading for many porous materials because it does not reflect (i) the magnitude, and (ii) the modality of the pore size dispersion; that is, how narrow the pore size distribution centers around one or several maxima. Only carbons with a very narrow pore size distribution, such as CNTs, some carbide-derived carbons (CDCs),[87] and many template-produced carbons,[88] exhibit a meaningful pore size "average," whereas most activated carbons or hierarchic porous materials exhibit a much broader distribution of pore sizes.

Depending on the carbon material and parameters such as synthesis conditions and possible postsynthesis treatment, the geometry of pores inside carbon particles may vary greatly. Abstracted, the pore shapes may be approximated as spherical, cylindrical, or slit shaped as the simplest geometries. Indeed, many templated carbons show cylindrical pores[89] and for most activated carbons we assume slit-shaped pores. More complex shapes are also possible: the space between dense nanoparticles shows pore walls with a positive curvature.[90]

Compared to the IUPAC definition, the literature on transport modeling in porous electrodes employs a different definition of the terms micro- and macropores (cf. Sections 15.2 through 15.6): Here, micropores are any pores inside the porous particles that constitute the electrode, and macropores are the (interparticle) void space between these porous particles. This distinction reflects the large difference in size between the large interparticle and small intraparticles pores. In film electrodes composed of porous carbon powders, the size, number, and magnitude (i.e., pore volume) of such macropores depend on the film preparation and are governed by parameters such as particle size, use of polymeric binder, and film compaction.

In the following sections, we focus on the influence of the properties of carbon particles on CDI performance, and thus the pore structure of the particles becomes important. To describe the pore structure in detail, we adhere to the IUPAC definition of pores which is commonly used for porous carbons (see Figure 15.12). If not defined otherwise, in this section we refer to pores inside porous carbon particles (i.e., intraparticle pores). Only where explicitly noted, we refer to interparticle pores (i.e., between particles). We also prefer the term intraparticle instead of interstitial[91] as the latter is a common term in crystallography and refers to atomic lattice defects.

For making CDI electrodes, we cannot use single carbon particles but need a film composed of such particles. Commonly, film electrodes for CDI are prepared similar to electrodes used for energy storage devices: carbon powders are mixed with a polymeric binder (usually ~10 wt%) and often some conductive additive like carbon black (CB). The components are thoroughly mixed, rolled, and dried, or directly cast on the current collector; the latter may be a graphite foil to avoid corrosion in saline water. A very elegant way to avoid both the use of polymeric binder and possible wash-out of powder particles is the use of a porous monolith,[92] an interwoven array of (porous) fibers,[93] or carbon felts and fabrics.[94,95] When such structured electrodes are either not available or undesired (e.g., because of cost considerations), a film electrode made from a carbon powder can either be protected by a membrane (see also MCDI) or mechanically stabilized by a porous separator which enables saline water flow.

15.7.2 Carbon Materials

15.7.2.1 Activated Carbons and Activated Carbon Cloths

Among porous carbons, activated carbons (ACs) stand out, because they are by far the most often used and usually the most cost-efficient material for many applications. Today, AC is the most common material found in water treatment systems and its use is already documented in early studies on CDI in the 1960/1970s.[4] ACs are derived from natural sources such as coconut shells, wood, coal, resins, or other organic precursors (including starch) and the combination of a high SSA (1000–2500 m^2/g) and low costs (~50 ¢/kg) makes this material particularly attractive for widespread commercial applications.[96] Resin-derived AC can be synthesized to form beads, fibers, or monoliths while most other ACs are usually powders composed of micrometer-sized particles. The variety of precursor materials and synthesis conditions (especially the kind and extent of either chemical or physical activation) translate to an equally great variety in pore structures and surface chemistry of the resulting AC material. Thus, a detailed description of the properties, most importantly the pore size distribution, is mandatory for any comprehensive understanding and comparison of the desalination capacity of AC electrodes. In the following sections, we will discuss the use of AC for desalination from selected studies.

A detailed study on ACs was presented by Zou et al.[97] where conventional AC, KOH-activated AC, and TiO_2/AC composites were compared. The starting material was mesoporous AC with a SSA of 932 m^2/g and an average pore size of 4.2 nm. KOH activation was carried out so that the wetting behavior of the hydrophobic AC was improved by introducing hydrophilic surface groups to the system; this procedure yielded AC with an SSA of 889 m^2/g and an average pore size of 4.2 nm so that the SSA and pore size average remained largely unchanged after the activation compared to the starting material. TiO_2 coating resulted in a slightly lower SSA (851 m^2/g) and a somewhat smaller average pore diameter (4.1 nm). Both KOH-activated and TiO_2-infiltrated/coated samples had a higher CDI performance: +5% and +10%, respectively, compared to the starting material. For KOH-treated materials, the improved performance was explained by the more hydrophilic nature of the material resulting in improved wetting. This is in line with a study by Ahn et al.[98] which indicates that charge transfer related to surface functional groups may also contribute to improved performance. It is still an open question whether this aspect can explain the measured performance and how long-term stability will be affected by the presence of such surface functional groups which, over time, may chemically degrade.

There are several further studies related to metal oxide modification of AC materials. The influence of Ti–O modification of activated carbon cloth (ACC) has been investigated by Ryoo et al.[99,100] For these studies, ACC was derived from a phenolic precursor and the SSA of the material decreased from 1980 to 1180 m^2/g after loading with titanium(IV)-butoxide. After synthesis, chemical analysis confirmed the presence of titanium and oxygen groups, possibly in a tetrahedral configuration as indicated by x-ray photoelectron spectroscopy. Such Ti–O-decorated ACC was found to show a higher salt electrosorption which was explained by the additional charge transfer during the change of the oxidation state of titanium. In contrast, no significant change in the CDI capacity was found for metal-oxide coatings of Si–O, Zr–O, and Al–O.[35]

Instead of utilizing metal oxide coatings, it is possible to modify the surface of ACC with an SSA of 1440 m^2/g by oxidation (HNO_3) or reduction (hydrogen annealing).[41] To mitigate the charge efficiency limiting effect of co-ion adsorption at the point of zero charge (PZC), the anode and cathode were chemically treated in such a fashion that both electrodes would have a different PZC. Although a reduced carbon electrode was found to be unstable and quickly re-oxidized in saline water, an asymmetric three electrode cell with one oxidized electrode, one untreated electrode, and one reference electrode showed a higher charge efficiency compared to a symmetric cell consisting of two untreated ACC electrodes.

A symmetric CDI cell with chemically modified ACC was studied by Oh et al.[79] Oxidation of the carbon material with KOH or HNO_3 slightly decreased the specific surface area (−16% and −5%

for KOH and HNO_3 treatment, respectively) but a higher salt electrosorption capacity was observed, although the pore size average remained unchanged (± 1 Å). The difference in salt adsorption capacity may be related to the activation of carbon, that is, with the opening of closed pores and the creation of new small micropores. The chemical treatment also improved the electrosorption kinetics.

15.7.2.2 Ordered Mesoporous Carbons

In contrast to the disordered arrangement of micropores in most ACs, ordered mesoporous carbons (OMC) show a highly periodic hexagonal or cubic arrangement of mesopores which may improve the transport of salt ions through the pore network. OMCs can be derived via soft or hard templating and the literature on OMCs is extensive including several review articles (e.g., Refs. 88, 101). For hard templating, a template, such as a zeolite or ordered mesoporous silica, is infiltrated with a carbon precursor which then is carbonized, and in a final step, the initial template is chemically removed (e.g., with hydrofluoric acid) yielding OMC powder. The other approach, soft templating, is a relatively new method for the synthesis of OMC powder and thin films and it involves, for example, the self-assembly of triblock copolymers and the thermal removal of the latter; hence, the only solid phase left in the end will be carbon which retains the ordered porous feature of the template. Depending on the synthesis conditions and possible activation, OMCs can have a very high SSA (>3000 m²/g), but more typical SSA values are in the range between 750 and 1500 m²/g. Thus, while the pore arrangement is very different, the SSA of OMC is comparable to many ACs.

OMCs synthesized from a modified sol–gel process with an SSA between 950 and 1594 m²/g and an average pore size between 3.3 and 4.0 nm were studied by Li et al.[102] Compared to AC, these OMCs had a significantly better CDI performance. Even the OMC with the lowest SSA (950 m²/g) showed a much better desalination capacity than AC with a comparably large SSA (845 m²/g). This is in agreement with a study by Zou et al.[17] which showed that AC with an SSA of 968 m²/g performs not as well as OMC with an SSA of 844 m²/g. One explanation is that the disordered arrangement of small AC micropores obstructs fast ion transport so that a certain percentage of the total SSA may not fully contribute to the dynamic process of salt electrosorption; in contrast, almost the entire SSA of OMCs with their transport-optimized pore network may participate in the salt immobilization which results in an improved CDI performance. With little differences in the porosity between two high SSA OMCs in the study by Li et al.[102] (1491 versus 1594 m²/g and 3.7 versus 3.3 nm average pore size), the sample with the lower SSA had a significantly higher desalination capacity. This may be related to the carbon structure and the electrical conductivity: cyclic voltammetry shows that the OMC with lower resistance also yields the better CDI performance.

15.7.2.3 Carbide-Derived Carbons

Unlike ACs, carbide-derived carbons (CDCs) can be synthesized in such a way that they only exhibit extremely narrowly distributed micropores and no mesopores; but, unlike OMCs, pores are not arranged in an ordered fashion. CDCs are most commonly produced by etching of carbide powders in dry chlorine gas at elevated temperatures (200–1200°C), but they can also be derived from monoliths, fibers, or thin films. The chlorine treatment followed by subsequent hydrogen annealing to remove residual chlorine compounds yields an SSA of typically between 1200 and 2000 m²/g but activation may increase the SSA to values of up to 3200 m²/g (cf. Ref. 87 for a review). Recently, the CDI capacity of CDCs derived from titanium carbide (i.e., TiC-CDC) has been investigated, and a positive relation between capacity and the volume of pores smaller than ~1 nm was suggested.[47] This study suggests that the pore volume associated with ion accessible micropores is particularly attractive for CDI, and not, as it may be implied from the studies on OMC, the volume of mesopores.

15.7.2.4 Carbon Aerogels

Another group of carbons are carbon aerogels (CA)[103] which combine a moderate SSA (typically 400–1100 m²/g, but also up to 1700 m²/g)[104] with high electrical conductivity (25–100 S/cm) and a low total density (<0.1 g/cm^3; see Ref. 82). So far, CAs have been synthesized in the form of powders, small beads, thin films, or monoliths and they are composed out of a network of rather dense carbon nanoparticles; thus, the low density is not related to a low skeletal density of the carbon particles but to their spacious arrangement. Most of the total SSA is usually related to interparticles pores (mesopores), but depending on the synthesis conditions there may also be micropores that are related to intraparticles porosity.[105] The latter may range from only 10 m^2/g or less, to more than 600 m^2/g.[106] Typical diameters of the rather round carbon nanoparticles range from 3 to 30 nm and, thus, the resulting pore network is characterized by pore walls with a positive curvature. Such a pore geometry is different from ACs, OMCs, and CDCs, which show predominantly pores with a negative surface curvature, and it may improve the charge storage capacity per unit area when compared to carbons with a negative pore curvature.[90,107,108]

A number of publications are dedicated to the electrosorption from mixed ionic solutions rather than investigating an single salt solution (such as NaCl). For instance, Farmer et al.[109,110] have studied CAs with an SSA of 600–800 m^2/g for CDI in mixed ionic solutions. A decay in salt electrosorption capacity of only ≈5% during a period of several months was observed when using deaerated electrolytes. The electrosorption of a variety of anions and cations (Na$^+$, K$^+$, Mg^{2+}, Rb$^+$, Br$^-$, Cl$^-$, SO$_4^{2-}$, NO$_3^-$) from natural river water was studied in Ref. 91 for CA paper with an SSA of 400–590 m^2/g and average pore sizes between 4 and 9 nm. It was found that monovalent ions with smaller (hydrated) ion size were preferentially electrosorbed by CA electrodes. Farmer et al. also show that the presence of organic components in the water reduces the lifetime of CA CDI electrodes.

Considering the low mechanical stability of CA as a result of the very low bulk density, pasterolling of CA with silica gel was studied by Yang et al.[104] as a method to improve the mechanical properties. Different carbon-to-silica mass ratios (100:0, 75:25, 50:50, 25:75) were investigated and a slight increase in CDI performance was observed when adding the silica gel.

15.7.2.5 CNTs and Graphene

CNTs and graphene are inherently exohedral carbons which have, in case of graphene only and in case of CNT mostly, an ion accessible outer surface. This is in contrast with, for example, AC, where almost the entire surface area is within the particles in the form of internal porosity. However, even when composed of 1D (CNT) or 2D (graphene) nanomaterials, the resulting CDI electrodes are still by nature three-dimensional and will also have a significant interparticle porosity; graphene flakes may also be wrinkled to form slit pores even within one sheet. Also, the presence of metal catalysts in the CNT material may be a detriment to the EC stability of the resulting CDI electrodes and parasitic side reactions can occur.

The most commonly found CNTs are multi-walled CNTs. Zhang et al.[78] studied multi-walled CNTs from catalytic decomposition of methane with surface areas between 50 and 129 m^2/g and the sample with the highest SSA had the largest CDI capacity. In contrast to other porous carbons, these SSA values are very moderate and as a result, the CDI performance is much smaller compared to AC, as shown by Wang et al.[111] Electrodes for CDI composed of multi-walled CNTs (SSA: 47–129 m^2/g)[33] and multi-walled CNT/AC composites[112] have been studied and for pure CNTs, the CDI capacity increased with the amount of SSA. A series of CNT/AC ratios (from 1:0 to 0:1) were studied and it was shown that the CDI capacity continuously decreased as the CNT content was increased. Yet, the energy efficiency improved at high CNT loadings as the device conductivity benefits from a high CNT content considering the moderate conductivity of AC.

Single-walled CNTs (455 m^2/g, average pore size: 4.8 nm) have a higher CDI capacity than double-wall CNTs (415 m^2/g, average pore size: 5.1 nm)[113] and both CNT varieties show a better desalination performance compared to AC with a significantly larger SSA (999 m^2/g, average pore size: 2.1 nm); yet, they are inferior to OMCs (SSA: 1491 m^2/g, average pore size: 3.7 nm).

Freestanding electrodes composed of either multi-walled CNTs or composites of multi-walled CNTs and polyacrylic acid (PAA) were obtained by Nie et al.[114] via electrophoretic deposition.

Employing PAA was shown to improve the CDI performance as it acts as a cation exchange membrane. Another composite electrode composed of PANI (polyaniline) and single-walled CNTs showed a slight increase in CDI performance (~10%) compared to films composed of pure single-walled CNTs. This improvement in desalination was not only explained by the increase in the mesopore volume but may also be related with the different EC properties of the PANI-CNT composite material.[115]

Pure graphene electrodes have been investigated by Li et al.[36] and their CDI electrodes consisted of graphene flakes (or AC for comparison) added to graphite powder and polymeric binder in a 72:20:8 weight ratio. Initially, the graphene SSA was 222 m^2/g, but this value decreased to ≤50 m^2/g after treatment with sulfuric/nitric acid (for comparison: the AC had an SSA of 990 m^2/g).

15.7.2.6 Carbon Black

CB consists of spherical and usually dense carbon nanoparticles with a low SSA (usually below 120 m^2/g)[116] and because of their high electrical conductivity, they are a common conductive additive to film electrodes composed of porous carbons. Indeed, as shown in Refs. 37 and 117, adding CB to CDI film electrodes made from AC significantly improves the salt removal in saline electrolytes with 670, and 1000 ppm NaCl. The very low SSA, however, limits the CDI performance of electrodes purely composed of CB particles, as shown in Refs. 6, 42, and 43.

15.7.2.7 Overview of Performance Data of Carbon Electrodes for CDI

In Table 15.1, we summarize reported data from the literature for the important CDI property of salt adsorption per gram of electrode material. Here, data are given as function of salinity and cell voltage, per gram of both electrodes combined. The experiment in all cases is done in a symmetric cell with the two electrodes of the same mass and material. As can be read from Table 15.1, reported numbers vary in a large range between ~0.7 and ~15 mg/g of adsorbed salt per gram of the electrodes. For scientific studies on CDI electrodes, our suggestion is to always report the measured value of salt adsorption either in mg/g or in mol/g. We suggest that a proper CDI electrode should have an adsorption of at least 8 mg/g at a cell voltage of 1.2 V.

In Table 15.1, it can be observed that experiments done in the BM method with low starting salinities score very low on salt adsorption. This may be because during adsorption, the concentration in the system drops further, and there is a limited amount of salt in the system. Thus, we suggest to do CDI experiments either using the SP-method, or—when using the BM method—at a sufficiently large initial salinity, such that the salinity does not drop by over 50%. In Figure 2 of Ref. 12, a dependence of absorbed amount of salt is observed as function of salt concentrations. Especially below 5 mM (~292 mg/L) salt adsorption decreases. Therefore, we suggest varying the salinity in the experimental program instead of using only a single value for salinity.

Table 15.2 presents a few selected data for electrode mass densities. Adsorptions reported in Table 15.1 can be multiplied by these numbers to obtain the adsorption per milliliter of both electrodes.

15.8 OUTLOOK

On the materials side, further research efforts have to identify the most appropriate pore size and pore size distribution. Excellent desalination obtained with microporous carbons has revised the textbook wisdom that only mesopores would yield a high CDI performance. But studies on OMC and other mesoporous materials also show that such larger pores and a mass-transport-optimized pore architecture clearly help to mitigate diffusional limitations. CDI, in the end, is a dynamic, flow-through process and materials science and system engineering need to factor this aspect in. Clearly, the future of CDI materials belongs to hierarchical porous materials which not necessarily but most probably will be carbon-based—at least for the time being. However, cost considerations are equally important for the widespread success and implementation of CDI technologies: CDI, in the end will have to be an affordable mass technology to truly make a transformative impact.

TABLE 15.1
Overview of Salt Adsorption Performance Reported for Different Electrode Materials Applied for CDI[a]

First Author/Journal/ Publication Year	Carbon Material	Experimental Conditions				Salt Adsorption [mg/g]
		Initial Salt Concentration [mg/L]	Cell Voltage [V]	Carbon Content [%]	Operational Mode	
J. C. Farmer/	Carbon aerogel	~50	1.2	nd	BM CDI	1.4
J. Electrochem. Soc./1996	Carbon aerogel	~500	1.2	nd	BM CDI	2.9
M. W. Ryoo/*Water Research*/2003	Ti-O-activated carbon cloth	~5844	1.0	nd	BM CDI	4.3
K. Dai/*Materials Letters*/2005	Multi-walled CNTs	~3000	1.2	nd	BM CDI	1.7
X. Z. Wang/*Electrochem. Solid-St. Let.*/2006	CNTs- nanofibers	~110	1.2	100	BM CDI	3.3
L. Zou/*Water Research*/2008	Ordered mesoporous carbon	~25	1.2	78	BM CDI	0.68
L. Li/*Carbon*/2009	Ordered mesoporous carbon	~50	0.8	78	BM CDI	0.93
H. Li/*Journal of Materials Chemistry*/2009	Graphene	~25	2.0	100	BM CDI	1.8
Y. J. Kim/*Sep. Purif. Tech.*/2010	Activated carbon	~200	1.5	nd	SP CDI	3.7
	Activated carbon	~200	1.5	nd	SP 0-MCDI	5.3
R. Zhao/*Journal of Physical Chemistry Letters*/2010	Commercial activated Carbon electrode	~292	1.2	nd	SP CDI	10.6[b]
		~1170	1.4	nd		13.0[b]
H. Li/*En. Sci. & Techn.*/2010	Graphene-like nanoflakes	~25	2.0	80	BM CDI	1.3
H. Li/*J. Electroanal. Chem.*/2011	Single-walled CNTs	~23	2.0	70	BM CDI	0.75
P. M. Biesheuvel/*Journal of Colloid and Interface Science*/2011	Commercial activated carbon electrode	~292	1.2	nd	SP CDI	10.5
		~292	1.2	nd	SP 0-MCI	12.8
		~292	1.2	nd	SP r-MCI	14.2
J. Yang/*Desalination*/2011	MnO₂-activated carbon	~25	1.2	nd	BM CDI	4.6
G. Wang/*Electrochimica Acta*/2012	Carbon nanofiber webs	~95	1.6	100	BM CDI	4.6
B. Jia/*Chemical Physics Letters*/2012	Sulfonated graphite nanosheet	~250	2.0	72	BM CDI	8.6
D. Zhang/*Journal of Materials Chemistry*/2012	Graphene-CNT	~29	2.0	90	BM CDI	1.4
H. Li/*Journal of Materials Chemistry*/2012	Reduced graphene oxide-activated carbon	~50	1.2	nd	BM CDI	2.9
Z. Peng/*Journal of Materials Chemistry*/2012	Ordered mesoporous carbon-CNT	~46	1.2	80	BM CDI	0.63
M. E. Suss/*Energy & Env. Sci.*/2012	Carbon aerogel monoliths	~2922	1.5	100	BM CDI	9.6

continued

TABLE 15.1 (continued)
Overview of Salt Adsorption Performance Reported for Different Electrode Materials Applied for CDI[a]

First Author/Journal/ Publication Year	Carbon Material	Experimental Conditions				Salt Adsorption [mg/g]
		Initial salt Concentration [mg/L]	Cell Voltage [V]	Carbon Content [%]	Operational Mode	
Z. Wang/ Desalination/2012	Reduced graphite oxidate—resol	~65	2.0	80	BM CDI	3.2
S. Porada/ACS Applied Materials & Interfaces/2012	Activated carbon (Norit DLC Super 50)	~292	1.2	85	SP CDI	6.9
		~292	1.4	85		8.4
		~292	1.2	85		12.4
	Carbide-derived carbon	~292	1.4	85		14.9

[a] All experiments use NaCl solutions. Reported numbers for adsorption give mg of NaCl adsorption per gram of both electrodes combined. SP, single-pass; BM, batch model; nd, no data. For SP, the given salinity is the inflow salinity. 0-MCDI uses ion exchange membranes with ion release at zero cell voltage; r-MCDI also uses membranes, but the ion release is carried out at reversed voltage.

[b] After electrode mass correction by 10.6/8.5 g/g.

TABLE 15.2
Volumetric Densities of a Few Selected Functional CDI Electrodes

Carbon cloth electrode[41]	0.27 g/mL
Carbon aerogel electrode[16]	0.33 g/mL
Carbide-derived carbon electrode[47]	0.47 g/mL
Activated carbon (AC) electrode[47]	0.5 g/mL
Commercial AC electrode[12,40]	0.55–0.58 g/mL

The next major challenge to improve CDI systems will also involve advanced electrode design. Present developments show a trend toward fiber mats or cloths (comparable to their application in redox flow cells) and freestanding wires which have already shown their potential. Future systems may include monolithic foams or self-assembled hierarchic porous thin films. In this line of thought, the next generation of CDI electrode design may use asymmetry in regards to SSA, film thickness, or PZC. The ability to utilize material composites may not only yield more efficient CDI systems, but may also help to abate detrimental aspects, such as biofouling or irreversible EC material degradation.

ACKNOWLEDGMENTS

Part of this work was carried out in the TTIW-cooperation framework of Wetsus, Centre of Excellence for Sustainable Water Technology. Wetsus is funded by the Dutch Ministry of Economic Affairs, the European Union Regional Development Fund, the Province of Friesland, the City of Leeuwarden, and the EZ/Kompas program of the "Samenwerkingsverband Noord-Nederland." The authors are grateful to Ran Zhao and Raul Rica for scientific discussions and support.

REFERENCES

1. Anderson, M. A. et al., *Electrochimica Acta*, 55, 3845, 2010.
2. Biesheuvel, P. M. *Journal of Colloid and Interface Science*, 332, 258, 2009.
3. Blair, J. W. et al., *Advances in Chemistry Series*, 27, 206, 1960.
4. Johnson, A. M. et al., *The Electrosorb Process for Desalting Water*, U.S. Dept. of the Interior, Washington, 1970.
5. Johnson, A. M. et al., *Journal of the Electrochemical Society*, 118, 510, 1971.
6. Oren, Y. et al., *Journal of the Electrochemical Society*, 125, 869, 1978.
7. Caudle, D. D. et al., *Electrochemical Demineralization of Water with Carbon Electrodes*, U.S. Dept. of the Interior, Washington, 1966.
8. Lee, J.-B. et al., *Desalination*, 196, 125, 2006.
9. Li, H. et al., *Water Research*, 42, 4923, 2008.
10. Biesheuvel, P. M. et al., *Journal of Membrane Science*, 346, 256, 2010.
11. Kim, Y.-J. et al., *Water Research*, 44, 990, 2010.
12. Biesheuvel, P. M. et al., *Journal of Colloid and Interface Science*, 360, 239, 2011.
13. Bouhadana, Y. et al., *Desalination*, 268, 253, 2011.
14. Zhao, R. et al., *Energy & Environmental Science*, 5, 9520, 2012.
15. Bouhadana, Y. et al., *The Journal of Physical Chemistry C*, 115, 16567, 2011.
16. Suss, M. E. et al., *Energy & Environmental Science*, 5, 9511, 2012.
17. Zou, L. et al., *Water Research*, 42, 2340, 2008.
18. Yang, J. et al., *Desalination*, 276, 199, 2011.
19. Lee, J. W. et al., *Applied Chemistry for Engineering*, 21, 265, 2010.
20. Biesheuvel, P. M. et al., *Physical Review E*, 83, 061507, 2011.
21. Biesheuvel, P. M. et al., *Physical Review E*, 81, 031502, 2010.
22. Biesheuvel, P. M. et al., *Russian Journal of Electrochemistry*, 48, 580, 2012.
23. Zhao, R. et al., *Journal of Colloid and Interface Science*, 384, 38, 2012.
24. Rouquerol, J. et al., *Pure and Applied Chemistry*, 66, 1739, 1994.
25. Bard, J. et al., Excess charge and capacitance, in *Electrochemical Methods: Fundamentals and Applications*, Harris, D., Ed. John Wiley & Sons, Inc., New York, 2001.
26. Posey, F. A. et al., *Journal of the Electrochemical Society*, 113, 176, 1966.
27. de Levie, R. *Electrochimica Acta*, 8, 751, 1963.
28. de Levie, R. *Electrochimica Acta*, 9, 1231, 1964.
29. Jung-Ae, L. et al., *Desalination*, 238, 37, 2009.
30. Reinhoudt, H. R. et al., *Water purification device* EPO (EP2253592), 2008.
31. Reinhoudt, H. R. et al., *A method for preparing a coated current collector, a coated current collector and an apparatus for de-ionizing water comprising such current collector* EPO (EP2253592), 2009.
32. Chang, L. M. et al., *Desalination*, 270, 285, 2011.
33. Dai, K. et al., *Materials Letters*, 59, 1989, 2005.
34. Lee, J.-B. et al., *Desalination*, 237, 155, 2009.
35. Leonard, K. C. et al., *Electrochimica Acta*, 54, 5286, 2009.
36. Li, H. et al., *Environmental Science & Technology*, 44, 8692, 2010.
37. Park, K.-K. et al., *Desalination*, 206, 86, 2007.
38. Porada, S. et al., *Electrochimica Acta*, 75, 148, 2012.
39. Farmer, J. C. et al., *Energy & Fuels*, 11, 337, 1997.
40. Zhao, R. et al., *Journal of Physical Chemistry Letters*, 1, 205, 2010.
41. Cohen, I. et al., *Journal of Physical Chemistry C*, 115, 19856, 2011.
42. Oren, Y. et al., *Journal of Applied Electrochemistry*, 13, 473, 1983.
43. Oren, Y. et al., *Journal of Applied Electrochemistry*, 13, 489, 1983.
44. Porada, S. et al., *The Journal of Physical Chemistry Letters*, 3, 1613, 2012.
45. Li, H. et al., *Chemical Physics Letters*, 485, 161, 2010.
46. Huang, Z.-H. et al., *Langmuir*, 28, 5079, 2012.
47. Porada, S. et al., *ACS Applied Materials & Interfaces*, 4, 1194, 2012.
48. Lee, J.-H. et al., *Desalination*, 258, 159, 2010.
49. Wang, M. et al., *New Journal of Chemistry*, 34, 1843, 2010.
50. Avraham, E. et al., *Electrochimica Acta*, 56, 441, 2010.
51. Avraham, E. et al., *Journal of the Electrochemical Society*, 158, 168, 2011.
52. Jeon, B. G. et al., *Desalination*, 274, 226, 2011.

53. Sigalov, S. et al., *Carbon*, 50, 3957, 2012.
54. Mani, A. et al., *Physical Review E*, 84, 061504, 2011.
55. Biesheuvel, P. M. *Journal of Colloid and Interface Science*, 355, 389, 2011.
56. von Helmholtz, H. L. F. *Ann. Phys. Chem.*, 89, 211, 1853.
57. von Helmholtz, H. L. F. *Ann. Phys. Chem.*, 7, 337, 1879.
58. Gouy, G. et al., *Seances Acad. Sci.*, 149, 654, 1909.
59. Gouy, G. *J. Phys.*, 9, 45, 1910.
60. Chapman, D. L. *Philos. Mag*, 25, 475, 1913.
61. Stern, O. *Z. Elektrochem. Angew. Phys. Chem.*, 30, 508, 1924.
62. Bazant, M. Z. et al., *Physical Review E*, 70, 021506, 2004.
63. Yang, K.-L. et al., *Langmuir*, 17, 1961, 2001.
64. Ying, T.-Y. et al., *Journal of Colloid and Interface Science*, 250, 18, 2002.
65. Hou, C.-H. et al., *Journal of Colloid and Interface Science*, 302, 54, 2006.
66. Yang, K.-L. et al., *Journal of Chemical Physics*, 117, 8499, 2002.
67. Huang, J. et al., *Chemistry—A European Journal*, 14, 6614, 2008.
68. Feng, G. et al., *ACS Nano*, 4, 2382, 2010.
69. Biesheuvel, P. M. et al., *Journal of Physical Chemistry C*, 113, 5636, 2009.
70. Kastening, B. et al., *Electrochimica Acta*, 50, 2487, 2005.
71. Philipse, A. et al., *Journal of Physics-Condensed Matter*, 23, 2011.
72. Arafat, H. A. et al., *Langmuir*, 15, 5997, 1999.
73. Grahame, D. C. *Chemical Reviews*, 41, 441, 1947.
74. Bazant, M. Z. et al., *SIAM Journal on Applied Mathematics*, 65, 1463, 2005.
75. Brogioli, D. et al., *Energy & Environmental Science*, 4, 772, 2011.
76. Wilson, J. R. *Demineralization by Electrodialysis*, Butterworths Scientific, London, 1960.
77. Li, H. B. et al., *Journal of Materials Chemistry*, 19, 6773, 2009.
78. Zhang, D. et al., *Materials Chemistry and Physics*, 97, 415, 2006.
79. Oh, H. J. et al., *Thin Solid Films*, 515, 220, 2006.
80. Li, J. et al., *Journal of Power Sources*, 158, 784, 2006.
81. Peng, Z. et al., *Journal of Physical Chemistry C*, 115, 17068, 2011.
82. Yoram, O. *Desalination*, 228, 10, 2008.
83. Nightingale, E. R. *The Journal of Physical Chemistry*, 63, 1381, 1959.
84. Rouquerol, J. et al., *Pure and Applied Chemistry*, 66, 1739, 1994.
85. Sing, K. S. W. et al., *Pure and Applied Chemistry*, 57, 603, 1985.
86. Nikitin, A. et al., Nanostructured carbide-derived carbon, in *Encyclopedia of Nanoscience and Nanotechnology*, Nalwa, H. S., Ed. American Scientific Publishers, CA, 2004, 553.
87. Presser, V. et al., *Advanced Functional Materials*, 21, 810, 2011.
88. Zhai, Y. et al., *Advanced Materials*, 23, 4828, 2011.
89. Korenblit, Y. et al., *ACS Nano*, 4, 1337, 2010.
90. Huang, J. et al., *Journal of Materials Research*, 25, 1525, 2010.
91. Gabelich, C. J. et al., *Environmental Science & Technology*, 36, 3010, 2002.
92. Chmiola, J. et al., *Science*, 328, 480, 2010.
93. Hung, K. et al., *Journal of Power Sources*, 193, 944, 2009.
94. Jost, K. et al., *Energy and Environmental Science*, 4, 5060, 2011.
95. Presser, V. et al., *Advanced Energy Materials*, 1, 422, 2011.
96. Marsh, H. *Activated Carbon*, Elsevier, Boston, MA, 2006.
97. Zou, L. et al., *Desalination*, 225, 329, 2008.
98. Ahn, H.-J. et al., *Materials Science and Engineering: A*, 449–451, 841, 2007.
99. Ryoo, M. W. et al., *Journal of Colloid and Interface Science*, 264, 414, 2003.
100. Ryoo, M. W. et al., *Water Research*, 37, 1527, 2003.
101. Kyotani, T. et al., Carbide derived carbon and templated carbons, in *Carbon Materials for Electrochemical Energy Storage Systems*, Beguin, F., and Frackowiak, E., Eds, CRC Press/Taylor & Francis, Boca Raton, FL, 2009, 77.
102. Li, L. et al., *Carbon*, 47, 775, 2009.
103. Pekala, R. W. et al., *Journal of Non-Crystalline Solids*, 145, 90, 1992.
104. Yang, C.-M. et al., *Desalination*, 174, 125, 2005.
105. Moreno-Castilla, C. et al., *Carbon*, 43, 455, 2005.
106. Yoshizawa, N. et al., *Journal of Non-Crystalline Solids*, 330, 99, 2003.
107. Feng, G. et al., *Physical Chemistry Chemical Physics*, 13, 1152, 2011.

108. Pech, D. et al., *Nat Nano*, 5, 651, 2010.
109. Farmer, J. C. et al., The use of capacitive deionization with carbon aerogel electrodes to remove inorganic contaminants from water, in *Low Level Waste Conference*, Orlando, 1995.
110. Farmer, J. C. et al., *Journal of the Electrochemical Society*, 143, 159, 1996.
111. Wang, X. Z. et al., *Electrochemical and Solid-State Letters*, 9, E23, 2006.
112. Dai, K. et al., *Chemical Engineering Science*, 61, 428, 2006.
113. Zou, L. Developing nano-structured carbon electrodes for capacitive brackish water desalination, in *Expanding Issues in Desalination*, Ning, R. Y., Ed. INTECH, Janeza Trdine, Croatia, 2011.
114. Nie, C. et al., *Electrochimica Acta*, 66, 106, 2012.
115. Yan, C. et al., *Desalination*, 290, 125, 2012.
116. Stoeckli, F. et al., *Carbon*, 40, 211, 2002.
117. Nadakatti, S. et al., *Desalination*, 268, 182, 2011.
118. Kim, Y.-J. et al., *Separation and Purification Technology*, 71, 70, 2010.
119. Li, H. et al., *Journal of Electroanalytical Chemistry*, 653, 40, 2011.
120. Wang, G. et al., *Electrochimica Acta*, 69, 65, 2012.
121. Jia, B. et al., *Chemical Physics Letters*, 548, 23, 2012.
122. Zhang, D. et al., *Journal of Materials Chemistry*, 22, 14696, 2012.
123. Li, H. et al., *Journal of Materials Chemistry*, 22, 15556, 2012.
124. Peng, Z. et al., *Journal of Materials Chemistry*, 22, 6603, 2012.
125. Wang, Z. et al., *Desalination*, 299, 96, 2012.

16 Carbon Nanotubes for Photoinduced Energy Conversion Applications

Ge Peng, Sushant Sahu, Mohammed J. Meziani, Li Cao, Yamin Liu, and Ya-Ping Sun

CONTENTS

16.1 INTRODUCTION

Photoinduced energy conversion processes are at the center of many green and renewable energy technologies.[1–7] Materials widely used for photo-energy conversion are classic semiconductors, especially those at the nanoscale, such as silicon nanoparticles, semiconductor quantum dots (QDs), or nanorods.[3,4,8–15] Since the discovery of fullerenes,[16] carbon nanomaterials have been widely pursued for their photon-harvesting and photoinduced redox characteristics relevant to energy conversion applications.[17–20] In fact, fullerene derivatives are among the most popular materials used in organic or related photovoltaic devices.[21–27]

Carbon nanotubes (CNTs) are also photoactive, with high optical absorption cross-sections (for both one- and multiphoton excitations),[28–31] rich excited state properties,[32–36] and efficient photoinduced charge-transfer processes.[35,37,38] CNTs have been investigated extensively for potential applications in photo-energy conversions.[39–42] For example, CNTs have been integrated successfully into organic photovoltaic devices as part of the photoactive layer, and as highly prospective materials to replace counter electrodes and transparent conductive oxide (TCO) layers. The first use of CNTs as electron acceptors in bulk-heterojunction solar cells was reported in 2002, when researchers blended single-walled carbon nanotubes (SWCNTs) with polythiophenes and observed an increase in the photocurrent by two orders of magnitude.[43] In another report, the use of CNTs in dye-sensitized solar cells (DSSCs) has doubled the efficiency of such photoelectrochemical solar cells.[44] It has also been demonstrated that SWCNT can be an efficient photovoltaic diode device material[45,46] with impressive electrical characteristics, including the generation of multiple electron–hole pairs when light is focused on an individual CNT. The current trend is to use a new kind of all carbon photovoltaic device, in which all the functional layers are made of graphitic nanomaterials.[41,47–49] For example, solar cells made with C_{60}/SWCNTs/reduced graphene oxide as the active layer and an additional evaporated C_{60} blocking layer have yielded a power conversion efficiency of 0.21%, and a significant increase to 0.85% was reported when replacing C_{60} with the better absorber C_{70}.[41] Recently, Strano and coworkers have demonstrated a polymer-free carbon-based photovoltaic device capable of harnessing light in the infrared region, which relies on exciton dissociation at the SWCNT/C_{60} interface.[49] In their device assembly, only highly purified single chirality (6,5) semiconducting SWCNTs were used, allowing for the distinction between intrinsic losses and those caused by impurities in SWCNT chirality. Although device efficiency was limited (~0.10%), it is interesting to note that it is comparable to many polymer-SWCNT bulk-heterojunction (BHJ) devices. They also found that the presence of 20 wt% of a second chirality of semiconducting SCWCNT (6,4) could result in more than 30 times decrease in power conversion efficiency. This effect clearly demonstrates the negative effects of multi-chirality present in an SWCNT active layer on photovoltaic device performance.

CNTs also offer distinctive properties that are not found in fullerenes, such as the very broad spectral coverage with band-gap electronic transitions well into the near-IR spectral region,[50] the ballistic electron transport that may facilitate charge transfer and/or separation over an extended distance,[51–53] and CNTs can be used as linear semiconductors of extremely large aspect ratios.[52,53] Unlike fullerenes, CNTs are capable of accommodating structural defects, which on one hand might be negative to some of the desired properties but on the other present excellent and possibly unique opportunities, from the functionalization for much improved dispersion or solution-phase processing to defect-derived photoluminescence and to the manipulation of photoinduced redox processes.[31,54–57]

In recent years, graphene has emerged to become a center of attention in the field of carbon nanomaterials.[58–62] However, when considering the unique properties of graphene, there are still distinct advantages for the use of CNTs in energy conversion applications. In this chapter, we will highlight the fundamental optical and electronic properties of CNTs, including their optical absorption, band-gap and defect-derived photoluminescence emissions, and photoinduced charge separation and transfers. The effect of their geometric structures, such as diameter and chiral indices, dispersion and chemical functionalization on these properties will be also discussed. Also, we will review the potential of CNTs as building blocks in solar energy conversion applications with a focus on the recent progress of water splitting, CO_2 conversion, and BHJ devices and DSSCs, along with some discussion on the challenges and perspectives in these exciting fields.

16.2 OPTICAL ABSORPTION

CNTs have unique cylindrical hollow structures of extremely large aspect ratios,[63] and one may conceptually visualize their structures by taking single or multiple graphene sheets and rolling them into seamless tubes.[64–66] Depending on the number of the graphene layers, CNTs are further categorized into SWCNT and multiple-walled (MWCNT) CNTs. Owing to this quasi-one-dimensionality, the electronic and optical properties of CNTs are very sensitive to their geometric structure, such as diameter and chiral indices, providing the possibility to tune these properties within a wide range of band gaps from metallic to semiconducting. The situation in MWCNTs is complicated as their electronic properties are determined by the contribution of all individual shells which have different structures. In general, because of the synthesis, MWCNTs usually show a higher defect concentration than SWCNTs. Often, measurements of optical absorption of MWCNT can be difficult and yield inaccurate results[67] and the few studies that have been carried out on CNTs have almost exclusively involved SWCNTs.[64,70–73]

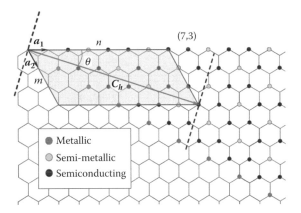

FIGURE 16.1   The conceptual SWCNT formation by rolling up a graphene sheet. As an example, the dashed lines represent the two edges that will merge in the rolling up of a (7,3) semiconducting SWCNT. (Adapted from Liu, X. et al., *Phys. Rev. B: Condens. Matter*, 66, 045411, 2002.)

For SWCNTs, the rolling of a graphene sheet will need to match carbon atoms on the edges, which can be described by a chiral vector, C_h, consisting of two primitive vectors ($C_h = na_1 + ma_2$), to match the graphene carbon atoms from edge to edge (Figure 16.1).[66,68,69] The chiral vector, also commonly referred to as the chiral index (n,m) (or chirality, helicity), uniquely defines the diameter (d) and chiral angle (θ) of an SWCNT:

$$d = \frac{\sqrt{3}a_{c-c}}{\pi}\sqrt{n^2 + nm + m^2} \tag{16.1}$$

$$\theta = \tan^{-1}[\sqrt{3}m/(2n + m)] \tag{16.2}$$

where a_{c-c}(~0.142 nm) is the nearest-neighbor C–C distance.

Depending on its chiral vector, an SWCNT can either be semiconducting or metallic (including semi- or quasi-metallic), which is often referred to as "metallicity." When $n-m \neq 3q$ (q is an integer), the electronic density of states (DOS) in the SWCNT exhibits a significant band gap near the Fermi level, and the nanotube is thus semiconducting; when $n-m = 3q$, the conductance and valence bands in the SWCNT overlap, and the nanotube becomes metallic (or semi-metallic when $n \neq m$). Statistically, there are twice as many ways for rolling graphene sheet into a semiconducting SWCNT as ways for rolling the same sheet into a metallic SWCNT. Therefore, the semiconducting-to-metallic nanotube ratio of 2:1 should generally be expected in an as-grown mixture of SWCNTs.

The electronic structures of both semiconducting and metallic SWCNTs are characterized by several pairs of van Hove singularities in the DOS, inducing unique optical features of SWCNTs, as illustrated in Figure 16.2.[64,70–73] Each van Hove singularity is labeled with the index of the sub-band to which it belongs. The peaks related to transitions between the first and second pairs of DOS singularities in semiconducting nanotubes, designated as S_{11} and S_{22}, are observed at about 0.9 and 1.5 eV, whereas the position of the feature because of the transitions between the first pair of DOS singularities in metallic nanotubes, designated as M_{11}, is about 2 eV. The optical absorption peaks are superimposed on a significant background, which is attributed to the tail of the inter-band π-plasmon resonance that exhibits a broad peak in the optical absorption at around 5 eV.[74] The π-plasmon represents a collective oscillation of π-electrons polarized along the tube axis, which is induced by a π–π* inter-band excitation, and its experimentally observed energy range agrees well with values predicted by theory for bundled SWCNTs.[75]

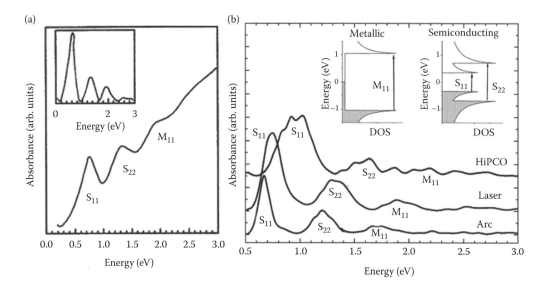

FIGURE 16.2 (a) Optical absorption spectrum of an SWCNT sample with an average diameter of 1.3 nm. The inset shows the same spectrum after subtraction of the surface π-plasmon. (Adapted from Liu, X. et al., *Phys. Rev. B: Condens. Matter*, 66, 045411, 2002.) (b) Absorption spectra of films of purified HiPCO, purified laser, and soluble Arc produced SWCNTs after baseline correction. The inset shows the corresponding interband transitions in the DOS diagram showing van Hove singularities and the levels/bands involved. (Adapted from Hamon, M.A. et al., *J. Am. Chem. Soc.*, 123, 11292, 2001.)

For the evaluation of electronic and optical features in SWCNTs, the nanotubes must be dispersed very well to minimize inter-nanotube quenching effects.[76–78] In this regard, O'Connell et al.[76] developed a method to isolate individual SWCNTs in surfactant micelles through strong ultrasonic treatment and ultra-high-speed centrifugation, yielding relatively short tube fragments (~80–200 nm in length). Optical absorption spectra of these isolated nanotubes exhibited more resolved and sharpened near-infrared absorption peaks. Within each transition region, the observed peaks were attributed to individual tubes, and the variation in the peak intensity originated from differences in both the relative abundance and the absorption cross-section of the tube species. Spectra of samples containing many species, including bundled tubes, typically showed broad, undifferentiated optical absorption features arising from strongly overlapped transitions, rather than sharp, resolved absorption bands.

Experimental work on the implication of chemical functionalization of SWCNTs on the electronic and optical properties has also been investigated.[79,80] It has been found that with an increasing degree of functionalization, the extended π-network is increasingly disrupted, leading to a significant electronic perturbation in the nanotubes. At lower functionalization degrees, the S_{11}, S_{22}, and M_{11} transitions can still be discerned in the UV–vis–NIR spectrum but with less intensity. For example, such behavior has been reported for oxidized SWCNTs,[76] and for SWCNTs modified by the addition of nitrenes.[81] At higher functionalization degrees, the S_{11}, S_{22}, and M_{11} features were no longer visible, as demonstrated for nanotube ozonolysis,[82,83] fluorination,[84,85] as well as the addition of various types of organic residues.[86–92] The weakening or loss of all of the inter-band transitions in both the semiconducting and metallic tubes was attributed to the saturation of the delocalized partial C–C double bonds by the attachment of functional groups, which leads to new bonds at the expense of the π-electrons in the highest occupied molecular orbitals (HOMOs) in the nanotubes.

Metallic and semiconducting SWCNTs have also been readily characterized by resonance Raman spectroscopy.[64,93–95] The radial breathing mode (RBM) in the Raman spectrum of SWCNTs is useful for determining the diameter and the (n, m) values of the nanotubes. Based on the RBM bands, electronic properties of the nanotubes can be predicted by using the so-called Kataura plots. The Raman band of SWCNTs centered at around 1580 cm^{-1} (G-band) also exhibits a feature at approximately 1540 cm^{-1} that is characteristic of metallic SWCNTs. The G-band can be deconvoluted to allow a determination on the relative proportions of metallic and semiconducting species.

16.3 PHOTOEXCITED STATE PROPERTIES

16.3.1 PHOTOLUMINESCENCE EMISSIONS

CNTs were found to display both band-gap fluorescence and defect-derived photoluminescence emissions, which are obviously different in origin, but complementary in certain properties. A shared requirement between the two kinds of emissions is that they are both highly sensitive to the nanotube dispersion. For the band-gap emission, the dispersion is often assisted by the use of surfactants or polymers with the CNTs and also ultra-high-speed centrifugation.[76–78,96–101] The functionalization is effective in the exfoliation of nanotube bundles to the level of individual nanotubes and very thin bundles, but it is hardly applicable to the investigation of band-gap fluorescence. For example, isolated SWCNTs in surfactant micelles, reported by O'Connell et al.,[76] displayed a series of band-gap fluorescence peaks in the near infrared (~800–1600 nm) that were attributed to fluorescence across the band gap of semiconducting nanotubes, as illustrated in Figure 16.3. The data showed that the sample contained many light-emitting species, with each displaying one dominant transition in this spectral range and a very small Stokes shift between its absorption and emission peaks. SWCNT emission was observed exclusively for E_{11} transitions and not for E_{22} or higher transitions, a result that is in accordance with the predictions of Kasha's rule that molecular electronic luminescence originates entirely from the lowest-lying electronic state within a spin multiplicity manifold.[102] Clearly, the many distinct spectral features in the E_{11} region correspond to different (n,m) species of semiconducting SWCNTs in the structurally heterogeneous sample. The mapping of SWCNTs with various diameters by using fluorescence spectroscopy has been accomplished, as shown in Figure 16.3. Such distinct features cannot be achieved with SWCNT bundles because of strong luminescence quenching by neighboring metallic tubes. Recently, there were reports on the detection of band-gap fluorescence for suspended SWCNTs in an ambient environment[99] and also for nanotubes produced by the laser ablation method.[96,97,100,101]

An accurate quantum yield value for the band-gap fluorescence is still being determined or decided, with current numbers ranging from 0.001% to 1.5%, presumably depending on several factors, such as nanotubes diameter (d), diameter distribution, the degree of nanotube bundling/aggregation, and surface doping or chemical treatment.[76–78,96–98,103,104] For example, it has been reported that the quantum efficiency of SWCNTs from the arc-discharge production ($d \sim 1.5$ nm) is weaker in band-gap fluorescence because the average diameter is larger than that of SWCNTs produced by following the HiPco method ($d \sim 0.7$–1.2 nm).[96] The band-gap emission is more prominent in the small-diameter nanotubes with an upper diameter limit of 1.5 nm. For example, the observed quantum yield of SWCNTs from the laser ablation production ($d \sim 1.4$ nm) is of the order of 1×10^{-5}; this is two orders of magnitude lower than that of the HiPco nanotubes. The band-gap emission was also found to be suppressed strongly after chemical functionalization and in the line of this observation, the emission spectrum of dispersed SWCNTs after the acid treatment was found to be weak and poorly structured.[96,105] In another investigation, Cognet et al.[105] reported reversible stepwise quenching of individual SWCNTs by acid and irreversible quenching of individual SWCNT exposed to diazonium salts.

Initially, both SWCNTs and MWCNTs on their surface modification or functionalization were found to be strongly luminescent in homogeneous organic or aqueous solution, exhibiting broad luminescence emission bands in the visible and well extending into the near-IR region.[54] In that report, Sun et al. found strong defect-derived photoluminescence emissions in CNTs purified by nitric acid treatment and then functionalized with amino or other polymers or oligomers, such as

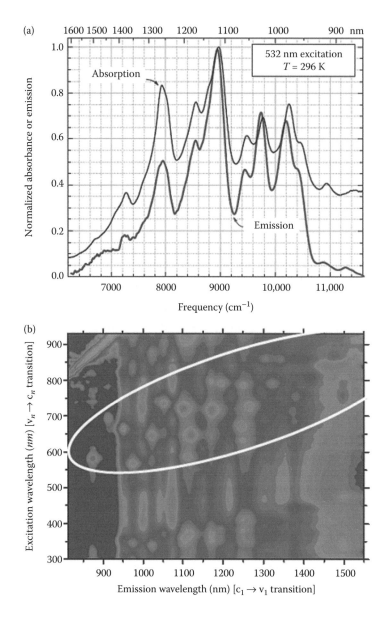

FIGURE 16.3 (a) Overlaid absorption and emission spectra of a sample of HiPco nanotubes in SDS/D_2O suspension. The emission was excited by a pulsed laser at 532 nm. (Adapted from O'Connell, M. et al., *Science*, 297, 593, 2002.) (b) Contour plot of fluorescence intensity versus excitation and emission wavelengths for a sample of SWCNTs suspended in SDS and deuterium oxide. (Adapted from Bachilo, S.M. et al., *Science*, 298, 2361, 2002.)

polyethylene glycol (PEG_{1500N}) or an aminopolymer poly(propionylethylenimine-*co*-ethylenimine) (PPEI-EI) (Figure 16.4), where the functionalization targeted and passivated selectively defects on the nanotube surface. The luminescence excitation spectra of these functionalized CNTs monitored at different emission wavelengths were consistent with the broad UV–vis absorption spectra. However, the emission spectra were strongly dependent on excitation wavelengths in a progressive fashion. The excitation wavelength dependence indicated the presence of significant inhomogeneity or a distribution of emitters in the sample (nanotubes of different diameters) or emissive excited states (trapping sites of different energies).[54,106–108] Defect-derived luminescence has also

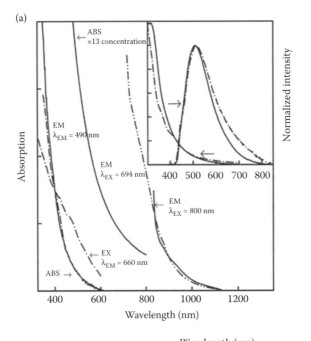

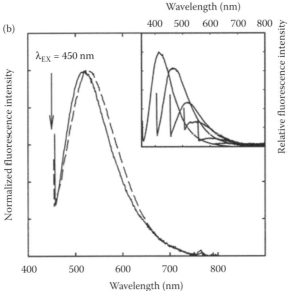

FIGURE 16.4   (a) Absorption (ABS), luminescence (EM), and luminescence excitation (EX) spectra of the PPEI-EI-MWCNT in room-temperature chloroform. Inset: A comparison of absorption and luminescence (440 nm excitation) spectra of PPEI-EI-MWCNT (solid line) and PPEI-EI-SWCNT (dashed line) in homogeneous chloroform solutions at room temperature. (Adapted from Riggs, J. E. et al., *J. Am. Chem. Soc.*, 122, 5879, 2000.) (b) Luminescence emission spectra (normalized, 450 nm excitation) of PPEI-EI-SWCNT (solid line) and PEG$_{1500N}$-SWCNT (dashed) in aqueous solution. Inset: the spectra of PPEI-EI-functionalized SWCNT excited at 350, 400, 450, 500, 550, 600 nm (intensities shown in relative quantum yields). (Adapted from Lin, Y. et al., *J. Phys. Chem. B*, 109, 14779, 2005.)

been observed in other well-functionalized CNT samples of diverse functional groups.[54,55,106–108] For example, Guldi et al. reported that the luminescence was associated with CNT samples from different production methods, including laser ablation and arc discharge, and with heavily oxidized nanotubes.[108] Similarly, Wong et al. reported strong visible luminescence from CNTs that are functionalized with Wilkinson's catalyst.[106]

The observed defect-derived luminescence quantum yields are generally high. As shown in Figure 16.4, for example, the luminescence quantum yields of PPEI-EI-functionalized SWCNTs (PPEI-EI-SWCNT) and PEG_{1500N}-functionalized SWCNTs (PEG_{1500N}-SWCNT) at 450 nm excitation are 4.5% and 3%, respectively.[55] Generally speaking, the luminescence quantum yields of SWCNTs and MWCNTs are of the same order of magnitude. The luminescence decays of functionalized nanotubes are relatively fast and nonexponential, with average lifetimes of the order of a few nanoseconds. The nonexponential nature of the luminescence decays is consistent with the presence of multiple emissive entities in the sample and the observed excitation wavelength dependence of luminescence.

Mechanistically, Sun et al. suggested that the broad visible luminescence could be attributed to the presence of passivated surface defects on CNTs, which serve as trapping sites for the excitation energy.[54] The passivation as a result of the surface modification and functionalization with oligomeric and polymeric species stabilizes the emissive sites in their competition with other excited state deactivation pathways.[54,55,107]

There was also other experimental evidence suggesting that the defect-derived luminescence is sensitive to the degree of functionalization and dispersion of the CNTs. The higher observed luminescence quantum yield is generally associated with better functionalized CNTs, as supported by results from the experiments of repeated functionalization and the defunctionalization of functionalized CNTs.[55,109] For example, Lin et al.[55] demonstrated this sensitivity by comparing nonfunctionalized and functionalized SWCNTs, with the former dispersed in dimethylformamide (DMF) with the assistance of polyimide under sonication and the latter functionalized with polyimide and dissolved in DMF. At the same equivalent nanotube content, the two solutions had comparable optical density at the same excitation wavelength (450 nm). However, the luminescence measurements of the two solutions revealed that the latter was much more luminescent than the former (Figure 16.5). This strongly supports the conclusion that the dispersion of CNTs plays a critical role in their luminescence.

Despite the extensive effort on the elucidation of the two kinds of emissions, there are still significant technical and mechanistic issues for both. For example, the accurate determination of the

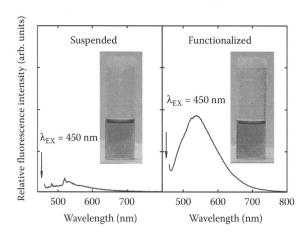

FIGURE 16.5 Luminescence emission spectra and pictures from SWCNTs dispersed with the aid of polyimide in DMF (left) and the PI-NH$_2$-SWCNT in DMF solution (right). The nanotube and polymer contents in the two samples were comparable. (Adapted from Lin, Y. et al., *J. Phys. Chem. B*, 109, 14779, 2005.)

quantum yield for the band-gap fluorescence remains difficult because of the wavelength region, whereas the nature and properties of the emissive excited states for the defect-derived luminescence require further investigation.

16.3.2 CHARGE SEPARATION AND TRANSFERS

Studies of photoinduced electron transfer processes of CNTs hybridized with electron-donating or other electron-accepting molecules have provided evidence for donor–acceptor interactions.[33,37,110–125] Such studies have played a key role in utilizing CNTs in solar energy conversion devices. In fact, it is well known that fast transfer of photoinduced charges from the donor to the acceptor level is very essential for an efficient photovoltaic device.[126] If the electron is not transferred within few femtoseconds in a photovoltaic device, the photogenerated exciton will decay to the ground state, emitting photoluminescence and resulting in a device with poor efficiency. Depending on the redox properties of its counterpart, CNT in the composites can act as either electron acceptor or electron donor. For example, CNTs act as electron acceptors in combination with porphyrins, phthalocyanines, tetrathiafulvalene (TTF) derivatives, semiconducting nanoparticles, and conjugated polymers,[33,37,110–122] or as an electron donors when combined with fullerenes.[123–125] In these systems, the donors or acceptors have been linked to nanotubes by either covalent or noncovalent coupling. In both approaches, the removal of impurities from CNTs must be done carefully because the use of strong oxidation agents (such as a mixture of sulfuric acid and hydrogen peroxide) hinders the photogeneration of free carriers at the donor–acceptor interface.

The most widely examined photoactive components bound to CNTs have been porphyrin molecules, because they are excellent visible light-harvesting chromophores, electron donors, and are also photostable with tunable redox properties.[33,36,110–116] For example, Sun et al. first developed an efficient covalent tethering of SWCNTs with porphyrin through the esterification of SWCNT-bound carboxylic groups and explored their photoexcited state properties via steady-state and time-resolved fluorescence methods (Figure 16.6).[112] Interestingly, the rates and efficiencies of the excited state transfer were found to depend on the length of the tether that links the porphyrins with the SWCNTs. Intramolecular excited state energy-transfer quenching of porphyrin fluorescence by the tethered nanotube occurred in the sample with a longer tether, but no fluorescence quenching was observed in the sample with a shorter tether. The observed fluorescence decays before and after the attachment to SWCNTs were essentially the same resembling static quenching behavior. Not only SWCNT but also MWCNT when covalently linked with porphyrin molecules can act as electron acceptors.[113]

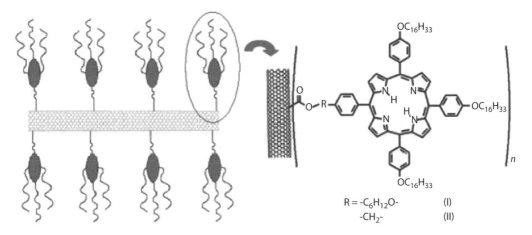

$R = -C_6H_{12}O-$ (I)

$-CH_2-$ (II)

FIGURE 16.6 Structure of porphyrin decorated SWCNTs I and II. (Adapted from Li, H. et al., *Adv. Mater.*, 16, 896, 2004.)

The same strategy was successful in integrating very strong electron donors like tetrathia-fulvalene (TTF), extended TTF, and ruthenium(II) bipyridine, which all supported the occurrence of photoinduced electron transfer processes.[120–122,127,128] Similarly, in most of these studies inserting spacers of different length was found to control the rate of the electron transfer. For example, the lifetime of the charge-separated states (typically of the order of several hundreds of nanoseconds) was appreciably extended when longer linkers and π-extended TTF were used.[122] Another popular concept was to utilize CNT networks as support to anchor light-harvesting metallic and/or semiconductor nanoparticles, which endowed CNTs with specific optoelectronic activities. Research efforts along these lines include organizing these nanoparticles on oxidized CNTs with the aid of crosslinking agents, such as thiol–amine and amine–carboxylic acid cross-linking agents.[129–132] For example, the interactions and the charge-transfer efficiencies between QDs (such as CdS, CdSe, and CdTe) and CNTs have been evaluated by studying the changes in the photoluminescence and using transient absorption (TA) spectroscopy.[133] The photoluminescence behaviors of these nanohybrids were found to be strongly dependent on how the QDs were attached to the CNTs. However, in most of these studies, the QD emission upon visible light excitation was strongly quenched when it was bound to the CNT, suggesting a fast electron transfer to the nanotubes.[133]

Overall, the absorption and fluorescence of these complexes showed that the CNTs are efficient electron acceptors, thus, paving the way to construct novel photovoltaic devices and light-harvesting systems using various configurations.[134,135] For example, porphyrin, ruthenium(II) bipyridine, and QDs when linked to SWCNTs allowed their integration for DSSCs.[134,135]

Considerable efforts have also been dedicated to understand the photoinduced electron transfers at the interface between conjugated polymers and CNTs because of their known potential in organic photovoltaic applications. Among the polymers, poly(3-hexylthiophene) (P3HT), poly(p-phenylene vinylene), and polyindenofluorene are important because of their solubility, high conductivity, and photoluminescence property, which can be tuned in their nanocomposites prepared by different procedures. An important consideration when associating CNTs with these polymers is to avoid a high density of structural defects to preserve the electronic structure. Defects on the CNT surfaces are known to markedly disrupt electron transfer from photoexcited polymer to the CNTs.[35] A versatile approach for the solubility of these nanocomposites often involved the wrapping of CNTs with polymer through noncovalent stacking interaction by sonication and ultracentrifugation cycles in different solvents.[103,136] Electron diffraction studies of these nanocomposites showed that the polymer backbone forms crystalline polymer monolayers wrapped around the nanotube to yield highly ordered polymer:SWCNT nanocomposites. These studies also showed that polymer crystallinity increases with the amount of SWCNT in the composite, and this could explain the improvement of exciton diffusion as well as of charge mobility in these nanocomposites.[137] In these nanocomposites, high photoluminescence quenching of polymers was often observed even in the presence of very minor quantities of CNTs, indicating the occurrence of photoinduced electron transfer. For example, based on photoluminescence quenching, Ferguson et al. demonstrated conclusively that charge transfer occurs in addition to the previously observed energy transfer mechanism.[138] They selectively excited either P3HT or SWCNTs in composite films and probed the free carrier generation with time-resolved microwave conductivity (TRMC). In this study, it was shown that photoexcitation of the polymer gives rise to long-lived carriers, due to spatial separation of the charges across the donor–acceptor interface. However, no charge transfer to the polymer was observed after photon absorption by the SWCNTs, and the carriers were short lived because of their confinement inside the tube along when we consider the electron acceptor character of SWCNTs. Detailed spectroscopic analyses in solution showed an intense redshift for the SWCNT transition energies, consistent with the formation of a type II (staggered band line-up) heterojunction at the P3HT/SWCNT interface,[136,139,140] where the charge separation occurs. Nicholas et al. showed that photoexcitation of P3HT forming a monolayer around an SWCNT leads to ultrafast (~430 fs) electron transfer between the materials using a combination of femtosecond spectroscopy.[141] The addition of excess P3HT leads to long-term charge separation in which free polarons remained separated at room temperature. Their time-correlated single-photon counting studies showed that the photoluminescence decay of P3HT was shortened from 1.7 to 0.43 ps when 1% of SWCNTs individually wrapped with P3HT were added to a blend.

Recent theoretical and experimental studies of P3HT/SWCNT interfaces demonstrated that in fact only semiconducting SWCNTs with a small diameter form favorable type-II heterojunctions with P3HT (Figure 16.7).[48,142,143] Metallic SWCNTs act as recombination sites either by electron transfer from the LUMO of the conductive polymers to empty states in the conduction band of metallic SWCNTs or by electron transfer from the valence band of metallic SWCNTs to the HOMO of P3HT.[142,143] Time-resolved microwave conductivity experiments revealed that the long-lived carrier population can be significantly increased up to three orders of magnitude by incorporating highly enriched semiconducting SWCNTs into semiconducting polymer composites. Recently, the utilization of semiconducting SWCNTs coated with an ordered P3HT layer was found to enhance the charge separation and electron transport in the active layer of BHJ solar cells.[143] A strong photoluminescence quenching was observed for the P3HT/s-SWCNT nanofilaments than in the case of the random P3HT/s-SWCNT mixture spin cast from solution (Figure 16.7), suggesting that well-ordered P3HT layers in the nanofilament configuration form an intimate contact with the semiconducting SWCNTs and can improve the dissociation of excitons generated in the polymer.

However, it is also energetically possible for CNTs to act as electron donors in donor–acceptor nanohybrids when combined with good electron acceptors exhibiting excellent light-harvesting ability. For example, such trend was successfully demonstrated by D'Souza et al. in a self-assembled

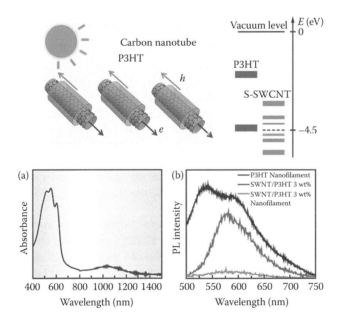

FIGURE 16.7 Upper: (left) Schematic image of a P3HT/semiconducting SWCNT nanofilament, and (right) Band alignment diagram for their interfaces. The positions of the SWCNT energy levels were determined based on the semiconducting SWCNT diameter distribution and corresponding energy gap values. The data used for SWCNT are derived from LDA density functional theory calculations, and slightly underestimate the nanotube electronic gaps and thus the band offsets. HOMO and LUMO band widths of 0.3 eV were used for P3HT. Lower: (a) Optical photoabsorption of a thin layer of P3HT/semiconducting SWCNT nanofilaments with SWCNT concentration 3 wt%, showing contributions to the absorbance from P3HT in the visible and from SWCNTs in the infrared. (b) Photoluminescence intensity measurements for different P3HT/semiconducting SWCNT morphologies. The nanofilament morphology (bottom curve) shows stronger quenching of the photoluminescence in the visible part of the spectrum compared to a simple mixture of P3HT and semiconducting SWCNT (middle curve). (Adapted from Ren, S. et al., *Nano Lett.*, 11, 5316, 2011.)

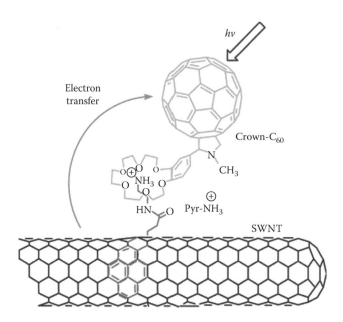

FIGURE 16.8 A scheme illustrating the methodology adopted for building the SWCNT-C_{60} nanohybrids using alkyl ammonium-functionalized pyrene (Pyr-NH_3^+) and benzo-18-crown-6-functionalized fullerene, crown-C_{60}. (Adapted from D'Souza, F. et al., *J. Am. Chem. Soc.*, 129, 15865, 2007.)

SWCNT-C_{60} hybrids, in which SWCNT acted as an electron donor and the fullerene as an electron acceptor (Figure 16.8).[123] Toward this attempt, SWCNTs were first noncovalently functionalized using alkyl ammonium functionalized pyrene (Pyr-NH_3^+) to form SWCNT/Pyr-NH_3^+ hybrids, which was further utilized to complex with benzo-18-crown-6 functionalized fullerene, crown-C_{60}, via ammonium-crown ether interactions to yield stable SWCNT/pyrene-NH_3^+/crown-C_{60} nanohybrids. Studies of steady-state and time-resolved fluorescence and nanosecond transient absorption revealed efficient quenching of the singlet excited state of C_{60} in the nanohybrids and confirmed electron transfer as the quenching mechanism resulting in the formation of $SWCNT^{+}$/pyrene- NH_3^+/crown-$C60^{-}$ charge-separated states (with a relatively longer lifetime of the order of 100 ns). The rates of charge separation, k_{CS}, and charge recombination, k_{CR}, were found to be 3.46×10^9 and 1.04×10^7 s^{-1}, respectively.

A similar conclusion was also derived by Guldi et al. when studying sapphyrin (a pentapyrrolic "expanded porphyrin" macrocycle)-functionalized SWCNT complexes obtained through donor–acceptor stacking interactions.[125] These complexes showed rapid decay of sapphyrin-excited states, something that does not occur for sapphyrin solutions in the absence of nanotubes. Upon photoexcitation of the nanotube material, the presence of a signal ascribable to the sapphyrin radical anion was observed at 840 nm in the transients, a finding that is consistent with electron transfer from the nanotubes to the sapphyrin. The same group recently reported that water-soluble perylenediimide dye with reasonably sized π-system behaved as a strong electron acceptor when mutually interfaced with semiconducting SWCNTs.[144] Detailed spectroscopic studies of these nanohybrids confirmed the occurrence of distinct ground and excited-state interactions and the formation of radical ion pair states within a few picoseconds. Such donor–acceptor nanohybrids may prove useful in the generation of nanotube-based electron transfer ensembles for light-harvesting and photovoltaic applications.

The combination of CNT with donors or acceptors upon illumination seems to give a fast charge separation and a slow charge recombination, which is expected to open up opportunities to a new generation of donor–acceptor nanohybrids. The long lifetimes of the charge-separated species make these systems excellent candidates for the fabrication of photovoltaic devices.

16.4 PHOTOINDUCED ENERGY CONVERSION

Numerous efforts have been made to take advantage of the outstanding optoelectronic and structural properties of CNTs in solar energy conversion, such as water splitting, CO_2 conversion, and organic solar cells. Highlighted below are some interesting recent contributions and achievements dealing with the above needs, along with some discussion on the challenges and perspectives in this exciting field.

16.4.1 CNTs in Photocatalytic Water Splitting and CO_2 Conversion

The past few years saw a renewed interest in photocatalytic water splitting and conversion of CO_2.[2–4,145] The aim is to find new material systems that enable conversion efficiencies beyond 10% which has been set as target for a commercially viable catalyst by the U.S. Department of Energy.[1] The majority of research focuses on the design of suitable materials (typically semiconductors) that are able to efficiently harvest sunlight for the generation of electrons. In general, when a semiconductor is illuminated with photons having energies greater than that of the band gap, electron–hole pairs are generated and separated in a space-charge layer in the material. The pairs, if not recombined, travel to the surface of a semiconductor, and split water to produce oxygen and hydrogen, or reduce CO_2 to yield hydrocarbons (e.g., alcohols). A schematic representation of the principle for these photocatalytic reactions is given in Figure 16.9. To produce hydrogen from water or to reduce CO_2 to fuel, the electrons in the conduction band must have a potential that is more negative than the redox potential of H^+/H_2 (0 eV versus normal hydrogen electrode (NHE)) or CO_2/CH_3OH (~0.03 eV versus NHE) to provide the driving force for the reaction. However, water oxidation occurs when the hole potential is more positive than the redox potential of O_2/H_2O (+1.23 eV versus NHE). On this basis, the minimum band-gap energy required to drive the reaction is 1.23–1.27 eV, which corresponds to absorption of solar photons of wavelengths below about 1000 nm. In practice, the ideal minimum band-gap energy should be higher (close to about 1.35 eV) because of energy losses associated with the over-potentials required for the two chemical reactions and driving force for charge carrier transportation.[3]

For a combination of cost, nontoxicity, chemical stability, and performance, nanoscale TiO_2 has been one of the most attractive photocatalysts among conventional semiconductors. However, the use of nanoscale titania remains limited in terms of the requirement for UV excitation and generally low conversion efficiencies.[4,146] The major drawback in using TiO_2 is its large band-gap

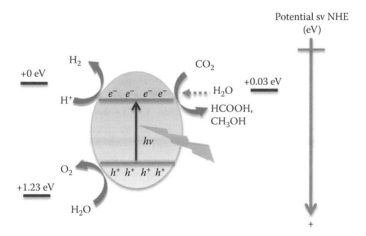

FIGURE 16.9 A schematic representation showing the basic overall principle in photocatalytic water splitting and conversion of CO_2.

energy (3.2 eV), which limits its photocatalytic activity to photons with wavelengths below 385 nm, leading to only approximately 4% effectiveness of the solar radiation, while 43% of light in visible region is wasted. Semiconductors with small band gaps, such as CdS (bandgap of ~2.2 eV), have been successfully used as visible-light photocatalysts because of their high activity and sufficiently negative flat-band potential. However, there are still some major problems such as the serious photocorrosion under long-term light irradiation, which hinders its broad applications. The other major limiting factors that also need to be dealt with are the high charge recombination rate of the photogenerated electrons and holes, and low surface area, which is detrimental for the forward reaction and adsorbance of reactants. In response to these limitations, several approaches have been employed by incorporating suitable atoms such as N and S into their nanostructures, loading nanoscale co-catalysts (e.g., Cu, Ag, Pt, Ru) or other semiconductors on their surfaces, and changing their nanoscale geometrical shapes (e.g., rods, tubes, wires, sheets).[2–4] In addition, to take full advantage of these nanocatalysts, their immobilization and dispersion have also been a major focus as it offers several advantages, such as increasing the specific surface area and facilitating the catalyst reusability.[3,4,147,148] For example, the agglomeration of the nanoparticles has been a critical problem associated with several photocatalyst platforms.[3,147,148]

As an emerging material, CNTs have attracted considerable attention for photocatalytic applications because of their special structures, excellent electronic and mechanical properties, high surface area (~1600 $m^2 g^{-1}$), chemical inertness, and stability. In fact, several works have recently been directed toward investigating CNT as a scaffold to anchor semiconductor nanoparticles and assist in promoting selectivity and efficiency of the photocatalytic process. It has been reported that CNTs may have three main roles in the composite photocatalysts. First, CNTs as an electron acceptor could induce an efficient charge transfer and retard the charge recombination.[149,150] Second, as a photosensitizer, CNTs could expand the visible-light absorption of the photocatalyst and enhance the visible light use efficiency.[151,152] Third, the presence of CNTs could help to enlarge the specific surface area of the photocatalysts, leading to their higher adsorptive ability and their protection from photocorrosion.[153]

Earlier, Ou et al. impregnated anatase (TiO_2) particles with small Ni clusters and used them as a catalyst to grow MWCNTs via chemical vapor deposition (CVD) at 550°C.[154] The hybrid materials were found to be active for H_2 evolution from a methanol/water solution under visible light illumination. The addition of organic alcohols as reducing agents considerably enhanced the reaction efficiency by preventing gasous H_2 and O_2 from recombining on the surface of TiO_2. In contrast to the completely inactive TiO_2-Ni catalyst, the addition of 4.4 wt% MWCNTs produced significant amounts of hydrogen, with a reaction rate of 38 μmol/(h·g) (5 μmol/(h·g) in pure water). Increasing the amount of CNTs above 4.4 wt%, however, led to a decrease in the absorption of light and, thus, in the activity of water splitting. This effect was explained in terms of MWCNTs acting as a photosensitizer, which made the absorption of the catalyst cover the whole range of the UV–vis spectrum. In another report, obvious improvement of the photocatalytic activity for hydrogen generation was achieved by assembling functionalized MWCNTs with Eosin Y dye and Pt and using triethanolamine (TEOA) as the electron donor.[155] The MWCNTs were treated with HNO_3 to form -COOH and -OH groups, which provided anchoring sites for Eosin Y. The highest hydrogen generation rate of 3.06 mmol/(h·g) and the apparent quantum yield of 12.14% were reached when the pH value was 7, the mass ratio of Eosin Y:MWCNT was 5:4, the Pt-loading content was 1 wt%, and the wavelength of incident light was longer than 420 nm. In a similar study, the same photocatalytic system was used but instead of Pt, mixed metal oxides CuO/NiO were used as efficient active center of H_2 evolution.[156] A rate of H_2 evolution of approximately 1.0 mmol/(h·g) was achieved under optimal conditions.

MWCNT–TiO_2 nanocomposites were also synthesized hydrothermally via the direct growth of titania nanoparticles on the surface of functionalized MWCNTs and then loaded with Pt for H_2 production from splitting water.[151] A quantity of 1 wt% Pt loaded 5 wt% MWCNT–TiO_2 nanocomposites showed the highest hydrogen generation rate (235.1 or 8092.5 μmol/(h·g), respectively) with TEOA as the electron donor under visible light ($\lambda > 420$ nm) or full spectral irradiation from a xenon lamp, whereas no capacity to split water was found on the Pt-loaded pristine TiO_2 and MWCNTs. In a similar study, hydrothermal technique was used to synthesize MWCNTs/CdS nanocomposites, and it was found that 10 wt% MWCNTs/CdS had the maximum photocatalytic H_2 production up to 2882.8 µmol after 10 h of photoreaction and better photostability than pure CdS, which was 2.25 times less.[157] In both studies, the enhanced photocatalytic activity was attributed to the excellent light absorption and charge separation on the interfaces between the modified MWCNTs and TiO_2 or CdS. In a separate report, Wang et al.[158] studied the effect of the mass ratio of CdS/CNT in the nanocomposites on the photocatalytic activities. Significant band-gap narrowing was observed due to the incorporation of CNT into CdS, indicating the strong interactions between CdS and CNT. The optimum rate of H_2 evolution for these CdS–CNT nanocomposites was obtained with a mass ratio of 1:0.05, which was found to be approximately four times higher than that of CdS alone. To further enhance the catalytic effect of CNT/CdS, Park et al. explored the ternary hybrids of CdS, MWCNT, and metal catalyst under visible light in the presence of electron donor (Na_2S and Na_2SO_3).[159] When hybridized with CdS and Pt, acid-treated CNT had the largest amount of hydrogen production (up to 3276 µmol/g). In such ternary hybrids, Pt, Ni, and Ru were found to be effective in catalyzing proton/water. Other metals (Pd, Au, Ag, Cu) showed very low activities with the following order: Pt > Ni > Ru > Pd > Au > Ag > Cu. The positive effect of SWCNTs on TiO_2 as photocatalyst for water splitting has been demonstrated by Ahmmad et al.[160] A drastic synergy effect was found with an increase in the amount of H_2 gas by a factor of ca. 400 simply by mixing SWCNTs with TiO_2.

Besides TiO_2 and CdS, other semiconductors with visible-light responses have also been investigated for water splitting by other researchers. For example, Yu et al.[161] demonstrated enhanced photocatalytic hydrogen production by MWCNTs modified $Cd_{0.1}Zn_{0.9}S$ prepared by hydrothermal treatment (Figure 16.10). The optimal MWCNT loading was determined to be about 0.25 wt% and the corresponding H_2-production rate using Na_2S and Na_2SO_3 as sacrificial agents was 1563.2 µmol/h·g with an apparent quantum efficiency of 7.9% at 420 nm even without any noble metal co-catalysts, exceeding that of pure $Cd_{0.1}Zn_{0.9}S$ by more than 3.3 times, and also higher than that (2.16%) of CNTs/CdS nanocomposites reported before. The composites also exhibited good stability and recycling performance of photocatalytic H_2 production.

Chai et al.[162] have reported a hydrothermal synthesis of MWCNTs/$ZnIn_2S_4$ composites and examined their photocatalytic H_2 production under visible-light irradiation. In their study, 3 wt% MWCNTs/$ZnIn_2S_4$ composite reached the maximum photocatalytic hydrogen production rate (about 684 µmol/h) with an apparent quantum efficiency as high as 23.3% under 420 nm light irradiation. Recently, MWCNTs and metal-free graphitic carbon nitride (g-C_3N_4) composites have been

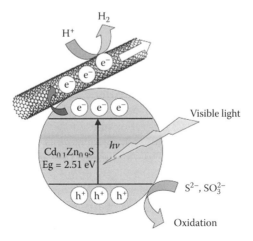

FIGURE 16.10 Proposed mechanism for the enhanced electron transfer in the CNT/$Cd_{0.1}Zn_{0.9}S$ composites under visible-light irradiation. (Adapted from Yu, J., Yang, B. and Cheng, B., *Nanoscale*, 4, 2670, 2012.)

demonstrated to be a good visible light photocatalyst for H_2 evolution in aqueous methanol solutions.[163] Mesoporous g-C_3N_4 was found to possess unique semiconductor properties with a narrow band-gap energy of 2.7 eV along with an open crystalline pore wall and a large surface area (about 200 m^2/g). The optimal MWCNT content was determined to be 2.0 wt% and corresponding H_2 evolution rate was 7.58 μmol/h which was about 3.7-fold higher than that of pure g-C_3N_4. It was suggested that MWCNTs have a higher capture electron capability and can promote electron transfer from g-C_3N_4 toward their surface, leading to the improvement of photocatalytic performance.

Very recently, the enrichment of larger diameter semiconducting SWCNTs was found to be useful for improvement in the performance of photocatalytic water splitting.[164] A CO_2-assisted arc-discharge method was developed to directly synthesize enriched semiconducting SWCNTs with content of >90%, and a majority having diameters of >1.5 nm. Results showed that enriched semiconducting SWCNTs combined with TiO_2 have much better photocatalytic enhancement. The optimal enriched semiconducting SWCNTs content was determined to be 10 wt% and corresponding H_2 evolution rate was 2.0 μmol/h. It was proposed that more allowable transitions from van Hove singularities of semiconducting SWCNTs between the conduction and valance band of TiO_2 make semiconducting SWCNTs more energetically favorable for the charge transfer and, thus, improve the water-splitting efficiency.

In contrast to water splitting, only one example has been reported for CO_2 photoconversion. Although CO_2 photoconversion has a similar mechanism, it requires two to eight electrons to reduce one CO_2 molecule to the desired product. In other words, more free electrons are required in the photocatalyst, which is often accompanied by a dramatic increase in the recombination rate. Xia et al.[165] studied the reduction of CO_2 with H_2O using MWCNTs-supported TiO_2 that were prepared by both sol–gel and hydrothermal methods. In using the sol–gel method, the MWCNTs were coated with anatase nanoparticles, and by the hydrothermal method, rutile nanorods were uniformly deposited on the MWCNTs. The selectivity of the product depended on the method used in material preparation: formic acid was obtained from hydrothermal synthesis and ethanol was produced from sol–gel synthesis.

These original studies provided a valid proof of concept for the successful integration of CNTs within a photocatalytic system and their ability to enhance H_2 production and CO_2 reduction, although the improvements seen so far are quite modest. The results demonstrate that the unique features of CNT make it an excellent supporting material for semiconductor nanoparticles as well as an electron collector and transporter to separate photogenerated electron–hole pairs. There exists still a large range of research opportunities to further enhance the performance in carefully designed hybrid assemblies. Fundamental research on exploring how electrons and holes move and react in these photocatalytic systems will also provide a better understanding of the factors that control photocatalytic activity.

16.4.2 CNTs in Solar Cells

The direct conversion of solar radiation into electricity using photovoltaic devices is currently viewed as an urgent component to meet our demand for clean energy. Generally, this process includes three successive procedures: (i) generation of excitons (photogenerated electron and hole pair) induced by photon absorption by semiconductor materials with the proper electronic band gap, (ii) separation of the electrons and holes from each other, and (iii) their migration through donor and acceptor molecules to the electrodes, generating a photocurrent in an external circuit (Figure 16.11).[166] The wavelength dependence of the power density of ambient sunlight on the Earth's surface is shown in Figure 16.11. The challenge for photovoltaic cell designers is to optimize energy conversion of this incident solar flux to electrons in an external circuit. The theoretical maximum efficiency from a Schockley–Queisser analysis for a solar cell was reported to be about 31%.[167] Advances in this technology continue to bring to the market different semiconductor materials and configurations to achieve thinner applications and improved efficiency

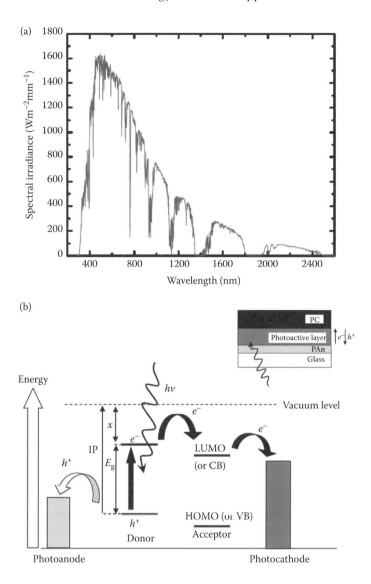

FIGURE 16.11 (a) The standard air mass 1.5 (AM1.5) global solar spectrum. Derived from the ASTMG173–03 Tables obtained from the Renewable Resource Data Center website at http://rredc.nrel.gov/solar/spectra/am1.5/. (b) Energy levels and the harvesting of energy from a photon for an acceptor–donor interface within a photoactive layer of a PV cell. The electron affinity and ionization potential are shown as χ and IP, respectively. LUMO and HOMO are the lowest unoccupied molecular orbital and highest occupied molecular orbital, respectively. CB and VB represent the conduction and valence bands, respectively. PC and PAn are photocathode and photoanode, respectively. A schematic of the PV cell design for which the above diagram applies is also shown. (Adapted from Saunders, B.R. et al., *Adv. Colloid and Interf. Sci.*, 138, 1, 2008.)

in converting light into electricity (Figure 16.12). Currently, the best single crystal silicon with a band gap of 1.1 eV can be used to make a photovoltaic device with about 25% efficiency,[168] but such systems suffer from high cost of manufacturing and installation. Cheaper solar cells (Figure 16.12) can be made from other materials, but their efficiency needs to be enhanced for making them practically viable. Among all the alternative technologies to silicon-based solar cells, organic photovoltaics have attracted significant attention because of their low costs, flexibility, and light weight ability. Most of the research focus for organic photovoltaics has been on

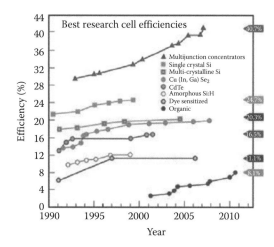

FIGURE 16.12   Year-wise progress of different solar cell technologies. (From National Renewable Energy Laboratory.)

either BHJ devices or DSSCs.[22,169–188] Currently, there are at least four major fundamental aspects in moving these organic photovoltaics toward commercial applications, including low charge mobility, lack of absorbance in the red/NIR spectral range, poor chemical stability, and excitonic character of photocarrier generation.

Recent trends suggest that the successful utilization of CNTs, in particular SWCNTs, could potentially overcome many of the above-mentioned deficiency and lead to the realization of high-efficiency and low-cost solar cells. SWCNTs exhibit several unique properties that make them possibly attractive candidates.[52,53,66] Remarkably, they offer a wide range of band gaps to match the solar spectrum, enhanced optical absorption, a charge mobility of the order of 10^5 cm^2/V s for individual nanotubes,[189] and approximately 60 cm^2/V s for CNT films.[190] This provides an opportunity for charge carrier dissociation and transport because of their large surface area (above 400 m^2/g) and electron-accepting properties. In addition, they also exhibit good chemical stability, and can be easily doped or functionalized; thus, tuning the Fermi level to the favorable position when forming heterojunction with semiconductors.[191–193] In recent years, various strategies and designs for the integration of CNTs in organic photovoltaics have been developed. These include the direct integration of CNTs inside the BHJ and DSSC photoactive layers, and their use as highly prospective materials to replace counter electrodes and TCO layers.

16.4.2.1   CNTs in Bulk Heterojunction

BHJ typically consists of an intimate blend of two semiconductors, an electron-donating conjugated polymer, strongly light absorbing, and usually a soluble fullerene derivative with electron-accepting properties, placed as the active layer between two electrodes (Figure 16.13).[137] With suitable offsets between the energy levels of the highest occupied molecular orbital (HOMO) and lowest unoccupied molecular orbital (LUMO), the blend forms a donor–acceptor system which is suitable to separate photogenerated electron–hole pairs at the interface. The probability that an exciton reaches the interface and dissociates is high if the length scale of the network features is consistent with the exciton diffusion length (10–30 nm). One of the most widely examined material systems consisted of poly(3-hexylthiophene) (P3HT) as the electron donor and 6,6-phenyl-C$_{61}$-butyric acid methyl ester (PCBM) as the electron acceptor. Such a combination has led to power conversion efficiencies between 6% and 9%,[25,26,194] although efficiencies exceeding 10% should be possible according to theoretical calculations.[195] Nevertheless, the efficiency in this kind of devices was found to be limited by the small charge mobility, due to the hopping transport, and the clustering of fullerenes. To prevent this problem, CNTs have often been proposed as an alternative to the fullerenes.

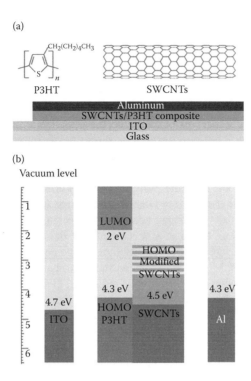

FIGURE 16.13 (a) Chemical structure of P3HT and SWCNTs, and schematic representation of a photovoltaic device. (b) Potential-energy diagram of the SWCNTs/P3HT system relative to vacuum level. (Adapted from Geng, J. and Zeng, T., *J. Am. Chem. Soc.*, 128, 16827, 2006.)

A common strategy to implement CNTs in BHJs involved the dispersion of CNTs in a solution of an electron donating conjugated polymer, such as P3HT or poly(3-octylthiophene) (P3OT) (Figure 16.13). The conjugated polymers act as the photoactive material and CNTs act as the acceptor of the dissociated electrons or holes and their transport path. These blends are then spin-coated onto a transparent conductive electrode with a thicknesses that varies from 60 to 120 nm. These conductive electrodes are usually glass covered with indium tin oxide (ITO) and a 40 nm sublayer of poly(3,4-ethylenedioxythiophene) (PEDOT) and poly(styrenesulfonate) (PSS). The latter two help to smooth the ITO surface, decreasing the density of pinholes and stifling current leakage that occurs along shunting paths. Through thermal evaporation or sputter coating, a 20–70 nm thick layer of aluminum and sometimes an intermediate layer of lithium fluoride are then applied on the photoactive material.

SWCNTs were first proposed as electron acceptor in the BHJ cells by Kymakis et al. who spin casted composite films of P3OT and 1 wt% SWCNTs onto transparent ITO-coated quartz substrates.[43] Compared with the pristine P3OT device, the composite exhibited current densities several orders of magnitude higher under AM1.5 (100 mW/cm²) illumination. The power efficiency of the blend device was dramatically increased from 2.5×10^{-5} to 0.04% with respect to the pristine one. The same authors were able to reach a maximum of 0.22% power efficiency after annealing the device to a temperature higher than the glass transition temperature of the polymer (120°C for 5 min).[196] The improved device performance was attributed to the introduction of internal polymer–nanotube junctions, which allowed for a better charge carrier transport in the polymer matrix and more effective charge separation and collection.

Doping a higher percentage of SWCNTs to the polymer matrix was often avoided because of the possibility that it may cause short circuits as the CNT lengths are comparable to the total thickness of photovoltaic films. The design of new donor materials with optimized properties (chemical stability and extended light absorption) was also found to improve the photovoltaic performances. For example, Lanzi et al. employed new thiophene derivatives bearing a u-methoxy-functionalized hexamethylene side chain as donor, blended with SWCNTs (3 wt%) as the acceptor.[197] This new class of polythiophenes is more soluble and resistant to atmospheric oxygen and led to solar cells with an efficiency of 0.53%. MWCNTs and SWCNTs integrated into the photoactive material were investigated, too, but most of the results indicate a very low power efficiency in the cells.[198–201] This poor performance has been, in part, associated with the presence of metallic SWCNTs that short circuit the cell (reducing the shunt resistance), impurities (mostly metal catalyst residue), SWCNT aggregation, and low-charge carrier mobility in the polymer matrix. As mentioned earlier, metallic SWCNTs lack a true band gap and can, therefore, act as an efficient recombination pathway for excitons in polymer-SWCNT blends, lowering charge-separation efficiency. By contrast, semiconducting SWCNTs display ultrafast photoinduced energy and charge transfer with conjugated polymers such as P3HT, and they can act as efficient acceptors at the interface with P3HT.[48,138,140–142,202,203] Through theoretical calculation, Kanai et al. proposed that the ground state interaction between P3HT and metallic SWCNTs substantially redistributes charge density, resulting in a potential well for holes on the metallic SWCNT and electrons on the polymer, thus seriously hindering charge-separation efficiency.[140] Experimentally, Holt et al. produced blends enriched with either metallic or semiconducting SWCNTs dispersed in a P3HT, and presented conclusive evidence that charge separation is significantly enhanced by at least threefold as the concentration of metallic SWCNTs is reduced, suggesting that composites of P3HT and semiconducting SWCNTs may be promising as the active layer in organic photovoltaic devices.[142] Recently, Ren et al. used enriched semiconducting SWCNTs coated with an ordered P3HT layer to enhance the charge separation and transport in the active layer and demonstrated an AM1.5 efficiency of 0.72%.[143] Contrary to previous prediction, both semiconducting and metallic SWCNTs were also found to function as efficient hole acceptors probably because of the heavy p-doping of P3HT, making selective omission of metallic SWCNT unnecessary. In this regard, Dissanayake et al. reported a surprising external quantum efficiency exceeding 90% for a millimeter-scale P3HT/SWCNTs device, irrespective of whether semiconducting or metallic SWCNTs were used.[204] Instead of BHJs, they used a bilayer heterojunction, depositing on a quartz substrate horizontally aligned arrays of SWCNTs (metallic and/or semiconducting ones) and directly contacting each of them at the extremes by metal electrodes (Al or Pd). Then, P3HT was deposited onto nanotubes and top-contacted by an ITO layer. Another parameter that was also found to crucially influence the charge and exciton transport in the active layer was the SWCNT diameter. The optimal conditions suggested a nanotube diameter between 1.3 and 1.5 nm.[141,205]

Recently, C_{60}-encapsulated semiconducting SWCNTs were used to fabricate $p–n$ heterojunctions with n-type Si.[206] The Si substrate was found to dominate the power-conversion efficiency of SWCNT/Si solar cells at wavelengths less than 1100 nm, whereas semiconducting SWCNTs played a critical role in transforming the infrared light (1550 nm) into the electrical energy. Although the power efficiency was low (~0.01%), the performance of this device was found to be much better than that observed in solar cells fabricated by C_{60}-encapsulated SWCNTs containing both metallic and semiconducting SWCNTs. It was found that when the light photon energy exceeds two times the band-gap energy of semiconducting SWCNTs, the efficiency suddenly increases, suggesting the occurrence of multiple exciton generation.

The other major issue limiting the device performance of CNT solar cells is the bundling/aggregation of nanotubes during polymer dispersion, which reduces the amount of interface available for charge separation, alters the properties of the isolated tubes, and allows only for low concentrations of SWCNTs to be achieved. Nogueira et al. covalently functionalized SWCNT with thiophene moieties to improve their dispersion in a P3OT matrix. The best BHJ solar cell was obtained using 5 wt% of the modified SWCNTs and showed an efficiency of 0.184%.[207] The same group further improved the efficiency to 1.48% under 15.5 W/m² light by using a polybithiophene layer between fluorine-doped tin-oxide (FTO) and a P3OT/SWCNT composite to collect and transport the holes and also avoid nanotube contact with the FTO.[208] To supply additional dissociation sites, Pal et al.

physically blended functionalized MWCNTs into P3HT polymer and then deposited a C_{60} layer to create a P3HT-MWCNT-C_{60} double-layered device.[209] However, the power efficiency was still relatively low at 0.01% under 100 mW/cm² white illumination. Recently, Ham et al.[210] reported a high charge transfer ability on nano-planar heterojunction structures consisting of highly oriented and isolated SWCNTs grown by CVD parallel to the substrate on which was deposited a P3HT layer. Their photovoltaic efficiencies per nanotube ranging from 3% to 3.8% were attributed to the absence of CNT aggregation.

To expand the absorption range of BHJs to the infrared region, several groups anchored QDs, such as CdSe and PbSe, onto the sidewalls of SWCNTs by functionalized thiol groups.[211,212] For example, the organic photovoltaic devices based on blend films of poly(vinyl carbazole)/PbSe QD-grafted SWCNTs exhibited an incident photon-to-current conversion efficiency of 2.6% upon infrared light illumination, which was two times larger than the control device without SWCNTs.[212]

SWCNTs or MWCNTs also served as efficient charge transporters, as a scaffold for fullerenes in conducting polymer-BHJ cells. Under these conditions, the electrons captured by C_{60} molecules are transferred to SWCNTs or MWCNTs, facilitating fast transport rather than slow hopping between C_{60} molecules. The current density of such device was greatly increased when 0.1 wt% MWCNTs were introduced to an 1:1 P3HT:PCBM solution; as a consequence, the power efficiency of the devices was improved to around 2.0%,[213] and this efficiency was nearly three times larger than that of the device without MWCNT addition. In another report, a photovoltaic device from C_{60}-modified SWCNTs and P3HT was made by first microwave irradiating a mixture of SWCNT–water solution and C_{60} solution in toluene, followed by adding P3HT.[214] The best power conversion efficiency of 0.57% under solar irradiation (95 mW/cm²) was achieved on the cell annealed at 120°C for 10 min. Sadhu et al. recently improved the efficiency of P3HT:PCBM to 2.5% by functionalizing the MWCNTs with thiophene groups.[215] The improved efficiency was attributed to the well-dispersed MWCNTs, resulting in an extension of the exciton dissociation volume and charge transport properties through the nanotube percolation network in P3HT/MWCNT, PCBM/carbon nanotube (CNT), or both phases. The highest power conversions reported to date with the use of CNTs were obtained by depositing an SWCNT layer on the cathodic side, either between the ITO and the PEDOT:PSS or between the PEDOT:PSS and the photoactive blend in a modified ITO/PEDOT:PSS/P3HT:PCBM/Al solar cell (Figure 16.14).[216] SWCNTs were deposited by dip coating from a hydrophilic suspension, after an initial nondestructive argon plasma treatment to achieve a power conversion efficiency of 4.9%, compared to 4% without SWCNTs.

Another effective way to further enhance the charge mobility of BHJ cells has been through the use of nitrogen- or boron-doped CNTs in active layers, as highly selective electron- or hole-transport enhancement materials.[217] The work function of N- and B-doped MWCNTs was found to match well with the LUMO of PCBM and the HOMO of P3HT, respectively, because of the excess or lack of electrons created during the doping. In particular, the incorporation of 1.0 wt% B-MWCNTs in P3HT/PCBM BHJ solar cells resulted in balanced electron and hole transport, and accomplished a power conversion efficiency improvement from 3.0% (conventional control cells without CNTs) to 4.1%.[217]

16.4.2.2 CNTs in Dye-Sensitized Solar Cells

DSSCs have emerged as an alternative to solid-state $p–n$ junction photovoltaic devices, primarily because of its potential to harvest visible light quite efficiently and generate electricity on resource abundant raw materials and energy-saving device processing.[173,175,176,218,219] DSSCs operate in terms of a process (Figure 16.15) that is similar in many respects to photosynthesis, the process by which green plants generate chemical energy from sunlight. Central to these cells is a thick film of nanometer-sized wide-band-gap semiconductor (such as TiO_2 or ZnO) that provides a large surface area for the adsorption of light-harvesting organic dye molecules. Upon illumination, the photoexcited dye injects an electron into the conduction band of the semiconductor, which carries the electron to the transparent electrodes.[175,176] This process is accompanied by a reduction of the oxidized dye

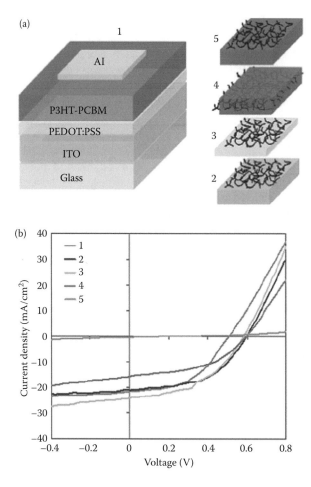

FIGURE 16.14   (a) Schematic of various BHJ solar cells, in which SWCNTs were precisely placed at different hierarchical levels in the device structure. (b) *J–V* curves for these cells (structures from 1 to 5) under AM 1.5 G illumination. (Adapted from Chaudhary, S. et al., *Nano Lett.*, 7, 1973, 2007.)

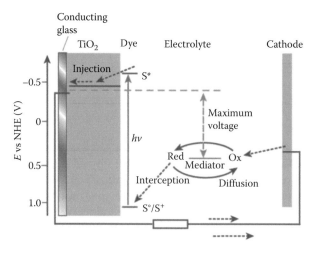

FIGURE 16.15   Schematic illustration for the operation of the DSSCs. (Adapted from Grätzel, M., *Nature*, 414, 338, 2001.)

by a redox-active electrolyte usually iodide/triiodide (I^-/I^{3-}), which transports the resulting positive charge to the platinized counter-electrode.

In terms of quantum efficiency, the maximum attainable efficiency has remained in the range of 11–12%,[220] which are far lower than that predicted theoretically (~20%).[221,222] Several strategies have been proposed to boost the performance of DSSCs, including efficient light harvesting in the visible region, the promotion of electron transport from the adsorbed dyes to the TiO_2 electrode to avoid charge recombination, and efficient hole transport to the counter electrode through the liquid or solid electrolytes. To this end, CNTs have been demonstrated to be promising materials in addressing some of these issues by incorporating them in TiO_2 active layers to improve charge separation and transport and directly using them as counter electrode in replacement of conventional platinum.

CNT networks have been used in TiO_2 active layers, with the purpose of promoting the charge collection and transport in TiO_2 films.[134,223–228] However, the enhancement of device performance is still limited, due to the insufficient contact between TiO_2 and the CNTs and the serious CNT aggregation. Lee et al.[228] fabricated a DSSC using TiO_2-coated MWCNTs, produced by sol–gel method, achieving an increase of 50% in the conversion efficiency with respect to conventional DSSC. Kamat et al. demonstrated the beneficial role of SWCNT network in improving the charge separation, as the rate of back electron transfer between the oxidized sensitizer (Ru(III)) and the injected electrons becomes slower in the presence of the nanotube scaffold (Figure 16.16).[134] It was found that the photoinduced electrons in TiO_2 survive about 50% longer when embedded within the SWCNT network, thus, reducing the charge recombination rate. The authors also noticed an improvement in the photocurrent generation, but this improvement was found to be neutralized at a lower photovoltage as the apparent Fermi level of the TiO_2 and SWCNT composite becomes more positive than that of pristine TiO_2. The beneficial aspect of CNT dispersion was demonstrated by Zhang et al., who improved the dispersion property of CNTs by introducing oxygen-containing groups using O_2 plasma, and then incorporating them within a TiO_2 matrix.[229] The resulting composite provided a better attachment of TiO_2 particles and greater degree of dye adsorption and lower levels of charge recombination. The authors demonstrated a conversion efficiency of 6.34% in DSSCs, 75% higher than that in conventional TiO_2-based devices.

For further improvement of the photoanode performance in DSSCs, it is highly desirable to use semiconducting SWCNTs with a suitable position of conduction band that would facilitate electron transport from mesoscopic TiO_2 particles to ITO electrode through the SWCNTs. Recently, Belcher et al. developed a new method to disperse and template compact core/shell SWCNT/TiO_2 nanocomposites using a genetically engineered M13 virus (Figure 16.17).[230] Using this method, SWCNTs

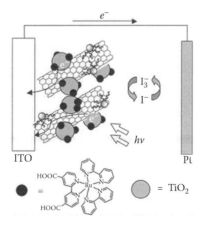

FIGURE 16.16 Schematic illustration for photocurrent generation in the DSSC using the ITO/SWCNT–TOAB/TiO_2/Ru(II) electrode. (Adapted from Umeyamaa, T. and Imahori, H., *Energy Environ. Sci.*, 1, 120, 2008; Brown, P. R. et al., *J. Phys. Chem.* C, 112, 4776, 2008.)

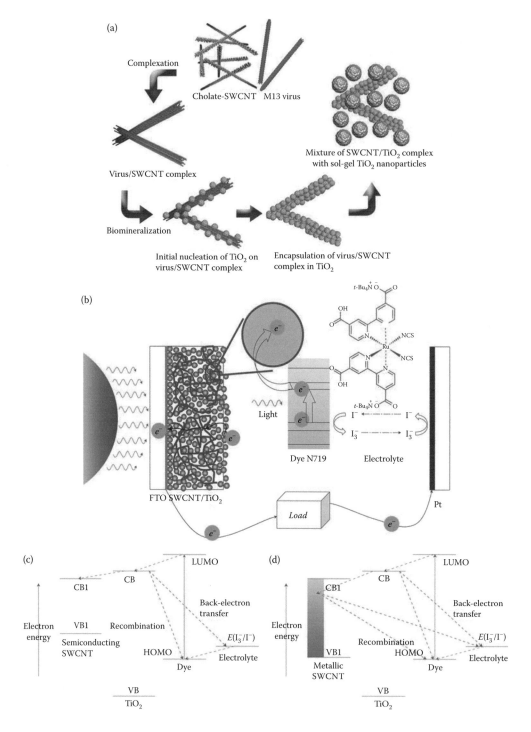

FIGURE 16.17 Schematic diagram of virus-enabled SWCNT/TiO₂ DSSCs. (a) Process of virus/SWCNT complexation, and biomineralization of TiO₂ on the surface of the virus/SWCNT complex. (b) Scheme of DSSCs incorporating the SWCNT/TiO₂ complex. Energy diagrams of DSSCs incorporating semiconducting SWCNTs (c) and metallic SWCNTs (d). The dye absorbs photons and generates electron–hole pairs, and instant charge separation then occurs at the dye/TiO₂ interface. Semiconducting SWCNTs improve electron collection at FTO electrodes, whereas incorporation of metallic SWCNTs results in recombination and back reaction. (Adapted from Dang, X. et al., *Nat. Nanotech.*, 6, 377, 2011.)

were stabilized without surfactants and surface modifications, and their electronic properties were preserved. By using these nanocomposites as photoanodes in DSSCs, the authors demonstrated that well-dispersed semiconducting SWCNTs can be used to improve the power conversion efficiency of DSSCs to 10.6%. They also demonstrated that metallic and semiconducting SWCNTs affect device performance in opposite ways, and aggregation states of SWCNTs affect device performance, guiding further studies in incorporating SWCNTs in photovoltaic devices.

A large number of studies have focused on the use of CNTs as effective counter electrodes because of their good catalytic activity, good conductivity, high surface area, environmental stability, flexibility, and comparatively lower cost.[42,231–243] Although Pt has been the preferred material for counter electrode, its high cost and tendency to degrade over time when in contact with the commonly used electrolytes are serious drawbacks for large-scale and cost-effective DSSC fabrication. In recent developments, it has been proposed that higher performing counter electrodes can be achieved with defect-rich CNT species and vertically aligned CNTs.[236–241,243] For example, Lee et al. found that defect-rich edge planes of bamboo-like-structure MWCNTs can ensure low charge-transfer resistance and an improved fill factor.[232] The device performed well and achieved an efficiency of 7.67%, compared to 7.83% for the Pt-based cell. The high catalytic activity has been attributed to the high surface area and the defect-rich basal plane of CNTs. It has also been debated that this may also be attributed to residual catalyst particles located in the nanotubes.[244]

Aligned CNTs have proved to be a very useful material because electrons can directly transfer through individual tubes, instead of hopping between the different carbon structures within this film which significantly increases their abilities for charge separation and transport. For example, it was found that aligned MWCNT sheet in DSSCs has efficiencies higher than the randomly dispersed CNT film and comparable with platinum.[241] The resulting cell showed an efficiency of 6.6%, compared with 5.27% of the platinum-based cell under the same conditions. In another study, an efficiency of 6.05% was achieved by using vertically aligned MWCNTs grown on graphene paper.[240] With respect to MWCNTs, even higher catalytic activity can be achieved with SWCNTs because of their higher surface area. For example, gel-coated SWCNTs-DSSCs exhibited good stability and reached an efficiency of 8% in conjunction with iodine electrolyte.[237] More recently, vertically aligned SWCNTs have been successfully implemented as efficient counter electrode in DSSCs, featuring notably improved electrocatalytic activity toward a noncorrosive thiolate/disulfide redox shuttle over conventional Pt counter electrodes (Figure 16.18).[245] Any iodine-based electrolyte was

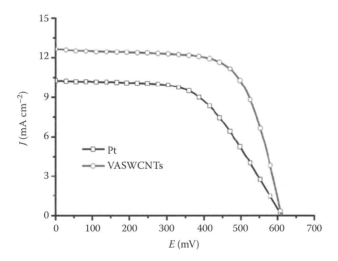

FIGURE 16.18 Photocurrent–voltage characteristics of DSCs using vertically aligned SWCNTs or Pt counter electrodes under 1 sun illumination (AM 1.5 G, 100 mW cm^{-2}). (Adapted from Hao, F. et al., *Sci. Rep.*, 2, 368, 2012.)

avoided because it tends to corrode metallic current collectors and has the tendency to absorb light in the visible wavelengths, which means fewer photons can be utilized. The device with vertically aligned SWCNTs counter electrode demonstrated power conversion efficiency up to 5.25%, lower than the DSSC record of 11% with iodine electrolytes and a platinum electrode, but significantly higher than 3.49% for a control test that combined the new electrolyte with a conventional Pt electrode.

To further improve the performance of the counter electrode in DSSCs, it is highly desirable to use enriched long and smaller diameter metallic SWCNTs.[246] The long tubes should reduce the inter-tube connections, increasing the conductivity of the cathode and small diameter to enhance the chemical reactivity. The use of these high conductive metallic SWCNTs would improve the rate transfer of the electrons arriving from the external circuit to the redox system.

16.4.2.3 CNTs in Transparent Electrodes

ITO has been the predominant material used to create transparent conductive electrodes in organic photovoltaic devices of BHJs and DSSCs, and other optoelectronic devices (e.g., displays, touch screens, organic light emitting diodes (OLEDs)) due to its superior combination of environmental stability, relatively low electric resistivity, and high optical transparency (70 Ω/sq at 90% transmittance).[247,248] However, this material suffers from significant deficiencies, including processing expense involving high temperature deposition, limited availability of indium, incompatibility with plastic substrates, and lack of flexibility, which make it completely unsuited for future flexible electronic devices.[249–252] Thus, the use of CNTs as potential replacements of ITO transparent electrodes, especially those requiring high flexibility, has been a research endeavor of great interest ever since the discovery of SWCNTs.[253–261] The huge intrinsic charge mobility of an individual CNT, the high optical transparency over a wide range of frequencies,[262] and the amenability to solution processing are attractive enough to design electrically conductive, flexible, transparent, and low-cost networks.

Recent investigations have demonstrated organic photovoltaics incorporating SWCNT films as a conductive transparent anode, with the primary barrier to greater efficiencies being the relatively high sheet resistance of the SWCNT film.[42,259,260,263] There has always been a trade-off between conductivity and transparency as the film thickness increases, the sheet resistance and optical transmittance decrease. To meet minimum industry standards, a nearly ideal SWCNT-based transparent conductive film should have a sheet resistance of <50 Ω/sq at 85% transparency, or 100 Ω/sq at 90% (generally taken at 550 nm).[264] Current state-of-the-art SWCNT transparent conductive films have sheet resistances of the order of 300 Ω/sq at 90% transmittance and 100 Ω/sq at 80% transmittance with some variations depending on the methods used.[42,260] In spite of their great potential, the widespread use of SWCNTs in electronic materials remains stalled by issues with unbundling and purifying them to obtain much greater precision over their electronic properties. For example, it is well known that the high electrical conductivity of SWCNTs is associated only with metallic SWCNTs, which generally represent the minority fraction in the mixture with the available synthesis methods for SWCNTs. In addition, experimental evidence has pointed out that the conductivity in nanotube films is also dominated by resistance at the tube–tube junctions.[265–269] Indeed, the contact resistance at metallic-semiconducting junctions is three orders of magnitude higher than that at the metallic–metallic junctions.[270] One of the common methods employed to overcome this obstacle has been chemical doping and functionalization of the films prior to device fabrication.[265,267,271–275] In particular, the adsorption of electron-withdrawing species was found to both lower the SWCNT film sheet resistance and bleach the primary peaks in the optical absorption spectrum, thereby increasing the film transparency.[265,267,272,273] The nanotube films' post-fabrication were often doped via treatment with a strong acid HNO_3, $SOCl_2$, and $AuCl_3$ to metallize semiconducting nanotubes and to decrease the inter-tube resistance at the junctions (the mitigation of Schottky barrier), thus significantly enhancing the electrical conductivity in the films.[267,271,273–275] For example, Feng et al. demonstrated that an acid treatment of few-walled CNTs (FWCNTs) can produce both a 10-fold decrease in resistance and a 1.5-fold increase in transparency.[276] Their FWCNTs ($T = 70\%$, $R_s = 86$ Ω/sq)/

P3HT:PCBM/Al device showed a maximum efficiency up to 0.61%, which was comparable to ITO (0.68%) with almost identical operation. Hecht et al. prepared CNT transparent conducting film using a chlorosulfonic acid dispersion of CVD-grown few-walled CNTs (from Unidym, Inc.) (predominantly single-walled and double-walled tubes grown by CVD with greater than 95% purity) and measured a sheet resistance of 60 Ω/sq at a transmittance of 90.9%, almost matching the performance of ITO films on plastic substrates.[277] In another interesting work, Barnes et al. prepared organic photovoltaic devices by spray deposited oxidized-SWCNTs electrodes giving lower resistance and increased power conversion efficiency from 3.5% to 4.1% with respect to ITO-based devices.[273] Although enhanced optoelectronic performance was achieved, these doped films are generally less stable thermally and chemically, often degrading in performance over time.[256,278] Other studies have also recognized the importance of the geometry of aligned CNTs to further improve thermal conductivity (TC) performance. Vertically aligned CNTs were found to provide a larger electrical and thermal conductivity along the tube axis than randomly dispersed ones.[262,279]

Nevertheless, the demonstrated effect of metallization does point to the great potential of transparent conductive films from separated metallic SWCNTs where there is no need for metallization.[280] Metallic SWCNTs have extremely high electrical conductivity (estimated theoretically as high as 10^6 S/cm)[281] with a ballistic propagation of electrons, largely free from scattering over a distance of thousands of atoms. With their resistance approaching the theoretical lower limits,[282] metallic nanotubes may, in principle, carry an electrical current density of 4×10^9 A/cm^2, which is more than 1000 times greater than that in metals such as copper.[283] Sun et al. recently highlighted major achievements in the development of postproduction separation methods, which are now capable of harvesting separated metallic SWCNTs from different production sources, and their use in transparent electrodes in various device nanotechnologies.[261,284–290] According to a direct comparison by Sun et al., the electrical conductive performance in the films from the separated metallic SWCNTs was consistently much better than that in the films from as-purified SWCNTs.[285–289] The results thus obtained suggested that at approximately 80% transmittance (550 nm) the surface resistivity of the film from metallic SWCNTs was less than 100 Ω/sq (Figure 16.19),[285] a performance level already competitive to that of ITO coatings for some applications.

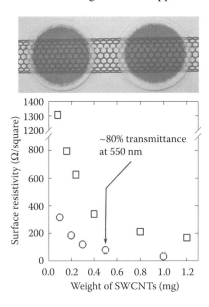

FIGURE 16.19 (a) Films of SWCNTs on alumina filters from the as-purified sample (left) and the separated metallic fraction (right). (b) A direct comparison of surface resistivity values in films of SWCNTs (fabricated via vacuum filtration) from the as-purified sample (□) and the separated metallic fraction (O). (Adapted from Lu, F. et al., *Chem. Phys. Lett.*, 497, 57, 2010.)

The separated metallic SWCNTs were also found to enhance transparent conductive performance in composite films with conductive polymers, particularly the poly(3,4-ethylenedioxythiophene): poly(styrenesulfonate) (PEDOT:PSS) blend as it is optically transparent in the visible spectral region (Figure 16.20).[286] In such composites, the conductive polymer blend served the function of dispersion agents, so no surfactants were necessary in the film fabrication. In the work by Wang et al., suspensions of nanotubes (enriched metallic or nonseparated SWCNTs) in DMSO were mixed with aqueous PEDOT:PSS in various compositions, and the resulting mixtures were sprayed onto an optically transparent substrate.[286] The sheet resistance results demonstrated that the composite films with enriched metallic SWCNTs were consistently and substantially better in performance than those with nonseparated SWCNTs (and both better than films with neat PEDOT:PSS). Aqueous PEDOT:PSS is not as effective as commonly used surfactants in the dispersion of SWCNTs, which

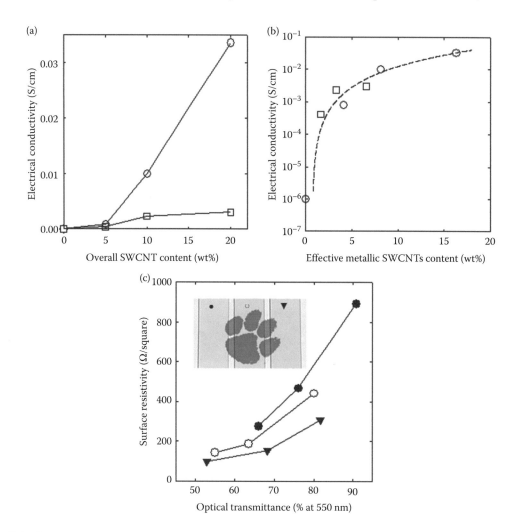

FIGURE 16.20 Electrical conductivity results of P3HT/SWCNT composite films depending on (a) different amounts of pre-separation (□) and separated metallic (○) nanotube samples, and (b) their corresponding effective metallic SWCNT contents in the films (dashed line: the best fit in terms of the percolation theory equation). (c) Surface resistivity results of PEDOT:PSS/SWCNT films on glass substrate with the same 10 wt% nanotube content (○: pre-separation purified sample and ▼: separated metallic SWCNTs; and for comparison, ●: blank PEDOT:PSS without nanotubes) but different film thickness and optical transmittance at 550 nm. Shown in the inset are representative films photographed with tiger paw print as background. (Adapted from Wang, W. et al., *J. Am. Chem. Soc.*, 130, 1415, 2008.)

is negative to the performance of the resulting nanocomposite films. Appropriate structural modifications to PEDOT:PSS for derivatives of improved dispersion characteristics may be pursued to optimize the composite films. It should be pointed out that the use of conductive polymers in transparent electrodes may yield other benefits not reflected in the performance of low surface resistivity. For example, to adjust and improve the interfacial work function in electronic devices such as OLEDs, PEDOT:PSS is often coated as an additional thin layer on a nanotube film or ITO coating in transparent electrodes.[291,292]

For photovoltaic applications, Tyler et al. recently used SWCNTs sorted via the density-gradient ultracentrifugation (DGU) method as transparent anode material and showed that a composition in metallic SWCNTs >70% affords a power conversion efficiency 50-fold higher than monodisperse semiconducting SWCNTs thin films (Figure 16.21).[293] Again, this result points out the necessity to use highly enriched metallic CNTs as electrodes with very low resistance.[294,295]

There is sufficient evidence to validate, in principle, the long-held expectation that metallic SWCNTs may ultimately be used in transparent electrodes, or at least alternatives to the ITO technology. In practice, however, many technical issues from materials (separated metallic SWCNTs)

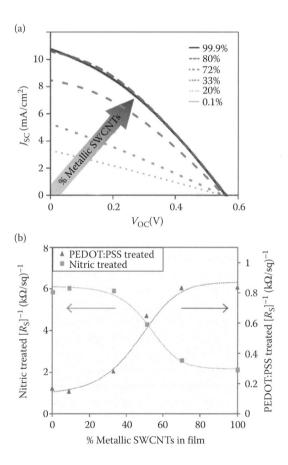

FIGURE 16.21 (a) *J–V* curves of representative devices show a trend of increasing performance with metallic SWCNT content, and nearly identical performance in devices fabricated with 80% and 99.9% metallic SWCNT anodes. (b) The sheet resistance of SWCNT films, plotted here in the inverse, shows a clear dependence on metallic SWCNT content. After nitric acid treatment (left axis), doped semiconducting films have lower sheet resistance than metallic films. After PEDOT:PSS treatment (right axis), the sheet resistance of all films is increased, but the metallic films show greater conductivity than semiconducting films. (Adapted from Tyler et al., *Adv. Energy Mater.*, 1, 785, 2011.)

to fabrication have yet to be addressed. Beyond transparent electrodes, metallic SWCNTs may find other applications in which extremely high electrical conductivity and excellent optical properties are both required, or even some in which optical transparency is not necessary.

16.5 CONCLUSIONS AND PERSPECTIVES

During the past several years, exciting progress has been made in the research on the integration of CNTs into photo-energy conversion systems as exemplified in the solar-driven water splitting for H_2 fuels, photoreduction of CO_2, and in the uniquely configured light-weight and low-cost photovoltaic devices of BHJs and DSSCs. SWCNTs in particular have shown a rich, unique, and useful array of optical and electronic properties. For example, each structural species of semiconducting nanotube displayed not only a set of intense and distinct absorption transitions ranging from near-infrared to ultraviolet wavelengths, but also well-defined fluorescent band-gap photoluminescence in the near infrared. Their combination with electron acceptors or donors favored a fast and long charge separation and a slow charge recombination, by providing a high electric field at the polymer/nanotube interfaces and allowing the electrons to transport efficiently along their length. CNTs are also capable of accommodating structural defects, which present in many cases unique opportunities, from the functionalization for much-improved dispersion or processing to defect-derived photoluminescence and to the manipulation of photoinduced redox processes.

Because of these unique features, CNTs have demonstrated great potential for the emerging energy conversion applications by playing several roles. For example, when converting water into H_2, or converting CO_2 into hydrocarbon fuels, besides providing a high surface area support and immobilization for semiconductor nanoparticles, their presence induced enhanced photocatalytic activity through one or all of the three primary mechanisms: minimization of electron–hole recombination, band-gap tuning/photosensitization, and provision of high-quality adsorptive active sites. Their incorporation into organic photovoltaic devices as components of the photoactive layer and as alternative materials for counter electrodes and transparent conductive films have also been very promising, enabling the development of cheaper, flexible, and more robust devices.

Although performance remains the major barrier for CNTs, the prospect for improvement is great in every step of CNT integration, from growth, purification, separation, dispersion, alignment, doping, to coating. One of the most important areas to explore is chiral selective growth and separation of SWCNTs. Metallic SWCNTs are more suited to replace counter electrodes and TCO layers because their conductivity is much higher than that of the mixture, whereas semiconducting SWCNTs are highly desired for the BHJ and DSSC photoactive layers because they display ultrafast photoinduced charge transfer with conjugated polymers. The separated SWCNTs will likely continue to find use in a variety of energy conversion devices, though higher metallic and semiconducting purity seems necessary for more high-end applications. Significant improvement may also be achieved by developing better dispersion and alignment processes and new stable dopants. High level of dispersion and vertical alignment will allow for even more surface space and for CNT–CNT junction resistance reduction, resulting in greater solar absorption and charge collection and therefore high power efficiency. Besides all these parameters, systematic evaluations of the device performance as a function of various film processing parameters such as concentrations within the composite, active layer film thickness, and film annealing temperature, as well as deeper understandings of the structure–function relationship are still lacking and need to be addressed before fully realizing the extraordinary properties of CNTs in energy conversion systems.

Finally, the recent emergence of graphene nanosheets and related materials may offer other great opportunities for the development of carbon tube–sheet hybrid nanotechnologies, which may speed up their applications for advanced energy conversion.

ACKNOWLEDGMENTS

Financial support from the Air Force Office of Scientific Research (AFOSR) through the program of Dr. Charles Lee is gratefully acknowledged. G.P. was on leave from Ningbo University of Technology in Ningbo, China (with a visiting scholarship jointly funded by CSC-China and Zhejiang Province). M.J.M. was a summer visiting faculty at Clemson supported by funds from NSF (0967423) and the Palmetto Academy program of the South Carolina Space Grant Consortium. L.C. was supported by a *Susan G. Komen for the Cure* Postdoctoral Fellowship.

REFERENCES

1. Green, M. A., *Third Generation Photovoltaics: Advanced Solar Energy Conversion*, Springer-Verlag, Berlin, 2006.
2. Kamat, P.V., *J. Phys. Chem. C*, 111, 2834, 2007.
3. Chen, X. et al., *Chem. Rev.*, 110, 6503, 2010.
4. Roy, S. C. et al., *ACS Nano*, 4, 1259, 2010.
5. Navarro, R.M. et al., *Energy Environ. Sci.*, 2, 35, 2009.
6. Tran, P. D. et al., *Energy Environ. Sci.*, 5, 5902, 2012.
7. Kubacka, A., Fernandez-García, M., and Colon, G., *Chem. Rev.*, 112, 1555, 2012.
8. Kang, Z.H., Liu, Y., and Lee, S.T., *Nanoscale*, 3, 777, 2011.
9. Peng, K.Q. and Lee, S.T., *Adv. Mater.*, 23, 198–215, 2011.
10. Huynh, W. U., Dittmer, J. J., and Alivisatos, A. P., *Science*, 295, 2425, 2002.
11. Kamat, P. V., *J. Phys. Chem. C*, 112, 18737, 2008.
12. Mora-Sero, I. and Bisquert, J., *J. Phys. Chem. Lett.*, 1, 3046, 2010.
13. Kramer, I. J. and Sargent, E. H., *ACS Nano*, 5, 8506, 2011.
14. Nozik, A. J. et al., *Chem. Rev.*, 110, 6873, 2010.
15. Zhou, H. et al., *Energy Environ. Sci.*, 5, 6732, 2012.
16. Kroto, H. W. et al., *Nature*, 318, 162, 1985.
17. Kamat, P. V., *Nano Today*, 1, 20, 2006.
18. Guldi, D. M. et al., Angew. *Chem. Int. Ed.*, 44, 2015, 2005.
19. Zhu, H. et al., *Sol. Energy Mater. Sol. Cells*, 93, 1461, 2009.
20. Liming Dai et al., *Small*, 8, 1130, 2012.
21. D'Souza, F. and Ito, O., *Chem. Soc. Rev.*, 41, 86, 2012.
22. Dennler, G., Scharber, M.C., and Brabec, C.J., *Adv. Mater.*, 21, 1323, 2009.
23. Clarke, T.M. et al., *Chem. Rev.*, 110, 6736, 2010.
24. Gunes, S., Neugebauer, H., and Sariciftci, N.S., *Chem. Rev.*, 107, 1324, 2007.
25. He, Y. and Li, Y., *Phys. Chem. Chem. Phys.*, 13, 1970, 2011.
26. Green, M.A. et al., *Prog. Photovolt: Res. Appl.*, 20, 12, 2012.
27. He, Z. et al., *Adv. Mater.*, 23, 4636, 2011.
28. Barros, E.B. et al., *Phys. Rev. B*, 73, 241406, 2006.
29. Wang, X. et al., *J. Phys. Chem. C*, 114, 20941, 2010.
30. Berciaud, S., Cognet, L., and Lounis, B., *Phys. Rev. Lett.*, 101, 077402, 2008.
31. Sun, Y.-P. et al., In: *NanoScience in Biomedicine*, Shi, D., Ed., Springer-Verlag and Tsinghua University Press, New York, 2009, Ch. 6.
32. Korovyanko, O. J. et al., *Phys. Rev. Lett.*, 92, 017403, 2004.
33. Guldi, D.M. et al., *Acc. Chem. Res.*, 38, 871, 2005.
34. Guldi, D. M. et al., *Chem. Soc. Rev.*, 35, 471, 2006.
35. Sgobba, V. and Guldi, D. M., *Chem. Soc. Rev.*, 38, 165, 2009.
36. D'souza, F. and Ito, O., *Chem. Commun.*, 4913, 2009.
37. Robel, I., Bunker, B. A., and Kamat, P. V., *Adv. Mater.*, 17, 2458, 2005.
38. Yang, C. et al., *Phys. B*, 338, 366, 2003.
39. Umeyamaa, T. and Imahori, H., *Energy Environ. Sci.*, 1, 120, 2008.
40. Cataldo, S. et al., *Energy Environ. Sci.*, 5, 5919, 2012.
41. Tung, V. C. et al., *Energy Environ. Sci.*, 5, 7810, 2012.
42. Brennan, L. J. et al., *Adv. Energy Mater.*, 1, 472, 2011.
43. Kymakis, E. and Amartunga, G. A., *Appl. Phys. Lett.*, 80, 112, 2002.
44. Kongkanand, A., Domínguez, R. M., and Kamat, P. V., *Nano Lett.*, 7, 676, 2007.

45. Gabor, N. M. et al., *Science*, 325, 1367, 2009.
46. Lee, J. U., *Appl. Phys. Lett.*, 87, 073101, 2005.
47. Tomokazu U. et al., *Adv. Mater.*, 22, 1767, 2010.
48. Bindl, D.J. et al., *Nano Lett.*, 11, 455, 2011.
49. Jain, R.M. et al., *Adv. Mater.*, 24, 4436, 2012.
50. Kataura, H. et al., *Synth. Met.*, 103, 2555, 1999.
51. Javey, A. et al., *Nature*, 424, 654, 2003.
52. Dresselhaus, M.S., Dresselhaus, G., and Avouris, P., *Carbon Nanotubes: Synthesis, Structure, Properties, and Applications*, Springer, Berlin, Germany, 2001.
53. Reich, S., Thomsen, C., and Maultzsch, J., *Carbon Nanotubes: Basic Concepts and Physical Properties*, VCH, Weinheim, Germany, 2004.
54. Riggs, J. E. et al., *J. Am. Chem. Soc.*, 122, 5879, 2000.
55. Lin, Y. et al., *J. Phys. Chem. B*, 109, 14779, 2005.
56. Zhou, B. et al., *Springer Ser. Fluoresc.*, 4, 363, 2008.
57. Guldi, D. M. et al., *Chem. Commun.*, 10, 1130, 2003.
58. Geim, A. K. and Novoselov, K. S., *Nat. Mater.*, 6, 183, 2007.
59. Geim, A. K., *Science*, 324, 1530, 2009.
60. Novoselov, K. S. et al., *Science*, 306, 666, 2004.
61. Kamat, P.V., *J. Phys. Chem. Lett.*, 2, 242, 2011.
62. Xiang, Q., Yu, J., and Jaroniec, M., *Chem. Soc. Rev.*, 41, 782, 2012.
63. Iijima, S. and Ichihashi, T., *Nature*, 354, 56, 1991.
64. Saito, R., Dresselhaus, G., and Dresselhaus, M. S., *Physical Properties of Carbon Nanotubes, Silverstein*, Imperial College Press, London, 1998.
65. Ajayan, P. M., *Chem. Rev.*, 99, 1787, 1999.
66. Dresselhaus, M. S., Dresselhaus, G., and Eklund, P. C., *Science of Fullerenes and Carbon Nanotubes*, Academic Press, San Diego, 1996.
67. Raffaelle, R.P. et al., *Mater. Sci. Eng., B*, 116, 233, 2005.
68. Lin, Y. et al., In: *Carbon Nanotechnology: Recent Developments in Chemistry, Physics, Material Science and Device Applications*, Dai, L., Ed., Elsevier, Netherlands, 2006, p. 255.
69. Dai, H., Acc. *Chem. Res.*, 35, 1035, 2002.
70. Liu, X. et al., *Phys. Rev. B: Condens. Matter*, 66, 045411, 2002.
71. Hamon, M.A. et al., *J. Am. Chem. Soc.*, 123, 11292, 2001.
72. Odom, T. W. et al., *J. Phys. Chem. B*, 104, 2794, 2000.
73. Niyogi, S. et al., *Acc. Chem. Res.*, 35, 1105, 2002.
74. Lauret, J.S. et al., *Phys. Rev. Lett.*, 90, 057404, 2003.
75. Shyu, F.L. and Lin, M.F., *Phys. Rev. B*, 62, 8508, 2000.
76. O'Connell, M. et al., *Science*, 297, 593, 2002.
77. Bachilo, S.M. et al., *Science*, 298, 2361, 2002.
78. Graff, R.A. et al., *Adv. Mater.*, 17, 980, 2005.
79. Burghard, M., *Surf. Sci. Rep.*, 58, 1, 2005.
80. Singh, P. et al., *Chem. Soc. Rev.*, 38, 2214, 2009.
81. Holzinger, M. et al., *J. Am. Chem. Soc.*, 125, 8566, 2003.
82. Cai, L.T. et al., *Chem. Mater.*, 14, 4235, 2002.
83. Banerjee, S. and Wong, S. S., *J. Phys. Chem. B*, 106, 12144, 2002.
84. Boul, P.J., et al. *Chem. Phys. Lett.*, 310, 367, 1999.
85. Zhao, W. et al., *J. Phys. Chem. B*, 106, 293, 2002.
86. Umek, P. et al., *Chem. Mater.*, 15, 4751, 2003.
87. Ying, Y.M. et al., *Org. Lett.*, 5, 1471, 2003.
88. Peng, H. Q. et al., *J. Am. Chem. Soc.*, 125, 15174, 2003.
89. Bahr, J. L. et al., *J. Am. Chem. Soc.*, 123, 6536, 2001.
90. Georgakilas, V. et al., *J. Am. Chem. Soc.*, 124, 760, 2002.
91. Dyke, C.A. and Tour, J.M., *J. Am. Chem. Soc.*, 125, 1156, 2003.
92. Wu, W. et al., *Macromolecules*, 36, 6286, 2003.
93. Rao, C. N. R. and Govindaraj, A., *Nanotubes and Nanowires: RSC. Nanoscience & Nanotechnology Series*, Royal Society of Chemistry, Cambridge, 2005.
94. Dresselhaus, M.S. et al., *J. Phys. Chem. C*, 111, 17887, 2007.
95. Strano, M.S., *J. Am. Chem. Soc.*, 125, 16148, 2003.
96. Lebedkin, S. et al., *J. Phys. Chem. B*, 107, 1949, 2003.

97. Lebedkin, S. et al., *New J. Phys.*, 5,140, 2003.
98. Jones, M. et al., *Phys. Rev. B*, 71,115426, 2005.
99. Lefebvre, J. et al., *Phys. Rev. B*, 69, 075403, 2004.
100. Hennrich, F. et al., *J. Phys. Chem. B*, 109, 10567, 2005.
101. Arnold, K. et al., *Nano Lett.*, 12, 2349, 2004.
102. Kasha, M., *Disc. Faraday Soc.*, 9, 14, 1950.
103. Nish, A. et al., *Nat. Nanotech.*, 2, 640, 2007.
104. Crochet, J., Clemens, M., and Hertel, T., *J. Am. Chem. Soc.*, 129, 8058, 2007.
105. Cognet, L. et al., *Science*, 316, 1465, 2007.
106. Banerjee, S. and Wong, S. S., *J. Am. Chem. Soc.*, 124, 8940, 2002.
107. Sun, Y.-P. et al., *Chem. Phys. Lett.*, 351, 349, 2002.
108. Guldi, D.M. et al., *Chem Commun.*, 1130, 2002.
109. Sun, Y.-P. et al., *Acc. Chem. Res.*, 35, 1096, 2002.
110. Chitta, R. and D'Souza, F., *J. Mater. Chem.*, 18, 1440, 2008.
111. Campidelli, S. et al., *J. Phys. Org. Chem.*, 19, 531, 2006.
112. Li, H. et al., *Adv. Mater.*, 16, 896, 2004.
113. Baskaran, D. et al., *J. Am. Chem. Soc.*, 127, 6916, 2005.
114. Guldi, D.M. et al., *Adv. Mater.*, 17, 871, 2005.
115. Guldi, D.M. et al., *Chem. Commun.*, 2034, 2004.
116. Saito, K. et al, *J. Phys. Chem. C*, 111, 1194, 2007.
117. Herranz, M. A. et al., *J. Am. Chem. Soc.*, 130, 66, 2008.
118. Ballesteros, B. et al., *J. Am. Chem. Soc.*, 129, 5061, 2007.
119. Ago, H. et al., *Phys. Rev. B*, 61, 2286, 2000.
120. Segura, J. L. and Martin, N., *Angew. Chem., Int. Ed.*, 40, 1372, 2001.
121. Martin, N. et al., *Acc. Chem. Res.*, 40, 1015, 2007.
122. Herranz, M. A. et al., *Angew. Chem. Int. Ed.*, 45, 4478, 2006.
123. D'Souza, F. et al., *J. Am. Chem. Soc.*, 129, 15865, 2007.
124. Hecht, D. S. et al., *Nano Lett.*, 6, 2031, 2006.
125. Boul, P. J. et al., *J. Am. Chem. Soc.*, 129, 5683, 2007.
126. Rait, S. et al., *Sol. Energy Mater. Sol. Cells*, 91, 757, 2007.
127. Lee, T. Y. and Yoo, J. B., *Diamond Rel. Mater.*, 14, 1888, 2005.
128. Jang, S. R., Vittal, R., and Kim, K. J., *Langmuir*, 20, 9807, 2004.
129. Liu, J. et al., *Science*, 280, 1253, 1998.
130. Zamudio, A. et al., *Small*, 2, 346, 2006.
131. Zanella, R. et al., *J. Phys. Chem. B*, 109, 16290, 2005.
132. Sheeney-Haj-Ichia, L., Basnar, B., and Willner, I., *Angew. Chem. Int. Ed.*, 44, 78, 2005.
133. Guldi, D.M. et al., *J. Am. Chem. Soc.*, 128, 2315, 2006.
134. Brown, P. R. et al., *J. Phys. Chem. C*, 112, 4776, 2008.
135. Berger, S. et al., *J. Appl. Phys.*, 105, 094323, 2009.
136. Schuettfort, T., Nish, A., and Nicholas, R. J., *Nano Lett.*, 9, 3871, 2009.
137. Geng, J. and Zeng, T., *J. Am. Chem. Soc.*, 128, 16827, 2006.
138. Ferguson, A.J. et al., *J. Phys. Chem. Lett.*, 1, 2406, 2010.
139. Stranks, S.D. et al. *ACS Nano*, 6, 6058, 2012.
140. Kanai, Y. and Grossman, J. C., *Nano Lett.*, 8, 908, 2008.
141. Stranks, S. D. et al., *Nano Lett.*, 11, 66, 2011.
142. Holt, J.M. et al., *Nano Lett.*, 10, 4627, 2010.
143. Ren, S. et al., *Nano Lett.*, 11, 5316, 2011.
144. Oelsner, C. et al., *J. Am. Chem. Soc.*, 133, 4580, 2011.
145. Li, K., Martin, D., and Tang, J., *Chin. J. Catal.*, 32, 879, 2011.
146. Usubharatana, P. et al., *Ind. Eng. Chem. Res.*, 45, 2558, 2006.
147. Pathak, P. et al., *Green Chem.*, 7, 667, 2005.
148. Pathak, P. et al., *Chem. Commun.*, 1234, 2004.
149. Woan, K., Pyrgiotakis, G., and Sigmund, W., *Adv. Mater.*, 21, 2233, 2009.
150. Kongkanand, A. and Kamat, P. V., *ACS Nano*, 1, 13, 2007.
151. Dai, K. et al., *Nanotechnology*, 20, 125603, 2009.
152. Wang, W. D. et al., *J. Mol. Catal. A: Chem.*, 235, 194, 2005.
153. Ma, L. L., *Nanotechnology*, 19, 115709, 2008.
154. Ou, Y. et al., *Chem. Phys. Lett.*, 429, 199, 2006.

155. Li, Q., Chen, L., and Lu, G., *J. Phys. Chem. C*, 111, 11494, 2007.
156. Kang, S-Z. et al., *Appl. Surf. Sci.*, 258, 6029, 2012.
157. Peng, T. et al., *Energy Fuels*, 25, 2203, 2011.
158. Ye, A. et al., *Catal. Sci. Technol.*, 2, 969, 2012.
159. Kim, Y. K. and Park, H., *Energy Environ. Sci.*, 4, 685, 2011.
160. Ahmmad, B. et al., *Catal. Commun.*, 9, 1410, 2008.
161. Yu, J., Yang, B., and Cheng, B., *Nanoscale*, 4, 2670, 2012.
162. Chai, B., *Dalton Trans.*, 41, 1179, 2012.
163. Ge, L. and Han, C., *Appl. Catal. B: Environ.*, 117, 268, 2012.
164. Li, N., *Carbon*, 49, 5132, 2011.
165. Xia, X. H., *Carbon*, 45, 717, 2007.
166. Saunders, B.R. et al., *Adv. Colloid and Interf. Sci.*, 138, 1, 2008.
167. Shockley, W. and Queisser, H., *J. Appl. Phys.*, 32, 123, 1961.
168. Green, M.A. et al., *Prog. Photovolt: Res. Appl.*, 19, 565, 2011.
169. Sun, S.-S. and Sariciftci, N.S., *Organic Photovoltaics: Mechanisms, Materials, and Devices (Optical Engineering)*, CRC, Boca Raton, FL, 2005.
170. Hiramoto, M., Fujiwara H., and Yokokawa M., *J. Appl. Phys.*, 72, 3781, 1992.
171. Peumansn, P., Uchida S., and Forrest S. R., *Nature*, 425, 158, 2003.
172. Xue, J. et al., *Appl. Phys. Lett.*, 84, 3013, 2004.
173. O'Regan, B. and Grätzel, M., *Nature*, 353, 737, 1991.
174. Bach, U. et al., *Nature*, 395, 583, 1998.
175. Hagfeldt, A. and Grätzel, M., *Acc. Chem. Res.*, 33, 269, 2000.
176. Grätzel, M., *Nature*, 414, 338, 2001.
177. Grätzel, M., *Pure Appl. Chem.*, 73, 459, 2001.
178. Watsonson, D.F. and Meyer, G.J., *Annu. Rev. Phys. Chem.*, 56, 119, 2005.
179. Anderson, N.A. and Lian, T., *Annu. Rev. Phys. Chem.*, 56, 491, 2005.
180. Lewis, N.S., *Inorg. Chem.*, 44, 6900, 2005.
181. Duncan, W.R. and Prezhdo, O.V., *Annu. Rev. Phys. Chem.*, 58, 143, 2007.
182. Li, C. et al., *Chem. Rev.*, 110, 6817, 2010.
183. Gendron, D. and Leclerc, M., *Energy Environ. Sci.*, 4, 1225, 2011.
184. Gonçalves, L.M. et al., *Energy Environ. Sci.*, 1, 655, 2008.
185. Ning, Z.J., Fu, Y., and Tian, H., *Energy Environ. Sci.*, 3, 1170, 2010.
186. Dennler, G. and Brabec C.J., *Organic Photovoltaics: Materials, Device Physics, and Manufacturing Technologies*, Wiley-VCH, Weinheim, 2008.
187. Hatton, R.A., Miller, A.J., and Silva, S.R.P., *J. Mater. Chem.*, 18, 1183, 2008.
188. Sgobba, V. and Guldi, D.M., *J. Mater. Chem.*, 18, 153, 2008.
189. Durkop, T. et al., *Nano Lett.*, 4, 35, 2004.
190. Xiao, K. et al., *Appl. Phys. Lett.*, 83, 150, 2003.
191. Sankapal, B.R., Setyowati, K., and Chen, J., *Appl. Phys. Lett.*, 91, 173103, 2007
192. Lee, R.S. et al., *Nature*, 388, 255, 1997.
193. Chen, J. et al., *Science*, 282, 95, 1998.
194. Service, R.F., *Science*, 332, 293, 2011.
195. Scharber, M.C. et al., *Adv. Mater.*, 18, 789, 2006.
196. Kymakis, E., *J. Phys. D: Appl. Phys.*, 39, 1058, 2006.
197. Lanzi, M., Paganin, L., and Caretti, D., *Polymer*, 49, 4942, 2008.
198. Miller, A.J., Hatton R.A., and Silva, S.R.P., *Appl. Phys. Lett.*, 89, 123115, 2006.
199. Kymakis, E., Alexandrou, I., and Amaratunga, G. A. J., *J. Appl. Phys.*, 93, 1764, 2003.
200. Landi, B.J. et al., *Progr. Photovolt.: Res. Appl.*, 13, 165, 2005.
201. Kazaoui, S. et al., *J. Appl. Phys.*, 98, 084314, 2005.
202. Kymakis, E. and Amaratunga, G.A.J., *Rev. Adv. Mat. Sci.*, 10, 300, 2005.
203. Bindl, D.J., Safron, N.S., and Arnold, M.S. *ACS Nano*, 4, 5657, 2010.
204. Dissanayake, N.M. and Zhong, Z., *Nano Lett.*, 11, 286, 2010.
205. Zhao, Y., Liao, A., and Pop, E., *IEEE Electron. Device Lett.*, 30, 1078, 2009.
206. Hatakeyama, R., *Appl. Phys. Lett.*, 97, 013104, 2010.
207. Nogueira, A. F. et al., *J. Phys. Chem. C*, 111, 18431, 2007.
208. Partyk, R.L., *Phys. Stat. Sol. (RRL)*, 1, R43, 2007.
209. Pradhan, B., Batabyal, S.K., and Pal, A.J., *Appl. Phys. Lett.*, 88, 093106/1, 2006.
210. Ham, M.H. et al., *ACS Nano*, 4, 6251, 2010.

211. Landi, B. et al., *Sol. Energy Mater. Sol. Cells*, 87, 733, 2005.
212. Cho, N. et al., *Adv. Mater.*, 19, 232, 2007.
213. Berson, S., *Adv. Funct. Mater.*, 17, 3363, 2007.
214. Li, C., *J. Mater. Chem.*, 17, 2406, 2007.
215. Sadhu, V., *Nanotechnology*, 22, 265607, 2011.
216. Chaudhary, S. et al., *Nano Lett.*, 7, 1973, 2007.
217. Lee, J.M. et al., *Adv. Mater.*, 23, 629, 2011.
218. Hagfeldt, A. et al., *Chem. Rev.*, 110, 6595, 2010.
219. Ardo, S. and Meyer, G.J., *Chem. Soc. Rev.*, 38, 115, 2009.
220. Yella, A. et al., *Science*, 334, 629, 2011.
221. Snaith, H.J. *Adv. Funct. Mater.*, 20, 13, 2010.
222. Frank, A.J., Kopidakis, N., and Lagemaat, J.V., *Coord. Chem. Rev.*, 248, 1165, 2004.
223. Sawatsuk, T. et al., *Diam. Rel. Mater.*, 18, 524, 2009.
224. Yu, J., Fan, J., and Cheng, B., *J. Power Source*, 196, 7891, 2011.
225. Lin, W.J., Hsu, C.T., and Tsai, Y.C., *J. Colloid Interf. Sci.*, 358, 562, 2011.
226. Yen, C. et al., *Nanotechnology*, 19, 375305, 2008.
227. Kim, S.L. et al., *J. Appl. Electrochem.*, 36, 1433, 2006.
228. Lee, T.Y., Alegaonkar, P.S., and Yoo, J.-B., *Thin Solid Films*, 515, 5131, 2007.
229. Zhang, S. et al., *J. Phys. Chem. C*, 115, 22025, 2011.
230. Dang, X. et al., *Nat. Nanotech.*, 6, 377, 2011.
231. Trancik, J.E., Barton, S.C., and Hone, J., *Nano Lett.*, 8, 982, 2008.
232. Lee, W.J. et al., *ACS Appl. Mater. Interf.*, 1, 1145, 2009.
233. Nam, J.G. et al., *Scr. Mater.*, 62, 148, 2010.
234. Dong, P. et al., *ACS Appl. Mater. Interf.*, 3, 3157, 2011.
235. Hsieh, C.T., Yang, B.H., and Lin, J.Y., *Carbon*, 49, 3092, 2011.
236. Han, J. et al., *ACS Nano*, 4, 3503, 2010.
237. Mei, X. et al., *Nanotechnology*, 21, 395202, 2010.
238. Huang, S.Q. et al., *Adv. Mater.*, 23, 4707, 2011.
239. Lee, K.S. et al., *Chem. Commun.*, 47, 4264, 2011.
240. Li, S. S. et al., *Adv. Energy Mater.*, 1, 486, 2011.
241. Yang, Z. et al., *Adv. Mater.*, 23, 5436, 2011.
242. Suzuki, K. et al., *Chem. Lett.*, 32, 28, 2003.
243. Tao. C. et al., *Nano Lett.*, 12, 2568, 2012.
244. Šljukic, B., Banks, C.E., and Compton, R.G., *Nano Lett.*, 6, 1556, 2006.
245. Hao, F. et al., *Sci. Rep.*, 2, 368, 2012.
246. Bonaccorso, F., *Int. J. Photoenergy*, 727134, 2010.
247. Gordon, R.G., *MRS Bull.*, 25, 52, 2000.
248. Kumar, A. and Zhou, C., *ACS Nano*, 4, 11, 2010.
249. Hecht, D. S., Hu, L., and Irvin, G., *Adv. Mater.*, 23, 1482, 2011.
250. Lagemaat, J. V. et al., *Appl. Phys. Lett.*, 88, 233503, 2006.
251. Saran, N. et al., *J. Am. Chem. Soc.*, 126, 4462, 2004.
252. Nguyen T. P. and De Vos, S. A., *Appl. Surf. Sci.*, 221, 330, 2004.
253. Hu, L., Hecht, D. S., and Gruner, G., *Nano Lett.*, 4, 2513, 2004.
254. Kaempgen, M., Duesberg, G. S., and Roth, S., *Appl. Surf. Sci.*, 252, 425, 2005.
255. Geng, H. Z. et al., *J. Am. Chem. Soc.*, 129, 7758, 2007.
256. Jackson, R. et al., *Adv. Funct. Mater.*, 18, 2548, 2008.
257. Li, L. et al., *Nano Lett.*, 5, 2472, 2006.
258. Ou, E. C. et al., *ACS Nano*, 3, 2258, 2009.
259. Rowell, M. W. et al., *Appl. Phys. Lett.*, 88, 233506, 2006.
260. Sasaki, Y. et al., *Appl. Phys. Express*, 4, 085102, 2011.
261. Lu, F. et al., *Langmuir*, 27, 4339, 2011.
262. Wu, Z. et al., *Science*, 305, 1273, 2004.
263. Pasquier, A. D., *Appl. Phys. Lett.*, 87, 203511, 2005.
264. De, S. et al., *ACS Nano*, 3, 714, 2009.
265. Blackburn, J. L. et al., *ACS Nano*, 2, 1266, 2008.
266. Hong, S. and Myung, S., *Nat. Nanotech.*, 2, 207, 2007.
267. Jackson, R. K et al., *ACS Nano*, 4, 1377, 2010.
268. Grüner, G. et al., *Appl. Phys. Lett.*, 89, 133112, 2006.

269. Nirmalraj, P.N. et al., *Nano Lett.*, 9, 3890, 2009.
270. Fuhrer, M. S. et al., *Science*, 288, 494, 2000.
271. Yang, S. B. et al., *Nanoscale*, 3, 1361, 2011.
272. Tenent, R. C. et al., *Adv. Mater.*, 21, 3210, 2009.
273. Barnes, T. M., *Appl. Phys. Lett.*, 96, 243309, 2010.
274. Scardaci, V., *Appl. Phys. Lett.*, 97, 023114, 2010.
275. Kim, K. K., *J. Am. Chem. Soc.*, 130, 12757, 2008.
276. Feng, Y. Y., *Appl. Phys. Lett.*, 94, 123302, 2009.
277. Hecht, D. S. et al., Nanotechnology, 22, 075201, 2011.
278. Shim, B. S. et al., ACS Nano, 4, 3725, 2010.
279. Berber, S., Kwon, Y.-K., and Tomanek, D., *Phys. Rev. Lett.*, 84, 4613, 2000.
280. Miyata, Y. et al., *Phys. Chem. C*, 112, 3591, 2008.
281. McEuen, P. L. and Park, J. Y., *MRS Bull.*, 29, 272, 2004.
282. Liang, W. et al., *Nature*, 411, 665, 2001.
283. Hong, S. and Myung, S., *Nat. Nanotechnol.*, 2, 207, 2007.
284. Li, H. et al., *J. Am. Chem. Soc.*, 126, 1014, 2004.
285. Lu, F. et al., *Chem. Phys. Lett.*, 497, 57, 2010.
286. Wang, W. et al., *J. Am. Chem. Soc.*, 130, 1415, 2008.
287. Sun, Y.-P., U.S. *Patent* 7374685, May 20, 2008.
288. Rahy, A. et al., *J. Appl. Surf. Sci.*, 255, 7084, 2009.
289. Wang, W. et al., I: *Encyclopedia of Inorganic Chemistry, Nanomaterials: Inorganic and Bioinorganic Perspectives*, Lukehart, C. M., and Scott. R. A., Eds., John Wiley & Sons, Chichester, UK, 2008, p. 169.
290. Yang, F et al., *J. Phys. Chem. C*, 116, 6800, 2012.
291. Zhang, D. et al., *Nano Lett.*, 6, 1880, 2006.
292. Wang, Y. et al., *Adv. Mater.*, 20, 4442, 2008.
293. Tyler et al., *Adv. Energy Mater.*, 1, 785, 2011.
294. Zheng, M. et al., *Nat. Mater.*, 2, 338, 2003.
295. Giordani, S. et al., *J. Phys. Chem. B*, 110, 15708, 2006.

Index

A

AA, *see* Ascorbic acid (AA)
ABSW equation, *see* Achar–Brindley–Sharp–
    Wendeworth equation (ABSW equation)
AC, *see* Alternating current (AC)
ACC, *see* Activated carbon cloth (ACC)
Achar–Brindley–Sharp–Wendeworth equation
    (ABSW equation), 360
Activated carbon (AC), 440, 453
    metal oxide modification, 453
    resin-derived, 453
    sources, 453
Activated carbon cloth (ACC), 453–454
AD-OLC, *see* Arc discharge onion-like carbon (AD-OLC)
AE, *see* Algebraic equation (AE)
Aerogels, carbon, 454–455
AES, *see* Atomic emission spectroscopy (AES)
AFM, *see* Atomic force microscopy (AFM)
Alchothermal reaction, *see* Solvothermal reaction
Algebraic equation (AE), 438
Aligned CNT electrodes, 200; *see also* Nonaligned CNT
        electrodes
    amino-dextran immobilization, 201, 202
    CP–CNT, 201, 202, 203
    fabrication process, 201
    H$_2$O–plasma etching, 201
    microfabrication, 200, 201
Alkaline phosphatase (ALP), 211
ALP, *see* Alkaline phosphatase (ALP)
α-SiC, 119
Alternating current (AC), 208, 222
Ambipolar field effect, 24
Amino-dextran immobilization, 201, 202
3-Aminopropyltriethoxysilane (APTES / APTS) , 31, 203
Ammonothermal reaction, *see* Solvothermal reaction
Angiogenesis, 228
Apoptosis, 228
APTES, *see* 3-Aminopropyltriethoxysilane (APTES/APTS)
APTS, *see* 3-Aminopropyltriethoxysilane (APTES/APTS)
Arc discharge onion-like carbon (AD-OLC), 384;
        see also Onion-like carbon (OLC)
    surface functionalization, 386
    TGA of, 385
Ascorbic acid (AA), 196
Atomic emission spectroscopy (AES), 125
Atomic force microscopy (AFM), 123, 222
Average crystallite size, 265
Aziridothymidine (AZT), 189
AZT, *see* Aziridothymidine (AZT)

B

Ball milling, 4
Batch-mode method (BM method), 425; *see also*
        Capacitive deionization (CDI)

simplified dynamic CDI transport model, 437–441
BET, *see* Brunauer–Emmett–Teller (BET)
β-SiC, 119
BHJ, *see* Bulk-heterojunction (BHJ)
Biomass, 408; *see also* Carbons
    chemical reactions of, 410
    hydrothermal carbonization, 409–410
    SEM images, 409
Biomolecule detection, 227
Biot number, 445
Bisadducts isomers of a general fulleropyrrolidine, 49
Bistable rotaxanes, 60
BM method, *see* Batch-mode method (BM method)
BN, *see* Boron nitride (BN)
Boron nitride (BN), 4
Bottom-up method, 251
Boudouard equation, 358
Brackish water, 420; *see also* Water desalination
Brunauer–Emmett–Teller (BET), 318
Bulk-heterojunction (BHJ), 464, 480; *see also* Carbon
        nanotubes (CNTs)
    solar cells, 484

C

CA, *see* Carbon aerogels (CA)
Capacitive deionization (CDI), 420, 456; *see also*
        Capacitive deionization electrode;
        Electrochemical analysis; Electrostatic
        double-layer (EDL); Porous electrode theory;
        Simplified dynamic CDI transport model;
        Water desalination
    activated carbon cloths, 453–454
    carbide-derived carbons, 454
    carbon aerogels, 454–455
    carbon black, 456
    carbon material selection, 450
    CDI and MCDI, 449
    charge and ion storage models, 431
    CNTs and graphene, 455–456
    commercial CDI system, 421
    conceptual approaches, 427
    CV vs. CC operation, 426–427
    cycle, 420
    electrochemical experimentation vs. desalination,
        422–423
    experimental approaches, 422
    flow-through capacitor technology, 423
    geometries for CDI testing, 423–425
    isotherm-based modeling, 428–429
    limit cycle, 424
    materials for, 449
    MCDI and IEM barrier role, 429
    mechanism of, 420
    modeling, 428
    operational voltage windows concept, 428

O

P

For Product Safety Concerns and Information please contact our
EU representative GPSR@taylorandfrancis.com Taylor & Francis
Verlag GmbH, Kaufingerstraße 24, 80331 München, Germany